Lecture Notes in Artificial Intelligence 9077

Subseries of Lecture Notes in Computer Science

More information about this series at http://www.springer.com/series/1244

Tru Cao · Ee-Peng Lim
Zhi-Hua Zhou · Tu-Bao Ho
David Cheung · Hiroshi Motoda (Eds.)

Advances in Knowledge Discovery and Data Mining

19th Pacific-Asia Conference, PAKDD 2015
Ho Chi Minh City, Vietnam, May 19–22, 2015
Proceedings, Part I

 Springer

Editors
Tru Cao
Ho Chi Minh City University of Technology
Ho Chi Minh City
Vietnam

Ee-Peng Lim
Singapore Management University
Singapore
Singapore

Zhi-Hua Zhou
Nanjing University
Nanjing
China

Tu-Bao Ho
Japan Advanced Institute of Science and
   Technology
Nomi City
Japan

David Cheung
The University of Hong Kong
Hong Kong
Hong Kong SAR

Hiroshi Motoda
Osaka University
Osaka
Japan

ISSN 0302-9743   ISSN 1611-3349   (electronic)
Lecture Notes in Artificial Intelligence
ISBN 978-3-319-18037-3   ISBN 978-3-319-18038-0   (eBook)
DOI 10.1007/978-3-319-18038-0

Library of Congress Control Number: 2015936624

LNCS Sublibrary: SL7 – Artificial Intelligence

Springer International Publishing AG Switzerland is part of Springer Science+Business Media
(www.springer.com)

Preface

After ten years since PAKDD 2005 in Ha Noi, PAKDD was held again in Vietnam, during May 19–22, 2015, in Ho Chi Minh City. PAKDD 2015 is the 19th edition of the Pacific-Asia Conference series on Knowledge Discovery and Data Mining, a leading international conference in the field. The conference provides a forum for researchers and practitioners to present and discuss new research results and practical applications.

There were 405 papers submitted to PAKDD 2015 and they underwent a rigorous double-blind review process. Each paper was reviewed by three Program Committee (PC) members in the first round and meta-reviewed by one Senior Program Committee (SPC) member who also conducted discussions with the reviewers. The Program Chairs then considered the recommendations from SPC members, looked into each paper and its reviews, to make final paper selections. At the end, 117 papers were selected for the conference program and proceedings, resulting in the acceptance rate of 28.9%, among which 26 papers were given long presentation and 91 papers given regular presentation.

The conference started with a day of six high-quality workshops. During the next three days, the Technical Program included 20 paper presentation sessions covering various subjects of knowledge discovery and data mining, three tutorials, a data mining contest, a panel discussion, and especially three keynote talks by world-renowned experts.

PAKDD 2015 would not have been so successful without the efforts, contributions, and supports by many individuals and organizations. We sincerely thank the Honorary Chairs, Phan Thanh Binh and Masaru Kitsuregawa, for their kind advice and support during preparation of the conference. We would also like to thank Masashi Sugiyama, Xuan-Long Nguyen, and Thorsten Joachims for giving interesting and inspiring keynote talks.

We would like to thank all the Program Committee members and external reviewers for their hard work to provide timely and comprehensive reviews and recommendations, which were crucial to the final paper selection and production of the high-quality Technical Program. We would also like to express our sincere thanks to the following Organizing Committee members: Xiaoli Li and Myra Spiliopoulou together with the individual Workshop Chairs for organizing the workshops; Dinh Phung and U Kang with the tutorial speakers for arranging the tutorials; Hung Son Nguyen, Nitesh Chawla, and Nguyen Duc Dung for running the contest; Takashi Washio and Jaideep Srivastava for publicizing to attract submissions and participants to the conference; Tran Minh-Triet and Vo Thi Ngoc Chau for handling the whole registration process; Tuyen N. Huynh for compiling all the accepted papers and for working with the Springer team to produce these proceedings; and Bich-Thuy T. Dong, Bac Le, Thanh-Tho Quan, and Do Phuc for the local arrangements to make the conference go smoothly.

We are grateful to all the sponsors of the conference, in particular AFOSR/AOARD (Air Force Office of Scientific Research/Asian Office of Aerospace Research and Development), for their generous sponsorship and support, and the PAKDD Steering

Committee for its guidance and Student Travel Award and Early Career Research Award sponsorship. We would also like to express our gratitude to John von Neumann Institute, University of Technology, University of Science, and University of Information Technology of Vietnam National University at Ho Chi Minh City and Japan Advanced Institute of Science and Technology for jointly hosting and organizing this conference. Last but not least, our sincere thanks go to all the local team members and volunteering helpers for their hard work to make the event possible.

We hope you have enjoyed PAKDD 2015 and your time in Ho Chi Minh City, Vietnam.

May 2015

Tru Cao
Ee-Peng Lim
Zhi-Hua Zhou
Tu-Bao Ho
David Cheung
Hiroshi Motoda

Organization

Honorary Co-chairs

Phan Thanh Binh   Vietnam National University, Ho Chi Minh City,
                               Vietnam
Masaru Kitsuregawa   National Institute of Informatics, Japan

General Co-chairs

Tu-Bao Ho   Japan Advanced Institute of Science and
                               Technology, Japan
David Cheung   University of Hong Kong, China
Hiroshi Motoda   Institute of Scientific and Industrial Research,
                               Osaka University, Japan

Program Committee Co-chairs

Tru Hoang Cao   Ho Chi Minh City University of Technology,
                               Vietnam
Ee-Peng Lim   Singapore Management University, Singapore
Zhi-Hua Zhou   Nanjing University, China

Tutorial Co-chairs

Dinh Phung   Deakin University, Australia
U. Kang   Korea Advanced Institute of Science and
                               Technology, Korea

Workshop Co-chairs

Xiaoli Li   Institute for Infocomm Research, A*STAR,
                               Singapore
Myra Spiliopoulou   Otto-von-Guericke University Magdeburg,
                               Germany

Publicity Co-chairs

Takashi Washio   Institute of Scientific and Industrial Research,
                               Osaka University, Japan
Jaideep Srivastava   University of Minnesota, USA

Proceedings Chair

Tuyen N. Huynh   John von Neumann Institute, Vietnam

Contest Co-chairs

Hung Son Nguyen   University of Warsaw, Poland
Nitesh Chawla   University of Notre Dame, USA
Nguyen Duc Dung   Vietnam Academy of Science and Technology,
                                       Vietnam

Local Arrangement Co-chairs

Bich-Thuy T. Dong   John von Neumann Institute, Vietnam
Bac Le   Ho Chi Minh City University of Science, Vietnam
Thanh-Tho Quan   Ho Chi Minh City University of Technology,
                                       Vietnam
Do Phuc   University of Information Technology, Vietnam
                                       National University at Ho Chi Minh City,
                                       Vietnam

Registration Co-chairs

Tran Minh-Triet   Ho Chi Minh City University of Science,
                                       Vietnam
Vo Thi Ngoc Chau   Ho Chi Minh City University of Technology,
                                       Vietnam

Steering Committee

Chairs

Tu-Bao Ho (Chair)   Japan Advanced Institute of Science and
                                       Technology, Japan
Ee-Peng Lim (Co-chair)   Singapore Management University, Singapore

Treasurer

Graham Williams   Togaware, Australia

Members

Tu-Bao Ho	Japan Advanced Institute of Science and Technology, Japan (Member since 2005, Co-chair 2012–2014, Chair 2015–2017, Life Member since 2013)
Ee-Peng Lim (Co-chair)	Singapore Management University, Singapore (Member since 2006, Co-chair 2015–2017)
Jaideep Srivastava	University of Minnesota, USA (Member since 2006)
Zhi-Hua Zhou	Nanjing University, China (Member since 2007)
Takashi Washio	Institute of Scientific and Industrial Research, Osaka University, Japan (Member since 2008)
Thanaruk Theeramunkong	Thammasat University, Thailand (Member since 2009)
P. Krishna Reddy	International Institute of Information Technology, Hyderabad (IIIT-H), India (Member since 2010)
Joshua Z. Huang	Shenzhen Institutes of Advanced Technology, Chinese Academy of Sciences, China (Member since 2011)
Longbing Cao	Advanced Analytics Institute, University of Technology, Sydney, Australia (Member since 2013)
Jian Pei	School of Computing Science, Simon Fraser University, Canada (Member since 2013)
Myra Spiliopoulou	Otto-von-Guericke-University Magdeburg, Germany (Member since 2013)
Vincent S. Tseng	National Cheng Kung University, Taiwan (Member since 2014)

Life Members

Hiroshi Motoda	AFOSR/AOARD and Institute of Scientific and Industrial Research, Osaka University, Japan (Member since 1997, Co-chair 2001–2003, Chair 2004–2006, Life Member since 2006)
Rao Kotagiri	University of Melbourne, Australia (Member since 1997, Co-chair 2006–2008, Chair 2009–2011, Life Member since 2007)
Huan Liu	Arizona State University, USA (Member since 1998, Treasurer 1998–2000, Life Member since 2012)

Ning Zhong   Maebashi Institute of Technology, Japan
                                (Member since 1999, Life member since 2008)
Masaru Kitsuregawa   Tokyo University, Japan (Member since 2000,
                                Life Member since 2008)
David Cheung   University of Hong Kong, China (Member since
                                2001, Treasurer 2005–2006,
                                chair 2006–2008, Life Member since 2009)
Graham Williams   Australian National University, Australia
                                (Member since 2001, Treasurer since 2006,
                                Co-chair 2009–2011, Chair 2012–2014,
                                Life Member since 2009)
Ming-Syan Chen   National Taiwan University, Taiwan, ROC
                                (Member since 2002, Life Member since 2010)
Kyu-Young Whang   Korea Advanced Institute of Science and
                                Technology, Korea (Member since 2003,
                                Life Member since 2011)
Chengqi Zhang   University of Technology, Sydney, Australia
                                (Member since 2004, Life Member since 2012)

Senior Program Committee Members

Arbee Chen   National Chengchi University, Taiwan
Bart Goethals   University of Antwerp, Belgium
Charles Ling   University of Western Ontario, Canada
Chih-Jen Lin   National Taiwan University, Taiwan
Dacheng Tao   University of Technology, Sydney, Australia
Dou Shen   Baidu, China
George Karypis   University of Minnesota, USA
Haixun Wang   Google, USA
Hanghang Tong   City University of New York, USA
Hui Xiong   Rutgers Univesity, USA
Ian Davidson   University of California Davis, USA
James Bailey   University of Melbourne, Australia
Jeffrey Yu   The Chinese University of Hong Kong, Hong Kong
Jian Pei   Simon Fraser University, Canada
Jianyong Wang   Tsinghua University, China
Jieping Ye   Arizona State University, USA
Jiuyong Li   University of South Australia, Australia
Joshua Huang   Shenzhen Institutes of Advanced Technology,
                                Chinese Academy of Sciences, China
Kyuseok Shim   Seoul National University, Korea
Longbing Cao   University of Technology, Sydney, Australia
Masashi Sugiyama   University of Tokyo, Japan
Michael Berthold   University of Konstanz, Germany

Ming Li	Nanjing University, China
Ming-Syan Chen	National Taiwan University, Taiwan
Min-Ling Zhang	Southeast University, China
Myra Spiliopoulou	Otto-von-Guericke-University Magdeburg, Germany
Nikos Mamoulis	University of Hong Kong, Hong Kong
Ning Zhong	Maebashi Institute of Technology, Japan
Osmar Zaiane	University of Alberta, Canada
P. Krishna Reddy	International Institute of Information Technology, Hyderabad, India
Peter Christen	Australian National University, Australia
Sanjay Chawla	University of Sydney, Australia
Takashi Washio	Institute of Scientific and Industrial Research, Osaka University, Japan
Vincent S. Tseng	National Cheng Kung University, Taiwan
Wee Keong Ng	Nanyang Technological University, Singapore
Wei Wang	University of California at Los Angeles, USA
Wen-Chih Peng	National Chiao Tung University, Taiwan
Xiaofang Zhou	University of Queensland, Australia
Xiaohua Hu	Drexel University, USA
Xifeng Yan	University of California, Santa Barbara, USA
Xindong Wu	University of Vermont, USA
Xing Xie	Microsoft Research Asia, China
Yanchun Zhang	Victoria University, Australia
Yu Zheng	Microsoft Research Asia, China

Program Committee Members

Aijun An	York University, Canada
Aixin Sun	Nanyang Technological University, Singapore
Akihiro Inokuchi	Kwansei Gakuin University, Japan
Alfredo Cuzzocrea	ICAR-CNR and University of Calabria, Italy
Andrzej Skowron	University of Warsaw, Poland
Anne Denton	North Dakota State University, USA
Bettina Berendt	Katholieke Universiteit Leuven, Belgium
Bin Zhou	University of Maryland, Baltimore County, USA
Bing Tian Dai	Singapore Management University, Singapore
Bo Zhang	Tsinghua University, China
Bolin Ding	Microsoft Research, USA
Bruno Cremilleux	Université de Caen Basse-Normandie, France
Carson K. Leung	University of Manitoba, Canada
Chandan Reddy	Wayne State University, USA
Chedy Raissi	Inria, France
Chengkai Li	The University of Texas at Arlington, USA

Chia-Hui Chang	National Central University, Taiwan
Chiranjib Bhattacharyya	Indian Institute of Science, India
Choochart Haruechaiy	National Electronics and Computer Technology Center, Thailand
Chun-Hao Chen	Tamkang University, Taiwan
Chun-hung Li	Hong Kong Baptist University, Hong Kong
Clifton Phua	NCS, Singapore
Daoqiang Zhang	Nanjing University of Aeronautics and Astronautics, China
Dao-Qing Dai	Sun Yat-Sen University, China
David Taniar	Monash University, Australia
David Lo	Singapore Management University, Singapore
De-Chuan Zhan	Nanjing University, China
Dejing Dou	University of Oregon, USA
De-Nian Yang	Academia Sinica, Taiwan
Dhaval Patel	Indian Institute of Technology, Roorkee, India
Dinh Phung	Deakin University, Australia
Dragan Gamberger	Ruđer Bošković Institute, Croatia
Du Zhang	California State University, Sacramento, USA
Duc Dung Nguyen	Institute of Information Technology, Vietnam
Enhong Chen	University of Science and Technology of China, China
Fei Liu	Carnegie Mellon University, USA
Feida Zhu	Singapore Management University, Singapore
Florent Masseglia	Inria, France
Geng Li	Oracle Corporation, USA
Giuseppe Manco	Università della Calabria, Italy
Guandong Xu	University of Technology, Sydney, Australia
Guo-Cheng Lan	Industrial Technology Research Institute, Taiwan
Gustavo Batista	University of São Paulo, Brazil
Hady Lauw	Singapore Management University, Singapore
Harry Zhang	University of New Brunswick, Canada
Hiroshi Mamitsuka	Kyoto University, Japan
Hong Shen	University of Adelaide, Australia
Hsuan-Tien Lin	National Taiwan University, Taiwan
Hua Lu	Aalborg University, Denmark
Hui Wang	University of Ulster, UK
Hung Son Nguyen	University of Warsaw, Poland
Hung-Yu Kao	National Cheng Kung University, Taiwan
Irena Koprinska	University of Sydney, Australia
J. Saketha Nath	Indian Insitiute of Technology, India
Jaakko Hollmén	Aalto University, Finland
Jake Chen	Indiana University–Purdue University Indianapolis, USA

James Kwok   Hong Kong University of Science and Technology,
                             China
Jason Wang   New Jersey Science and Technology University,
                             USA
Jean-Marc Petit   Université de Lyon, France
Jeffrey Ullman   Stanford University, USA
Jen-Wei Huang   National Cheng Kung University, Taiwan
Jerry Chun-Wei Lin   Harbin Institute of Technology Shenzhen,
                             China
Jia Wu   University of Technology, Sydney, Australia
Jialie Shen   Singapore Management University, Singapore
Jiayu Zhou   Samsung Research America, USA
Jia-Yu Pan   Google, USA
Jin Soung Yoo   Indiana University–Purdue University
                             Indianapolis, USA
Jingrui He   IBM Research, USA
Jinyan Li   University of Technology, Sydney, Australia
John Keane   University of Manchester, UK
Jun Huan   University of Kansas, USA
Jun Gao   Peking University, China
Jun Luo   Huawei Noah's Ark Lab, Hong Kong
Jun Zhu   Tsinghua University, China
Junbin Gao   Charles Sturt University, Australia
Junjie Wu   Beihang University, China
Junping Zhang   Fudan University, China
K. Selcuk Candan   Arizona State University, USA
Keith Chan   Hong Kong Polytechnic University, Hong Kong
Khoat Than   Hanoi University of Science and Technology,
                             Vietnam
Kitsana Waiyamai   Kasetsart University, Thailand
Krisztian Buza   Semmelweis University, Budapest, Hungary
Kun-Ta Chuang   National Cheng Kung University, Taiwan
Kuo-Wei Hsu   National Chengchi University, Taiwan
Latifur Khan   University of Texas at Dallas, USA
Ling Chen   University of Technology, Sydney, Australia
Lipo Wang   Nanyang Technological University, Singapore
Manabu Okumura   Japan Advanced Institute of Science and
                             Technology, Japan
Marco Maggini   Università degli Studi di Siena, Italy
Marian Vajtersic   University of Salzburg, Austria
Marut Buranarach   National Electronics and Computer Technology
                             Center, Thailand
Mary Elaine Califf   Illinois State University, USA
Marzena Kryszkiewicz   Warsaw University of Technology, Poland

Masashi Shimbo	Nara Institute of Science and Technology, Japan
Meng Chang Chen	Academia Sinica, Taiwan
Mengjie Zhang	Victoria University of Wellington, New Zealand
Michael Hahsler	Southern Methodist University, USA
Min Yao	Zhejiang University, China
Mi-Yen Yeh	Academia Sinica, Taiwan
Muhammad Cheema	Monash University Australia
Murat Kantarcioglu	University of Texas at Dallas, USA
Ngoc-Thanh Nguyen	Wrocław University of Technology, Poland
Nguyen Le Minh	Japan Advanced Institute of Science and Technology, Japan
Pabitra Mitra	Indian Institute of Technology Kharagpur, India
Patricia Riddle	University of Auckland, New Zealand
Peixiang Zhao	Florida State University, USA
Philippe Lenca	Télécom Bretagne, France
Philippe Fournier-Viger	University of Moncton, Canada
Qian You	Amazon, USA
Qingshan Liu	NLPR Institute of Automation, Chinese Academy of Sciences, China
Raymond Chi-Wing Wong	Hong Kong University of Science and Technology, Hong Kong
Richi Nayak	Queensland University of Technology, Australia
Rui Camacho	Universidade do Porto, Portugal
Salvatore Orlando	University of Venice, Italy
Sanjay Jain	National University of Singapore, Singapore
See-Kiong Ng	Institute for Infocomm Research, A*STAR, Singapore
Shafiq Alam	University of Auckland, New Zealand
Sheng-Jun Huang	Nanjing University of Aeronautics and Astronautics, China
Shoji Hirano	Shimane University, Japan
Shou-De Lin	National Taiwan University, Taiwan
Shuai Ma	Beihang University, China
Shu-Ching Chen	Florida International University, USA
Shuigeng Zhou	Fudan University, China
Silvia Chiusano	Politecnico di Torino, Italy
Songcan Chen	Nanjing University of Aeronautics and Astronautics, China
Tadashi Nomoto	National Institute of Japanese Literature, Japan
Takehisa Yairi	University of Tokyo, Japan
Tetsuya Yoshida	Nara Women's University, Japan
Toshihiro Kamishima	National Institute of Advanced Industrial Science and Technology, Japan

Tuyen N. Huynh	John von Neumann Institute, Vietnam
Tzung-Pei Hong	National University of Kaohsiung, Taiwan
Van-Nam Huynh	Japan Advanced Institute of Science and Technology, Japan
Vincenzo Piuri	Università degli Studi di Milano, Italy
Wai Lam The	Chinese University of Hong Kong, Hong Kong
Walter Kosters	Universiteit Leiden, The Netherlands
Wang-Chien Lee	Pennsylvania State University, USA
Wei Ding	University of Massachusetts Boston, USA
Wenjie Zhang	University of New South Wales, Australia
Wenjun Zhou	University of Tennessee, Knoxville, USA
Wilfred Ng	Hong Kong University of Science and Technology, Hong Kong
Wu-Jun Li	Nanjing University, China
Wynne Hsu	National University of Singapore, Singapore
Xiaofeng Meng	Renmin University of China, China
Xiaohui (Daniel) Tao	University of Southern Queensland, Australia
Xiaoli Li	Institute for Infocomm Research, A*STAR, Singapore
Xiaowei Ying	Bank of America, USA
Xin Wang	University of Calgary, Canada
Xingquan Zhu	Florida Atlantic University, USA
Xintao Wu	University of Arkansas, Arkansas
Xuan Vinh Nguyen	University of Melbourne, Australia
Xuan-Hieu Phan	University of Engineering and Technology–Vietnam National University, Hanoi, Vietnam
Xuelong Li	University of London, UK
Xu-Ying Liu	Southeast University, China
Yang Yu	Nanjing University, China
Yang-Sae Moon	Kangwon National University, Korea
Yasuhiko Morimoto	Hiroshima University, Japan
Yidong Li	Beijing Jiaotong University, China
Yi-Dong Shen	Chinese Academy of Sciences, China
Ying Zhang	University of New South Wales, Australia
Yi-Ping Phoebe Chen	La Trobe University, Australia
Yiu-ming Cheung	Hong Kong Baptist University, Hong Kong
Yong Guan	Iowa State University, USA
Yonghong Peng	University of Bradford, UK
Yue-Shi Lee	Ming Chuan University, Taiwan
Zheng Chen	Microsoft Research Asia, China
Zhenhui Li	Pennsylvania State University, USA
Zhiyuan Chen	University of Maryland, Baltimore County, USA
Zhongfei Zhang	Binghamton University, USA
Zili Zhang	Deakin University, Australia

External Reviewers

Ahsanul Haque	University of Texas at Dallas, USA
Ameeta Agrawal	York University, Canada
Anh Kim Nguyen	Hanoi University of Science and Technology, Vietnam
Arnaud Soulet	Université François Rabelais, Tours, France
Bhanukiran Vinzamuri	Wayne State University, USA
Bin Fu	University of Technology, Sydney, Australia
Bing Tian Dai	Singapore Management University, Singapore
Budhaditya Saha	Deakin University, Australia
Cam-Tu Nguyen	Nanjing University, China
Cheng Long	Hong Kong University of Science and Technology, Hong Kong
Chung-Hsien Yu	University of Massachusetts Boston, USA
Chunming Liu	University of Technology, Sydney, Australia
Dawei Wang	University of Massachusetts Boston, USA
Dieu-Thu Le	University of Trento, Italy
Dinusha Vatsalan	Australian National University, Australia
Doan V. Nguyen	Japan Advanced Institute of Science and Technology, Japan
Emmanuel Coquery	Université Lyon1, CNRS, France
Ettore Ritacco	ICAR-CNR, Italy
Fan Jiang	University of Manitoba, Canada
Fang Yuan	Institute for Infocomm Research A*STAR, Singapore
Fangfang Li	University of Technology, Sydney, Australia
Fernando Gutierrez	University of Oregon, USA
Fuzheng Zhang	University of Science and Technology of China, China
Gensheng Zhang	University of Texas at Arlington, USA
Gianni Costa	ICAR-CNR, Italy
Guan-Bin Chen	National Cheng Kung University, Taiwan
Hao Wang	University of Oregon, USA
Heidar Davoudi	York University, Canada
Henry Lo	University of Massachusetts Boston, USA
Ikumi Suzuki	National Institute of Genetics, Japan
Jan Bazan	University of Rzeszów, Poland
Jan Vosecky	Hong Kong University of Science and Technology, Hong Kong
Javid Ebrahimi	University of Oregon, USA
Jianhua Yin	Tsinghua University, China
Jianmin Li	Tsinghua University, China
Jianpeng Xu	Michigan State University, USA
Jing Ren	Singapore Management University, Singapore
Jinpeng Chen	Beihang University, China

Jipeng Qiang	University of Massachusetts Boston, USA
Joseph Paul Cohen	University of Massachusetts Boston, USA
Junfu Yin	University of Technology, Sydney, Australia
Justin Sahs	University of Texas at Dallas, USA
Kai-Ho Chan	Hong Kong University of Science and Technology, Hong Kong
Kazuo Hara	National Institute of Genetics, Japan
Ke Deng	RMIT University, Australia
Kiki Maulana Adhinugraha	Monash University, Australia
Kin-Long Ho	Hong Kong University of Science and Technology, Hong Kong
Lan Thi Le	Hanoi University of Science and Technology, Vietnam
Lei Zhu	Huazhong University of Science and Technology, China
Lin Li	Wuhan University of Technology, China
Linh Van Ngo	Hanoi University of Science and Technology, Vietnam
Loc Do	Singapore Management University, Singapore
Maksim Tkachenko	Singapore Management University, Singapore
Marc Plantevit	Université de Lyon, France
Marian Scuturici	INSA de Lyon, CNRS, France
Marthinus Christoffel du Plessis	University of Tokyo, Japan
Md. Anisuzzaman Siddique	Hiroshima University, Japan
Min Xie	Hong Kong University of Science and Technology, Hong Kong
Ming Yang	Binghamton University, USA
Minh Nhut Nguyen	Institute for Infocomm Research A*STAR, Singapore
Mohit Sharma	University of Minnesota, USA
Morteza Zihayat	York University, Canada
Mu Li	University of Technology, Sydney, Australia
Naeemul Hassan	University of Texas at Arlington, USA
NhatHai Phan	University of Oregon, USA
Nicola Barbieri	Yahoo Labs, Spain
Nicolas Béchet	Université de Bretagne Sud, France
Nima Shahbazi	York University, Canada
Pakawadee Pengcharoen	Hong Kong University of Science and Technology, Hong Kong
Pawel Gora	University of Warsaw, Poland
Peiyuan Zhou	Hong Kong Polytechnic University, Hong Kong
Peng Peng	Hong Kong University of Science and Technology, Hong Kong
Pinghua Gong	University of Michigan, USA

Qiong Fang	Hong Kong University of Science and Technology, Hong Kong
Quan Xiaojun	Institute for Infocomm Research A*STAR, Singapore
Riccardo Ortale	ICAR-CNR, Italy
Sabin Kafle	University of Oregon, USA
San Phyo Phyo	Institute for Infocomm Research A*STAR, Singapore
Sang The Dinh	Hanoi University of Science and Technology, Vietnam
Shangpu Jiang	University of Oregon, USA
Shenlu Wang	University of New South Wales, Australia
Shiyu Yang	University of New South Wales, Australia
Show-Jane Yen	Ming Chuan University, Taiwan
Shuangfei Zhai	Binghamton University, USA
Simone Romano	University of Melbourne, Australia
Sujatha Das Gollapalli	Institute for Infocomm Research A*STAR, Singapore
Swarup Chandra	University of Texas at Dallas, USA
Syed K. Tanbeer	University of Manitoba, Canada
Tenindra Abeywickrama	Monash University, Australia
Thanh-Son Nguyen	Singapore Management University, Singapore
Thin Nguyen	Deakin University, Australia
Tiantian He	Hong Kong Polytechnic University, Hong Kong
Tianyu Kang	University of Massachusetts Boston, USA
Trung Le	Deakin University, Australia
Tuan M. V. Le	Singapore Management University, Singapore
Xiaochen Chen	Google, USA
Xiaolin Hu	Tsinghua University, China
Xin Li	University of Science and Technology, China
Xuhui Fan	University of Technology, Sydney, Australia
Yahui Di	University of Massachusetts Boston, USA
Yan Li	Wayne State University, USA
Yang Jianbo	Institute for Infocomm Research A*STAR, Singapore
Yang Mu	University of Massachusetts Boston, USA
Yanhua Li	University of Minnesota, USA
Yanhui Gu	Nanjing Normal University, China
Yathindu Rangana Hettiarachchige	Monash University, Australia
Yi-Yu Hsu	National Cheng Kung University, Taiwan
Yingming Li	Binghamton University, USA
Yu Zong	West Anhui University, China
Zhiyong Chen	Singapore Management University, Singapore
Zhou Zhao	Hong Kong University of Science and Technology, Hong Kong
Zongda Wu	Wenzhou University, China

Contents – Part I

Social Networks and Social Media

Maximizing Friend-Making Likelihood for Social Activity
Organization. 3
    Chih-Ya Shen, De-Nian Yang, Wang-Chien Lee, and Ming-Syan Chen

What Is New in Our City? A Framework for Event Extraction Using
Social Media Posts 16
    Chaolun Xia, Jun Hu, Yan Zhu, and Mor Naaman

Link Prediction in Aligned Heterogeneous Networks 33
    Fangbing Liu and Shu-Tao Xia

Scale-Adaptive Group Optimization for Social Activity Planning 45
    Hong-Han Shuai, De-Nian Yang, Philip S. Yu, and Ming-Syan Chen

Influence Maximization Across Partially Aligned Heterogenous
Social Networks 58
    Qianyi Zhan, Jiawei Zhang, Senzhang Wang, Philip S. Yu,
    and Junyuan Xie

Multiple Factors-Aware Diffusion in Social Networks. 70
    Chung-Kuang Chou and Ming-Syan Chen

Understanding Community Effects on Information Diffusion 82
    Shuyang Lin, Qingbo Hu, Guan Wang, and Philip S. Yu

On Burst Detection and Prediction in Retweeting Sequence. 96
    Zhilin Luo, Yue Wang, Xintao Wu, Wandong Cai, and Ting Chen

#FewThingsAboutIdioms: Understanding Idioms and Its Users
in the Twitter Online Social Network 108
    Koustav Rudra, Abhijnan Chakraborty, Manav Sethi, Shreyasi Das,
    Niloy Ganguly, and Saptarshi Ghosh

Retweeting Activity on Twitter: Signs of Deception 122
    Maria Giatsoglou, Despoina Chatzakou, Neil Shah,
    Christos Faloutsos, and Athena Vakali

Resampling-Based Gap Analysis for Detecting Nodes with High
Centrality on Large Social Network 135
    Kouzou Ohara, Kazumi Saito, Masahiro Kimura, and Hiroshi Motoda

Classification

Double Ramp Loss Based Reject Option Classifier.   151
Naresh Manwani, Kalpit Desai, Sanand Sasidharan,
and Ramasubramanian Sundararajan

Efficient Methods for Multi-label Classification   164
Chonglin Sun, Chunting Zhou, Bo Jin, and Francis C.M. Lau

A Coupled k-Nearest Neighbor Algorithm for Multi-label Classification   176
Chunming Liu and Longbing Cao

Learning Topic-Oriented Word Embedding for Query Classification.   188
Hebin Yang, Qinmin Hu, and Liang He

Reliable Early Classification on Multivariate Time Series
with Numerical and Categorical Attributes.   199
Yu-Feng Lin, Hsuan-Hsu Chen, Vincent S. Tseng, and Jian Pei

Distributed Document Representation for Document Classification   212
Rumeng Li and Hiroyuki Shindo

Prediciton of Emergency Events: A Multi-Task Multi-Label Learning
Approach.   226
Budhaditya Saha, Sunil Kumar Gupta, and Svetha Venkatesh

Nearest Neighbor Method Based on Local Distribution for Classification . . .   239
Chengsheng Mao, Bin Hu, Philip Moore, Yun Su, and Manman Wang

Immune Centroids Over-Sampling Method for Multi-Class Classification . . .   251
Xusheng Ai, Jian Wu, Victor S. Sheng, Pengpeng Zhao, Yufeng Yao,
and Zhiming Cui

Optimizing Classifiers for Hypothetical Scenarios.   264
Reid A. Johnson, Troy Raeder, and Nitesh V. Chawla

Repulsive-SVDD Classification   277
Phuoc Nguyen and Dat Tran

Centroid-Means-Embedding: an Approach to Infusing Word Embeddings
into Features for Text Classification   289
Mohammad Golam Sohrab, Makoto Miwa, and Yutaka Sasaki

Machine Learning

Collaborating Differently on Different Topics: A Multi-Relational Approach
to Multi-Task Learning   303
Sunil Kumar Gupta, Santu Rana, Dinh Phung, and Svetha Venkatesh

Multi-Task Metric Learning on Network Data 317
Chen Fang and Daniel N. Rockmore

A Bayesian Nonparametric Approach to Multilevel Regression 330
Vu Nguyen, Dinh Phung, Svetha Venkatesh, and Hung H. Bui

Learning Conditional Latent Structures from Multiple Data Sources 343
Viet Huynh, Dinh Phung, Long Nguyen, Svetha Venkatesh, and Hung H. Bui

Collaborative Multi-view Learning with Active Discriminative Prior
for Recommendation 355
Qing Zhang and Houfeng Wang

Online and Stochastic Universal Gradient Methods for Minimizing
Regularized Hölder Continuous Finite Sums in Machine Learning 369
Ziqiang Shi and Rujie Liu

Context-Aware Detection of Sneaky Vandalism on Wikipedia Across
Multiple Languages. 380
Khoi-Nguyen Tran, Peter Christen, Scott Sanner, and Lexing Xie

Uncovering the Latent Structures of Crowd Labeling 392
Tian Tian and Jun Zhu

Use Correlation Coefficients in Gaussian Process to Train Stable
ELM Models 405
Yulin He, Joshua Zhexue Huang, Xizhao Wang, and Rana Aamir Raza

Local Adaptive and Incremental Gaussian Mixture for Online Density
Estimation 418
Tianyu Qiu, Furao Shen, and Jinxi Zhao

Latent Space Tracking from Heterogeneous Data with an Application
for Anomaly Detection 429
Jiaji Huang and Xia Ning

A Learning-Rate Schedule for Stochastic Gradient Methods to Matrix
Factorization 442
Wei-Sheng Chin, Yong Zhuang, Yu-Chin Juan, and Chih-Jen Lin

Applications

On Damage Identification in Civil Structures Using Tensor Analysis 459
Nguyen Lu Dang Khoa, Bang Zhang, Yang Wang, Wei Liu, Fang Chen, Samir Mustapha, and Peter Runcie

Predicting Smartphone Adoption in Social Networks   472
Le Wu, Yin Zhu, Nicholas Jing Yuan, Enhong Chen, Xing Xie,
and Yong Rui

Discovering the Impact of Urban Traffic Interventions Using Contrast
Mining on Vehicle Trajectory Data.   486
Xiaoting Wang, Christopher Leckie, Hairuo Xie,
and Tharshan Vaithianathan

Locating Self-collection Points for Last-mile Logistics using Public
Transport Data   498
Huayu Wu, Dongxu Shao, and Wee Siong Ng

A Stochastic Framework for Solar Irradiance Forecasting Using Condition
Random Field.   511
Jin Xu, Shinjae Yoo, Dantong Yu, Hao Huang, Dong Huang,
John Heiser, and Paul Kalb

Online Prediction of Chess Match Result.   525
Mohammad M. Masud, Ameera Al-Shehhi, Eiman Al-Shamsi,
Shamma Al-Hassani, Asmaa Al-Hamoudi, and Latifur Khan

Learning of Performance Measures from Crowd-Sourced Data
with Application to Ranking of Investments.   538
Greg Harris, Anand Panangadan, and Viktor K. Prasanna

Hierarchical Dirichlet Process for Tracking Complex Topical Structure
Evolution and its Application to Autism Research Literature   550
Adham Beykikhoshk, Ognjen Arandjelović, Svetha Venkatesh,
and Dinh Phung

Automated Detection for Probable Homologous Foodborne Disease
Outbreaks   563
Xiao Xiao, Yong Ge, Yunchang Guo, Danhuai Guo, Yi Shen,
Yuanchun Zhou, and Jianhui Li

Identifying Hesitant and Interested Customers for Targeted Social
Marketing   576
Guowei Ma, Qi Liu, Le Wu, and Enhong Chen

Activity-Partner Recommendation.   591
Wenting Tu, David W. Cheung, Nikos Mamoulis, Min Yang,
and Ziyu Lu

Iterative Use of Weighted Voronoi Diagrams to Improve Scalability
in Recommender Systems   605
Joydeep Das, Subhashis Majumder, Debarshi Dutta,
and Prosenjit Gupta

Novel Methods and Algorithms

Principal Sensitivity Analysis. 621
  Sotetsu Koyamada, Masanori Koyama, Ken Nakae, and Shin Ishii

SocNL: Bayesian Label Propagation with Confidence. 633
  Yuto Yamaguchi, Christos Faloutsos, and Hiroyuki Kitagawa

An Incremental Local Distribution Network for Unsupervised Learning 646
  Youlu Xing, Tongyi Cao, Ke Zhou, Furao Shen, and Jinxi Zhao

Trend-Based Citation Count Prediction for Research Articles. 659
  Cheng-Te Li, Yu-Jen Lin, Rui Yan, and Mi-Yen Yeh

Mining Text Enriched Heterogeneous Citation Networks. 672
  Jan Kralj, Anita Valmarska, Marko Robnik-Šikonja, and Nada Lavrač

Boosting via Approaching Optimal Margin Distribution 684
  Chuan Liu and Shizhong Liao

o-HETM: An Online Hierarchical Entity Topic Model for News Streams . . . 696
  Linmei Hu, Juanzi Li, Jing Zhang, and Chao Shao

Modeling User Interest and Community Interest in Microbloggings:
An Integrated Approach. 708
  Tuan-Anh Hoang

Minimal Jumping Emerging Patterns: Computation and Practical
Assessment 722
  Bamba Kane, Bertrand Cuissart, and Bruno Crémilleux

Rank Matrix Factorisation 734
  *Thanh Le Van, Matthijs van Leeuwen, Siegfried Nijssen,
  and Luc De Raedt*

An Empirical Study of Personal Factors and Social Effects on Rating
Prediction 747
  Zhijin Wang, Yan Yang, Qinmin Hu, and Liang He

Author Index 759

Contents – Part II

Opinion Mining and Sentiment Analysis

Emotion Cause Detection for Chinese Micro-Blogs Based on ECOCC
Model 3
   Kai Gao, Hua Xu, and Jiushuo Wang

Parallel Recursive Deep Model for Sentiment Analysis 15
   Changliang Li, Bo Xu, Gaowei Wu, Saike He, Guanhua Tian,
   and Yujun Zhou

Sentiment Analysis in Transcribed Utterances 27
   Nir Ofek, Gilad Katz, Bracha Shapira, and Yedidya Bar-Zev

Rating Entities and Aspects Using a Hierarchical Model 39
   Xun Wang, Katsuhito Sudoh, and Masaaki Nagata

Sentiment Analysis on Microblogging by Integrating Text and Image
Features 52
   Yaowen Zhang, Lin Shang, and Xiuyi Jia

TSum4act: A Framework for Retrieving and Summarizing Actionable
Tweets during a Disaster for Reaction 64
   Minh-Tien Nguyen, Asanobu Kitamoto, and Tri-Thanh Nguyen

Clustering

Evolving Chinese Restaurant Processes for Modeling Evolutionary
Traces in Temporal Data 79
   Peng Wang, Chuan Zhou, Peng Zhang, Weiwei Feng, Li Guo,
   and Binxing Fang

Small-Variance Asymptotics for Bayesian Nonparametric Models
with Constraints 92
   Cheng Li, Santu Rana, Dinh Phung, and Svetha Venkatesh

Spectral Clustering for Large-Scale Social Networks via a Pre-Coarsening
Sampling Based NystrÖm Method 106
   Ying Kang, Bo Yu, Weiping Wang, and Dan Meng

pcStream: A Stream Clustering Algorithm for Dynamically Detecting
and Managing Temporal Contexts. 119
   Yisroel Mirsky, Bracha Shapira, Lior Rokach, and Yuval Elovici

Clustering Over Data Streams Based on Growing Neural Gas   134
  Mohammed Ghesmoune, Mustapha Lebbah, and Hanene Azzag

Computing and Mining ClustCube Cubes Efficiently   146
  Alfredo Cuzzocrea

Outlier and Anomaly Detection

Contextual Anomaly Detection Using Log-Linear Tensor Factorization.   165
  Alpa Jayesh Shah, Christian Desrosiers, and Robert Sabourin

A Semi-Supervised Framework for Social Spammer Detection.   177
  *Zhaoxing Li, Xianchao Zhang, Hua Shen, Wenxin Liang,
  and Zengyou He*

Fast One-Class Support Vector Machine for Novelty Detection   189
  Trung Le, Dinh Phung, Khanh Nguyen, and Svetha Venkatesh

ND-Sync: Detecting Synchronized Fraud Activities   201
  *Maria Giatsoglou, Despoina Chatzakou, Neil Shah, Alex Beutel,
  Christos Faloutsos, and Athena Vakali*

An Embedding Scheme for Detecting Anomalous Block Structured
Graphs.   215
  Lida Rashidi, Sutharshan Rajasegarar, and Christopher Leckie

A Core-Attach Based Method for Identifying Protein Complexes
in Dynamic PPI Networks   228
  Jiawei Luo, Chengchen Liu, and Hoang Tu Nguyen

Mining Uncertain and Imprecise Data

Mining Uncertain Sequential Patterns in Iterative MapReduce   243
  Jiaqi Ge, Yuni Xia, and Jian Wang

Quality Control for Crowdsourced POI Collection   255
  *Shunsuke Kajimura, Yukino Baba, Hiroshi Kajino,
  and Hisashi Kashima*

Towards Efficient Sequential Pattern Mining in Temporal Uncertain
Databases.   268
  Jiaqi Ge, Yuni Xia, and Jian Wang

Preference-Based Top-k Representative Skyline Queries on Uncertain
Databases.   280
  Ha Thanh Huynh Nguyen and Jinli Cao

Cluster Sequence Mining: Causal Inference with Time and Space
Proximity under Uncertainty   293
   Yoshiyuki Okada, Ken-ichi Fukui, Koichi Moriyama,
   and Masayuki Numao

Achieving Accuracy Guarantee for Answering Batch Queries
with Differential Privacy   305
   Dong Huang, Shuguo Han, and Xiaoli Li

Mining Temporal and Spatial Data

Automated Classification of Passing in Football   319
   Michael Horton, Joachim Gudmundsson, Sanjay Chawla,
   and Joël Estephan

Stabilizing Sparse Cox Model Using Statistic and Semantic Structures
in Electronic Medical Records   331
   Shivapratap Gopakumar, Tu Dinh Nguyen, Truyen Tran,
   Dinh Phung, and Svetha Venkatesh

Predicting Next Locations with Object Clustering and Trajectory
Clustering   344
   Meng Chen, Yang Liu, and Xiaohui Yu

A Plane Moving Average Algorithm for Short-Term Traffic Flow
Prediction   357
   Lei Lv, Meng Chen, Yang Liu, and Xiaohui Yu

Recommending Profitable Taxi Travel Routes Based on Big Taxi
Trajectories Data   370
   Wenxin Yang, Xin Wang, Seyyed Mohammadreza Rahimi,
   and Jun Luo

Semi Supervised Adaptive Framework for Classifying Evolving Data
Stream   383
   Ahsanul Haque, Latifur Khan, and Michael Baron

Feature Extraction and Selection

Cost-Sensitive Feature Selection on Heterogeneous Data   397
   Wenbin Qian, Wenhao Shu, Jun Yang, and Yinglong Wang

A Feature Extraction Method for Multivariate Time Series Classification
Using Temporal Patterns   409
   Pei-Yuan Zhou and Keith C.C. Chan

Scalable Outlying-Inlying Aspects Discovery via Feature Ranking   422
  Nguyen Xuan Vinh, Jeffrey Chan, James Bailey, Christopher Leckie,
  Kotagiri Ramamohanarao, and Jian Pei

A DC Programming Approach for Sparse Optimal Scoring   435
  Hoai An Le Thi and Duy Nhat Phan

Graph Based Relational Features for Collective Classification   447
  Immanuel Bayer, Uwe Nagel, and Steffen Rendle

A New Feature Sampling Method in Random Forests for Predicting
High-Dimensional Data   459
  Thanh-Tung Nguyen, He Zhao, Joshua Zhexue Huang, Thuy Thi Nguyen,
  and Mark Junjie Li

Mining Heterogeneous, High Dimensional, and Sequential Data

Seamlessly Integrating Effective Links with Attributes for Networked
Data Classification   473
  Yangyang Zhao, Zhengya Sun, Changsheng Xu, and Hongwei Hao

Clustering on Multi-source Incomplete Data via Tensor Modeling
and Factorization   485
  Weixiang Shao, Lifang He, and Philip S. Yu

Locally Optimized Hashing for Nearest Neighbor Search   498
  Seiya Tokui, Issei Sato, and Hiroshi Nakagawa

Do-Rank: DCG Optimization for Learning-to-Rank in Tag-Based Item
Recommendation Systems   510
  Noor Ifada and Richi Nayak

Efficient Discovery of Recurrent Routine Behaviours in Smart Meter Time
Series by Growing Subsequences   522
  Jin Wang, Rachel Cardell-Oliver, and Wei Liu

Convolutional Nonlinear Neighbourhood Components Analysis for Time
Series Classification   534
  Yi Zheng, Qi Liu, Enhong Chen, J. Leon Zhao, Liang He,
  and Guangyi Lv

Entity Resolution and Topic Modelling

Clustering-Based Scalable Indexing for Multi-party Privacy-Preserving
Record Linkage   549
  Thilina Ranbaduge, Dinusha Vatsalan, and Peter Christen

Efficient Interactive Training Selection for Large-Scale Entity Resolution. . . .   562
  Qing Wang, Dinusha Vatsalan, and Peter Christen

Unsupervised Blocking Key Selection for Real-Time Entity Resolution   574
  Banda Ramadan and Peter Christen

Incorporating Probabilistic Knowledge into Topic Models.   586
  Liang Yao, Yin Zhang, Baogang Wei, Hongze Qian, and Yibing Wang

Learning Focused Hierarchical Topic Models with Semi-Supervision
in Microblogs.   598
  Anton Slutsky, Xiaohua Hu, and Yuan An

Predicting Future Links Between Disjoint Research Areas Using
Heterogeneous Bibliographic Information Network.   610
  Yakub Sebastian, Eu-Gene Siew, and Sylvester Olubolu Orimaye

Itemset and High Performance Data Mining

CPT+: Decreasing the Time/Space Complexity of the Compact Prediction
Tree   625
  *Ted Gueniche, Philippe Fournier-Viger, Rajeev Raman,
  and Vincent S. Tseng*

Mining Association Rules in Graphs Based on Frequent Cohesive
Itemsets.   637
  *Tayena Hendrickx, Boris Cule, Pieter Meysman, Stefan Naulaerts,
  Kris Laukens, and Bart Goethals*

Mining High Utility Itemsets in Big Data   649
  Ying Chun Lin, Cheng-Wei Wu, and Vincent S. Tseng

Decomposition Based SAT Encodings for Itemset Mining Problems.   662
  Said Jabbour, Lakhdar Sais, and Yakoub Salhi

A Comparative Study on Parallel LDA Algorithms in MapReduce
Framework.   675
  *Yang Gao, Zhenlong Sun, Yi Wang, Xiaosheng Liu, Jianfeng Yan,
  and Jia Zeng*

Distributed Newton Methods for Regularized Logistic Regression   690
  Yong Zhuang, Wei-Sheng Chin, Yu-Chin Juan, and Chih-Jen Lin

Recommendation

Coupled Matrix Factorization Within Non-IID Context.   707
  Fangfang Li, Guandong Xu, and Longbing Cao

Complementary Usage of Tips and Reviews for Location
Recommendation in Yelp.   720
    Saurabh Gupta, Sayan Pathak, and Bivas Mitra

Coupling Multiple Views of Relations for Recommendation   732
    Bin Fu, Guandong Xu, Longbing Cao, Zhihai Wang, and Zhiang Wu

Pairwise One Class Recommendation Algorithm   744
    Huimin Qiu, Chunhong Zhang, and Jiansong Miao

RIT: Enhancing Recommendation with Inferred Trust.   756
    Guo Yan, Yuan Yao, Feng Xu, and Jian Lu

Author Index   769

Social Networks and Social Media

Maximizing Friend-Making Likelihood for Social Activity Organization

Chih-Ya Shen[1]([✉]), De-Nian Yang[2], Wang-Chien Lee[3], and Ming-Syan Chen[1,2]

[1] Research Center for Information Technology Innovation,
Academia Sinica, Taipei, Taiwan
{chihya,mschen}@citi.sinica.edu.tw
[2] Institute of Information Science, Academia Sinica, Taipei, Taiwan
dnyang@iis.sinica.edu.tw
[3] Department of Computer Science and Engineering,
The Pennsylvania State University, University Park, USA
wlee@cse.psu.edu

Abstract. The social presence theory in social psychology suggests that computer-mediated online interactions are inferior to face-to-face, in-person interactions. In this paper, we consider the scenarios of organizing in person friend-making social activities via online social networks (OSNs) and formulate a new research problem, namely, Hop-bounded Maximum Group Friending (HMGF), by modeling both existing friendships and the likelihood of new friend making. To find a set of attendees for socialization activities, HMGF is unique and challenging due to the interplay of the group size, the constraint on existing friendships and the objective function on the likelihood of friend making. We prove that HMGF is NP-Hard, and no approximation algorithm exists unless $P = NP$. We then propose an error-bounded approximation algorithm to efficiently obtain the solutions very close to the optimal solutions. We conduct a user study to validate our problem formulation and perform extensive experiments on real datasets to demonstrate the efficiency and effectiveness of our proposed algorithm.

1 Introduction

With the popularity and accessibility of online social networks (OSNs), e.g., Facebook, Meetup, and Skout[1], more and more people initiate friend gatherings or group activities via these OSNs. For example, more than 16 millions of events are created on Facebook each month to organize various kinds of activities[2], and more than 500 thousands of face-to-face activities are initiated in Meetup[3]. The activities organized via OSNs cover a wide variety of purposes, e.g., friend gatherings, cocktail parties, concerts, and marathon events. The wide spectrum of these activities shows that OSNs have been widely used as a convenient means for initiating real-life activities among friends.

---

[1] http://www.skout.com/
[2] http://newsroom.fb.com/products/
[3] http://www.meetup.com/about/

© Springer International Publishing Switzerland 2015
T. Cao et al. (Eds.): PAKDD 2015, Part I, LNAI 9077, pp. 3–15, 2015.
DOI: 10.1007/978-3-319-18038-0_1

On the other hand, to help users expand their circles of friends in the cyberspace, friend recommendation services have been provided in OSNs to suggest candidates to users who may likely become mutual friends in the future. Many friend recommendation services employ link prediction algorithms, e.g., [10,11], to analyze the features, similarity or interaction patterns of users in order to derive potential future friendship between some users. By leveraging the abundant information in OSNs, link prediction algorithms show high accuracy for recommending online friends in OSNs.

As social presence theory [16] in social psychology suggests, computer-mediated online interactions are inferior to face-to-face, in-person interactions, off-line friend-making activities may be favorable to their on-line counterparts in cyberspace. Therefore, in this paper, we consider the scenarios of organizing face-to-face friend-making activities via OSN services. Notice that finding socially cohesive groups of participants is essential for maintaining good atmosphere for the activity. Moreover, the function of making new friends is also an important factor for the success of social activities, e.g., assigning excursion groups in conferences, inviting attendees to housewarming parties, etc. Thus, for organizing friend-making social activities, both activity organization and friend recommendation services are fundamental. However, there is a gap between existing activity organization and friend recommendation services in OSNs for the scenarios under consideration. Existing activity organization approaches focus on extracting socially cohesive groups from OSNs based on certain cohesive measures, density, diameter, of social networks or other constraints, e.g., time, spatial distance, and interests, of participants [5–8]. On the other hand, friend recommendation services consider only the *existing friendships* to recommend potential new friends for an individual (rather than finding a group of people for engaging friend-making). We argue that in addition to themes of common interests, it is desirable to organize friend-making activities by mixing the "potential friends", who may be interested in knowing each other (as indicated by a link prediction algorithm), with existing friends (as lubricators). To the best knowledge of the authors, the following two important factors, 1) the existing friendship among attendees, and 2) the potential friendship among attendees, have not been considered simultaneously in existing activity organization services. To bridge the gap, it is desirable to propose a new activity organization service that carefully addresses these two factors at the same time.

In this paper, we aim to investigate the problem of selecting a set of candidate attendees from the OSN by considering both the existing and potential friendships among the attendees. To capture the two factors for activity organization, we propose to include the likelihood of making new friends in the social network. As such, we formulate a new research problem to find groups with tight social relationships among existing friends and potential friends (i.e., who are not friends yet). Specifically, we model the social network in the OSN as a heterogeneous social graph $G = (V, E, R)$ with edge weight $w : R \rightarrow (0, 1]$, where V is the set of individuals, E is the set of *friend edges*, and R is the set of *potential friend edges* (or potential edges for short). Here a friend edge (u, v) denotes that individuals u and v are mutual friends, while a potential edge $[u', v']$

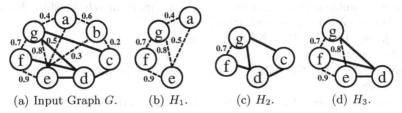

(a) Input Graph G.   (b) H_1.   (c) H_2.   (d) H_3.

Fig. 1. Illustrative Example

indicates that individuals u' and v' are likely to become friends (the edge weight $w[u', v']$ quantifies the likelihood). The potential edges and the corresponding edge weights can be obtained by employing a link prediction algorithm in friend recommendation.

Given a heterogeneous social graph $G = (V, E, R)$ as described above, we formulate a new problem, namely, *Hop-bounded Maximum Group Friending (HMGF)*, to find a group that 1) maximizes the likelihood of making new friends among the group, i.e., the group has the highest ratio of total potential edge weight to group size, 2) ensures that the social tightness, i.e., hop count on friend edges in G between each pair of individuals is small, and 3) is a sufficiently large group, i.e., too small a group may not work well for socialization activities.

Figure 1 illustrates the social graph and the interplay of the above factors. Figure 1(a) shows a social graph, where a dash line, e.g., $[a, b]$ with weight 0.6, is a potential edge and a solid line, e.g., (c, d), is a friend edge. Figure 1(b) shows a group $H_1:\{a, e, f, g\}$ which has many potential edges and thus a high total weight. However, not all the members of this group have common friends as social lubricators. Figure 1(c) shows a group $H_2:\{c, d, f, g\}$ tightly connected by friend edges. While H_2 may be a good choice for gathering of close friends, the goal of friend-making in socialization activities is missed. Finally, Figure 1(d) shows $H_3:\{d, e, f, g\}$ which is a better choice than H_1 and H_2 for socialization activities because each member of H_3 is within 2 hops of another member via friend edges in G. Moreover, the average potential edge weight among them is high, indicating members are likely to make some new friends.

Processing HMGF to find the best solution is very challenging because there are many important factors to consider, including hop constraint, group size and the total weight of potential edges in a group. Indeed, we prove that HMGF is an NP-Hard problem with no approximation algorithm. Nevertheless, we prove that if the hop constraint can be slightly relaxed to allow a small error, there exists a 3-approximation algorithm for HMGF. Theoretical analysis and empirical results show that our algorithm can obtain good solutions efficiently.

The contributions made in this study are summarized as follows.

- For socialization activity organization, we propose to model the existing friendship and the potential friendship in a heterogeneous social graph and formulate a new problem, namely, Hop-bounded Maximum Group Friending (HMGF), for finding suitable attendees. To our best knowledge, HMGF is the first problem that considers these two important relationships between attendees for activity organization.

- We prove that HMGF is NP-Hard and there exists no approximation algorithm for HMGF unless $P = NP$. We then propose an approximation algorithm, called MaxGF, with a guaranteed error bound for solving HMGF efficiently.
- We conduct a user study on 50 users to validate our argument for considering both existing and potential friendships in activity organization. We also perform extensive experiments on real datasets to evaluate the proposed algorithm. Experimental results manifest that HMGF can obtain solutions very close to the optimal ones, very efficiently.

The rest of this paper is organized as follows. Section 2 formulates HMGF and proves it NP-Hard with no approximation algorithm. Section 3 reviews the related works, and Section 4 details the algorithm design. Section 5 reports a user study and experimental results. Section 6 concludes this paper.

2   Problem Formulation

Based on the description of heterogeneous social graph described earlier, here we formulate the *Hop-bounded Maximum Group Friending (HMGF)* tackled in this paper. Given two individuals u and v, let $d_G^E(u, v)$ be the shortest path between u and v via friend edges in G. Moreover, given $H \subseteq G$, let $w(H)$ denote the total weight of potential edges in H and let *average weight*, $\sigma(H) = \frac{w(H)}{|H|}$ denote the average weight of potential edges connected to each individual in H^4. HMGF is formulated as follows.

Problem: Hop-bounded Maximum Group Friending (HMGF).
Given: Social network $G = (V, E, R)$, hop constraint h, and size constraint p.
Objective: Find an induced subgraph $H \subseteq G$ with the maximum $\sigma(H)$, where $|H| \geq p$ and $d_G^E(u, v) \leq h, \forall u, v \in H$.

Efficient processing of HMGF is very challenging due to the following reasons: 1) The interplay of the total weight $w(H)$ and the size of H. To maximize $\sigma(H)$, finding a small H may not be a good choice because the number of edges in a small graph tends to be small as well. On the other hand, finding a large H (which usually has a high $w(H)$) may not lead to an acceptable $\sigma(H)$, either. Therefore, the key is to strike a good balance between the graph size $|H|$ and the total weight $w(H)$. 2) HMGF includes a hop constraint (say $h = 2$) on friend edges to ensure that every pair of individuals is not too distant socially from each other. However, selecting a potential edge $[u, v]$ with a large weight $w[u, v]$ may not necessarily satisfy the hop constraint, i.e., $d_G^E(u, v) > h$ which is defined based on existing friend edges. In this case, it may not always be a good strategy to prioritize on large-weight edges in order to maximize $\sigma(H)$, especially when u and v do not share a common friend nearby via the friend edges.

---

[4] Note that $\sigma(H) = 0$ if $H = \varnothing$.

In the following, we prove that HMGF is NP-Hard and *not approximable* within any factor. In other words, there exists no approximation algorithm for HMGF.

Theorem 1. *HMGF is NP-Hard and there is no approximation algorithm for HMGF unless $P = NP$.*

Proof. Due to the space constraints, we prove this theorem in the full version of this paper (available online [1]).

The above theorem manifests that HMGF has no approximation algorithm. Nevertheless, we show that HMGF becomes approximable if a small error h is allowed in the hop constraint. More specifically, in Section 4, we first propose an error-bounded approximation algorithm for HMGF, which returns a solution with guaranteed $\sigma(H)$, while $d_G^E(u, v)$ for any two vertices u and v in H may exceed h but is always bounded by $2h$. Afterward, we present a post-processing procedure to tailor the solution for satisfying the hop constraint.

3 Related Work

Extracting dense subgraphs or social cohesive groups among social networks is a natural way for selecting a set of close friends for a gathering. Various social cohesive measures have been proposed for finding dense social subgraphs, e.g., diameter [2], density [3], clique and its variations [4]. Although these social cohesive measures cover a wide range of application scenarios, they focus on deriving groups based only on existing friendship in the social network. In contrast, the HMGF studied in this paper aims to extract groups by considering both the existing and potential friendships for socialization activities. Therefore, the existing works mentioned above cannot be directly applied to HMGF tackled in this paper.

Research on finding a set of attendees for activities based on the social tightness among existing friends [5–9] have been reported in the literature. Social-Temporal Group Query [5] checks the available times of attendees to find the social cohesive group with the most suitable activity time. Geo-Social Group Query [6,7] extracts socially tight groups while considering certain spatial properties. The willingness optimization for social group problem in [8] selects a set of attendees for an activity while maximizing their willingness to participate. Finally, [9] finds a set of compatible members with tight social relationships in the collaboration network. Although these works find suitable attendees for activities based on existing friendship among the attendees, they ignore the likelihood of making new friends among the attendees. Therefore, these works may not be suitable for socialization activities discussed in this paper.

Link prediction analyzes the features, similarity or interaction patterns among individuals in order to recommend possible friends to the users [10–14]. Link prediction algorithms employ different approaches including graph-topological features, classification models, hierarchical probabilistic model, and linear algebraic methods. These works show good prediction accuracy for friend recommendation in social networks. In this paper, to estimate the likelihood of how individuals may potentially become friends in the future, we employ link prediction algorithms for deriving the potential edges among the individuals.

To the best knowledge of the authors, there exists no algorithm for activity organization that considers both the existing friendship and the likelihood of making new friends when selecting activity attendees. The HMGF studied in this paper examines the social tightness among existing friends and the likelihood of becoming friends for non-friend attendees. We envisage that our research result can be employed in various social network applications for activity organization.

4 Error-Bounded Approximation Algorithm for HMGF

4.1 Algorithm Description

To tackle HMGF, a naive approach is to enumerate all possible combinations of vertices, and extracts the subgraph H with the maximum $\sigma(H)$ following the hop and group size constraints. However, this approach is computationally expensive and thus not applicable for a large-scale social network. To efficiently answer HMGF, we propose an algorithm, called *MaxGF*, which is a 3-approximation algorithm with a guaranteed error bound h. MaxGF limits the search space of candidate solutions by dividing the graph into different hop-bounded subgraphs such that their sizes are much smaller than $|V|$. Then, it employs a greedy app-roach on the hop-bounded subgraphs to iteratively remove the vertices that are inclined to generate a small $\sigma(H)$. Specifically, we define the *incident weight* of a vertex v in an induced subgraph $H \subseteq G$ as $\tau_H(v)$, where $\tau_H(v) = \sum_{u \in H} w[v, u]$, i.e., the incident weight of v is the total weight of the potential edges incident to v in H. By carefully examining the incident weights of the vertices, we can remove from the hop-bounded subgraph those vertices that contribute no gain in the objective function. Moreover, we propose an effective pruning strategy for trimming redundant search. Finally, a post-processing procedure is proposed to ensure that the returned solution follows the hop constraint.

The pseudo code of MaxGF is presented in Algorithm 1. Basically, to obtain the hop-bounded subgraphs, MaxGF sorts the vertices in terms of their incident weights and iteratively selects a vertex v with the maximum incident weight from G as a reference vertex. A hop-bounded subgraph H_v is constructed from v by including every vertex u with at most h hops from v on the friend edges, i.e., $H_v = \{u | d_G^E(u, v) \leq h\}$. Moreover, if $|H_v| < p$, it is no longer necessary to examine H_v because any subgraph in H_v will never be a feasible solution due to the size constraint. Therefore, redundant search space is effectively pruned.

In addition, another pruning condition is also proposed to further prune the resulted subgraph H_v. Let S^{APX} denote the best solution obtained so far. If half of the maximum incident weight among the vertices u in H_v, i.e., $(1/2) \cdot \max_{u \in H_v} \tau_{H_v}(u)$, does not exceed $\sigma(S^{APX})$, there will never be any solution

---

Algorithm 1. MaxGF

Input: Social graph $G = (V, E, R)$, hop constraint h, and size constraint p

1: $U \leftarrow G,\ S^{APX} \leftarrow \varnothing$
2: **while** $U \neq \varnothing$ **do**
3:   $v \leftarrow \arg\max_{u \in U} \tau_G(u),\ U \leftarrow U - \{v\}$
4:   let H_v be the induced subgraph of G with vertices as $\{u \mid d_G^E(u, v) \leq h\}$
5:   **if** $|H_v| < p$ or $\frac{1}{2} \cdot \max_{u \in H_v} \tau_{H_v}(u) \leq \sigma(S^{APX})$ **then**
6:     continue;
7:   let $S_1 \leftarrow H_v$
8:   **for** $i \leftarrow 1$ to $|H_v|$ **do**
9:     $\hat{v}_i \leftarrow \arg\min_{u \in S_i} \tau_{S_i}(u)$
10:    $S_{i+1} \leftarrow S_i - \{\hat{v}_i\}$
11:   let S_v^* be the S_i with the maximum $\sigma(S_i)$ where $|S_i| \geq p$
12:   **if** $\sigma(S_v^*) > \sigma(S^{APX})$ **then**
13:     $S^{APX} \leftarrow S_v^*$
14: PostProcessing(S^{APX})
15: **output** S^{APX}

---

better than $\sigma(S^{APX})$ in H_v. The reason is that the average weight $\sigma(H)$ of any subgraph $H \subseteq H_v$ must satisfy the following inequality,

$$\sigma(H) = \frac{\sum_{t \in H} \tau_H(t)}{2|H|} \leq \frac{\max_{u \in H_v} \tau_{H_v}(u) \cdot |H|}{2|H|} = \frac{1}{2} \cdot \max_{u \in H_v} \tau_{H_v}(u).$$

Therefore, if $\frac{1}{2} \cdot \max_{u \in H_v} \tau_{H_v}(u) \leq \sigma(S^{APX})$ holds, there exists no subgraph in H_v with the average weight larger than $\sigma(S^{APX})$, and H_v can be pruned.

Next, MaxGF starts to find the solution in H_v with the maximized average weight, which includes $|H_v|$ steps. Let S_{i+1} denote the subgraph after removing a vertex $\hat{v}_i$ from S_i in step i. That is, we set $S_1 = H_v$ initially, and at each step i afterwards, S_{i+1} is the subgraph $S_i - \{\hat{v}_i\}$. During each step i, $\hat{v}_i$ is selected as the vertex which has the lowest incident weight in S_i, i.e., $\hat{v}_i = \arg\min_{u \in S_i} \tau_{S_i}(u)$. This is based on the intuition that excluding vertices with low incident weights is more inclined to increase the average weight of the the remaining subgraph. Then, $\hat{v}_i$ and its incident potential edges are removed from S_i and the remaining graph is S_{i+1}. Then, S_{i+1} is processed in the next step $i+1$. The above procedure ends until S_i is empty.

To maximize the objective function $\sigma(H) = \frac{w(H)}{|H|}$, after a hop-bounded subgraph H_v is processed, S_v^* is extracted as the subgraph S_i with the maximum $\sigma(S_i)$ in H_v where $|S_i| \geq p$. If $\sigma(S_v^*) > \sigma(S^{APX})$, we replace S^{APX} with S_v^*. Then, we continue to extract the next vertex v' for examining the corresponding hop-bounded subgraph $H_{v'}$ until all vertices have been examined. Afterward, a post-processing procedure (detailed in Section 4.3) is employed on the best solution obtained in the algorithm, i.e., S^{APX}, to ensure that the hop constraint is satisfied and to further maximize $\sigma(S^{APX})$. Finally, S^{APX} is output as the solution.

4.2   Theoretical Bound

In the following, given the hop-bounded subgraph H_v, we first prove that there exists a subgraph $F \subseteq H_v$ such that $3 \cdot w(F)$ is an upper bound of the total potential edge weight of the optimal solution to the HMGF instance on H_v. Then, we prove that for each H_v, the average weight of S_v^* obtained in the algorithm, i.e., $\sigma(S_v^*)$, is at least $\frac{1}{3}$ the average weight of the optimal solution of HMGF on H_v. Finally, based on the properties of the hop-bounded subgraph and S^{APX}, we prove that the proposed algorithm is a 3-approximation algorithm with guaranteed error bound to HMGF.

Let S_v^{OPT} denote the optimal solution of the HMGF instance on H_v with $\sigma(S_v^{OPT}) > 0$, we first prove that the largest subgraph F in H_v, where $\tau_F(u) \geq \frac{2}{3}\sigma(S_v^{OPT})$, $\forall u \in F$, is not an empty graph.

Lemma 1. *The largest subgraph $F \subseteq H_v$, where $\tau_F(u) \geq \frac{2}{3}\sigma(S_v^{OPT})$, $\forall u \in F$, is not an empty graph.*

Proof. The proof is presented in the online version [1].

With the existence of F proven above, we now derive an upper bound of the total potential edge weight of S_v^{OPT}, i.e., $w(S_v^{OPT})$, according to $w(F)$.

Lemma 2. $w(S_v^{OPT})$ *is upper bounded by* $3 \cdot w(F)$, *i.e.,* $3 \cdot w(F) > w(S_v^{OPT})$.

Proof. The proof is presented in the online version [1].

Then, with the properties derived above, we turn our attention to analyzing MaxGF proposed in this section. In MaxGF, given H_v and when we are iteratively extracting $\hat{v}_i$ which has the minimum incident weight in S_i, if $\hat{v}_i$ is the first extracted vertex such that $\hat{v}_i \in F$ (i.e., step i is the earliest step such that $\hat{v}_i \in F$), then we have the following lemma.

Lemma 3. *Given H_v in MaxGF, if step i is the earliest step where the extracted $\hat{v}_i$ from S_i is in F, then $\tau_{S_i}(u) \geq \frac{2}{3}\sigma(S_v^{OPT})$, $\forall u \in S_i$. Moreover, $F = S_i$.*

Proof. The proof is presented in the online version [1].

We combine the results obtained above, and derive the bound on $\sigma(S_v^*)$, where S_v^* is the group S_i which has the maximum $\sigma(S_i)$ among all S_i with $|S_i| \geq p$ obtained by MaxGF in H_v. Please note that Lemma 3 proves that during the steps of extracting $\hat{v}_i$ from S_i, there exists $\hat{v}_i$ with $\tau_{S_i}(\hat{v}_i) \geq \frac{2}{3}\sigma(S_v^{OPT})$.

Theorem 2. *Given H_v in MaxGF, let i be the earliest step such that $\hat{v}_i$ satisfies $\tau_{S_i}(\hat{v}_i) \geq \frac{2}{3}\sigma(S_v^{OPT})$, then $\sigma(S_v^*) \geq \frac{1}{3}\sigma(S_v^{OPT})$.*

Proof. The proof is presented in the online version [1].

Finally, let S^{OPT} denote the optimal solution of HMGF on G, the following theorem proves that the solution obtained by MaxGF, i.e., S^{APX}, has $\sigma(S^{APX})$ at least $\frac{1}{3} \cdot \sigma(S^{OPT})$, and the error is bounded by h.

Theorem 3. *MaxGF returns the solution S^{APX} with $\sigma(S^{APX}) \geq \frac{\sigma(S^{OPT})}{3}$ and $d_G^E(u,v) \leq 2 \cdot h, \forall u, v \in S^{APX}$.*

Proof. The proof is presented in the online version [1].

4.3   Post Processing and Time Complexity

A post-processing procedure is designed to tailor S^{APX} for meeting the hop constraint and further maximizing the average weight. More specifically, given S^{APX} obtained in the algorithm, we first define the notion of *boundary vertices*. A vertex u in S^{APX} is a boundary vertex if there exists at least one other vertex v in S^{APX} such that the shortest path from u to v via friend edges contains more than h edges. Let B denote the set of boundary vertices. MaxGF includes the following adjustment steps in the post-processing procedure. 1) Expand: a vertex $v \in (V \backslash S^{APX})$ can be added into S^{APX} if adding v does not increase $|B|$ and increases $\sigma(S^{APX})$. We give priority to the v which maximizes $\sigma(S^{APX} \cup \{v\})$. 2) Shrink: given a boundary vertex $u \in B$, u can be safely removed if after removing u from S^{APX}, $|B|$ decreases but $\sigma(S^{APX})$ does not. We give priority to the u that maximizes $\sigma(S^{APX} - \{u\})$. Please note that the above post-processing procedure minimizes $\max_{u,v \in S^{APX}} d_G^E(u,v)$ while increasing $\sigma(S^{APX})$. Therefore, after post processing, the performance and error bounds in Theorem 3 still hold.

Time Complexity. The time complexity of MaxGF is $O(|V| \log |V| \cdot (|E| + |R|))$. The detailed analysis is presented in the online version of this paper [1].

5   Experimental Results

We implement HMGF in Facebook and invite 50 users to participate in our user study. Each user, given 12 test cases of HMGF using her friends in Facebook as the input graph, is asked to solve the HMGF cases, and compare her results with the solutions obtained by MaxGF. In addition to the user study, we evaluate the performance of MaxGF on two real social network datasets, i.e., FB [15] and the MS dataset from KDD Cup 2013[5]. The FB dataset is extracted from Facebook with 90K vertices, and MS is a co-author network with 1.7M vertices. We extract the friend edges from these datasets and identify the potential edges with a link prediction algorithm [11]. The weight of a potential edge is ranged within (0,1]. Moreover, we compare MaxGF with two algorithms, namely, Baseline and DkS [3]. Baseline finds the optimal solution of HMGF by enumerating all the subgraphs satisfying the constraints, while DkS is an $O(|V|^{1/3})$-approximation algorithm for finding a p-vertex subgraph $H \subseteq G$ with the maximum density on $E \cup R$ without considering the potential edges and the hop constraint. The algorithms are implemented in an IBM 3650 server with Quadcore Intel X5450 3.0 GHz CPUs. We measure 30 samples in each scenario. In the following, Fea-Ratio and ObjRatio respectively denote the ratio of feasibility (i.e., the portion

---

[5] https://www.kaggle.com/c/kdd-cup-2013-author-paper-identification-challenge/data

12   C.-Y. Shen et al.

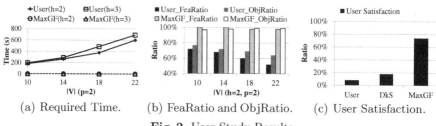

Fig. 2. User Study Results

of solutions satisfying the hop constraint) and the ratio of $\sigma(H)$ in the solutions obtained by MaxGF or DkS to that of the optimal solution.

5.1   User Study

Figure 2 presents the results of the user study. Figure 2(a) compares the required time for users and MaxGF to solve the HMGF instances. Users need much more time than MaxGF due to challenges brought by the hop constraint and tradeoffs in potential edge weights and the group size, as explained in Section 2. As $|V|$ or h grows, users need more time because the HMGF cases become more complicated. Figure 2(b) compares the solution feasibility and quality among users and MaxGF. We employ Baseline to obtain the optimal solutions and derive FeaRatio and ObjRatio accordingly. The FeaRatio and ObjRatio of users are low because simultaneously considering both the hop constraint on friend edges and total weights on potential edges is difficult for users. As shown, users' FeaRatio and ObjRatio drop when $|V|$ increases. By contrast, MaxGF obtains the solutions with high FeaRatio and ObjRatio. In Figure 2(c), we ask each user to compare her solutions with the solutions obtained by MaxGF and DkS, to validate the effectiveness of HMGF. 74% of the users agree that the solution of MaxGF is the best because HMGF maximizes the likelihood of friend-making while considering the hop constraint on friend edges at the same time. By contrast, DkS finds the solutions with a large number of edges, but it does not differentiate the friend edges and potential edges. Therefore, users believe that the selected individuals may not be able to socialize with each other effectively.

5.2   Performance Evaluation

Baseline can only find the optimal solutions of small HMGF cases since it enumerates all possible solutions. Therefore, we first compare MaxGF against Baseline and DkS on small graphs randomly extracted from FB. Figure 3(a) compares the execution time of the algorithms by varying the size of input graph. Since Baseline enumerates all the subgraphs H with $|H| \geq p$, the execution time grows exponentially. The execution time of MaxGF is very small because the hop-bounded subgraphs and the pruning strategy effectively trim the search space. Figures 3(b) and 3(c) present the FeaRatio and ObjRatio of the algorithms, respectively. MaxGF has high ObjRatio because MaxGF iteratively removes vertices with low incident

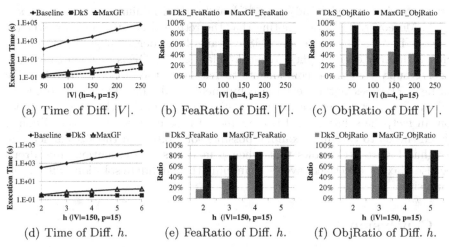

(a) Time of Diff. $|V|$.   (b) FeaRatio of Diff. $|V|$.   (c) ObjRatio of Diff $|V|$.

(d) Time of Diff. h.   (e) FeaRatio of Diff. h.   (f) ObjRatio of Diff. h.

Fig. 3. Comparisons with Optimal Solutionss

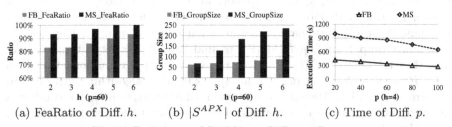

(a) FeaRatio of Diff. h.   (b) $|S^{APX}|$ of Diff. h.   (c) Time of Diff. p.

Fig. 4. Experimental Results on Different Datasets

weights from each hop-bounded subgraph H_v, and extracts the solution S^{APX} with maximized $\sigma(S^{APX})$ among different subgraphs in different H_v to strike a good balance on total edge weights and group sizes as describe in Section 2. Moreover, the high FeaRatio and ObjRatio also indicate that the post-processing procedure effectively restores the hop constraint and maximizes the average weight accordingly. By contrast, DkS does not consider the hop constraint and different edge types in finding solutions and thus generates the solutions with smaller FeaRatio and ObjRatio.

Figures 3(d)-(f) compare execution time, FeaRatio and ObjRatio again but by varying h. When h increases, the execution time of MaxGF grows slowly because the pruning strategy avoids examining the hop-bounded subgraphs that do not lead to a better solution. The FeaRatio and ObjRatio of MaxGF with different h are high because MaxGF employs hop-bounded subgraphs to avoid generating solutions with large hop distances on friend edges, and the post-processing procedure effectively restores the hop constraint and maximizes the objective function.

Figure 4 compares MaxGF in different datasets, i.e., FB and MS. Figures 4(a) and 4(b) present the FeaRatio and the solution group sizes with different h. As h increases, MaxGF on both datasets achieves a higher FeaRatio due to the post-processing procedure adjusts S^{APX} and further minimizes $d_G^E(u, v), \forall u, v \in S^{APX}$. Moreover, it is worth noting that the returned group sizes grow when h increases in MS. This is because MS contains large densely connected components with large edge weights. When h is larger, MaxGF is inclined to extract larger groups from these components to maximize the objective function. By contrast, FB does not have large components and MaxGF thereby tends to find small groups to reduce the group size for maximizing the objective function. In fact, the solutions in FB are almost the same with different h. Finally, MaxGF needs to carefully examine possible solutions with the sizes at least p, and thus Figure 4(c) shows that when p increases, the execution time drops because MaxGF effectively avoids examining the candidate solutions with small group sizes.

6   Conclusion

To bridge the gap between the state-of-the-art activity organization and friend recommendation in OSNs, in this paper, we propose to model the individuals with existing and potential friendships in OSNs for friend-making activity organization. We formulate a new research problem, namely, Hop-bonded Maximum Group Friending (HMGF), to find suitable activity attendees. We prove that HMGF is NP-Hard and there exists no approximation algorithms unless $P = NP$. We then propose an approximation algorithm with guaranteed error bound, i.e., MaxGF, to find good solutions efficiently. We conduct a user study and extensive experiments to evaluate the performance of MaxGF, where MaxGF outperforms other relevant approaches in both solution quality and efficiency.

References

1. Shen, C.-Y., Yang, D.-N., Lee, W.-C., Chen, M.-S.: Maximizing Friend-Making Likelihood for Social Activity Organization (2015). arXiv:1502.06682
2. Wasserman, S., Faust, K.: Social Network Anlysis: Methods and Applications. Cambridge University Press (1994)
3. Feige, U., Kortsarz, G., Peleg, D.: The Dense k-Subgraph Problem. Algorithmica (2001)
4. Mokken, R.: Cliques, Clubs and Clans. Quality and Quantity: International Journal of Methodology (1979)
5. Yang, D.-N., Chen, Y.-L., Lee, W.-C., Chen, M.-S.: On Social-Temporal Group Query with Acquaintance Constraint. VLDB (2011)
6. Yang, D.-N., Shen, C.-Y., Lee, W.-C., Chen, M.-S.: On socio-spatial group query for location-based social networks. In: KDD (2012)
7. Zhu, Q., Hu, H., Xu, J., Lee, W.C.: Geo-Social Group Queries with Minimum Acquaintance Constraint (2014). arXiv:1406.7367v1
8. Shuai, H.-H., Yang, D.-N., Yu, P.S., Chen, M.-S.: Willingness Optimization for Social Group Activity. VLDB (2014)
9. Surian, D., Liu, N., Lo, D., Tong, H., Lim, E.-P., Faloutsos, C.: Recommending people in developers' collaboration network. In: WCRE (2011)

10. Kashima, H., Abe, N: A parameterized probabilistic model of network evolution for supervised link prediction. In: ICDM (2006)
11. Liben-Nowell, D., Kleinberg, J.: The Link Prediction Problem for Social Networks. Journal of the American Soceity for Information Science and Technology (2007)
12. Clause, A., Moore, C., Newman, M.: Hierarchical Structure and the Prediction of Missing Links in Network. Nature (2008)
13. Kunegis, J., Lommatzsch, A.: Learning spectral graph transformations for link prediction. In: ICML (2009)
14. Leung, C., Lim, E.-P., Lo, D., Weng, J.: Mining interesting link formation rules in social networks. In: CIKM (2010)
15. Viswanath, B., Mislove, A., Cha, M., Gummadi, K.: On the evolution of userinter-action in facebook. In: WOSN (2009)
16. Short, J., Williams, E., Christie, B.: The Social Psychology of Telecommunications. Wiley, London (1976)

What Is New in Our City? A Framework for Event Extraction Using Social Media Posts

Chaolun Xia[1]([✉]), Jun Hu[1], Yan Zhu[1], and Mor Naaman[2]

[1] Rutgers University, 57 US Highway 1, New Brunswick, NJ 08901, USA
cx28@cs.rutgers.edu
[2] Jacobs Institute, Cornell Tech, 111 8th Ave., New York, NY 10011, USA

Abstract. Post streams from public social media platforms such as Instagram and Twitter have become precious but noisy data sources to discover what is happening around us. In this paper, we focus on the problem of detecting and presenting local events in real time using social media content. We propose a novel framework for real-time city event detection and extraction. The proposed framework first applies bursty detection to discover candidate event signals from Instagram and Twitter post streams. Then it integrates the two posts streams to extract features for candidate event signals and classifies them into true events or noise. For the true events, the framework extracts various information to summarize and present them. We also propose a novel method that combines text, image and geolocation information to retrieve relevant photos for detected events. Through the experiments on a large dataset, we show that integrating Instagram and Twitter post streams can improve event detection accuracy, and properly combining text, image and geolocation information is able to retrieve more relevant photos for events. Through case studies, we also show that the framework is able to report detected events with low spatial and temporal deviation.

Keywords: Data mining · Social media · Event extraction

1 Introduction

With the growing popularity of mobile devices and applications, more and more people are sharing their moments with their friends and the public through mainstream social media platforms such as Facebook (Instagram) and Twitter. A recent report[1] shows that Instagram now has more than 200 Million monthly-active-users (MAU) and these users upload more than 1.5 billion photos and videos per month. Twitter has even larger traffic and popularity, 255 Million MAUs and 15 billion tweets per month.

Although a dominating proportion of posts from such social media platforms are about users' personal life [19], such as emotional feeling, opinions, food, travel and even self-portraits, there are still considerable amount of posts recording

---

[1] http://jennstrends.com/instagram-statistics-for-2014/

© Springer International Publishing Switzerland 2015
T. Cao et al. (Eds.): PAKDD 2015, Part I, LNAI 9077, pp. 16–32, 2015.
DOI: 10.1007/978-3-319-18038-0_2

what are happening in our city. A user may upload photos of a fashion show to her Instagram account, or talk about emergency or crime on Twitter. These valuable social media posts have made it possible for researchers and developers to accurately detect and present local events in real time. Such techniques will benefit various users, ranging from government officers, journalists, to tourists and residents, etc. For example, a system quickly reporting fire or car accidents can help the local police to make a timely response to the emergency; detecting entertaining events in real time and representing them to nearby tourists or residents can provide opportunities in social engagement.

However, the problem of detecting and representing local events in real time from social media data streams remains challenging. First, event-related social media data are sparse, although the volume of posts from any popular social media platform is large. Second, most current research focuses on Twitter [2], while there are various types of social media platforms which can potentially contribute to the detection problem. However, the problem to choose or combine multiple data sources to detect events is challenging due to the heterogeneity of posts from different data sources. Third, after detecting events, to represent events with the most relevant posts is still challenging due to the noisy posts stream with heterogeneous content including image, text and geolocation.

To address these challenges, we propose a novel framework in this paper. This framework first detects candidate event signals from Instagram and Twitter post streams, and then extract features to classify whether an event signal is a true event or noise. Finally, it summarizes the detected event by retrieving relevant photos and topics and then estimating the occurrence time and location. Besides the proposed framework, our contributions also include that we analyze different methods to integrate Instagram and Twitter post streams, and experimentally show that they improve the detection accuracy. To our best knowledge, we are the first to integrate Instagram and Twitter posts to detect events in real time. For event summarization, we propose a method to retrieve relevant photos, which utilizes image content, text and geolocation information. Finally, we conduct case studies to show that our framework has low spatial and temporal deviation for detected events.

The rest of this paper is constructed as follows. Section 2 reviews the previous works. In Section 3, we formally define the local event detection problem. We introduce the detailed methodology and our system framework in Section 4. We analyze and discuss our experiment results in Section 5, and conclude our work in Section 6.

2   Related Work

There have been plenty of research regarding detecting events or news. They can be categorized according to several aspects, including types of events, data sources and methods [2].

Prior to detecting events from social media streams, [12][13][14][16] detect events from traditional media data. As a seminar work to this problem, [16] uses an infinite-state automaton to model the term frequency in documents, and considers the burst, for example, the significant change in term frequency, as potential events. [13][14] detect events by modeling feature burst with spectral analysis and Gaussian mixture respectively. [12] heuristically identifies bursty terms and then groups these terms to discover potential events. Since all of them use information which has been existing for long time as data sources, their systems can hardly produce events in real time.

The introduction of social media platform brings new opportunities and challenges to the problem of event detection, and plenty of methods have been proposed. Inspired by the idea of detecting bursty feature, EDCoW [33] uses wavelet theory to model the signal of words and capture their bursts to detect events. [28] monitors bursty topics instead of unigram or tweet-segments [21]. Similarly, [7] detects events by discovering trending topics. Modeling trending topics can produce real-time detection, however, it is not applicable to our scenario of detecting and locating events since trending topics are usually weak signals for small scale event, and it has large error to estimate the location of trending topics [1][15][17]. Other than detecting the trending topics, based on influential theories of emotions, [32] automatically assigns a single tweet with an emotional label which is neutral or comes from one of the 6 Ekman's emotions. Then they monitor the sudden change of tweets' emotions in countries as the signals to detect events. All of the above works consider burst of certain features as signals of potential events. They model different bursty features including n-gram, terms, topics and emotions, and the common idea behind is absorbed into our framework.

Some detection frameworks are specific-event driven, that is, assigning a specific event type to each detection task. TEDAS [22] was proposed to detect crime and disaster related events from twitter stream. Earthquake center and typhoon trajectory have been successfully estimated in [30]. Besides disasters, [11][20] use twitter posts to detect local festivals by monitoring the movements of crowds. Twitterstand [31] classifies tweets as news and non-news to detect news events. Different from these methods, our proposed framework is not restricted to any event type.

Considering the data source, most of the previous works collect data from Twitter posts [11][20][22]. We put two data collectors in Instagram and Twitter monitoring and collecting useful information from the live post streams from these two social media platforms. Our previous work [35] uses Instagram posts to detect events with high accuracy. Unlike them, in this paper we use the posts from both of the two popular OSNs together to detect events.

After detecting events, retrieving relevant content to represent existing events is a challenging problem. Focusing on Twitter content, [3][5] extracts tweets and topics for known events. [25] generates a journalistic summary of a sport event using status updates from Twitter. Including Twitter, [4] retrieves social media content across YouTube and Flickr for existing events. In [9], photo tags are used to detect events and then retrieve photos based on tags to represent an event. It does not use the rich image content but heavily rely on user generated tags that are not always reliable [27]. Although the work in [27] combines photo content and tags to detect events, it needs to discover landmarks first and then detect events around the landmarks. To be different, our work does not rely on landmark discovery, thus we can detect more general events. As to retrieving images, most existing methods rank images based on certain similarity measurements to a specific query, a keyword or image. In large scale applications, approximate nearest neighbor algorithms and hashing method [18][24] are widely explored. However, these methods are not directly applicable to our problem since we do not have a specific query. Instead, our query is an entire event that consists of noisy photos, text and geolocations. In our system, we observe that for a true event, the images that are relevant to it usually share common patterns. While other irrelevant images are considered as noise, which are usually independent and randomly distributed.

3   Problem Definition

Following [32][35], we define an event as *a real world activity that occurs during time period T within a geographical area L*. To detect such events in real time, we define a framework as follows. The framework takes the real-time streams of posts from Instagram and Twitter as the input. The system is expected to output detected events in sequence. For each detected event, we extract its related content namely the set of related images, topics (a set of keywords), the estimated occurrence location, and the estimated occurrence time.

4   System Framework and Methodology

In this section we introduce our architecture, each component and methods. The system framework is shown in Figure 1. Given a fixed geographical region L from which we want to detect events in real time, first we distribute event sensors over the entire region. Each event sensor is designed to be independently responsible for discovering events in a single sub-region l. In other words, we divide the entire region, i.e. New York City in this paper, into k sub-regions, $L = \{l_1, ..., l_k\}$. For each sub-region l, we allocate an event sensor, which has three components, Event Signal Discovery, Event Signal Classification and Event Summarization. Although more advanced methods that divide an entire region to sub-regions according to topic distribution [1][17] or population density [20] might improve the overall performance, in this paper we do not focus on this problem, and we divide the entire New York City into $N \times M$ grids of equal size.

The architecture of our system is shown in figure 1. The Event Signal Discovery component takes the input of Instagram and Twitter data streams in real time and outputs candidate event signals. The Event Signal Classification component takes candidate event signals as input, extracts various features for them, and finally outputs event signals which are classified as true events. The Event Summarization component selects the most relevant content, including photos and text to represent the event. Besides, it produces the estimated occurrence location and time of the event.

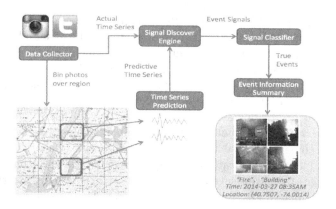

Fig. 1. Architecture of our system framework

4.1 Event Signal Discovery

The Event Signal Discovery component contains 3 sub-components, data stream collector, time series estimator and bursty detector. The motivation behind follows the general idea of modeling bursts [16] of certain features as potential events. Unlike other papers which model the sudden change of emotions [32], the movements of crowds [20] or the trending topics/terms/n-grams [21][33][7], we adopt the method in [34][35] which considers the abnormal increase of social media posts as the potential signal of events. This is because we observe that the change in the number of posts is sensitive to event occurrence, especially to the occurrence of small-scale local events.

The data stream collectors keep collecting posts from Twitter and Instagram in real time. We only collect posts containing geo-location information. In order to find bursty signals, an event sensor monitors the change in the number of posts in a sub-region l. A time series of the post number is constructed for l. We use t to denote the time, and $v_l(t)$ denotes the post number at l and within t. In practice, t stands for a time period and its window length in our experiments is 15 minutes.

The time series estimator is implemented by Gaussian Process Regressor (GPR) [29]. We use GPR due to its great performance in modeling various time series data such as stock prices [29]. Due to limited space, the detail of GPR is available in our previous work [35].

Once the GPR model is built on the historical data, it is able to predict the number of posts at l during t, as $\hat{v}_l(t)$ for any given t. When the data stream collector gathers the true number of posts in a sub-region l at t, we compare the actual number of posts, i.e. $v_l(t)$, with the predicted number of posts, i.e. $\hat{v}_l(t)$. If there is a large deviation between these two numbers, this signal is marked as a potential event signal. Following bursty detection, we are only interested in when the predicted number of posts is larger than the actual number of posts. Typically, we define an abnormality score as $[\hat{v}_l(t) - v_l(t)]/\hat{\sigma}(t)$. $\hat{\sigma}(t)$ is the predictive standard deviation given by GPR. It indicates the confidence of the prediction and a smaller $\hat{\sigma}(t)$ indicates stronger confidence. If the abnormality score in sub-region l during time t exceeds a given threshold, the Event Signal Discovery component outputs a candidate event $e(l,t)$ that stands for the set of all the Instagram and Twitter posts that are posted during time t and within location l.

4.2 Event Signal Classification

Once Event Signal Discovery component produces a candidate event signal $e(l,t)$, the Event Signal Classification component first extracts features from $e(l,t)$ and classifies it as true or false by a supervised learning model. Since a candidate event signal (shortened as candidate event) $e(l,t)$ is a set of Instagram and Twitter posts bounded by location l and time t, we can extract various types of features from them. Based on these features, the classifier determines whether $e(l,t)$ represents a true event or not. Note that, even if there is an event at location l and time t, not all the posts in $e(l,t)$ is related to that event. Thus we will choose relevant posts to represent the event which is discussed in Section 4.3. At this step, we focus on extracting robust features from the Instagram and Twitter post streams.

Feature Fusion. Before design specific features for candidate events, we first model the fusion of Instagram and Twitter posts. We previously assume when the number of total posts (including Instagram and Twitter) bounded by location l and time t suddenly increases, some event $e(l,t)$ may happen. However, we do not know which data source, Instagram or Twitter, records this event, or both. This is caused by the heterogeneity of Instagram and Twitter posts and users. Although they are both popular social media, their users have different habits and interests. Instagram is more about recording personal life and daily activity while Twitter is considered as an influential news media [19]. Thus, it is expected that some events are recorded by only one data source while some are recorded by both. We can either extract features from Instagram posts or Twitter posts only, or from both of them. In this paper, we consider two methods to fuse two data sources for feature extraction and classification.

The first fusion method is to integrate Instagram and Twitter posts at data level, i.e. before feature extraction. In this way, we need to consider a Twitter post and a Instagram post as homogeneous. For each event signal $e(l,t)$, we extract its features vector $\mathbf{x}_e$ from all the posts during time period t within location l. This method mitigates the sparsity problem of geo-tagged posts, and it is expected to benefit the classification of small-scale events with a few of posts in total.

The second method is to integrate Instagram and Twitter posts at feature level, i.e. after separate feature extraction. We extract feature vector $\mathbf{x}_e^I$ from Instagram posts and extract feature vector $\mathbf{x}_e^T$ from Twitter posts respectively, and then concatenate them to form the final feature vector $\mathbf{x}_e$. Note that by this method, the size of feature vector $\mathbf{x}_e$ is nearly doubled compared to the first method. The benefit of this method is that, we can extract different features from Twitter and Instagram, and further incorporate other inhomogeneous data sources.

Feature Extraction. To represent an event signal $e(l, t)$, we extract four types of features from all the posts bounded by l and t, namely topic features, emotional features, spatial features and social features. Formally, we use $P_e = \{p_1, ..., p_n\}$ to denote the set of posts associated with the event $e(l, t)$ and $n = |P_e|$. Note that, here we do not extract feature from a single post, instead, we extract features from the set of posts P_e associated to event signal $e(l, t)$.

First we extract five topic features from posts' text, i.e. photo captions and/or tweets. We first build a background topic distribution θ_B for location l. We use word unigram language model to represent the topic distribution of the background posts, i.e. all the posts during last 24 hours within l. We also build the event topic distribution θ_E for all the posts in P_e in the same way. We calculate (1) the total number of words that are in θ_E but missing in θ_B. A novel word, which has never appeared before, may indicate something new. We calculate (2) the average KL-divergence $[KL(\theta_B || \theta_E) + KL(\theta_E || \theta_B)]/2$. We expect that the topic distribution changes when there are true events. We also compute (3) average number of hashtags in p_i, (4) the average text length of p_i and (5) the average frequency of the 3 most frequent words in P_e.

The second type is emotional features. Inspired by [32] in which the authors experimentally prove when there are large event occurring, user emotions on Twitter change. In order to capture emotional changes, we compute the number of emotion-related punctuations and words from P_e: (1-2) the number of exclamations and question marks respectively and (3-8) the total number of words from P_e categorized to each of the six Ekman's emotions [32] respectively. Similar to topic features, we construct a background emotion-related word and punctuation count vector E_B, and take the deviation between E_B and the (1-8) features as the (9-16) features, indicating the change of emotion with location l and time t.

The third type is geolocation features. They are (1) mean and (2) standard deviation of pairwise post geo-distance, i.e. $dist(p_i, p_j) \forall p_i, p_j \in P_e$; and (3) the entropy [35] of the spatial distribution of all posts in P_e. The intuition behind these features is that we observe that when there is an event, the event-related posts tend to form a cluster. Similarly, we also compute these features from the background posts, and take the corresponding difference from (1-3) features as (4-6) features.

The fourth class includes a social feature. We compute the average number of mentioned users, i.e. @Alex of all posts in P_e. We finally extract 28 features of four categories in total. We also normalize the topic and emotional features by text length.

Fig. 2. Five sampled Instagram photos from a detected Knicks NBA game event in NYC. From journalists' perspective, the first three images are considered representative to summarize the event. Although the last two images were uploaded at the stadium and their captions are also about game, they are not informative for describing this event. Typically, the forth photo has a user privacy issue.

4.3   Event Summarization

In this section, we introduce our methods to summarize a detected event. A candidate event signal that consists of a number of Instagram and Twitter posts bounded by time period t and location l is classified as a true event or not. Provided that the classifier in Section 4.2 judges an event signal is a true event, we still do not know what the event is, a concert or a car crash, because the event classifier in this framework is designed to be general, i.e. independent of event type or scale.

Therefore we summarize an event from 4 aspects: topics, photos, occurrence location and occurrence time. Extracting topics from user generated posts is well studied [3][4][5][8]. Thus it is not our focus in this paper, and we use existing methods to select keywords from tweets and photo captions as the topics of an event. Besides, it is straightforward to estimate the occurrence location and time of an event in our framework. Due to their simplicity, the methods are discussed together with performance in the experiments, Section 5.4. Here we only cover the method we proposed to retrieve relevant photos for a detected event. Since tweets are seldom associated with photos, we only retrieve photos from Instagram posts to represent an event.

We need to retrieve photos because not all the Instagram post bounded by time period t and Location l are related to that event. For example, an Instagram photo was uploaded near a fire accident event, but the image content is about beers. Besides, we observe that users frequently upload self-portraits or food in events, which are not helpful for other users to understand the event. Moreover, some photos involve user privacy issue as shown in Figure 2. Therefore, we need to select relevant and representative photos to visually summarize an event. For simplicity, we name them **event-related** photos.

Our proposed method is based on the following observations. We observe that event-related photos usually share similar image content. For example, photos related to a fire accident usually record smoke, fire or the police. We also observe similarity of text associated to event-related photos. For example, users are likely to use "fire" or "smoke" to describe the photo related to a fire. Besides image content and text, most events occur in a fixed place, such as NBA matches, thus event-related photos tend to geographically form a cluster in the event center. Different from event-related photos, we observe that, for most noisy photos, their image content and text share very limited similarity to each other, and they are not necessarily close to the geographic center. However, we also observe outliers. For example, a user uploads a photo of her food during a NBA game and write a caption "A wonderful NBA game! # Knicks!" to the photo. In this example, when we compute the relevance score of the photo by only considering its geolocation and/or text, we find many popular text algorithms consider this photo highly relevant to the event. But when we compute its relevance score based on image content, the relevance diminishes.

Inspired by the above observations, for each photo x in an event e, we individually compute the image content relevance score (to the event e) given the image content of x only, as $s_c(x, e)$, the text relevance score given the text of x only, as $s_t(x, e)$, and the geolocation relevance score given the geolocation of x only, as $s_l(x, e)$. Note that, here P_e denotes the set of photos associated to event e, and thus $x \in P_e$. After that, we linearly combine the three individual relevance scores into the finalized relevance score $s(x, e)$ in Eq (1) that denotes how the photo x is overall relevant to the event e.

$$s(x, e) = a_c s_c(x, e) + a_t s_t(x, e) + a_l s_l(x, e) \tag{1}$$

a_c, a_t and a_l are the weights for the three independent relevance scores. Conventionally, we specify $a_c + a_t + a_l = 1$ and $a_t, a_c, a_l \geq 0$. The weights are the marginal effects of individual relevance score contributed to the overall relevance score. Intuitively, the larger a weight is, the larger positive impact that the corresponding single relevance score has on selecting event-related photos. Since we model retrieving event-related photos as an unsupervised ranking problem, the choices of a_c, a_t and a_l are discussed through experiments. We introduce the models to compute the three individual relevance scores as follows.

Image Content Relevance Model. To compute the image relevance score function $s_c(x, e)$, we use color histogram and GIST features [26] as image descriptor. These two image features are known for effectively describing discriminative scene characteristics. Our image relevance ranking method is adapted from an unsupervised image outlier removal method [23].

For a detected event e, its corresponding posts set is denoted as $P_e = \{\mathbf{x}_i \| \mathbf{x}_i \in \mathbb{R}^d, i = 1, 2, ..., n\}$. Since each post is always associated with an image, here we use the same notation for a post (x) and its image ($\mathbf{x}$ in vector space). For each image $\mathbf{x}_i$, we learn a scoring function $s(\mathbf{x}_i)$ to manifest its relevance to the event:

$$\min_{s \in \mathcal{H}, y_i \in \{t^+, t^-\}} \sum_{i=1}^{n} (s(\mathbf{x}_i) - y_i)^2 + \alpha s^T L s - \frac{2\beta}{n - n^-} \sum_{y_i > 0} s(\mathbf{x}_i) \tag{2}$$

Note that, the value of $s(\mathbf{x}_i)$ is exactly the image relevance score $s_c(x, e)$ in Eq (1). n^- is the number of posts which is considered as irrelevant to the events,

L is the graph Laplacian matrix, computed from the k nearest neighbor graph. We construct the neighborhood graph G by defining the affinity matrix W as:

$$W_{ij} = \begin{cases} exp(-\frac{dist(x_i,x_j)}{\sigma^2}), & if \ \mathbf{x}_i \in \mathcal{N}_k(\mathbf{x}_j) \ or \ \mathbf{x}_j \in \mathcal{N}_k(\mathbf{x}_i) \\ 0 \ , otherwise \end{cases} \quad (3)$$

$dist(,)$ is the Euclidean distance, $N_k(\mathbf{x}_i)$ is the set of k-nearest neighbors of $\mathbf{x}_i$, and σ is the bandwidth parameter. L in Eq (2) is the graph Laplacian matrix of G, computed as $L = D - W$, where D is a diagonal matrix with diagonal elements defined as $D_{ii} = \sum_{j=1}^{n} W_{ij}$. α and β are two model parameters balancing the regularization of graph Laplacian term and the effect of pushing the average positive example away from the margin.

During optimization, we do not have any label supervision on whether an image is event-related or not, therefore we are essentially solving an unsupervised learning problem: y_i is unknown in our optimization problem. As suggested by Eq (2), we treat y_i as a variable which is softly labeled as t^+ or t^- during optimization. Following the experiment results in [23], we dynamically set (t^+, t^-) as $(\sqrt{\frac{n^-}{n-n^-}}, \sqrt{\frac{n-n^-}{n^-}})$, where n^- is updated in each iteration. Eq (2) is minimized by alternating optimization: iterating between fixing y to minimize s and fixing s to minimize y, until convergence. The first subproblem, fixing y to minimize s is achieved by solving a constrained eigenvalue problem with a closed form solution. The other subproblem, fixing s to minimize y, is achieved via sorting and sweeping cut to find an optimal threshold. Throughout our experiment, the similarity between any two posts is measured in Gaussian kernel space. Finally, we use the score $s(\mathbf{x}_i)$ as the image content relevance score for images $\mathbf{x}_i$, i.e. $s_c(x, e)$.

Text and Relevance Model. We directly use the method in [3][5] to compute the relevance score of a photo's text to the event. For each photo's text, we represent it by a character n-gram language model where each photo's text is converted to a large and sparse vector. Then we compute the textual centroid $\mathbf{c}_t$ of these photos as $\mathbf{c}_t = \frac{1}{n}\sum_{i=1}^{n} \mathbf{x}_i$ where $\mathbf{x}_i$ denotes the i-th photo's text (in vector space) of the event. According to [3][5], the text relevance of a photo $\mathbf{x}$ to the event e could be computed by the closeness of $\mathbf{x}$ to the centroid $\mathbf{c}_t$. Thus, we compute $s_t(\mathbf{x}, e)$ as the cosine similarity between $\mathbf{x}$ and $\mathbf{c}_t$.

Geolocation Relevance Model. Each photo is associated with a coordinate (u, v) which denotes its latitude and longitude respectively. Similarly, we compute the geographical centroid c_l of the event as $(\frac{1}{n}\sum_{i=1}^{n} u_i, \frac{1}{n}\sum_{i=1}^{n} v_i)$ where u_i and v_i respectively denote the latitude and longitude of the i-th photo of the event. Thus, the geolocation relevance score $s_l(x, e)$ can be computed by the earth surface distance between x and c_l.

Interpretations and Advantages. In the above method, we extract image, text and geolocation features from a photo and the event to compute the three relevance scores separately, and finally linearly combine them into the overall relevance score. An alternative method is not to compute the individual relevance scores separately. Instead, it extracts the same image, text and geolocation features but concatenates all the three feature vectors into a longer feature vector, and then apply a unified model to assign relevance score to the photo given its event. However, such a unified model has problems caused by the heterogeneous characteristics of image, text and geolocation information. First, the dimensions of the three feature vectors are largely different. An efficient text representation is n-gram model which transfers a piece of text to a large and sparse feature vector. However, the geolocation information is efficiently represented as a feature vector in two dimensions only. Thus, if we just simply concatenate them without robust feature selection, the geolocation information is easily overwhelmed in such feature space. Second, the hypotheses of relevance (or similarity or closeness) are semantically different in the three aspects. In modeling the relevance of image content, many previous researches find the similarity between photos defined in Gaussian kernel space is proper. While in modeling geolocation closeness, the earth surface distance is naturally the best. Thus, it is not ideal to model all the three types of information in a unified distance space.

5   Experiments

In this section, we first introduce the dataset and parameter setting. Then we evaluate event detection accuracy by Instagram and Twitter post streams. We also evaluate the event-related photo retrieval. Finally, we sample detected true events to evaluate the temporal latency and spatial accuracy by case studies.

5.1   Dataset and Setting

We use Twitter APIs and Instagram APIs to crawl geo-tagged posts in New York City. Each crawled Instagram post (shortened as photo) is associated with an image, a text, a pair of coordinates, created time and other information. Each crawled tweet is associated with a non-empty text, a pair of coordinates, created time and other information. From 2012-12 to 2014-06, we collected $12,453,448$ geo-tagged tweets and $31,188,195$ geo-tagged photos.

Event Classification Annotation. We use crowdsourcing to accomplish this labeling task: given an event signal $e = (l_e, t_e)$ and its associated posts, it is labeled based on whether there is a true event during time period t_e within location l_e. We first used Amazon Mechanical Turk to label the discovered event signals and then invited three journalists from a local newsroom in New York city to calibrate the labeling to guarantee our dataset is as correct as possible. We sampled 1945 events signals with associated posts to label. As a result, we get 1084 events signals with valid and confident labeling. Among them, 477 events signals are labeled as true events while the other 607 are labeled as false events (noise).

Image Relevance Annotation. To evaluate our proposed relevant photo retrieving method, we randomly select 153 true events which contain at least 8 photos[2]. For each event, we label all its photos (if the number of photos in an event exceeds 35, we sample 35 photos). We tried to use a third-party crowdsourcing to label, but find out their correctness is largely below our expectation. Therefore we trained an independent user, and ask the user to label whether a photo is relevant to a given event based on these criteria, 1 for relevant, 0.5 for partially relevant and 0 for irrelevant. For example, we consider self-portraits and food as irrelevant. We also give the user the location and time information and topics of the event to facilitate labeling. On average, an event has 24.7 photos, and 39.4% of its photos are labeled as relevant, 5.8% are partially relevant and the rest, i.e. 54.8% are irrelevant.

Parameter Setting. To monitor the entire New York City, we divide NYC into $25 * 25$ sub-regions (0.45 square kilometers for each geo-region). We also turned the window size t in Gaussian Process Regressor to be 15 minutes. A reasonably long time interval will lead to large detecting latency while a tiny interval will cause the decrease of the detection accuracy since there may be very few posts during a tiny time interval. The experiments on the choice of these parameters are in our previous work [35].

5.2   Detection Accuracy

In this section, we evaluate the performance of Event Signal Classification. Before extract text-related features, we preprocess posts' text by NLTK [6]. We remove stopwords, non-English characters and urls. We also separate capitalized and concatenated words, such as from "ILoveThisGame" to "I", "love", "this" and "game". Then we use 10-fold cross-validation to evaluate the effectiveness of feature extraction and fusion with standard classifiers. To avoid the variance caused by different classifiers, we run all the experiments with three popular and representative supervised classifiers, Support Vector Machine (SVM), Logistic Regression (LR) and Random Forest (RF).

We show the evaluation (on test data) to the event signal classifiers with different settings in Table 1. In the setting of the Instagram-only method, we discard all Twitter posts. We only extract features from Instagram posts and train all the three classifiers with Instagram data. Then we discard all Instagram posts but extract features and train the classifiers from Twitter data only, as the Twitter-only method. From Table 1, we can find that if we just use a single data source to classify the candidate events, Instagram data outperforms Twitter data. Furthermore, we evaluate the event classifiers on integrated data with two fusion methods. We find that the classifiers trained with the two fusion methods, no matter in data-level or feature-level, both outperform the classifiers trained on a single data source. We investigate results in detail and conclude this

---

[2] In some special cases, we find there is more than one true event simultaneously recorded in a candidate event signal. We consider them as true events but do not label their photos to avoid ambiguity.

Table 1. Event Signal Classification Results

Features	Classifier	Precision	Recall	F score	Overall Accuracy
	SVM	0.833	0.740	0.784	82.01%
Instagram-only	LR	0.855	0.719	0.781	82.28%
	RF	0.793	0.780	0.786	81.36%
	SVM	0.845	0.675	0.751	80.25%
Twitter-only	LR	0.815	0.681	0.742	79.15%
	RF	0.761	0.719	0.739	77.67%
	SVM	0.876	0.755	0.811	84.50%
Data-level Fusion	LR	0.866	0.759	0.809	84.22%
	RF	0.830	**0.849**	**0.839**	**85.70%**
	SVM	**0.883**	0.746	0.809	84.50%
Feature-level Fusion	LR	0.856	0.774	0.813	84.31%
	RF	0.835	0.836	0.836	85.51%

improvement is caused by the following reasons. First, although we have plenty of geo-tagged posts, in certain sub-regions, we still encounter severe data sparsity problem. Either fusion method brings us more valuable data to mitigate this problem. Second, small-scale events whose weak signals are easily overwhelmed in noisy content. But when we find the weak signals in both data sources, our system are more confident to detect them. Due to the limited length, the evaluation of feature importance is not included. In short, by investigating the weights in Logistic Regression, topic and spatial features are far more discriminative than the emotional and social features.

5.3 Relevant Photo Retrieval

To evaluate the efficiency of the relevant photo retreiving method in Eq (1), we use Normalized Discounted Cumulative Gain (NDCG@k) in Eq (4) as the metric. For each event, we rank its photos decreasingly by overall relevance scores in Eq (1), and then compute NDCG@k for the ranking.

$$NDCG_k = \frac{1}{z_n} \sum_{i=1}^{k} \frac{2^{r_i} - 1}{log_2(i+1)} \qquad (4)$$

z_n is a normalization factor. r_i is the actual relevance score of the ranked i-th photo, and it is given by our labeler, 1, 0.5 or 0. k is a free parameter to control the number of ranked photos to compute NDCG@k. To reduce the variance caused by k, we compute the NDCG@k for k from 1 to 10. We compare the combined relevance model in Eq (1) with three baselines, image content relevance model, text relevance model and geolocation relevance model introduced in Section 4.3. As shown in Table 2, we have the following observations. First, among all the three baselines, the relevance model based on text information works the best. Second, the combination of all the three single relevance models constantly performance better than any of the three single relevance models. Third, by grid search, we empirically find that around $(a_c, a_t, a_l) = (\frac{1}{4}, \frac{1}{2}, \frac{1}{4})$,

Table 2. NDCG@k of Relevance Models over k

Relevance Model	NDCG@k									
	k=1	2	3	4	5	6	7	8	9	10
Image Content	0.517	0.505	0.495	0.486	0.476	0.465	0.459	0.451	0.445	0.439
Text	0.516	0.547	0.558	0.563	0.572	0.568	0.565	0.562	0.556	0.548
Geolocation	0.334	0.342	0.355	0.366	0.383	0.393	0.402	0.404	0.406	0.405
Combined	**0.652**	**0.656**	**0.642**	**0.639**	**0.629**	**0.622**	**0.612**	**0.598**	**0.593**	**0.582**

the combined relevance score reaches the maximal on our labeled dataset. This implies the importance of each factor's contribution to the overall relevance.

5.4 Spatial and Temporal Deviation

In this paper we focus on local event detection in real time, thus we also evaluate the detecting deviation of spatial and temporal factors of events. The detecting deviation of the spatial factor of an event is the geographical distance between the coordinates where the event actually occurred and the coordinates our framework estimated for the event. Similarly, the detecting deviation of the temporal factor of an event is the time period between when the event actually occurred and the time our framework estimated for the event. However, since it is expensive to manually collect accurate spatial and temporal information of an event, we choose 20 events to evaluate their spatial deviation, and 5 events to evaluate their temporal deviation as case study.

We first evaluate spatial deviation of detected events. Many events are held dynamically in a wide region, e.g. New Year parade in China town and marathon, thus we are unable to track all the areas associated with that event. Therefore we only consider events that take place in a fixed area, such as fire accident and basketball games. For each event, we find the name of the associated place, and then take the coordinates of the associated place from Google Maps as the actual event coordinates, in a pair of longitude and latitude. On the other hand, our system calculates the geographic center of of all Instagram and Twitter posts related to that event, as the estimated coordinates of the event. More specifically, the estimated longitude is the arithmetic mean of the longitudes of all related posts, the same for estimated latitude. Then a spatial deviation, i.e. spherical distance, is calculated between the estimated coordinates and the actual coordinates of an event. On average the estimated coordinates of these 20 events are 104.46 meters far from their actual coordinates with a standard deviation of 37.75. Table 3 shows the results for 5 example events.

Similarly, here we evaluate the temporal deviation of detected events. To acquire the exact knowledge of when events occurred, we manually check with websites, the police or news reports for their actual occurrence time. Meanwhile, we take the time of the earliest post among all posts related to that event as the estimated time. We report the time interval between actual time and estimated time of an event as its temporal deviation. Table 4 shows the results of 5 detected

Table 3. Spatial Deviation of Detected Events

Event Name	Actual Location (lat, lon)	Deviation(m)
NBA Knicks Game	40.750733, -73.992743	63.1
Trey Songz Concert	40.751222, -73.994749	109.7
Christmas Eve	40.758250, 73.981217	173.1
Fire Accident	40.750369, 73.992726	106.2
Car Crash	40.739061, -74.001488	78.1

Table 4. Temporal Deviation of Detected Events

Event Name	Date	Estimated Time	Actual Time	Time Interval
NBA Knicks game	Nov 30 2012	19:11pm	19:30pm	19 mins **in advance**
Boxing Cotto vs Trout	Dec 01 2012	20:07pm	21:00pm	53 mins **in advance**
Fire in West Village	Nov 17, 2013	10:40am	10:26am	14 mins later
Fire in 34st	Mar 27, 2014	08:44am	08:35am	9 mins later
Car Crash	Feb 12 2014	08:26am	05:45am	3 hours later

events as examples. We can find that "NBA Knick Game" and "Boxing: Cotto vs Trout", which are two planned events, are detected prior to the actual event time. This is because as more people arrived to the stadium in advance and started to post about the coming events, our system detected the local unusual increasing trends before the game actually started. For "Fire in West Village", "Fire in 34 St", which are two emergencies, our event responded 14 minutes and 9 minutes after the events happened respectively. Notice that for the "Car Crash" event, our system responded 3 hours later. This failure is probably because it happened at 5:45AM, when most of local residents were still sleeping. In this case, few related posts can be detected at the early stage of this event.

6 Conclusion and Future Work

In this paper, we proposed a general framework for real-time event detection from Instagram and Twitter post streams. Our proposed system uses three components to discover and classify the events. Then we can extract high-level knowledge from detected events. Extensive experiments on NYC social media data show the promising results. Based on our general framework, a lot of future work can be investigated to potentially boost the performance. For example, we plan to further study how to adaptivity divide sub-regions in the city based on their topic distributions [10]. Also, more sophisticated feature fusion approaches for event knowledge extraction can be investigated.

Acknowledgments. This work was partially supported by a Magic Grant from the Brown Institute for Media Innovation, and the National Science Foundation grants Numbers 1054177 and 1017845. Any opinions, findings, and conclusions or recommendations expressed in this material are those of the author(s) and do not necessarily reflect the views of the National Science Foundation.

References

1. Ahmed, A., Hong, L., Smola, A.J.: Hierarchical geographical modeling of user locations from social media posts. In: WWW, pp. 25–36 (2013)
2. Atefeh, F., Khreich, W.: A survey of techniques for event detection in twitter. Computational Intelligence (2013)
3. Becker, H., Chen, F., Iter, D., Naaman, M., Gravano, L.: Automatic identification and presentation of twitter content for planned events. In: ICWSM (2011)
4. Becker, H., Iter, D., Naaman, M., Gravano, L.: Identifying content for planned events across social media sites. In: WSDM, pp. 533–542 (2012)
5. Becker, H., Naaman, M., Gravano, L.: Selecting quality twitter content for events. In: ICWSM (2011)
6. Bird, S.: Nltk: The natural language toolkit. In: Proceedings of the ACL Workshop on Effective Tools and Methodologies for Teaching Natural Language Processing and Computational Linguistics (2002)
7. Budak, C., Georgiou, T., El Abbadi, D.A.A.: Geoscope: Online detection of geo-correlated information trends in social networks. In: Proceedings of the VLDB Endowment (2013)
8. Chakrabarti, D., Punera, K.: Event summarization using tweets. In: ICWSM (2011)
9. Chen, L., Roy, A.: Event detection from flickr data through wavelet-based spatial analysis. In: CIKM, pp. 523–532 (2009)
10. Flatow, D., Naaman, M., Xie, K.E., Volkovich, Y., Kanza, Y.: On the accuracy of hyper-local geotagging of social media content. In: WSDM (2015)
11. Fujisaka, T., Lee, R., Sumiya, K.: Discovery of user behavior patterns from geo-tagged micro-blogs. In: Proceedings of the 4th International Conference on Uniq-uitous Information Management and Communication, ICUIMC, pp. 36:1–36:10 (2010)
12. Fung, G.P.C., Yu, J.X., Yu, P.S., Lu, H.: Parameter free bursty events detection in text streams. In: VLDB, pp. 181–192 (2005)
13. He, Q., Chang, K., Lim, E.-P.: Analyzing feature trajectories for event detection. In: SIGIR, pp. 207–214 (2007)
14. He, Q., Chang, K., Lim, E.-P., Zhang, J.: Bursty feature representation for clus-tering text streams. In: SDM (2007)
15. Hong, L., Ahmed, A., Gurumurthy, S., Smola, A.J., Tsioutsiouliklis, K.: Discover-ing geographical topics in the twitter stream. In: WWW, pp. 769–778 (2012)
16. Kleinberg, J.: Bursty and hierarchical structure in streams. In: KDD, pp. 91–101 (2002)
17. Kling, C.C., Kunegis, J., Sizov, S., Staab, S.: Detecting non-gaussian geographical topics in tagged photo collections. In: WSDM (2014)
18. Kulis, B., Grauman, K.: Kernelized locality-sensitive hashing for scalable image search. In: ICCV, pp. 2130–2137 (2009)
19. Kwak, H., Lee, C., Park, H., Moon, S.: What is twitter, a social network or a news media? In: WWW, pp. 591–600 (2010)
20. Lee, R., Wakamiya, S., Sumiya, K.: Discovery of unusual regional social activities using geo-tagged microblogs. World Wide Web 14(4), 321–349 (2011)
21. Li, C., Sun, A., Datta, A.: Twevent: Segment-based event detection from tweets. In: CIKM, pp. 155–164 (2012)
22. Li, R., Lei, K.H., Khadiwala, R., Chang, K.C.-C.: Tedas: a twitter based event detection and analysis system. In: ICDE, pp. 1273–1276 (2012)

23. Liu, W., Hua, G., Smith, J.R.: Unsupervised one-class learning for automatic outlier removal. In: CVPR (2014)
24. Liu, W., Wang, J., Kumar, S., Chang, S.-F.: Hashing with graphs. In: ICML, pp. 1–8 (2011)
25. Nichols, J., Mahmud, J., Drews, C.: Summarizing sporting events using twitter. In: Proceedings of the 2012 ACM International Conference on Intelligent User Interfaces, pp. 189–198 (2012)
26. Oliva, A., Torralba, A.: Modeling the shape of the scene: A holistic representation of the spatial envelope. International Journal of Computer Vision **42**, 145–175 (2001)
27. Papadopoulos, S., Zigkolis, C., Kompatsiaris, Y., Vakali, A.: Cluster-based landmark and event detection on tagged photo collections. IEEE Multimedia (2010)
28. Parikh, R., Karlapalem, K.: Et: Events from tweets. In: WWW Companion, pp. 613–620 (2013)
29. Rasmussen, C.E., Williams, C.K.I.: Gaussian Processes for Machine Learning (Adaptive Computation and Machine Learning) (2005)
30. Sakaki, T., Okazaki, M., Matsuo, Y.: Earthquake shakes twitter users: real-time event detection by social sensors. In: WWW, pp. 851–860 (2010)
31. Sankaranarayanan, J., Samet, H., Teitler, B.E., Lieberman, M.D., Sperling, J.: Twitterstand: news in tweets. In: Proceedings of the 17th ACM SIGSPATIAL International Conference on Advances in Geographic Information Systems, pp. 42–51. ACM (2009)
32. Valkanas, G., Gunopulos, D.: How the live web feels about events. In: CIKM, pp. 639–648 (2013)
33. Weng, J., Lee, B.-S.: Event detection in twitter. In: ICWSM (2011)
34. Xia, C., Schwartz, R., Xie, K., Krebs, A., Langdon, A., Ting, J., Naaman, M.: Citybeat: Real-time social media visualization of hyper-local city data. In: WWW Companion, pp. 167–170 (2014)
35. Xie, K., Xia, C., Grinberg, N., Schwartz, R., Naaman, M.: Robust detection of hyper-local events from geotagged social media data. In: Proceedings of the 13th Workshop on Multimedia Data Mining in KDD (2013)

Link Prediction in Aligned Heterogeneous Networks

Fangbing Liu[1,2] and Shu-Tao Xia[1,2(✉)]

[1] Graduate School at Shenzhen, Tsinghua University, Beijing, China
[2] Tsinghua National Laboratory for Information Science and Technology,
Beijing, China
lfb13@mails.tsinghua.edu.cn, xiast@sz.tsinghua.edu.cn

Abstract. Social networks develop rapidly and often contain heterogeneous information. When users join a new social network, recommendation affects their first impressions on this social network. Therefore link prediction for new users is significant. However, due to the lack of sufficient active data of new users in the new social network (target network), link prediction often encounters the cold start problem. In this paper, we attempt to solve the user-user link prediction problem for new users by utilizing data in a similar social network (source network). In order to bridge the two networks, three categories of local features related to single edge and one category of global features associated with multiple edges are selected. The Aligned Factor Graph (*AFG*) model is proposed for prediction, and *Aligned Structure Algorithm* is used to reduce the factor graph scale and keep the prediction performance at the same time. Experiments on two real social networks, i.e., Twitter and Foursquare show that *AFG* model works well when users leave little data in target network.

Keywords: Link prediction · Heterogeneous network · Aligned factor graph model

1 Introduction

In recent years, Social networks have become part of our life. When users join a new social network, their first impressions are very important to keep them active in this network. Thus how to predict future links for new users according to the current snapshot of the network is significant.

Link prediction can be seen as a classification problem. A classifier trained with simple topology features such as the number of common neighbors and the Adamic/Adar measure can successfully identify missing links in social networks [1]. Weak ties and interaction activities can also be useful for inference [2,3].

This research is supported in part by the Major State Basic Research Development Program of China (973 Program, 2012CB315803), the National Natural Science Foundation of China (61371078), and the Research Fund for the Doctoral Program of Higher Education of China (20130002110051).

© Springer International Publishing Switzerland 2015
T. Cao et al. (Eds.): PAKDD 2015, Part I, LNAI 9077, pp. 33–44, 2015.
DOI: 10.1007/978-3-319-18038-0_3

Actually, nodes in a social network often have abundant attributes such as time and location [4]. In addition, geographic distance has been shown to play an important role in creating new social connections [5]. In [6,7], network structure and node attributes are used simultaneously to improve prediction performance.

Many of the current studies mainly focus on a single social network. However, sometimes data in one network is not sufficient to train a good classifier. In particular, when users join a new network (target network), link prediction will encounter the cold start problem [8]. But if we can use data from another network (source network), the prediction performance should be better intuitively. In general, there are two ways to utilize the source network to help prediction, one is based on transfer learning through different feature spaces and the other is based on the factor graph. In the transfer learning method, items having both features in source space and target space are utilized [9]. The factor graph method uses the phenomenon that different social networks obey common rules such as triad social balance and triad status balance [10,11].

The works [9–11] focus on information transfer between *two different types of networks*. A widespread phenomenon is that some social networks are similar to each other except for some specific services. Users often have accounts in multiple social networks to enjoy distinctive services. Networks connected by accounts of same users are aligned networks. Link prediction for new users in aligned networks is first discussed in [12]. Though rich features are used for training, the important fact that user-user relationships affect each other is ignored.

In this paper, we study the link prediction problem for new users in target network from a new perspective. And the aligned source network is utilized to solve the cold start problem. Our method can get good prediction performance estimated by Area Under Curve (Auc) and Accuracy (Acc). The contributions can be summarized as follows:

- Three categories of local features and one category of global features are selected, which describe the social networks accurately and reflect the edges interaction. These features play an important role in improving prediction performance.
- An Aligned Factor Graph (AFG) model is proposed to solve the link prediction problem for new users in target network, making full use of a similar source network. It performs well when we encounter the cold start situation. In addition, in order to control the scale of the factor graph and guarantee an efficient inference, *Aligned Sturcture Algorithm* is used in building model.
- Experiments on two real social networks - Twitter and Foursquare are carried out and results show that AFG model improves prediction performance by utilizing source network data when compared with the Basic Factor Graph (BFG) model. And AFG model performs better than $SCAN$-PS model [12].

This paper is organized as follows: Section 2 gives basic definitions and related works in link prediction. Meanwhile, BFG model is introduced. Section 3 is our prediction method. Aligned Factor Graph (AFG) model is proposed and the *Aligned Structure Algorithm* is demonstrated. Besides, the parameter learning algorithm and feature selection are discussed. Section 4 includes some experimental results and analysis. Section 5 is the conclusion.

2   Preliminaries and Related Works

Definition 1. *(Aligned Heterogeneous Networks [12]): Let V_i^k be the set of the same kind of nodes in network G_k, $V_k = \cup_i V_i^k$ is the set of different kinds of nodes. $E_k = \cup_i E_i^k$ is the set of different kinds of edges. Let f be a one-to-one mapping between user $u_i^s \in U_s \subseteq V_s$ in the source network and user $u_j^t \in U_t \subseteq V_t$ in the target network, if $\exists(F = \cup_{i,j} f(u_i^s, u_j^t)) \neq \emptyset$, then network $G_s = (V_s, E_s)$, $G_t = (V_t, E_t)$ are called aligned heterogeneous networks. The link (u_i^s, u_j^t) is called an anchor link and all these links form the set of anchor links E_A.*

Definition 2. *(Edge Descriptor): An edge e_{ij} can be described as $d_{ij} = (l_{ij}, p_{ij})$, where l_{ij} is the edge label belonging to $\{0, 1\}$, p_{ij} is the probability that e_{ij} having this label. $l_{ij} = 0$ means the edge does not exist.*

Definition 3. *(Triad Social Balance [14]): Undirected edges between three users form a triad. It is social balanced if three or one edge exists.*

Definition 4. *(Triad Status Balance [15]): Directed edges between three users form a triad. It is status balanced if three edges are not in a directed cycle.*

There are many works which use source network to help prediction in the target network. In [16], relationship prediction is studied under space feature transfer learning framework and inter-domain edges are enhanced by discovering new edges and strengthening existing ones. In [17,18], domain connection sparsity and data non-consistent problem are studied .

The prediction method based on factor graph concern triad features transfer between two different networks [10] and the *BFG* model [11] is used. Friend recommendation problem is solved by limiting friend candidates in two hops to keep factor graph in bearable scale [11]. And parameters of triad features are the same in source and target networks during training.

A factor graph [19] is defined as a bipartite graph containing variable nodes and factor nodes. In the *BFG* model, the user-user relationship e_{ij} between user u_i and u_j is mapped to a variable node v_{ij} in the factor graph, while variable nodes connecting to the same factor node reflect the interactive influence between relationships' formation. A simple explanation for *BFG* model is shown in Fig. 1.

3   Social Network Prediction

We try to solve the link prediction problem for new users in target network by utilizing data of aligned heterogeneous source network. Firstly, we extend the *BFG* model to the Aligned Factor Graph (*AFG*) model. Besides, the *Aligned Structure Algorithm* is used for controlling factor graph scale when building the model. Secondly, the parameter inference framework is proposed. Thirdly, a detailed parameter learning algorithm is studied. Fourthly, new user links are inferred by maximizing an objective function. At last, both local and global features used in prediction are given.

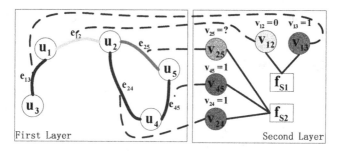

Fig. 1. BFG model. The first layer is observations, and the second layer is a factor graph. Each observation corresponds to a variable node. Black edge in the first layer means friend relationship exists between two users and the corresponding variable nodes's state is 1. Yellow edge indicates no friendship and variable node state is 0. Red edge represents unobserved relationship and variable node's state is ?, i.e., unknown.

3.1 The Aligned Factor Graph Model

The AFG model is also a two layer model. The first layer is composed of two observations deriving from source network G_s and target network G_t. The second layer is a factor graph containing two aligned parts $FG = \{FG_s \cup FG_t\}$. A more intuitive description of AFG model is shown in Fig. 2. The two networks in the first layer are fully aligned networks. Relationships between each pair of users are taken into consideration. Thus we can also find one-to-one mappings between variable nodes in FG_s and FG_t. Moreover, if variable node v_j^t's state is unknown in FG_t, the structure of v_j^t must be the same with the structure of the corresponding variable node v_i^s in FG_s. Local features belonging to variable nodes in the second layer can be got according to the corresponding edges' attributes in the first layer. Global features belonging to factor nodes are drawn from the edge cycles in the first layer, determining the factor graph structure.

Building AFG model efficiently is important for prediction. Firstly, the first layer observations can be got easily given G_s and G_t. Secondly, states of all variable nodes in the second layer are determined according to the observations. $state = 1$ and $state = 0$ variable nodes are state-known variable nodes while $state =?$ variable nodes belong to the state-unknown set. Thirdly, for combinations of state-known variable nodes satisfying global features defined in section 3.5, we build a factor node and connect it with the variable nodes in this combination. Fourthly, take the state-unknown variable nodes into consideration. If we build a factor node for each combination of variable nodes, the factor graph scale will be too large and the complexity will be too high. However, if we build a factor node and connect it with variable nodes randomly, the prediction performance will decrease. In this paper, Algorithm 1 is used to determine the accurate structures of state-unknown variable nodes.

3.2 Parameters Inference Framework

The first layer can be built given source network $G_s = (U_s, E_s, A_s)$ and target network $G_t = (U_t, E_t, A_t)$, where U_s, U_t are the sets of users, E_s, E_t are the sets

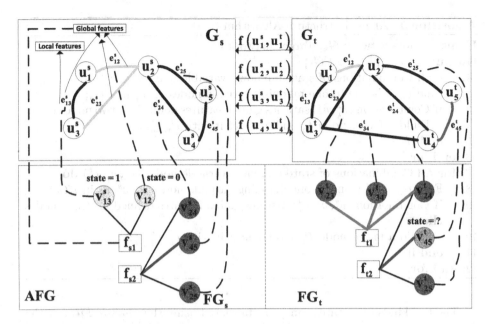

Fig. 2. AFG model. The second layer has an aligned structure. Edge color meanings are the same as Fig. 1. Red variable node v_{45}^t is connected to the factor node f_{t2} to keep aligned structure with FG_s as purple lines show. However, $v_{23}^t, v_{34}^t, v_{24}^t$ can have different structures from FG_s, shown in green lines, because their states are known. $v_{23}^s, v_{34}^s, v_{24}^s$ are not connected with a factor node in FG_s.

of user-user relationships, A_s, A_t are the sets of local attribute vectors belonging to edges. According to the observations in the first layer, a factor graph $FG = \{FG_s, FG_t\} = \{V_s, F_s, EF_s, V_t, F_t, EF_t\}$ in the second layer can be established, where V_s, V_t are the sets of variable nodes, F_s, F_t are the sets of factor nodes and EF_s, EF_t are the edge sets. In network G_t (so does G_s), each $e_{ij}^t \in E_t$ is associated with an attribute vector $a_{ij}^t \in A_t$ and is mapped to a variable node $v_{ij}^t \in FG_t$. e_{ij}^t has an edge descriptor $d_{ij}^t = (l_{ij}^t, p_{ij}^t)$ related to v_{ij}'s state and marginal probability. As all edge descriptors in G_s are known while only part of edge descriptors in G_t are known, the link prediction problem can be described as maximizing the following probability

$$P(D_t, D_s|G_s, G_t) = \prod_{ij} f_l(v_{ij}^s, a_{ij}^s) g_c(v_{ij}^s, G(v_{ij}^s)) \prod_{pq} f_l(v_{pq}^t, a_{pq}^t) g_c(v_{pq}^t, G(v_{pq}^t)) \quad (1)$$

where G_t is the target network, G_s is the source network. D_t and D_s are the sets of edge descriptors in G_t, G_s.

The state of a variable node is affected by two features

- $f_l(v_{ij}, a_{ij})$: local feature, it describes how local attributes influence the friend relationship formation between user u_i and u_j.
- $g_c(v_{ij}, G(v_{ij}))$: global feature, it describes how two or three edges interact in forming the relationship. $G(v_{ij})$ is the set of variable nodes connecting to the same factor node with v_{ij}.

Algorithm 1. Aligned Structure Algorithm

Input: Source network G_s, Target network G_t
Output: $FG = \{FG_s \cup FG_t\}$
1: **for all** Combinations of state-unknown variable nodes (v_p^t, v_q^t) **do**
2:    Find v_p^t, v_q^t's one-to-one mapping variable nodes v_i^s, v_j^s in FG_s;
3:    **if** Combination (v_i^s, v_j^s) satisfies global features defined for two nodes **then**
4:       Build a factor node f_n^t and connect it with v_p^t, v_q^t;
5:    **end if**
6: **end for**
7: **for all** Combinations of state-unknown variable nodes (v_p^t, v_q^t, v_r^t) **do**
8:    Find v_p^t, v_q^t, v_r^t's one-to-one mapping variable nodes v_i^s, v_j^s, v_k^s in FG_s;
9:    **if** Combination (v_i^s, v_j^s, v_k^s) satisfies global features defined for three nodes
      then
10:       Build a factor node f_n^t and connect it with v_p^t, v_q^t, v_r^t;
11:    **end if**
12: **end for**

The two kinds of features can be instantiated using the *Markov Field* or the *Bayesian Theory*. In this paper, the *Hammersley-Clifford Theorem* [20] is used and the two probabilities are defined as

$$f_l(v_{ij}, a_{ij}) \quad = \frac{1}{Z_1} \times \exp\{\sum_k \alpha_k r_k(a_{ij}^k)\} \tag{2}$$

$$g_c(v_{ij}, G(v_{ij})) = \frac{1}{Z_2} \times \exp\{\sum_c \sum_d \beta_d h_d(G(v_{ij}))\} \tag{3}$$

where Z_1, Z_2 are the normalization factors, k is the local attribute index, a_{ij}^k represents the kth attribute in attribute vector a_{ij}. $G(v_{ij})$ is the set of variable nodes concerning v_{ij} and $|G(v_{ij})| = c$. If three edges affect each other, then $c = 3$. r_k is the kth local feature function. For example, it can be a function calculating common neighbor number. h_d is the dth global feature function. For instance, if a triad is social balanced, $h_d = 1$. α_k, β_d are the weights of features.
Then the joint probability defined by Eq. (1) can be written as

$$P(D_t, D_s | G_s, G_t) = \frac{1}{Z} \times \prod_{ij} \prod_{pq} \exp\{\sum_k \alpha_k(r_k(a_{pq}^{kt}) +$$

$$r_k(a_{ij}^{ks})) + \sum_c \sum_d \beta_d\{h_d(G(v_{pq}^t)) + h_d(G(v_{ij}^t))\}\} \tag{4}$$

where Z is normalization factor. Thus, the source and target networks union objective function is

$$O(\theta) = \log P(D_t, D_s | G_s, G_t)$$

$$= \sum_k \alpha_k \{ \sum_{p=1}^{|U_{new}^t|} \sum_{q=1}^{|U_{all}^t|} (r_k(a_{pq}^{kt})) + \sum_{m=1}^{|U_{new}^s|} \sum_{n=1}^{|U_{all}^s|} (r_k(a_{mn}^{ks})) \} \tag{5}$$

$$+ \sum_c \sum_d \beta_d \{ \sum_{p=1}^{|U_{new}^t|} \sum_{q=1}^{|U_{all}^t|} h_d(G(v_{pq}^t)) + \sum_{m=1}^{|U_{new}^s|} \sum_{n=1}^{|U_{all}^s|} h_d(G(v_{mn}^s)) \} - \log Z$$

Algorithm 2. Learning Algorithm

Input: *Learning Rate* η
Output: *Model Parameters* θ
1: **repeat**
2:      Calculate $E_{p(D_{tu}|D_s,D_{tl},G_s,G_t)}(r_k(a_n^{st}))$, $E_{p(D_{tu}|D_{tl},D_s,G_s,G_t)}(h_d(G(v_m^{st})))$ using the LBP algorithm;
3:      Calculate $E_{p(D_s,D_t|G_s,G_t)}(r_k(a_n^{st}))$, $E_{p(D_t,D_s|G_s,G_t)}(h_d(G(v_m^{st})))$ using the LBP algorithm;
4:      Calculate gradient according to Eqs. (6) and (7);
5:      Update parameter set θ with learning rate
6:      $\theta_{new} = \theta_{old} - \eta \times \frac{\partial O(\theta)}{\partial \theta}$
7: **until** Converage
8: Output θ

---

where U_{new}^t is the set of new users, U_{all}^t is the set of all users in G_t (so does G_s). We try to find parameter set $\theta = (\alpha, \beta)$ that maximizing the objective function.

3.3 Learning Algorithm

In order to solve the objective function, the gradient decent algorithm is used. As Z is the normalization factor, all variable nodes' likelihoods in the factor graph need to be calculated including the state-unknown variable nodes. The gradients of parameters are calculated as follows

$$\frac{\partial O(\theta)}{\partial \alpha_k} = E_{p(D_{tu}|D_{tl},D_s,G_s,G_t)}(r_k(a_n^{st})) - E_{p(D_s,D_t|G_s,G_t)}(r_k(a_n^{st})) \tag{6}$$

$$\frac{\partial O(\theta)}{\partial \beta_d} = E_{p(D_{tu}|D_{tl},D_s,G_s,G_t)}\{h_d(G(v_m^{st}))\} - E_{p(D_t,D_s|G_s,G_t)}\{h_d(G(v_m^{st}))\} \tag{7}$$

where a_n^{st} is local attribute vector associating with variable node v_n^{st} in AFG model's second layer and v_m^{st} is the mth variable node. D_{tu} is the set of unknown descriptors and D_{tl} is the set of known descriptors. $E_{p(D_{tu}|D_{tl},D_s,G_s,G_t)}(r_k(a_n^{st}))$ is the expectation of the local function given all known descriptors of edges, while $E_{p(D_s,D_t|G_s,G_t)}(r_k(a_n^{st}))$ is the expectation given the estimated model. As the factor graph has different topology, it is hard to directly calculate the second part. In this paper, we use Loopy Belief Propagation (LBP) [21] to approximate the gradients. With LBP, the marginal probilities of different states of variable nodes can be calculated. After this, we sum over all nodes to obtain the gradient. The detailed algorithm is shown in Algorithm 2.

3.4 New User Link Inference

Model parameters θ can be got through learning. Then new user link inference problem is defined as finding the descriptors that maximizing the probability

$$O(D_{tu}) = P(D_{tu}|D_{tl}, D_s, G_s, G_t, \theta) \tag{8}$$

where D_{tu} is the set of unknown edge descriptors, D_{tl} is the set of known edge descriptors. The *LBP* algorithm is used to compute the marginal probability of each variable node v_i^{st} in factor graph. And we choose the state $l_i^{st} \in \{0,1\}$ with larger marginal probability $p_i^{st} = \max\{p(0|\theta), p(1|\theta)\}$ as v_i^{st}'s state. The edge descriptor corresponding to variable node v_i^{st} is $d_i^{st} = (l_i^{st}, p_i^{st})$.

Time cost is also very important when applying the prediction framework. If n users exist in social network, building *AFG* costs $\mathcal{O}(n^3)$ time. Parameter learning complexity is $\mathcal{O}(cnt)$, t is the number of iterations, c is a constant. Thus, the whole prediction algorithm can be finished in polynomial time.

3.5   Feature Selection

Table 1 is a list of all local features. As the networks we study are heterogenous and contain different types of data, three categories of local features can be selected, namely topology feature, location feature and time feature. s stands for source user and t stands for target user of an edge. FI_t is the set of users who follow user t and F_t is the set of users whom user t follows. Loc_t is location vector of user t, each element is the user's visited number of this location. Tim_t is time vector of user t with length 24, corresponding to the 24 hours of a day.

Taking the interactive effects of edges into consideration, one category of global features is drawn. We find that more than 90% triads in our data set are triad social balanced and triad status balanced. According to this observation, we choose the global features in Table 2.

Table 1. Local features

Category	Feature Name	Definition
Topology	BoolOpinionLeader	0 *or* 1
	InDegree	$|FI_s|, |FI_t|$
	OutDegree	$|F_s|, |F_t|$
	TotalDegree	$|FI_s \cup F_s|, |FI_t \cup F_t|$
	NumCommonNeighbor	$|(FI_s \cup F_s) \cap (FI_t \cup F_t)|$
	NumTotalNeighbor	$|(FI_s \cup F_s) \cup (FI_t \cup F_t)|$
	SimAdamic	$\sum_{i \in (FI_s \cup F_s) \cap (FI_t \cup F_t)} \{1/\log|FI_i \cup F_i|\}$
	SimJaccard	$|(FI_s \cup F_s) \cap (FI_t \cup F_t)|/|(FI_s \cup F_s) \cup (FI_t \cup F_t)|$
Location	LocationDis	$(\sum_i (Loc_{s,i} - Loc_{t,i})^2)^{1/2}$
	LocationCosine	$(Loc_s \cdot Loc_t)/(\| Loc_s \| \cdot \| Loc_t \|)$
	LocationJaccard	$|Loc_s \cap Loc_t|/|Loc_s \cup Loc_t|$
Time	TimeDis	$(\sum_i (Tim_{s,i} - Tim_{t,i})^2)^{1/2}$
	TimeCosine	$(Tim_s \cdot Tim_t)/(\| Tim_s \| \cdot \| Tim_t \|)$
	TimeExtendJaccard	$(Tim_s \cdot Tim_t)/(|Tim_s|^2 + |Tim_t|^2 - Tim_s \cdot Tim_t)$

Table 2. Global features

Feature Name	Definition
CommonSourceUser	0 or 1
CommonTargetUser	0 or 1
SocialBalance	0 or 1
SocialStatus	0 or 1

4 Experiment

4.1 Experiment Settings and Results

We use Twitter and Foursquare data sets and the same method as [12] to divide data for 5-cross-validation. Firstly, we randomly choose 1000 users to form two fully aligned networks. Secondly, 20% of users are chosen as new users. Thirdly, all existing friend relationship edges related to new users are put into an existing link set, equivalent number of non-existing friend relationship edges are put into non-existing link set. Fourthly, both the existing link set and the non-existing link set are divided into five parts. Fifthly, if the old users' information is used, just keep balance when expanding the two link sets. The ratio of new users' data used for training is defined as user novelty. Ratio 0.0 means brand-new users. All relationships related to new users are sampled according to the setting novelty.

Table 3. Experiment settings and results

	Target	Source	Model	Baseline	Auc ↑	Acc ↑
Group1	Twitter	None	BFG	TRAD	−3%	2%
	Twitter	Foursquare	AFG	SCAN-PS	10%	11%
	Twitter	Foursquare	AFG	BFG	36%	31%
Group2	Foursquare	None	BFG	TRAD	−15%	−9%
	Foursquare	Twitter	AFG	SCAN-PS	7%	3%
	Foursquare	Twitter	AFG	BFG	32%	25%

Two groups of experiments are carried out in this paper. Traditional Link Prediction *(TRAD)* and Supervised Cross Aligned Networks Link Prediction with Personalized Sampling *(SCAN-PS)* proposed in [12] are used as baseline methods. *SCAN-PS* merges features extracted from the anchor link in source network to expand the feature vector of corresponding link in target network to train a classifier. *Auc* and *Acc* are the performance evaluation criteria. Detailed comparative models and main results are shown in Table 3.

4.2 Performance Analysis

Fig. 3 is the results of first group experiments. Twitter is the target network, Foursquare is the source network in this group experiments.

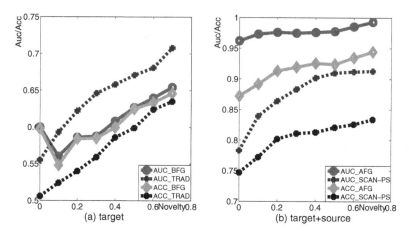

Fig. 3. Source network: Foursquare, target network: Twitter. It shows how Auc and Acc change with the user novelty. 1) In (a), BFG model has a higher Acc in average while $TRAD$ model performs well under Auc. 2) In (b), AFG model improves Auc by about 10% and Acc by about 11% than $SCAN$-PS model. 3) AFG model in (b) improves Auc by 36% and Acc by 31% in average compared with BFG model in (a).

- In Fig. 3(a), BFG model performs worse than $TRAD$ model in Auc because there are many state-unknown variable nodes in target network, the factor graph structure can not be decided accurately. Inaccurate factor graph structure decreases BFG model performance, but it has no effect on $TRAD$ model, which only makes use of the local features. As Twitter is follow-follow network, users having most fans play important role in network formation. That is the reason why we get high Auc and Acc when user novelty is 0.0.
- In Fig. 3(b), AFG model performs better than $SCAN$-PS model both in Auc and Acc. That is because source network information expands the training set and *Aligned Sturcture Algorithm* determines accurate factor graph structure.
- AFG model uses Foursquare to help new user link prediction in Twitter while BFG model only use Twitter data. Comparing AFG' performance in Fig. 3(b) and BFG's performance in Fig. 3(a), we find that AFG model can make full use of source network to improve the prediction performance.

Though target network and source network are similar, they also have own characteristics. Foursquare provides location based service while Twitter provides Tweet service. In order to prove that AFG model is suitable to solve new user link prediction problem in similar aligned networks regardless of their positions, we use Foursquare as target network and Twitter as source network in second group experiments. And the results of second group experiments are shown in Fig. 4.

- In Fig. 4(a), BFG model performs worse than $TRAD$ model, keeping the same trend with Fig. 3(a) for the same reason.

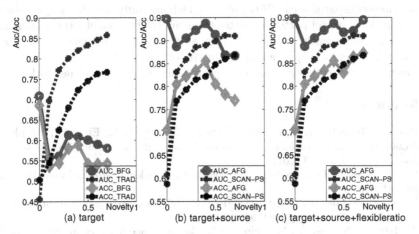

Fig. 4. Source network: Twitter, target network: Foursquare. It shows how Auc and Acc change with the user novelty. 1) In (a), BFG model performs worse than $TRAD$ model. 2) In (b), AFG model performs better than $SCAN$-PS model when user novelty is less than 0.5. 3) In (c), AFG model improves the Auc by 7% and Acc by 3% in average than $SCAN$-PS model. 4) AFG model in (c) improves Auc by 32% and Acc by 25% in average compared with BFG model in (a).

- In Fig. 4(b), the AFG model performance increases gradually before user novelty reaches 0.5, then its performance decreases. That is because the Twitter part in union training set contains noise. Only 6.5% users in Twitter have location data [12]. We use the corresponding users' location data in Foursquare on condition that the user-location links are in the training set of Foursquare part. The location data replacement causes noise, though the training set is expanded.
- In Fig. 4(c), AFG model performs better than $SCAN$-PS model both in Auc and Acc. Balance between training data amount and low data noise is achieved by using different ratio of the source network data as user novelty changes. This method improves the performance compared with curves in Fig. 4(b) when user novelty exceeds 0.5.

5   Conclusion

The link prediction problem for new users is studied in this paper. Recommendations for new users have significant influence on their keeping active in this social network. However, the cold start problem is often encountered. The AFG model is proposed to utilize data from a similar source network to help prediction in target network. Three categories of local features and one category of global features are put forward for training. The *Aligned Structure Algorithm* is brought up to reduce the scale of the factor graph and keep high prediction accuracy when building the model. Experiments on Twitter and Foursquare show that AFG model can make full use of source network data to improve prediction performance compared with *BFG* model, which can only use the target network data. And *AFG* model performs better than *SCAN-PS* model. *Auc* is increased by 10% and *Acc* is increased by 11% in average when Foursquare is source network and Twitter is target network. On the other hand, 7% *Auc* and 3% *Acc* improvements are achieved when swapping positions of the two networks.

References

1. Fire, M., Tenenboim, L., Lesser, O., Puzis, R., Rokach, L., Elovici, Y.: Link prediction in social networks using computationally efficient topological features. In: SocialCom, pp. 73–80 (2011)
2. Lü, L., Zhou, T.: Link prediction in weighted networks: The role of weak ties. EPL 89(1) (2010)
3. Xiang, R., Neville, J., Rogati, M.: Modeling relationship strength in online social networks. In: WWW, pp. 981–990 (2010)
4. Scellato, S., Noulas, A., Mascolo, C.: Exploiting place features in link prediction on location-based social networks. In: SIGKDD, pp. 1046–1054 (2011)
5. Allamanis, M., Scellato, S., Mascolo, C.: Evolution of a location-based online social network: analysis and models. In: IMC, pp. 145–158 (2012)
6. Backstrom, L., Leskovec, J.: Supervised random walks: predicting and recommending links in social networks. In: WSDM, pp. 635–644 (2011)
7. Davis, D., Lichtenwalter, R., Chawla, N.V.: Supervised methods for multi-relational link prediction. Social Network Analysis and Mining **3**(2), 127–141 (2013)
8. Leroy, V., Cambazoglu, B.B., Bonchi, F.: Cold start link prediction. In: SIGKDD, pp. 393–402 (2010)
9. Dai, W., Chen, Y., Xue, G.R., Yang, Q., Yu, Y.: Translated learning: Transfer learning across different feature spaces. In: NIPS, pp. 353–360 (2008)
10. Tang, J., Lou, T., Kleinberg, J.: Inferring social ties across heterogenous networks. In: WSDM, pp. 743–752 (2012)
11. Dong, Y., Tang, J., Wu, S., Tian, J., Chawla, N.V., Rao, J., Cao, H.: Link prediction and recommendation across heterogeneous social networks. In: ICDM, pp. 181–190 (2012)
12. Zhang, J., Kong, X., Yu, P.S.: Predicting social links for new users across aligned heterogeneous social networks. In: ICDM, pp. 1289–1294 (2013)
13. Liu, L., Tang, J., Han, J., Jiang, M., Yang, S.: Mining topic-level influence in heterogeneous networks. In: CIKM, pp. 199–208 (2010)
14. Zheng, X., Zeng, D., Wang, F.Y.: Social balance in signed networks. Information Systems Frontiers 1(19) (2014)
15. Leskovec, J., Huttenlocher, D., Kleinberg, J.: Signed networks in social media. In: SIGCHI, pp. 1361–1370 (2010)
16. Cremonesi, P., Tripodi, A., Turrin, R.: Cross-domain recommender systems. In: ICDMW, pp. 496–503 (2011)
17. Tang, J., Wu, S., Sun, J., Su, H.: Cross-domain collaboration recommendation. In: SIGKDD, pp. 1285–1293 (2012)
18. Lu, Z., Pan, W., Xiang, E.W., Yang, Q., Zhao, L., Zhong, E.: Selective transfer learning for cross domain recommendation. In: SDM pp. 641–649 (2013)
19. Frey, B.J., Kschischang, F.R., Loeliger, H.A., Wiberg, N.: Factor graphs and algorithms. In: Allerton, pp. 666–680 (1997)
20. Hammersley, J.M., Clifford, P.: Markov fields on finite graphs and lattices (1971)
21. Yedidia, J.S., Freeman, W.T., Weiss, Y.: Bethe free energy, Kikuchi approximations, and belief propagation algorithms. In: NIPS (2001)

Scale-Adaptive Group Optimization for Social Activity Planning

Hong-Han Shuai[1][✉], De-Nian Yang[2], Philip S. Yu[3], and Ming-Syan Chen[1,2]

[1] Graduate Institute of Communication Engineering,
National Taiwan University, Taipei, Taiwan
{hhshuai,dnyang}@iis.sinica.edu.tw
[2] Institute of Information Science, Academia Sinica, Taipei, Taiwan
mschen@cc.ee.ntu.edu.tw
[3] Department of Computer Science, University of Illinois at Chicago, Chicago, USA
psyu@uic.edu

Abstract. Studies have shown that each person is more inclined to enjoy a group activity when 1) she is interested in the activity, and 2) many friends with the same interest join it as well. Nevertheless, even with the interest and social tightness information available in online social networks, nowadays many social group activities still need to be coordinated manually. In this paper, therefore, we first formulate a new problem, named Participant Selection for Group Activity (PSGA), to decide the group size and select proper participants so that the sum of personal interests and social tightness of the participants in the group is maximized, while the activity cost is also carefully examined. To solve the problem, we design a new randomized algorithm, named Budget-Aware Randomized Group Selection (BARGS), to optimally allocate the computation budgets for effective selection of the group size and participants, and we prove that BARGS can acquire the solution with a guaranteed performance bound. The proposed algorithm was implemented in Facebook, and experimental results demonstrate that social groups generated by the proposed algorithm significantly outperform the baseline solutions.

1 Introduction

Studies have shown that two important factors are usually involved in a person's decision to join a social group activity: (1) interest in the activity topic or content, and (2) social tightness with other attendees [5,8]. For example, if a person who appreciates jazz music has complimentary tickets for a jazz concert in Rose Theatre, she is inclined to invite her friends or friends of friends who are also jazzists. However, even the information on the two factors is now available online, the attendees of most group activities still need to be selected manually, and the process will be tedious and time-consuming, especially for a large social activity, given the complicated social link structure and the diverse interests of potential attendees.

© Springer International Publishing Switzerland 2015
T. Cao et al. (Eds.): PAKDD 2015, Part I, LNAI 9077, pp. 45–57, 2015.
DOI: 10.1007/978-3-319-18038-0_4

Recent studies have explored community detection, graph clustering and graph partitioning to identify groups of nodes mostly based on the graph structure [1]. The quality of an obtained community is usually measured according to its internal structure, together with its external connectivity to the rest of the nodes in the graph [7] . Those approaches are not designed for activity planning because it does not consider the interests of individual users along with the cost of holding an activity with different numbers of participants. An event which attracts too few or too many attendees will result in unacceptable loss for the planner. Therefore, it is important to incorporate the preference of each potential participant, their social connectivity, and the activity cost during the planning of an activity.

With this objective in mind, a new optimization problem is formulated, named *Participant Selection for Group Activity (PSGA)*. The problem is given a cost function related to the group size and a social graph G, where each node represents a potential attendee and is associated with an interest score that describes the individual level of interest. Each edge has a social tightness score corresponding to the mutual familiarity between the two persons. Since each participant is more inclined to enjoy the activity when 1) she is interested in the activity, and 2) many friends with the same interest join as well, the *preference* of a node v_i for the activity can be represented by the sum of its interest score and social tightness scores of the edges connecting to other participants, while the *group preference* is sum of the total interest scores of all participants and the social tightness scores of the edges connecting to any two participants. Moreover, the *group utility* here is represented by the group preference subtracted by the *activity cost* (ex. the expense in food and siting), which is usually correlated to the number of participants.[1] The objective of PSGA is to determine the best group size and select proper participants, so that the group utility is maximized. In addition, the induced graph of the set F of selected participants is desired to be a connected component, so that each attendee is possible to become acquainted with another attendee according to a social path[2].

One possible approach to solving PSGA is to examine every possible combination on every group size. However, this enumeration approach of group size k requires the evaluation of C_k^n candidate groups, where n is the number of nodes in G. Therefore, the number of group size and attendee combinations is $O(2^n)$, and it thereby is not feasible in practical cases. Another approach is to incrementally construct the group using a greedy algorithm that iteratively tries each group size and sequentially chooses an attendee that leads to the largest increment in group utility at each iteration. However, greedy algorithms are inclined to be trapped in local optimal solutions. To avoid being trapped in local optimal

---

[1] Different weighted coefficients can be assigned to the group utility and activity cost according to the corresponding scenario.

[2] For some group activities, it is not necessary to ensure that F leads to a connected subgraph, and those scenarios can be handled by adding a virtual node v connecting to every other node in G, and choosing v in F for PSGA always creates a connected subgraph in $G \cup \{v\}$, but F may not be a connected subgraph in G.

solutions, randomized algorithms have been proposed as a simple but effective strategy to solve problems with large instances [12].

A simple randomized algorithm is to randomly choose multiple start nodes initially. Each start node is considered as a partial solution, and a node neighboring the partial solution is randomly chosen and added to the partial solution at each iteration later. Nevertheless, this simple strategy has three disadvantages. Firstly, a start node that has the potential to generate final solutions with high group utility does not receive sufficient computational resources for randomization in the following iterations. More specifically, each start node in the randomized algorithm is expanded to only one final solution. Thus, a good start node will usually fail to generate a solution with high group utility since it only has one chance to randomly generate a final solution. The second disadvantage is that the expansion of the partial solution does not differentiate the selection of the neighboring nodes. Each neighboring node is treated equally and chosen uniformly at random for each iteration. Even this issue can be partially resolved by assigning the selection probability to each neighboring node according to its interest score and the social tightness of incident edges, this assignment will lead to the greedy selection of neighbors and thus tends to be trapped in local optimal solutions as well. The third disadvantage is that the linear scanning of different group sizes is not computationally tractable for real scenarios as an online social network contains an enormous number of nodes.

Keeping the above observations in mind, we propose a randomized algorithm, called *Budget-Aware Randomized Group Selection (BARGS)*, to effectively select the start nodes, expand the partial solutions, and estimate the suitable group size. The computational budget represents the target number of random solutions. Specifically, *BARGS* first selects a group size limit k_{max} in accordance with the cost function[3]. Afterward, m start nodes are selected, and neighboring nodes are properly added to expand the partial solution iteratively, until k_{max} nodes are included, while the group size corresponding to the largest group utility is acquired finally. Each start node in *BARGS* is expanded to multiple final solutions according to the assigned budget. To properly invest the computational budgets, each stage of *BARGS* invests more budgets on the start nodes and group sizes that are more inclined to generate good final solutions, according to the sampled results from the previous stages. Moreover, the node selection probability is adaptively assigned in each stage by exploiting the cross entropy method. In this paper, we show that our allocation of computation budgets is the optimal strategy, and prove that the solution acquired by *BARGS* has a guaranteed performance bound.

The rest of this paper is organized as follows. Section 2 formulates PSGA and surveys related works. Sections 3 explains *BARGS* and derives the performance bound. User study and experimental results are presented in Section 4, and we conclude this paper in Section 5.

---

[3] For instance, if the largest capacity of available stadiums for a football game is $20,000$, k_{max} is set as $20,000$.

2   Preliminary

2.1   Problem Definition

Given a social network $G = (V, E)$, where each vertex $v_i \in V$ and each edge $e_{i,j} \in E$ are associated with an interest score η_i and a social tightness score $\tau_{i,j}$ respectively, we study a new optimization problem for finding a set F of vertices which maximizes the *group utility* $U(F)$, i.e.,

$$U(F) = \sum_{v_i \in F} \left(\eta_i + \sum_{v_j \in F : e_{i,j} \in E} \pi_{i,j} \right) - \beta C(|F|), \tag{1}$$

where F with $|F| \leq k_{max}$ is a connected subgraph in G to encourage each attendee to be acquainted with another attendee with at least one social path in F, C is a non-negative activity cost function based on the number of attendees, and β is a weighted coefficient between the preference and cost. For each node v_i, let $\eta_i + \sum_{v_j \in F : e_{i,j} \in E} \pi_{i,j}$ denote the *preference* of node v_i on the social group activity[4]. PSGA is very challenging due to the tradeoff between interest, social tightness, and the cost function, while the constraint assuring that F is connected also complicates this problem because it is no longer able to arbitrarily choose any nodes from G. Indeed, we show that PSGA is NP-hard in [15].

2.2   Related Works

A recent line of study has been proposed to find cohesive subgroups in social networks with different criteria, such as cliques, n-clubs, k-core, and k-plex. Saríyüce et al. [14] proposed an efficient parallel algorithm to find a k-core subgraph, where every vertex is connected to at least k vertices in the subgraph. Xiang et al. [16] proposed a branch-and-bound algorithm to acquire all maximal cliques that cannot be pruned during the search tree optimization. Moreover, finding the maximum k-plexes was comprehensively discussed in [11]. On the other hand, community detection and graph clustering have been exploited to identify the subgraphs with the desired structures [1]. The quality of a community is measured according to the structure inside the community and the structure between the community and the rest of the nodes in the graph, such as the density of local edges, deviance from a random null model, and conductance [7]. Nevertheless, the above models did not examine the interest score of each user and the social tightness scores between users, which have been regarded as crucial factors for social group activities. Moreover, the activity cost for the group is not incorporated during the evaluation.

   In addition to dense subgraphs, social groups with different characteristics have been explored for varied practical applications. Expert team formation in social

---

[4] Different weights λ and (1-λ) can be assigned to the interest scores and social tightness such that $U(F) = \sum_{v_i \in F}(\lambda_i \eta_i + (1 - \lambda_i) \sum_{v_j \in F : e_{i,j} \in E} \tau_{i,j}) - \beta C(|F|)$. λ_i can be set directly by a user or according to the existing model [18]. The impacts of different λ will be studied later in Section 4.

networks has attracted extensive research interest. The problem of constructing an expert team is to find a set of people possessing the required skills, while the communication cost among the chosen friends is minimized to optimize the rapport among the team members to ensure efficient operation. Communication costs can be represented by the graph diameter, the size of the minimum spanning tree, and the total length of the shortest paths [9]. Finding influential event organizers who can influence largest number of attendees to join the event is studied [6]. By contrast, minimizing the total spatial distance with R-Tree from the group with a given number of nodes to the rally point is also studied [17]. Nevertheless, this paper focuses on a different scenario that aims at identifying a group with the most suitable size according to the activity cost, while those selected participants also share the common interest and high social tightness.

3   Algorithm Design for PSGA

To solve PSGA, a baseline approach is to incrementally constructing the solution by sequentially choosing and adding a neighbor node that leads to the largest increment in the group preference until k_{max} people are selected. Afterward, we derive the group utility for each k by incorporating the activity cost, $1 \leq k \leq k_{max}$, and extract the group size k^* with the maximum group utility. The theoretical analysis of greedy algorithm is presented in [15] due to the space constraint.

The greedy algorithm, despite the simplicity, the search space of the greedy algorithm is limited and thus tends to be trapped in a local optimal solution, because only a single sequence of solutions is explored. To address the above issues, this paper proposes a randomized algorithm $BARGS$ to randomly choose m start nodes[5]. $BARGS$ leverages the notion of Optimal Computing Budget Allocation (OCBA) [3] to systematically generate the solutions from each start node, where the start nodes with more potential to generate the final solutions with large group utility will be allocated with more budgets (i.e., expanded to more final solutions). In addition, since each start nodes can generate the final solutions with different group sizes, the size with larger group utility will be associated with more budgets as well (i.e., generated more times). Specifically, $BARGS$ includes the following two phases.

1) Selection and Evaluation of Start Nodes and Group Sizes: This phase first selects m start nodes according to the summation of the interest scores and social tightness scores of incident edges. Each start node acts as a seed to be expanded to a few final solutions. At each iteration, a partial solution, which consists of only a start node at the first iteration or a connected set of nodes at each iteration afterward, is expanded by randomly selecting a node neighboring to the partial solution, until $k_{\max}$ nodes are included. The group utility of each intermediate and final solution is evaluated to optimally allocate different computational budgets to different start nodes and different group sizes in the next phase.

---

[5] The impact of m will be studied in Section 4.

2) *Allocation of Computational Budgets:* This phase is divided into r stages[6], while each stage shares the same total computational budget. In the first stage, the computational budget allocated to each start node is determined by the sampled group utility in the first phase. In each stage afterward, the computational budget allocated to each start node is adjusted by the sampled results in the previous stages. Note that each node can generate different numbers of final solutions with different group sizes. The sizes with small group utility sampled in the previous stages will be associated with smaller computational budgets in the current stage. Therefore, if the activity cost is a convex cost function, the cost increases more significantly as the group size grows, and *BARGS* tends to allocate smaller computational budgets and thus generates fewer final solutions with large group sizes.

During the expansion of the partial solutions, we differentiate the probability to select each node neighboring to a partial solution. One intuitive way is to associate each neighboring node with a different probability according to the sum of the interest scores and social tightness scores on the incident edges. Nevertheless, this assignment is similar to the greedy algorithm as it limits the scope to only the local information, making it difficult to generate a final solution with large group utility. By contrast, *BARGS* exploits the cross entropy method [13] according to sampled results in the previous stages in order to optimally assign a probability to the edge incident to a neighboring node.

Due to the space constraint, the detailed pseudocode is presented in [15]. In the following, we first present how to optimally allocate the computational budgets to different start nodes and different group sizes. Afterward, we exploit the cross entropy method to differentiate the neighbor selection during the expansion of the partial solutions. The performance bound and illustrative example of the proposed algorithm are provided in the full version [15].

Allocation of Computational Budgets. Similar to the baseline greedy algorithm, allocating more computational budgets to a start node v_i with larger group preference (i.e., $\sum_{v_i \in F}(\eta_i + \sum_{v_j \in F: e_{i,j} \in E} \pi_{i,j})$) examines only the local information and thus is difficult to generate the solution with large group utility. Therefore, to optimally allocate the computational budgets for each start node and each group size, we first define the solution quality as follows.

Definition 1. *The solution quality, denoted by Q, is defined as the maximum group utility of the solution generated from the m start nodes among all sizes.*

For each stage t of phase 2 in *BARGS*, let $N_{i,k,t}$ denote the computational budgets allocated to the start node v_i with size k in the t-th stage. In the following, we first derive the optimal ratio of the computational budgets allocated to any two start nodes v_i and v_j with group size k and l, respectively. Let two random variables $Q_{i,k}$ and $Q^*_{i,k}$ denote the sampled group utility of any solution

---

[6] The detailed settings of the parameters of the algorithm, such as m, r, α, and β are presented in [15].

and the maximal sampled group utility of a solution for start node v_i with size k, respectively. If the activity cost is not considered, according to the central limit theorem, $Q_{i,k}$ follows the normal distribution when $N_{i.k}$ is large, and it can be approximated by the uniform distribution in $[c_{i,k}, d_{i,k}]$ as analyzed in OCBA [3], where $c_{i,k}$ and $d_{i,k}$ denote the minimum and maximum sampled group utility in the previous stages, respectively. On the other hand, when the activity cost is considered, the cumulative distribution function is shifted by $C(k)$, and it still follows the same distribution. Therefore, we have the following lemma.

Lemma 1. *Assume that $d_{j,l} \geq c_{i,k}$, the probability that the solution generated from the start node v_i with size k is better than the solution generated from the start node v_j with size l, i.e., $P(Q_{i,k}^* \leq Q_{j,l}^*)$, is at least $\frac{1}{2}(\frac{d_{j,l} - c_{i,k}}{d_{i,k} - c_{i,k}})^{N_{i,k}}$.*

Proof. Due to the space constraint, the detailed proof is presented in [15].

Let v_b and k_b^* denote the best start node and best activity size for v_b, respectively. The ratio between $N_{i,k,t}$ and $N_{j,l,t}$ equals $P(Q_{i,k}^* \geq Q_{b,k_b^*}^*) : P(Q_{j,l}^* \geq Q_{b,k_b^*}^*)$, which is optimal as shown in OCBA [3]. However, the computational costs for different group sizes are not the same, e.g., the computational cost of the total group utility for size 1 is much smaller than the computational cost for size 100. Since the computational complexity of adding a node to a partial solution of size $k - 1$ is $O(k)$, we derive the ratio of the computational budgets between $N_{i,k,t}$ and $N_{j,l,t}$ as follows.

$$\frac{N_{i,k,t}}{N_{j,l,t}} = \frac{\frac{1}{k} \cdot P(Q_{i,k}^* \geq Q_{b,k_b^*}^*)}{\frac{1}{l} \cdot P(Q_{j,l}^* \geq Q_{b,k_b^*}^*)}. \tag{2}$$

Note that if the allocated computational budgets for a start node is 0 in the t-th stage, we prune off the start node in the any stage afterward. Moreover, when we generate a solution with group size k, the solutions from size 1 to size $k - 1$ are also generated as well. Therefore, to avoid generating an excess number the solutions with small group sizes, it is necessary to relocate the computation budgets. Let $\hat{N}_{i,k,t}$ denote the reallocated budget of start node v_i with size k in the t-th stage. *BARGS* derives $N_{i,k,t}$ as follows.

$$\hat{N}_{i,k,t} = \max(0, N_{i,k,t} - \sum_{l > k} \hat{N}_{i,l,t}). \tag{3}$$

Specifically, after deriving $N_{i,k,t}$ with Eq. 2, *BARGS* derives $\hat{N}_{i,k,t}$ from $k = k_{max}$ to 1. Initially, $\hat{N}_{i,k_{\max},t} = N_{i,k_{\max},t}$. Afterward, for $k = k_{max} - 1$, if $N_{i,k_{max}-1,t}$ is equal to $N_{i,k_{max},t}$, it is not necessary to generate additional solutions with size $k_{max} - 1$ since they have been created during the generation of the solutions with size k_{max}. In this case, $\hat{N}_{i,k_{max}-1,t}$ is 0. Otherwise, *BARGS* sets $\hat{N}_{i,k_{max}-1,t} = N_{i,k_{max}-1,t} - \hat{N}_{i,,k_{max},t}$. The above process repeats until $k = 1$. Since the number of solutions with size k is still $N_{i,k,t}$, the computational budget allocation is still optimal as shown in Eq. 2.

Neighboring Node Differentiation. To effectively differentiate neighbor selection, *BARGS* exploits the cross entropy method [13] to achieve importance sampling by adaptively assigning a different probability to each neighboring node from the sampled results in previous stages.

Definition 2. *A Bernoulli sample vector, denoted as* $X_{i,k,q} = \langle x_{i,k,q,1}, ..., x_{i,k,q,j},$ $..., x_{i,k,q,n} \rangle$, *is defined to be the q-th sample vector from start node* v_i, *where* $x_{i,k,q,j}$ *is 1 if node* v_j *is selected in the q-th sample and 0 otherwise.*

Take start node v_i with size k as an example, after collecting $N_{i,k,1}$ samples $X_{i,k,1}, X_{i,k,2}, ..., X_{i,k,q}, ..., X_{i,k,N_{i,k,1}}$ generated from start node v_i, *BARGS* calculates the total group utility $U(X_{i,k,q})$ for each sample and sorts them in the descending order, $U_{(1)} \geq ... \geq U_{(N_{i,k,1})}$. Let $\gamma_{i,k,1}$ denotes the group utility of the top-ρ performance sample, i.e. $\gamma_{i,k,1} = U_{(\lceil \rho N_{i,k,1} \rceil)}$. With those sampled results, we set the selection probability $p_{i,k,t+1,j}$ of every node v_j in iteration $t + 1$ for the partial solution expanded from node v_i by fitting the distribution of top-ρ performance samples as follows.

$$p_{i,k,t+1,j} = \frac{\sum_{q=1}^{N_{i,k,t}} I_{\{U(X_{i,k,q}) \geq \gamma_{i,k,t}\}} x_{i,k,q,j}}{\sum_{q=1}^{N_{i,k,t}} I_{\{U(X_{i,k,q}) \geq \gamma_{i,k,t}\}}}, \quad (4)$$

where $I_{\{U(X_{i,k,q}) \geq \gamma_{i,k,t}\}}$ is 1 if the group utility of sample $X_{i,k,q}$ is no smaller than a threshold $\gamma_{i,k,t} \in \mathbb{R}$, and 0 otherwise. Intuitively, the neighbor that tends to generate a better solution will be assigned a higher selection probability. As shown in [13], the above probability assignment scheme has been proved to be optimal from the perspective of cross entropy. Eq. 4 minimizes the Kullback-Leibler cross entropy (KL) distance between node selection probability and the distribution of top-ρ performance samples, such that the performance of random samples in the $(t + 1)$-th stage is guaranteed to be closest to the top-ρ performance samples in the t-th stage. Due to the space constraint, the illustrative example and theoretical results are provided in the full version [15].

Time Complexity of BARGS. The time complexity of *BARGS* contains two parts. The first phase selects m start nodes with $O(E + n + m \log n)$ time, where $O(E)$ is to sum up the interest and social tightness scores, $O(n + m \log n)$ is to build a heap and extract m nodes with the largest sum. Afterward, the second phase of *BARGS* includes r stages, and each stage allocates the computational resources with $O(m)$ time and generates $O(\frac{T}{r})$ new partial solutions with at most k_{max} nodes for all start nodes. Therefore, the time complexity of the second phase is $O\left(r(m + \frac{T}{r} k_{max})\right) = O(k_{max}T)$, and *BARGS* therefore needs $O(E + m \log n + k_{max}T)$.

4   Experimental Results

We implement *BARGS* in Facebook and invite 50 people from various communities, e.g., schools, government, technology companies, and businesses to join our user study. We compare the solution quality and running time of manual coordination and *BARGS* for answering PSGA problems, to evaluate the need of an automatic group recommendation service. Each user is asked to plan 5 social activities with the social graphs extracted from their social networks in Facebook. The interest scores follow the power-law distribution with the exponent as 2.5 according to the recent analysis [4] on real datasets. The social tightness score between two friends is derived according to the number of common friends, which represents the proximity interaction [2], and the probability of negative weights [10]. Then, the weighted coefficient λ on social tightness scores and interest scores and the weighted coefficient β on group preference and activity cost in Footnote 4 are set as the average value specified by the 50 people, i.e., $\lambda = 0.527$ and $\beta = 0.514$. Most importantly, after the scores are returned by the above renowned models, each user is allowed to fine-tune the two scores by themselves. In addition to the user study, two real datasets are evaluated in the experiment. The first dataset is crawled from Facebook with $90,269$ users in the New Orleans network[7]. The second dataset is crawled from DBLP dataset with $511,163$ nodes and $1,871,070$ edges.

In this paper, the activity cost is modelled by a piecewise linear function, which can approximate any non-decreasing functions. We set the activity cost according to the auditorium cost and other related cost in Duke Energy Center[8].

$$
C(k) = \begin{cases} 400 - k & \text{if } 0 \le k \le 100. \\ 850 - k & \text{if } 100 < k \le 600. \\ 2200 - k & \text{if } 600 < k \le 1750. \end{cases}
$$

We compare deterministic greedy (*DGreedy*), randomized greedy (*RGreedy*), and *BARGS* in an HP DL580 server with four Intel E7-4870 2.4 GHz CPUs and 128 GB RAM. *RGreedy* first chooses the same m start nodes as *BARGS*. At each iteration, *RGreedy* calculates the preference increment of adding a neighboring node v_j to the intermediate solution V_S obtained so far for each neighboring node, and sums them up as the total preference increment. Afterward, *RGreedy* sets the node selection probability of each neighbor as the ratio of the corresponding preference increment to the total preference increment, similar to the concept in the greedy algorithm. Notice that the computation budgets represent the number of generated solutions. With more computation budgets, *RGreedy* generates more solutions of group size k_{max}, examines the group utility by subtracting the activity cost from group size 1 to k_{max}, and selects the group with maximum group utility. It is worth noting that *RGreedy* is computationally intensive and not scalable to support a large group size because it is necessary to sum up the interest scores and social tightness scores during the selection of a node neighboring to each partial solution. Therefore, we can only present the results of *RGreedy* with small group sizes. Due to the space constraint, detailed experimental results of the DBLP dataset are presented in [15].

---

[7] http://socialnetworks.mpi-sws.org/data-wosn2009.html
[8] http://www.dukeenergycenterraleigh.com/uploads/venues/rental/5-rateschedule.pdf

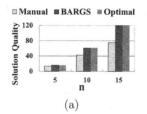 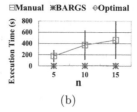 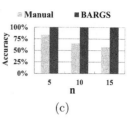

Fig. 1. Results of user study

The default m in the experiment is set as n/k_{max} since n/k_{max} groups can be acquired from a network with n nodes if each group has k_{max} participants. The default cross-entropy parameters ρ and α are set as 0.3 and 0.99 as recommended by the cross-entropy method [13]. Since *BARGS* natively supports parallelization, we also implemented them with OpenMP for parallelization, to demonstrate the gain in parallelization with more CPU cores.

4.1   User Study

Figures 1(a)-(c) compare manual coordination and *BARGS* in the user study. In addition, the optimal solution is also derived with the enumeration method since the network size is very small. Figures 1(a) and (b) present the solution quality and execution time with different network sizes. The result indicates that the solutions obtained by *BARGS* are identical to the optimal solutions, but users are not able to acquire the optimal solutions even when $n = 5$. As n increases, the solution quality of manual coordination degrades rapidly. We also compare the accuracy of selecting the optimal group size in Figure 1(c). As n increases, it becomes more difficult for a user to correctly identify the optimal size, while *BARGS* can always select the optimal one. Therefore, it is desirable to deploy *BARGS* as an automatic group recommendation service, especially to address the need of a large group in a massive social network nowadays.

4.2   Performance Comparison and Sensitivity Analysis

Figure 2(a) compares the execution time of *DGreedy*, *RGreedy*, and *BARGS* by sampling different numbers of nodes from Facebook data. *DGreedy* is always the fastest one since it is a deterministic algorithm and generates only one final solution, whereas *RGreedy* requires more than 10^5 seconds. The results of *RGreedy* do not return in 2 days as n increases to 10000. To evaluate the performance of *BARGS* with multi-threaded processing, Figure 2(b) shows that we can accelerate the processing speed to 7.2 times with 8 threads. The acceleration ratio is slightly lower than 8 because OpenMP forbids different threads to write at the same memory position at the same time. Therefore, it is expected that *BARGS* with parallelization is promising to be deployed as a value-added *cloud service*.

In addition to the running time, Figure 2(c) compares the solution quality of different approaches. The results indicate that *BARGS* outperforms *DGreedy*

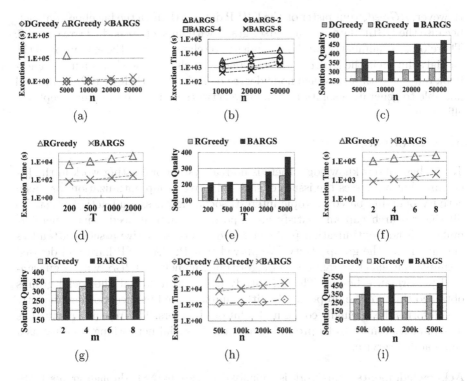

Fig. 2. Experimental results on Facebook and DBLP datasets

and *RGreedy*, especially under a large n. The group utility of *BARGS* is 45% better than the one from *DGreedy* when $n = 50000$. On the other hand, *RGreedy* outperforms *DGreedy* since it has a chance to jump out of the local optimal solution.

Figures 2(d) and (e) compare the execution time and solution quality of two randomized approaches under different total computational budgets, i.e., T. As T increases, the solution quality of *BARGS* increases faster than that of *RGreedy* because it can optimally allocate the computation resources. Even though the solution quality of *RGreedy* is closer to *BARGS* in some cases, *BARGS* is much faster than *RGreedy* by an order of 10^{-2}.

Figures 2(f) and (g) present the execution time and solution quality of *RGreedy* and *BARGS* with different numbers of start nodes, i.e., m. The results show that the solution quality in Figure 2(g) is almost the same as m increases, demonstrating that it is sufficient for m to be set as a value smaller than $\frac{n}{k_{max}}$ as recommended by OCBA [3]. The running time of *BARGS* for $m = 2$ is only 60% of the running time for $m = 4$ as shown in Figure 2(f), while the solution quality remains almost the same.

BARGS is also evaluated on the DBLP dataset. Figures 2(h) and (i) show that *BARGS* outperforms *DGreedy* by 50% and *RGreedy* by 26% in solution quality when $n = 500000$. *BARGS* is still faster than *RGreedy* by an order of 10^{-2}.

However, *RGreedy* runs faster on the DBLP dataset than on the Facebook dataset, because the DBLP dataset is a sparser graph with an average node degree of 3.66. Therefore, the number of candidate nodes to be chosen during the expansion of the partial solution in the DBLP dataset increases much more slowly than in the Facebook dataset with an average node degree of 26.1. Nevertheless, *RGreedy* is still not able to generate a solution for a large network size n due to its unacceptable efficiency.

5 Conclusion

To the best of our knowledge, there is no real system or existing work in the literature that addresses the issues of scale-adaptive group optimization for social activity planning based on topic interest, social tightness, and activity cost. To fill this research gap and satisfy an important practical need, this paper formulated a new optimization problem called PSGA to derive a set of attendees and maximize the group utility. We proved that PSGA is NP-hard and devised a simple but effective randomized algorithms, namely *BARGS*, with a guaranteed performance bound. The user study demonstrated that the social groups obtained through the proposed algorithm implemented in Facebook significantly outperforms the manually configured solutions by users. This research result thus holds much promise to be profitably adopted in social networking websites as a value-added service.

Acknowledgments. This work is supported in part by NSF through grants CNS-1115234, and OISE-1129076.

References

1. Brandes, U., Delling, D., Gaertler, M., Goerke, R., Hoefer, M., Nikoloski, Z., Wagner, D.: On modularity clustering. IEEE Transactions on Knowledge and Data Engineering **20**, 172–188 (2008)
2. Chaoji, V., Ranu, S., Rastogi, R., Bhatt, R.: Recommendations to boost content spread in social networks. In: WWW, pp. 529–538 (2012)
3. Chen, C.H., Yucesan, E., Dai, L., Chen, H.C.: Efficient computation of optimal budget allocation for discrete event simulation experiment. IIE Transactions **42**(1), 60–70 (2010)
4. Clauset, A., Shalizi, C.R., Newman, M.E.J.: Power-law distributions in empirical data. SIAM **51**(4), 661–703 (2009)
5. Deutsch, M., Gerard, H.B.: A study of normative and informational social influences upon individual judgment. JASP **51**(3), 291–301 (1955)
6. Feng, K., Cong, G., Bhowmick, S.S., Ma, S.: In search of influential event organizers in online social networks. In: Proc. SIGMOD (2014)
7. Gleich, D.F., Seshadhri, C.: Vertex neighborhoods, low conductance cuts, and good seeds for local community methods. In: KDD, pp. 597–605 (2012)
8. Kaplan, M.F., Miller, C.E.: Group decision making and normative versus informational influence: Effects of type of issue and assigned decision rule. JPSP **53**(2), 306–313 (1987)

9. Kargar, M., An, A.: Discovering top-k teams of experts with/without a leader in social networks. In: CIKM, pp. 985–994 (2011)
10. Leskovec, J., Huttenlocher, D., Kleinberg, J.: Signed networks in social media. In: CHI (2010)
11. McClosky, B., Hicks, I.V.: Combinatorial algorithms for max k-plex. Journal of Combinatorial Optimization (2012)
12. Mitzenmacher, M., Upfal, E.: Probability and computing: Randomized algorithms and probabilistic analysis. Cambridge University Press (2005)
13. Rubinstein, R.Y.: Combinatorial optimization, cross-entropy, ants and rare events. In: Uryasev, S., Pardalos, P.M. (eds.) Stochastic Optimization: Algorithms and Applications, pp. 304–358. Kluwer Academic (2001)
14. Saríyüce, A.E., Gedik, B., Jacques-Silva, G., Wu, K.-L., Çatalyüreks, U.V.: Streaming algorithms for k-core decomposition. VLDB **6**(5), 433–444 (2013)
15. Shuai, H.-H., Yang, D.-N., Yu, P.S., Chen, M.-S.: Scale-adaptive group optimization for social activity planning. In: CoRR (1502.06819) (2015)
16. Xiang, J., Guo, C., Aboulnaga, A.: Scalable maximum clique computation using mapreduce. In: ICDE, pp. 74–85 (2013)
17. Yang, D.N., Shen, C.Y., Lee, W.C., Chen, M.S.: On socio-spatial group query for location-based social networks. In: KDD, pp. 949–957 (2012)
18. Ye, M., Liu, X., Lee, W.C.: Exploring social influence for recommendation - a probabilistic generative model approach. In: SIGIR, pp. 671–680 (2012)

Influence Maximization Across Partially Aligned Heterogenous Social Networks

Qianyi Zhan[1]([✉]), Jiawei Zhang[2], Senzhang Wang[3], Philip S. Yu[2,4], and Junyuan Xie[1]

[1] National Laboratory for Novel Software Technology,
Nanjing University, Nanjing 210023, China
`zhanqianyi@gmail.com, jyxie@nju.edu.cn`
[2] University of Illinois at Chicago, Chicago, IL, USA
`{jzhan9,psyu}@uic.edu`
[3] Beihang University, Beijing 100191, China
`szwang@buaa.edu.cn`
[4] Institute for Data Science, Tsinghua University, Beijing 100084, China

Abstract. The influence maximization problem aims at finding a subset of seed users who can maximize the spread of influence in online social networks (OSNs). Existing works mostly focus on one single homogenous network. However, in the real world, OSNs (1) are usually heterogeneous, via which users can influence each others in multiple channels; and (2) share common users, via whom information could propagate across networks.

In this paper, for the first time we study the influence maximization problem in multiple partially aligned heterogenous OSNs. A new model, multi-aligned multi-relational network influence maximizer (M&M), is proposed to address this problem. M&M extracts multi-aligned multi-relational networks (MMNs) from aligned heterogeneous OSNs based on a set of inter and intra network social meta paths. Besides, M&M extends traditional linear threshold (LT) model to depict the information diffusion across MMNs. In addition, M&M, which selects seed users greedily, is proved to achieve a $(1 - \frac{1}{e})$-approximation of the optimal solution. Extensive experiments conducted on two real-world partially aligned heterogeneous OSNs demonstrate its effectiveness.

1 Introduction

Witnessing the rapid growth of online social networks, viral marketing (i.e., influence maximization) in social networks has attracted much attention of data mining community in the last decade [5,7,10]. Traditional viral marketing problem aims at selecting the set of seed users to maximize the awareness of ideas or products merely based on the *social connections* among users in *one single social network* [3,8,11]. However, in the real world, social networks usually contain heterogeneous information [18–20], e.g., various types of nodes and complex links, via which users are extensively connected and have multiple channels to influence each other [9].

© Springer International Publishing Switzerland 2015
T. Cao et al. (Eds.): PAKDD 2015, Part I, LNAI 9077, pp. 58–69, 2015.
DOI: 10.1007/978-3-319-18038-0_5

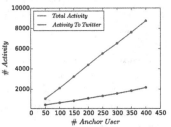

(a) anchor users' reposting   (b) cross-network reposted activities

Fig. 1. Cross-network information propagation analysis

Meanwhile, as studied in [13, 20], users nowadays are usually involved in multiple social networks simultaneously to enjoy more social network services. The shared users across multiple social networks are named as *anchor users* [13]. *Anchor users* exist widely in the real world. Via these anchor users, influence can propagate not only within but also across social networks [16]. To support such a claim, we investigate the partially aligned network dataset studied in this paper (i.e., Twitter and Foursquare) and the results are given in Fig. 1. In Fig. 1(a), we randomly sample a subset of anchor users from Foursquare and observe that 409 out of 500 (i.e., 81.8%) sampled users have reposted their activities (e.g., tips, location checkins, etc.) to Twitter. Meanwhile, the activities reposted by these 409 anchor users only account for a small proportion of their total activities in Foursquare, as shown in Fig. 1(b).

In this paper, we study the influence maximization problem across multiple partially aligned heterogenous social networks simultaneously. This is formally defined as the *Aligned Heterogeneous network Influence maximization (AHI)* problem. The *AHI* problem studied in this paper is very important and has extensive concrete applications in real-world social networks, e.g., *cross-community* [1] even *cross-platform* [16] *product promotion* [17] and *opinion diffusion* [2].

To help illustrate the *AHI* problem, we give an example in Fig. 2, where Fig. 2-A shows the two partially aligned heterogeneous input networks. To conduct viral marketing in the input networks and solve the *AHI* problem, we first extract multiple influence channels (i.e., multi-relations) among users with the heterogeneous information (e.g., traditional follow links, retweet, location checkins, as well as anchor links, etc.) and then select the optimal seed user set based on the constructed multi-relational network, as shown in Fig. 2-B.

The AHI problem is a novel problem and totally different from conventional works on information diffusion and influence maximization, including:(1) traditional viral marketing problems in one single homogeneous social network [6, 12, 17], like the Twitter network shown in Fig. 2-C; (2) topic diffusion in heterogeneous information networks [9], which explores information diffusion in one single multi-relational network (e.g., the Twitter network in Fig. 2-D); and

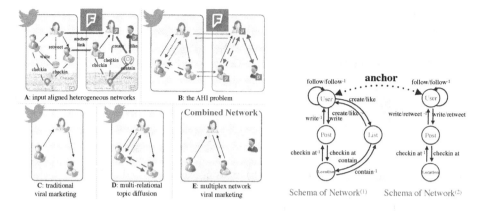

Fig. 2. Comparison of related works   **Fig. 3.** Network schema

(3) influence maximization in multiplex social networks [16], which studies information maximization problem across multiple homogeneous social networks by simply combining multiple networks into one single homogeneous network (e.g., the network shown in Fig. 2-E). In paticular, [16] assumes that the shared users will propagate all the information reaching them to the other network, which is unrealistic and severely violates our observation in Fig. 1(b). Different from all these related works, in the *AHI* problem: (1) the social networks are heterogeneous [18]; (2) multiple social networks [20] are studied simultaneously, where the different heterogeneous networks may have different structures or network schema as shown in Fig. 3; and (3) social networks studied in this paper are partially aligned by *anchor links* [20] instead of being simply merged together.

Addressing the *AHI* problem is very difficult due to the following challenges:

- *Information Diffusion in Heterogeneous Networks*: Users in heterogeneous networks are extensively connected with each other by different types of links and information can diffuse among users via different channels. Modeling information diffusion in heterogeneous social networks is very challenging.
- *Cross-Network Information Propagation*: Via the anchor links, information can propagate across networks. Modeling inter-network information diffusion remains an open problem.
- *NP-hard*: The *AHI* problem is proved to be *NP-hard*, which cannot be solved in polynomial time.

To address the above challenges, a new model <u>M</u>ulti-aligned <u>M</u>ulti-relational network influence maximizer (M&M) is proposed in this paper. M&M first extracts multi-aligned multi-relational networks with the heterogeneous information across the input OSN based on a set of inter and intra network social meta paths [18,20]. M&M extends the traditional Linear Threshold (LT) model to depict the information propagation within and across these multi-aligned multi-relational networks. Based on the extended diffusion model, the influence function which maps seed user set to the number of activated users is proved to be both *monotone* and *submodular*. Thus the greedy algorithm used in M&M, which selects seed users greedily at each step, is proved to achieve a $(1 - \frac{1}{e})$-approximation of the optimal result.

The remaining parts of this paper are organized as follows. We formulate the studied problem in Section 2. In Sections 3-4, we introduce the proposed M&M method. Experiments are given in Section 5. Finally, we introduce the related works in Section 6 and conclude the paper in Section 7.

2   Problem Formulation

In this paper, we will follow the definitions of concepts *"anchor user"*, *"heterogeneous networks"*, *"aligned networks"*, ect., proposed in [20]. Based on the definitions of these terminologies, the *AHI* problem can be formulated as follows:

AHI: Given two partially aligned networks [20] $G^{(1)}$ and $G^{(2)}$ together with the undirected anchor link set $\mathcal{A}$ [13] between $G^{(1)}$ and $G^{(2)}$, the user sets of $G^{(1)}$ and $G^{(2)}$ can be represented as $\mathcal{U}^{(1)}$ and $\mathcal{U}^{(2)}$ respectively. Let $\sigma(\cdot) : \mathcal{Z} \to \mathbb{R}, \mathcal{Z} \subset \mathcal{U}^{(1)} \cup \mathcal{U}^{(2)}$ be the *influence function* [12] which maps the seed user set $\mathcal{Z}$ to the number of users influenced by users in $\mathcal{Z}$. The *AHI* problem aims at selecting the optimal set $\mathcal{Z}^*$ which contains d seed users to maximize the propagation of information across the networks, i.e., $Z^* = \arg\max_{\mathcal{Z} \subseteq \mathcal{U}^{(1)} \cup \mathcal{U}^{(2)}} \sigma(\mathcal{Z})$.

3   Proposed Model

In this section, we will introduce the method M&M in details. M&M can extract multi-aligned multi-relation networks (MMNs) based on a set of inter and intra network social meta paths. The traditional LT model is extended in M&M to depict the information propagation across MMNs.

3.1   Multi-aligned Multi-relational Networks Extraction

We utilize the meta paths [18, 20] defined based on the *network schema* to extract multi-aligned multi-relational networks with the heterogeneous information in aligned networks.

Definition 1. *Network Schema:* For the given network G, its network schema can be defined as $S_G = (O, R)$ with O and R denoting the set of node types and link types in G.

For the partially aligned input networks shown in Fig. 2-A. We note that the network schemas of the two networks are different, so the heterogeneous networks cannot be simply merged together as in the homogeneous case [16]. Based on the network schema, we can represent the diffusion channels as a set of intra and inter network social meta paths that are defined as follows.

Definition 2. *Intra-network Social Meta Path:* *An intra-network social meta path $\mathcal{P}$, based on the given network schema $S_G = (O, R)$, is denoted as $\mathcal{P} = O_1 \xrightarrow{R_1} O_2 \xrightarrow{R_2} \cdots \xrightarrow{R_{k-1}} O_k (k > 1)$ where $O_i \in O, i \in \{1, 2, \cdots, k\}$ and $R_i \in R, i \in \{1, 2, \cdots, k-1\}$. In addition, $O_1 \cdots O_k = User \in O$ as we are mainly concerned about meta paths connecting users, i.e., social meta paths [20].*

Definition 3. *Inter-network Social Meta Path:* *Given two partially aligned heterogenous networks $G^{(1)}$ and $G^{(2)}$ with network schemas $S_{G^{(1)}} = (O^{(1)}, R^{(1)})$ and $S_{G^{(2)}} = (O^{(2)}, R^{(2)})$, $\mathcal{Q} = O_1 \xrightarrow{R_1} O_2 \xrightarrow{R_2} \cdots \xrightarrow{R_{k-1}} O_k (k > 1)$ can be defined to be an inter-network social meta path between $G^{(1)}$ and $G^{(2)}$, where $O_i \in O^{(1)} \cup O^{(2)}, i \in \{1, 2, \cdots, k\}$, $R_i \in R^{(1)} \cup R^{(2)} \cup \{Anchor\}, i \in \{1, 2, \cdots, k-1\}$ and Anchor is the anchor link type. Furthermore, $O_1 = User \in O^{(1)}$, $O_k = User \in O^{(2)}$, and $\exists m \in \{1, 2, \cdots, k-1\}$ such that $R_m = \{Anchor\}$.*

In both Foursquare and Twitter, users can follow other users and check-in at locations, forming two intra-network influence channels among users. Meanwhile, (1) in Foursquare, users can create/like lists containing a set of locations; (2) while in Twitter, users can retweet other users' tweets, both of which will form an intra-network influence channel among users in Foursquare and Twitter respectively. The set of intra network social meta paths considered in this paper as well as their physical meanings are listed as follows:

intra-network social meta paths in Foursquare

(1) *follow*: User $\xrightarrow{follow^{-1}}$ User

(2) *co-location checkins*: User $\xrightarrow{checkin}$ Location $\xrightarrow{checkin^{-1}}$ User

(3) *co-location via shared lists*: User $\xrightarrow{create/like}$ List $\xrightarrow{contain}$ Location $\xrightarrow{contain^{-1}}$ List $\xrightarrow{create/like^{-1}}$ User

intra-network social meta paths in Twitter

(1) *follow*: User $\xrightarrow{follow^{-1}}$ User

(2) *co-location checkins*: User $\xrightarrow{checkin}$ Location $\xrightarrow{checkin^{-1}}$ User

(3) *contact via tweet*: User $\xrightarrow{write}$ Tweet $\xrightarrow{retweet}$ Tweet $\xrightarrow{write^{-1}}$ User

Users can diffuse information across networks via the anchor links formed by anchor users. This can be abstracted as *inter-network social meta path*: User $\xrightarrow{Anchor}$ User. By taking the inter-network meta paths into account, the studied problem becomes even more complex due to the fact that non-anchor users in both networks can also be connected via intra- and inter-network meta paths. As a result, the number of social meta path instances grows mightily.

Each meta path defines an influence propagation channel among linked users. If linked users u, v are connected by only intra-network meta path, we say u has *intra-network relation* to v, otherwise there is *inter-network relation* between them. Based on these relations, we can construct multi-aligned multi-relational networks (e.g., the network shown in Fig. 2-B) for the aligned heterogeneous networks (e.g., the networks shown in Fig. 2-A). The formal definition of multi-aligned multi-relational networks is given as follows:

Definition 4. *Multi-Aligned Multi-Relational Networks:* For two given heterogenous networks $G^{(1)}$ and $G^{(2)}$, we can define the multi-aligned multi-relational network constructed based on the above intra and inter network social meta paths as $M = (U, E, R)$, where $U = U^{(1)} \cup U^{(2)}$ denote the user nodes in the MMNs M. Set E is the set of links among nodes in U and element $e \in E$ can be represented as $e = (u, v, r)$ denoting that there exists at least one link (u, v) of link type $r \in R = R^{(1)} \cup R^{(2)} \cup \{Anchor\}$, where $R^{(1)}$, $R^{(2)}$ are the intra-network link types of networks $G^{(1)}$, $G^{(2)}$ and the inter-network Anchor link between $G^{(1)}$ and $G^{(2)}$ respectively.

3.2 Influence Propagation in Multi-aligned Multi-relational Networks

In this subsection, we will extend the traditional *linear threshold* (LT) model to handle the information diffusion across the multi-aligned multi-relational networks (MMNs).

In traditional *linear threshold* (LT) model for single homogeneous network $G = (V, E)$, user $u_i \in V$ can influence his neighbor $u_k \in \Gamma_{in}(u_i) \subseteq V$ according to weight $w_{i,k} \geq 0$ ($w_{i,k} = 0$ if u_i is *inactive*), where $\Gamma_{in}(u_i)$ represents the users following u_i (i.e., set of users that u_i can influence) and $\sum_{u_k \in \Gamma_{in}(u_i)} w_{i,k} \leq 1$. Each user, e.g., u_i, is associated with a *static threshold* θ_i, which represents the minimal required influence for u_i to become *active*.

Meanwhile, based on the MMNs $M = (U, E, R)$, the weight of each pair of users with different diffusion relations is estimated by pathsim [18]. Formally, the intra-network (inter-network) diffusion weight between user u and v with relation $i(j)$ is defined as:

$$\phi^i_{(u,v)} = \frac{2|P^i_{(u,v)}|}{|P^i_{(u,)}| + |P^i_{(,v)}|}, \quad \psi^j_{(u,v)} = \frac{2|Q^j_{(u,v)}|}{|Q^j_{(u,)}| + |Q^j_{(,v)}|},$$

where $P^i_{(u,v)}(Q^j_{(u,v)})$ is the set of intra-network (inter-network) diffusion meta paths instances, starting from u and ending at v with relation $i(j)$. $|\cdot|$ denotes the size of the set. Thus, $P^i_{(u,)}(Q^j_{(u,)})$ and $P^i_{(,v)}(Q^j_{(,v)})$ means the number of meta path instances with users u, v as the starting and ending users, respectively.

Based on the traditional LT model, influence propagates in discrete steps in the network. In step t, all *active* users remain *active* and *inactive* user can be *activated* if the received influence exceeds his threshold. Only activated users at step t can influence their neighbors at step $t+1$ and the activation probability for user v in one network (e.g., $G^{(1)}$) with intra-network relation i and inter-network relation j can be represented as $g^{(1)}_{v,i}(t + 1)$ and $h^{(1)}_{v,j}(t + 1)$ respectively:

$$g^{(1)}_{v,i}(t + 1) = \frac{\sum_{u \in \Gamma_{in}(v,i)} \phi^i_{(u,v)} \varphi(u, t)}{\sum_{u \in \Gamma_{in}(v,i)} \phi^i_{(u,v)}}, \quad h^{(1)}_{v,j}(t + 1) = \frac{\sum_{u \in \Gamma_{in}(v,j)} \phi^j_{(u,v)} \varphi(u, t)}{\sum_{u \in \Gamma_{in}(v,j)} \phi^j_{(u,v)}}$$

where $\Gamma_{in}(v, i), \Gamma_{in}(v, j)$ are the neighbor sets of user v in relations i and j respectively and $\varphi(u, t)$ denotes if user u is activated at timestamp t. Note that anchor user $v^{(1)}$ is activated does not mean that $v^{(2)}$ is activated at the same time, but $v^{(2)}$ will get influence from $v^{(1)}$ via anchor link.

By aggregating all kinds of intra-network and inter-network relations, we can obtain the integrated activation probability of $v^{(1)}$ [9]. Here logistic function is used as the aggregation function.

$$p_v^{(1)}(t+1) = \frac{e^{\sum_{(i)} \rho_i^{(1)} g_{v,i}^{(1)}(t+1) + \sum_{(j)} \omega_j^{(1)} h_{v,j}^{(1)}(t+1)}}{1 + e^{\sum_{(i)} \rho_i^{(1)} g_{v,i}^{(1)}(t+1) + \sum_{(j)} \omega_j^{(1)} h_{v,j}^{(1)}(t+1)}},$$

where $\rho_i^{(1)}$ and $\omega_j^{(1)}$ denote the weights of each relation in diffusion process, whose value satisfy $\sum_{(i)} \rho_i^{(1)} + \sum_{(j)} \omega_j^{(1)} = 1$, $\rho_i^{(1)} \geq 0$, $\omega_j^{(1)} \geq 0$. Similarly, we can get activation probability of a user $v^{(2)}$ in $G^{(2)}$.

4   Influence Maximization Problem in M&M model

In this section, we will first analyze the influence maximization problem based on M&M model, and then provide M&M Greedy algorithm for seed users selection.

4.1   Analysis of Influence Maximization Problem

Kempe et al. [12] proved traditional influence maximization problem is a NP-hard for LT model, but the objective function of influence $\sigma(\mathcal{Z})$ is *monotone* and *submodular*. Based on these properties, the greedy approximation algorithms can achieve an approximation ratio of $1 - 1/e$.

With the above background knowledge, we will show that the influence maximization problem under the M&M model is also NP-hard and prove the influence spread function $\sigma(\mathcal{Z})$ is monotone and submodular.

Theorem 1. *Influence Maximization Problem across Partially Aligned Heterogenous Social Networks(AHI) is NP-hard.*

Proof: The AHI problem can be easily mapped to "Vertex Cover" problem which is NP-complete. Thus AHI problem is NP-hard.

Theorem 2. *For the M&M model, the influence function $\sigma(\mathcal{Z})$ is monotone.*
Proof: Given the existing seed user sets $\mathcal{Z}$, let z be a seed user selected in this round. Since the weights of multi-relation are nonnegative, adding a new seed user z will not decrease the number of influenced users, i.e.,$\sigma(\mathcal{Z} + z) \geq \sigma(\mathcal{Z})$. Therefore the influence spread function is monotone for the given M&M model.

Theorem 3. *For the M&M model, the influence function $\sigma(\mathcal{Z})$ is submodular.*
Proof: It can be proved with the live-edge path method proposed in [12] very easily. The detail is omitted due to space limitation.

Algorithm 1. M&M Greedy Algorithm for AHI problem

Input: $G^{(1)}$, $G^{(2)}$, anchor user matrix $A_{n_{(1)} \times n_{(2)}}$, d

Output: seed set Z

1: initialize $Z =$, seed index $i = 0$;
2: get network schema $S_G^{(1)}$ and $S_G^{(2)}$, get user set $U = U^{(1)} \cup U^{(2)}$;
3: **for** $v = 0$ to $|U|$ **do**
4:      extract intra and inter network diffusion meta paths of v;
5: **end for**
6: calculate relations' diffusion strength $\phi_{(u,v)}$ and $\psi_{(u,v)}$;
7: define activation probability vector $P^{(1)}$, $P^{(2)}$ and calculate their initial value;
8: **while** $i < d$ **do**
9:      **for** $u \in U \setminus Z$ **do**
10:          using Monte Carlo method to estimate u's marginal gain $M_u = \sigma(Z \cup \{u\}) - \sigma(Z)$ based on users' activation probability;
11:      **end for**
12:      select $z = \arg \max_{u \in U \setminus Z} M_u$
13:      $Z = Z \cup \{z\}$
14:      update users' activation probability in $P^{(1)}$, $P^{(2)}$ and $i = i + 1$.
15: **end while**

4.2   Greedy Algorithm for AHI problem

Since the influence function is monotone and submodular based on the M&M model, step-wise greedy algorithms which select the users who can lead to the maximum increase of influence can achieve a $(1 - \frac{1}{e})$-approximation of the optimal result. Algorithm 1 is a greedy algorithm to solve the AHI problem based on M&M model.

5   Experiment

5.1   Experiment Preparation

In this part, we will introduce the dataset and baselines used in the experiments.

Dataset Description: The partially aligned heterogeneous network dataset used in the experiment are Foursquare and Twitter, The statistics of the two datasets are given in Table 1. For more detailed information about the dataset as well as its crawling methods, please refer to [13].

Baselines: We use following methods as baselines:

Table 1. Properties of the Heterogeneous Social Networks

	property	network	
		Twitter	**Foursquare**
# node	user	500	500
	post	741,529	7,504
	location	34,413	6,300
# link	friend/follow	5,341	2,934
	write	741,529	7,504
	locate	40,203	7,504

- *The* M&M *method* (M&M): M&M is the method proposed in this paper, which can select seed users greedily from the extracted MMNs. Depending on from which network to select the seed users, different variants of M&M are compared: (1) M&M (which selects seed users from both Foursquare and Twitter), (2) M&M-Foursquare (selecting only from Foursquare), and (3) M&M-Twitter (selecting only from Twitter).
- *Lossless method for multiplex networks (LCI)*: Method LCI is the influence maximization method proposed for multiplex networks in [16], which selects seed users from the merged network as shown in Fig. 2-E.
- *Greedy method for single heterogenous network (Greedy)*: Based on a multi-relational network (as shown in Fig. 2-D), method Greedy selects seed users who can lead to the maximum influence gain within one single network. Similar to M&M, Greedy also has two variants: Greedy-Foursquare and Greedy-Twitter.
- *Seed Selection method based on traditional LT model(LT)*: Based on one single homogeneous network (e.g., Fig. 2-C), LT selects seed users who can lead to the maximum influence gain. Two variants of method LT, LT-Foursquare and LT-Twitter, are compared in the experiments.

5.2 Experiment Setup

Based on the input aligned heterogeneous networks, the MMNs are extracted based on a set of intra and inter network social meta paths. The influence score among users in each relation is used to calculate the aggregated activation probability with the logistic function. For simplicity, the weights of all relations (both intra and inter network) are set to be equal (i.e., 0.25 in this paper). The thresholds of users are randomly select from the uniform distribution within range [0,1]. The number of selected seed users is selected from $\{5, 10, \cdots, 50\}$. To simulate different partially aligned networks, we randomly sample the anchor links from the networks with different anchor ratios: $\{0.3, 0.6\}$, where 0.3 denotes that 30% anchor links are preserved while the remaining 70% are removed.

To evaluate the performance of all comparison methods, the number of finally activated users by the seed users is counted as the evaluation metric in the experiments, where anchor users are counted at most once. For example, for an anchor user u (whose accounts in Foursquare and Twitter are $u^{(1)}$ and $u^{(2)}$ respectively), if neither $u^{(1)}$ nor $u^{(2)}$ is activated, then u will not be counted as the activated user (i.e., 0); otherwise u will be counted as one activated user finally (i.e., 1).

5.3 Experiment Results

The experiment results are given in Fig.4, where the anchor ratios of (a) and (b) are 30% and 60% respectively.

As shown in both figures, the number of influenced users will increase as more users are added as the seed users. M&M outperforms all the baselines consistently.

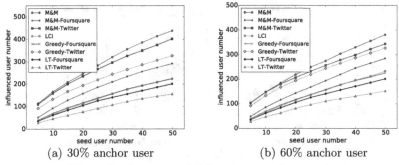

(a) 30% anchor user   (b) 60% anchor user

Fig. 4. Performance of different comparison methods

By comparing M&M with M&M-Foursquare and M&M-Twitter, we observe that M&M can perform better than both M&M-Foursquare and M&M-Twitter in both Fig. 4(a)- 4(b). It demonstrates that selecting seed users globally (i.e., both of the networks) can achieve better results than the method selecting seed users locally (i.e., either Foursquare or Twitter).

Compared with LCI, M&M can outperform LCI with significant advantages. For example, in Fig. 4(a) with seed user set size 20, seed users selected by M&M can activate 246 users, which is 117% larger than the 113 users activated by the seed users selected by LCI. Similar results can be observed for other seed user set sizes in both Fig. 4(a) and Fig. 4(b). As a result, M&M which selects seed users from MMNs can perform better than LCI which selects seed users from combined multiplex networks.

Furthermore, by comparing M&M with Greedy methods (both Greedy-Foursquare and Greedy-Twitter) and LT methods (both LT-Foursquare and LT-Twitter), M&M can always achieve better performance for different seed user set sizes and anchor ratios in Fig. 4(a)-4(b). In summary, selecting seed users based on cross-network information propagation model can select better seed user sets than those merely based on intra-network information propagation models.

5.4   Parameter Analysis

To study the effects of anchor ratio parameter, we compare the performance of all these comparison methods achieved at anchor ratio 0.3 and 0.6, whose results are shown in Fig. 5, where Fig. 5(a)-5(b) correspond to the seed user set sizes 5 and 50 respectively. We abbreviate M&M-Twitter and M&M-Foursquare as M&M-T and M&M-F, while Greedy is abbreviated as G.

By comparing the performance of all the comparison methods achieved with different anchor ratios in Fig. 5(a)-5(b), we observe that Greedy-Foursquare, Greedy-Twitter, LT-Foursquare and LT-Twitter can perform exactly the same with different ratios, as these comparison methods are all based on intra-network information propagation models.

However, the M&M methods can influence more users in aligned networks with lower anchor ratio, e.g., 0.3. With lower anchor ratio, less information can

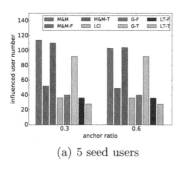 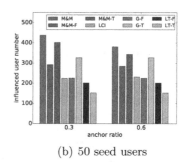

(a) 5 seed users   (b) 50 seed users

Fig. 5. Influence diffusion range with different anchor user ratio

propagate across networks. However, with lower anchor ratio, more users will be non-anchor users. According to the evaluation metric introduced in Subsection 5.2, anchor users' accounts in multiple aligned networks will be counted at most once in the results, which is the reason why M&M can perform a little better for networks with anchor ratio 0.3 than those with anchor ratio 0.6.

6   Related Work

Influence maximization problem as a popular research topic recent years was first proposed by Domingos et al. [6]. It was first formulated as an optimization problem in [12]. Since then a considerable number of work focused on speeding up the seed selection algorithms. CELF in [14] is faster 700 times than original Greedy method, and Chen designed heuristic algorithms for both IC model [4] and LT model [5]. Some other papers extended information diffusion models and provided efficient algorithms [3]. However almost all existing work studied influence maximization problem only for one single network. Nguyen et al. [16] studied the least cost influence problem across multiplex networks.

As to another related topic, information diffusion study, heterogenous and multi-relational networks became an increasingly hot topic [18,19]. Tang et al. [15] proposed a generative graphical model to mine topic-level influence strength with both link and textual information. Gui et al. [9] proposed models by considering weighted combination of different types of relations. While all these work focused on one network.

7   Conclusion

In this paper, we study the novel problem of influence maximization across partially aligned heterogeneous social networks. To solve this problem, we propose multi-aligned multi-relation network based on intra and inter network meta paths to model information diffusion process. Greedy algorithm is proposed to select seed users in multiple heterogenous networks. Extensive experiments conducted on two real OSNs verify the effectiveness of the proposed algorithm. We believe that our work will not only advance the research on influence maximization problem, but also benefit many real-world applications.

Acknowledgments. This work is supported in part by NSFC(61375069, 61403156), NSF through grants CNS-1115234, and OISE-1129076, Google Research Award, and the Pinnacle Lab at Singapore Management University.

References

1. Belak, V., Lam, S., Hayes, C.: Towards maximising cross-community information diffusion. In: ASONAM (2012)
2. Chen, W., Collins, A., Cummings, R., et al.: Influence maximization in social networks when negative opinions may emerge and propagate - microsoft research. In: SDM (2011)
3. Chen, W., Wang, C., Wang, Y.: Scalable influence maximization for prevalent viral marketing in large-scale social networks. In: KDD (2010)
4. Chen, W., Wang, Y., Yang, S.: Efficient influence maximization in social networks. In: KDD (2009)
5. Chen, W., Yuan, Y., Zhang, L.: Scalable influence maximization in social networks under the linear threshold model. In: ICDM (2010)
6. Domingos, P., Richardson, M.: Mining the network value of customers. In: KDD (2001)
7. Du, N., Song, L., Gomez-Rodriguez, M., Zha, H.: Scalable influence estimation in continuous-time diffusion networks. In: NIPS (2013)
8. Goyal, A., Lu, W., Lakshmanan, L.: Celf++: optimizing the greedy algorithm for influence maximization in social networks. In: WWW (2011)
9. Gui, H., Sun, Y., Han, J., Brova, G.: Modeling topic diffusion in multi-relational bibliographic information networks. In: CIKM (2014)
10. He, X., Song, G., Chen, W., Jiang, Q.: Influence blocking maximization in social networks under the competitive linear threshold model. In: SDM (2012)
11. Jiang, Q., Song, G., Cong, G., Wang, Y., Si, W., Xie, K.: Simulated annealing based influence maximization in social networks. In: AAAI (2011)
12. Kempe, D., Kleinberg, J., Tardos, É.: Maximizing the spread of influence through a social network. In: KDD (2003)
13. Kong, X., Zhang, J., Yu, P.: Inferring anchor links across multiple heterogeneous social networks. In: CIKM (2013)
14. Leskovec, J., Krause, A., Guestrin, C., et al.: Cost-effective outbreak detection in networks. In: KDD (2007)
15. Liu, L., Tang, J., Han, J., Jiang, M., Yang, S.: Mining topic-level influence in heterogeneous networks. In: CIKM (2010)
16. Nguyen, D., Zhang, H., Das, S., Thai, M., Dinh, T.: Least cost influence in multiplex social networks: model representation and analysis. In: ICDM (2013)
17. Richardson, M., Domingos, P.: Mining knowledge-sharing sites for viral marketing. In: KDD (2002)
18. Sun, Y., Han, J., Yan, X., Yu, P., Wu, T.: Pathsim: meta path-based top-k similarity search in heterogeneous information networks. In: VLDB (2011)
19. Tang, J., Lou, T., Kleinberg, J.: Inferring social ties across heterogenous networks. In: WSDM (2012)
20. Zhang, J., Yu, P., Zhou, Z.: Meta-path based multi-network collective link prediction. In: KDD (2014)

Multiple Factors-Aware Diffusion
in Social Networks

Chung-Kuang Chou$^{(\boxtimes)}$ and Ming-Syan Chen

Department of Electrical Engineering, National Taiwan University, Taipei, Taiwan
ckchou@arbor.ee.ntu.edu.tw, mschen@cc.ee.ntu.edu.tw

Abstract. Information diffusion is a natural phenomenon that information propagates from nodes to nodes over a social network. The behavior
that a node adopts an information piece in a social network can be
affected by different factors. Previously, many diffusion models are proposed to consider one or several fixed factors. The factors affecting the
adoption decision of a node are different from one to another and may
not be seen before. For a different scenario of diffusion with new factors,
previous diffusion models may not model the diffusion well, or are not
applicable at all. In this work, our aim is to design a diffusion model
in which factors considered are flexible to extend and change. We further propose a framework of learning parameters of the model, which
is independent of factors considered. Therefore, with different factors,
our diffusion model can be adapted to more scenarios of diffusion without requiring the modification of the diffusion model and the learning
framework. In the experiment, we show that our diffusion model is very
effective on the task of activation prediction on a Twitter dataset.

Keywords: Social networks · Diffusion models

1 Introduction

Information diffusion in social networks has been an active research field in about
a decade. It is a natural phenomenon that information propagates from nodes
to nodes over a social network, which acts like an epidemic. There are many
applications on information diffusion, such as promoting an idea more effectively
[5,10] , blocking adverse opinions [3,11], or identifying information flows [13]
in a network. A well-known problem therein is called *influence maximization*,
formulated by Kempe et al. [10]. The problem of influence maximization is to
find a group of target nodes to be convinced of an idea initially to maximize
the spread size, i.e. the number of nodes adopting the idea, on a given diffusion
model.

To model how information diffuses in a network, researchers have proposed
various diffusion models from different aspects. In these diffusion models, the
Independent Cascading (IC) model and the Linear Threshold (LT) Model [10]

© Springer International Publishing Switzerland 2015
T. Cao et al. (Eds.): PAKDD 2015, Part I, LNAI 9077, pp. 70–81, 2015.
DOI: 10.1007/978-3-319-18038-0_6

have been widely employed in many applications and have several variants [1]. Both models consider the influence strength of a neighbor. The key difference is that, for a node turning to adopt an idea, IC considers only the influence from exactly one activated neighbor with uncertainty, while LT considers the collaborative influence contribution from all activated neighbors. In other words, IC places importance on which neighbor tries to affect the node to adopt the idea whereas LT thinks highly on the overall influence contribution from neighbors. Nevertheless, the real world is so complicated that a simple concern is hard to capture such complexity. Many factors probably affect the decision of adoption. For example, an idea that has been adopted by most people will have more chance to influence somebody [9] and an idea is harder to be adopted by someone as time passes. Moreover, a person may have different strength of interests in different topics [1]. However, the factors considered by previous diffusion models in social networks are all fixed. For a different scenario of diffusion with new factors, previous diffusion models may not model the diffusion well, or are not applicable at all. Therefore, one usually has to propose a new diffusion model for modeling diffusion of a specific scenario better by considering new factors.

To design a diffusion model for different factors one by one and to propose the corresponding algorithms, e.g. parameters learning and influence maximization, both become tedious. In this work, we aim to design a diffusion model which can consider multiple factors flexibly and further propose a framework of learning parameters of the model, related to information transmission likelihood between nodes and adoption prediction of a node. To the best of our knowledge, no existing work has the same sight. Specifically, we propose a Multiple-Factors Aware Diffusion (MFAD) model which is able to consider multiple factors flexibly that may affect adoption behaviors. MFAD is a two-stage propagation model. In the first stage, called *influence transmission*, an activated node u tries to influence its inactivated neighbor v with a probability. If the influence of u is successfully transmitted to v, in the second stage, called *adoption decision*, v decides whether it becomes activated based on its considerations, predicted via its related classification model trained on historical adoption information. Unfortunately, due to the limitation of observation in the real world, only positive instances are available to train classifiers , which is hard to achieve good performance. We further design a mechanism to get unlabeled instances to help train nodes' classifiers and propose the learning framework to learn the classifiers and transmission probabilities between nodes. Our contributions are summarized as follows.

1. Our proposed MFAD model is flexible to extend and change factors since we employ a classification approach to predicting the adoption behavior of a node.
2. Our proposed learning framework is independent of factors considered and we show the learning framework is effective in the experiment.
3. Due to the limitation of observation on diffusion in the real world, to predict adoption behaviors is hard to reach good results. We explicitly tackle this issue by learning nodes' classifiers for adoption decision with only positive and unlabeled instances.

The remaining of the paper is organized as follows. We next review the related work in Section 2. We introduce our diffusion model in Section 3 and how to learn the model in Section 4. In Section 5, we conduct experiments on a Twitter dataset. Finally, we conclude in Section 6.

2   Related Work

In this section, we briefly review the related works on diffusion models in social networks and learning parameters of diffusion models.

Diffusion models interpret how information spreads within a network. As mentioned above, IC and LT are two classical models and have been widely employed since they were connected to the influence maximization problem [10]. Recently, more factors of diffusion are explored in the literature like [1], [9] and [14], to name a few. N. Barbieri et al. [1] extend the IC model to consider topic distribution of items. T.-A. Hoang and E.-P. Lim [9] propose a model considering three factors, user virality, user susceptibility and item virality. Moreover, S.A. Myers et al. [14] explore not only internal influence from activated nodes in a network, but also external influence outside the network. However, as discussed above, the factors considered by these diffusion models are all fixed. For a different scenario of diffusion with new factors, these models may not model the diffusion well, or are not applicable at all.

Although diffusion models in social networks have been proposed for a long time, algorithms to learn parameters of a diffusion model are proposed recently. For example, K. Saito et al. [16] first propose a learning method for the IC model. The following up works mainly propose learning methods for their own diffusion models [1]. Moreover, A. Goyal et al. [7] propose the Credit Distribution model that directly estimates spread size from diffusion data without learning influence probabilities between nodes. Since we aim to design a learning framework that is independent of the diffusion factors considered, the above results do not apply to our scenario.

3   Proposed Model

In the work, our aim is to design a diffusion model which considers multiple factors flexibly for information propagation. We propose the Multiple Factors-Aware Diffusion (MFAD) model in the section.

Given a social graph $G = (V, E)$, where V is the node set and E is the edge set composed of directed edges without multiple edges and self-loops, let $p_{u,v}$ denote the probability of node u to successfully transmit influence to node $v \in u$'s out-neighbors $N^{out}(u)$ after u is activated by an item i and let $f_v(x)$ denote a probabilistic classifier for node v where $f_v(x)$ considers multiple predefined features, i.e. factors, that affect the tendency of u to adopt item i after v is exposed to i and x is the feature vector of the exposure. Note that a nonprobabilistic classification model, the outputs of which can be transformed to the probabilistic outputs [15], is applicable to $f_v(x)$, e.g. SVM with Platt scaling. The Multiple Factors-Aware Diffusion (MFAD) model is defined as follows.

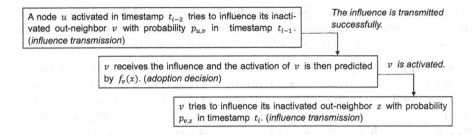

Fig. 1. The successful activation process of MFAD

Definition 1. *Multiple Factors-Aware Diffusion (MFAD) model. The propagation of diffusion starts with initial seed nodes $S_0 \subseteq V$ which have adopted item i. In the first timestamp t_1, each $u \in S_0$ tries to influence u's out-neighbors which are not activated by item i. The successful activation probability for $v \in N^{out}(u)$ is calculated as $p_{u,v} f_v(x)$. The activated nodes in t_1 are denoted as S_1. In timestamp t_2, new activated nodes in S_1 try to activate their inactivated out-neighbors in the same process as the above. The process runs iteratively. If S_j is empty in timestamp t_j, the diffusion terminates. Note that the spread size can be expressed as $|\cup_{0 \le i \le j-1} S_i|$ and when a node becomes activated, it never turns to be inactivated.*

MFAD is a two-stage propagation model. An illustration of the successful activation process in MFAD is shown as Figure 1. In the first stage, called *influence transmission*, a node u tries to influence each $v \in N^{out}(u)$ which is not activated with probability $p_{u,v}$ in timestamp t_{i-1}, where u is activated in timestamp t_{i-2}. However, the influence successfully transmitted over the edge does not directly activate a node. Our model has the following stage, called *adoption decision*. In the second stage, if the influence of u is successfully transmitted with probability $p_{u,v}$ to v previously, v receives this influence and decides whether it becomes activated based on its considerations, predicted via its classification model $f_v(x)$. If v is activated, v will try to influence its neighbors in the next timestamp t_i. Consider a news diffusing in an online social network, e.g. Facebook and Twitter. A user u posted a message about the news recently. Due to the ranking mechanism designed by Facebook or since there are too many messages, a friend v of u may not see the message, which is modeled by influence transmission. Moreover, even if the friend v reads the message, v may consider whether to share or reply to the message based on several concerns, e.g. v's interests, the importance of the news, which is modeled by adoption decision. In contrast to the traditional diffusion models, MFAD is flexible to consider multiple factors and considers more in a microscopic view for information propagation.

4   Two-Stage Learning

In this section, we propose a two-stage learning framework for MFAD since MFAD is a two-stage propagation model. In the first stage, the classifier of each node is trained, which corresponds to adoption decision, while the transmission probability between two connected nodes is estimated in the second stage, which corresponds to influence transmission. We first introduce the observed data.

Observed Data. A propagation trace consists of activation records. Each record $\{u, v, i, t\}$ represents that node u adopts an item i at timestamp t and the adoption is caused by u's in-neighbor v. If u is actually a seed of the item in the observed data, v does not exist and is set to be NIL. In reality, we usually do not observe that a node fails to influence others. In other words, we would only have positive instances for training classifiers directly from the propagation trace. We next discuss how to learn classifiers of nodes in such a situation.

4.1   Learning Classifiers of Nodes

Due to the limitation of observation in the real world, only positive instances are available to train binary classifiers of each nodes. However, a classifier trained on only positive instances is hard to achieve good performance. In fact, *unlabeled* instances can provide more information for learning, e.g. feature distribution, and can be generated via observing that an inactivated out-neighbor of an activated node does not turn activated in the next timestamp. We use the term *unlabeled* instead of *negative* since an unlabeled instance may be positive or negative due to the limitation of observation. Thus, the task of the first-stage learning becomes training classifiers by using positive and unlabeled instances. In the literature [6,12], the problem is called *positive and unlabeled learning*. Among previous work on positive and unlabeled learning, C. Elkan and K. Noto [6] provide a principled way to assigning weights to unlabeled instances. Based on their work [6], we construct a framework to learn nodes' classifiers for MFAD. In the following, we first describe how to obtain unlabeled instances from observed positive records and then describe how to train a node's classifier based on positive and unlabeled instances.

Obtaining Unlabeled Instances. With observed positive records, we can analyze the whole propagation trace to get positive instances for training nodes' classifiers with ease. However, negative instances are hard to obtain due to two main reasons: (1) an item does not successfully be exposed to a node from its in-neighbor; (2) the observation window for a node is not long enough. Fortunately, we can generate unlabeled instances to help train a node's classifier.

Assume that we have the complete propagation trace which consists of positive records in the format of $(u, s, i, t, o = 1)$ where the binary variable o indicates whether a record is an observed positive record in the trace or not. If a node u adopts an item i from s at timestamp t, we can observe u's out-neighbors who haven't adopted i from timestamp t. For an out-neighbor v of u, if v does not adopt i in the complete propagation trace, we generate an *unlabeled record* $(v, u, i, t', o = 0)$, where t' is the end observation time of propagation trace. However, the approach is with high cost and does not work if the size of the trace is extremely large. For a program to sequentially trace the positive records in

Algorithm 1. Unlabeled Record Generation

Require: graph $G = (V, E)$, complete propagation trace $\mathbb{D}$ sorted in chronological
    order, time window T
Ensure: unlabeled records
 1: Let M be a table to memorize possible unlabeled instances
 2: **for all** $(u, s, i, t, 1)$ in $|\mathbb{D}|$ **do**
 3:    **while** M contains a record $(u, *, i, *, 0)$ **do**
 4:       remove the record from M
 5:    **for all** v in u's out-neighbors **do**
 6:       **if** v hasn't adopted i **then**
 7:          insert an unlabeled record $(v, u, i, t, 0)$ into M
 8:    **while** the timestamp of the oldest record in M is less than $t - T$ **do**
 9:       output and remove the record from M
10: **while** $|M| > 0$ **do**
11:    output and remove the record from M

chronological order, it has to memorize all items that a node has not adopted
in order to generate unlabeled records at the end of tracing, which is impossible for a single machine with limited memory size. Otherwise, multiple scans are
needed, which incurs many disk I/O operations and therefore is time-consuming.

In a more general way, we propose Algorithm 1 to trace positive records to
generate unlabeled records. Let T be the time window to observe whether an
out-neighbor v of u, for a positive record $(u, s, i, t, 1)$, adopts i before $t + T$.
The main idea of the algorithm is that if v does not adopt i before $t + T$, the
algorithm generates an unlabeled record $(v, u, i, t, 0)$. Although the pseudo code
of Algorithm 1 is written in a batch way, it is easy to adapt it to process positive
records coming in a streaming way. Note that a feature vector x, i.e. an instance,
is generated at the same time when a positive record is traced or an unlabeled
record is generated in order to capture the state of an exposure.

Training a Node's Classifier. With the above approach, for a node u, we
can obtain positive instances $\mathbb{P}(u)$ and unlabeled instances $\mathbb{U}(u)$ for training
u's classifier in order to predict the adoption tendency of an instance. Given
an instance x, the goal is to predict $p(a = 1|x)$, where a is a binary random
variable to indicate whether the instance is positive $(a = 1)$ or negative $(a = 0)$.
Recall that o is a binary variable to indicate an instance is observed $(o = 1)$ or
unlabeled $(o = 0)$. In the lemma derived in [6], $p(a = 1|x) = p(o = 1|x)/c$, where
$c = p(o = 1|a = 1)$ is a constant value[1]. Based on the lemma, they [6] further
reach the result on how to give weights to instances rigorously as the following.
The weight of a positive instance is still unit, while an unlabeled instance have
two copies, where one copy is a positive instance with weight $p(a = 1|x, o = 0)$
and the other copy is a negative instance with weight $1 - p(a = 1|x, o = 0)$.

---

[1] Due to the space limit, we omit the details of the lemma. If readers are interested
  in the lemma and the corresponding results, please refer to [6].

Note that $p(a=1|x,o=0)=\frac{1-c}{c}\frac{p(o=1|x)}{1-p(o=1|x)}$. Thus, our task here becomes three subtasks: (1) to learn $p(o=1|x)$, (2) estimate c and (3) learn $p(a=1|x)$.

Specifically, for a node u, (1) we first train a nontraditional classifier $g_u(x)=p_u(o=1|x)$ on $\mathbb{P}(u)-\mathbb{V}(u)$ and $\mathbb{U}(u)$, where the instances in $\mathbb{V}(u)$ are randomly selected from $\mathbb{P}(u)$, which is reserved as a validation set to estimate c. (2) Next, c is estimated as $\frac{1}{|\mathbb{V}(u)|}\sum_{x\in\mathbb{V}(u)}g_u(x)$ according to [6]. (3) Finally, we construct positive instances $\mathbb{P}'(u)$ and negative instances $\mathbb{N}(u)$ to train a traditional classifier $f_u(x)=p_u(a=1|x)$ for the node u. $\mathbb{P}'(u)$ contains the instances in $\mathbb{P}(u)$ and copies from the instances in $\mathbb{U}(u)$ with each weight $p_u(a=1|x,o=0)$. $\mathbb{N}(u)$ consists of copies from the instances in $\mathbb{U}(u)$ with each weight $1-p_u(a=1|x,o=0)$. In the experiments, we use the logistic regression to train both $g_u(x)$ and $f_u(x)$ since its output probability is well-calibrated [6] by applying the above way.

4.2   Learning the Transmission Probability

With the above $\mathbb{P}(v)$, $\mathbb{U}(v)$ and the trained classifier $f_v(x)$ for each node $v\in V$, we now describe how to learn transmission probability between two connected nodes. Let $\mathbb{D}=\bigcup_{v\in V}(\mathbb{P}(v)\cup\mathbb{U}(v))$ denote the dataset for learning transmission probabilities. An instance $x\in\mathbb{D}$ is in the format of $(f_1,f_2,...,f_m)_{[u,v,o,t,i]}$ where u is the node that tries to activate v by item i before time t ($x_o=1$) or at time t ($x_o=0$), o is a binary variable to indicate whether v is activated during the observation in Algorithm 1 and $f_1,f_2,...,f_m$ are factors of adoption, calculated in the same time of running Algorithm 1 for the exposure. An instance $x\in\mathbb{D}$ is unlabeled if $x_o=0$; otherwise, x is positive.

We train the MFAD model via maximizing the likelihood of $\mathbb{D}$ in the MFAD model. Let D_s denote the data of node s in $\mathbb{D}$, i.e. $\mathcal{D}_s=\{x\in\mathbb{D}|x_v=s\}$, and let Θ denote all parameters of the MFAD model to learn, i.e. all transmission probabilities, and $\Theta_s=\{p_{q,s}|q\in N^{in}(s)\}$ denote transmission probabilities between node s and its in-neighbors $N^{in}(s)$. Assuming adoptions between nodes are independent, the complete data log-likelihood can be expressed as follows.

$$\mathcal{L}(\Theta;\mathbb{D})=\log\prod_{s\in V}\mathcal{L}(\Theta_s;\mathcal{D}_s) \qquad (1)$$

Note that since we want to learn transmission probability between two connected nodes, we exclude an instance x, x_u of which is NIL, from $\mathbb{D}$. Moreover, since $f_s(x)$ for each node s has trained in the above, the data likelihood of each $\mathcal{D}_s$ is only related to Θ_s. To maximize Eq.(1) is equal to maximizing each $\mathcal{L}(\Theta_s;\mathcal{D}_s)$,i.e.

$$\forall s\in V,\max_{\Theta_s}\log\mathcal{L}(\Theta_s;\mathcal{D}_s). \qquad (2)$$

In reality, the diffusion happens in a continuous time space, while MFAD is a discrete time-based diffusion model. We include time constraints Δ^+ and Δ^- to decide the validity of an instance. Let $\mathcal{D}^+=\{x\in\mathbb{D}|x_o=1\}$ and $\mathcal{D}^-=\{x\in\mathbb{D}|x_o=0\}$. We define $\mathcal{D}_s^+$ and $\mathcal{D}_s^-$ as follows.

$$\mathcal{D}_s^+=\{x\in\mathcal{D}_s|x_o=1\wedge\exists y\in\mathcal{D}^+(y_v=x_u\wedge y_i=x_i\wedge 0\leq x_t-y_t\leq\Delta^+)\} \quad (3)$$

$$\mathcal{D}_s^- = \{x \in \mathcal{D}_s | x_o = 0 \wedge \exists y \in \mathcal{D}^+(y_v = x_u \wedge y_i = x_i \wedge 0 \le x_t - y_t \le \Delta^-)\} \quad (4)$$

Note that for an instance $x \in \mathcal{D}^+$, $x_u \in N^{in}(x_v)$ should hold, where $N^{in}(v)$ is v's in-neighbor set. The data likelihood of node s is then defined as

$$\mathcal{L}(\Theta_s; \mathcal{D}_s) = \prod_{q \in N^{in}(s)} \prod_{x \in \mathcal{D}_{q,s}^+} (p_{q,s} f_s(x)) \prod_{q \in N^{in}(s)} \prod_{x \in \mathcal{D}_{q,s}^-} (1 - p_{q,s} f_s(x)), \quad (5)$$

where

$$f_s(x) = p_s(a = 1 | x), \mathcal{D}_{q,s}^+ = \{x \in \mathcal{D}_s^+ | x_u = q\}$$
$$\text{and } \mathcal{D}_{q,s}^- = \{x \in \mathcal{D}_s^- | x_u = q\}. \quad (6)$$

The data log-likelihood of node s is:

$$\log \mathcal{L}(\Theta_s; \mathcal{D}_s) = \sum_{q \in N^{in}(s)} [\sum_{x \in \mathcal{D}_{q,s}^+} \log(p_{q,s} f_s(x)) + \sum_{x \in \mathcal{D}_{q,s}^-} \log(1 - p_{q,s} f_s(x))] \quad (7)$$

To find $p_{q,s}$ by maximizing the above log likelihood, let $\frac{\partial \log \mathcal{L}(\Theta_s; \mathcal{D}_s)}{\partial p_{q,s}} = 0$:

$$\sum_{x \in \mathcal{D}_{q,s}^+} \frac{1}{p_{q,s}} + \sum_{x \in \mathcal{D}_{q,s}^-} \frac{-f_s(x)}{1 - p_{q,s} f_s(x)} = 0 \quad (8)$$

Since no closed form solution for Eq.(8) exists, we employ the Brent's algorithm [2]. The Brent's algorithm uses a combination of golden section search and successive parabolic interpolation. For an initial good guess $p_{q,s}^0$ in order to converge fast, we apply the first order Taylor series to approximate $\frac{-f_s(x)}{1 - p_{q,s} f_s(x)}$ at $p_{q,s} = 0$ as $-f_s(x) - f_s(x)^2 p_{q,s}$. Thus, the Eq. (8) becomes

$$\sum_{x \in \mathcal{D}_{q,s}^+} \frac{1}{p_{q,s}} + \sum_{x \in \mathcal{D}_{q,s}^-} [-f_s(x) - f_s(x)^2 p_{q,s}] = 0 \quad (9)$$

and by some mathematical manipulation we get $\hat{p}_{q,s} = \frac{-C + \sqrt{C^2 + 4BD}}{2D}$, where $B = |\mathcal{D}_{q,s}^+|$, $C = \sum_{x \in \mathcal{D}_{q,s}^-} f_s(x)$ and $D = \sum_{x \in \mathcal{D}_{q,s}^-} f_s(x)^2$. Note that D should be a real number greater than 0 and obviously, B and C are non-negative real numbers. In some situation, $\hat{p}_{q,s}$ will not be a valid probability value, the value $\frac{(3 - \sqrt{5})}{2}$ suggested in the Brent's algorithm [2] is used instead.

5 Experiments

In the section, we conduct experiments on a Twitter dataset to evaluate the effectiveness on activation prediction. We first describe the setup.

Table 1. Data Statistics

T	training instances			testing instances		
	positive	unlabeled	all	positive	unlabeled	all
		positive negative			positive negative	
3	17,197	274,575 236,807	528,579	115,955	80,243 15,288	211,486
6	17,197	270,467 236,807	524,471	114,463	78,788 15,940	209,191
12	17,197	263,750 236,807	517,754	112,063	76,496 17,001	205,560
24	17,197	252,274 236,807	506,278	108,180	72,864 18,893	199,937
48	17,197	233,820 236,807	487,824	101,010	66,049 20,705	187,764
96	17,197	204,919 236,807	458,923	88,807	54,793 24,520	168,120

5.1 Setup

Dataset. We use the real dataset collected from Twitter by L. Weng et al. [17]. We use standard preprocessing steps, similar to the steps used in [1] to clean diffusion data. However, in order to obtain enough size of training data for training nodes' classifiers, we allow multiple activation records of the same item for a node. After the preprocessing, each node has at least 20 activation records, i.e. retweets and tweets with hashtags in Twitter, and each item, i.e. hashtags, are adopted by at least 20 nodes. Moreover, there is no isolated node left. The remaining social graph consists of $24,045$ nodes and $871,745$ directed edges. The remaining activation records contain $8,427$ different items and the number of all activation records is $1,105,316$. The dataset spans from March 23 to April 25 in 2012, approximately one month long.

Factors. We first define some notations. Let $deg_{in}(v)$ and $deg_{out}(v)$ denote v's in-degree and out-degree in the graph. For an item i, we use $t_{glo}(i)$ to denote the earliest time in which i is adopted by some node in the data and $t_{loc}(i, v, t)$ to denote the earliest time in which i is exposed to v by an in-neighbor of v that adopts i before time t. Let $node_{glo}(i, t)$ denote all nodes activated by item i before time t and $node_{loc}(i, t, v)$ to represent in-neighbors of v which are activated by item i before time t. For a directed edge from u to v, we use $ratio_{from}(v, u, t)$ to denote the ratio that v's adoptions are caused by u and $ratio_{same}(v, u, t)$ to denote the ratio that v's adopted items are the same as u's adopted items before time t.

For an instance $x = (f_1, f_2, ..., f_m)_{[u,v,o,t,i]}$, where node u tries to activate v by item i before time t ($x_o = 1$) or at time t ($x_o = 0$), the features $f_1, f_2, ..., f_m$ are composed of three types, structure-based, time-based and history-based features. Structure-based features include $deg_{in}(u)$, $deg_{out}(u)$, $deg_{in}(v)$, $deg_{out}(v)$ and the number of common neighbors between u and v. Time-based features are $t - t_{glo}(i)$, $t - t_{loc}(i, v, t)$ and $t(i, u) - t$, where $t(i, u)$ is the time that node u adopts the item and if it is unavailable in the dataset, we assume $t(i, u) - t = 0$. The first two are able to reflect global and local freshness. The last one is to measure the adoption latency. History-based features are $|node_{glo}(i, t)|$, $|node_{loc}(i, t, v)|$, $ratio_{from}(v, u, t)$ and $ratio_{same}(v, u, t)$. Thus, there are $m = 12$ features in total for training a node's classifier in the experiment.

Instances. We use the approach introduced in Section 4.1 to generate positive and unlabeled instances from the dataset with different T for the most active 100 nodes, measured by the number of positive records in the whole dataset. The statistics of training and testing instances are summarized as Table 1. In both training and testing sets, we exclude instance x, x_u of which is NIL, since we want to learn transmission probability between two connected nodes for diffusion models. Note that the positive instances for testing consist of positive instances and unlabeled positive instances, while the negative instances for testing consist of unlabeled negative instances. For each T, we use the earliest 20% instances as the training set. From the latest 80% instances, the testing set only contains instances related to $p_{u,v}$ that is trained in the training data for MFAD. Thus, the satisfied testing instances are not too many. Moreover, the number of unlabeled instances for training is much more than the labeled positive instances since the earliest 20% time ($\sim$ 6.6 days) is relative short and when a node u adopts an item i at time $t(i,u)$ but u's in-neighbors all adopt i before $t(i,u) - T$, $|N^{in}(u)|$ unlabeled positive instances will be generated.

Methods. We include the following three methods to predict activations of nodes: (1) the logistic regression directly trained on positive and unlabeled instances (LOGIST), which is the classical approach, (2) our proposed learning framework for the MFAD model (MFAD) and (3) the independent cascading model (IC). Note that only MFAD and IC are diffusion models, while LOGIST is a classification algorithm only and cannot be applied to other applications on diffusion, e.g. influence maximization [10]. We select the IC model instead of the LT model since IC is also a probabilistic diffusion model. The parameters of IC are inferred by the maximum-likelihood estimation conducted in the same approach of Section 4.2. For two connected nodes u and v, the influence probability $p_{u,v}$ is $\frac{|\mathcal{D}_{u,v}^+|}{|\mathcal{D}_{u,v}^+| + |\mathcal{D}_{u,v}^-|}$ for IC. While training the nodes' classifiers for both LOGIST and MFAD, the class imbalance problem is encountered, i.e. skewed class distribution. We use SMOTE[4], which doubles the size of the minority class, and then apply SpreadSubsample [8] to undersample instances of the majority class to balance the class distribution. Moreover, we set time constraints Δ^+ and Δ^- in Eq. (3) and (4) as the same value of T. All methods are implemented in Java with Weka [8] and executed in a PC with an Intel i7 3.4GHz CPU. The running time of a run of MFAD for the same T does not exceed 2 hours, including time for sampling, training 100 nodes' classifiers and learning related transmission probabilities.

Metrics. We use four metrics, precision, recall, F-Measure and accuracy, to measure the results of activation prediction, based on true positive (TP), false positive (FP), true negative (TN) and false negative (FN) instances. Precision is $\frac{TP}{TP+FP}$. Recall is $\frac{TP}{TP+FN}$. F-Measure is $\frac{2 \times precision \times recall}{precision+recall}$ and accuracy is defined as $\frac{TP+TN}{TP+FP+TN+FN}$.

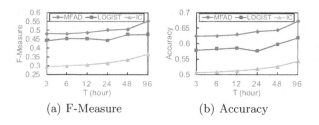

(a) F-Measure   (b) Accuracy

Fig. 2. Overall Results

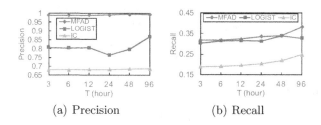

(a) Precision   (b) Recall

Fig. 3. Results of Components of F-Measure

5.2   Results

The overall results of activation prediction are shown in Figure 2. Our MFAD
outperforms the other two methods significantly in the overall metrics, F-Measure
and accuracy. F-Measure in Fig. 2(a) concerns mainly on true positive, false positive and false negative instances, while accuracy in Fig. 2(b) takes true negative
instances into consideration. F-Measure is suitable for the scenario of retrieval
of activated nodes whereas accuracy is more suitable for the scenario of spread
estimation. MFAD works great for both scenarios. For the components of F-
Measure, MFAD is very effective in precision as shown in Fig. 3(a), which means
the size of false positive instances is much smaller than those of LOGIST and
IC. The recalls of MFAD and LOGIST are close to each other but much better
than that of IC as shown in Fig. 3(b). Moreover, as T increases, F-Measure and
accuracy become better for all methods since the positive unlabeled instances
are fewer and thus more positive instances are available for training classifiers.

In summary, MFAD is the best method to predict activation of nodes among
three methods. Most importantly, MFAD is a diffusion model and therefore
can simulate how information diffuses whereas LOGIST cannot. IC is also a
diffusion model, but it cannot reflect the state of an exposure precisely and thus
do not model the diffusion well. Although there is an extension of IC in [1],
called TIC, to consider the topic factor, the dataset does not have the detailed
textual information of tweets and hence we do not include TIC in the experiment.
Nevertheless, our MFAD model can consider the topic factor by defining new
features for nodes' classifiers with ease, which does not require the modifications
of the learning framework.

6   Conclusions

In this work, we propose the model of Multiple-Factors Aware Diffusion Model (MFAD) which explicitly models *influence transmission* and *adoption decision* and considers multiple factors flexibly that may affect adoption behaviors. The learning framework of MFAD is independent of factors considered and is effective as shown in the experiment. Therefore, MFAD has more flexibility and can be applied to different scenarios for different purposes with ease. In the future, we will design influence maximization algorithms for MFAD.

Acknowledgments. This work is in part supported by MOST of Taiwan (103-2221-E-001-038-MY2).

References

1. Barbieri, N., Bonchi, F., Manco, G.: Topic-aware social influence propagation models. In: ICDM, pp. 81–90 (2012)
2. Brent, R.P.: Algorithms for minimization without derivatives. Prentice-Hall, Englewood Cliffs (1973)
3. Budak, C., Agrawal, D., El Abbadi, A.: Limiting the spread of misinformation in social networks. In: WWW, pp. 665–674 (2011)
4. Chawla, N.V., Bowyer, K.W., Hall, L.O., Kegelmeyer, W.P.: Smote: Synthetic minority over-sampling technique. J. Artif. Int. Res. **16**(1), 321–357 (2002)
5. Chen, W., Wang, C., Wang, Y.: Scalable influence maximization for prevalent viral marketing in large-scale social networks. In: KDD, pp. 1029–1038 (2010)
6. Elkan, C., Noto, K.: Learning classifiers from only positive and unlabeled data. In: KDD, pp. 213–220 (2008)
7. Goyal, A., Bonchi, F., Lakshmanan, L.V.S.: A data-based approach to social influence maximization. Proc. VLDB Endow. **5**(1), 73–84 (2011)
8. Hall, M., Frank, E., Holmes, G., Pfahringer, B., Reutemann, P., Witten, I.H.: The weka data mining software: An update. SIGKDD Explor. Newsl. **11**(1), 10–18 (2009)
9. Hoang, T.A., Lim, E.P.: Virality and susceptibility in information diffusions. In: ISWSM (2012)
10. Kempe, D., Kleinberg, J., Tardos, E.: Maximizing the spread of influence through a social network. In: KDD, pp. 137–146 (2003)
11. Kimura, M., Saito, K., Motoda, H.: Blocking links to minimize contamination spread in a social network. ACM TKDD 3(2), 9:1–9:23 (2009)
12. Lee, W.S., Liu, B.: Learning with positive and unlabeled examples using weighted logistic regression. In: ICML (2003)
13. Lin, C., Mei, Q., Han, J., Jiang, Y., Danilevsky, M.: The joint inference of topic diffusion and evolution in social communities. In: ICDM, pp. 378–387 (2011)
14. Myers, S.A., Zhu, C., Leskovec, J.: Information diffusion and external influence in networks. In: KDD, pp. 33–41 (2012)
15. Niculescu-mizil, A., Caruana, R.: Predicting good probabilities with supervised learning. In: ICML, pp. 625–632 (2005)
16. Saito, K., Nakano, R., Kimura, M.: Prediction of information diffusion probabilities for independent cascade model. In: KES, pp. 67–75 (2008)
17. Weng, L., Menczer, F., Ahn, Y.Y.: Virality prediction and community structure in social networks. Nature Scientific Report 3(2522) (2013)

Understanding Community Effects
on Information Diffusion

Shuyang Lin[1]([✉]), Qingbo Hu[1], Guan Wang[3], and Philip S. Yu[1,2]

[1] Department of Computer Science,
University of Illinois at Chicago, Chicago, IL, USA
slin38@uic.edu
[2] Institute for Data Science, Tsinghua University, Beijing, China
[3] LinkedIn, Mountain View, CA, USA

Abstract. In social network research, community study is one flourishing aspect which leads to insightful solutions to many practical challenges. Despite the ubiquitous existence of communities in social networks and their properties of depicting users and links, they have not been explicitly considered in information diffusion models. Previous studies on social networks discovered that links between communities function differently from those within communities. However, no information diffusion model has yet considered how the community structure affects the diffusion process.

Motivated by this important absence, we conduct exploratory studies on the effects of communities in information diffusion processes. Our observations on community effects can help to solve many tasks in the studies of information diffusion. As an example, we show its application in solving one of the most important problems about information diffusion: the influence maximization problem. We propose a community-based fast influence (CFI) model which leverages the community effects on the diffusion of information and provides an effective approximate algorithm for the influence maximization problem.

1 Introduction

For many years, community study is one of the hot topics in social network research. Studies in this area offer insightful solutions to many classic problems of social network research, such as network evolution [14], recommendation system [19], and expert finding [2]. Communities can be potentially helpful for studies on diffusion of information in social networks. Previous studies found that links between communities function differently from those within communities: friends in the same communities have stronger links, but weaker links between friends in different communities are crucial in the diffusion of novel information, because these links provide more useful information to people [1, 7, 12].

Some key problems in the studies of information diffusion have been found difficult to solve by traditional information diffusion methods. Studies on the community structure of social networks may bring new ideas for solving these

© Springer International Publishing Switzerland 2015
T. Cao et al. (Eds.): PAKDD 2015, Part I, LNAI 9077, pp. 82–95, 2015.
DOI: 10.1007/978-3-319-18038-0_7

problems. For example, one of the key problems, the influence maximization problem for the independent cascade model has been proved to be NP-hard [15]. By considering the effects of community studies on the diffusion of information, we can easily come up with some intuitive heuristics to solve that problem more efficiently. For example, we may utilize the community homophily to quickly estimate the influence of users. We may also select seed nodes from different communities to minimize the overlap and maximize the influence.

However, few existing work explicitly studied the effects of communities on the diffusion of information, or use these effects to solve diffusion-related problems. Motivated by this important absence, we introduce the first exploratory study on the effects of communities on information diffusion processes. By analyzing real-world datasets, we study the diffusion of information with communities. We first observe the action homophily of communities, and then introduce the concept of role-based homophily of communities, which consists of influencee role homophily and influencer role homophily. We discover that the role-based homophily is significantly stronger than the action homophily.

Our findings on community effects can lead to insightful solutions to many problems in information diffusion studies. As an example application of these findings, we propose an approximate solution for the influence maximization problem. We design a community-based fast influence (CFI) model based on the influencee role homophily of communities. The CFI model applies a community clustering method to social networks, and makes aggregations on users' roles as influencees. Influence maximization algorithm based on the CFI model can efficiently select seed nodes to maximize the influence. The main contributions are summarized in the following:

1. We conduct quantitative analyses on real-world datasets to explore the effects of community on the diffusion of information.

2. We get valuable findings about the community effects from quantitative analyses. We introduce the concept of role-based homophily of communities. These understandings can bring new insights to the studies of information diffusion.

3. We show an example application of our findings on the influence maximization problem. We propose a community-based fast influence (CFI) model, and an efficient approximate influence maximization algorithm based on that model.

2   Related work

Information Diffusion Problem. Several models have been proposed for the information diffusion processes. The independent cascade (IC) model and its variants are most widely used information diffusion models [10,15,16,21]. The basic idea of the IC model is: if a node in a social network becomes active, it can make its neighbors active with a probability, and for each node the attempts of its neighbors to activate it are independent. The influence maximization problem has been defined for the IC model and a few other information diffusion models [15]. Given an IC model, the problem is to select a seed set with k nodes so that the expected number of active nodes are maximized. This problem has been proved to be NP-hard. The first solution to it is a greedy algorithm that repeatedly invokes a computational expensive sampling method [15]. Heuristic algorithms and optimized versions of the greedy algorithm have been proposed in previous works [3,4,17]. Work in [22] proposed an heuristic algorithm which finds influencers from communities. Different from that work, our proposed model is built on observations on real data and baed on a substantially different idea. A recent work in [8] defined a group-based version of the influence maximization problem. The predefined groups studied in that work were not conceptually equivalent to the communities studied in this paper.

Community Detection. Community detection in social networks has been studied for years. Varieties of algorithms have been proposed. A good survey is available in [18]. We are not going to discuss the varieties of existing community detection methods, except for those that are related to our work in this paper. Modularity-based methods are a major class of community detection methods. Among these methods, the fast greedy method [5] is frequently used for community detection on large-scale networks. In [20], Rosvall et al. proposed the infomap method. Substantially different from modularity-based methods, the infomap method is based on flows carried by networks [20]. The SHRINK algorithm in [13] is another algorithm that is related to our work. It is a parameter-free hierarchical network clustering algorithm that combines the advantages of density-based clustering and modularity optimization methods. Work in [23] utilized social influence modeling methods in the detection of communities.

3   Preliminary

3.1   Notations

A **social network** $G = \{V, E\}$ is a directed graph with a node set V and an edge set E. A node $v_i \in V$ represents a user in the social network, while a directed edge $e_{ij} \in E$ represents a link from v_i to v_j.

A **community** C in the social network G is a subset of the node set V. We consider non-overlapping communities in this paper. In other words, we consider the partition of V into a set of communities $\mathcal{C} = \{C_i\}_{i=1}^{m}$. Each user in the network should belong to exactly one of the communities in $\mathcal{C}$. Given a graph G, a **community detection** algorithm divides the graph G into a set of communities $\mathcal{C}$. There are a lot of different community detection algorithms. Generally, a good community detection algorithm finds a partition, so that (1) each community is a relatively independent compartment of the graph, and (2) nodes in the same community tend to have dense links between each other.

We follow the definition of **information diffusion process** in the IC model [15]. An information diffusion process starts with a set of seed nodes that are active at the first place. Active nodes can activate their out-neighbors in the social network. Once a node is activated, it becomes active and can never become inactive again. It is quite often for real applications that the information diffusion processes cannot be directly observed. For example, we may observe that a person got infected by influenza, but we do not know from whom he got infected. We define a **cascade** $O = \{(v_1, t_1), \ldots, (v_m, t_m)\}$ as the set of user actions during an information diffusion process. An action (v_i, t_i) in O represents that the user v_i becomes active at time t_i. In this paper, we focus on the scenario that the information diffusion processes are not directly observed, but a set of cascades is observed.

3.2 Datasets

Foursquare[9]. In this dataset, nodes represent users of the Foursquare website, while edges represent friendship relations. Actions are defined by check-ins of users. Each cascade corresponds to a location. When a user checks in at a location for the first time, she becomes active for the corresponding cascade. This dataset contains 18,107 users, 245,034 friendship relations, and 476,482 actions of 43,063 cascades.

DBLP. In this dataset, nodes represent authors, while edges represent co-author relations. We extract a subgraph of the DBLP network with authors and papers in the areas of data mining and machine learning. We define cascades by terms (defined by bi-grams) in the titles of papers. When an author has a paper with a certain term in the title for the first time, he becomes active for the corresponding cascade. This dataset contains 6,896 users, 111,044 friendship relations, and 1,655,778 actions of 162,904 cascades.

4   Observations

In this section, we explore the community effects on information diffusion processes via analyses on real-world datasets. We first identify communities in social networks, and then study cascades with respect to these communities.

4.1 Identifying Communities for Information Diffusion

Communities in social networks can be defined in many different ways. To understand the effects of communities on the information diffusion in general, we apply two different community detection algorithms to the two networks, and conduct community effect analyses for both algorithms.

The two community detection methods that we use to identify communities are the fast greedy (FG) method [5], and the infomap (IM) method [20]. The FG method is based on the well-adopted idea of modularity maximization, while the IM method is a flow-based method, which is essentially different from the modularity maximization methods. We choose these two methods because (1) they are all widely-used community detection methods that prove to be efficient and accurate, and (2) they are based on substantially different ideas. Both methods are implemented in the igraph network analysis package [6].

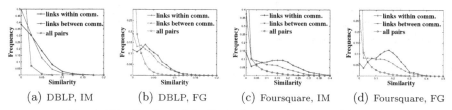

(a) DBLP, IM   (b) DBLP, FG   (c) Foursquare, IM   (d) Foursquare, FG

Fig. 1. Distribution of similarity between actions of user pairs

4.2   Action Homophily of Communities

We first look into the effects that communities have on the actions of users. We construct a vector for each user to keep the action information of that user, and then compare the vectors between pairs of users. We check whether the users who belong to the same community are more likely to have similar actions.

Given a set of cascades $\mathcal{O} = \{O_1, \ldots, O_m\}$, we define an action vector $\mathbf{a_i}$ for each user v_i, where $\mathbf{a_i} = (a_{i0}, \ldots, a_{im})$. If the user v_i has an action in the cascade O_j, we set a_{ij} to 1. Otherwise, we set a_{ij} to 0. For each pair of users v_i and v_j, we calculate the cosine similarity between the action vectors $\mathbf{a_i}$ and $\mathbf{a_j}$, and then study the distribution of similarity. We consider three different cases here: (1) There is an edge e_{ij} between v_i and v_j, and v_i and v_j belong to the same community; (2) There is an edge e_{ij} between v_i and v_j, but v_i and v_j belong to different communities; (3) v_i and v_j is an arbitrary pair of nodes, may or may not having an edge between them. For each case, we plot the distributions of similarity, and check whether there is any difference between the distributions.

Figure 1 shows the distributions of similarity in two datasets, with two sets of communities in each dataset. In each setting, we observe a similar discrepancy among the three distributions: comparing with linked pairs in different communities, linked pairs in the same communities have larger similarity; comparing with arbitrary pairs, linked pairs have much larger similarity. Intuitively, friends in the same communities tend to have stronger link between each other, and they have more chances to influence each other indirectly via common friends. The results are quite consistent for different community detection methods.

4.3   Role-Based Homophily of Communities

We have observed the action homophily of communities. However, although the similarity between linked pairs in the same communities is relative larger than the similarity in the other two cases, it is still quite small (typically, less than 0.3). In this section, we introduce the role-based homophily of communities, and show that the role-based homophily is more significant than the action homophily.

With a set of cascades $\mathcal{O}$, we build a support matrix S for the influence between users in the social networks. The element at the i-th row and the j-th column of the matrix S is the number of potential influences from the user v_i to the user v_j. We say there is a potential influence from v_i to v_j, if both of them have actions in the same cascade, and the time of v_i's action is earlier

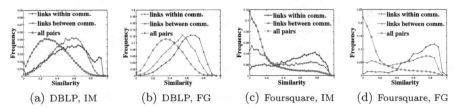

Fig. 2. Distribution of similarity between influencer feature vectors of user pairs

than the time of v_j's action. Formally, it is defined as: $s_{ij} = |\{O_k \in \mathcal{O} \mid v_i, v_j \in V(O_k) \wedge t_i^{O_k} < t_j^{O_k}\}|$, where $V(O_k)$ is the set of users that has an action in the cascade O_k, and $t_i^{O_k}$ is the time of v_i's action in the cascade O_k.

We define $\mathbf{s}_{i*}$, the i-th row of S, as the **influencer feature vector** of v_i, and $\mathbf{s}_{*i}$, the i-th column of S, as the **influencee feature vector** of v_i. The influencer feature vector $\mathbf{s}_{i*}$ captures the influence that v_i has on other users in the social networks, while the influencee feature vector $\mathbf{s}_{*i}$ captures the influence from other users to the user v_i.

Similar to what we did in Section 4.2, we calculate the cosine similarity between the influencer/influencee feature vectors, and compare the distributions. Figures 2 and 3 show the comparison of distributions of influencer feature vector and influencee feature vector, respectively. Similar to Figure 1, comparing with the other two cases, the similarity is larger for the case that users are linked and are in the same communities. There are a few new observations that are interesting:

First, distributions of similarity between influencer/influencee feature vector (Figures 2 and 3) show significantly larger discrepancy than the distributions of similarity of action vector (Figure 1). This observation suggests that for users in the same communities the role-based homophily is much stronger than the action homophily. The effect of community in the information diffusion process is better reflected by the roles that users play in the information diffusion process, rather than the results of information diffusion process (whether being active or inactive for a cascade).

Second, for friends in the same community, the similarity value of influencer and influencee feature vectors (typically larger than 0.5) is larger than the similarity of action vectors (typically less than 0.3). It suggests that aggregation on the influencer/influencee feature vectors of users without significant loss of accuracy is more feasible.

Third, the influencee-based homophily is more significant than the influencer-based homophily, especially for the FG algorithm. This is easy to understand by the following example: professors and students in the same research lab have similar behaviors as influencees, because when a cascade reaches anyone in the lab, it is very likely that cascade will reach everyone in the lab quickly, but professors are probably much stronger influencers than students.

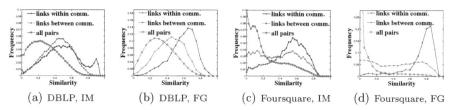

(a) DBLP, IM	(b) DBLP, FG	(c) Foursquare, IM	(d) Foursquare, FG

Fig. 3. Distribution of similarity between influencee feature vectors of user pairs

5   Community-Based Fast Influence Model

Based on the observations in Section 4, we are able to design an efficient influence model which makes use of the community effects. The community-based fast influence (CFI) model we propose in this section is an approximate model for the IC model. The whole framework has three components, namely influence decoupling, community detection, and influence maximization.

5.1   Influence Decoupling

An intuitive way to construct an approximate information diffusion model based on community effects is to consider each community as a "super-node" and make information propagates through "super-edges" between "super-nodes". The coarse-grained information diffusion model in [8] is based on a similar idea.

Although this intuitive model is simple and seems reasonable, it may not work for our task here. When we consider a community as a "super-node", we have to aggregate users' roles as influencers as well as users' roles as influencees. This may cause a problem: the influence maximization problem requires us to determine how influential each user is and find the set of seed nodes that maximizes the influence. When we aggregate the roles of users as influencers, we lose the necessary information for solving the influence maximization problem.

To avoid this problem, the CFI model considers the roles of users as influencers and influencees separately. To be specific, we split each node v_i in the network G into an influencer node v_i^{out} and influencee node v_i^{in}, and transform the network into a bipartite graph G_b. In the graph G_b, there is an edge from v_i^{out} to v_j^{in} if and only if the edge e_{ij} exists in the original graph G. We call this transformation from the original network G to the bipartite graph G_b **influence decoupling**. The left part of Figure 4(a) shows an example of influence decoupling. In the original graph G, there is an edge from v_4 to v_1. Correspondingly, there is an edge from v_4^{out} to v_1^{in} in G_b. The result of influence decoupling is that we can apply the community-based aggregation to the influencee nodes only.

If we apply the original IC model directly to the decoupled graph G_b, we will end up with cascades with only two levels, i.e. only the nodes that are direct out-neighbors of the seed nodes can become active. This problem can be approximately solved by the community detection and the aggregation of influencee nodes. Instead of limiting influence to direct out-neighbors, the CFI

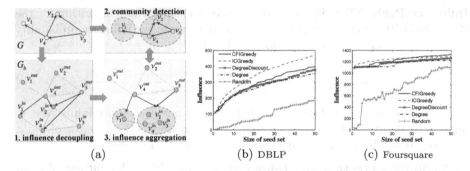

Fig. 4. (a) Inference of the CFI model. (b)-(c) Influence of different sizes of seed set

model allows users to have influence on the communities that their direct out-neighbors belong to. Notice that we do omit the indirect influence from a user to the nodes that are neither his out-neighbor nor in the same communities as his out-neighbor. This is indeed a trade-off between the accuracy and efficiency, but the loss of accuracy is actually negligible. This is because the influence between nodes in different communities are smaller than the influence between nodes in the same communities, and indirect influence are generally very small. We will also show by experiment that the CFI model is a good enough approximation to the original IC model.

5.2 Identifying Communities

We now discuss the community detection algorithm. As we have discussed in the last section, users in the same community should be similar influencees. To identify communities so that users in the same communities are similar as influencees, we design an agglomerative clustering algorithm. It starts with clusters with single users, and iteratively merges clusters together based on similarity between clusters. As shown in Figure 4(a), the clustering procedure is conducted on the original graph G, but the similarity is defined by users' roles as influencees, and the communities detected by the algorithm will finally be applied to the influencee nodes in the decoupled graph G_b.

Similarity. The similarity between two clusters is defined as the cosine similarity between their incident influence probability vectors. Let $p_{i \to j}$ be the probability that v_i influences v_j directly or indirectly (i.e. the probability that v_j becomes active if v_i is the single seed node). For a cluster $C = \{v_{i_1}, \ldots v_{i_{n_C}}\}$, we define the influence that user v_j on C as:

$$q_{j \to C} = \begin{cases} \frac{1}{n_c} \sum_{k=1}^{n_C} p_{j \to i_k} & \text{if } e_{j,i_k} \in E \text{ for some } i_k \in C \\ 0 & \text{otherwise.} \end{cases} \quad (1)$$

where n_C is the number of nodes in the cluster C.

We define incident influence probability vectors of community C as $\mathbf{q}_C = (q_{1 \to C}, \ldots q_{n \to C})$, and the similarity between two clusters C_1 and C_2 as $sim(C_1, C_2) = \mathbf{q}_{C_1} \cdot \mathbf{q}_{C_2} / (\|\mathbf{q}_{C_1}\| \|\mathbf{q}_{C_2}\|)$.

Influence Probability Estimation. Similar to the learning algorithm for the IC model in [10], given a set of cascades $\mathcal{O}$, we estimate the influence probability $p_{i \to j}$ from cascades by the equation as follows:

$$\hat{p}_{i \to j} = \frac{s_{ij}}{s_i} = \frac{\mid \{O_k \in \mathcal{O} \mid v_i, v_j \in V(O_k) \wedge t_i^{O_k} < t_j^{O_k}\} \mid}{\mid \{O_k \in \mathcal{O} \mid v_i \in V(O_k)\} \mid} \tag{2}$$

Since that v_i becomes active earlier than v_j does not necessarily imply that v_i directly or indirectly influences v_j, $\hat{p}_{i \to j}$ is not an unbiased estimator of $p_{i \to j}$. However, it is still a good enough estimator for the CFI model.

Community Detection and Influence Aggregation. The purpose of community detection in the CFI model learning is to aggregate users who play similar roles as influencees, while keep the accuracy of the original IC model. To serve this purpose, we adopt a community detection strategy that is similar to the algorithm in [13]. By iteratively merging clusters into larger one, we get a sequence of super-graph $G_0, G_1, G_2, \ldots$. Each node in these super-graphs corresponds to a cluster. The algorithm starts with graph G_0, in which each cluster contains a single user. In each step t, we find from G_t connected subgraphs that contain similar nodes, and merge these subgraphs to generate a new super-graph G_{t+1}. We repeat these steps, until the similarity between any two neighbors in G_t are below a threshold θ.

Let $\mathcal{C}^{(t)} = \{C_1, \ldots C_{m^{(t)}}\}$ be the set of clusters at the t-th iteration. We say two clusters C_1 and C_2 are neighbors, if there exist $v_i \in C_1$ and $v_j \in C_2$, s.t. edge e_{ij} or e_{ji} exists. For a pair of connected clusters, we say they are a **mutually most similar pair (ms-pair)** with similarity ϵ (denoted by $C_1 \leftrightarrow^\epsilon C_2$), if $\epsilon = sim(C_1, C_2) = \max_{C_i \in N(C_1)} sim(C_1, C_i) = \max_{C_i \in N(C_2)} sim(C_2, C_i)$, where $N(C_i)$ is the set of neighbors of C_i.

We define a **ms-subgraph** as a maximal connected subgraph of G_t that are connected by ms-pairs. Formally, a graph D is a ms-subgraph of G_t with similarity ϵ if and only if (1) for any two nodes $C_i, C_j \in D$, there exist a path $< C_i, C_1 \ldots C_k, C_j >$ in D, s.t. $C_i \leftrightarrow^\epsilon C_1, C_1 \leftrightarrow^\epsilon C_2, \ldots, C_{k-1} \leftrightarrow^\epsilon C_k, C_k \leftrightarrow^\epsilon C_j$; (2) for any nodes $C_i \notin D$ and $C_j \in D$, C_i and C_j are not a ms-pair. By this definition, the graph can be partitioned into ms-subgraphs (some ms-subgraphs may contain only one single node). By merging ms-subgraphs into new nodes, the original super-graph can be reduced into a smaller super-graph.

At the iteration t, we first find out all the ms-subgraphs of G_t, and then merge each ms-subgraph D that contains more than one nodes and has similarity $\epsilon \geq \theta$ into a new node. The new node is a neighbor to any node that was a neighbor of any node in D, and the similarity between the new nodes and its neighbors need to be recalculated. The algorithm stops when the similarity between each linked nodes are less than the threshold θ, and the clusters at that point of time are taken as communities.

5.3   CFI-Based Influence Maximization Algorithm

In this subsection, we show how we can design a CFI-based algorithm for the influence maximization of the IC model. The influence maximization problem is defined as follows: Given an IC model and an integer $k > 0$, find a set of k seed nodes, so that the influence of the seed nodes is maximized. The standard method to solve this problem is a greedy algorithm [15]. It starts with finding one seed node that maximizes the influence, and then adds a second node to the seed node set so that the increase of influence is maximized. In this way, it repeatedly adds nodes to the seed node set, until it gets k seed nodes. This greedy algorithm is very time-consuming, because in each step it uses the Monte Carlo method to evaluate all the remaining nodes. Optimized versions of the greedy algorithm have been proposed in [17] and [11], and heuristic algorithms have been proposed in [3]. These algorithms also use sampling for the evaluation of nodes. We can get a new heuristic algorithm based on the CFI model. This new heuristic algorithm does not involve random sampling, so it is faster than the existing algorithms, especially when the number of seed nodes k is large.

The CFI-based influence maximization algorithm also adopts a greedy framework. In each step t, the node that can maximize the influence increase is selected. The problem is how we can estimate the influence increase using the CFI model. When $t = 1$, the problem is reduced to estimating the influence of each single node. Let $\mathcal{C} = \{C_1, \ldots C_m\}$ be the set of communities in the CFI model. We estimate the influence of a user v_i as $Inf(\{v_i\}) = \sum_{C \in \mathcal{C}} n_c q_{i \to C}$, where n_c is the number of users in the community C.

Once we select the node with the greatest influence to be the first seed node, we cannot simply select the second most influential node to be the second seed node, because the nodes activated by the first node and the second node may overlap. We need to deduct the number of nodes that has already been activated by the first node. To do that, we decrease the number of nodes from each community by the estimated influence of the first node v_{i_1}. Formally, we let $n_c^1 = n_c - n_c q_{i_1 \to C}$, which is the expected number of nodes in the community C that are not activated by the influence of v_{i_1}, and then we select the second node v_{i_2} by maximizing the increase of influence: $\Delta Inf(v) = Inf(\{v_{i_1}, v\}) - Inf(\{v_{i_1}\}) = \sum_{C \in \mathcal{C}} n_c^1 q_{i_2 \to C}$. For $t = 3, \ldots, k$, we can repeat the above step to select $v_{i_3}, \ldots, v_{i_k}$. Generally, we select v_{i_t} by maximizing $\sum_{C \in \mathcal{C}} n_c^{t-1} q_{i_t \to C}$, where $n_c^{t-1} = n_c^{t-2} - n_c^{t-2} q_{i_{t-1} \to C}$.

6 Experiment

6.1 Experiment Setup

We use the DBLP and Foursquare networks described in Section 3 for the experiment. For each network, we construct an IC model by assigning diffusion probability $1 - e^{-0.01c}$ to each edge. A similar method for model construction has been used in [4]. For the DBLP network, c is the number of papers coauthored by the two authors. For the Foursquare network, c is the number of locations that both users visited. We do not construct the ground-truth models by learning the diffusion probabilities directly from the actions in the datasets because we want to avoid the inaccuracy caused by model learning algorithm.

We then sample each ground truth IC model to get 5,000 cascades, each with 10 seed nodes and use the sampled cascades to learn the CFI model. For the baselines, since they are all based on the IC model, we directly apply the influence maximization algorithms on the ground truth model. We evaluate the influence of seed nodes by sampling the ground truth model 10,000 times to get the average number of active nodes. We compare the following algorithms:

- **CFIGreedy** The CFI-based influence maximization algorithm with $\theta = 0.4$.
- **ICGreedy** The greedy influence maximization of IC model with the CELF++ optimization [11]. We take a sample size of $10,000$ to estimate the influence.
- **Degree** The heuristic algorithm that selects the nodes with the largest weighted degree. The weighted degree of a node is the sum of the diffusion probabilities over the out-going edges.
- **DegreeDiscount** The degree discount heuristics based on the degree heuristics [4]. The basic idea is to discount the degree for users whose friends have been selected as seed nodes.
- **Random** Randomly selecting seed nodes.

6.2   Results

Effectiveness Results for Influence Maximization. First, we present the effectiveness results of the influence maximization algorithms in terms of the number of seed nodes. We test the effectiveness of each algorithm with increasing number of seed nodes. The results of the DBLP and Foursquare datasets are illustrated in Figures 4(b) and 4(c), respectively. In each case, we illustrate the number of seed nodes on the X-axis, and the influence of seed nodes on the Y-axis. For the Foursquare data, $CFIGreedy$ performs worse than $ICGreedy$ when the size of the seed set is small, but does better than $ICGreedy$ when the size is greater than 25. This is a very interesting observation. Although the CFI model is designed to be an approximate model for the IC model, the greedy algorithm of the CFI model does not necessarily performs worse than the greedy algorithm of the IC model. This is because the CFI model considers the community structure of social networks, and the consideration of community structure may favor combinations of seed nodes that cover more communities. For the DBLP dataset, $CFIGreedy$ is less effective than $ICGreedy$. However, the difference is not very significant, especially when we consider the fact that $CFIGreedy$ is significantly faster. Besides, $CFIGreedy$ consistently outperforms $DegreeDiscount$, $Degree$ and $Random$. Notice that although $DegreeDiscount$ is a simple heuristic method, previous work showed that it is a very effective method that nearly matches the performance of $ICGreedy$ [4].

Efficiency Results for Influence Maximization. We also tested the efficiency of influence maximization methods with varying number of seed nodes. The efficiency results for the DBLP and Foursquare datasets are illustrated in Figures 5(a) and 5(b), respectively. The X-axis denotes the number of seed nodes, whereas the Y-axis denotes the running time. Since heuristics as $Random$,

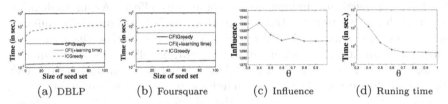

Fig. 5. (a)-(b) Running time with different sizes of seed set; (c)-(d) Effects of θ

Degree, and *DegreeDiscount* are obviously very fast, we only show the running time of *CFIGreedy* and *ICGreedy*. As illustrated in the figures, influence maximization based on *CFIGreedy* is several orders of magnitudes faster than *ICGreedy* with CELF++ optimization. We also add together the time spent on the learning of the CFI model and the running time of *CFIGreedy* to get a total time for the influence maximization on the CFI model, and illustrate the total time as "CFI(+learning time)" in the figures. Even when the learning time is added, the total running time for the CFI model is still significantly smaller the running time of *ICGreedy*. For example, for the DBLP dataset, it takes *ICGreedy* 9,079 seconds to find 60 seed nodes, while the total running time of the CFI model is 34 seconds. Notice that, in real applications, the IC models also need to be learned from user actions, and the running time of *ICGreedy* should also be added with the learning time of the IC model.

Parameter Sensitivity. Finally, we tested the sensitivity of the CFI-based influence maximization with the clustering threshold θ. Figure 5(c) shows the influence of seed nodes selected by the CFI model with varying θ. We illustrate the value of θ on the X-axis, and the influence of seed nodes on the Y-axis. Figure 5(d) shows the total running time of influence maximization with varying θ. We illustrate the value of θ on the X-axis, and the total running time of the influence maximization (the running time of model learning plus the running time of *CFIGreedy*) on the Y-axis. In each case, the number of seed nodes is set to 50. We show the results on the Foursquare dataset, while similar trends are observed on the DBLP dataset. When the threshold θ decreases, the running time increases, because the agglomerative clustering takes more steps when θ is smaller. It is an interesting observation that the influence does not monotonically increases when θ decreases. When θ is too large, the size of communities are very small, so the CFI model omits too much indirect influence. When θ is too small, the users in the same community do not have enough similarity between each other. Both cases cause loss of accuracy. Nevertheless, we notice that the variation of influence is not significant. The Y-axis of Figure 5(c) does not start at 0. When θ varies from 0.3 to 1.0, the variation of influence is within $\pm 1.5\%$.

7   Conclusion

In this paper, we explore the effects of communities on the information diffusion processes. We quantitatively analyze the real-world information diffusion datasets to get insightful findings on the community effects. As an application of these findings, we propose the CFI model, which is substantially different from existing approximate algorithms. Experiment shows that the CFI-based influence maximization algorithm can get comparable effectiveness as influence maximization algorithms based on the IC model, but is significantly faster. Our work sheds light on the effects of communities in the diffusion of information, and brings a new idea to the approximation of information diffusion processes.

Acknowledgments. This work is supported in part by NSF through grants CNS-1115234, and OISE-1129076, Google Research Award, and the Pinnacle Lab at Singapore Management University.

References

1. Bakshy, E., Rosenn, I., Marlow, C., Adamic, L.: The role of social networks in information diffusion. In: WWW (2012)
2. Balog, K., Azzopardi, L., de Rijke, M: Formal models for expert finding in enterprise corpora. In: SIGIR (2006)
3. Chen, W., Wang, C., Wang, Y.: Scalable influence maximization for prevalent viral marketing in large-scale social networks. In: KDD (2010)
4. Chen, W., Wang, Y.: Efficient influence maximization in social networks. In: KDD (2009)
5. Clauset, A., Newman, M.E.J., Moore, C.: Finding community structure in very large networks. Phys. Rev. E **70**(6 Pt 2), 066111 (2004)
6. Csardi, G., Nepusz, T.: The igraph software package for complex network research. Inter. Journal, Complex Systems, 1695 (2006)
7. Eagle, N., Macy, M., Claxton, R.: Network diversity and economic development. Science **328**(5981), 1029–1031 (2010)
8. Eftekhar, M., Ganjali, Y., Koudas, N.: Information cascade at group scale. In: KDD (2013)
9. Gao, H., Tang, J., Liu, H.: Exploring social-historical ties on location-based social networks. In: ICWSM (2012)
10. Goyal, A., Bonchi, F., Lakshmanan, L.V.: Learning influence probabilities in social networks. In: WSDM (2010)
11. Goyal, A., Lu, W., Lakshmanan, L.V.: Celf++: optimizing the greedy algorithm for influence maximization in social networks. In: WWW (2011)
12. Granovetter, M.S.: The Strength of Weak Ties. The American Journal of Sociology **78**(6), 1360–1380 (1973)
13. Huang, J., et al.: Shrink: a structural clustering algorithm for detecting hierarchical communities in networks. In: CIKM (2010)
14. Jin, E.M., Girvan, M., Newman, M.E.J.: Structure of growing social networks. Phys. Rev. E **64**, 046132 (2001)
15. Kempe, D., Kleinberg, J.: Maximizing the spread of influence through a social network. In: KDD (2003)
16. Lappas, T., Terzi, E., Gunopulos, D., Mannila, H.: Finding effectors in social networks. In: KDD (2010)
17. Leskovec, J., Krause, A., Guestrin, C., Faloutsos, C., Vanbriesen, J., Glance, N.: Cost-effective outbreak detection in Networks. In: KDD (2007)

18. Newman, M.E.J.: Modularity and community structure in networks. PNAS **103**(23), 8577–8582 (2006)
19. Reddy, P.K., Kitsuregawa, M., Sreekanth, P., Rao, S.S.: A graph based approach to extract a neighborhood customer community for collaborative filtering. In: Bhalla, S. (ed.) DNIS 2002. LNCS, vol. 2544, pp. 188–200. Springer, Heidelberg (2002)
20. Rosvall, M., Axelsson, D., Bergstrom, C.T.: The map equation. The European Physical Journal Special Topics **178**(1), 13–23 (2009)
21. Saito, K., Kimura, M., Ohara, K., Motoda, H.: Learning continuous-time information diffusion model for social behavioral data analysis. In: Zhou, Z.-H., Washio, T. (eds.) ACML 2009. LNCS, vol. 5828, pp. 322–337. Springer, Heidelberg (2009)
22. Wang, Y., Cong, G., Song, G., Xie, K.: Community-based greedy algorithm for mining top-K influential nodes in mobile social networks. In: KDD (2010)
23. Zhou, Y., Liu, L.: Social influence based clustering of heterogeneous information networks. In: KDD (2013)

On Burst Detection and Prediction
in Retweeting Sequence

Zhilin Luo[1], Yue Wang[2], Xintao Wu[3]($\boxtimes$), Wandong Cai[4], and Ting Chen[5]

[1] Shanghai Future Exchange, Shanghai, China
luo.zhilin@shfe.com.cn
[2] University of North Carolina at Charlotte, Charlotte, USA
ywang91@uncc.edu
[3] University of Arkansas, Fayetteville, USA
xintaowu@uark.edu
[4] Northwestern Polytechnical University, Xi'an, China
caiwd@nwp.edu.cn
[5] Northeastern University, Boston, USA
tingchen@ccs.neu.edu

Abstract. Message propagation via retweet chain can be regarded as a social contagion process. In this paper, we examine burst patterns in retweet activities. A burst is a large number of retweets of a particular tweet occurring within a certain short time window. The occurring of a burst indicates the original tweet receives abnormally high attentions during the burst period. It will be imperative to characterize burst patterns and develop algorithms to detect and predict bursts. We propose the use of the Cantelli's inequality to identify bursts from retweet sequence data. We conduct a comprehensive empirical analysis of a large microblogging dataset collected from the Sina Weibo and report our observations of burst patterns. Based on our empirical findings, we extract various features from users' profiles, followship topology, and message topics and investigate whether and how accurate we can predict bursts using classifiers based on the extracted features. Our empirical study of the Sina Weibo data shows the feasibility of burst prediction using appropriately extracted features and classic classifiers.

1 Introduction

Microblogging, such as Twitter and Sina Weibo, has attracted a huge number of users and becomes increasingly popular. In Twitter, a user can tweet any message within 140-character limit or share pictures, follow any interesting users, and comment or retweet messages that she received from her followees. A tweet can reach the immediate followers of the owner user and can further reach other users when retweeted by some followers. Hence the retweeting mechanism empowers users to spread their ideas beyond the research of the original tweet's followers. Message propagation via retweet chain can be regarded as a social contagion process.

T. Cao et al. (Eds.): PAKDD 2015, Part I, LNAI 9077, pp. 96–107, 2015.
DOI: 10.1007/978-3-319-18038-0_8

In this paper, we examine burst patterns in retweet activities. A burst is a large number of retweets of a particular tweet occurring within a certain short time window. We can consider a burst as a spike and its duration is often short as compared to the surrounding non-burst durations. The occurring of a burst indicates the original tweet receives abnormally high attentions during the burst period. As a result, it will be imperative to characterize burst patterns and develop algorithms to detect and predict bursts.

Many tweets receive little interests in their life cycle and have no burst at all. The propagation of those no-burst tweets often experiences only two stages: low growth and long extinction. However, for tweets that receive significant attention and spread widely in microblogging sites, their propagation often experience eruption, continuance and extinction. Some tweets have a distribution with single-burst whereas other tweets have a distribution with multi-burst. The single-burst indicates that the original tweet receives wide intensive attention in its short period of eruption and then fades away gradually without raising any further significant attention. On the contrary, some tweets may receive intensive attention in several different periods of times during their life cycle due to some triggering event. As a result, they have a distribution with multi-burst. The multi-burst is often characterized by slowly alternating phases of near steady state behavior and rapid spikes. The propagation of multi-burst tweets has five cyclic stages: eruption, continuance, decay, dormant and reflourish, within their (often long) life durations.

In this paper, we propose the use of the Cantelli's inequality to identify bursts from retweet sequence data. We treat bursts as outliers (i.e., significantly different from the average) in retweet sequence data. We then conduct a comprehensive empirical study of burst pattern using the Sina Weibo data and examine various factors, including tweet users and topics, that may have effects on burst. We extract various features from users' profiles, followship topology, and message topics and investigate whether and how accurate we can predict bursts using classifiers based on the extracted features.

2 Burst Characterization

We define the life duration of a particular tweet as the time period from when it was originally posted to when it was lastly retweeted. We convert the retweet frequency information of a given tweet into a time series where each value indicates the number of occurrence of retweets during the time window. The size of the time window could be minutes, hours, or even days dependent on the application. Formally, denote t_i the ith time window after the original tweet is posted, and x_{t_i} the number of retweets in the ith time window. The retweet time series is defined by $X = x_{t_1}, x_{t_2}, \ldots, x_{t_n}$. A burst is a large number of retweets occurring within certain time windows of the tweet's life duration. We define the **burst duration** of the tweet as the total number of time windows for which all bursts last.

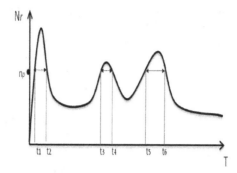

Fig. 1. The retweeting time series of a tweet

Figure 1 shows an example of a retweet series with three bursts: t_1 to t_2 for the first burst, t_3 to t_4 for the second burst, and t_5 to t_6 for the third burst. The retweeting sequence often has a much longer non-burst duration which includes the period before the the first burst, the period between two consecutive bursts, and the period after the last burst. We propose the use of the Cantelli's inequality to identify those bursts.

Theorem 1. *(Cantelli's inequality) Let X be a random variable with finite expected value μ and finite non-zero variance σ^2. Then for any real number λ*

$$Pr(X - \mu \geq \lambda) \begin{cases} \leq \frac{\sigma^2}{\sigma^2 + \lambda^2} & if \quad \lambda > 0, \\ \geq 1 - \frac{\sigma^2}{\sigma^2 + \lambda^2} & if \quad \lambda < 0. \end{cases} \qquad (1)$$

The Cantelli's inequality is a generalization of Chebyshev's inequality in the case of a single tail. When $\lambda > 0$, we have $Pr(X - \mu \geq \lambda\sigma) \leq \frac{1}{1+\lambda^2}$. We treat as outliers those values that are more than λ standard deviations σ away from the mean μ. The number of outliers are no more than $1/(1 + \lambda^2)$ of the distribution values. Those outliers form the bursts in the retweeting sequence. In our paper, we set the λ value as 2.

3   Empirical Evaluation of Burst Patterns

We conduct an empirical study using the WISE 2012 Challenge Data [1]. The WISE 2012 Challenge is based on a dataset collected from the Sina Weibo, one of the most popular Microblogging service in China. In the data, content of tweets are removed and some tweets are annotated with events. For each event, the terms that are used to identify the event are given. Each tweet includes the basic information such as time, user ID, message ID, mentions (user IDs appearing in tweets), retweet paths, and whether containing links. The followship network is also provided. The data set contains 5,636,858 users with 46,584,914 original tweets being retweeted by 190,920,026 times.

---

[1] http://www.wise2012.cs.ucy.ac.cy/challenge.html

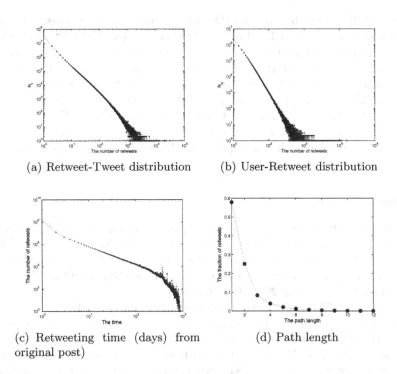

(a) Retweet-Tweet distribution   (b) User-Retweet distribution

(c) Retweeting time (days) from original post)   (d) Path length

Fig. 2. Retweet distribution from WISE 12 Challenge data (200M tweets)

3.1 An Overview of Retweet Patterns

Our preliminary findings are summarized below.

- Figure 2(a) shows the distribution of the number of retweets that each tweet receives. We observe from the figure that most of the original tweets receive less than 10 retweets in their lifetime, while a small number of tweets receive hundreds or even thousands retweets, e.g., the largest number of retweets from a single tweet reaches 34,096 in the data set.
- Figure 2(b) shows the distribution of the number of retweets that each user receives. Tweets authored by a small number of influential users (e.g., celebrities, actors, stars) are very popular and receive most retweets. For example, the top 100 most influential users receive 46,094,478 retweets in total, about 24.2% of all retweets; while the top 1000 most influential users receive 86,501,021 retweets, about 45.3% of all retweets.
- Figure 2(c) shows the distribution of retweeting time of each retweet. We can observe that most retweets occur in a very short period of time after the tweet's posting. For example, 81.8% of the tweets would not be retweeted any more after the first day and only 6.28% of the tweets would last for more than two days. However, a small number of tweets would still be retweeted even after 100 days.

– Figure 2(d) shows the path length distribution of each retweet. The path length of a retweet is defined as the number of hops of the retweeting user away from the original user who posts the tweet. For example, given a retweeting sequence of $A \to B \to C \to D$, user A's tweet is retweeted by user B, then retweeted by user C through user B, and finally retweeted by user D through user C. The path length of the retweet by user C is 2 while the path length of the last retweet by user D is 3. We can observe from the figure that 57.9% of the retweets have one single hop and 98.6% of the retweets are within five hops, matching the concept of *six degrees of separation* in social networks.

3.2 Burst Pattern

In the previous section, we found that some popular tweets are widely retweeted and their retweets last a long time after their posting. On the contrary, a majority of tweets would not be retweeted any more shortly after their posting. In this section, we focus on those tweets that have been retweeted more than 100 times in the data set. We extract 207,259 such tweets.

For those 207,259 popular tweets (each receiving more than 100 retweets), the majority (68.71%) include only one single burst, which often occurs in the first day when the original tweet is posted. 12.84% tweets have no burst and 18.45% of tweets have multi-burst. Tweets with multi-burst often have longer path lengths and longer active duration time than tweets with single or no burst, as shown in Table 1. Among the tweets with multi-burst, there are 31,300 tweets with two bursts and 2,782 tweets with more than 4 bursts. The maximum number of bursts is 17 in the data set.

Table 1. The burst distribution and the average path length of 207,259 original tweets (each receiving more than 100 retweets)

Burst	Number of tweets	Ratio(%)	Avg. of path length	Avg. of life duration (days)
No	26620	12.84	2.03	4.78
Single	142406	68.71	2.09	10.27
Multi	38233	18.45	2.32	18.87

Different Bursts. We examine whether the average path length of retweets occurred in each burst period is different. Our conjecture is that for a tweet authored by user A, its retweets occurred in the first burst are more from user A's immediate followers and retweets occurred in later bursts are more from A's indirect followers. Our findings show that the average path length of retweets in the first burst is shorter than that in following bursts. Specifically, the average path lengths for the first four bursts in our data set are 2.08, 2.29, 2.92, and 2.99 respectively, which validates our conjecture.

We further examine the path length distribution of retweets occurred in each burst. Each curve in Figure 3 shows the fraction of retweets for each path length value from 1 to 10. For retweets occurred in the first burst, 45.6% of retweets

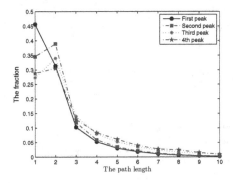

Fig. 3. Path length distribution of retweets occurred in each peak

have the path length of 1 and 30.6% of retweets have the path length of 2. However, for retweets occurred in the second burst, 39.3% of retweets have the path length of 2, which are more than the number of retweets (35.1%) with the path length of 1. This shows that the second burst is mainly caused by non-immediate followers of original users who post the tweets. We have the similar phenomenon for the third burst and the fourth burst.

Burst Pattern vs. Topics. We examine whether topics of original tweets have effects on burst patterns. We extract four hot topics, i.e., house price, xiao mi release, family violence of Li Yang, and case of running fast car in Hei Bei university. We denote them as *House*, *Xiaomi*, *Li Yang*, and *He Bei*, respectively. For those tweets with no assigned topic, we group them in the *Unknown* category.

Table 2. The comparison of different topics

Topic	Avg. of path length.	Avg. burst duration(days)	Avg. life duration(days)
Unknown	1.90	2.95	15.85
House	2.66	3.05	22.75
Xiao Mi	2.62	3.39	16.32
Li Yang	2.89	3.71	21.20
He Bei	2.97	3.64	28.15

Table 2 shows the general comparison of different categories in terms of path length, burst duration, and life duration. We can see that the tweets from the *Unknown* category have shorter path length, shorter burst duration, and shorter life duration time than the tweets with known topics. This indicates tweets with some particular hot topics are often widely propagated in the microblogging site.

Figure 4 shows the path length distribution for each topic category under study. The curve of *No Topic* (aka, *Unknown*) is significantly different from other curves corresponding to known topic categories. We can observe that the proportions of retweets from *Unknown* category with path length 1 and 2 are 45% and 38% respectively, which are much higher than the corresponding proportions for retweets with known topics.

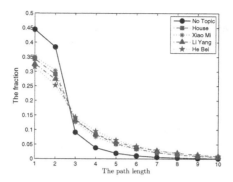

Fig. 4. Path length distribution of retweets of each topic

Table 3. The information of different types of users

Users	Avg. of path length.	Avg. burst duration(days)	Avg. life duration(days)
Top 100	1.86	2.73	10.63
Top 100-1000	2.17	3.05	18.34
Normal	3.04	3.77	28.46

Burst Pattern vs. Users. We examine whether different types of users who post tweets have effects on burst pattern. Table 3 shows the general comparison of three types of users: top 100, top 100-1000, and normal users. We define the top 100 users as those who rank among the top 100 in terms of the total number of retweets each user receives. We can see there are significant differences in terms of path length, peak time, and duration time among three types of users. Tweets from the top 100 most influential users have much shorter path lengths, burst duration, and tweet life duration than the top 100-1000 and normal users.

Figure 5 shows the path length distribution for each type of user under study. We can observe that the proportions of retweets of those tweets authored by the

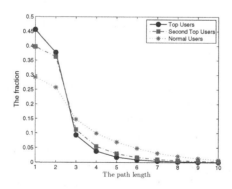

Fig. 5. Path length distribution of retweets from three types of users

top 100 users with path length 1 and 2 are 46% and 39%, respectively, which are much higher than the corresponding proportions for tweets from the top 100-1000 and normal users. This phenomena shows that the top 100 most influential users can propagate their messages more quickly in the microblogging site than other users.

4 Burst Prediction

We are interested in the following prediction problem: given a tweet with known information about its content, its user profile, the followship topology, and the observed retweet sequence in the first 12 hours, can we predict whether the tweet will have multi-burst in the future of its life cycle.

One challenge here is what kind of features we can extract from the known information and how useful they are for burst prediction. In our study, we extract 178 features from the a-priori known information of a tweet (i.e., its topics, user profile, followship topology, and its observed retweet sequence in the first 12 hours). The extracted features can be roughly grouped into two main classes: user-related and tweet-related.

In the user-related class, we extract features from the profile of the user who posts the original tweet. For example, we extract the number of his immediate followees, the number of his two-hop followees, the number of tweets the user has authored, the average number of retweets received in the first 12 hours for all his tweets, and the numbers of tweets with no, single, and multiple bursts.

In the tweet-related class, we extract the features such as the tweet's post time, first retweeting time, the presence/absence of hot topics in the tweet, the presence/absence of hot topics in its retweets, the presence/absence of @users in the tweet, the presence/absence of @users in its retweets, the number of retweets containing @users and the number of @users in its retweets, etc. For each tweet, we also build a retweet tree from its observed retweet sequence in the first 12 hours and extract features such as the maximum width, the maximum height, the number of retweet users, and the average path length.

In our experiment, we exclude from the Sina Weibo dataset those records in which the original tweets' user ID could not be found in the followship network. Finally, we build a training data set with 30,084 tweets with no multi-burst and 30,030 tweets with multi-burst.

We run a suite of 7 classifiers: Logistic Regression (LR), Random Forest(RF), Decision Tree (DT), Naive Bayes (NB), Support Vector Machine (SVM), Stochastic Gradient Descent (SGD), and k-Nearest Neighbor (kNN). We take the 10 fold cross-validation for each classifier. The accuracy result is shown in Figure 6. We can observe that Random Forest, Decision Tree, k-Nearest Neighbor, and Logistic Regression achieve good prediction results in terms of accuracy (higher than 72%).

We then analyze the effect of each feature on prediction. We take the logistic regression coefficient as the effect. The regression coefficients represent the change in the logit for each unit change in the feature. The larger the absolute value of the coefficient is, the more effect the feature takes. Formally, we can

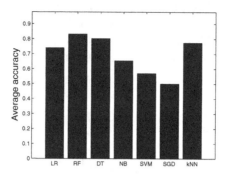

Fig. 6. Accuracy of Classifiers: Logistic Regression (LR), Random Forest(RF), Decision Tree (DT), Naive Bayes (NB), Support Vector Machine (SVM), Stochastic Gradient Descent (SGD), and k-Nearest Neighbor (kNN)

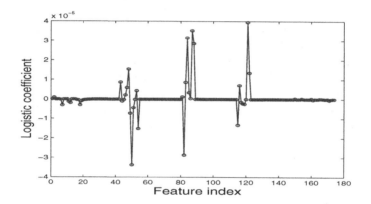

Fig. 7. Logistic Coefficient of Features

use the likelihood ratio test or the Wald statistic to assess the significance of an individual feature. Our results show that there are only 20 features with relatively large coefficient values. Figure 7 plots the logistic regression coefficient for each feature where X-axis represents different features and Y-axis shows each feature's coefficient value. We list top 5 most significant features in Table 4. We can see that the average number of retweets with path length 1 of the user's all

Table 4. Top 5 most significant features (PL1 denotes path length 1)

Index	Meaning	Coefficient
121	Avg no of PL1 retweets of user's all tweets	3.95E-05
87	Avg no of PL1 retweets (first 12h) of user's no-burst tweets	3.51E-05
50	Avg no of retweets (first 12h) of user's multi-burst tweets	-3.37E-05
84	Avg no of retweets (first 12h) of user's no-burst tweets	3.14E-05
82	Avg no of retweets of user's no-burst retweets	-2.86E-05

tweets is the most significant feature with the coefficient value 3.95E-05. In our future work, we will conduct detailed correlation analysis and examine prediction performance after removing those redundant features.

5   Related Work

Examining retweet behavior has been an active research area recently [7–9,12, 13]. For example, the authors in [7] studies the coverage prediction of retweets, i.e., what is the number of times that a particular message posted by a user will be retweeted. In [13], the authors examine various factors such as user, message, and time and propose a factor graph model to predict whether a user will retweet a message. The authors in [9] study why people retweet and examine the anti-homophily phenomenon. In [8], the authors examine the use of log-linear modeling to identify multi-way interactions between retweet and various features such as power ratio, link structure and users' profile information. In [12], the authors analyze the ways in which hashtags spread on twitter and show widely-used hashtags on different topics spread significantly different.

Change detection models [1,4] provide a standard approach to detecting deviations from baseline. Usually we assume the mean and variance of a distribution representing normal behavior and the mean and variance of another distribution representing behavior that is abnormal. We can measure deviations from normal using the generalized likelihood ratio. For example, in [4], the authors assume both distributions are Gaussian with the same variance and the change is reflected in the mean of the observations. In this context, they apply the generalized likelihood ratio to score changes from baseline.

Techniques for finding burst patterns in data streams have also been presented in [6,11,15,16]. In [6], the authors examine bursty structure in temporal text streams (e.g., emails or blogs). They examine how frequency words change over time. The burstiness of words is defined as those words with significantly higher frequency than others. They propose to model the stream using an infinite-state automaton, in which bursts appear naturally as state transitions. In [16], the authors examine point monitoring and aggregate monitoring in time series data streams and design a new structure, called the Shifted Wavelet Tree, for elastic burst monitoring. In [15], the authors propose a family of data structures based on the Shifted Binary Tree for elastic burst detection and develop a heuristic search algorithm to find an efficient structure given the input. In [11], the authors study how to detect, characterize and classify bursts in user query logs of large scale e-commerce systems. The authors build several models that continually detect newer bursts with minimal computation and provide a mechanism to rank the identified bursts based on a number of factors such as burst concentration, burst intensity and burst interestingness. They also propose several quantities to rank bursts including duration of burst, mass of burst, arrival rate for burst, span ratio, momentum of burst, and concentration of burst, and apply unsupervise learning techniques to classify the bursts based on their patterns. Although extensive work has been done in related fields for mining various temporal patterns, we notice that very little work has been done to detect and predict interesting burst patterns from large-scale retweet sequence data.

Message propagation can be regarded as a social contagion process. There has been research on rumor propagation [5,10,14]. In [14], the authors study the dynamics of an epidemic-like model for the spread of a rumor on a small-world network. In [10], the authors study the dynamics of a generic rumor model on complex scale-free topologies and investigate the impact of the interaction rules on the efficiency and reliability of the rumor process. In [5], the authors apply the susceptible-infectious-recovered and susceptible-infectious-susceptible models to study the spreading process in complex networks. However, we notice that very little work has been done to detect and predict burst patterns.

6   Conclusion

In this paper, we have proposed the use of the Cantelli's inequality to identify bursts from retweet sequence data. With the use of the Cantelli's inequality, we do not need to assume the distribution of the retweet sequence data and can still identify bursts efficiently. We conducted a complete empirical study of burst pattern using Sina Weibo data and examined what factors would affect burst. We extracted various features from users' profiles, followship topology, and message topics and investigated whether and how accurate we can predict bursts using various classifiers based on the extracted features. Our empirical evaluation results show the burst prediction is feasible with appropriately extracted features and classifiers.

In our future work, we will investigate various regression analysis methods [3] on extracted features to predict when a tweet produces its first burst as well as following bursts. We will analyze the bursts to see what their causality was by matching external events that might have caused the bursts. In our future work, we will also study how to classify bursts based upon their shapes, durations, and derived burst characteristics. We will examine various burst characteristics such as burst concentration, burst intensity and burst interestingness. We will study how the window size affects burst detection and categorization. Finally, we will study the use of topic modeling [2] to analyze tweet content and automatically identify the topics of every tweet.

Acknowledgments. The authors would like to thank anonymous reviewers for their valuable comments and suggestions. This work was supported in part by U.S. National Science Foundation (CCF-1047621), U.S. National Institute of Health (1R01GM103309), and the Chancellor's Special Fund from UNC Charlotte.

References

1. Basseville, M., Nikiforov, I., et al.: Detection of abrupt changes: theory and application, vol. 104. Prentice Hall, Englewood Cliffs (1993)
2. Blei, D., Ng, A., Jordan, M.: Latent dirichlet allocation. JMLR **3**, 993–1022 (2003)

3. Cohen, J., Cohen, P.: Applied multiple regression/correlation analysis for the behavioral sciences. Lawrence Erlbaum (1975)
4. Curry, C., Grossman, R., Locke, D., Vejcik, S., Bugajski, J.: Detecting changes in large data sets of payment card data: a case study. In: KDD, pp. 1018–1022. ACM (2007)
5. Kitsak, M., Gallos, L., Havlin, S., Liljeros, F., Muchnik, L., Stanley, H., Makse, H.: Identification of influential spreaders in complex networks. Nature Physics 6(11), 888–893 (2010)
6. Kleinberg, J.: Bursty and hierarchical structure in streams. Data Mining and Knowledge Discovery 7(4), 373–397 (2003)
7. Luo, Z., Wang, Y., Wu, X.: Predicting retweeting behavior based on autoregressive moving average model. In: Wang, X.S., Cruz, I., Delis, A., Huang, G. (eds.) WISE 2012. LNCS, vol. 7651, pp. 777–782. Springer, Heidelberg (2012)
8. Luo, Z., Wu, X., Cai, W., Peng, D.: Examining multi-factor interactions in microblogging based on log-linear modeling. In: ASONAM (2012)
9. Macskassy, S.A., Michelson, M.: Why do people retweet? anti-homophily wins the day! In: ICWSM (2011)
10. Moreno, Y., Nekovee, M., Pacheco, A.: Dynamics of rumor spreading in complex networks. Physical Review E 69(6), 066130 (2004)
11. Parikh, N., Sundaresan, N.: Scalable and near real-time burst detection from ecommerce queries. In: KDD, pp. 972–980. ACM (2008)
12. Romero, D.M., Meeder, B., Kleinberg, J.: Differences in the mechanics of information diffusion across topics: idioms, political hashtags, and complex contagion on twitter. In: WWW, pp. 695–704. ACM (2011)
13. Yang, Z., Guo, J., Cai, K., Tang, J., Li, J., Zhang, L., Su, Z.: Understanding retweeting behaviors in social networks. In: CIKM, pp. 1633–1636. ACM (2010)
14. Zanette, D.: Dynamics of rumor propagation on small-world networks. Physical Review E 65(4), 041908 (2002)
15. Zhang, X., Shasha, D.: Better burst detection. In: ICDE, pp. 146–146. IEEE (2006)
16. Zhu, Y., Shasha, D.: Efficient elastic burst detection in data streams. In: KDD, pp. 336–345. ACM (2003)

#FewThingsAboutIdioms: Understanding Idioms and Its Users in the Twitter Online Social Network

Koustav Rudra[1]([⊠]), Abhijnan Chakraborty[1], Manav Sethi[1], Shreyasi Das[1], Niloy Ganguly[1], and Saptarshi Ghosh[2,3]

[1] Department of CSE, Indian Institute of Technology Kharagpur, Kharagpur, India
koustav.rudra@cse.iitkgp.ernet.in
[2] Max Planck Institute for Software Systems, Kaiserslautern, Germany
[3] Department of CST, Indian Institute of Engineering
Science and Technology Shibpur, Howrah, India

Abstract. To help users find popular topics of discussion, Twitter periodically publishes 'trending topics' (trends) which are the most discussed keywords (e.g., hashtags) at a certain point of time. Inspection of the trends over several months reveals that while most of the trends are related to events in the off-line world, such as popular television shows, sports events, or emerging technologies, a significant fraction are *not* related to any topic / event in the off-line world. Such trends are usually known as *idioms*, examples being #4WordsBeforeBreakup, #10thingsI-HateAboutYou etc. We perform the first systematic measurement study on Twitter idioms. We find that tweets related to a particular idiom normally do not cluster around any particular topic or event. There are a set of users in Twitter who predominantly discuss idioms – common, not-so-popular, but active users who mostly use Twitter as a conversational platform – as opposed to other users who primarily discuss topical contents. The implication of these findings is that within a single online social network, activities of users may have very different semantics; thus, tasks like community detection and recommendation may not be accomplished perfectly using a single universal algorithm. Specifically, we run two (link-based and content-based) algorithms for community detection on the Twitter social network, and show that idiom oriented users get clustered better in one while topical users in the other. Finally, we build a novel service which shows trending idioms and recommends idiom users to follow.

1 Introduction

Twitter is now considered more of an 'information network' than a social network [6] and almost the entire focus of the research community has been on 'topical' content in Twitter, such as tweets / hashtags related to sports or technology or emergency situations in the off-line world [2]. However, a closer inspection of the Twitter *trending topics* ('trends' in short) – keywords periodically declared

© Springer International Publishing Switzerland 2015
T. Cao et al. (Eds.): PAKDD 2015, Part I, LNAI 9077, pp. 108–121, 2015.
DOI: 10.1007/978-3-319-18038-0_9

Table 1. Percentage of Twitter trends collected over ten months, and classified into nine different categories that were identified by human volunteers (details in Section 2). Also given are few examples of trends.

Category	%	Example trends
Entertainment	33%	#5sosonKiis, #IWishICould, #Austinonidol
Sports	30%	#argentinavsholanda, #lakers, #bravsger
Idioms	9%	**#WhenIWasATeenager, #FactsaboutMe, I get angry when**
Technology	8%	#iphone6, #galaxy4, AppleWatch, ios8
Politics	5%	#tcot, #pjnet, #obama, #gaza
Business	5%	#amazon, #AlibabaIPO, #FedReserve
Religion	3%	#EidMubarak, #jesus, #citrt
Health	2%	#Ebola, #Who, #breastcancer
Others	5%	#garlicparmpizza, #filipino, cheesecake, pizza is healthy

by Twitter as being the most discussed at that point in time – indicates some exceptions to this view, and provides the motivation for the present study.

We collected US trends over a duration of 10 months (January – October, 2014) using the Twitter API at 15-minute intervals. This gave about 18,500 distinct trending topics during this period. We then developed a classifier *Odin*[1] and classified the trends into multiple categories such as sports, entertainment, technology etc. – these broad categories were identified by human volunteers (details in Section 2). Table 1 shows the distribution of the trends in the different broad categories. While most of the categories are topical and related to events in the off-line world, it can be observed that a special category, known as *idioms*[2], regularly becomes trending. Examples of idioms include #4WordsBeforeBreakup, #11ThingsAboutYou, and apparently these are not related to any topic or event in the off-line world.

The frequent presence of such trends is intriguing – it raises the question whether their dynamics as well as the users discussing such trends are similar to those of the topical counterparts. To understand the dynamics, we collected tweets related to hundreds of idioms and the users who discuss them, and conducted a detailed measurement study. We find that the tweets containing idioms are mainly conversational in nature; for instance, they hardly contain URLs. On investigating the users who post the tweets (the idiom-users), we find that they are mostly general and active Twitter users, as opposed to being popular experts / celebrities who usually drive topics such as politics and entertainment. The idiom-users maintain close friendships among themselves and interact on diverse issues with their friends. Thus, the study unfurls that hidden within the

---

[1] Named after the God of Wisdom according to Norse mythology; details in Section 2.

[2] In this study, we follow the definition of idioms given by [13] – an idiom is a keyword representing a conversational theme on Twitter, consisting of a concatenation of at least two common words which does not include names of people, places or music albums etc.

information network of Twitter, there is a *social network* of users who regularly have "non-topical" conversations among themselves.

Such an inference has far-reaching implications. It essentially means that multiple dominant dynamics are present in the same social network – so the standard tasks like community detection, recommendation, and so on, *cannot* be done using a one-parameter-fits-all approach. An algorithm to identify (recommend) topical groups might fail to identify (recommend) idiom-users. To test this proposition, we run two community detection algorithms – one identifying topical groups [2] and the other, Infomap [14] which detects communities using link structure. We find that the idiom-users are well identified by Infomap while the topical groups are better identified by [2]. This establishes that different approaches for tasks such as clustering may have different utilities in a heterogeneous online social network. Further, considering that all existing recommender services are specifically meant to recommend topical experts, we develop a service *Idiomatic* where one can easily follow popular idiom-users, see the recent and past trending idioms and post tweets using them.

2   Classification of Trends

In order to perform a large scale study on idioms and the trends related to topics / events in the off-line world, we built an automatic classifier *Odin*, to distinguish particular trends based on whether they are idioms or related to some topic. Note that some prior studies [7,21] have also attempted to classify trends (not necessarily into the same categories found by Odin), utilizing the textual contents of the tweets containing the trends. However, tweets (restricted to 140 characters) often contain informal language and abbreviations which potentially results in lower classification accuracy [21]. Hence, we adopt a different approach that combines both tweets and related web documents and uses several web-based knowledge engines to perform the classification. Odin classifies a given trend following the steps presented below.

2.1   Preprocessing

Segmentation: Trends often consist of multiple words [13] recognizing which is easy for multi-word phrases and hashtags written in CamelCase style (e.g., #WorldCupSoccer), but is very difficult for trends which simply have the words juxtaposed without any separation (e.g., #everythingididntsay). Since it is important to identify the individual words which make up a trend in order to understand its topic, trends need to be segmented into the component words. Odin follows a modified version of the Viterbi Algorithm [1], which uses a model of word distribution to calculate the most probable character sequence forming a word. Odin computes the word distribution from Google n-gram corpus (*https://books.google.com/ngrams*).

Given a trend, Odin segments the trend into its constituent words based on this calculated probability estimates (details omitted for brevity).

Categorization of Related Web Documents: Odin searches different Web search engines (e.g., Google, Bing) with the segmented trend, to get a large set of web-pages relevant to the given trend. Often the tweets containing the trend have URLs, which become another source for getting related web-pages.[3] For a given trend, Odin collects all the web-pages pointed from the tweets and returned by the search engines; and then a set of category keywords are extracted for these collected web-pages using the NLP-based AlchemyAPI web service (www. alchemyapi.com).

Entity Extraction and Categorization: Sometimes the trend contains names of people, organisations or locations (e.g., #EMABiggestFansJustinBieber) detecting which might give a clear idea on the category of the trend. Similarly, the web documents and the tweets associated with a particular trend have many such named entities present in them. Odin extracts such entities using AlchemyAPI and then queries Freebase (www.freebase.com) to know the 'notable type' of such named entities (e.g., according to Freebase, notable type for 'Justin Bieber' is '/music/artist').

2.2 Classification

At the end of preprocessing steps, for a given trend, Odin collects the categories of the related web documents and the notable types of the related named entities. Treating the number of web documents and named entities in the various categories as features, Odin uses a Support Vector Machine (SVM) classifier with Radial Basis Function kernel to classify a particular trend into one of the 9 categories shown in Table 1.

Training Data Preparation: For creation of training data, three human volunteers (regular users of Twitter, who are not authors of this paper) were asked to manually inspect 700 distinct trends collected during the first two weeks of January 2014 (along with tweets containing these trends), and classify the trends into different categories. The volunteers identified the *nine broad categories* shown in Table 1, such as Entertainment, Sports, Technology, Idioms (following the definition of idioms in [13]). Out of the 700 trends, all three volunteers agreed upon a particular category for 575 trends. We created the training data considering this unanimous categorization as the ground truth.

Classification Performance: Standard 10-fold cross validation on the data of the 575 trends showed that Odin attains 77.15% accuracy in predicting trend categories, which is good considering that it is a complex nine-class classification task.

---

[3] URLs leading to social media sites like Facebook, Twitter, Instagram, are ignored, since these pages usually do not have much content to help the topic categorization process.

Table 2. Statistics of data collected

Property	Idiom	Sports	Entertainment	Technology
Number of trends	150	150	150	150
Total #tweets containing the trends (millions)	6.205	6.787	6.967	6.105
Mean #tweets per trend	41,369	45,257	46,455	40,721
Total #distinct users posting the trend (millions)	2.74	2.71	1.90	1.75
Mean #distinct users per trend	18,315	18,098	12,725	11,705

3   Dataset

Since most of the Twitter trends were related to the three topics *entertainment*, *sports*, and *technology* (see Table 1), we decided to focus on idioms and trends related to these three topics; the trends related to any of these three topics are collectively referred to as '*topical trends*'. For each of the trends belonging to the four selected categories, we collected as many tweets containing the trend as possible using the Twitter search API. To get a better understanding about the trends, in our analysis as presented in later sections, we used only those trends for which we were able to collect more than 30,000 tweets. To maintain uniformity across categories, we finally selected a set of 150 trends related to each of the categories (the actual distribution is stated in Table 1).

For each of the 600 selected trends, we further collected detailed statistics about all the users (including their profile details, social links and recently posted tweets) who posted a tweet containing any of the selected trends. Table 2 summarizes the statistics of the data collected for the trends of the four categories.

4   Comparing Idioms and Topical Trends

In this section, we compare how idioms and topical trends are discussed in the Twitter social network, and the users who discuss them frequently.

4.1   How Trends Are Discussed in Twitter

We first analyze how the trends of different categories are, in general, discussed in Twitter. For a given trend t, we consider all tweets containing t, and measure what percentage of these tweets contain other hashtags (apart from t itself), and URLs.

Figure 1 shows mean values of the percentage of tweets containing other hashtags and URLs, where the mean values are computed over all trends of a particular category. Statistical measures like two sample KS-test and Mann-Whitney U test with significance level 0.05 show that there is a significant difference in the distribution of the mean values among the four categories. Expectedly, we find that the topical trends are much more likely to be accompanied by other hashtags and URLs

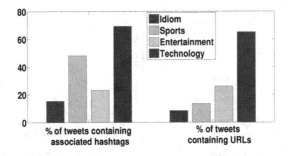

Fig. 1. Comparing topical trends and idioms: Percentage of tweets which contain (i) other hashtags (apart from the trend under consideration), and (ii) URLs. All values averaged over all trends of a particular category.

related to the corresponding event in the off-line world. For instance, the sports-related trend #LIVvCHE (referring a match between the two English soccer clubs Liverpool and Chelsea) is accompanied by the hashtag #Torres which indicates a player who is a part of the match. On the other hand, Twitter-specific idioms are very seldom accompanied by other hashtags since they are not related to external websites or news-stories in the off-line world.

We also observed the timeline evolution of trends, i.e., how they start getting tweeted and become popular in Twitter. Expectedly, most topical trends emerge as a result of some related event in the off-line world, such as a sports or musical event, or a socio-political incident / issue. In case of idioms, an interesting pattern observed is that many idioms *initially* propagate along with hashtags related to some specific event in the off-line world. For example, the idiom '#MyFavouriteActor' first appeared with the hashtag '#PeoplesChoice' (related to the People's Choice awards), while the idiom '#SexRequirements' initially appeared with the health-related hashtag '#FitnessPromo'. These idioms, however, follow their independent path with users innovating interesting comments and thus making them popular.

4.2 Characterising Users Interested in Various Categories

In order to understand the nature of the users who are interested in promoting particular types of trends, we identify sets of users who are interested in the different categories (sports / technology / entertainment / idioms), and compare various characteristics of these users.

Identifying Users Interested in a Certain Category: To identify users who are interested in a certain category, we identify those users who *frequently* discuss trends of that category. For a particular category, we initially consider all the users who have posted at least one tweet on a trend in that category. We rank the users based on the number of different trends in that category on which they have posted at least one tweet. Subsequently, for each category, the 10,000 users

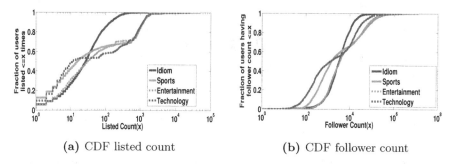

(a) CDF listed count   (b) CDF follower count

Fig. 2. Distribution of Listed & Follower count of four categories (idiom, sports, entertainment, technology) of users

who have posted most number of distinct trends in that category (according to our dataset) are considered. Since our objective is to identify users who are genuinely interested in trends of a certain category, we next attempt to verify whether the users selected above frequently discuss trends on that category. For this, we collected the 3,200 most recent tweets for each of the selected users, by crawling their time-line through the Twitter API, and used our classifier *Odin* to classify the hashtags contained in these tweets, to check what fraction of these hashtags were related to that category. For instance, for a certain user u included among the top 10,000 users who posted on most sports-related trends in our dataset, we checked whether a significant fraction of all hashtags included in u's recent tweets were related to sports. Additionally, Opencalais (www.opencalais. com) tool is used to identify the topic of each tweet present in the timeline of a user. We included a user in the final selected set for a category, if at least 30% of the hashtags and 70% of the tweets posted by her (among her recent tweets) were judged to be related to that category.

In this way, we finally identified a set of 5,000 users who are genuinely interested in each of the four categories. We refer to these sets of users as *idiom-users*, *sports-users*, *entertainment-users*, and *technology-users*. The rest of this section studies the characteristics of these sets of users.

Popularity of the Users: We start by checking the popularity of the users interested in the various categories. We use two standard metrics of popularity of users in the Twitter social network [3,4] – (i) follower-count, i.e., the number of followers of a given user, and (ii) listed-count, i.e, the number of Twitter Lists a given user is included in.[4] Both metrics resulted in very similar observations. Figure 2 shows the distribution of the listed-count and follower-count values of users who predominantly discussed the trends in the four categories.

We observe an interesting trend. Almost *all* idiom-users are relatively less popular – 65% of the idiom-users have listed-count values in the range 0–40. In contrast, a significant fraction of the users who predominantly discuss the topical

---
[4] Lists are a feature by which a user can group together accounts on a common theme [4,16].

Table 3. Characterizing the users who frequently discuss trends in each category – top 5 words appearing in (i) the user-bio in the profile, (ii) Lists in which the users are members, and (iii) tweets posted by the users.

Idiom-users			Sports-users			Entertainment-users			Technology-users		
Bio	Lists	Tweets	Bio	Lists	Tweets	Bio	Lists	Tweets	Bio	Lists	Tweets
life	faves	friend	sports	wwe	game	5sos	band	show	news	social	iphone
love	ily	people	football	wrestling	season	justin	album	music	tech	media	ios
fun	luke	hobby	wrestling	sports	team	bands	music	video	tech	tech	android
cool	nigg	niall	wwe	chelsea	nfl	ariana	youtubers	photo	oracle	tweet	google
harry	styles	school	soccer	cricket	football	luke	idols	album	software	business	apple

trends(sports, entertainment, technology) are very popular users, which includes experts from their respective fields. The above statistics lead to some interesting insights. There seems to be two very distinct types of users who dominantly participate in discussions on topics related to the off-line world (e.g., sports, entertainment, technology) – (i) popular users who are experts on these topics (e.g., researchers, sports-persons, journalists, musicians), and (ii) the common masses who are interested in these topics. This agrees with findings in recent research studies [2,20]. In sharp contrast, users who dominantly participates in idioms are mostly common masses.

How the Users are Described: We next focus on how the users who are interested in various categories describe themselves, and how they are described by others. To infer the characteristics of a given user u, we refer to two sources – (i) the bio of u, which is a short description written by the user to describe herself, and (ii) the name and description of Twitter Lists in which u is included as a member – this indicates how other users (those who created the Lists and added u as a member) describe u [16,18]. For a given category, we consider the bio (or List names and descriptions) of all the 5000 users chosen for this category (as described above), and find the words which occur in the bio (or Lists) of most number of these users.[5]

Table 3 shows the top 5 words which appear in the bio and Lists of the users for each category. As expected, the users for the topical categories (sports, entertainment, technology) are characterized by words related to the topics. For instance, sports-users are described by 'wrestling', 'wwe', entertainment-users are identified by '5sos', 'justin', and technology-users by 'social', 'tech'. On the other hand, the idiom-users are mostly described by words related to day-to-day conversation and positive sentimental words such as 'love', 'life', 'faves', 'ily (i love you)' and so on.

Content Posted by the Users: We next focus on the content (tweets) posted by the users. Similar to the previous analysis, we consider the set of tweets posted by the users interested in a certain category, and find out the most frequent words in the tweets. Table 3 shows the top 5 words posted by users in each category – we again find that while the sports-users, technology-users

---

[5] The bio and List-names are pre-processed using standard techniques such as case-folding, removal of a common set of stopwords, and so on.

and entertainment-users mostly post words related to the corresponding topics, the idiom-users mostly use conversational words and phrases related to musical events, celebrities etc.

4.3 Studying the Interactions Among the Users

We now investigate how the users in the four groups interact among themselves. In Twitter, the primary ways by which a user u can interact with another user v are (i) u can subscribe to the content posted by v by following v, or by following a List which has v as its member, and (ii) u can @mention v in her tweet.

Analysing Interaction Networks: We construct two types of interaction networks among the users. The first is a *subscription network* where a directed link $u \to v$ indicates that user (node) u subscribes to the content posted by user v. The second is a *mention network* where the link $u \to v$ indicates that user u has @mentioned v.

To quantify the level of interaction among the users, we measure two structural properties of the subscription and mention network – (i) *density*, which measures what fraction of all links which can be present in a network, are actually present, and (ii) *reciprocity*, which indicates what fraction of the directed links are reciprocated, i.e., both the links $u \to v$ and $v \to u$ exist in the network. The importance of reciprocity is that if two users share a reciprocal link, then the two users are *mutual friends* with a higher probability (as compared to the chance of a fan subscribing to a celebrity, but the celebrity not reciprocating).

Table 4 shows the reciprocity and density of the mention and subscription networks among different groups of users. We find that the density of the subscription network among the idiom-users is significantly higher compared to that for the sports-users, entertainment-users, and technology-users. Also, the reciprocity is significantly higher for both the subscription network and the mention network for idiom-users, indicating that a large fraction of the interactions are between mutual friends. These observations indicate that, just like users interested in a common topic (sports, entertainment or technology), the idiom-users form their own group; in fact, they subscribe to / mention one another much more frequently than the topical groups of users.

Note that the density of the mention networks are comparable for all the user-groups. This is because, as observed earlier, the sports-users, technology-users, and entertainment-users contain a large number of common (less popular)

Table 4. Reciprocity and density of the mention and subscription networks among different groups of users

User-group	Mention Network		Subscription Network	
	Reciprocity	Density	Reciprocity	Density
Idiom	21.88%	0.0012	49.57%	0.0221
Sports	14.67%	0.0017	10.19%	0.0030
Entertainment	13.40%	0.0010	13.76%	0.0058
Technology	13.91%	0.0011	4.87%	0.0025

users and a few popular celebrities, and most of the @mentions result from the common users mentioning the celebrities. For instance, a significant fraction of the @mentions among technology-users are directed towards @twitter and a few software companies. However, the reciprocities are lower for the topical groups, since the celebrities do *not* mention the common users. On the other hand, most of the mentions among the idiom-users (who have similar popularity) come from conversations among mutual friends, leading to high reciprocity. In fact, as much as 62.5% of the mentions among the idiom-users are between two users who share a *reciprocal* link in the corresponding subscription network (i.e., are likely to be mutual friends), where as this percentage is less than 35% for the topical user-groups.

Nature of Conversations Among the Users: Finally, we analyze the nature of the conversations among the users of the same group. Specifically, when a user retweets or mentions another user *in the same group*, we check whether the hashtags used in the tweets are related to the common topic of interest of the users. For instance, among the hashtags which a sports-user retweets or mentions to another sports-user along with the tweets, we check what fraction of such hashtags are related to sports. For this, we use our classifier *Odin* to classify hashtags present in the tweets where a user mentions or retweets another user from the same-group. The results are shown in Table 5. More than 74% of the hashtags that are mentioned / retweeted among the sports-users, entertainment-users, and technology-users are related to the corresponding common topic of interest of that user-group. In sharp contrast, only about 25% of the hashtags that are exchanged among idiom-users are idioms. This again shows that idiom-users are not a focused topical group rather they engage themselves in diverse issues.

4.4   Type of User-Groups and Their Identifiability

Our analyses reveal that the group of users interested in Twitter-specific idioms has very different characteristics compared to the groups of users interested in topics such as technology, sports and entertainment. In this section, we attempt to explain the differences and their implications on identifiability of the groups.

Explaining Group Formation: Formation of user-groups in a social network has been a long-standing topic of research in sociology, and several theories have been proposed to explain their formation [8,11,19]. According to the well-accepted *common identity and common bond theory* [5,10,12], there are two

Table 5. Percentage of hashtags (present in tweets) where a user of a certain group mentions or retweets another user of the same group, which are related to the topic of interest of that user-group

User-group	Idioms	Sports	Entertainment	Technology
% of topical hashtags in retweeted tweets	22.83%	78.47%	81.63%	79.57%
% of topical hashtags in mentioned tweets	25.74%	74.58%	77.12%	78.54%

primary types of groups. In *identity-based groups*, people join the group due to their interest in a well-defined common theme (topic), whereas *bond-based groups* are driven by personal social relations (bonds) among the members, and may be characterized by the absence of any common topic of discussion. As a result, bond-based groups have higher reciprocity among the members than identity-based groups. Also, the discussions in bond-based groups tend to vary widely and cover multiple subjects, while in identity-based groups, they tend to be related to the common topic of interest of the group.

The above analyses on the four user-groups show that, as expected, the users interested in a common topic like sports, entertainment or technology form identity-based groups, with fewer interactions (@mentions) among friends, and most of the discussions among the members being related to the topic of common interest (Table 5). On the other hand, the idiom-users group is characterised by relatively higher levels of personal interactions with mutual friends, and a relatively small fraction of the conversation among the friends is related to their common topic i.e. idioms. Hence, the idiom-users form a bond-based social community within Twitter, in which they discuss their personal topics of interest as well as conversational matters.

Identifiability of the Groups: The differences in the nature of various user-groups can have significant impact on the identifiability of the groups. To demonstrate this, we used two algorithms for detecting groups in the Twitter social network, and checked how well they could identify the idiom-users group and the topical groups.

(i) We used the well-known Infomap community detection algorithm [14] on the Twitter subscription network among all the users spanning the four user-groups. Then we enumerated the number of different communities identified by Infomap, where the members in any of the four user-groups are distributed. Table 6 (second row) shows that the topical groups were scattered into significantly higher number of Infomap communities, as compared to the idiom-users group.

(ii) Bhattacharya et al. [2] proposed a methodology to identify *topical communities* in Twitter (comprising of users who are experts on a topic or interested in the topic). We used this method to check the number of distinct topical communities a member in our dataset is placed. We found that, on average, a user in the idiom-users group is placed in many more topical communities, than a user in the sports-users / entertainment-users and technology-users groups (Table 6, last row).

Table 6. (i) Number of communities identified by Infomap, into which a user-group is scattered, (ii) average number of topical groups assigned per user by the topical group identification approach developed in [2]

User-group	Idioms	Sports	Entertainment	Technology
Nos. communities	107	284	272	281
Nos. groups assigned per user	9	2	2	3

These observations reveal that within Twitter, there exist two different kinds of network structure – one is an information network, and other one is social communication network. Any community detection method which considers only one facet of the network might not be able to identify all the communities accurately.

5 Idiomatic: Service for Idiom Lovers

As stated earlier, the focus of the research community has been entirely on the topical content discussed in Twitter, such as identifying experts on various topics [4,17]. However, for a user who is interested in idioms (idiom lover), there is no existing service to recommend whom she could follow to know interesting idioms being discussed in Twitter. Hence, we have developed *Idiomatic* (`http://cse.iitkgp.ac.in/resgrp/cnerg/idiomatic`), a service where one can easily follow popular idiom-users (ranked according to the number of idioms they post), have a quick look at recent and past trending idioms (classified by an enhanced version of the Odin classifier presented in Section 2 from continuous stream of trending topics collected at 15 minute intervals), and post tweets using idioms.

To evaluate the quality of the recommended idiom-users, we used human feedback since relevance of user-profiles to a certain topic / theme is subjective in nature. The evaluators were shown the most recent 100 tweets of the idiom-users, and were asked to judge whether the user appears to be an active idiom-user or not. 15 human volunteers individually judged the top 20 idiom-users shown by the service. Out of the top 20 users, 18 were judged as active idiom-user by *all* the evaluators, and even the remaining two users were judged as active idiom-users by majority of the evaluators.

6 Related Work

The present study focuses on the characteristics of Twitter idioms, identifying users who actively participate in idioms, and understanding the social behaviour of the groups of these users. Some prior studies on trending topics in Twitter have focused on classification of the trends [9,21], whereby the presence of idioms [21] is identified. However, there has been little effort in analyzing the characteristics of idioms, and of the users who post the idioms. To our knowledge, the only prior study which attempted to compare idioms with trends related to events in the off-line world is by Naaman et al. [9], where they used different features like content, interaction etc. to classify the trends. However, they did not attempt to analyze the users who discuss such idioms.

Also note that there have been prior attempts to distinguish between bond-based and identity-based groups in online social networks (see Section 4.4). For instance, [15] classified chats among users on a text-based communication platform into two categories – on-topic chats which are on a common topic (identity-based) and off-topic chats where people chatted on a variety of topics (bond-based). More recently, [2] identified a large number of topical groups in

Twitter, comprising of users who are experts or seekers of information on various topics, and showed that these groups are essentially identity-based. In this work, we explored the nature of the groups among the idiom-users, and found that they reveal bond-based characteristics.

7 Conclusion

The popular perception of the research community is that, there are two parts of Twitter – one interesting part where participants read and post a wide variety of topical tweets, and another part which comprises of pointless babble and is hence unimportant and uninteresting. However, in our study, we find that these pointless babbles, even though not related to any off-line event, frequently become trending in Twitter due to participation of large number of common masses. These users form bond-based groups among themselves to discuss their personal interests – idioms and some other forms of fun and gossip. This study has several implications, e.g., for community detection in social networks. Keeping in mind the popularity of idioms, we developed a whom-to-follow recommendation service where idiom lovers can easily find trending idioms and users who post idioms actively and frequently.

Acknowledgments. The authors thank the anonymous reviewers for their valuable suggestions which immensely helped us to improve the paper. K. Rudra was supported by a fellowship from Tata Consultancy Services, and S. Ghosh was supported by a fellowship from the Alexander von Humboldt Foundation.

References

1. Berardi, G., Esuli, A., Marcheggiani, D., Sebastiani, F.: ISTI@TREC Microblog Track 2011: Exploring the Use of Hashtag Segmentation and Text Quality Ranking. In: NIST TREC (2011)
2. Bhattacharya, P., Ghosh, S., Kulshrestha, J., Mondal, M., Zafar, M.B., Ganguly, N., Gummadi, K.P.: Deep Twitter Diving: Exploring Topical Groups in Microblogs at Scale. In: ACM CSCW (2014)
3. Cha, M., Haddadi, H., Benevenuto, F., Gummadi, K.P.: Measuring User Influence in Twitter: The Million Follower Fallacy. In: Proc. AAAI ICWSM (May 2010)
4. Ghosh, S., Sharma, N., Benevenuto, F., Ganguly, N., Gummadi, K.: Cognos: crowdsourcing search for topic experts in microblogs. In: Proc. ACM SIGIR (2012)
5. Grabowicz, P.A., Aiello, L.M., Eguiluz, V.M., Jaimes, A.: Distinguishing topical and social groups based on common identity and bond theory. In: Proc. ACM WSDM (2013)
6. Kwak, H., Lee, C., Park, H., Moon, S.: What is Twitter, a social network or a news media? In: Proc. World Wide Web Conference (WWW) (2010)
7. Lee, K., Palsetia, D., Narayanan, R., Patwary, M.M.A., Agrawal, A., Choudhary, A.: Twitter Trending Topic Classification. In: Proc. IEEE International Conference on Data Mining Workshops (2011)
8. McMillan, D., Chavis, D.: Sense of community: A definition and theory. Journal of Community Psychology **14**(1), 6–23 (1986)

9. Naaman, M., Becker, H., Gravano, L.: Hip and trendy: Characterizing emerging trends on Twitter. Journal of the American Society for Information Science and Technology **62**(5), 902–918 (2011)
10. Prentice, D.A., Miller, D.T., Lightdale, J.R.: Asymmetries in attachments to groups and to their members: Distinguishing between common-identity and common-bond groups. Personality and Social Psychology Bulletin **20**(5), 484–493 (1994)
11. Chakraborty, A., Ghosh, S., Ganguly, N.: Detecting overlapping communities in folksonomies. In: Proc. ACM Hypertext Conference (2012)
12. Ren, Y., Kraut, R., Kiesler, S.: Applying Common Identity and Bond Theory to Design of Online Communities. Organization Studies **28**(3), 377–408 (2007)
13. Romero, D.M., Meeder, B., Kleinberg, J.: Differences in the mechanics of information diffusion across topics: idioms, political hashtags, and complex contagion on twitter. In: Proc. World Wide Web Conference (WWW), pp. 695–704 (2011)
14. Rosvall, M., Bergstrom, C.T.: Maps of random walks on complex networks reveal community structure. PNAS **105**, 1118–1123 (2008)
15. Sassenberg, K.: Common bond and common identity groups on the Internet: Attachment and normative behavior in on-topic and off-topic chats. Group Dynamics Theory Research And Practice **6**(1), 27–37 (2002)
16. Sharma, N., Ghosh, S., Benevenuto, F., Ganguly, N., Gummadi, K.: Inferring Who-is-Who in the Twitter Social Network. In: Proc. WOSN Workshop (2012)
17. Twitter Help Center — Twitter's suggestions for who to follow. `https://support.twitter.com/articles/227220-twitter-s-suggestions-for-who-to-follow`
18. Wagner, C., Liao, V., Pirolli, P., Nelson, L., Strohmaier, M.: It's not in their tweets: Modeling topical expertise of twitter users. In: Proc. ASE/IEEE SocialCom (2012)
19. Chakraborty, A., Ghosh, S.: Clustering hypergraphs for discovery of overlapping communities in folksonomies. In: Dynamics on and of Complex Networks, vol. 2. Springer (2013)
20. Wu, S., Hofman, J.M., Mason, W.A., Watts, D.J.: Who says what to whom on Twitter. In: Proc. World Wide Web Conference (WWW) (2011)
21. Zubiaga, A., Spina, D., Fresno, V., Martínez, R.: Real-Time Classification of Twitter Trends. Journal of the American Society for Information Science and Technology (2014)

Retweeting Activity on Twitter: Signs of Deception

Maria Giatsoglou[1]([✉]), Despoina Chatzakou[1], Neil Shah[2],
Christos Faloutsos[2], and Athena Vakali[1]

[1] Informatics Department, Aristotle University of Thessaloniki, Thessaloniki, Greece
{mgiatsog,deppych,avakali}@csd.auth.gr
[2] School of Computer Science, Carnegie Mellon University, Pittsburgh, USA
{neilshah,christos}@cs.cmu.edu

Abstract. Given the re-broadcasts (i.e. retweets) of posts in Twitter, how can we spot fake from genuine user reactions? What will be the tell-tale sign — the connectivity of retweeters, their relative timing, or something else? High retweet activity indicates influential users, and can be monetized. Hence, there are strong incentives for fraudulent users to artificially boost their retweets' volume. Here, we explore the identification of fraudulent and genuine retweet threads. Our main contributions are: (a) the discovery of *patterns* that fraudulent activity seems to follow (the "TRIANGLES" and "HOMOGENEITY" patterns, the formation of micro-clusters in appropriate feature spaces); and (b) "RTGEN", a realistic generator that mimics the behaviors of both honest and fraudulent users. We present experiments on a dataset of more than 6 million retweets crawled from Twitter.

1 Introduction

Can we spot patterns in fake retweeting behavior? When a large number of Twitter users re-broadcast a given post, should we attribute this burst of activity to organic, genuine expression of interest or rather to a fraudulent, paid contract? Twitter is arguably the most popular micro-blogging site and one of the first sites forbidden by authoritarian regimes. High-quality tweets are re-broadcasted (*retweeted*) by many users, indicating that their authors are influential. Since such influence can be monetized via per-click advertisements, Twitter hosts many fraudsters trying to falsely create the impression of popularity by artificially generating a high volume of retweets for their posts. In our work, we observe a thriving ecosystem of spammers, content advertisers, users paying for content promotion, bots disguised as regular users promoting content and humans retweeting for various incentives. Such content is at best vacuous, but often spammy or malicious and detracts from Twitter content's credibility and honest users' experiences.

Despite previous efforts on Twitter fraudsters' activity [8,17,18], the different manifestations of fake retweets have not been adequately studied. Previous approaches focus mainly on specific URL broadcasting, instead of retweet threads, and rely on temporal and textual features to identify bots [5,11]. Fraudsters on Twitter, though, constantly evolve and adopt advanced techniques to obscure their activities. The identification of patterns associated with "fake"

© Springer International Publishing Switzerland 2015
T. Cao et al. (Eds.): PAKDD 2015, Part I, LNAI 9077, pp. 122–134, 2015.
DOI: 10.1007/978-3-319-18038-0_10

retweet activity is, thus, crucial for spotting retweet threads and their authors as fraudulent. This work's primary goal is to distinguish organic from fake retweet activity and the informal problem definition we address is

Informal Problem 1 (RETWEET-THREAD LEVEL).
Given: *the connectivity network (who-follows-whom); the i-th tweet of user; and the retweet activity (IDs and timestamps of the users that retweeted it)*
Find: *features of the retweet activity*
To determine *whether the activity is organic or not.*

Here, we focus on identifying features and patterns in relation to the connectivity and temporal behavior of retweeters that will allow the classification of the motive behind retweet threads as driven by users' genuine reactions to tweeted content, or resulting from a paid contract. We also aim at spotting users who are suspicious of long-term spam activity, but manage to evade suspension from Twitter by using *camouflage*.

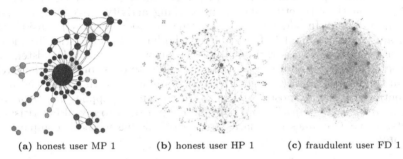

   (a) honest user MP 1   (b) honest user HP 1   (c) fraudulent user FD 1

Fig. 1. CONNECTIVITY: Retweeter networks for retweet threads of size (a) 117, (b) 1132, (c) 336. Dense connections in (c) indicate the TRIANGLES pattern. Retweeter networks of honest and fake activities can be distinguished by several other patterns (e.g. DEGREES, HOMOGENEITY). In the depicted networks, a double edge indicates a reciprocal relationship and a node's size is relative to its degree.

The contributions of this work are the following:

- **Patterns:** Our proposed approach, RTSCOPE, identifies multiple patterns that we found indicative of fraudulent behavior by analyzing the retweeter networks of Twitter accounts. For example, in one class of fraudulent accounts, all accounts follow each other and thus have an excessive number of triangles ("TRIANGLES" pattern) — see Figure 1. It is important that these patterns can be detected based on partial snapshots of the fraudsters' relationship network. Moreover, other fraudsters retweet concurrently within a fixed time from each-other in lockstep fashion, with little variation ("HOMOGENEITY" pattern).
- **Generator:** Based on our analysis, we provide RTGEN, a data generator which produces (ID, timestamp) pairs mimicking traces of fraudulent as well as organic retweet activity. The significance of RTGEN is highlighted by the difficulty of obtaining real world organic and fraudulent retweeting data for experimentation, due to the lack of a standard dataset and the strict policies of social network APIs.
- **Reproducibility:** We share an (anonymized) version of our dataset and RTGEN's code at: http://oswinds.csd.auth.gr/project/RTSCOPE.

2   Related Work

Related work mainly spans: *anomaly detection* in social networks and *fraud* on Twitter.

Anomaly detection and fraud detection in social networks has led to several methods: NetProbe [13] identifies fraud on eBay using belief propagation. MalSpot [12] uses tensor decomposition for computer network intrusion detection. Copy-Catch [1] spots lockstep behavior in Facebook Page Like patterns. [6] leverages spectral analysis to reveal various types of lockstep behavior in social networks.

Fraud on Twitter: [18] analyzes the relationships of criminal accounts inside and outside of the criminal account community to infer types of accounts which serve as criminal supporters. [2] proposes a classification method relying on tweeting behavior, tweet content and account properties for computing the likelihood of an unknown user being a human, bot or cyborg. [16] shows the strong classification and prediction performance of temporal features for distinguishing between account types. However, all these works address the detection of spammers based on their tweeting and/or networking activity, instead of the fake retweeting problem. In addition, most existing methods (e.g. [17]) consider the typical and out-dated model of a fraudster who has uniform posting frequency and a followers-to-followees ratio close to 1 — nowadays, many fraudsters are more sophisticated. [5] addresses a problem similar to ours, but uses the URLs found in tweets instead of retweet threads in conjunction with a time and user-based entropy to classify posting activity and content. [9] applies *disparity*, also known as *inverse participation ratio* [3], on Twitter data to reveal favoritism in retweets. Table 1 outlines the characteristics of existing methods compared to RTScope.

Table 1. RTScope comparison against alternatives

	[5]	[18]	[2]	[16]	**RTScope**
Can be applied for individual retweet chains	✓				✓
Can operate without timestamps		✓			✓
Independent of tweet content	✓			✓	✓
Exploits network topology		✓			✓
Detects bot activity	✓		✓	✓	✓

3   Background on Fake Retweet Thread Detection

Our intitial intuition is that a large proportion of "fake" retweets originate from bot accounts or human accounts which employ the use of automated software. This implies the existence of similarity in the temporal behavior of the individual retweeters, due to the posting (and retweeting) scheduling capabilities of automation tools. We also expect that it is highly probable that fraudulent retweeters of a given user will operate concurrently in lockstep fashion. This is indicative of collaboration between spammers or a contract between the author and a third party for a purchase of retweets. To study the retweeting activity in terms of time and retweeting users, given a user u_m (*author*) we represent the ith tweet posted by u_m with $tw_{m,i}$ as a tuple $(u_m, t_{m,i})$, where $t_{m,i}$ is the tweet's creation time. Then, a retweet thread is defined as follows:

Definition 1 (Retweet thread). *Given an author u_m and a tweet $tw_{m,i}$, a retweet thread $R_{m,i}$ is defined as the set of all tweets that retweeted $tw_{m,i}$.*

We hypothesize that certain types of fraudulent retweet threads are generated by users with abnormal connectivity in terms of their *follow* relationships in Twitter. An example of such *abnormal connectivity* would be a much denser network of fraudulent (compared to honest) retweeters, corresponding to a group of fraudsters following each other in an attempt to maintain reputability. To validate our hypothesis on the importance of connectivity as a feature, we consider the following two types of relationship networks:

Definition 2 (Relationship networks). *Given a retweet thread $R_{m,i}$ we define the "R-A" and "R" networks as the induced networks of:*
"R-A" network *author u_m and all retweeters of $tw_{m,i}$;*
"R" network *all retweeters of $tw_{m,i}$ minus zero-degree nodes, i.e. retweeters who are disconnected from the rest.*

We highlight the fact that the considered network types are partial snapshots of the complete Twitter followers network, since we operate under the constraint of limited visibility. Constraining the followers network to specific subgraphs is important given that the massive size of the Twitter network poses computational burdens to the application of graph algorithms for pattern detection.
We then formulate two versions of the fake retweet detection problem.

*Problem 1 (*RETWEET-THREAD LEVEL*).*

Given: a tweet $tw_{m,i}$ and a retweet thread $R_{m,i}$,
Identify: whether $R_{m,i}$ is organic.

*Problem 2 (*USER LEVEL*).*

Given: a user u_m, a set of tweets $tw_{m,i}$ and their induced retweet threads,
Identify: whether u_m is a spammer.

The RETWEET-THREAD LEVEL problem addresses the detection of single instances of fraud, thus is suitable for "occasional" fraudsters (who occasionally purchase retweets or are paid to participate in promotions, but otherwise exhibit normal activity) and *promiscuous* professional spammers (their fake retweet threads can be spotted without additional data on their past activities). The USER LEVEL problem addresses also the detection of more *cautious* spammers, whose retweet threads are not suspicious on their own, but they reveal suspicious recurring patterns when they are jointly analyzed.

4 Dataset and Preliminary Observations

We examine our hypotheses on a dataset comprising several retweet threads of honest and fraudulent Twitter users. RTSCOPE requires *complete* retweet threads, i.e. with no gaps in the tuples representing a tweet's retweets. Due to Twitter Streaming API's constraint of allowing access to only a sample of the published posts, our need for *complete* retweet threads and the lack of a relevant (labeled) dataset, we manually selected a set of *target* users and tracked all their posts and retweets for a given time period.

We selected target user accounts based on two approaches. The first involved the examination of a 2-day sample of the Twitter timeline, followed by the identification of the users who had posted the most retweeted tweets, and those who posted tweets containing keywords heavily used in spam campaigns (e.g. casino, followback). The second approach was based on "Twitter Counter"[1], a web application publishing lists that rank Twitter users based on criteria such as their number of followers and tweets, and involved the selection of users based on their posting frequency and influence (i.e. we kept only users who posted several posts per week and had received more than 100 retweets on some of their recent posts). We manually labeled target users as "fraudulent" (FD) if (a) inspection of their tweets' content led to the discovery of spammy links to external web pages, spam-related terms, and repetitive posts with the same promotions, or (b) their profile information was clearly fabricated. We labeled the rest of target users (of different popularity scales for the sake of diversity) as "honest" and further divided them into high-, medium- and low-popularity (HP, MP, LP, respectively), using the cut-offs of $>100K$ followers for HP and $< 10K$ followers for LP. We monitored the initial set of target users for 30 days and eliminated those who had all their posts retweeted less than 50 times. Then, we reinforced the remaining dataset with an extra number of similarly selected users, and collected data for an additional 60-days period. At the end of this period, we again pruned users using the same filtering criterion. Overall, this process left a total number of 24 users in the dataset, of which 11 honest (5 HP, 4 MP, and 2 LP) and 13 fraudulent, while after the end of the monitoring period we identified that 4 of our fraudulent users had been suspended by Twitter. Table 2 shows the activity characteristics for the dataset's honest and fraudulent users. For the reproducibility of our results, we make available an anonymized version of our dataset at http://oswinds.csd.auth.gr/project/RTSCOPE.

Table 2. Activity statistics per user class

Type	# Tweets	# Original tweets	# Retweeted tweets	# Retweets
honest	35,179	18,706	13,261	708,814
fraudulent	92,520	50,536	27,809	5,330,407
BOTH	127,699	69,242	41,070	6,039,221

From our data collection and preliminary analysis, we make two main observations:

Observation 1 (Variety). *Fraudsters have various behaviors in terms of their posting frequency and timing.*

Specifically, some fraudsters are *hyperactive*, posting many tweets (> 100 per day); others are more *subtle*, posting few tweets per day, while sometimes mixing original posts with retweets to other users' posts, implying some type of cooperation (half of our dataset's FD users are *hyperactive*). We also noticed that some FD users often produced (resembling) honest posts along with fraudulent ones. This may indicate the existence of "occasional" fraudsters, or intended *camouflage* practiced by "professional" fraudsters.

---

[1] http://twittercounter.com/

Observation 2 (FF imbalance). *Despite earlier reports of success, the followers-to-followees ratio (FF) is uninformative for several fraudsters.*

The reasoning behind this observation is that although previous works considered fraudsters with a similar number of followers and followees, we found that some fraudsters maintain a high FF ratio (in our dataset, only two FD users have a ratio close to 1, while for the rest it ranges in 1.3 - 2061). Further complicating the problem, hijacked accounts have honest followers and followees with "normal" FF ratio (significantly different from 1).

Given the various types of fraudulent behavior types and inefficacy of the commonly used FF ratio, what additional features can we use to spot fake retweets? This is exactly the focus of RTSCOPE, which is described next.

5   RTScope: Discovery of Retweeting Activity Patterns

In this section we propose RTSCOPE and present the results of its application on our dataset. RTSCOPE includes a series of tests that address:

- the RETWEET-THREAD LEVEL problem (1), namely: *ConR*, connectivity analysis of "R" and "R-A" relationship networks (Sect. 5.1);
- the USER LEVEL problem (2), namely: *RAct*, detection of retweeters' activation patterns across a given user's posts (Sect. 5.2), and *ASum*, inspection of the activity summarization features per retweet thread (Sect. 5.3).

The most significant features involved in each test are summarized in Table 3. We note here that in this approach only the *ASum* features require the retweets' timestamps, which, in some cases, may be hard to obtain, or easy for the fraudsters to manipulate.

Table 3. Signs and explanations of suspicious retweeting activity

Feature Category	Alias	Description	Fraud Sign
		RETWEET-THREAD LEVEL	
Retweeters' connectivity	ConR1	Number of triangles (TRIANGLES)	Excessive
	ConR2	Distribution of degrees (DEGREES)	Non power-law
Activity summarization features	ASum1	Activated followers ratio (ENTHUSIASM)	High
	ASum2	IQR (=spread) of interarrival times (MACHINE-GUN)	Low
		USER LEVEL	
Retweeters' activation pattern	RAct	Distr. of # retweets (HOMOGENEITY)	Homogeneous
Activity summarization features	ASum3	Formation of microclusters (REPETITION)	Yes

5.1   Retweeter Networks Connectivity: TRIANGLES and DEGREES Patterns

To study the connectivity between the retweeters of a given tweet, we selected a sample of the largest retweet threads for each user in the dataset, identified their follower relations via the Twitter API and generated the "R" and "R-A" graphs[2].

---

[2] Due to the hard limits of Twitter API in terms of requesting information on users' relations, it was impossible to generate the "R" networks for all retweet threads of the dataset.

Interestingly, we observed that for some retweet threads of fraudulent users there were no connections between the retweeters, whereas for others, none of the retweeters was connected to the author. These phenomena were mostly observed in the context of *occasional fraudsters*. However, we noticed that in these cases, a significant (more than 20%) percentage of the original retweeters were suspended some time afterwards, thus affecting the remaining users' connectivity. For the rest of the retweet threads (of fraudulent and honest users) the percentage of suspended retweeters was less than 10%.

The connectivity analysis of the "R" and "R-A" networks led to Observation 3. Next, we discuss the details of our analysis approach and findings.

Observation 3 (CONNECTIVITY). *"R" and "R-A" networks of honest and fraudulent users differ substantially and exhibit the* TRIANGLES, DEGREES *and* SATELLITE *patterns, on which we elaborate below:*

TRIANGLES: Some fraudulent users have a very well connected network of retweeters, resulting in many triangles in their "R" network. The triangles vs. degree plots of fraudsters often exhibit power-law behavior with high (1.1-2.5) slope. Figure 2 shows that honest users (top row, (a)-(c)) have "R" networks with <100 and often 0 triangles. Conversely, the "R" networks of fraudulent users (bottom row, (d)-(f)) are near-cliques with almost the maximum count of triangles for each node ($(d-1)(d-2)/2$ for a node of degree d).

Such networks are probably due to several bot accounts created by a script and made to follow each other in botnet fashion.

DEGREES: Honest users have "R-A" and "R" networks with power-law degree distribution (Figure 3(a)) while fraudulent ones deviate (Figure 3(b)). The spike at degree ≈ 30 for the latter, agrees with the botnet hypothesis.

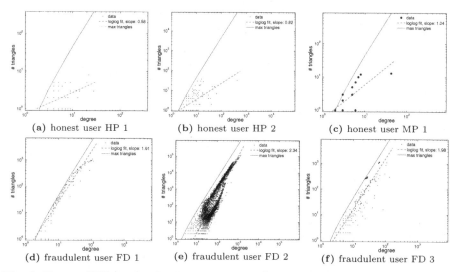

Fig. 2. Dense "R" networks for fraudsters (TRIANGLES pattern): log-log scatter plots of the number of triangles vs. degree, for each node of selected users' "R" networks. Red line indicates maximum number of triangles ($\approx$ degree^2 for a clique). Dashed green line denotes the least squares fit. Honest users (top) have fewer triangles and smaller slope than fraudsters (bottom).

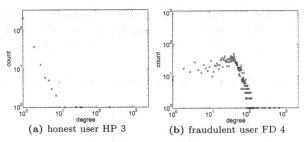

(a) honest user HP 3   (b) fraudulent user FD 4

Fig. 3. Fraudsters disobey the degree power-law (DEGREES pattern): log-log scatter plots of count of nodes with degree deg_i vs. degree deg_i for "R" networks of selected users. Honest users, depicted in (a), tend to follow power-law behavior; fraudsters, depicted in (b), do not.

SATELLITE: In honest "R-A" networks, the author has many "satellites", i.e. retweeters that follow him, and no other retweeters. The fraction s of such satellite nodes is $0.1 < s < 0.9$ for honest users, but $s < 0.001$ for many fraudulent users.

5.2   Retweet Activity Frequency: FAVORITISM and HOMOGENEITY Patterns

Given a target user's posts, what is the distribution of retweets across the retweeters? Do most retweets originate from a specific set of *dedicated* users, or are they distributed uniformly across all the user's connections?

To investigate this distibution, we use the *disparity* measure which quantifies, given a finite number of instances (in our case, retweets), the number of different states or subsets these instances can be distributed into. With respect to a given target user, the number of instances corresponds to the total number of retweets, while a given state is the number of retweets made by a single user. Disparity reveals whether the retweeting activity spreads *homogeneously* over a set of users, or if it is strongly *heterogeneous*, in the sense that it is skewed towards a small set of very active *dedicated* retweeters.

Given target user u_i and a retweet thread size of k, generated by u_j for $j = 1 \ldots k$ retweeters, we examine disparity with respect to the total retweeting activity of these k users. We define the number of retweets made from user j to user i as r_{ij}, and the total number of retweets from u_j users as $SR = \sum_{j=1}^{k} r_{ij}$. Then, we consider that the number of retweets r_{ij} defines the *state* of user u_j, ranging from $r_{ij} = 1$ to $r_{ij} = SR$.

Definition 3 (Disparity). *The disparity of retweeting activity with respect to author u_i and a retweet thread size k is defined as:*

$$Y(k, i) = \sum_{j=1}^{k} \left(\frac{r_{ij}}{SR}\right)^2 \tag{1}$$

In the case that there exists more than one retweet thread of size k, we simply take the average of the $Y(k, i)$ values over retweet threads.

To give an intuition of disparity, we provide two extreme examples of activity distribution: (a) the *homogeneous*, where all users are in the same state (i.e. they have the same r_{ij} value), and (b) the *super-skewed*, where there exists some user u_l who is at a state of much larger value compared to the rest — that is, $r_{il} \simeq SR$, whereas for $j \neq l$, $r_{ij} = q << SR$. The disparities for these situations are derived as follows:

Lemma 1. *The disparity $Y_h(k, i)$ for the homogeneous activity distribution obeys*

$$Y_h(k, i) = \sum_{j=1}^{k} (\frac{r_{ij}}{SR})^2 = \sum_{j=1}^{k} (\frac{1}{k})^2 = \frac{1}{k} \tag{2}$$

Lemma 2. *The disparity for the super-skewed activity distribution is given by:*

$$Y_{ss}(k, i) = \sum_{j=1}^{k} (\frac{r_{ij}}{SR})^2 = (\frac{r_{il}}{SR})^2 + \sum_{j, j \neq l} (\frac{b}{SR})^2 \simeq 1, \tag{3}$$

thus it is independent of the retweet thread's size k.

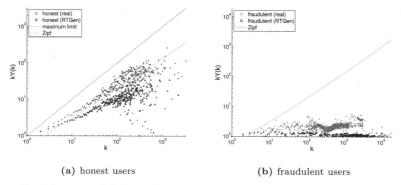

(a) honest users   (b) fraudulent users

Fig. 4. Fraudsters exhibit uniform retweet disparity. (FAVORITISM and HOMO-GENEITY patterns): log-log scatter plots of $kY(k, i)$ vs. k for real and simulated retweets of (a) honest users and (b) fraudulent users. Magenta (green) line corresponds to the *super-skewed* case of eq. 3 (the realistic Zipf distribution of Lemma 3). Black triangles correspond to RTGEN retweet threads for: honest-like, in (a) and fraudulent-like, in (b).

Figure 4 exhibits the relation between $Y(k, i)$ and k averaged over all honest (Figure 4a) and fraudulent users (Figure 4b). We observe that $kY(k, i)$ for honest users appears to have exponential relationship to k, with an exponent of less than 1 (from equation 3). Fraudulent users' activity is fundamentally different and is close to the homogeneous case, where $kY(k, i) = 1$. The most homogeneous behavior is encountered at large values of k which correspond to heavily promoted tweets, whereas less homogeneity is encountered for small retweet threads, likely for camouflage-related reasons.

We try to approximate the relationship between disparity and k under the hypothesis that the different states r_{ij} of users u_j for $j = 1 \ldots k$ follow a Zipf distribution. If we sort the different r_{ij} states by decreasing order of magnitude, we can express the jth frequency $p_j = \frac{r_{ij}}{SR}$ as $p_j = \frac{1}{j \times \ln(1.78*k)}$[15]. Then, we derive the following lemma:

Lemma 3. *The disparity of a Zipf distribution is given by:* $Y_{Zipf}(k,i) \simeq \frac{k-1}{k \times \ln^2(1.78*k)}$

Proof. As per equation 1, the disparity of the Zipf distribution can be approximated by:

$$Y_{Zipf}(k,i) \simeq \int_{j=1}^{k} \left(\frac{1}{j \times \ln(1.78*k)} \right)^2$$

$$= \frac{1}{\ln^2(1.78*k)} \int_{j=1}^{k} \frac{1}{j^2} = \frac{k-1}{k \times \ln^2(1.78*k)} \qquad \square$$

Figure 4a depicts the k-$kY_{Zipf}(k,i)$ relation with a green line, which is a good fit for honest users' behavior (FAVORITISM pattern). Conversely, fraudulent users' disparity is characteristic of a zero slope (HOMOGENEITY pattern), as indicated by Figure 4b.

Observation 4 (FAVORITISM). *The disparity of retweeting activity to honest users' posts can be modeled under the hypothesis that the participation of users to retweets follows a Zipf law.*

Observation 5 (HOMOGENEITY). *The disparity of retweeting activity to fraudulent users' posts can be modeled under the hypothesis that the participation of users to retweets is homogeneous.*

5.3 Activity Summarization Features: MACHINE-GUN, ENTHUSIASM and REPETITION Patterns

We further extracted the following temporal and popularity (ASum) features with respect to the retweet threads included in the datasets:

- *ratio of activated followers*, i.e. author's followers who retweeted;
- *response time*, i.e. time elapsed between the tweet's posting and its first retweet;
- *lifespan*, i.e. time elapsed between the first and the last (observed) retweet, constrained to 1 month to remove bias with respect to later tweets;
- *Arr-IQR*, i.e inter-quartile range of interarrival times for retweets.

Figure 5a depicts the scatterplot of activated followers ratio vs. response time for retweet threads of all target users. Interestingly, several red points of users suspected of fraud are clearly separated from honest users' retweet threads due to their high or low response time and high *activated followers ratio*. In addition, the consideration of various feature combinations can be useful for identifying fake retweet threads. Figure 5b, which depicts the scatter plot of the Arr-IQR vs. lifespan for retweets of all target users' retweet threads, indicates that several retweet threads of the same fraudulent users tend to exhibit similar values for these features, resulting in the formation of dense microclusters of points. For example, the cluster appearing at the figure's bottom-left side is created from retweet threads whose author is fraudulent user FD 5.

From this analysis, we draw several additional observations.

Observation 6 (ENTHUSIASM). *Followers of fraudulent retweeters have a high infection probability.*

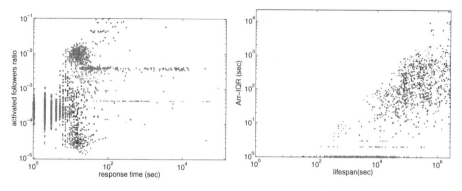

(a) Activated followers ratio vs. Response time

(b) Arr-QR vs. Lifespan

Fig. 5. Dense microclusters formed by fraudsters. (ENTHUSIASM, MACHINE-GUN and REPETITION patterns): log-log scatter plots of ASum features for all target users - each point is a retweet thread, each author has a different glyph. HP, MP, LP users are in blue, green, cyan, and fraudsters are in red.

Observation 7 (MACHINE-GUN). *Fraudsters retweet all at once, or with similar time-delay.*

Observation 8 (REPETITION). *Groups of fake retweet threads exhibit the same values in terms of response time, Arr-IQR and activated followers ratio, forming microclusters.*

6   RTGen Generator

We propose RTGEN, a generator that simulates the retweeting activity of honest and fraudulent users, highlight its properties, and present its results with respect to disparity.

Algorithm 1 outlines the process for the simulation of the retweeting behavior over a network $G(V, E)$, where V_i is the set of users and $E_{i,j}$ is the set of directed *who-follows-whom* relationships between them. In our model, a given user u_i from the set V_i is considered a *candidate* for retweeting if u_i follows either the *author* or another user who has already retweeted (an activated user). Each run of the generator involves the selection of a random user and the simulation of the tweet forwarding process for N tweet events. More specifically, in the first simulation, the *author* of a tweet is randomly selected, and the author's followers become candidate retweeters. Each candidate is then added to a list of activated users with a given retweeting probability. This process is executed recursively until all activated users' followers have been examined and there are no more candidate users. Then, RTGEN continues with the next simulation. Each simulation (tweet) is characterized by a varying *interestingness* value representing the infection probability given the significance of the tweet's content.

RTGEN simulates the scenarios of honest and fraudulent retweeting behavior by forming hypotheses on the underlying graph and the users' inclination to retweet. In specific, based on the discovered TRIANGLES and DEGREES patterns, RTGEN uses a Kronecker graph [10] to simulate honest users' networks and a dense Erdös-Rényi graph [4] for fraudsters' networks. Moreover, RTGEN assumes the same infection probability for all fraudulent users, based on the ENTHUSIASM

Data: $G(V, E)$ = Examined network, N = number of simulations, b =
            interestingness in $[B_1,..., B_n]$
Result: *activatedUsers* : activated nodes $\in V$ per simulation
author $\leftarrow$ user randomly selected from V;
sim $\leftarrow$ 1 ;
while *sim* $\leq N$ **do**
    initialInterestingness $\leftarrow$ pick an interestingness b from B_i ;
    candidateUsers $\leftarrow$ authors' followers ;
    for *each user in candidateUsers* **do**
        followers $\leftarrow$ take followers of candidateUsers ;
        for *each follower f in followers* **do**
            if *f not in activated users* **then**
                calculate retweet probability $bUser_f$;
                add f to *activatedUsers* with probability $bUser_f$;

    sim $\leftarrow$ *sim* + 1 ;

Algorithm 1. Pseudocode for RTGEN

and REPETITION patterns. Conversely, honest users have different activation rates depending on the tweet's interestingness, topics of interest and limited attention. For generality, we follow the *weighted cascade model* [7] and assume that user u_i's infection probability is inversely proportional to the number of followers. This lowers the retweeting probability for users with a large number of followers, simulating limited attention and content competition. For organic retweet thread simulation, the probability $bUser_v$ of user v is thus taken as:

$$P_{honest}(v, i) = b_i * (1/|f_v|) \qquad (4)$$

where $b_i \in [B_1, ..., B_n]$ is the tweet's interestigness in the i_{th} simulation sim_i and $|f_v|$ is the number of followers for user v. Respectively, for the fake retweet thread case:

$$P_{fraudulent}(v, i) = b_i \qquad (5)$$

where, here, b_i is randomly selected between two probability values $[B_1, B_2]$. B_1 represents *camouflage* retweeting activity, and B_2 represents *fake* retweeting activity, with B_2 being much higher than B_2 (in our experiments by an order of magnitude).

RTGEN was applied on: (a) a Kronecker graph of 500k nodes, 14M edges (generated with a parameter matrix $\left(\begin{smallmatrix} 0.9999 & 0.5542 \\ 0.5785 & 0.2534 \end{smallmatrix}\right)$ [14]), and (b) an Erdös-Rényi graph of 10k nodes, 1M edges, for 10 users and 100 simulations. Based on the simulation results, we calculated the disparity for each author and k-sized retweet thread and averaged the disparity values separately for honest and fraudulent authors. Figure 4 depicts the relation between disparity and k for each class of users, which emulate those derived from real Twitter data.

7 Conclusions

Fake retweet behavior incentivized by monetary and social benefits negatively impacts the credibility of content and the perception of honest users on Twitter. In this work, we focus on spotting fake from organic retweet behavior, as well as identifying the fraudsters to blame by carefully extracting features from the activity of their retweeters. Specifically, our main contributions are:

– **Patterns:** We discovered several patterns (RTSCOPE) for characterizing various types of fraud: e.g. the "TRIANGLES" pattern reveals strong connectivity in retweeter networks, the "HOMOGENEITY" pattern indicates uniform retweet disparity.
– **Generator:** We propose RTGEN, a scalable, realistic generator which produces both organic and fraudulent retweet activity using the weighted cascade model. RTGEN can be useful for experimentation and evaluation scenarios where actual, labeled retweet data are missing.

References

1. Beutel, A., et al.: CopyCatch: stopping group attacks by spotting lockstep behavior in social networks. In: WWW, pp. 119–130. ACM (2013)
2. Chu, Z., et al.: Who is Tweeting on Twitter: Human, Bot, or Cyborg? ACSAC, 21–30 (2010)
3. Derrida, B., et al.: Statistical Properties of Randomly Broken Objects and of Multivalley Structures in Disordered Systems. Journal of Physics A: Mathematical and General **20**(15), 5273–5288 (1987)
4. Erdos, P., et al.: On the evolution of Random Graphs. Publ. Math. Inst. Hungary. Acad. Sci. **5**, 17–61 (1960)
5. Ghosh, R., et al.: Entropy-based classification of 'retweeting' activity on twitter. In: KDD Workshop on Social Network Analysis (SNA-KDD) (2011)
6. Jiang, M., Cui, P., Beutel, A., Faloutsos, C., Yang, S.: Inferring strange behavior from connectivity pattern in social networks. In: Tseng, V.S., Ho, T.B., Zhou, Z.-H., Chen, A.L.P., Kao, H.-Y. (eds.) PAKDD 2014, Part I. LNCS, vol. 8443, pp. 126–138. Springer, Heidelberg (2014)
7. Kempe, D., et al.: Maximizing the spread of influence through a social network. In: Proceedings of the Ninth ACM SIGKDD International Conference on Knowledge Discovery and Data Mining, KDD 2003, pp. 137–146. ACM, New York (2003)
8. Kurt, T., et al.: Suspended Accounts in Retrospect: an Analysis of Twitter Spam. IMC, 243–258 (2011)
9. Kwak, H., et al.: What is Twitter, a Social Network or a News Media? In: WWW, pp. 591–600 (2010)
10. Leskovec, J., et al.: Kronecker Graphs: An Approach to Modeling Networks. JMLR **11**, 985–1042 (2010)
11. Lin, P.-C., et al.: A Study of Effective Features for Detecting Long-surviving Twitter Spam Accounts. ICACT 841 (2013)
12. Mao, H.-H., Wu, C.-J., Papalexakis, E.E., Faloutsos, C., Lee, K.-C., Kao, T.-C.: MalSpot: Multi^2 malicious network behavior patterns analysis. In: Tseng, V.S., Ho, T.B., Zhou, Z.-H., Chen, A.L.P., Kao, H.-Y. (eds.) PAKDD 2014, Part I. LNCS, vol. 8443, pp. 1–14. Springer, Heidelberg (2014)
13. Pandit, S., et al.: Netprobe: a fast and scalable system for fraud detection in online auction networks. In: WWW, pp. 201–210. ACM (2007)
14. Rao, A., et al.: Modeling and Analysis of Real World Networks using Kronecker Graphs. Project report (2010)
15. Schroeder, M.: Fractals, Chaos, Power Laws, 6th edn. W. H. Freeman, New York (1991)
16. Tavares, G., et al.: Scaling-Laws of Human Broadcast Communication Enable Distinction between Human, Corporate and Robot Twitter Users. PLoS ONE **8**(7), e65774 (2013)
17. Wu, X., Feng, Z., Fan, W., Gao, J., Yu, Y.: Detecting marionette microblog users for improved information credibility. In: Blockeel, H., Kersting, K., Nijssen, S., Železný, F. (eds.) ECML PKDD 2013, Part III. LNCS, vol. 8190, pp. 483–498. Springer, Heidelberg (2013)
18. Yang, C., et al.: Analyzing spammers' social networks for fun and profit: a case study of cyber criminal ecosystem on twitter. In: WWW, pp. 71–80 (2012)

Resampling-Based Gap Analysis for Detecting Nodes with High Centrality on Large Social Network

Kouzou Ohara[1]([✉]), Kazumi Saito[2], Masahiro Kimura[3], and Hiroshi Motoda[4,5]

[1] Department of Integrated Information Technology, Aoyama Gakuin University, Kanagawa, Japan
ohara@it.aoyama.ac.jp
[2] School of Administration and Informatics, University of Shizuoka, Shizuoka, Japan
k-saito@u-shizuoka-ken.ac.jp
[3] Department of Electronics and Informatics, Ryukoku University, Shiga, Japan
kimura@rins.ryukoku.ac.jp
[4] Institute of Scientific and Industrial Research, Osaka University, Osaka, Japan
[5] School of Computing and Information Systems, University of Tasmania, Hobart, Australia
motoda@ar.sanken.osaka-u.ac.jp, hmotoda@utas.edu.au

Abstract. We address a problem of identifying nodes having a high centrality value in a large social network based on its approximation derived only from nodes sampled from the network. More specifically, we detect gaps between nodes with a given confidence level, assuming that we can say a gap exists between two adjacent nodes ordered in descending order of approximations of true centrality values if it can divide the ordered list of nodes into two groups so that any node in one group has a higher centrality value than any one in another group with a given confidence level. To this end, we incorporate confidence intervals of true centrality values, and apply the resampling-based framework to estimate the intervals as accurately as possible. Furthermore, we devise an algorithm that can efficiently detect gaps by making only two passes through the nodes, and empirically show, using three real world social networks, that the proposed method can successfully detect more gaps, compared to the one adopting a standard error estimation framework, using the same node coverage ratio, and that the resulting gaps enable us to correctly identify a set of nodes having a high centrality value.

Keywords: Gap analysis · Error estimation · Resampling · Node centrality

1 Introduction

Recently, social media such as Facebook, Digg, Twitter, etc. becomes an extremely popular communication tool on a global scale, and generates large-scale social networks on the web. Such networks allow us to share a wide variety of topics

© Springer International Publishing Switzerland 2015
T. Cao et al. (Eds.): PAKDD 2015, Part I, LNAI 9077, pp. 135–147, 2015.
DOI: 10.1007/978-3-319-18038-0_11

that have been posted on social media because those topics can rapidly and widely spread through the networks. Thus, in recent years, social media plays an important role as information infrastructure, and social networks constructed on it have been extensively investigated from various angles [4,8].

In such social network analysis, we can get an insight into some features of a given network by using the node centrality [1,3,5,7,14], which characterizes nodes in the network based on its topology. Typical ones include the degree, closeness, and betweenness centralities. Some of them such as the degree centrality are based only on the information of neighboring nodes of a target node, but some others are also on global structure of a network. For example, to compute the betweenness centrality, we have to enumerate paths between arbitrary node pairs, which is computationally very expensive. Since a social network on the web can easily grow in size, it is crucial to efficiently compute values of such a centrality to analyze a large network.

To this kind of problem on scalability, sampling-based approaches have been proposed so far [6,10,11], which investigate sampling methods that can obtain better approximations of true centrality values. Those methods are roughly categorized into uniform sampling, non-uniform sampling, and traversal/walk-based sampling. In contrast to them, we proposed a framework that ensures the accuracy of the approximations under uniform sampling [13], in which we estimated the approximation error referred to as resampling error by considering all possible partial networks of a fixed size that are generated by resampling nodes according to a given coverage ratio and approximated centrality values derived from them. It is empirically shown that the resampling-based framework provides a tighter approximation error with a higher confidence level than the traditional standard error in statistics under a given sampling ratio.

Unlike these existing approaches, in this paper, we consider detecting a set of nodes having a high centrality value only from approximations derived from sampled nodes with an adequate confidence level, instead of trying to accurately estimate the centrality value itself. We are interested in such nodes because they tend to play an important role for information diffusion on the network. To this end, we consider a list of nodes in descending order of the approximate centrality value, and devise an algorithm to efficiently detect gaps that exist between two adjacent nodes in the list. Here, we say a gap, or a boundary exists between two adjacent nodes in the list if it can divide the ordered list of nodes into two groups so that any node belonging to one group has a higher centrality value than any node in another group with a given confidence level. We incorporate confidence intervals of true centrality values for each node to detect such gaps, and adopt the above resampling-based estimation framework to estimate the confidence intervals as accurately as possible. The results of extensive experiments on three real world social networks demonstrate that using the resampling error for detecting gaps outperforms using the standard error in terms of the number of gaps detected, and that the resulting gaps allow us to correctly identify nodes having a high centrality value.

2   Resampling-Based Estimation Framework

In this section, according to the work [13], we revisit the resampling-based framework for estimating an approximation error with a given confidence level and its application to computing the node centrality.

2.1   General Framework

Let S be a set of objects such that $|S| = L$, and f a function that assigns a value to each object $s \in S$. Then, the problem we address is estimating the average μ over the set of entire values $\{f(s) \mid s \in S\}$ only from its arbitrary subset of partial values $\{f(t) \mid t \in T \subset S\}$. Let $\mu(T)$ be the partial average over a subset T whose number of elements is N, i.e., $\mu(T) = (1/N)\sum_{t \in T} f(t)$. Then, we consider using this partial average $\mu(T)$ as an approximate solution of the true average μ and estimating an expected approximation error $RE(N)$, referred to as resampling error, which is the difference between μ and $\mu(T)$, with respect to the number of elements N, if L is too large to compute μ. Given $\mathcal{T} \subset 2^S$ that is a family of subsets of S such that $|T| = N$ for $T \in \mathcal{T}$, the resampling error $RE(N)$ is defined as follows:

$$RE(N) = \sqrt{\langle (\mu - \mu(T))^2 \rangle_{T \in \mathcal{T}}} = \sqrt{\binom{L}{N}^{-1} \sum_{T \in \mathcal{T}} \left(\mu - \frac{1}{N}\sum_{t \in T} f(t) \right)^2} = C(N)\sigma,$$

(1)

where the factor $C(N) = \sqrt{(L-N)/((L-1)N)}$ and $\sigma = \sqrt{L^{-1}\sum_{s \in S}(f(s) - \mu)^2}$ is the standard deviation. Note that since the estimation error of Equation (1) is regarded as the standard deviation with respect to the number of elements N, we can claim from a statistical viewpoint that for a given subset T such that $|T| = N$, and its partial average value $\mu(T)$, the probability that $|\mu(T) - \mu|$ is larger than $1.96 \times RE(N)$, is less than 5%. In other words, the range of $\mu(T) \pm 1.96 \times RE(N)$ is regarded as the 95% confidence interval of μ.

On the other hand, we can consider a standard approach to this problem that is based on the i.i.d. (independently identical distribution) assumption. More specifically, for a given subset T that has N elements, that is, $T = \{t_1, \cdots, t_N\}$, it is assumed that each element $t \in T$ is independently selected according to some distribution $p(t)$ such as an empirical distribution $p(t) = 1/L$. Then, the standard error $SE(N)$ based on this assumption is defined as follows:

$$SE(N) = \sqrt{\langle (\mu - \mu(T))^2 \rangle} = \sqrt{\sum_{t_1 \in S} \cdots \sum_{t_N \in S} \left(\mu - \frac{1}{N}\sum_{n=1}^{N} f(t_n) \right)^2 \prod_{n=1}^{N} p(t_n)} = D(N)\sigma,$$

(2)

where $D(N) = 1/\sqrt{N}$ and σ is the standard deviation.

It is noted that the difference between Equations (1) and (2) is only their coefficient terms, $C(N)$ and $D(N)$, and that $C(N) \leq D(N)$, $C(L) = 0$ and $D(L) \neq 0$. Namely, $RE(N) \leq SE(N)$ for any N, and $RE(N)$ becomes 0 when

$N = L$, but not $SE(N)$. Note that the true standard deviation σ is needed in both Equations (1) and (2), but in practice, we can use, instead of σ, the standard deviation σ' that is derived from a subset S' ($\subset S$) such that $|S'| = L'$ is small enough to compute σ' within a reasonable time if $|S|$ is too large to compute σ, which is just the case where sampling is needed.

2.2 Application to Node Centrality Estimation

Next, we present the way to apply the above estimation framework to node centrality estimation of a social network that is represented as a directed graph $G = (V, E)$, where V and E ($\subset V \times V$) are the sets of all the nodes and the links in the network, respectively. Here, we consider two node centrality measures, the closeness centrality and the betweenness centrality as in [13].

The closeness $cls_G(u)$ of a node u on a graph G is defined as

$$cls_G(u) = \frac{1}{(|V| - 1)} \sum_{v \in V, v \neq u} \frac{1}{spl_G(u, v)}, \tag{3}$$

where $spl_G(u, v)$ stands for the shortest path length from u to v in G, and we set $spl_G(u, v) = \infty$ when node v is unreachable from node u on G. Intuitively, a node u has a high value for this closeness centrality if a large number of nodes are reachable from u within relatively short path lengths. A standard technique for computing $cls_G(u)$ of each node $u \in V$ is the burning algorithm [12] whose computational complexity is $O(|E|)$. Thus, it takes a large amount of computation time for a huge social network consisting of millions of nodes. To apply the above estimation framework to the computation of an approximation of the closeness centrality $cls_G(u)$ of each node $u \in V$, we instantiate the set of objects S and the function f to this problem. In fact, we consider $S_u = V \setminus \{u\}$ as the set S and $f_u(v) = 1/spl_G(u, v)$ as the function f, and thereby can calculate a partial average value $cls_G(u; T)$ from an arbitrary subset $T \subset S_u \cup \{u\}$ and its approximation error, $RE(u; |T|)$ and $SE(u; |T|)$, according to the above framework.

Next, the betweenness $btw_G(u)$ of a node u on a graph G is defined as

$$btw_G(u) = \frac{1}{(|V| - 1)(|V| - 2)} \sum_{v \in V, v \neq u} \left(\sum_{w \in V, w \neq u, w \neq v} \frac{nsp_G(v, w; u)}{nsp_G(v, w)} \right), \tag{4}$$

where $nsp_G(v, w)$ is the number of the shortest paths from v to w in G, and $nsp_G(v, w; u)$ is the number of the shortest paths from v to w that pass through node u. Thus, the betweenness of a node u becomes high if a large number of shortest paths between two nodes pass through node u. The Brandes algorithm [2] is a standard technique for computing $btw_G(u)$ of each node $u \in V$ and its computational complexity is $O(|E|)$. Thus, it requires a large amount of computation time for a large social network, too. Again, we consider instantiating S and f of the above estimation framework for computing an approximation of the betweenness centrality $btw_G(u)$. More specifically, we regard the expression inside the large parentheses in Equation (4) as a function $btw_G(u; v)$, the betweenness of node u that restricts its starting node to v. Then, by considering $S_u = V \setminus \{u\}$ and $f_u(v) = btw_G(u; v)/(|V|-2)$, we can calculate a partial average value $btw_G(u; T)$ from an arbitrary subset $T \subset S_u \cup \{u\}$ and its estimation error, $RE(u; |T|)$ and $SE(u; |T|)$, based on the above estimation framework.

3   Gap Detection Method

In this section, we consider the way to detect a set of nodes having a high centrality value with a given confidence level based only on centrality values estimated from a subset of nodes in a network. First of all, we formally define the problem we address here. For a network $G(V, E)$, let $\mu_G(v)$ be the true value of a certain centrality measure for node $v \in V$, $\mu_G(v; T)$ be its estimation derived only from a subset of nodes $T \subseteq V$, and $\sigma(v; |T|)$ be its approximation error such as $RE(v; |T|)$ and $SE(v; |T|)$. In addition, given a node v, let $V_H(v; T) = \{u \in V; \mu_G(u; T) \geq \mu_G(v; T)\}$ and $V_L(v; T) = \{w \in V; \mu_G(w; T) < \mu_G(v; T)\}$ be disjoint partitions of V with respect to $\mu_G(v; T)$. Then, incorporating the confidence interval estimation in statistics, the problem can be defined as finding out all nodes $v \in V$ that satisfy the following inequality for $\forall u \in V_H(v; T)$ and $\forall w \in V_L(v; T)$:

$$\mu_G(u; T) - z(\alpha) \cdot \sigma(u; |T|) > \mu_G(w; T) + z(\alpha) \cdot \sigma(w; |T|) \qquad (5)$$

where $0 < \alpha < 1$ and $z(\alpha)$ is the upper $\alpha/2$ critical value of the standard normal distribution. In other words, $\mu_G(u) > \mu_G(w)$ holds for $\forall u \in V_H(v; T)$ and $\forall w \in V_L(v; T)$ with the confidence level $C = 100(1 - \alpha)\%$. Here, the upper half set $V_H(v; T)$ is a set that we want to identify, and we say that a gap exists between v and $v' \in \arg\max_{w \in V_L(v;T)} \mu_G(w; T)$. It is obvious that a straightforward approach to this problem requires the computational complexity of $O(|V|^3)$ because it has to check $|V_H(v; T)||V_L(v; T)|$ pairs of nodes for each v, which is not acceptable when a given social network is very large.

To cope with this, we first consider a lower error bound of $V_H(v; T)$ and an upper error bound of $V_L(v; T)$, respectively defined as $LB(V_H(v; T); \alpha) = \min_{u \in V_H(v)}(\mu_G(u; T) - z(\alpha)\sigma(u; |T|))$ and $UB(V_L(v; T); \alpha) = \max_{w \in V_L(v)}(\mu_G(w; T) + z(\alpha)\sigma(w; |T|))$. Hereafter, for simplicity, $LB(V_H(v; T); \alpha)$ and $UB(V_L(v; T); \alpha)$ are denoted by $LB(V_H(v); T, \alpha)$ and $UB(V_L(v); T, \alpha)$, respectively. Then, we focus on the fact that the above problem is reduced to finding all nodes $v \in V$ that satisfy the relation $LB(V_H(v); T, \alpha) > UB(V_L(v); T, \alpha)$ for given α. Since both $LB(V_H(v); T, \alpha)$ and $UB(V_L(v); T, \alpha)$ can be simultaneously computed for arbitrary $v \in V$ by making only one pass through V, the total computational complexity becomes $O(|V|^2)$, which is smaller than $O(|V|^3)$, but it is still hard to find all of such nodes when the size of a network gets larger.

Thus, we further consider an ordered list $(v_1, v_2, \cdots, v_{|V|})$ of nodes in V resulted from sorting them in descending order of the value of $\mu_G(v; T)$, i.e., $\mu_G(v_i; T) \geq \mu_G(v_{i+1}; T)$ for $i \in \{1, \cdots, |V| - 1\}$. Then, $LB(V_H(v_k); T, \alpha)$ is recursively defined as $LB(V_H(v_k); T, \alpha) = \min(LB(V_H(v_{k-1}); T, \alpha), \mu_G(v_k; T) - z(\alpha)\sigma(v_k; |T|))$. As well, $UB(V_L(v_k); T, \alpha)$ is defined as $UB(V_L(v_k); T, \alpha) = \max(UB(V_L(v_{k+1}); T, \alpha), \mu_G(v_{k+1}; T) + z(\alpha)\sigma(v_{k+1}; |T|))$. Considering these definitions, we can compute $LB(V_H(v); T, \alpha)$ and $UB(V_L(v); T, \alpha)$ for every node $v \in V$ by making only one pass, each, through the list $(v_1, v_2, \cdots, v_{|V|})$, respectively, which implies that we can detect all gaps by making two passes through the ordered list. More specifically, in the first pass, referred to as the forward step, we compute $LB(V_H(v_k); T, \alpha)$ varying k from 1 to $|V| - 1$, and then, in the second pass called the backward step, we compute $UB(V_L(v_k); T, \alpha)$ and detect a gap if $LB(V_H(v_k); T, \alpha) > UB(V_L(v_k); T, \alpha)$ holds varying k from $|V|$ to 2. The computational complexity of this method is governed by that of its sorting process, and thus becomes $O(|V| \log |V|)$, which enables the practical gap analysis even for a large social network. The procedure is summarized as follows:

1. $A \leftarrow \emptyset$, $LB(V_H(v_1); T, \alpha) = \mu_G(v_1; T) - z(\alpha)\sigma(v_1; |T|))$, and $UB(V_L(v_{|V|}); T, \alpha) = 0$;
2. (Forward step) For $k = 2$ to $|V| - 1$,
   $LB(V_H(v_k); T, \alpha) = \min(LB(V_H(v_{k-1}); T, \alpha), \mu_G(v_k; T) - z(\alpha)\sigma(v_k; |T|))$;
3. (Backward step) For $k = |V| - 1$ to 2,
   (a) $UB(V_L(v_k); T, \alpha) = \max(UB(V_L(v_{k+1}); T, \alpha), \mu_G(v_{k+1}; T) + z(\alpha)\sigma(v_{k+1}; |T|))$;
   (b) $A \leftarrow A \cup \{v_k\}$ if $LB(V_H(v_k); T, \alpha) > UB(V_L(v_k); T, \alpha)$;
4. Output A, and terminate.

We consider three kinds of methods by adopting different definitions of the estimated error $\sigma(v; |T|)$, which are $\sigma(v; |T|) = 0$, $\sigma(v; |T|) = SE(v; |T|)$, and $\sigma(v; |T|) = RE(v; |T|)$. We refer to these methods as the naive, SE, and RE method, respectively. Note that the naive method assumes $\mu_G(v; T) = \mu_G(v)$. Thus, it determines that there exists a gap between nodes v_k and v_{k+1} for every k such that $\mu_G(v_k; T) \neq \mu_G(v_{k+1}; T)$. On the other hand, since $SE(v; |T|)$ overestimates the approximation error of $\mu_G(v; T)$ compared to $RE(v; |T|)$, the number of gaps detected by the SE method becomes less than that by the RE method. For more details, we empirically compare these methods through experiments on real world social networks as described below.

4 Experiments

4.1 Datasets

We empirically evaluated the three gap detection methods described in the previous section on three datasets of real world networks that are represented as directed graphs. The first dataset is a network extracted from a Japanese blog service site "Ameba"[1], which has $56,604$ nodes representing blogs in "Ameba" and $734,737$ directed links among them. Each directed link is constructed from

---

[1] http://www.ameba.jp/

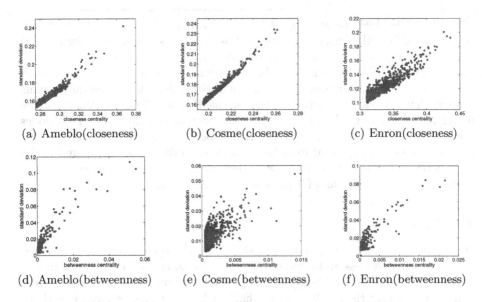

(a) Ameblo(closeness)   (b) Cosme(closeness)   (c) Enron(closeness)

(d) Ameblo(betweenness)   (e) Cosme(betweenness)   (f) Enron(betweenness)

Fig. 1. Centrality values and their standard deviations of the top $1,000$ nodes in descending order of the true value of each centrality in the Ameblo, Cosme, and Enron networks

blog u to blog v if blog u is registered as a favorite one in blog v. We refer to this network as the Ameblo network. The second one is a network extracted from a Japanese word-of-mouth communication site for cosmetics, "@cosme"[2], consisting of $45,024$ nodes representing its users and $351,299$ directed links, in which a link (u, v) means that user v registers user u as her favorite one. We refer to this directed network as the Cosme network. The last one is a network derived from the Enron Email Dataset [9], which has $19,603$ nodes and $210,950$ links. In this network, a node is an email address that appears in the dataset as either a sender or a recipient, while a directional link (u, v) between two email addresses u and v means that u sent an email to v. We refer to this directed network as the Enron network. These three networks are not very huge, but large enough to investigate the basic performance of the three methods from various angles. We thus simply use the standard deviation σ derived from S to compute the resampling and standard errors.

Figures 1(a) to 1(c) show the top 1,000 nodes in descending order of true value of the closeness centrality in the Ameblo, Cosme, and Enron networks, respectively, while Figures 1(d) to 1(f) show the top 1,000 nodes in descending order of true value of the betweenness centrality for the same three networks. We only plotted the top 1,000 nodes because we are interested in nodes having high centrality values. In each figure, the horizontal axis indicates the values of corresponding centrality, and the vertical axis shows its standard

^{2} http://www.cosme.net/

deviation defined as $\sigma_{\mu_G}(u) = \sqrt{(|V| - 1)^{-1} \sum_{v \in V, v \neq u} (f_u(v) - \mu_G(u))^2}$, where $\mu_G(u)$ sands for either $cls_G(u)$ or $btw_G(u)$, and $f_u(v)$ is $1/spl_G(u, v)$ for $cls_G(u)$ and $btw_G(u; v)/(|V| - 2)$ for $btw_G(u)$. From these figures, we can observe that higher-ranked nodes in each centrality measure are distinguishable from each other in every network because of their distinctive values of the centrality, while it looks hard to do the same for lower-ranked nodes. This tendency can be found more clearly in the plots for the betweenness centrality in which nodes are scattered over a larger area. From these observations, we can expect that it is harder to detect gaps that exist between lower-ranked nodes compared to the ones between higher-ranked nodes and that more gaps can be detected for the betweenness centrality than for the closeness centrality.

4.2   Results

We applied the naive, SE, and RE methods to the three networks mentioned above for the closeness and betweenness centralities, and examined the number of gaps they detected and how many gaps among them were correct. A correct gap is the one that the resulting upper half set $V_H(v_k; T)$ corresponds exactly to the true upper half set that is a set of the top k nodes in the descending order of the true centrality value. In this experiment, we adopted the confidence level of 95% ($\alpha = 0.05$) as a typical one and fixed it, while we varied the coverage $|T|/|V|$ from 0.01 to 1.00 by 0.01 points to see how the number of gaps detected changes according to the coverage. More precisely, we randomly sampled nodes from V without replacement, added it to the subset T one by one, and counted the number of gaps detected and the number of gaps correctly detected each time the coverage increases by 0.01. Since we are interested in nodes having a high centrality value, we considered only the top K nodes in descending order of the estimated value of the corresponding centrality at each coverage. We repeated this process $R = 1,000$ times and computed the average over them.

Figure 2 shows the results for the closeness centrality in the case of $K = 100$. The horizontal axis means the coverage, and the vertical axis means the number of gaps. The blue solid line and the red broken line represent the number of gaps detected and the number of gaps incorrectly detected by the corresponding method, respectively, which are defined as follows:

$$(\text{\# of gaps detected}) \quad \frac{1}{R} \sum_{r=1}^{R} \frac{|A(c, r)|}{|A_{nv}(c, r)|} \times K \tag{6}$$

$$(\text{\# of gaps incorrectly detected}) \quad \frac{1}{R} \sum_{r=1}^{R} \frac{|A(c, r) \setminus A^*(c, r)|}{|A_{nv}(c, r)|} \times K, \tag{7}$$

where $A(c, r)$ is the set of nodes corresponding to gaps, i.e., A in the algorithm in Section 3 detected by the respective method at coverage c in the r-th iteration, while $A^*(c, r)$ is the set of nodes correctly detected among them. It is noted that since some of the top K nodes may have the same estimation, these numbers

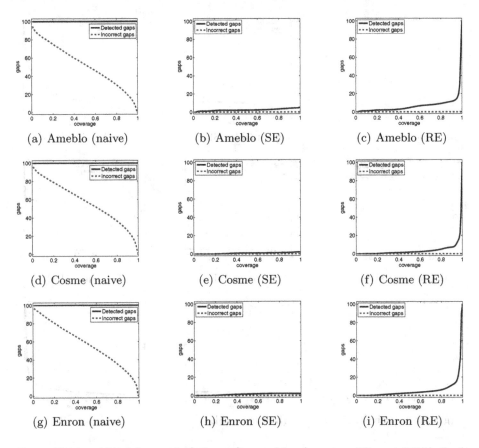

Fig. 2. Fluctuation of the number of gaps detected by the naive, SE, and RE methods as a function of the coverage for the top $K = 100$ nodes in descending order of the estimated value of the *closeness* centrality in the Ameblo, Cosme, and Enron networks

are normalized by the number of gaps detected by the naive method $|A_{nv}(c,r)|$ that corresponds to the number of node pairs v_i and v_{i+1} having different estimations. Thus, the blue solid line for the naive method always exhibits the best performance $(=K)$.

From these results, it is found that although the number of gaps incorrectly detected by the naive method decreases as the coverage becomes larger, it is much larger than the ones by the other two methods that are almost exactly 0. Whereas, the number of gaps detected either by the SE or RE method is very small compared to the one by the naive method. Especially, the number of gaps detected by the SE method increases only a very little even if the coverage becomes closer to 1.0. On the other hand, the number of gaps detected by th RE method is slightly larger than the one by the SE method while the coverage is small, but it rapidly increases at around $c = 0.9$ and finally becomes 100 while the number of gaps incorrectly detected remains almost 0. This difference

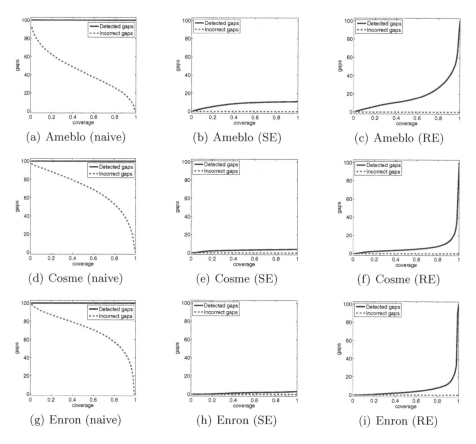

Fig. 3. Fluctuation of the number of gaps detected by the naive, SE, and RE methods as a function of the coverage for the top $K = 100$ nodes in descending order of the estimated value of the *betweenness* centrality in the Ameblo, Cosme, and Enron networks

comes from their nature that the resampling error $RE(v; |T|)$ converges to 0 as $|T|$ approaches to $|V|$, while the standard error $SE(v; |T|)$ does not. These tendencies are also observed in the results for the betweenness centrality shown in Fig. 3.

Next, we examined in the cases of $K = 10$ and $1,000$. Due to the page limitation, we will show only the results for the Ameblo network here, but we observed the same tendencies for the others. Figures 4 and 5 show the results for the closeness centrality and for the betweenness centrality, respectively. From Figs. 4(a) and 5(a), the number of gaps incorrectly detected by the naive method is relatively small compared to the results for $K = 100$ although it is still larger than the ones by the other methods that are almost 0 in this case, too. This is because the higher-ranked nodes in the true centrality value are distinguishable as shown in Fig. 1. Due to the same reason, the number of gaps detected either by the SE or RE method is relatively large compared to the case of $K = 100$.

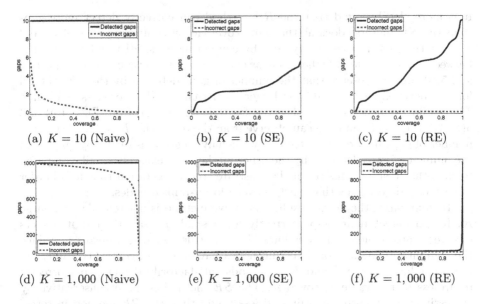

Fig. 4. Fluctuation of the number of gaps detected by the naive, SE, and RE methods as a function of the coverage for the top $K = 10$ and $K = 1,000$ nodes in descending order of the estimated value of the *closeness* centrality in the Ameblo network

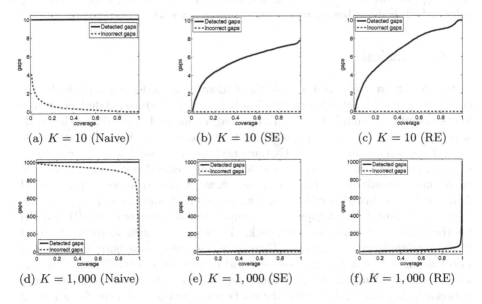

Fig. 5. Fluctuation of the number of gaps detected by the naive, SE, and RE methods as a function of the coverage for the top $K = 10$ and $K = 1,000$ nodes in descending order of the estimated value of the *betweenness* centrality in the Ameblo network

It is more clearly found that the RE method can correctly detect more gaps than the SE method does at the same coverage by comparing Figs. 4(b) and 4(c) for the closeness centrality, and by comparing Figs. 5(b) and 5(c) for the betweenness centrality. Furthermore, as expected above, by comparing Figs. 4(b) and 5(b), we can observe that the number of gaps detected by the SE method for the betweenness centrality is larger than that for the closeness centrality. The similar tendency can be observed for the RE method from Figs. 4(c) and 5(c). On the other hand, we can observe from the results for $K = 1,000$ that the number of gaps incorrectly detected by the naive method is relatively large, and the number of gaps detected by the other methods is relatively small, compared to the other results. This result demonstrates our expectation that it is harder to correctly detect gaps that exist between lower-ranked nodes.

To summarize the above results, the naive method is not reliable for a large K. It can detect many gaps correctly for a small K, say 10, but it detects incorrect gaps if the coverage is low. This is not desirable as a means to reduce the computational cost for detecting nodes having a high centrality value. On the other hand, the SE and RE methods satisfactorily detect gaps correctly regardless of the value of coverage. The SE method is more conservative by overestimating the error margin and less useful than the RE method in terms of the number of gaps detected at the same coverage. Note that although the number of gaps detected by the RE method is limited for a low coverage, the resulting gaps are more likely to appear between nodes having a high centrality value, which is desirable for us to detect important nodes in a network.

5   Conclusion

In this paper, we addressed a problem of identifying nodes having a high centrality value in a social network based only on its approximation derived from a limited number of sampled nodes. To this end, we focused on confidence intervals of true centrality value for each node, and considered detecting gaps that divide a set of nodes into two groups so that any node in one group has a higher centrality value than any one in another does with a given confidence level. To estimate confidence intervals as accurately as possible, we employed the resampling-based framework for estimation of the approximation error, and devised an algorithm that can efficiently detect gaps whose computational complexity is $O(|V|log|V|)$ for the number of nodes in a network, $|V|$, which is much less than $O(|V|^3)$ of the straightforward approach. Through extensive experiments on three real world social networks for the closeness and betweenness centralities, we empirically confirmed that the proposed method can correctly detect gaps that exist between high-ranked nodes with the confidence level of 95% even for a partial network whose coverage is small, say 0.2, and can detect more gaps compared to the one that uses the standard error to estimate confidence intervals at the same coverage ratio. Especially, the ratio of gaps incorrectly detected to the total number of detected gaps is almost 0 for both the methods. It is noted that the method we proposed is not only specific to identification of nodes having a high centrality value, but also applicable to any other estimation problems to which the resampling-based estimation framework is applicable. We believe that the conclusions obtained in this paper can generalize but we have yet to test out the proposed method in a broader setting and in different domains, too.

Acknowledgments. This work was partly supported by Asian Office of Aerospace Research and Development, Air Force Office of Scientific Research under Grant No. AOARD-13-4042, and JSPS Grant-in-Aid for Scientific Research (C) (No. 26330261).

References

1. Bonacichi, P.: Power and centrality: A family of measures. Amer. J. Sociol. **92**, 1170–1182 (1987)
2. Brandes, U.: A faster algorithm for betweenness centrality. Journal of Mathematical Sociology **25**, 163–177 (2001)
3. Brin, S., Page, L.: The anatomy of a large-scale hypertextual web search engine. Computer Networks and ISDN Systems **30**, 107–117 (1998)
4. Chen, W., Lakshmanan, L., Castillo, C.: Information and influence propagation in social networks. Synthesis Lectures on Data Management **5**(4), 1–177 (2013)
5. Freeman, L.: Centrality in social networks: Conceptual clarification. Social Networks **1**, 215–239 (1979)
6. Henzinger, M.R., Heydon, A., Mitzenmacher, M., Najork, M.: On near-uniform url sampling. The International Journal of Computer and Telecommunications Networking **33**(1–6), 295–308 (2000)
7. Katz, L.: A new status index derived from sociometric analysis. Sociometry **18**, 39–43 (1953)
8. Kleinberg, J.: The convergence of social and technological networks. Communications of ACM **51**(11), 66–72 (2008)
9. Klimt, B., Yang, Y.: The enron corpus: a new dataset for email classification research. In: Boulicaut, J.-F., Esposito, F., Giannotti, F., Pedreschi, D. (eds.) ECML 2004. LNCS (LNAI), vol. 3201, pp. 217–226. Springer, Heidelberg (2004)
10. Kurant, M., Markopoulou, A., Thiran, P.: Towards unbiased bfs sampling. IEEE Journal on Selected Areas in Communications **29**(9), 1799–1809 (2011)
11. Leskovec, J., Faloutsos, C.: Sampling from large graphs. In: Proceedings of the 12th ACM SIGKDD International Conference on Knowledge Discovery and Data Mining (KDD 2006), pp. 631–636 (2006)
12. Newman, M.E.J.: Scientific collaboration networks. ii. shortest paths, weighted networks, and centrality. Physical Review E 64, 016132 (2001)
13. Ohara, K., Saito, K., Kimura, M., Motoda, H.: Resampling-based framework for estimating node centrality of large social network. In: Džeroski, S., Panov, P., Kocev, D., Todorovski, L. (eds.) DS 2014. LNCS, vol. 8777, pp. 228–239. Springer, Heidelberg (2014)
14. Zhuge, H., Zhang, J.: Topological centrality and its e-science applications. Journal of the American Society of Information Science and Technology **61**, 1824–1841 (2010)

Classification

Double Ramp Loss Based Reject Option Classifier

Naresh Manwani[1]($\boxtimes$), Kalpit Desai[2],
Sanand Sasidharan[1], and Ramasubramanian Sundararajan[3]

[1] Data Mining Laboratory, GE Global Research, JFWTC, Whitefield,
Bangalore, India
{Naresh.Manwani,Sanand.Sasidharan}@ge.com
[2] Bidgely Technologies Pvt Ltd., Bangalore, India
kvdesai@gmail.com
[3] Sabre Airline Solutions, Bangalore, India
gs.ramsu@gmail.com

Abstract. The performance of a reject option classifiers is quantified using $0 - d - 1$ loss where $d \in (0, .5)$ is the loss for rejection. In this paper, we propose *double ramp loss* function which gives a continuous upper bound for $(0 - d - 1)$ loss. Our approach is based on minimizing regularized risk under the double ramp loss using *difference of convex programming*. We show the effectiveness of our approach through experiments on synthetic and benchmark datasets. Our approach performs better than the state of the art reject option classification approaches.

1 Introduction

The primary focus of classification problems has been on algorithms that return a prediction on every example. However, in many real life situations, it may be prudent to *reject* an example rather than run the risk of a costly potential misclassification. Consider, for instance, a physician who has to return a diagnosis for a patient based on the observed symptoms and a preliminary examination. If the symptoms are either ambiguous, or rare enough to be unexplainable without further investigation, then the physician might choose not to risk misdiagnosing the patient. He might instead ask for further medical tests to be performed, or refer the case to an appropriate specialist. The principal response in these cases is to "reject" the example. This paper focuses on learning a classifier with a reject option. From a geometric standpoint, we can view the classifier as being possessed of a decision surface as well as a rejection surface. The rejection region impacts the proportion of examples that are likely to be rejected, as well as the proportion of predicted examples that are likely to be correctly classified. A well-optimized classifier with a reject option is the one which minimizes the rejection rate as well as the mis-classification rate on the predicted examples.

Let $\mathbf{x} \in \mathbb{R}^p$ is the feature vector and $y \in \{-1, +1\}$ is the class label. Let $\mathcal{D}(\mathbf{x}, y)$ be the joint distribution of $\mathbf{x}$ and y. A typical *reject option classifier* is defined using a bandwidth parameter (ρ) and a separating surface $(f(\mathbf{x}) = 0)$.

© Springer International Publishing Switzerland 2015
T. Cao et al. (Eds.): PAKDD 2015, Part I, LNAI 9077, pp. 151–163, 2015.
DOI: 10.1007/978-3-319-18038-0_12

ρ is the parameter which determines the rejection region. Then a reject option classifier $h(f(\mathbf{x}), \rho)$ is formed as:

$$h(f(\mathbf{x}), \rho) = 1.\mathbb{I}_{\{f(\mathbf{x}) > \rho\}} + 0.\mathbb{I}_{\{|f(\mathbf{x})| \le \rho\}} - 1.\mathbb{I}_{\{f(\mathbf{x}) < -\rho\}} \tag{1}$$

where $\mathbb{I}_{\{A\}}$ is an indicator function which takes value 1 if predicate 'A' is true, else 0. The reject option classifier can be viewed as two parallel surfaces with the rejection area in between. The goal is to determine $f(\mathbf{x})$ as well as ρ simultaneously. The performance of this classifier is evaluated using L_{0-d-1} [8,12] which is

$$L_{0-d-1}(f(\mathbf{x}), y, \rho) = 1.\mathbb{I}_{\{yf(\mathbf{x}) < -\rho\}} + d.\mathbb{I}_{\{|f(\mathbf{x})| \le \rho\}} + 0.\mathbb{I}_{\{yf(\mathbf{x}) \ge -\rho\}} \tag{2}$$

In the above loss, d is the cost of rejection. If $d = 0$, then we will always reject. When $d > .5$, then we will never reject (because expected loss of random labeling is 0.5). Thus, we always take $d \in (0, .5)$.

To learn a reject option classifier, the expectation of $L_{0-d-1}(., ., .)$ with respect to $\mathcal{D}(\mathbf{x}, y)$ (*risk*) is minimized. Since $\mathcal{D}(\mathbf{x}, y)$ is fixed but unknown, the empirical risk minimization principle is used. The risk under L_{0-d-1} is minimized by *generalized Bayes discriminant* [4,8]. $h(f(\mathbf{x}), \rho)$ (Eq. (1)) is shown to be infinite sample consistent with respect to the generalized Bayes classifier [13].

Table 1. Convex surrogates for L_{0-d-1}

Loss Function	Definition
Generalized Hinge	$L_{\mathrm{GH}}(f(\mathbf{x}), y) = \begin{cases} 1 - \frac{1-d}{d} yf(\mathbf{x}), & \text{if } yf(\mathbf{x}) < 0 \\ 1 - yf(\mathbf{x}), & \text{if } 0 \le yf(\mathbf{x}) < 1 \\ 0, & \text{otherwise} \end{cases}$
Double Hinge	$L_{\mathrm{DH}}(f(\mathbf{x}), y) = \max[-y(1-d)f(\mathbf{x}) + H(d), -ydf(\mathbf{x}) + H(d), 0]$ where $H(d) = -d\log(d) - (1-d)\log(1-d)$

Since minimizing the risk under L_{0-d-1} is computationally cumbersome, convex surrogates for L_{0-d-1} have been proposed. *Generalized hinge loss L_{GH}* (see Table 1) is a convex surrogate for L_{0-d-1} [3,12]. It is shown that a minimizer of risk under L_{GH} is consistent to the generalized Bayes classifier [3]. *Double hinge loss L_{DH}* (see Table 1) is another convex surrogate for L_{0-d-1} [7]. Minimizer of the risk under L_{DH} is shown to be *strongly universally consistent* to the generalized Bayes classifier [7]. We observe that these convex loss functions have some limitations. For example, L_{GH} is a convex upper bound to L_{0-d-1} provided $\rho < 1 - d$ and L_{DH} forms an upper bound to L_{0-d-1} provided $\rho \in (\frac{1-H(d)}{1-d}, \frac{H(d)-d}{d})$ (see Fig. 1). Also, both L_{GH} and L_{DH} increase linearly in the rejection region instead of remaining constant. These convex losses can become unbounded for misclassified examples with the scaling of parameters of f. Moreover, limited experimental results are shown to validate the practical significance of these losses [3,7,12]. A non-convex formulation for learning reject

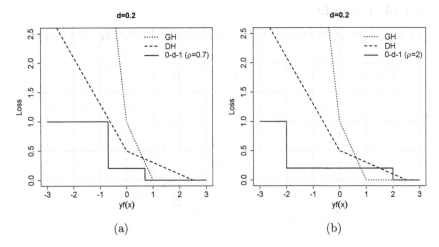

Fig. 1. L_{GH} and L_{DH} for $d = 0.2$. (a) For $\rho = 0.7$, both the losses upper bound the L_{0-d-1}. For $\rho = 2$, both the losses fail to upper bound L_{0-d-1}. L_{GH} and L_{DH} both increase linearly even in the rejection region than being flat.

option classifier is proposed in [5]. However, theoretical guarantees for the approach proposed in [5] are not known. While learning a reject option classifier, one has to deal with the overlapping class regions and outliers. SVM and other convex loss based approaches are less robust to label noise and outliers in the data [10]. It is shown that ramp loss based approach is more robust to noise [6].

Motivated by this, we propose *double ramp loss* (L_{DR}) which incorporates a different loss value for rejection. L_{DR} forms a continuous nonconvex upper bound for L_{0-d-1} and overcomes many of the issues of convex surrogates of L_{0-d-1}. To learn a reject option classifier, we minimize the regularized risk under L_{DR} which becomes an instance of difference of convex (DC) functions. To minimize it, we use DC programming approach [1]. The proposed method has following advantages: (1) the proposed loss L_{DR} gives a tighter upper bound to the L_{0-d-1}, (2) L_{DR} requires no constraint on ρ unlike L_{GH} and L_{DH}, (3) our approach can be easily kernelized for dealing with nonlinear problems.

The rest of the paper is organized as follows. In Section 2 we define the *double ramp loss* (L_{DR}). Then we discuss its properties and the proposed formulation based on L_{DR}. In Section 3 we derive the (L_{DR}) based reject option classifier learning algorithm. We present experimental results in Section 4. We conclude the paper with the discussion in Section 5.

2   Proposed Approach

Our approach for learning classifier with reject option is based on minimizing regularized risk under L_{DR} (double ramp loss).

2.1   Double Ramp Loss

Double ramp loss is defined as a sum of two ramp loss functions as follows:

$$L_{\mathrm{DR}}(f(\mathbf{x}), y, \rho) = \frac{d}{\mu}\Big[\big[\mu - yf(\mathbf{x}) + \rho\big]_+ - \big[-\mu^2 - yf(\mathbf{x}) + \rho\big]_+\Big]$$

$$+ \frac{(1-d)}{\mu}\Big[\big[\mu - yf(\mathbf{x}) - \rho\big]_+ - \big[-\mu^2 - yf(\mathbf{x}) - \rho\big]_+\Big] \quad (3)$$

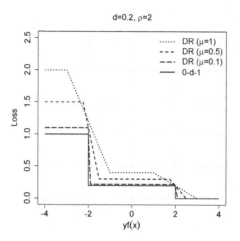

Fig. 2. L_{DR} and L_{0-d-1} : $\forall \mu \geq 0, \rho \geq 0$, L_{DR} is an upper bound for L_{0-d-1}

where $[a]_+ = \max(0, a)$. $\mu \in (0, 1]$ defines the slope of ramps in the loss[1]. Parameter ρ defines the width of the rejection region. Fig. 2 shows L_{DR} for $d = 0.2, \rho = 2$ for different μ.

Theorem 1. *(i)* $L_{DR} \geq L_{0-d-1}, \forall \mu > 0, \rho \geq 0$. *(ii)* $\lim_{\mu \to 0} L_{DR}(f(\mathbf{x}), \rho, y) = L_{0-d-1}(f(\mathbf{x}), \rho, y)$. *(iii)* In the rejection region, $yf(\mathbf{x}) \in (\rho - \mu^2, -\rho + \mu)$, $L_{DR}(f(\mathbf{x}), y, \rho) = d(1 + \mu)$, a const. *(iv)* $L_{DR} \leq (1 + \mu), \forall \rho \geq 0, d \geq 0$. *(v)* When $\rho = 0$, L_{DR} is same as μ-ramp loss ([11]). *(vi)* L_{DR} is a non-convex function of $(yf(\mathbf{x}), \rho)$.

The proof of Theorem 1 is omitted due to the space constraints. We see that L_{DR} does not put any restriction on ρ for it to be an upper bound of L_{0-d-1}.

2.2   Risk Formulation Using L_{DR}

Let $\mathcal{S} = \{(\mathbf{x}_n, y_n), \ n = 1 \dots N\}$ be the training dataset, where $\mathbf{x}_n \in \mathbb{R}^p$, $y_n \in \{-1, +1\}$, $\forall n$. As discussed, we minimize regularized risk under L_{DR} to find

---

[1] While L_{DR} is parametrized by μ and d as well, we omit them for the sake of notational consistency.

a reject option classifier. In this paper, we use l_2 regularization. Let $\Theta = [\mathbf{w}^T \quad b \quad \rho]^T$. Thus, for $f(\mathbf{x}) = (\mathbf{w}^T \phi(\mathbf{x}) + b)$, regularized risk under double ramp loss is

$$
\begin{aligned}
R(\Theta) &= \frac{1}{2}||\mathbf{w}||^2 + \frac{C}{\mu} \sum_{n=1}^{N} \Big\{ d\big[\mu - y_n f(\mathbf{x}_n) + \rho\big]_+ - d\big[-\mu^2 - y_n f(\mathbf{x}_n) + \rho\big]_+ \\
&\quad + (1-d)\big[\mu - y_n f(\mathbf{x}_n) - \rho\big]_+ - (1-d)\big[-\mu^2 - y_n f(\mathbf{x}_n) - \rho\big]_+ \Big\} \\
&= \frac{1}{2}||\mathbf{w}||^2 + \frac{C}{\mu} \sum_{n=1}^{N} \Big\{ d\big[\mu - y_n f(\mathbf{x}_n) + \rho\big]_+ + (1-d)\big[\mu - y_n f(\mathbf{x}_n) - \rho\big]_+ \\
&\quad - d\big[-\mu^2 - y_n f(\mathbf{x}_n) + \rho\big]_+ - (1-d)\big[-\mu^2 - y_n f(\mathbf{x}_n) - \rho\big]_+ \Big\}
\end{aligned}
$$

where C is regularization parameter. While minimizing $R(\Theta)$, no non-negativity condition on ρ is required due to the following lemma.

Lemma 1. *At the minimum of $R(\Theta)$, ρ must be non-negative.*

Proof. Let $\Theta' = (\mathbf{w}', b', \rho')$ minimizes $R(\Theta)$, where $\rho' < 0$. Thus $-\rho' > 0$. Consider $\Theta'' = (\mathbf{w}', b', -\rho')$ as another point.

$$
\begin{aligned}
R(\Theta') - R(\Theta'') &= \frac{C(1 - 2d)}{\mu} \sum_{n=1}^{N} \Big\{ -\big[\mu - y_n f(\mathbf{x}_n) + \rho'\big]_+ + \big[-\mu^2 - y_n f(\mathbf{x}_n) + \rho'\big]_+ \\
&\quad + \big[\mu - y_n f(\mathbf{x}_n) - \rho'\big]_+ - \big[-\mu^2 - y_n f(\mathbf{x}_n) - \rho'\big]_+ \Big\} \\
&= C(1 - 2d) \sum_{n=1}^{N} \Big\{ L_{ramp}(y_n f(\mathbf{x}_n) + \rho') - L_{ramp}(y_n f(\mathbf{x}_n) - \rho') \Big\}
\end{aligned}
$$

where $L_{ramp}(t) = \frac{1}{\mu}([\mu - t]_+ - [-\mu^2 - t]_+)$ is a monotonically non-increasing function of t [11]. Since $\rho' < 0$, thus, $y_n f(\mathbf{x}_n) + \rho' < y_n f(\mathbf{x}_n) - \rho'$, $\forall n$. This implies $L_{ramp}(y_n f(\mathbf{x}_n) + \rho') \geq L_{ramp}(y_n f(\mathbf{x}_n) - \rho')$, $\forall n$. Also $(1 - 2d) \geq 0$, since $0 \leq d \leq 0.5$. Thus $R(\Theta') - R(\Theta'') \geq 0$, which contradicts that Θ' minimizes $R(\Theta)$. Thus, at the minimum of $R(\Theta)$, ρ must be non-negative.

3   Solution Methodology

$R(\Theta)$ (Eq. (4)) is a nonconvex function of Θ. However, $R(\Theta)$ can be written as $R(\Theta) = R_1(\Theta) - R_2(\Theta)$, where $R_1(\Theta)$ and $R_2(\Theta)$ are convex functions of Θ.

$$
R_1(\Theta) = \frac{1}{2}||\mathbf{w}||^2 + \frac{C}{\mu} \sum_{n=1}^{N} \Big[d\big[\mu - y_n f(\mathbf{x}_n) + \rho\big]_+ + (1-d)\big[\mu - y_n f(\mathbf{x}_n) - \rho\big]_+ \Big]
$$

$$
R_2(\Theta) = \frac{C}{\mu} \sum_{n=1}^{N} \Big[d\big[-\mu^2 - y_n f(\mathbf{x}_n) + \rho\big]_+ + (1-d)\big[-\mu^2 - y_n f(\mathbf{x}_n) - \rho\big]_+ \Big]
$$

In this case, DC programming guarantees to find a local optima of $R(\Theta)$ [1]. In the simplified DC algorithm [1], an upper bound of $R(\Theta)$ is found using the convexity property of $R_2(\Theta)$ as follows.

$$R(\Theta) \leq R_1(\Theta) - R_2(\Theta^{(l)}) - (\Theta - \Theta^{(l)})^T \nabla R_2(\Theta^{(l)}) =: ub(\Theta, \Theta^{(l)}) \qquad (4)$$

where $\Theta^{(l)}$ is the parameter vector after $(l)^{th}$ iteration, $\nabla R_2(\Theta^{(l)})$ is a sub-gradient of R_2 at $\Theta^{(l)}$. $\Theta^{(l+1)}$ is found by minimizing $ub(\Theta, \Theta^{(l)})$. Thus, $R(\Theta^{(l+1)}) \leq ub(\Theta^{(l+1)}, \Theta^{(l)}) \leq ub(\Theta^{(l)}, \Theta^{(l)}) = R(\Theta^{(l)})$. Which means, in every iteration, the DC program reduces the value of $R(\Theta)$.

3.1   Learning Reject Option Classifier Using DC Programming

In this section, we will derive a DC algorithm for minimizing $R(\Theta)$. We initialize with $\Theta = \Theta^{(0)}$. Given $\Theta^{(l)}$, we find $\Theta^{(l+1)}$ as

$$\Theta^{(l+1)} \in \arg\min_{\Theta} \ ub(\Theta, \Theta^{(l)}) = \arg\min_{\Theta} \ R_1(\Theta) - \Theta^T \nabla R_2(\Theta^{(l)}) \qquad (5)$$

where $\nabla R_2(\Theta^{(l)})$ is the subgradient of $R_2(\Theta)$ at $\Theta^{(l)}$. We choose $\nabla R_2(\Theta^{(l)})$ as:

$$\nabla R_2(\Theta^{(l)}) = \sum_{n=1}^{N} \beta_n'^{(l)} [-y_n \phi(\mathbf{x}_n)^T \quad - y_n \quad 1]^T + \sum_{n=1}^{N} \beta_n''^{(l)} [-y_n \phi(\mathbf{x}_n)^T \quad - y_n \quad - 1]^T$$

where

$$\begin{cases} \beta_n'^{(l)} = \dfrac{Cd}{\mu} \mathbb{I}_{\{y_n(\phi(\mathbf{x}_n)^T \mathbf{w}^{(l)} + b^{(l)}) - \rho^{(l)} < -\mu^2\}} \\ \beta_n''^{(l)} = \dfrac{C(1-d)}{\mu} \mathbb{I}_{\{y_n(\phi(\mathbf{x}_n)^T \mathbf{w}^{(l)} + b^{(l)}) + \rho^{(l)} < -\mu^2\}} \end{cases} \qquad (6)$$

For $f(\mathbf{x}) = (\mathbf{w}^T \phi(\mathbf{x}) + b)$, we rewrite the upper bound minimization problem described in Eq. (5) as follows,

$$P^{(l+1)} = \min_{\Theta} \ R_1(\Theta) - \Theta^T \nabla R_2(\Theta^{(l)})$$

$$= \min_{\mathbf{w},b,\rho} \ \frac{1}{2} ||\mathbf{w}||^2 + \frac{C}{\mu} \sum_{n=1}^{N} \Big[d[\mu - y_n f(\mathbf{x}_n) + \rho]_+ + (1-d)[\mu - y_n f(\mathbf{x}_n) - \rho]_+ \Big]$$

$$+ \sum_{n=1}^{N} \beta_n'^{(l)} [y_n f(\mathbf{x}_n) - \rho] + \sum_{n=1}^{N} \beta_n''^{(l)} [y_n f(\mathbf{x}_n) + \rho]$$

We rewrite $P^{(l+1)}$ as

$$P^{(l+1)} = \min_{\mathbf{w},b,\boldsymbol{\xi}',\boldsymbol{\xi}'',\rho} \ \frac{1}{2} ||\mathbf{w}||^2 + \frac{C}{\mu} \sum_{n=1}^{N} [d\xi_n' + (1-d)\xi_n''] + \sum_{n=1}^{N} \beta_n'^{(l)} [y_n(\mathbf{w}^T \phi(\mathbf{x}_n) + b) - \rho]$$

$$+ \sum_{n=1}^{N} \beta_n''^{(l)} [y_n(\mathbf{w}^T \phi(\mathbf{x}_n) + b) + \rho]$$

$$\begin{aligned} s.t. \quad & y_n(\mathbf{w}^T \phi(\mathbf{x}_n) + b) \geq \rho + \mu - \xi_n', \quad \xi_n' \geq 0, \quad n = 1 \ldots N \\ & y_n(\mathbf{w}^T \phi(\mathbf{x}_n) + b) \geq -\rho + \mu - \xi_n'', \quad \xi_n'' \geq 0 \quad n = 1 \ldots N \end{aligned}$$

where $\boldsymbol{\xi}' = [\xi'_1 \; \xi'_2 \dots \xi'_N]^T$ and $\boldsymbol{\xi}'' = [\xi''_1 \; \xi''_2 \dots \xi''_N]^T$. The dual optimization problem $D^{(l+1)}$ of $P^{(l+1)}$ is as follows.

$$D^{(l+1)} = \min_{\boldsymbol{\gamma}',\boldsymbol{\gamma}''} \frac{1}{2} \sum_{n=1}^{N} \sum_{m=1}^{N} y_n y_m (\gamma'_n + \gamma''_n)(\gamma'_m + \gamma''_m) k(\mathbf{x}_n, \mathbf{x}_m) - \mu \sum_{n=1}^{N} (\gamma'_n + \gamma''_n)$$

$$s.t. \quad \begin{cases} -\beta'^{(l)}_n \leq \gamma'_n \leq \frac{Cd}{\mu} - \beta'^{(l)}_n & n = 1 \dots N \\ -\beta''^{(l)}_n \leq \gamma''_n \leq \frac{C(1-d)}{\mu} - \beta''^{(l)}_n & n = 1 \dots N \\ \sum_{n=1}^{N} y_n(\gamma'_n + \gamma''_n) = 0 \quad \sum_{n=1}^{N}(\gamma'_n - \gamma''_n) = 0 \end{cases}$$

where $\boldsymbol{\gamma}' = [\gamma'_1 \; \gamma'_2 \dots \dots \gamma'_n]^T$ and $\boldsymbol{\gamma}'' = [\gamma''_1 \; \gamma''_2 \dots \dots \gamma''_n]^T$ are dual variables. At the optimality of $P^{(l+1)}$, $\mathbf{w}$ can be found as $\mathbf{w} = \sum_{n=1}^{N} y_n(\gamma'_n + \gamma''_n)\phi(\mathbf{x}_n)$.

Since $P^{(l+1)}$ has quadratic objective and linear constraints, it holds strong duality with $D^{(l+1)}$. Solving $D^{(l+1)}$ is more useful as it can be easily kernelized for non-linear problems. Behavior of γ'_n and γ''_n under different cases is as follows.

$$\begin{cases} y_n(\mathbf{w}^T\phi(\mathbf{x}_n) + b) - \mu > \rho & \Rightarrow \gamma'_n = -\beta'^{(l)}_n; \;\; \gamma''_n = -\beta''^{(l)}_n \\ y_n(\mathbf{w}^T\phi(\mathbf{x}_n) + b) - \mu = \rho & \Rightarrow \gamma'_n \in \left(-\beta'^{(l)}_n, \frac{Cd}{\mu} - \beta'^{(l)}_n\right); \;\; \gamma''_n = -\beta''^{(l)}_n \\ y_n(\mathbf{w}^T\phi(\mathbf{x}_n) + b) - \mu \in (-\rho, \rho) & \Rightarrow \gamma'_n = \frac{Cd}{\mu} - \beta'^{(l)}_n; \;\; \gamma''_n = -\beta''^{(l)}_n \\ y_n(\mathbf{w}^T\phi(\mathbf{x}_n) + b) - \mu = -\rho & \Rightarrow \gamma'_n = \frac{Cd}{\mu} - \beta'^{(l)}_n; \;\; \gamma''_n \in \left(-\beta''^{(l)}_n, \frac{C(1-d)}{\mu} - \beta''^{(l)}_n\right) \\ y_n(\mathbf{w}^T\phi(\mathbf{x}_n) + b) - \mu < -\rho & \Rightarrow \gamma'_n = \frac{Cd}{\mu} - \beta'^{(l)}_n; \;\; \gamma''_n = \frac{C(1-d)}{\mu} - \beta''^{(l)}_n \end{cases}$$

3.2 Finding $b^{(l+1)}$ and $\rho^{(l+1)}$

To find $b^{(l+1)}$ and $\rho^{(l+1)}$, we consider $\mathbf{x}_n \in \text{SV}'^{(l+1)} \cup \text{SV}''^{(l+1)}$, where

$$\text{SV}'^{(l+1)} = \{\mathbf{x}_n \mid y_n(\phi(\mathbf{x}_n)^T \mathbf{w}^{(l+1)} + b^{(l+1)}) = \rho^{(l+1)} + \mu\}$$
$$\text{SV}''^{(l+1)} = \{\mathbf{x}_n \mid y_n(\phi(\mathbf{x}_n)^T \mathbf{w}^{(l+1)} + b^{(l+1)}) = -\rho^{(l+1)} + \mu\}$$

We already saw that

1. If $\mathbf{x}_n \in \text{SV}'^{(l+1)}$, then $\gamma'^{(l+1)}_n \in \left(-\beta'^{(l)}_n, \frac{Cd}{\mu} - \beta'_n(l)\right)$ and $\gamma''^{(l+1)}_n = -\beta''^{(l)}_n$
2. If $\mathbf{x}_n \in \text{SV}''^{(l+1)}$, then $\gamma'^{(l+1)}_n = \frac{Cd}{\mu} - \beta'^{(l)}_n$ and $\gamma''^{(l+1)}_n \in \left(-\beta''^{(l)}_n, \frac{C(1-d)}{\mu} - \beta''^{(l)}_n\right)$

We solve the system of linear equations corresponding to sets $\text{SV}'^{(l+1)}$ and $\text{SV}''^{(l+1)}$ for identifying $b^{(l+1)}$ and $\rho^{(l+1)}$.

3.3 Summary of the Algorithm

We fix $d \in [0, .5]$, $\mu \in (0, 1]$ and C and initialize the parameter vector Θ as $\Theta^{(0)}$. In any iteration (l), we find $\beta'^{(l)}_n, \beta''^{(l)}_n$, $n = 1 \dots N$ (see Eq. (6)). We solve

$D^{(l+1)}$ to find $\gamma'^{(l+1)}, \gamma''^{(l+1)}$. $\mathbf{w}^{(l+1)}$ is found as $\mathbf{w}^{(l+1)} = \sum_{n=1}^{N} y_n(\gamma_n'^{(l+1)} + \gamma_n''^{(l+1)})\phi(\mathbf{x}_n)$. We find $b^{(l+1)}$ and $\rho^{(l+1)}$ as described in Section 3.2. Thus, we have found $\Theta^{(l+1)}$. Using $\Theta^{(l+1)}$, we now find $\beta_n'^{(l+1)}, \beta_n''^{(l+1)}$, $n = 1 \dots N$. We repeat the above two steps until the parameter vector Θ changes significantly. More formal description of our algorithm is provided in Algorithm 1.

Algorithm 1. Learning Reject Option Classifier by Minimizing $R(\Theta)$

Input : $d \in [0, .5]$, $\mu \in (0, 1]$, $C > 0$, $\mathcal{S}$
Output : $\mathbf{w}^*, b^*, \rho^*$
Initialize $\mathbf{w}^{(0)}, b^{(0)}, \rho^{(0)}, l = 0$
repeat
   Compute $\beta_n'^{(l)} = \frac{Cd}{\mu} \mathbb{I}_{\{y_n(\phi(\mathbf{x}_n)^T \mathbf{w}^{(l)} + b^{(l)}) - \rho^{(l)} < -\mu^2\}}$
   $\beta_n''^{(l)} = \frac{C(1-d)}{\mu} \mathbb{I}_{\{y_n(\phi(\mathbf{x}_n)^T \mathbf{w}^{(l)} + b^{(l)}) + \rho^{(l)} < -\mu^2\}}$
   Find $\gamma'^{(l+1)}, \gamma''^{(l+1)}$ by solving $D^{(l+1)}$ described in Eq. (7)
   Find $\mathbf{w}^{(l+1)} = \sum_{n=1}^{N} y_n(\gamma_n'^{(l+1)} + \gamma_n''^{(l+1)})\phi(\mathbf{x}_n)$
   Find $b^{(l+1)}$ and $\rho^{(l+1)}$ by solving the system of linear equations corresponding to sets $\mathrm{SV}_1^{(l+1)}$ and $\mathrm{SV}_2^{(l+1)}$, where

$$\mathrm{SV}'^{(l+1)} = \{\mathbf{x}_n \mid y_n(\phi(\mathbf{x}_n)^T \mathbf{w}^{(l+1)} + b^{(l+1)}) = \rho^{(l+1)} + \mu\}$$
$$\mathrm{SV}''^{(l+1)} = \{\mathbf{x}_n \mid y_n(\phi(\mathbf{x}_n)^T \mathbf{w}^{(l+1)} + b^{(l+1)}) = -\rho^{(l+1)} + \mu\}$$

until convergence of $\Theta^{(l)}$

3.4 γ' and γ'' at the Convergence of Algorithm 1

At the convergence of Algorithm 1, let $\gamma_n'^*, \gamma_n''^*$, $n = 1 \dots N$ become the values of the dual variables. The behavior of $\gamma_n'^*$ and $\gamma_n''^*$ is described in Table 2. For any $\mathbf{x}_n$, only one of $\gamma_n'^*$ and $\gamma_n''^*$ can be nonzero. We observe that parameters $\mathbf{w}, b$ and ρ are determined by the points whose margin $(yf(\mathbf{x}))$ is in the range $[\rho - \mu^2, \rho + \mu] \cup [-\rho - \mu^2, -\rho + \mu]$. We call these points as *support vectors*. We also see that $\mathbf{x}_n$ for which $y_n f(\mathbf{x}_n) \in (\rho + \mu, \infty) \cup (-\rho + \mu, \rho - \mu^2) \cup (-\infty, -\rho - \mu^2)$, both $\gamma_n'^*, \gamma_n''^* = 0$. Thus, points which are correctly classified with margin at least $(\rho + \mu)$, points falling close to the decision boundary with margin in the interval $(-\rho + \mu, \rho - \mu^2)$ and points misclassified with a high negative margin (less than $-\rho - \mu^2$), are ignored in the final classifier. Thus, our approach not only rejects points falling in the overlapping region of classes, it also ignores potential outliers. We illustrate these insights through experiments on a synthetic dataset as shown in Fig. 3. 400 points are uniformly sampled from the square region $[0 \ 1] \times [0 \ 1]$. We consider the diagonal passing through the origin as the separating surface and assign labels $\{-1, +1\}$ to all the points using it. We changed the labels of 80 points inside the band (width=0.225) around the separating surface.

   Fig. 3 shows the reject option classifier learnt using the proposed method. We see that the proposed approach learns the rejection region accurately. We also observe that all of the support vectors are near the two parallel hyperplanes.

Table 2. Behavior of γ'^* and γ''^*

Condition	$\gamma_n'^* \in$	$\gamma_n''^* \in$
$y_n(\mathbf{w}^T\phi(\mathbf{x}_n)+b) \in (\rho+\mu, \infty)$	0	0
$y_n(\mathbf{w}^T\phi(\mathbf{x}_n)+b) = \rho+\mu$	$(0, \frac{Cd}{\mu})$	0
$y_n(\mathbf{w}^T\phi(\mathbf{x}_n)+b) \in [\rho-\mu^2, \rho+\mu)$	$\frac{Cd}{\mu}$	0
$y_n(\mathbf{w}^T\phi(\mathbf{x}_n)+b) \in (-\rho+\mu, \rho-\mu^2)$	0	0
$y_n(\mathbf{w}^T\phi(\mathbf{x}_n)+b) = -\rho+\mu$	0	$(0, \frac{C(1-d)}{\mu})$
$y_n(\mathbf{w}^T\phi(\mathbf{x}_n)+b) \in [-\rho-\mu^2, -\rho+\mu)$	0	$\frac{C(1-d)}{\mu}$
$y_n(\mathbf{w}^T\phi(\mathbf{x}_n)+b) \in (-\infty, -\rho-\mu^2)$	0	0

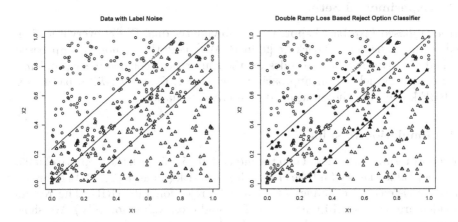

Fig. 3. Left figure shows that label noise affects points near the true classification boundary. Right figure shows reject option classifier learnt using L_{DR} based approach ($C = 100$, $\mu = 1$, $d = .2$). Filled *circles* and *triangles* represent the support vectors.

4 Experimental Results

We show the effectiveness of our approach by showing its performance on several datasets. We also compare our approach with the approach proposed in [7].

4.1 Dataset Description

We report experimental results on 1 synthetic datasets and 2 datasets taken from UCI ML repository [2].

1. **Synthetic Dataset :** Let f_1 and f_2 be two mixture density functions in $\mathbb{R}^2$ defined as follows:

$$f_1(\mathbf{x}) = 0.45\mathcal{U}([1,0] \times [1,1]) + 0.5\mathcal{U}([4,3] \times [0,1]) + 0.05\mathcal{U}([10,0] \times [5,5])$$
$$f_2(\mathbf{x}) = 0.45\mathcal{U}([0,1] \times [1,1]) + 0.5\mathcal{U}([9,10] \times [1,0]) + 0.05\mathcal{U}([0,10] \times [5,5])$$

where $\mathcal{U}(A)$ denotes the uniform density function with support set A. We sample 150 points independently each from f_1 and f_2. We label these points using the hyperplane with $\mathbf{w} = [1 \quad 0]^T$ and $b = 0$. We choose 10% of these points uniformly at random and flip their labels.

2. **Ionosphere Dataset [2]** : This dataset describes the problem of discriminating *good versus bad radars* based on whether they send some useful information about the Ionosphere. There are 34 variables and 351 observations.

3. **Parkinsons Disease Dataset [2]** : This dataset is used to discriminate people with Parkinsons disease from the healthy people. There are 195 feature vectors with each vector having 22 features.

4.2 Experimental Setup

In the proposed $L_{\mathbf{DR}}$ based approach, for solving the dual $D^{(l)}$ at every iteration, we have used the *kernlab* package [9] in **R**. We thank the authors of $L_{\mathbf{DH}}$ based method [7] for providing the codes for their approach. For nonlinear problems, we use RBF kernel. In our approach, we set $\mu = 1$. C and σ (width parameter for RBF kernel) are chosen using 10-fold cross validation.

4.3 Simulation Results

We report results for values of d in the interval $[0.05 \quad .5]$ with the step size of 0.05. For every value of d, we find the cross validation risk (under L_{0-d-1}), % accuracy on the non-rejected examples (Acc) and % rejection rate (RR). The results provided are based on 10 repetitions of 10-fold cross validation (CV). We show the average values and standard deviation (computed over the 10 repetitions).

We now discuss the experimental results. Fig. 4(a) shows the Synthetic dataset and the true classification boundary. Fig. 4(b) and (c) show the classifiers learnt using L_{DR} and L_{DH} based approaches respectively for $d = 0.2$. L_{DR} based approach accurately finds the true classification boundary as oppose to

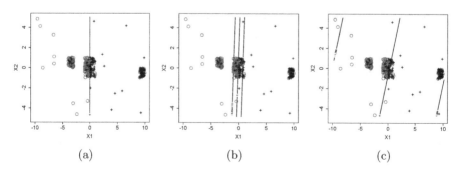

(a)   (b)   (c)

Fig. 4. (a) Synthetic Dataset and the true classification boundary. Reject option classifiers learnt using (b) proposed L_{DR} based approach for $d = 0.2$, (c) L_{DH} based approach for $d = 0.2$.

Table 3. Comparison results on Synthetic dataset (linear classifiers for both the approaches)

d	L_{DR} $(C = 2)$			L_{DH} $(C = 32)$		
	Risk	**RR**	**Acc(unrej)**	**Risk**	**RR**	**Acc(unrej)**
0.05	0.068±0.015	90.87±5.79	75.87±7.95	**0.05**	100	NA
0.1	0.138±0.023	70.35±12.18	79.05±6.87	**0.105**±0.002	95.53±1.69	77.20±6.06
0.15	**0.135**±0.003	65.41±5.06	89.66±0.90	0.136	72.77±0.23	90.56±0.66
0.2	**0.155**±0.006	43.18±4.31	88.56±0.75	0.17	72.67	90.36±1.44
0.25	**0.164**±0.014	32.13±8.43	87.97±1.42	0.204±0.003	66.5±1.7	91±0.74
0.3	**0.148**±0.012	13.23±7.52	87.67±0.69	0.197	46.73±0.14	89.37±0.32
0.35	**0.134**±0.005	4.57±1.80	87.68±0.23	0.21±0.002	43.33±0.65	90.02±0.38
0.4	**0.131**±0.003	1.51±0.56	87.29±0.30	0.21±0.006	31.17±1.26	87.41±0.55
0.45	**0.128**±0.002	0.86±0.45	87.45±0.25	0.265±0.008	9.13±1.1	75.58±0.98
0.5	**0.136**±0.01	0	86.41±0.99	0.297±0.004	0	70.27±0.44

Table 4. Comparison results on Ionosphere dataset (nonlinear classifiers using RBF kernel for both the approaches)

d	L_{DR} $(C = 2,\ \gamma = 0.125)$			L_{DH} $(C = 16,\ \gamma = 0.125)$		
	Risk	**RR**	**Acc(unrej)**	**Risk**	**RR**	**Acc(unrej)**
0.05	**0.025**±0.002	34.84±0.92	98.94±0.31	0.029	52.61±0.73	99.47±0.06
0.1	**0.027**±0.003	8.81±0.32	97.99±0.33	0.047±0.002	43.44±0.85	99.46±0.17
0.15	**0.039**±0.003	5.78±0.57	96.81±0.29	0.042±0.003	24.02±1.62	99.3±0.37
0.2	0.044±0.001	3.46±0.51	96.18±0.15	**0.04**±0.002	17.43±0.59	99.42±0.25
0.25	0.047±0.002	1.76±0.41	95.68±0.23	**0.046**±0.001	14.47±0.79	98.9±0.16
0.3	0.052±0.003	0.92±0.46	95.08±0.35	**0.051**±0.003	12.57±0.75	98.56±0.31
0.35	**0.051**±0.003	0.03±0.09	94.88±0.29	0.054±0.002	9.33±0.59	97.72±0.21
0.4	**0.051**±0.002	0	94.95±0.24	0.054±0.003	6.72±0.86	97.09±0.35
0.45	**0.054**±0.002	0	94.64±0.21	0.055±0.003	3.53±0.41	95.97±0.36
0.5	**0.054**±0.001	0	94.62±0.13	0.055±0.005	0	94.55±0.47

L_{DH} based approach. Also, the reject region found by L_{DR} based approach is the most ambiguous region unlike L_{DH} based approach which rejects almost all the points.

Table 3-5 show the experimental results on all the datasets. We observe the following:

1. We see that the proposed L_{DR} based method outperforms L_{DH} based approach in terms of the risk (expectation of L_{0-d-1}). For Synthetic dataset, except for $d = 0.05$ and 0.1, L_{DR} based method has lower CV risk. Similarly, for Ionosphere dataset, except for $d = 0.2, 0.25$ and 0.3, L_{DR} based method has lower CV risk. For Parkinsons dataset, L_{DR} based method has lower CV risk except for $d = 0.35$.

2. We also observe that L_{DR} based method outputs classifiers with significantly lesser rejection rate for all the datasets and for all values of d.

Thus, the proposed L_{DR} based approach outputs classifiers with lesser risk and lesser rejection rate compared to the L_{DH} based approach.

Table 5. Comparison results on Parkinsons Disease dataset (linear classifiers for both the approaches)

d	L_{DR} $(C = 32)$			L_{DH} $(C = 32)$		
	Risk	**RR**	**Acc(unrej)**	**Risk**	**RR**	**Acc(unrej)**
0.05	**0.031**±0.002	43.88±0.80	98.33±0.49	0.043±0.001	86.38±0.92	100
0.1	**0.051**±0.004	41.79±0.77	98.07±1.03	0.061±0.002	53.76±1.64	98.61±0.62
0.15	**0.071**±0.002	40.08±1.21	98.14±0.48	0.086±0.004	39.56±1.13	95.8±0.72
0.2	**0.095**±0.004	37.67±1.04	96.99±0.55	0.125±0.008	29.78±2.06	90.86±1.5
0.25	**0.133**±0.009	20.46±2.79	90.26±1.30	0.142±0.004	22.3±1.95	89.02±0.73
0.3	**0.129**±0.01	4.06±2.06	87.83±1.15	0.131±0.009	14.19±1.05	89.76±1.01
0.35	0.134±0.007	2.49±1.04	87.19±0.76	**0.133**±0.004	9.97±1.18	89.10±0.57
0.4	**0.131**±0.008	0.56±0.44	87.06±0.75	0.133±0.006	6.10±1.62	88.53±0.92
0.45	**0.133**±0.013	0.05±0.17	86.72±1.28	0.14±0.009	2.92±1.09	86.96±1.05
0.5	**0.133**±0.009	0	86.65±0.94	0.139±0.008	0	86.06±0.76

5   Conclusion and Future Work

In this paper, we have proposed a new loss L_{DR} (*double ramp*) for learning the reject option classifier. L_{DR} gives tighter upper bound for L_{0-d-1} compared to convex losses L_{DH} and L_{GH}. Our approach learns the classifier by minimizing the regularized *risk* under the double ramp loss which becomes an instance of DC optimization problem. Our approach can also learn nonlinear classifiers by using appropriate kernel function. Experimentally, we have shown that our approach works superior to L_{DH} based approach for learning reject option classifiers.

References

1. An, L.T.H., Tao, P.D.: Solving a class of linearly constrained indefinite quadratic problems by d.c. algorithms. Journal of Global Optimization **11**, 253–285 (1997)
2. Bache, K., Lichman, M.: UCI machine learning repository (2013)
3. Bartlett, P.L., Wegkamp, M.H.: Classification with a reject option using a hinge loss. Journal of Machine Learning Research **9**, 1823–1840 (2008)
4. Chow, C.K.: On optimum recognition error and reject tradeoff. IEEE Transactions on Information Theory **16**(1), 41–46 (1970)
5. Fumera, G., Roli, F.: Support Vector Machines with Embedded Reject Option. In: Lee, S.-W., Verri, A. (eds.) SVM 2002. LNCS, vol. 2388, pp. 68–82. Springer, Heidelberg (2002)
6. Ghosh, A., Manwani, N., Sastry, P.S.: Making risk minimization tolerant to label noise. CoRR, abs/1403.3610 (2014
7. Grandvalet, Y., Rakotomamonjy, A., Keshet, J., Canu, S.: Support vector machines with a reject option. In: NIPS, pp. 537–544 (2008)
8. Herbei, R., Wegkamp, M.H.: Classification with reject option. The Canadian Journal of Statistics **34**(4), 709–721 (2006)
9. Karatzoglou, A., Smola, A., Hornik, K., Zeileis, A.: kernlab - an S4 package for kernel methods in R. Journal of Statistical Software **11**(9), 1–20 (2004)

10. Manwani, N., Sastry, P.S.: Noise tolerance under risk minimization. IEEE Transactions on Systems, Man and Cybernetics: Part-B, 43, 1146–1151 (2013)
11. Ong, C.S., An, L.T.H.: Learning sparse classifiers with difference of convex functions algorithms. Optimization Methods and Software (ahead-of-print), 1–25 (2012)
12. Wegkamp, M., Yuan, M.: Support vector machines with a reject option. Bernaulli **17**(4), 1368–1385 (2011)
13. Yuan, M., Wegkamp, M.: Classification methods with reject option based on convex risk minimization. Journal of Machine Learning Research **11**, 111–130 (2010)

Efficient Methods for Multi-label Classification

Chonglin Sun[1]([✉]), Chunting Zhou[2], Bo Jin[1], and Francis C.M. Lau[2]

[1] School of Innovation Experiment, Dalian University of Technology, Dalian, China
chonglinsun@gmail.com, jinbo@dlut.edu.cn
[2] Department of Computer Science,
The University of Hong Kong, Hong Kong, China
chunting.violet.zhou@gmail.com, fcmlau@cs.hku.hk

Abstract. As a generalized form of multi-class classification, multi-label classification allows each sample to be associated with multiple labels. This task becomes challenging when the number of labels bulks up, which demands a high efficiency. Many approaches have been proposed to address this problem, among which one of the main ideas is to select a subset of labels which can approximately span the original label space, and training is performed only on the selected set of labels. However, these proposed sampling algorithms either require nondeterministic number of sampling trials or are time consuming. In this paper, we propose two label selection methods for multi-label classification (i) clustering based sampling (CBS) that uses deterministic number of sampling trials; and (ii) frequency based sampling (FBS) utilizing only label frequency statistics which makes it more efficient. Moreover, neither of these two algorithms needs to perform singular value decomposition (SVD) on label matrix which is used in previously mentioned approaches. Experiments are performed on several real world multi-label data sets with the number of labels ranging from hundreds to thousands, and it is shown that the proposed approaches achieve the state-of-the-art performance among label space reduction based multi-label classification algorithms.

Keywords: Classification · Clustering · Dimension reduction

1 Introduction

Multi-label classification [22] has been widely used in real world applications such as text categorization [11], image annotation [12,21], web advertising [2] and music categorization [18]. In these applications there are usually tens or hundreds of thousands of labels, while the number is still increasing. It is of great significance to perform such tasks with high efficiency. In multi-label classification, each sample can be assigned with a set of labels, which makes it a more challenging task. Many approaches have been proposed to improve its performance.

The traditional approach called binary relevance (BR) [1,16] is to train one classifier for each label being predicted independently. Despite its low training

© Springer International Publishing Switzerland 2015
T. Cao et al. (Eds.): PAKDD 2015, Part I, LNAI 9077, pp. 164–175, 2015.
DOI: 10.1007/978-3-319-18038-0_13

and testing efficiency, memory usage is also a bottleneck as the number of labels becoming larger. Recently, many approaches have been proposed either to exploit hierarchical label structures or conduct dimension reduction using label correlations. Proposed approaches for constructing hierarchical label structures [4,6] are usually converted to a very complex optimization problem mainly aiming at improving testing efficiency, but the training procedure is still very slow. In this paper, we focus on conducting label space dimension reduction by incorporating label correlations.

Label transformation and label selection are two main strategies used for label space dimension reduction. The main idea of label transformation is to transform the original label set into another small set of labels that is manageable for learning [8,13,17,20], i.e., project the original d-dimensional label vector into a k-dimensional vector ($k \ll d$) and then training is performed on the projected vectors. However, these transformed labels are usually difficult to learn. Label selection can overcome the limitation of label transformation, which just selects a small subset of labels from the original set and uses this selected set of labels for training [5,7]. These approaches are based on the assumption that non-selected labels could be correctly reconstructed from the selected ones. However, previously proposed approaches on label selection are either simple random sampling which requires nondeterministic number of sampling trials or computationally expensive. And singular value decomposition (SVD) is needed to be performed in almost all of these methods.

In this paper, two new label selection based methods for multi-label classification are proposed to alleviate these problems, clustering based sampling (CBS) and frequency based sampling (FBS). For CBS, labels are clustered before being selected. With the assumption that a label can faithfully construct other labels in the same cluster, we use K-means to group labels into k clusters (if k labels are to be selected), then sample one label from each cluster. This approach requires only k sampling trials if k labels are to be selected. FBS is an efficient method with high performance where only label frequency statistics is considered for label selection. Contrary to other label selection based multi-label classification approaches, FBS does not formulate label selection as a general column subset selection problem (CSSP). It is able to make better use of the property of label matrix: (1) sparse with each row containing only a few non-zero items; (2) containing values of only 0 or 1. Neither of the proposed two approaches needs to perform SVD or solve complex optimization problems. Experimental results on several real world multi-label datasets show that the proposed algorithms achieve state-of-the-art performance among label space dimension reduction based methods.

The rest of this paper is organized as follows. We first give a brief introduction to label selection and its related problems, and then two proposed label selection methods for multi-label classification are presented. Followed are the experimental results and analysis. Finally we conclude this paper.

Notations: The following notations will be used in this section and the rest of this paper. n, m, d are used to denote the number of samples, the dimension of features and the number of labels respectively. $X \in \mathbb{R}^{n \times m}$ denotes the training data, $Y \in \{0,1\}^{n \times d}$ is the corresponding label matrix. For a vector v, its transpose is denoted by v^T. For a matrix A, its transpose is denoted by A^T, $A^\dagger$ denotes the Moore-Penrose pseudo-inverse, $\|A\|_F$, $\|A\|_2$ denote the Frobenius norm and spectral norm respectively. Besides, we use $A^{(i)}$ to denote the ith row of A, $A_{(j)}$ to denote the jth column of A, and $A_{i,j}$ to denote the jth column of ith row of matrix A.

2   Label Selection

Label selection is a very efficient class of label space dimension reduction oriented approaches for multi-label classification. By selecting a small subset of labels and performing training on these selected labels, non-selected labels can be predicted using the selected ones. Obviously, this method is based on the assumption that non-selected labels can be faithfully constructed from the selected ones. Therefore, the label selection method is the key to classification performance. In this section, we will first give a brief introduction to label selection problem.

One method for label selection is to formulate it as a regularized least squares regression model [5]

$$\min_{W} \|Y - YW\|_F^2 + \lambda_1 \|W\|_{1,2} + \lambda_2 \|W\|_1 \ , \tag{1}$$

where $\|W\|_{1,2} = \sum_{i=1}^d \sqrt{\sum_{j=1}^d W_{i,j}^2}$, $\|W\|_1 = \sum_{i=1}^d \sum_{j=1}^d |W_{ij}|$. $W \in \mathbb{R}^{d \times d}$ is the coefficient matrix with only a few non-zero rows, and λ_1, λ_2 are the regularization parameters. The second term in (1) enforces joint group sparsity across the rows of W, and the third term is the traditional l_1-regularizer over the whole W. However, when the number of labels becomes large, problem (1) becomes computationally expensive.

An alternative is to treat label selection as one column subset selection problem (CSSP) [3,10]. For a matrix $A \in \mathbb{R}^{n \times d}$, CSSP aims at finding exactly k columns so that these selected columns can approximately span A. Concretely, we expect to find a column set C with size k such that $\left\| A - A_C A_C^\dagger A \right\|_F$ is minimized. A_C denotes the sub-matrix of the columns in set C of matrix A. Exact solution is impossible for its high computational complexity $O(d^k)$. Some approximate solution such as randomized sampling [3,7,10] has been proposed to alleviate this problem. In Bi & Kowk's [7] work, they proposed to select k columns from A with the probability for selecting the ith column being

$$p_i = \frac{1}{k} \left\| (V_k^T)_{(i)} \right\|_2^2 \ , \tag{2}$$

where $V_k \in \mathbb{R}^{n \times k}$, is the top k right singular vectors of partial SVD performed on A. p_i corresponds to the leverage score of $A_{(i)}$ on the best subspace of A. Our proposed algorithm is based on CSSP but with different calculation of p_i and different sampling procedures.

3   Algorithms

Similar to CSSP, the aim of the proposed algorithms is to select k columns from a given label matrix Y such that the reconstruction error is minimized. That is

$$\min_{C} \left\| Y - Y_C Y_C^\dagger Y \right\|_F \; , \tag{3}$$

where C is the selected set of k columns, and Y_C is the sub-matrix that contains k columns of Y. We will elaborate the proposed two column selection algorithms for multi-label classification, clustering based sampling (CBS) and frequency based sampling (FBS) in the following sections. Unlike previous proposed label selection algorithms, SVD is not needed in our proposed algorithms.

3.1   Clustering Based Sampling (CBS)

The main idea of CBS is to group labels (columns of label matrix) into k clusters, and then sample one label from each cluster. In order to cluster labels, we need first to generate embeddings for each label. In this work, we represent a label vector as

$$L^{(t)} = \frac{\sum_{i=1}^{n} X^{(i)} Y_{i,t}}{\sum_{i=1}^{n} Y_{i,t}} \; , \tag{4}$$

where $L \in \mathbb{R}^{d \times m}$ denotes the embedding matrix with each row denoting the vector of one label, and $L^{(t)}$ is an m-dimensional row vector which denotes the embedding of the tth label. Although this label embedding looks quite simple, experiments in Sect. 4 show that CBS performs well on several real world datasets. The full procedure is described in Algorithm 1.

Algorithm 1. Clustering Based Sampling for Multi-label Classification

---

1: Input: X, Y, k.
2: Compute the label embedding matrix L according to formula (4).
3: Cluster label embedding L using K-means to generate k clusters, clu_1, clu_2, ..., clu_k.
4: $C \leftarrow \emptyset$
5: **for** $i \leftarrow 1$ to k **do**
6:    Sample one label l from clu_i (For sampling algorithm, we use the method described in Algorithm 2, line 2-9)
7:    $C \leftarrow C \cup \{l\}$
8: **end for**
9: train a classifier $f(x)$ from $\{X^{(n)}, Y_C^{(n)}\}_{n=1}^{N}$
10: For a new test sample x, obtain its prediction $h = f(x)$, return $\hat{y}$ by rounding $h^T Y_C^\dagger Y$.

In Algorithm 1, we first obtain the label (column of label matrix) embeddings with a combination of sample features. Then K-means algorithm is used to group labels into k clusters. For labels with few occurrences, it is hard to obtain a good embedding for clustering. In our implementation, we just put these labels into the same cluster. Since we only sample one label from each cluster, exact k sampling trials are needed.

3.2 Frequency Based Sampling (FBS)

Most of the previously proposed column selection based algorithms for multi-label classification are formulated as a general CSSP problem, they did not utilize the unique property of label matrix Y: (1) they are usually sparse with each row containing only a few non-zero items and (2) the matrix contains values of only 0 or 1. And there are usually a lot of redundant labels in multi-label classification tasks with many labels. For example, in text categorization, one text can be categorized as *machine learning*, it also belongs to the category of *ML* which is short for *machine learning*. And samples with label *ML* is often a subset of samples with label *machine learning*. We want labels like *machine learning* to be selected. Based on this fact, we propose a frequency based sampling algorithm in which each label(column) is selected with a probability

$$p_j = \frac{\sum_{i=1}^n Y_{i,j}}{Z} \ , \ Z = \sum_{i=1}^n \sum_{j=1}^d Y_{i,j} \ , \tag{5}$$

where p_j is the probability of the jth column being selected. Intuitively, labels with higher frequency is assigned with higher sampling probability, and is more likely to be selected. The sampling procedure is described in Algorithm 2. From Algorithm 2, it is easy to see that the probability of selecting a label is proportional to its occurrence frequency.

Proposition 1. *The probability of the jth column being selected $p_j \geq \frac{1}{cn}$, for all $1 \leq j \leq d$, where c is a constant $(c \ll d)$.*

Proof. As the property of label matrix, each row has only a few non-zero terms, let c be the average number of non-zero terms in each row. The sum of all label occurrences is cn and each label (column) will occur at least once. Thus, $p_j \geq \frac{1}{cn}$.

Proposition 2. *The average sampling trials of selecting k different columns is $\Omega(n \log \frac{d}{d-k})$.*

Proof. Let p_j be the probability of the jth column being selected, and T_i be the expected number of sampling trials for sampling exactly i different columns, and C_i the set of selected columns with size i. Then we have the following formula

$$T_i = T_{i-1} + \frac{1}{\sum_{j \notin C_{i-1}} p_j} \ ,$$

$$T_0 = 0 \ , \ C_0 = \emptyset, \sum_{j=1}^d p_j = 1 \ , \ p_j > 0 (1 \leq j \leq d) \ , \tag{6}$$

Algorithm 2. Frequency Based Sampling for Multi-label Classification

---
1: Input: X, Y, k.
2: Compute the sampling probability p_j for each column according to formula (5).
3: $C \leftarrow \emptyset$
4: **while** $|C| < k$ **do**
5:     select a column from $\{1, 2, ..., d\}$ where the probability of selecting j^{th} column is p_j.
6:     **if** $j \notin C$ **then**
7:         $C \leftarrow C \cup \{j\}$
8:     **end if**
9: **end while**
10: train a classifier $f(x)$ from $\{X^{(n)}, Y_C^{(n)}\}_{n=1}^N$
11: For a new test sample x, obtain its prediction $h = f(x)$, return $\hat{y}$ by rounding $h^T Y_C^\dagger Y$.

Together with Proposition 1, we can obtain the lower bound of the second term of (6), $\sum_{j \notin C_{i-1}} p_j \geq \frac{d-i+1}{nc}$, the expected number of sampling trials

$$T_i \leq T_{i-1} + \frac{cn}{d-i+1} , \qquad (7)$$

From (7), we my obtain

$$T_k \leq cn \log \frac{d}{d-k} \qquad (8)$$

where c is a constant as described in Proposition 1. Because $k \ll d$, $\log \frac{d}{d-k} \ll 1$. Typically when $k = 0.1d$, $\log \frac{d}{d-k} = 0.152$, thus $cn \log \frac{d}{d-k} \ll cn$.

3.3 Prediction

The proposed two algorithms have two different column selection procedures, but they share the same prediction procedure as [7]. On prediction, a new test sample is first applied to the k learned classifiers to obtain a k-dimensional prediction vector h. And the d-dimensional prediction vector can be constructed as $\hat{y} = h^T Y_C^\dagger Y$.

3.4 Comparison with Other Methods

The proposed approaches are mainly compared with label selection based method: ML-CSSP [7] and label space transformation based methods: PLST [17], CPLST [8] for the reported high performance. The comparison are shown in Table 1. For our CBS method, it takes $O(nmd)$ in general to obtain label embeddings, however, because the label matrix Y is extremely sparse with only a few non-zero terms in each row, this complexity can be reduced to $O(nm)$. And the time complexity of K-means is $O(kdm)$. For our proposed FBS algorithm, only the frequency of each label is required which takes $O(nd)$. In order to sample k different columns, ML-CSSP requires $O(k \log k)$ sampling trials, our proposed CBS only uses k trials, and

FBS needs $nlog\frac{d}{d-k}$ trials. Besides, PLST and CPLST do not need to sample labels. Among these five methods, our proposed FBS is the most computationally efficient, and our CBS uses the minimum number of sampling trials. For easy comparison, the complexity of PLST and CPLST are also provided.

Table 1. Comparison of various algorithms

	time complexity	sampling trials
CBS (ours)	$O(nm) + O(kdm) + O(k)$	$O(k)$
FBS (ours)	$O(nd) + \Omega(n \log \frac{d}{d-k})$	$\Omega(n \log \frac{d}{d-k})$
ML-CSSP	$O(ndk) + O(k \log k)$	$O(k \log k)$
PLST	$O(ndk)$	-
CPLST	$O(\min\{nm^2, n^2m\}) + O(d^3)$	-

4   Experiments

In this section, experiments are conducted on a number of benchmark datasets[1] shown in Table 2. **cal500** is a dataset of human-generated musical annotations that describe 500 popular western musical tracks. Each song is annotated by a vocabulary of 174 tags [18]. **corel5k** is a set of PCD images. There are 371 words in total in the vocabulary and each image has 4-5 keywords [9]. **delicious** contains textual data of web pages along with their tags extracted from del.icio.us social bookmarking site [15]. **ESPGame** is a list of 100,000 images with English labels from the ESP Game. To make the training process efficient, a subset of the image dataset ESPGame is randomly selected and we pick up tags that occur at least twice within the subset. Each instance in ESPGame is represented with a 905-D feature vector [2] extracted with LIRe [14].

Table 2. Data sets used in the experiment

data sets	#samples	#features	#labels
cal500	502	68	174
corel5k	5000	499	374
delicious	16150	500	983
ESPGame	5000	905	1943

The proposed two methods FBS, CBS are compared with ML-CSSP [7], PLST [17] and CPLST [8]. All the compared methods are implemented in python

---

[1] These datasets are available at http://mulan.sourceforge.net and http://www.cs. utexas.edu/~grauman/courses/spring2008/datasets.htm

[2] 33-D Color Layout, 480-D Gabor, 40-D Edge Histogram, 256-D Color Histogram and 96-D FCTH.

with linear regression as the classifier. The number of selected labels k varies from $0.05d$ to $0.4d$. We just give the results of $k = 0.1d$ since there is no significant difference in relative performance as the value of k varies (But we provide the results with the variation of k on **cal500**).

For performance evaluation, we use two measures: RMSE defined in (10) as in [7], and micro-averaged area under precision-recall curve (AUPRC) [19]. Squared RMSE is proportional to the commonly used Hamming loss $\frac{1}{nd} \left\| Y - \widehat{Y} \right\|^2$. 10-fold cross-validation is performed.

$$RMSE = \frac{1}{\sqrt{n}} \left\| \widehat{Y} - Y \right\|_F . \tag{9}$$

Table 3. Testing RMSE obtained on several datasets (According to pairwise t-test with 95% confidence, number in square brackets indicates the rank)

data sets	cal500	corel5k	delicious	ESPGame
CBS	**4.94** ± 0.09[1]	**1.89** ± 0.02[1]	4.35 ± 0.02[2]	**2.38** ± 0.10[1]
FBS	**4.94** ± 0.09[1]	**1.90** ± 0.02[1]	4.34 ± 0.02[2]	2.49 ± 0.12[2]
ML-CSSP	4.95 ± 0.10[2]	1.92 ± 0.03[2]	4.38 ± 0.03[3]	2.50 ± 0.13[2]
PLST	4.97 ± 0.10[3]	1.91 ± 0.02[2]	**4.26** ± 0.03[1]	2.52 ± 0.12[3]
CPLST	5.01 ± 0.12[4]	1.92 ± 0.02[2]	**4.25** ± 0.03[1]	2.57 ± 0.15[4]

4.1 Accuracy

We compare testing accuracy of our proposed two methods with ML-CSSP [7], PLST [17] , CPLST [8]. The RMSE results of the five methods on several datasets are presented in Table 3. These results are obtained using pairwise t-test with 95% confidence. Our CBS achieves the best performance on 3 of the 4 datasets, which is best among the three approaches. Our FBS also obtains the best performance on 2 out of the 4 datasets, and its overall performance is competitive compared to ML-CSSP, PLST and CPLST. Both FBS and CBS achieve the state-of-the-art performance. We also calculate AUPRC, and results are shown in Table 4. And the variation of RMSE with the number of selected labels k on **cal500** are shown in Fig. 1.

Table 4. Testing AUPRC obtained on several datasets (According to pairwise t-test with 95% confidence, number in square brackets indicates the rank)

data sets	cal500	corel5k	delicious	ESPGame
CBS	**0.441** ± 0.03[1]	0.075 ± 0.01[5]	0.285 ± 0.02[3]	0.033 ± 0.003[4]
FBS	0.438 ± 0.03[2]	0.091 ± 0.01[3]	0.282 ± 0.03[5]	**0.067** ± 0.005[1]
ML-CSSP	0.437 ± 0.02[2]	0.088 ± 0.005[4]	0.283 ± 0.01[4]	0.061 ± 0.003[3]
PLST	0.439 ± 0.03[2]	0.098 ± 0.005[2]	0.301 ± 0.02[2]	**0.066** ± 0.005[1]
CPLST	0.426 ± 0.04[3]	**0.101** ± 0.01[1]	**0.310** ± 0.02[1]	0.063 ± 0.003[2]

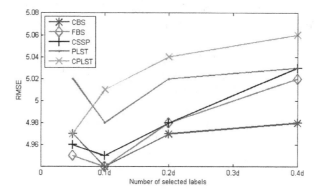

Fig. 1. Variation of Testing RMSE on **cal500** by selecting different number of labels

Table 5. Sampling trials obtained on several datasets

data sets	cal500	corel5k	delicious	ESPGame
CBS	$17 \pm 0[1]$	$37 \pm 0[1]$	$98 \pm 0[1]$	$194 \pm 0[1]$
FBS	$19 \pm 3[2]$	$51 \pm 10[2]$	$129 \pm 7[2]$	$307 \pm 20[2]$
ML-CSSP	$19 \pm 2[2]$	$57 \pm 7[3]$	$138 \pm 7[3]$	$310 \pm 19[2]$

4.2 Sampling Trials and Encoding Time

In this section, we compare sampling trials and encoding time of our CBS, FBS with ML-CSSP, PLST and CPLST.

The sampling trials of the 5 approaches on several datasets are shown in Table 5. As the theoretical analysis shown in Sect. 3.4, CBS uses the fewest number of trials on all datasets which is equal to the number of labels being selected (k). Although FBS and ML-CSSP uses a bit more trials, it is still far from their bound $\Omega(n \log \frac{d}{d-k})$ and $O(k \log k)$ respectively. And on average, FBS uses fewer sampling trials than ML-CSSP. Besides, PLST, CPLST do not have the process of sampling.

Table 6 shows the encoding time on several data sets. Our FBS achieves the best encoding efficiency, significantly faster than the other ones. From the result, we can see that CBS is less efficient than ML-CSSP, which is because our embedding approach leads to high dimensional embedding vectors which makes it slow for K-means, as is the case especially in the largest dataset. However, this is not a serious problem since we can embed labels into low dimensional vectors using other techniques to help accelerate the clustering. The efficiency of PLST is similar to ML-CSSP since they both need to perform SVD on label matrix. CPLST is the least efficient among the five methods because it has to perform SVD on a much larger matrix.

Table 6. Encoding time (in seconds) on several datasets (Number in square brackets indicates the rank)

data sets	cal500	corel5k	delicious	ESPGame
CBS	0.08[3]	0.31[2]	7.32[2]	54.74[5]
FBS	**0.01**[1]	**0.01**[1]	**0.07**[1]	**0.17**[1]
ML-CSSP	0.04[2]	0.56[3]	9.65[3]	17.68[3]
PLST	0.03[2]	0.58[3]	9.62[3]	15.62[2]
CPLST	0.03[2]	3.52[4]	46.78[4]	23.92[4]

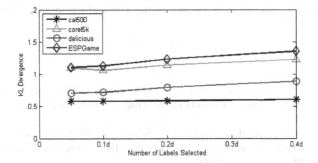

Fig. 2. KL divergence of distributions defined by FBS and ML-CSSP

4.3 Comparison of FBS and ML-CSSP

Although FBS uses a strategy to calculate the probability of sampling a label which is different from ML-CSSP as defined in (5), (2) respectively, FBS achieves comparative RMSE and AUPRC as ML-CSSP. In this section, we analyze a bit about the statistical property of the distribution of label selection given by FBS and ML-CSSP. For ML-CSSP, the probability of selecting a label varies with the number of labels to be selected, whereas it remains constant for FBS. We use KL divergence to measure the similarity between them

$$D_{KL}(p\|q) = \sum_i p(i) \log \frac{p(i)}{q(i)} \ , \tag{10}$$

where p denotes the distribution defined in (5), and q denotes the distribution in (2). The KL divergence on several data sets are shown in Fig. 2, it is small on all four data sets. Thus, the selection procedures in these two algorithms behave alike and obtain similar accuracies. However, FBS outperforms ML-CSSP in efficiency.

5 Conclusion

In this paper, we propose two efficient approaches for multi-label classification. Unlike previous proposed label selection methods, although we also formulate label selection as a column subset selection problem (CSSP), SVD is not needed in our methods. For our proposed clustering based sampling (CBS) method, only k sampling trials are needed for selecting k labels. And theoretical analysis and experimental results have demonstrated that our proposed frequency based sampling (FBS) method has the highest efficiency, and FBS is believed to make better use of the property of label matrix. Experiments on a number of real world datasets with many labels demonstrate that our proposed two algorithms achieve the state-of-the-art performance among recently proposed label space dimension reduction based multi-label classification algorithms.

Acknowledgments. The work is supported by the Natural Science Foundation of China (No. 51105052).

References

1. Allwein, E.L., Schapire, R.E., Singer, Y.: Reducing Multiclass to Binary: A Unifying Approach for Margin Classifiers. Journal of Machine Learning Research **1**, 113–141 (2001)
2. Beygelzimer, A., Langford, J., Lifshits, Y., Sorkin, G., Strehl, A.: Conditional probability tree estimation analysis and algorithm. In: Conference in Uncertainty in Artificial Intelligence (UAI) (2009)
3. Boutsidis, C., Mahoney, W.M., Drineas, P.: An improved approximation algorithm for the column subset selection problem. In: Proceedings of the 20th ACMSIAM Symposium on Discrete Algorithms, pp. 968–977 (2009)
4. Bengio, S., Weston, J. Grangier, D.: Label embedding trees for large multi-class tasks. Advances in Neural Information Processing Systems 23 (2010)
5. Balasubramanian, K., Lebanon, G.: The landmark selection method for multiple output prediction. In: Proceedings of the 29th International Conference on Machine Learning, pp. 983–990 (2012)
6. Bi, W., Kwok, J.T.: Mandatory leaf node prediction in hierarchical multilabel classification. Proceedings of the Neural Information Processing Systems (NIPS) (2012)
7. Bi, W., Kwok, J.T.: Efficient multi-label classification with many labels. In: Proceedings of the Thirtieth International Conference on Machine Learning (ICML) (2013)
8. Chen, Y.-N., Lin, H.-T.: Feature-aware label space dimension reduction for multi-label classification. Advances in Neural Information Processing Systems **25**, 1538–1546 (2012)
9. Duygulu, P., Barnard, K., de Freitas, J.F.G., Forsyth, D.: Object Recognition as Machine Translation: Learning a Lexicon for a Fixed Image Vocabulary. In: Heyden, A., Sparr, G., Nielsen, M., Johansen, P. (eds.) ECCV 2002, Part IV. LNCS, vol. 2353, pp. 97–112. Springer, Heidelberg (2002)
10. Drineas, P., Mahoney, M.W., Muthukrishnan, S.M.: Subspace Sampling and Relative-Error Matrix Approximation: Column-Based Methods. In: Díaz, J., Jansen, K., Rolim, J.D.P., Zwick, U. (eds.) APPROX 2006 and RANDOM 2006. LNCS, vol. 4110, pp. 316–326. Springer, Heidelberg (2006)
11. Dekel, O., Shamir, O.: Multiclass-Multilabel Learning when the Label Set Grows with the Number of Examples. In: Artificial Intelligence and Statistics (AISTATS) (2010)

12. Deng, J., Dong, W., Socher, R., Li, Li-Jia, Li, K., Li, Fei-Fei: Imagenet: A large-scale hierarchical image database. In: Conference on Computer Vision and Pattern Recognition (CVPR), pp. 248–255 (2009)
13. Hsu, D., Kakade, S.M., Langford, J., Zhang, T.: Multilabel prediction via compressed sensing. Advances in Neural Information Processing Systems **22**, 772–780 (2009)
14. Lux, M., S.A., Chatzichristofis: Lire: lucene image retrieval: an extensible java cbir library. MM (2008)
15. Tsoumakas, G., Katakis, I., Vlahavas, I.: Effective and Efficient Multilabel Classification in Domains with Large Number of Labels. In: Proceedings of the ECML/PKDD 2008 Workshop on Mining Multidimensional Data (2008)
16. Tsoumakas, G., Katakis, I., Vlahavas, I.: Mining multi-label data. Data Mining and Knowledge Discovery Handbook, pp. 667–685 (2010)
17. Tai, F., Lin, H.: Multilabel Classification with Principal Label Space Transformation. Neural Computation **24**(9), 2508–2542 (2012)
18. Turnbull, D., Barrington, L., Torres, D., Lanckriet, G.: Semantic annotation and retrieval of music and sound effects. IEEE Transactions on Audio, Speech and Language Processing **16**(2), 467–476 (2008)
19. Vens, C., Struyf, J., Schietgat, L., Džeroski, S., Blockeel, H.: Decision trees for hierarchical multi-label classification. Machine Learning **73**, 185–214 (2008)
20. Zhang, Y., Schneider, J.: Multi-label output codes using canonical correlation analysis. In: Proceedings of the 14th International Conference on Artificial Intelligence and Statistics, pp. 873–882 (2011)
21. Nguyen, C.-T., Zhan, D.-C., Zhou, Z.-H.: Multi-modal image annotation with multi-instance multi-label LDA. In: Proceedings of the 23rd International Joint Conference on Artificial Intelligence (IJCAI 2013), pp. 1558–1564 (2013)
22. Zhang, M.-L., Zhou, Z.-H.: A review on multi-label learning algorithms. IEEE Transactions on Knowledge and Data Engineering **26**(8), 1819–1837 (2014)

A Coupled k-Nearest Neighbor Algorithm for Multi-label Classification

Chunming Liu$^{(\boxtimes)}$ and Longbing Cao

AAI, University of Sydney Technology, Sydney, Australia
Chunming.Liu@student.uts.edu.au, LongBing.Cao@uts.edu.au

Abstract. ML-kNN is a well-known algorithm for multi-label classification. Although effective in some cases, ML-kNN has some defect due to the fact that it is a binary relevance classifier which only considers one label every time. In this paper, we present a new method for multi-label classification, which is based on lazy learning approaches to classify an unseen instance on the basis of its k nearest neighbors. By introducing the coupled similarity between class labels, the proposed method exploits the correlations between class labels, which overcomes the shortcoming of ML-kNN. Experiments on benchmark data sets show that our proposed Coupled Multi-Label k Nearest Neighbor algorithm (CML-kNN) achieves superior performance than some existing multi-label classification algorithms.

Keywords: Multi-label · Coupled · Classification · Nearest neighbor

1 Introduction

Although traditional single-label classification approaches have been proved to be successful in handling some real world problems, for the problems which the objects not fit the single-label rule, they may not work well, for example, in image classification, an image may contain several concepts simultaneously, such as beach, sunset and kangaroo. Such tasks are usually denoted as multi-label classification problems. In fact, a conventional single-label classification problem can simply be taken as a special case of the multi-label classification problem where there has only one label in the class label space. Multi-label classification problems exist in many domains, for example, in automatic text categorization, a document can associate with several topics, such as arts, history and Archeology; and in gene functional analysis of bio-informatics, a gene can belong to both metabolism and transcription classes; and in music categorization, a song may labeled as Mozart and sad.

In the last decades, there have been a variety of methods developed for multi-label classifications. These methods are generally grouped into two categories: One is the problem transformation methods and another is the algorithm adaptation methods. Problem transformation methods first transform the multi-label learning tasks into multiple single-label learning tasks, which are then handled

© Springer International Publishing Switzerland 2015
T. Cao et al. (Eds.): PAKDD 2015, Part I, LNAI 9077, pp. 176–187, 2015.
DOI: 10.1007/978-3-319-18038-0_14

by the standard single-label learning algorithms. Another approach is called algorithm adaptation method, which modifies existing single-label learning algorithms in order to extend its ability to handle multi-label data, such as ML-kNN [17], IBLR [7], BSVM [2], and BP-MLL [16].

Researchers have tried to extend the kNN concept to handle the multi-label classification problem, such as ML-kNN. ML-kNN applies maximum a posteriori principle for classification and ranking, and the likelihood is estimated by using the k nearest neighbors of an instance. Although simple and powerful, there are some shortcomings in its processing strategy. ML-kNN uses the popular binary relevance (BR) strategy [13], which may transfer the problem into many class-imbalance tasks, and then tend to degrade the performance of the classifiers. Another problem of it is the estimation of the posteriori may be affected by the facts that the instances with and without a particular label are typically highly imbalanced. Furthermore, its ignorance of the inter relationship between labels is another issue which limits its usage. Such relationship is described as a Coupled behavior in some previous research [4,6]. In [8,14], Can and Liu etc. analysis the coupling relationship on categorical data. These works all proved the effectiveness of considering the dependency between different attributes.

In this paper, we propose a novel kNN-based multi-label learning approach (CML-kNN for short) based on non-iidness [5]. The major contribution of this paper is summarized as follows:

- We propose a novel multi-label learning algorithm that based on lazy learning and the inner relationship between labels.
- We introduce a new coupled label similarity for multi-label kNN algorithm. Rather than only select the neighbors with a specific label, the coupled label similarity will include more similar neighbors in the process to overcome the problem of lacking neighbors with certain label.
- We extended the concept of the nearest neighbor in multi-label classification with coupled label similarity. Based on this extended nearest neighbors, we introduce a new frequency array strategy.

The structure of this paper is organized as follows. Section 2 briefly reviews the ML-kNN algorithm. Preliminary definitions are specified in Section 3.1. And section 3 gives a detailed description of the new algorithm we proposed. The experimental results are discussed in Section 4. Finally, the conclusion is discussed in Section 5.

2   ML-kNN

A number of multi-label learning methods are adapted from kNN [3,11,15,17]. ML-kNN, the first multi-label lazy learning approach, is based on the traditional kNN algorithm and the maximum a posteriori (MAP) principle [17].

The main idea of the ML-kNN approach is that an instance's labels depend on the number of neighbors that possess identical labels. Given an instance x with an unknown label set $L(x) \subseteq L$, ML-kNN first identifies the k nearest neighbors in the training data and counts the number of neighbors belonging to each class (i.e. a variable z from 0 to k). Then the maximum a posteriori principle is used to determine the label set for the test instance. The posterior probability of $l_i \in L$ is given by

$$P(l_i \in L(x)|z) = \frac{P(z|l_i \in L(x)) \cdot P(l_i \in L(x))}{P(z)} \tag{1}$$

where z is the number of neighbors belonging to each class ($0 \leq z \leq k$). Then, for each label $l_i \in L$, the algorithm builds a classifier h_i using the rule

$$h_i(x) = \begin{cases} 1 & P(l_i \in L(x)|z) > P(l_i \notin L(x)|z) \\ 0 & \text{otherwise} \end{cases} \tag{2}$$

where $0 \leq z \leq k$. If $h_i(x) = 1$, it means label l_i is in x's real label set, while 0 means it does not. The prior and likelihood probabilities in Eq. 1 are estimated from the training data set in advance.

ML-kNN has two inheriting merits from both lazy learning and MAP principle: One is the decision boundary can be adaptively adjusted due to the varying neighbors identified for each new instance, and another one is the class-imbalance issue can be largely mitigated due to the prior probabilities estimated for each class label. However, ML-kNN is actually a binary relevance classifier, because it learns a single classifier h_i for each label independently. In other words, it does not consider the correlations between different labels. The algorithm is often criticized because of this drawback.

3   Methodology

3.1   Problem Statement

We formally define the multi-label classification problem as this: Let X denotes the space of instances and $Y = \{l_1, \ldots, l_n\}$ denotes the whole label set where $|Y| = n$. $T = \{(x_1, L(x_1)), \ldots, (x_m, L(x_m))\}$ ($|T| = m$) is the multi-label training data set, whose instances are drawn identically and independently from an unknown distribution D. Each instance $x \in X$ is associated with a label set $L(x) \in Y$. The goal of our multi-label classification is to get a classifier $h : X \to Y$ that maps a feature vector to a set of labels, while optimizing some specific evaluation metrics.

3.2   Coupled Label Similarity

It is much easier for numerical data to calculate the distance or similarity, since the existing metrics such as Manhattan distance and Euclidean distance are mainly built for numeric variables, but the labels are categorical data. How to denote the similarity between them is a big issue. As we all know, matching and frequency [1] are the most common ways to measure the similarity of categorical data. Accordingly, two popular similarity measures are defined: For two categorical value v_i and v_j, the Overlap Similarity is defined as

$$\text{Sim\_Overlap}(v_i, v_j) = \begin{cases} 1, & \text{if } v_i = v_j \\ 0, & \text{if } v_i \neq v_j, \end{cases} \tag{3}$$

and the Frequency Based Cosine Similarity between two vectors V_i and V_j is defined as

$$\text{Sim\_Cosine}(V_i, V_j) = \frac{V_i \cdot V_j}{||V_i|| \, ||V_j||}. \tag{4}$$

The overlap similarity between two categorical values is to assign 1 if they are identical otherwise 0 if different. Further, for two multivariate categorical data points, the similarity between them will be proportional to the number of features in which they match. While for frequency based measures, they assume the different categorical values but with the same occurrence times as the same.

Hence, the Overlap measure and Frequency Based measure are too simplistic by just giving the equal importance to matches and mismatches. The co-occurrence information in categorical data reflects the interaction between features and can be used to define what makes two categorical values more or less similar. However, such co-occurrence information hasn't been incorporated into the existing similarity metrics.

To capture the inner relationship between categorical labels, we introduce an *Intra-Coupling Label Similarity (IaCLS)* and an *Inter-Coupling Label Similarity (IeCLS)* below to capture the interaction of two label values from two different labels.

Definition 1. *Given a training multi-label data set D and two different labels l_i and l_j ($i \neq j$), the label value is v_i^x, v_j^y respectively. The **Intra-Coupling Label Similarity** (IaCLS) between label values v_i^x and v_j^y of label l_i and l_j is formalized as:*

$$\delta^{Intra}(v_i^x, v_j^y) = \frac{RF(v_i^x) \cdot RF(v_j^y)}{RF(v_i^x) + RF(v_j^y) + RF(v_i^x) \cdot RF(v_j^y)}, \tag{5}$$

where $RF(v_i^x)$ and $RF(v_j^y)$ are the occurrence frequency of label value v_i^x and v_j^y in label l_i and l_j, respectively.

The Intra-coupling Label Similarity reflects the interaction of two different label values in the label space. The higher these similarities are, the closer such two values are. Thus, Equation (5) is designed to capture the label value similarity in terms of occurrence times by taking into account the frequencies of categories. Besides, since $1 \leq RF(v_i^x), RF(v_j^y) \leq m$, then $\delta^{Intra} \in [1/3, m/(m+2)]$.

In contrast to the Intra-Coupling, we also define an *Inter-Coupling Label Similarity* below to capture the interaction of two different label values according to the co-occurrence of some value (or discretized value group) from feature spaces.

Definition 2. *Given a training multi-label data set D and two different labels l_i and l_j $(i \neq j)$, the label value is v_i^x, v_j^y respectively. v_i^x and v_j^y are defined to be Inter-Coupling related if there exists at least one pair value (v_p^{zx}) or (v_p^{zy}) that occurs in feature a_z and labels of instance U_p. The **Inter-Coupling Label Similarity** (IeCLS) between label values v_i^x and v_i^y according to feature value v_p^z of feature a_z is formalized as:*

$$\delta^{Inter}(v_i^x, v_j^y | v_p^z) = \frac{\min\left(F(v_p^{zx}), F(v_p^{zy})\right)}{\max(RF(v_i^x), RF(v_j^y))}, \tag{6}$$

where $F(v_p^{zx})$ and $F(v_p^{zy})$ are the co-occurrence frequency count function for value pair v_p^{zx} or v_p^{zy}, and $RF(v_i^x)$ and $RF(v_j^y)$ is the occurrence frequency of related class label. v_p^z is the value in categorical feature a_z or the discretized value group in numerical feature a_z.

Accordingly, we have $\delta^{Ie} \in [0, 1]$. The Inter-Coupling Label Similarity reflects the interaction or relationship of two label values from label space but based on the connection to some other features.

Definition 3. *By taking into account both the Intra-Coupling and the Inter-Coupling, the **Coupled Label Similarity** (CLS) between two label values v_i^x and v_j^y is formalized as:*

$$CLS(v_i^x, v_j^y) = \delta^{Intra}(v_i^x, v_j^y) \cdot \sum_{k=1}^{n} \delta^{Inter}(v_i^x, v_j^y | v_k), \tag{7}$$

where v_i^x and v_j^y are the label values of label l_i and l_j, respectively. δ^{Intra} and δ^{Inter} are the intra-coupling label similarity (Eq. 5) and inter-coupling label similarity (Eq. 6), respectively. The n is the number of attributes and v_k denotes the values in the kth feature a_k.

Table 1. An Example of Multi-label Data

Instances	Label1	Label2	Label3	Label4
u_1	l_1			l_4
u_2			l_3	l_4
u_3	l_1		l_3	
u_4		l_2	l_3	
u_5		l_2	l_3	l_4

The *Coupled Label Similarity* defined in Eq. 7 reflects the interaction or similarity of two different labels. The higher the *CLS*, the more similar two labels be. In Table 1, for example, $CLS(l_1, l_4) = 0.33$, $CLS(l_1, l_3) = 0.25$, so in the data set, an instance with label l_4 is more similar or close to instances with label l_1 than those instances with label l_3 do. That is to say, label pair (l_1, l_4) is closer to each other than the label pair (l_1, l_3). For Table 1, we got the coupled label similarity array which showed in Table 2.

Table 2. CLS Array

	Label1	Label2	Label3	Label4
Label1	1.0	0	0.25	0.33
Label2	0	1.0	0.50	0.33
Label3	0.25	0.50	1.0	0.50
Label4	0.33	0.33	0.50	1.0

3.3 Extended Nearest Neighbors

Based on the Coupled Label Similarity, we introduce our extended nearest neighbors. Based on the similarity between labels, we can transfer a label set into a set with only a certain label, it also means a multi-label instance can be extended to a set of single-label. If we specify a basic label l_b, then any instance can be transformed into a set with only one label l_b. For example, in Table 1, instance u_5 has a label set of $\{l_2, l_3, l_4\}$, then according to the label similarity array Table 2, it can be transformed into $\{1 \cdot l_2, 0.5 \cdot l_2, 0.33 \cdot l_2\}$ if we choose label l_2 as the basic label. We can then call the original multi-label instance u_5 equals a single-label instance with a label of $\{1.83 \cdot l_2|l_2\}$.

Table 3. Extended Nearest Neighbors

instance	Extended Neighbors	To Label
u_5	$0 \cdot l_1 + 0.25 \cdot l_1 + 0.33 \cdot l_1$	l_1
u_5	$1 \cdot l_2 + 0.5 \cdot l_2 + 0.33 \cdot l_2$	l_2
u_5	$0.5 \cdot l_3 + 1 \cdot l_3 + 0.5 \cdot l_3$	l_3
u_5	$0.33 \cdot l_4 + 0.5 \cdot l_4 + 1 \cdot l_4$	l_4

If u_5 is the neighbor of some instance, when we consider the label l_2, the instance u_5 can be presented as an instance which contains $1 + 0.5 + 0.33 = 1.83$ label l_2, and vice versa, instance u_5 also presents there are $(1 - 1) + (1 - 0.5) + (1 - 0.33) = 1.17$ instances which not contain the label l_2, and there will have $(1.83 + 1.17 = 3 = |L(u_5)|)$. This is the basic idea when we finding our extended nearest neighbors.

3.4 Coupled ML-kNN

For the unseen instance x, lets $N(x)$ represents the set of its k nearest neighbors identified in data set D. For the j-th class label, CML-kNN chooses to calculate the following statistics:

$$C_j = Round(\sum_{i=1}^{k} \delta_{L_i^*|j}) \tag{8}$$

Where L_i is the label set of the i-th neighbor and $L_i \in N(x)$, and $\delta_{L_i^*|j}$ denotes the sum of the CLS values of the i-th neighbor's label set to the j-th label l_j, and $Round()$ is the rounding function.

Namely, C_j is a rounding number which records all the CLS value of all x's neighbors to label l_j.

Let H_j be the event that x has label l_j, and $P(H_j|C_j)$ represents the posterior probability that H_j holds under the condition that x has exactly C_j neighbors with label l_j. Correspondingly, $P(\neg H_j|C_j)$ represents the posterior probability that H_j doesn't hold under the same condition. According to the MAP rule, the predicted label set is determined by deciding whether $P(H_j|C_j)$ is greater than $P(\neg H_j|C_j)$ or not:

$$Y = \{l_j | \frac{P(H_j|C_j)}{P(\neg H_j|C_j)} > 1, 1 \le j \le q\} \tag{9}$$

According to the Bayes Theory, we have:

$$\frac{P(H_j|C_j)}{P(\neg H_j|C_j)} = \frac{P(H_j) \cdot P(C_j|H_j)}{P(\neg H_j) \cdot P(C_j|\neg H_j)} \tag{10}$$

Here, $P(H_j)$ and $P(\neg H_j)$ represents the prior probability that H_j holds and doesn't hold. Furthermore, $P(C_j|H_j)$ represents the likelihood that x has exactly C_j neighbors with label l_j when H_j holds, and $(P(Cj|\neg Hj))$ represents the likelihood that x has exactly C_j neighbors with label l_j when H_j doesn't hold.

When we count the prior probabilities, we integrated our coupled label similarity into the process:

$$P(H_j) = \frac{s + \sum_{i=1}^{m} \delta_{L_i^*|j}}{s \times 2 + m \times n}; \tag{11}$$
$$P(\neg H_j) = 1 - P(H_j);$$

where $(1 \le j \le n)$ and m is the records number in training set, and s is a smoothing parameter controlling the effect of uniform prior on the estimation which generally takes the value of 1 (resulting in Laplace smoothing).

Same as ML-kNN, for the j-th class label l_j, our CML-kNN maintains two frequency arrays α_j and β_j. As our method considers the other labels which have a similarity to a specific label, the frequency arrays will contain $k \times n + 1$ elements:

$$\alpha_j[r] = \sum_{i=1}^{m} \delta_{L_i^*|j} | C_j(x_i) = r \qquad (\delta_{L_i^*|j} \ge 0.5)$$
$$\beta_j[r] = \sum_{i=1}^{m} (n - \delta_{L_i^*|j}) | C_j(x_i) = r \quad (\delta_{L_i^*|j} < 0.5) \tag{12}$$

Where $(0 \le r \le k \times n)$. We take an instance with $\delta_{L_i^*|j} \ge 0.5$ as an instance which does have label j and we take an instance with $\delta_{L_i^*|j} < 0.5$ as an instance which doesn't have label j. Therefore, $\alpha_j[r]$ counts the sum of CLS values to label j of training examples which have label l_j and have exactly r neighbors with label l_j, while $\beta_j[r]$ counts the CLS to label j of training examples which don't have label l_j and have exactly r neighbors with label l_j. Afterwards, the likelihood can be estimated based on elements in α_j and β_j:

$$P(C_j|H_j) = \frac{s + \alpha_j[C_j]}{s \times (k \times n + 1) + \sum_{r=0}^{k \times n} \alpha_j[r]}$$

$$P(C_j|\neg H_j) = \frac{s + \beta_j[C_j]}{s \times (k \times n + 1) + \sum_{r=0}^{k \times n} \beta_j[r]} \quad (13)$$

$$(1 \leq j \leq n, 0 \leq C_j \leq k \times n)$$

Thereafter, by combing the prior probabilities (Eq.11) and the likelihoods (Eq.13) into Eq.(10), we will get the predicted label set in Eq.(9).

Algorithm 1. Coupled ML-kNN Algorithm

Input: An unlabeled instance x_t and a labeled dataset
  $T\{(x_1, L(x_1)), \ldots, (x_m, L(x_m))\}$, where $|T| = m$ and $|L| = n$
Output: The label set $L(x_t)$ of instance x_t
1: Calculate the CLS array $A(L)$ according to Eq.(7);
2: **for** $i = 1$ **to** m **do**;
3:     Identify the k nearest neighbors $N(x_i)$ for x_i
4: **end for**
5: **for** $j = 1$ **to** n **do**
6:     Calculate $P(H_j)$ and $P(\neg H_j)$ according to Eq.(11)
7:     Maintain the label-coupled frequency arrays α_j, β_j using Eq.(12)
8: **end for**
9: Identify the k nearest neighbors $N(x_t)$ for x_t
10: **for** $j = 1$ **to** n **do**
11:     Calculate the statistic C_j according to Eq.(8)
12: **end for**
13: **Return** the label set $L(x_t)$ of instance x_t according to Eq.(9)

3.5   Algorithm

Given an unknown test instance x_t, the algorithm determines the final label set of the instance. Algorithm 1 illustrates the main idea of our process. Our proposed CML-kNN contains of six main parts. a)Maintain the label similarity array; b)Finding the nearest neighbors for every instance in training set; c)Getting the prior probabilities and frequency arrays; d)Finding the nearest neighbors for the target instance; e)Calculate the statistics value; f)Calculate the result.

Firstly, we calculate the label similarity according to their inter-relationships and maintain the Coupled Label Similarity Array $A(L)$ from the training data set. Secondly, for every training instance, we identify its traditional k nearest neighbors. After that, for every different label, we calculate its prior probability which combined with CLS. Simultaneously, we expand the neighbors set for every instance to a new label-coupled neighbors set using the CLS, and calculate the frequency array for every label. After these works done, we identify the k neighbors of the test instance x_t. After applying CLS on this neighbor set and calculate the label statistics, we can finally get the predicted label set.

It is worth noting that our key idea is the label similarity, which tries to learn the label distance and then transfer any label into a specific label.

4 Experiments and Evaluation

4.1 Experiment Data

A total of eight commonly used multi-label data sets are tested for experiments in this study, and the statistics of the data sets are shown in Table 4. Given a multi-label data set $M = \{(x_i, L_i)|1 \leq i \leq q\}$, we use $|M|$, $f(M)$, $La(M)$, $F(M)$ to represent the number of instances, number of features, number of total labels, and feature type respectively. In addition, several multi-label statistics [9] are also shown in the Table. The Label cardinality $(LC(M))$ measures the average number of labels per example; the Label density $(LD(M))$ normalizes $LC(M)$ by the number of possible labels; the Distinct label sets $(DL(M))$ counts the number of distinct label combinations appeared in the data set; the Proportion of distinct label sets $(PDL(M))$ which normalizes $DL(M)$ by the number of instances. As shown in Table 4, eight data sets are included and are ordered by Label density $LD(M)$.

4.2 Experiment Setup

In our experiments, we compare the performance of our proposed CML-kNN with that some state-of-the-art multi-label classification algorithms: ML-kNN, IBLR and BSVM. All nearest neighbor based algorithms are parameterized by the size of the neighborhood k. We repeat the experiments with $k = 5, 7, 9$ respectively (odd number for voting), and use the Euclidean metric as the distance function when computing the nearest neighbors. For BSVM, models are learned via the cross-training strategy[2]. We also choose the BR-kNN as the basic algorithm to compare with. We perform 10-fold cross-validation three times on all the above data sets.

Table 4. Experiment Data Sets

Data Set	$|M|$	f(M)	La(M)	LC(M)	LD(M)	DL(M)	PDL(M)	F(M)
emotions	593	72	6	1.869	0.311	27	0.046	n
yeast	2417	103	14	4.237	0.303	198	0.082	n
image	2000	294	5	1.236	0.247	20	0.010	n
scene	2407	294	6	1.074	0.179	15	0.006	n
enron	1702	1001	53	3.378	0.064	753	0.442	c
genbase	662	1185	27	1.252	0.046	32	0.048	c
medical	978	1449	45	1.245	0.028	94	0.096	c
bibtex	7395	1836	159	2.402	0.015	2856	0.386	c

4.3 Evaluation Criteria

Multi-label classification requires different metrics than those used in traditional single-label classification. A lot of criteria have been proposed for evaluating the performance of multi-label classification algorithms [12]. In this paper, we use three popular evaluation criteria for multi-label classification: the **Hamming Loss**, the **One Error** and the **Average Precision**. The definitions of them can be found in [10].

4.4 Experiment Results

The experiment results are shown in Table 5 - Table 7. For each evaluation criterion, "↓" indicates "the smaller the better", while "↑" indicates "the bigger the better". And the numbers in parentheses denote the rank of the algorithms among the five compared algorithms.

The result tables indicate that CML-kNN and BSVM outperforms other algorithms significantly, which implies that exploiting the frequency of neighbors' label is effective, and especially for our CML-kNN, the improvement is significant compared to BR-kNN, that means incorporating the label relationship will greatly improve the BR strategy. Meanwhile, ML-kNN, IBLR and BR-kNN do not perform as well compared to the other algorithms. This implies that only exploiting the exact neighbor information is not sufficient, and the similar neighbor (correlations between labels) should also be considered.

Table 5. Experiment Result1 - Hamming Loss↓

	CML-kNN	BR-kNN	ML-kNN	IBLR	BSVM
emotions	0.189(1)	0.219(5)	0.194(2)	0.201(4)	0.199(3)
yeast	0.194(1)	0.205(5)	0.195(2)	0.198(3)	0.199(4)
image	0.157(1)	0.189(5)	0.172(2)	0.182(4)	0.176(3)
scene	0.078(1)	0.152(5)	0.084(2)	0.089(3)	0.104(4)
enron	0.061(4)	0.052(2)	0.052(2)	0.064(5)	0.047(1)
genbase	0.003(2)	0.004(3)	0.005(4)	0.005(4)	0.001(1)
medical	0.013(1)	0.019(4)	0.016(3)	0.026(5)	0.013(1)
bibtex	0.013(1)	0.016(4)	0.014(2)	0.016(4)	0.015(3)
AvgRank	**(1.50)**	4.13	2.38	4.00	2.50

Table 6. Experiment Result2 - One Error↓

	CML-kNN	BR-kNN	ML-kNN	IBLR	BSVM
emotions	0.244(1)	0.318(5)	0.263(3)	0.279(4)	0.253(2)
yeast	0.222(1)	0.235(4)	0.228(2)	0.237(5)	0.232(3)
image	0.267(1)	0.601(5)	0.319(3)	0.432(4)	0.314(2)
scene	0.197(1)	0.821(5)	0.219(2)	0.235(3)	0.251(4)
enron	0.308(3)	0.237(1)	0.313(4)	0.469(5)	0.245(2)
genbase	0.008(2)	0.012(5)	0.009(3)	0.011(4)	0.002(1)
medical	0.158(2)	0.327(4)	0.252(3)	0.414(5)	0.151(1)
bibtex	0.376(1)	0.631(5)	0.589(3)	0.576(2)	0.599(4)
AvgRank	**(1.50)**	4.25	2.88	4.00	2.38

Table 7. Experiment Result3 - Average Precision↑

	CML-kNN	BR-kNN	ML-kNN	IBLR	BSVM
emotions	0.819(1)	0.595(5)	0.799(3)	0.798(4)	0.807(2)
yeast	0.769(1)	0.596(5)	0.765(2)	0.759(3)	0.749(4)
image	0.824(1)	0.601(5)	0.792(3)	0.761(4)	0.796(2)
scene	0.885(1)	0.651(5)	0.869(2)	0.862(3)	0.849(4)
enron	0.591(3)	0.435(5)	0.626(2)	0.564(4)	0.702(1)
genbase	0.994(3)	0.992(4)	0.989(5)	0.994(2)	0.998(1)
medical	0.876(1)	0.782(4)	0.806(3)	0.686(5)	0.871(2)
bibtex	0.567(1)	0.329(5)	0.351(4)	0.476(3)	0.531(2)
AvgRank	**(1.50)**	4.75	3.00	3.50	2.25

Overall, our proposed CML-kNN outperforms all the compared methods on all three measures. The average ranking of our method on these data sets using three different metrics is the first one, with (1.50, 1.50, 1.50) respectively, while the second best algorithm, BSVM, only achieves (2.50, 2.38, 2.25). The BR-kNN performs the worst, which only achieves (4.13,4.25,4.75).

It is worth noting that although our proposed method runs the best on average, it does not mean that it is suitable for all kinds of data. For example, when used on data set "enron" and "genbase", the result is not as good as on other data sets. Sometimes it even got a worse result than BR-kNN. For example, when used on "enron" and evaluated by the Hamming Loss, our supposed CML-kNN only achieved a *4th* rank(0.061), while BR-kNN can get a second well result(0.052). The reason is because of the weak or loose connection between different labels in those data sets, and our extended neighbors may introduce more noisy information than useful information. But in terms of average performance, our method performs the best (the first rank).

5   Conclusions and Future Work

ML-kNN learns a single classifier h_i for each label l_i independently, so it is actually a binary relevance classifier. In other words, it does not consider the correlations between different labels. The algorithm is often criticized for this drawback. In this paper, we introduced a coupled label similarity, which explores the inner-relationship between different labels in multi-label classification according to their natural co-occupance. This similarity reflects the distance of the different labels. Furthermore, by integrating this similarity into the multi-label kNN algorithm, we overcome the ML-kNN's shortcoming and improved the performance. Evaluated over three commonly-used multi-label data sets and in terms of Hamming Loss, One Error and Average Precision, the proposed method outperforms ML-KNN, BR-kNN, IBLR and even BSVM. This result shows that our supposed coupled label similarity is appropriate for multi-label learning problems and can work more effectively than other methods.

Our future work will focus on expanding our coupled similarity to categorical multi-label data, and even mixed type multi-label data for which current numerical distance metrics is not suitable.

References

1. Boriah, S., Chandola, V., Kumar, V.: Similarity measures for categorical data: A comparative evaluation. red 30(2), 3 (2008)
2. Boutell, M.R., Luo, J., Shen, X., Brown, C.M.: Learning multi-label scene classification. Pattern Recognition 37(9), 1757–1771 (2004)
3. Brinker, K., Hüllermeier, E.: Case-based multilabel ranking. In: IJCAI, pp. 702–707 (2007)
4. Cao, L.: Coupling learning of complex interactions. Information Processing & Management (2014)
5. Cao, L.: Non-iidness learning in behavioral and social data. The Computer Journal 57(9), 1358–1370 (2014)
6. Cao, L., Ou, Y., Yu, P.S.: Coupled behavior analysis with applications. IEEE Transactions on Knowledge and Data Engineering 24(8), 1378–1392 (2012)
7. Cheng, W., Hüllermeier, E.: Combining instance-based learning and logistic regression for multilabel classification. Machine Learning 76(2–3), 211–225 (2009)
8. Liu, C., Cao, L., Yu, P.S.: Coupled fuzzy k-nearest neighbors classification of imbalanced non-iid categorical data. In: 2014 International Joint Conference on Neural Networks (IJCNN), pp. 1122–1129. IEEE (2014)
9. Read, J., Pfahringer, B., Holmes, G., Frank, E.: Classifier chains for multi-label classification. Machine Learning 85(3), 333–359 (2011)
10. Schapire, R.E., Singer, Y.: Boostexter: A boosting-based system for text categorization. Machine Learning 39(2–3), 135–168 (2000)
11. Spyromitros, E., Tsoumakas, G., Vlahavas, I.P.: An Empirical Study of Lazy Multilabel Classification Algorithms. In: Darzentas, J., Vouros, G.A., Vosinakis, S., Arnellos, A. (eds.) SETN 2008. LNCS (LNAI), vol. 5138, pp. 401–406. Springer, Heidelberg (2008)
12. Tsoumakas, G., Katakis, I.: Multi-label classification: An overview. International Journal of Data Warehousing and Mining (IJDWM) 3(3), 1–13 (2007)
13. Vembu, S., Gärtner, T.: Label ranking algorithms: A survey. In: Preference learning, pp. 45–64. Springer (2011)
14. Wang, C., Cao, L., Wang, M., Li, J., Wei, W., Ou, Y.: Coupled nominal similarity in unsupervised learning. In: Proceedings of the 20th ACM International Conference on Information and Knowledge Management, pp. 973–978. ACM (2011)
15. Wieczorkowska, A., Synak, P., Raś, Z.W.: Multi-label classification of emotions in music. In: Intelligent Information Processing and Web Mining, pp. 307–315. Springer (2006)
16. Zhang, M.L., Zhou, Z.H.: Multilabel neural networks with applications to functional genomics and text categorization. IEEE Transactions on Knowledge and Data Engineering 18(10), 1338–1351 (2006)
17. Zhang, M.L., Zhou, Z.H.: Ml-knn: A lazy learning approach to multi-label learning. Pattern Recognition 40(7), 2038–2048 (2007)

Learning Topic-Oriented Word Embedding for Query Classification

Hebin Yang[1,2](✉), Qinmin Hu[1,2], and Liang He[1,2]

[1] Department of Computer Science and Technology,
East China Normal University Shanghai, 200241 Shanghai, China
`yanghebin@outlook.com`, {`qmhu,lhe`}`cs.ecnu.edu.cn`
[2] Shanghai Key Laboratory of Multidimensional Information Processing,
East China Normal University, Shanghai 200241, China

Abstract. In this paper, we propose a topic-oriented word embedding approach to address the query classification problem. First, the topic information is encoded to generate query categories. Then, the user click-through information is also incorporated in the modified word embedding algorithms. After that, the short and ambiguous queries are enriched to be classified in a supervised learning way. The unique contributions are that we present four neural network strategies based on the proposed model. The experiments are designed on two open data sets, namely Baidu and Sogou, which are two famous commercial search companies. Our evaluation results show that the proposed approach is promising on both large data sets. Under the four proposed strategies, we achieve the high performance as 95.73% in terms of Precision, 97.79% in terms of the F1 measure.

Keywords: Query classification · Word embedding · Word2vec · Supervised learning

1 Introduction

How people who seek information interact with the search engine is an important research problem. What is the user intent? How to understand their information needs? A good way of starting to pick apart the puzzle is to classify the query types.

In order to understand what a user truly desires when searching for information, the intuitive idea is to ask each user what it is they are after. While this scenario may work for the offline mom and pop store, it definitely is not feasible for a search engine. Therefore, an automated approach, instead of the manual interaction, is necessary to be proposed.

Query classification is to map the queries to a list of predefined topic categories. However, most queries are short and ambiguous [14]. For example, according to our statistics in the experiments, we have: (1) around 20% queries contain three words; (2) queries having no more than four keywords are as frequent as almost 80%; (3) many keywords such as "java" have multiple aspects.

© Springer International Publishing Switzerland 2015
T. Cao et al. (Eds.): PAKDD 2015, Part I, LNAI 9077, pp. 188–198, 2015.
DOI: 10.1007/978-3-319-18038-0_15

Motivated by the work of Le [8], Socher [15] and Tang et. al. [17], we consider to represent the queries through the word embedding method. Then, the query embedding can be treated as the features to classify the queries.

The goal of learning word embedding is to associate each term in the vocabulary with a low-dimensional real-value vector [13]. Then, words with similar contexts are mapped to close vector space. Although existing word embedding, such as Word2Vec[10] and C&W[5] models, has yielded the state-of-the-art results in many natural language processing (NLP) tasks, current research does not show their effectiveness on query classification. Another problem for word embedding is that it models the syntactic contexts of words [17], but ignores the topic category information of queries. This results in misclassifying the queries into topic categories such that the categories lose their discriminative ability.

In this paper, we propose a topic-oriented word embedding (TOWE) approach to address the following problems as: (1) mapping short and ambiguous queries into right categories; (2) adopting explicit/implicit category information, especially the topic information of categories. The main contribution is that four neural network language models, namely as $TOWE_e$, $TOWE_{eu}$, $TOWE_i$ and $TOWE_{iu}$, are presented based on the Word2Vec model. Here $TOWE_e$ and $TOWE_{eu}$ encode explicit category information of categorised queries into word embedding as a supervised learning. $TOWE_i$ and $TOWE_{iu}$ incorporate user click-through information. Finally, we evaluate the proposed approach on two datasets which are widely published by two famous commercial search engine companies. The experimental results show that our approach is promising and outperforms Word2Vec[10].

The rest of the paper is organised as follows. In Section 2, we briefly review the related work on query classification , word embeddings and its applications. Then, we present our methodologies in Section 3, followed by the experimental design in Section 4. After that, we show our experimental results in Section 5 and discuss them in Section 6. Finally, we draw the conclusions and future work in Section 7.

2 Related Work

Here we mainly review the related studies on query classification, word embeddings and its applications.

2.1 Query Classification

Mapping web-user posted queries to a predefined taxonomy with a reasonable degree of accuracy is the heart of search engine and also particularly challenging. Since web-user queries are typically short, they yield few keywords features per query based on traditional bag-of-words representations.

In recent years, to overcome the problem of feature sparsity, many research efforts are devoted to enrich query features. The wining solution[14] of the 2005 KDD Cup associates a set of web pages by sending queries to search engines.

Then, the titles of Web pages, snippets and their contents are used as features to build classifiers based on document taxonomy.

However, with the coming of big data era, especially in the search engine domain, building a query classification system using queries expansion approaches are computationally infeasible, since fetching external text from large quantities of data is expected to be very time consuming and heavily depends on the quality of search engines.

To address the above problem, Broder et al.[4] propose to classify all of the web pages and then group the queries by voting methods using those pre-classified web pages. There is still a big problem which is to classify all of web pages in the search engine, since the huge number of web pages. If the predefined taxonomy has been changed, re-classify all of the web pages is required.

There are situations that queries are typically short, but terms in the vocabulary are extremely large. Therefore, large corpus of labeled training data is required for a query classification system[6]. But categorized queries are limited. Many studies focus on leveraging unlabeled queries to improve query classification performance using semi-supervised learning algorithm.

Beitzel et al.[1] exploit both labeled and unlabeled queries using several classification approaches. They emphasize on an application of computational linguistics, named selectional preference, to automatically generate some association rules. Then using those generated rules to label large numbers of training data from the vast amount of unlabeled web query logs. Li et al.[9] propose two semisupervised learning methods to infer the class memberships of unlabeled queries using click-through data.

2.2   Word Embeddings

How to represent text is central to many text classification tasks and determines the performance of tasks.

Bag-of-words representation is the most common and popular fixed-length vector representation for texts owing to its simplicity and efficiency. However, bag-of-words representation treats each term in vocabulary as an atomic unit. Such representation suffers from the problem of high dimensionality and sparsity. It considers very little in the semantic connection between words.

The problem is more acute in the query classification task. To overcome this shortcoming, numerous studies have been done to learn other word representations, such as Latent Semantic Analysis (LSA)[18] and Latent Dirichlet Allocation (LDA)[3].

With the revival of deep learning, word embedding, also referred as continuous distributed word representations, has been proved to be invaluable resource for many NLP tasks.

Word embedding is introduced by Hinton[7] to solve the problems of high dimensionality and sparsity in their work. Bengio et al.[2] propose a feed-forward neural network language model to predict the next word based on its previous contextual words. But the time complexity is very high.

Many works have been done to reduce the training time of the neural network language mode. Collobert and Weston[5] propose a ranking-type word embedding learning algorithm (C&W). Mikolov et al.[11] introduce the Recurrent neural network language models (RNNLMs). In 2013, Mikolov[10] et al. propose Word2Vec to learn word embedding.

Word embedding has achieved great success in many NLP tasks[19], such as Chinese word segmentation[16, 19], POS tagging[5], sentiment analysis[15], name entity recognition[5] etc.

Zhang et al.[19] propose a feature-based neural network language for learning feature embedding instead of human crafted feature and achieved the state-of-the-art result. Sun et al.[16] propose Radical-enhaced model based on C&W model to learn enhanced word embedding by exploiting Chinese word radical information and utilize neural-CRF to the Chinese word segmentation task.

In this paper, we focus on learning topic-oriented word embedding by exploiting the explicit topic information of categorized queries and implicit topic information of the very large click-through data. We represent queries through topic-oriented word embedding. Then, the query embedding is treated as features for the query classification task.

3   TOWE for Query Classification

In this section, the details of learning topic-oriented word embedding (TOWE) is proposed for web user query classification. Four neural networks based on Word2vec[1] are proposed to learn TOWE. In the following sections, we describe the Word2Vec model firstly and then present the details of the four proposed methods.

3.1   Word2Vec Model

The goal of Word2Vec is to associate each term in the vocabulary $w \in W$ with a unique d dimensional real value vector $v_w \in R^d$. Words with similar syntactic context are assigned to close vector space. We use the query "apple retina macbook air 13" as the demonstration.

Word2Vec predicts each term w in the vocabulary W based on its syntactic context c_w – the set of words in the window of size ws centered at w (w excluded). For $ws = 2$, the syntactic context of w is $c_w = \{w_{-2}, w_{-1}, w_1, w_2\}$. In our example the context of *macbook* are *apple*, *retina*, *air*, *13*. And the detail architecture of Word2Vec is shown in Figure 1(a). In this framework, every word in the syntactic context $w_i \in c_w$ is mapped to a unique vector. And the sum of the vectors is used as features for prediction Huffman code H_w of target word w. The negative maximum log likelihood loss function of softmax layer is

$$Loss_{w2v}(w, c_w) = - \sum_{j=1}^{|H_w|} \log \left(f_{d_j}^{w2v}(c_w)^{(1-d_j)} * (1 - f_{d_j}^{w2v}(c_w))^{d_j} \right) \qquad (1)$$

---

[1] In this paper, we utilize CBOW model because it performance better than the Skip-Gram model in our experiment.

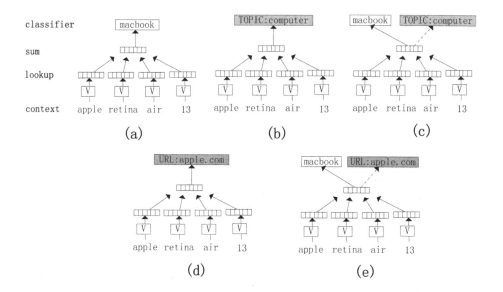

Fig. 1. The traditional Word2Vec CBOW model and our proposed neural network $(TOWE_e, TOWE_{eu}, TOWE_i$ and $TOWE_{iu})$ for learning topic-oriented word embedding

Where d_j is the jth Huffman code H_w value. And $f_{d_j}^{w2v}(c_w)$ is the predict probability of $d_j = 0$, which is calculated as given Equation (2). Where v_{c_w} is sum of vectors in context c_w which is calculated as $v_{c_w} = \sum_{w \in c_w} v_w$, and v_{d_j} is the parameter of huffman tree node. $g(x) = \frac{1}{1+e^{-x}}$ is the sigmoid function.

$$f_{d_j}^{w2v}(c_w) = g(v_{d_j} \cdot v_{c_w}) \qquad (2)$$

3.2 Topic-Oriented Word Embedding

In this section, we incorporate the explicit or implicit topic information into the Word2Vec model to learn topic-oriented word embedding. We develop four neural networks with different strategies to integrate the topic information of queries.

Model 1 ($TOWE_e$). The Word2Vec model does not capture the explicit topic category information of categorized queries. For this reason, explicit topic-oriented word embedding ($TOWE_e$) is proposed to integrate topic category information by predicting the topic category distribution of text based on input ngram. We assume that if the query q belongs to topic category k, the syntactic context c_w in the query q also belongs to topic category k. Therefore, we predict the topic category based on each categorized syntactic context.

Assuming there are K topic categories. We modify the top softmax layer by predicting its topic distribution instead of the target word. The detail neural network framework ($TOWE_e$) is shown in Figure 1(b). Let td^{c_w} be the topic distribution of c_w. If $c_w \in q$ and the category of q is k then $td_k^{c_w} = 1$, otherwise $td_k^{c_w} = 0$. Therefore the negative maximum log likelihood error of softmax layer is computed as:

$$Loss_{TOWE_t}(k, c_w) = -\sum_{k=1}^{K} \log \left(f_k^t(c_w)^{td_k^{c_w}} * (1 - f_k^t(c_w))^{(1-td_k^{c_w})} \right) \qquad (3)$$

Where $f^t(c_w)$ is the predicted topic distribution, and td^{c_w} is the standard topic distribution mentioned above.

Model 2 ($TOWE_{eu}$). The Word2Vec model learns word embedding by modeling syntactic contexts but disregard the topic category information of text. And $TOWE_e$ learn topic-oriented word embedding by exploiting explicit topic distribution of text but ignore the syntactic contexts of words. In this part, we develop a unified model ($TOWE_{eu}$). The details neural network framework is shown in Figure 1(c). And the loss function is computed as:

$$Loss_{TOWE_{eu}}(w, k, c_w) = \alpha \cdot Loss_{w2v}(w, c_w) + \gamma \cdot (1 - \alpha) \cdot Loss_{TOWE_e}(k, c_w) \quad (4)$$

Where $Loss_{w2v}(w, c_w)$ is the loss of the context part, $Loss_{TOWE_e}(k, c_w)$ is the loss of topic category part, and α is the linearly weights of the two parts. If context cw is labeled as k then $\gamma = 1$, otherwise $\gamma = 0$.

Model 3 ($TOWE_i$). Large corpus of categorized queries are not available for training $TOWE_e$ and $TOWE_{eu}$ in some case. In this section, implicit topic-oriented word embedding ($TOWE_i$) is proposed to conquer this problem. $TOWE_i$ learns topic information from large corpus of click-through data instead of categorized queries. Click-through data can be extracted easily form search log. Because the length of queries are variant, so we do not use entire queries as input. An assumption is made that if a user submit query q and click web page u in the corresponding search results, the syntactic context c_w in query q is likely to click url u. In $TOWE_i$ framework, the vector of syntactic context is used as features to predict the Huffman code of clicked URL. The detail of neural network framework architecture is shown in Figure 1(d). And the loss function is similar to Word2Vec, and is calculated as:

$$Loss_{TOWE_i}(url, c_w) = -\sum_{j=1}^{|H_{url}|} \log \left(f_{d_j}^i(c_w)^{(1-d_j)} * (1 - f_{d_j}^i(c_w))^{d_j} \right) \quad (5)$$

Where H_{url} is the Huffman code of clicked URL, d_j is the jth value of H_{url}. $f_{d_j}^i(c_w)$ is the predict value of d_j which is calculated same as Equation (2).

Model 4 ($TOWE_{iu}$). Model 3 $TOWE_i$ learns word embedding by modeling implicit topic relevance of click-through data but ignore the syntactic context of queries. So unified implicit topic-oriented word embedding ($TOWE_{iu}$) is proposed by linear combination. The detail of neural network architecture is shown in Figure 1(e). The loss function is:

$$Loss_{TOWE_{iu}}(w, url, c_w) = \beta \cdot Loss_{w2v}(w, c_w) + \xi \cdot (1 - \beta) \cdot Loss_{TOWE_i}(url, c_w) \quad (6)$$

Where β is the linear weight of two loss parts. And $\xi = 1$ if c_w click url, otherwise $\xi = 0$.

3.3 Query Embedding

We apply TOWE for query classification under a supervised learning framework. Instead of hand-crafting feature and feature enrichment, we represent queries based on word embedding. Owing to the short length and simple structure of queries, we represent a query by considering it as the sum of all words with ignoring word orders. This can be expressed by the following equation:

$$v_q = \sum_{w \in q} v_w \quad (7)$$

Where v_q is a the embedding of query q, and v_w is the embeddings of word in query q. We use this d dimension vector as features of query to predict its topic category. And linear SVM in Scikit-learn[12] is employed to build classifier.

4 Experimental Setup

To make a comprehensive comparison between our proposed TOWE models and the Word2Vec model, we utilize two real-word query datasets to evaluate the effectiveness of TOWE provided by Baidu[2] and Sogou[3], which are the most popular Chinese search engines. Same statistical information of Baidu and Sogou datasets is given in Table 1.

In Baidu dataset, we randomly sample 80% of categorized queries as training data and the remained 20% as test data. And exact category names are not provided in Baidu dataset. In Sogou dataset, predefined taxonomy and categorized queries are not available. So we manually label some queries which randomly select from query log for training and testing. At the same time, we also create a large corpus of categorized queries using rule based classifier for training $TOWE_e$ and $TOWE_{eu}$. Taxonomy and statistical information of Sogou dataset is shown in Table 2.

In this paper, we apply the standard precision, recall and F1 measure as evaluation metrics. And the open source toolkit Word2Vec[4] and four our proposed neural networks is used to train word embeddings. The same parameter settings are used for Word2Vec and TOWE models. The context window size is 5; the learning rate is set as 0.05.

---

[2] http://openresearch.baidu.com/
[3] http://www.sogou.com/labs/dl/q.html
[4] https://code.google.com/p/word2vec/

Table 1. Statistic of Baidu and Sogou datasets

	Baidu	Sogou
Total Query Number	10,787,584	4,240,938
Labeled Query Number	1,143,928	0
Topic Category Number	32	9
Train Number	915,388	6,846
Test Number	228,540	3,796
Click-Through Log Number	0	43,545,444
Average Word Length	3.3	2.7

Table 2. Taxonomy and statistics of Sogou dataset

Category	IT	Health	Sports	Military	Job	Education	Travel	Car	Finance
Train	908	500	877	120	982	977	510	982	989
Test	419	299	447	357	444	460	449	462	459
Rule Labeled	464,798	92,613	18,897	321,958	49823	223,917	9,310	80,676	57,784

5 Experimental Results

We present the experimental results in Table 3: (1) Bag-of-Word is the Traditional text representation; (2) Word2Vec, as a stat-of-the-art word embedding learning algorithm, is the baseline; (3) $TOWE_e$ is one of our proposed algorithms for learning word embedding by modeling explicit topic of labeled queries; (4) $TOWE_{eu}$ is another proposed word embedding learning algorithm, which learns

Table 3. Results of our four TOWE model, Word2Vec model and TF-IDF based Bag-of-Word query representation. "+label" means only categorized queries are used for training word embedding model.

Dataset	Model	Precision	Recall	F1
Baidu	Bag-of-Word	87.00%	100.0%	93.05%
	Word2Vec	82.03%	99.94%	90.10%
	Word2Vec + label	79.06%	99.90%	88.27%
	$TOWE_e$ + label	**95.17%**	99.90%	**97.47%**
	$TOWE_{eu}$	**95.73%**	99.94%	**97.79%**
Sogou	Bag-of-Word	71.80%	100.0%	83.58%
	Word2Vec	83.23%	99.30%	90.55%
	Word2Vec + label	77.63%	98.28%	86.74%
	$TOWE_e$ + label	82.66%	98.28%	89.80%
	$TOWE_{eu}$	**87.58%**	99.29%	**93.07%**
	$TOWE_i$	**86.76%**	99.29%	**92.60%**
	$TOWE_{iu}$	**88.17%**	99.30%	**93.40%**

both syntactic context information and explicit topic; (5) $TOWE_i$ learns word embedding by mimic user search behavior using large corpus of click-through data; (6) $TOWE_{iu}$ learns word embedding by modeling user search behavior and syntactic contexts of user submitted queries. (7) "+label" means that only categorized queries are used to train word embedding.

6   Analysis and Disscussion

6.1   TOWE *VS* Word2Vec

We compare topic-oriented word embedding ($TOWE_e$, $TOWE_{eu}$, $TOWE_i$ and $TOWE_{iu}$) with the baseline of Word2Vec by only using query embedding as features for query classification. We are comparable to Word2Vec model as it has achieved great success in many NLP tasks. And the embeddings of Word2Vec and TOWE are trained with same datasets and same parameters.

Table 3 shows the performance of query classification on the Baidu and Sogou datasets. From Table 3, we can see that the performance of TOWE is obviously better than Word2Vec as feature for query classification. The reason is that Word2Vec do not capture the topic information, resulting in that the words with different topics are mapped to neighboring word vectors space. The classification performance is affected since the discriminative ability of topic words are weakened when such word embeddings are fed as features. $TOWE_e$, $TOWE_{eu}$ effectively separate words with different topic category to different ends of the spectrum and perform better compare with Word2Vec model in both datasets. $TOWE_i$ and $TOWE_{iu}$ outperform Word2Vec model by exploiting more user click information from large corpus of click-through data.

6.2   Word Embedding *VS* Bag-of-Word

We compare word embedding with the traditional bag-of-word query representation for query classification. Classification performance is showed in Table 3. We can see that the performance of TOWE is obviously outperform bag-of-word and Word2Vec representation using same svm classifier with same parameters. Word2Vec model perform slightly worse than bag-of-word in Baidu dataset and better in Sogou dataset. TOWE capture the syntactic information of text but also the explicit topic and implicit topic information from categorized queries or large number of user-click-through data.

6.3   Effect of α in $TOWE_{eu}$

We tune the hyper-parameter α of $TOWE_{eu}$ model. As given in Equation (4), α is the weighing score of syntactic context loss part and explicit topic category loss part.

Figure 2(a) shows the precision of $TOWE_{eu}$ on query topic category classification on Baidu and Sogou datasets. The performance of $TOWE_{eu}$ is better when α in the range of $[0.3, 0.8]$. The $TOWE_{ue}$ model with $\alpha = 1$ stands for the Word2Vec model, which learns word embedding by modeling syntactic context of queries. The importance of topic information in learning word embedding for query classification can be verified by the sharp decline at $\alpha = 1$.

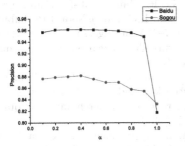
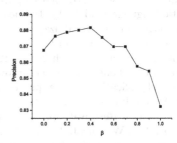

(a) Precision of $TOWE_{eu}$ with different hyper-parameter α

(b) Precision of $TOWE_{iu}$ with different hyper-parameter β

Fig. 2. Hyper-parameter tuning results

6.4 Effect of β in $TOWE_{iu}$

We also tune the hyper-parameter β of $TOWE_{iu}$ to learn TOWE for query classification. As gave in Equation (6), β is the weighing score of syntactic loss implicit topic loss.

Figure 2(b) shows the precision of $TOWE_{iu}$ on query classification with different β on Sogou test sets. We can see that $TOWE_{iu}$ performs better when β is in the range of $[0.3, 0.5]$. The model with $\beta = 1$ stands for the standard Word2Vec model. The sharp decline at $\beta = 1$ reflects the importance of implicit topic relevant information in learning word embeddings for query classification.

7 Conclusions and Future Work

We draw our conclusions as follows. First of all, we propose a topic oriented word embedding approach, configuring with four neural network strategies. Second, the features we learn for query classification is under a supervised learning framework, which makes the proposed model duplicable in other data sets. Third, compared to the traditional word embedding, we fully adopt the implicit and explicit topic information. Fourth, we suggest the parameters for the proposed TOWE model. Finally, our experiments confirm that TOWE with four strategies is successful on both data sets. Furthermore, we achieve the high performance as 95.73% in terms of Precision, 97.79% in terms of F1.

In the future, we will continue on learning word embedding with more information such advertiser information, URL category. And study how to composition more effective query embedding.

References

1. Beitzel, S.M., Jensen, E.C., Lewis, D.D., Chowdhury, A., Frieder, O.: Automatic classification of web queries using very large unlabeled query logs. ACM Transactions on Information Systems (TOIS) 25(2), 9 (2007)

2. Bengio, Y., Schwenk, H., Senécal, J.-S., Morin, F., Gauvain, J.-L.: Neural probabilistic language models. In: Holmes, D.E., Jain, L.C. (eds.) Neural Probabilisticlanguage Models. StudFuzz, vol. 194, pp. 137–186. Springer, Heidelberg (2006)
3. Blei, D.M., Ng, A.Y., Jordan, M.I.: Latent dirichlet allocation. the. Journal of Machine Learning Research 3, 993–1022 (2003)
4. Broder, A.Z., Fontoura, M., Gabrilovich, E., Joshi, A., Josifovski, V., Zhang, T.: Robust classification of rare queries using web knowledge. In: Proceedings of the 30th Annual International ACM SIGIR Conference on Research and Development in Information Retrieval, pp. 231–238. ACM (2007)
5. Collobert, R., Weston, J., Bottou, L., Karlen, M., Kavukcuoglu, K., Kuksa, P.: Natural language processing (almost) from scratch. The Journal of Machine Learning Research 12, 2493–2537 (2011)
6. Ganti, V., König, A.C., Li, X.: Precomputing search features for fast and accurate query classification. In: Proceedings of the Third ACM International Conference on Web Search and Data Mining, pp. 61–70. ACM (2010)
7. Hinton, G.E.: Learning distributed representations of concepts. In: Proceedings of the Eighth Annual Conference of the Cognitive Science Society, vol. 1, p. 12. Amherst, MA (1986)
8. Le, Q.V., Mikolov, T.: Distributed representations of sentences and documents. arXiv preprint arXiv:1405.4053 (2014)
9. Li, X., Wang, Y.-Y., Shen, D., Acero, A.: Learning with click graph for query intent classification. ACM Transactions on Information Systems (TOIS) 28(3), 12 (2010)
10. Mikolov, T., Chen, K., Corrado, G., Dean, J.: Efficient estimation of word representations in vector space. arXiv preprint arXiv:1301.3781 (2013)
11. Mikolov, T., Karafiát, M., Burget, L., Cernockỳ, J., Khudanpur, S.: Recurrent neural network based language model. In: INTERSPEECH, pp. 1045–1048 (2010)
12. Pedregosa, F., Varoquaux, G., Gramfort, A., Michel, V., Thirion, B., Grisel, O., Blondel, M., Prettenhofer, P., Weiss, R., Dubourg, V., Vanderplas, J., Passos, A., Cournapeau, D., Brucher, M., Perrot, M., Duchesnay, E.: Scikit-learn: Machine learning in Python. Journal of Machine Learning Research 12, 2825–2830 (2011)
13. Rei, L., Mladenic, D.: Learning semantic representations of words and their compositionality (2014)
14. Shen, D., Pan, R., Sun, J.-T., Pan, J.J., Wu, K., Yin, J., Yang, Q.: Query enrichment for web-query classification. ACM Transactions on Information Systems (TOIS), 24(3), 320–352 (2006)
15. Socher, R., Perelygin, A., Wu, J.Y., Chuang, J., Manning, C.D., Ng, A.Y., Potts, C.: Recursive deep models for semantic compositionality over a sentiment treebank. In: Proceedings of the Conference on Empirical Methods in Natural Language Processing (EMNLP), Citeseer, pp. 1631–1642 (2013)
16. Sun, Y., Lin, L., Tang, D., Yang, N., Ji, Z., Wang, X.: Radical-enhanced chinese character embedding. arXiv preprint arXiv:1404.4714 (2014)
17. Tang, D., Wei, F., Yang, N., Zhou, M., Liu, T., Qin, B.: Learning sentiment-specific word embedding for twitter sentiment classification. ACL (2014)
18. Zelikovitz, S., Marquez, F.: Transductive learning for short-text classification problems using latent semantic indexing. International Journal of Pattern Recognition and Artificial Intelligence 19(02), 143–163 (2005)
19. Zhang, M., Zhang, Y., Che, W., Liu, T.: Chinese parsing exploiting characters. ACL 1, 125–134 (2013)

Reliable Early Classification on Multivariate Time Series with Numerical and Categorical Attributes

Yu-Feng Lin[1], Hsuan-Hsu Chen[1], Vincent S. Tseng[2(⊠)], and Jian Pei[3]

[1] Department of Computer Science and Information Engineering,
National Cheng Kung University, Tainan, Taiwan, Republic of China
{aorborcord,tp6vm0vm4}@idb.csie.ncku.edu.tw
[2] Department of Computer Science, National Chiao Tung University,
Hsinchu, Taiwan, Republic of China
vtseng@cs.nctu.edu.tw
[3] School of Computing Science, Simon Fraser University Burnaby, Burnaby, BC, Canada
jpei@cs.sfu.ca

Abstract. Early classification on multivariate time series has recently emerged as a novel and important topic in data mining fields with wide applications such as early detection of diseases in healthcare domains. Most of the existing studies on this topic focused only on univariate time series, while some very recent works exploring multivariate time series considered only numerical attributes and are not applicable to multivariate time series containing both of numerical and categorical attributes. In this paper, we present a novel methodology named *REACT (Reliable EArly ClassificaTion)*, which is the first work addressing the issue of constructing an effective classifier on multivariate time series with numerical and categorical attributes in serial manner so as to guarantee stability of accuracy compared to the classifiers using full-length time series. Furthermore, we also employ the GPU parallel computing technique to develop an extended mechanism for building the early classifier efficiently. Experimental results on real datasets show that *REACT* significantly outperforms the state-of-the-art method in terms of accuracy and earliness, and the GPU implementation is verified to substantially enhance the efficiency by several orders of magnitudes.

Keywords: Early classification · Multivariate time series · Serial classifier · Numerical and categorical attributes · Shapelets · GPU

1   Introduction

Early classification, which refers to predict occurrences as early as possible, is an emerging subject in data mining with various time-sensitive applications such as health-informatics. For example, a retrospective study of clinical data from neonatal intensive unit found that abnormal heartbeat rate was significantly associated with sepsis in infants [9]. Monitoring the heartbeat time series and classifying them as early as possible may lead to earlier diagnosis and effective treatment.

The aim of early classification is naturally different than that of classic classification, which focuses only on accuracy without taking earliness into account. That is,

© Springer International Publishing Switzerland 2015
T. Cao et al. (Eds.): PAKDD 2015, Part I, LNAI 9077, pp. 199–211, 2015.
DOI: 10.1007/978-3-319-18038-0_16

early classifiers can keep similar accuracy which is comparable to classic classifiers, while they should also be able to predict the results at an earlier time. Several effective early classifiers have been proposed to make early prediction on univariate time series [18, 19], and these classifiers retained accuracy which was comparable to traditional classifiers [2, 4, 14]. However, to gain insights into the classification results in many applications, not only univariate time series but also multivariate time series need to be considered further.

To overcome the deficiency of the previous early classification methods that consider only univariate time series, *early classification on multivariate time series* has recently emerged as a novel and important topic of research [7, 8, 10]. The common idea of the existing methods is to extract *multivariate shapelets* as main features from all dimensions of time series with numerical attributes (or called numerical time series) that can manifest the target classes, where *shapelet* indicates a segment of numerical time series [8, 14, 20]. However, multivariate time series is usually composed of both numerical and categorical attributes in lots of real world data sets. For example, chronic asthmatic sufferers have to constantly observe not only vital signs and diagnostic records, but also environmental factors such as suspended particulates or humidity level. If the interactions between diagnostic records and environmental factors can be found, it is possible to predict the probability of asthmatic attack in advance using different variants of multivariate time series with numerical and categorical attributes. Moreover, Xing *et al.* [18, 19] argued that an early classifier should guarantee the stability of accuracy which was comparable to the classifier using full-length time series (defined as *serial* [18]), which can ensure an early classifier to be reliable and consistent.

In this paper, we propose a novel method for reliable early classification on *M*ulti-variate *T*ime *S*eries with *N*umerical and *C*ategorical attributes (abbreviated as *MTS-NC*). However, achieving such an aim is not an easy task with the following challenges: (I) Multivariate time series is heterogeneous and each variable has different characteristics with either numerical or categorical type. Hence, it is not easy to find the potential interactions/relations between different variables in *MTS-NC*. (II) It is not an easy task to build an early classifier being *serial* on *MTS-NC*. To the best of our knowledge, the *serial* property is designed for shapelets extracted from multivariate time series with numerical attribute [18], and it cannot be applied directly to categorical attributes. (III) Studying the tradeoff between earliness and accuracy of *REACT* on *MTS-NC* is not an easy task. In literatures [8, 18, 19], various measurements such as discrimination, frequency, earliness are employed to estimate the qualities of features for studying the tradeoff between earliness and accuracy. However, these criteria cannot be directly applied to *MTS-NC*, and they may be ineffective in obtaining the features satisfying these conditions. (IV) The proposed classifier has to efficiently extract features on *MTS-NC*. In the feature extraction of univariate time series with categorical attribute, the existing method [2] uses a two-phase approach by generating all frequent patterns and then selecting the discriminative patterns in different phases. However, the two-phase approach cannot be directly employed to generate patterns from *MTS-NC*, which might lead to a huge number of redundant patterns. In addition, in the feature extraction of univariate/multivariate time series with numerical attribute, discovering shapelets still has a higher computation overhead on existing methods [7, 8, 10].

To address all of the above challenges, this paper proposes a novel framework named *REACT* on *MTS-NC*. The major contributions of this work are shown below:

1. *REACT* incorporates the concept of heterogeneous multivariate time series with both numerical and categorical attributes into early classification to simultaneously consider numerical and categorical time series on construction of early classifier.
2. *REACT* constructs a reliable early classifier which is *serial* and guarantees the stability of accuracy compared to the classifier using full-length time series.
3. To avoid generating a huge number of features which may be redundant, we design a procedure of feature extraction in *REACT* named *MEG* (Mining Equivalence classes with shapelet Generators) based on the concept of *Equivalence Classes Mining* [12, 15]. *MEG* can efficiently and effectively generate the discriminative features. In addition, several strategies are proposed to prune the search space and reduce the number of redundant features in the processes of feature extraction.
4. Since discovering shapelet generators takes huge calculation operations, *REACT* incurs still high computation overhead. In view of this, we employ and integrate concepts of GPU technique of parallel computing [4] to propose a process of parallel *MEG* for substantially reducing the computational overhead of discovering shapelet generators.
5. We conduct an extensive empirical evaluation on several real datasets. The results show that *REACT* outperforms the state-of-the-art method in terms of f-score and earliness. In addition, the GPU implementation significantly runs faster than the baseline approach of building *REACT* by several orders of magnitudes.

The remainder of this paper is organized as follows. Section 2 introduces the background of early classification on multivariate time series. We then describe *REACT* in section 3. Experiments are reported in Section 4. Finally, we conclude our work and give prospective future work in Section 5.

2   Preliminaries and Related Work

2.1   Preliminaries

We introduce definitions and properties related to early classification on multivariate time series. For more details, readers can refer to [8, 14, 18, 19, 20].

Definition 1 (*MTS-NC*). A time series t is a set of readings of the form $< r_1, r_2, ..., r_{len(t)}>$, where $len(t)$ is the length of t and r_k is the k-th reading of t for all $1 \leq k \leq len(t)$. Given a time series $t = < r_1, r_2, ..., r_{len(t)}>$, t is called *categorical time series* if r_j is category for $1 \leq j \leq len(t)$. On the other hand, t is called *numerical time series* if r_i is number for $1 \leq i \leq len(t)$. A *MTS-NC* $mt = \{t_1, t_2, ..., t_n\}$ is composed of n time series, where t_x is a categorical/numerical time series, where $1 \leq x \leq n$. Let $C(mt)$ be a corresponding class label of *MTS-NC* mt. *Dataset of MTS-NC D* is a collection of mt and $C(mt)$, where $C(mt) \in$ class label set C. In addition, D_c is defined the subset of D carrying class label c, that is, $D_c = \{mt \mid mt \in D$ and $C(mt) = c\}$. Figure 1(a) shows an example of *MTS-NC*.

Definition 2 (Subsequence and super-sequence of time series). Given two *time series* $t = <r_1, r_2, ..., r_{len(t)}>$ and $t' = <r_1', r_2', ..., r_{len(t')}'>$, where $len(t') \leq len(t)$. We say that t' is a *subsequence* of time series t if there exists a sequence of integers $1 \leq z_1 < z_2 < ...z_{len(t')} \leq len(t)$ such that $r_i' = r_{zi}$ for all i, where $1 \leq i \leq len(t')$, denoted by $t' \sqsubseteq t$. On the other hand, t is a *super-sequence* of time series t'.

Definition 3 (Shapelet/Numerical feature). Given two numerical time series $nt1$ and $nt2 = <r_1, r_2, ..., r_{len(t)}>$, where $len(nt1) = len(nt2)$, we denote the *set of all distinct subsequences of time series* as $ST(t)_l = \{ ST(t)_{1,l}, ST(t)_{2,l}, ..., ST(t)_{n,l} \}$, and *normalized Euclidean distance* is defined by $dist(nt1, nt2) = \sqrt{\frac{1}{n}\sum_{i=1}^{len(nt1)}(nt1_i - nt2_i)^2}$, where $n = len(t) - l + 1$ and $1 \leq l < len(t)$. The *best matching distance* is denoted by $BMD(nt1, nt2) = minimum\{dist(nt1, ST(nt2)_{j,len(nt1)})| 1\leq j \leq len(nt2)-l+1\}$, where $len(nt1) = len(nt2)$. A *shapelet/numerical feature* is a pair (s, δ), where δ is a distance threshold, $s=<r_1, r_2, ..., r_{len(s)}>$, and $r_i \in \mathbb{N}$ for all i ($1 \leq i \leq len(s)$). A shapelet f is said to appear in a time series t, denoted by $f \sqsubseteq t$, if $BMD(s, t) \leq \delta$.

Definition 4 (Categorical feature). Let ct be a categorical time series, a *categorical feature* is a subsequence of length l extracted from ct and denotes by $f = <r_1, r_2,... , r_l>$, where $l \leq len(ct)$.

Definition 5 (Utility of a feature). Given a feature f and a dataset of *MTS-NC D* containing N instances and C different class labels, and assume that each class label c_i has n_i instances in D, where $1 \leq i \leq C$ and $N=\sum_{i=1}^{C} n_i$. The *entropy* of D is defined as $E(D) = -\sum_{i=1}^{C} \frac{n_i}{N} log \left(\frac{n_i}{N}\right)$. In addition, the *minimum prefix of t* is defined as the readings from the first reading to the ith reading, where f firstly appears in t for $1 \leq i \leq len(t)$, which is denoted as $minprefix(t, f)$. Its *Earliest Matching Time* is the time point of minimum prefix and denoted by $EMT(t, f)$. The *utility of feature* is defined as $U(f) = (E(D)-E(D_f))^\omega \times wsup(f)$, where $D_f = \{mt \mid mt \in D, f \sqsubseteq mt\}$ is the sub-dataset of mt where f appears in, and $wsup(f) = \frac{\sum_{f \sqsubseteq mt, mt \in D}\frac{1}{EMT(mt,f)}}{|D|}$ is the weighted support to measure frequency and earliness of features, in which $|D|$ is the number of instances in D. The parameter $\omega \geq 1$ determines the relative importance of information versus earliness and popularity.

SID	MiNCA	Class
1	`<b, c, a, b, a>` `<10, 20, 30, 15, 25>`	low
2	`<a, b, c, c, b>` `<30, 10, 50, 25, 5>`	high
3	`<b, c, a, b, a>` `<15, 15, 30, 10, 20>`	low
4	`<b, c, b, b, c>` `<30, 5, 45, 25, 10>`	high
5	`<c, b, b, a, c>` `<10, 10, 40, 25, 5>`	low

(a) *MTS-NC*

ID	Extracted Feature
#1	`<b, a>`
#2	`<c, b>`
#3	`(<10, 20>, 5)`
#4	`(<10, 50, 25>, 5)`
#5	`<b, c>`

Categorical Generator

Numerical Generator

(b) Extracted Features

SID	Encoded Sequence	Class
1	`<(#3, #5)^2, (#1, #3)^5>`	low
2	`<(#5)^3, (#4)^4, (#2)^5>`	high
3	`<(#3, #5)^2, (#1, #3)^5>`	low
4	`<(#5)^2, (#2)^3, (#4)^4, (#5)^5>`	high
5	`<(#1)^2, (#1, #4)^4>`	low

Earliest matching time

(c) Encoding Sequence

Fig. 1. Examples of *MTS-NC*, extracted feature, and encoding sequence

Definition 6 (Information gain and separation gap w.r.t. a shapelet). Given a dataset D and a shapelet $f = (s, \delta)$, the *information gain* of the split point δ is defined as $I(s, \delta) = E(D) - \frac{|D_f|}{|D|} E(D_f) - \frac{|D_n|}{|D|} E(D_n)$, where D_f is the sub-dataset of D in which all instances match f, and D_n is the remained time series removing D_f. The *separation gap* of the split point δ is computed as $G(s, \delta) = \frac{1}{|D_n|} \Sigma_{mt \in D_n} dist(mt, s) - \frac{1}{|D_f|} \Sigma_{mt \in D_f} dist(mt, s)$.

2.2   Related Works

Early classification on numerical time series aimed to classify a partial case only using the prefix of complete time series, which was first introduced by Diez *et al.* [4]. They simply used linear combination of available predicates of prefixes for classification. Xing *et al.* [18] then explored a feature based method for early classification on categorical time series. However, it had to discretize the time series when this method was applied to real-valued time series. In 2009, Xing *et al.* [19] proposed a novel nearest neighbor approach to tackle the problem of early classification on numeric time series. However, to gain insights into the classification results does not only be caused from univariate time series, but also multivariate time series. To overcome the deficiency, *early classification on multivariate time series with numerical attribute* has recently emerged as a novel and important topic of research [7, 8, 10]. In the existing frameworks, *multivariate shapelets* are extracted as candidates to build the early classifier, where *shapelet* is a segment of numerical time series [8, 14, 20]. However, multivariate time series is usually composed of numerical and categorical attributes in lots of real world data sets. Therefore, this paper simultaneously considers *MTS-NC* on construction of early classifier.

3   Methodology

In this section, we shall describe the proposed methodology named *REACT* (*Reliable EArly ClassificaTion*) on *M*ultivariate *T*ime *S*eries with *N*umerical and *C*ategorical attributes (abbreviated as *MTS-NC*). The framework of *REACT* is shown in Figure 2. We will introduce each process in the following subsections, and we discuss imbalance problem and implementation on GPUs in the last two subsections, respectively.

3.1   Feature Extraction

An equivalence class (abbreviated as EqC) was firstly introduced by Pasquier *et al.*[17], the maximal frequent itemsets in EqC are called *closed*, and the minimal frequent itemsets in EqC are called *generators*. Frequent closed patterns can form a concise and lossless representation of frequent itemsets, and they have been extensively studied [6, 13]. In addition, by Minimum Description Length principle, generators are preferable to closed patterns for model selection and classification [6, 13]. In [3, 12], authors gave discussions for the benefit of generators over closed patterns.

In the following paragraphs, we introduce how to extract categorical generators and shapelet generators from categorical and numerical time series, respectively.

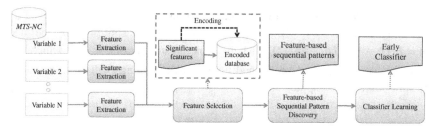

Fig. 2. The framework of *REACT*

● *Categorical Generators Extraction*

According to Definition 5, the attractive property of upper bound can be defined as below. Given a *MTS-NC* dataset D and a categorical feature f, the *sequential upper bound* of utility is computed as $SeqUB(f) = \frac{E(D)^{\omega}}{|D|} \sum_{mt \in D, f \sqsubseteq mt} \frac{1}{EMT(mt,f)+1}$. If the sequential upper bound of a categorical feature f is less than *min_utility*, the utility of super-sequence of f' must be less than *min_utility* [17].

Definition 7 (Extension timestamp of a categorical feature w.r.t a categorical time series). The *extension timestamp* of a categorical feature f w.r.t a categorical time series S is defined as $ET(f) = \{t \mid t = \text{matching time} + 1, t \leq len(S)\}$.

The essence of feature extraction of categorical generators is to check whether the patterns satisfy the stop conditions of extension or not. We first scan the projected database of prefix s once, and compute the exact utility for each item α that the extended pattern $(s{+}{+}\alpha)$ appears, where $(s{+}{+}\alpha)$ defines the *concatenation* of s and α. Initially the prefix is empty, and the projected database of empty is the original dataset. If there are some valid items in the projected database, we then identify the equivalence class for each valid item α and examine whether the categorical feature $(s{+}{+}\alpha)$ should be continued or not by Downward Closure Property of Non-Generator [17].

● *Shapelet Generators Extraction*

We adopt *best matching distance* as the similarity between shapelet f and time series t [8, 14, 20], i.e. $BMD(f, t)$. A time series t can be classified based on a shapelet $f = (s, \delta)$, once we find that the distance between t and f is no greater than the distance threshold δ. In addition, if several shapelets in the different classes satisfy the assumption, we select the first shapelet. Furthermore, to avoid existence of redundant patterns in the set of shapelets, generator mining is applied to shapelet extraction.

Definition 8 (Shapelet generator). A shapelet $f = (s, \delta)$ is called shapelet generator if (I) there is no shapelet $f' = (s', \delta')$ satisfying $U(f') \geq U(f)$, (II) the covered instances of s are no less than that of s', and (III) the distance between s and s' is no larger than δ in the same equivalence class.

Example 1. Given two shapelets $f_1 = (<10, 20>, 10)$ and $f_2 = (<15, 15, 30>, 10)$, which belong to the same equivalence class, and f_1 and f_2 have the same covered instances. In addition, f_1 has higher utility value because f_1 precedes f_2, and the distance between $<10, 20>$ and $<15, 15, 30>$ is 5 less than 10. Therefore, we refer to $<10, 20>$ as a subsequence of $<15, 15, 30>$, and f_1 is a generator in the equivalence class.

Procedure: *MEG*
Input : (1) *D* : *MTS-NC*; (2) *minLen* and *maxLen* : the minimal and maximal length of shapelet;
(3) *min_util* and *min_sup* : minimum utility and minimum support;
Output: The set of shapelet generators *SGs*;
01. **for** $i := 1$ to $|D|$ **do**
02. $S := i^{\text{th}}$ time series of *D*;
03.   **for** $j := 1$ to $|D|$ **do**
04. $S_{temp} := j^{\text{th}}$ time series of *D*;
05. $M_j :=$ statement between *S* and S_{temp};
06.   **for** $l := minLen$ to *maxLen* **do**
07.   **for** $s := 1$ to $|S|-l+1$ **do**
08.   **for** $k := 1$ to $|D|$ **do** $d[k] := BMD(S_{s,l}, M_k)$;
09. $(\delta, gain, gap, upper) :=$ best of $I(d)$ and $G(d)$;
10.   **if** $sup(S_{s,l}) \geq min\_sup$ **then**
11.   Identify equivalence class;
12.   Calculate exact utility $Utility(S_{s,l})$;
13.   **if** $Utility(S_{s,l}) \geq min\_util$ **then** $SGs := SGs \cup$ equivalence class of *s*;

Fig. 3. Pseudo code of procedure *MEG*

We show the procedure of Mining Equivalence classes with shapelet Generators (abbreviated as *MEG*) in Figure 3. The procedure scans dataset *D* once and gets distances from candidates to all time series (line 1-5). For the length of shapelet between *minLen* and *maxLen*, the best matching distance (which refers to Definition 3) are computed (line 6-8). After the distance thresholds are calculated, the information gain and separation gap are computed for each candidate (line 9), and the procedure then obtains the set of the supporting instances of shapelets to determine equivalence classes (line 10-12). If the utility of shapelet is no less than user-specified threshold, it is collected into the set *SGs* (line 13).

3.2 Feature Selection

As indicated by many existing associative classification [2, 4, 14], learning an optimal set of features for classification is very expensive and non-scalable. In this work, a greedy algorithm of feature selection works as below. The procedure first ranks the equivalence classes in descending order using their utility score, and then iterates over the features starting from the highest ranked one.

Step 1. We select the feature and remove all covered instances. Here, a feature *f* is said to cover an instance *mt* if *mt* matches *f* and their classes are the same.

Step 2. We then use the next highest ranked feature to see whether it covers any of the remained instances or not.

Step 3. If it covers some of them, then we select the feature and remove all instances that are covered. Otherwise, we discard it and proceed to next one.

Step 4. This process continues step 2 to step 4 until the set of extracted features or the remaining dataset is empty.

3.3   Feature-Based Sequential Pattern Discovery

In discovery of feature-based sequential patterns, we consider two kinds of combinations of features, sequential combination and simultaneous combination, to improve further the effectiveness of the classification model. Therefore, to discovery the relationship of the features mined from feature extraction, we associate each feature with a unique identifier, and then construct encoded sequence database composed of these identifiers. Figure 1(c) shows an example of encoded sequence database.

However, the encoded sequence dataset contains simultaneous event type, and there is more than one item at the same timestamp. It may increase the complexity of patter mining procedure. The *downward closure Property of Feature-based Generator* is thus proposed to modify for reducing the computational overhead.

Definition 9 (Simultaneous extension timestamp). The simultaneous extension timestamp of a feature-based sequential pattern P for an encoded sequence S is defined as $SET(P,S) = \{t \mid t = earliest\ matching\ time\ in\ S\}$.

Property 1 (Downward closure property of feature-based generator). Given a feature-based sequential pattern P_1, if $\exists$ a pattern P_2 such that the elements of $\{P_1\}$-projected database and $\{P_1\}$-projected database are the same, and P_2 is a sub-pattern of P_1, and then, any serial extension of P_1 are not generators. On the other hand, if $SET(P_1) = SET(P_2)$ for each instance and P_2 is a sub-pattern of P_1, then P_1 and all simultaneous extensions are not generators.

3.4   Serial Decision Tree

A tree-based classifier named *SDT* (Serial Decision Tree) is built with all extracted information from Feature-based Sequential Pattern Discovery, as shown in Figure 4. Similar to classical decision tree algorithm [2, 6, 17, 19], we select the attribute of the highest information gain as root of the tree, and determine the dominant class by maximizing confidence. The dataset is then divided into two sub-datasets, the first consist of the instances matching this feature and the other comprise of all remained instances. Once the root is constructed, the sub-trees of branches of the root can be constructed recursively. In addition, for each leaf node in *SDT*, we consider the stability of *error rate* between subspace and full-space. If the *error rate* of a node at time point i passes the user-specified threshold σ, then the MPL of the node is set to $i+1$. We make sure that all *error rates* at timestamp MPL+k for $k \geq 0$ are smaller than σ.

SID	Encoding sequence	$SDT(mt)^{full}$
1	$<(\#3)^1, (\#8)^2>$	low
2	$<(\#1, \#2, \#4, \#7)^1, (\#3)^2, (\#6)^3>$	high
3	$<(\#3, \#5)^1, (\#8)^2, (\#4, \#7)^3>$	low
4	$<(\#4, \#7)^1, (\#6)^2>$	high
5	$<(\#3)^1, (\#6)^2, (\#8)^3, (\#7)^4>$	low

(a) Encoding sequence

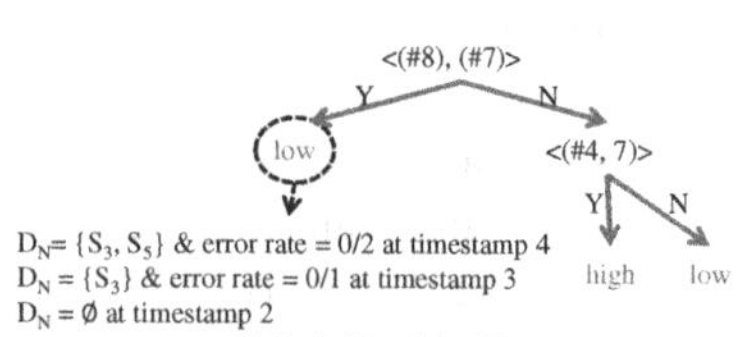

$<(\#8), (\#7)>$

low $<(\#4, 7)>$

$D_N = \{S_3, S_5\}$ & error rate = 0/2 at timestamp 4
$D_N = \{S_3\}$ & error rate = 0/1 at timestamp 3
$D_N = \emptyset$ at timestamp 2

high   low

(b) Serial Decision Tree

Fig. 4. Examples of encoding sequence and serial decision tree

Definition 10 (Error Rate of a leaf node). The *error rate* of a leaf node N in *SDT* is a ratio of difference of classification results between subspace formed by the prefix of length l and full-space, and is computed as $ER(N)^l = \frac{|\{mt|mt \in D_N, C(mt)^l \neq C(mt)^{full}\}|}{|D_N|}$, where D_N is the sub-dataset which N represents and $SDT(mt)$ is the class label of multivariate time series mt classified by SDT at time point l.

3.5   Imbalance Issue

To tackle the imbalance problem, we utilize the *ratio of sub-dataset* to instead of using the standard information gain and determine the discriminations of features, as shown in Definition 11.

Definition 11 (Ratio confidence and ratio entropy). Given a dataset D from C different classes and a sub-dataset D_f from D, the ratio of D_f for class c is defined as $ratio(D_f \to c) = \frac{|\{mt \in D_f|C(mt)=c\}|}{|\{mt \in D|C(mt)=c\}|}$, where $c \in C$. The ratio confidence of class c is the ability to manifest c, and defined as $(f \to c) = \frac{ratio(D_f \to c)}{\sum ratio(D_f \to c')}$. In addition, the ratio entropy of D_f is computed as $rE(D_f) = -\sum_{c \in C} \frac{ratio(D_f \to c)}{\sum ratio(D_f \to c')} \times \log(\frac{ratio(D_f \to c)}{\sum ratio(D_f \to c')})$.

Example 2. Given a pattern P and a dataset of 2-classes, in which the number of instances of majority class and minority class are 1733 and 72 respectively, and suppose all instances in this dataset match P. The confidence and entropy of P are 96% and 0.24 respectively. It is a highly discriminative pattern at first sight. As a matter of fact, this pattern always appears in this dataset. In this work, the discrimination is estimated by the modified formula such that $rconf(P \to c) = 50\%$ and $rE(D_P) = 1$.

3.6   Implementation on GPUs

The GPU implementation [4] employed in our study for parallel Mining Equivalence classes with shapelet Generators (abbreviated as parallel *MEG*) is shown in Figure 5. For each thread, parallel *MEG* loads all subsequences of time series in dataset to shared memory T_i and synchronizes all the threads (line 1-2). Then again, parallel *MEG* loads all subsequences to shared memory T_j and performs calculation of line 06-13 of algorithm *MEG* for each thread (line 3-6). Finally, parallel *MEG* returns the set of shapelet generators (line 7).

Procedure: Parallel *MEG*
Input : (1) *MTS-NC*; (2) The minimal and maximal length of shapelet; (3) Minimum utility and minimum support;
Output: The set of shapelet generators *SGs*;
01.   Thread s loads subsequence of time series T_i to shared memory $T_i[s]$;
02.   syncthreads();
03.   Thread s loads subsequence of time series T_j to shared memory $T_j[s]$;
04.   syncthreads();
05.   Get M_j := statement between $T_i[s]$ and $T_j[s]$;
06.   Perform calculation of line 06 to 13 of procedure *MEG*
07.   Return *SGs;*

Fig. 5. Pseudo code of Algorithm Parallel *MEG*

4   Experimental Evaluation

We use several real-world datasets to evaluate the performance of the proposed classifier. All experiments were performed on a computer with a four-core Intel Xeon host CPU at 2.40GHz with 96GB of memory, and this computer combined an NVIDIA Fermi C2075 GPU with 448 cores at 1.15GHz, 64KB shared memory per GPU multiprocessor, 64KB constant memory, and 6GB global memory. All algorithms are implemented in Java language and the GPU code is implemented in CUDA C++.

The experiments were performed on several real-world datasets: drug response [3], robot execution failures [1], ECG [16], wafer [16] and asthma [11]. Table 1 shows the characteristics of the datasets in the experiments. For evaluating the performance of the proposed model, we first employ 5-fold cross validation (CV) to divide into training and testing dataset, and then generate 20 runs of 5-fold CVs to calculate the average results. We compare five versions of the algorithm named as follows: *REACT*, *REACT*-Full (*REACT* with full-length time series), MSD ([8], the only study addressed to early classification with interpretability on multivariate time series), MSD-Full (MSD with full-length time series), and 1NN-Full (full-length 1NN which is strongly suggested by a comparison of dozens of time series classification algorithm on various datasets [5]). The similarity measures of numerical time series for MSD and 1NN-Full are Euclidean distance.

For shapelet extraction, we set *minLen* = 1 and *maxLen* to be 50% of the maximum length if length > 30. Otherwise, *maxLen* is set to equal to the maximum length. The results are under the best parameter setting for each dataset. In the results, we report the average of f-score, applicability and earliness. The *average f-score* is computed as $Avg.\ f\text{-}score = \frac{1}{|C|}\sum_{c'\in C}\frac{2\times precision(c')\times recall(c')}{precision(c')+recall(c')}$, where $Precision(c') = \frac{TP}{TP+FP}$ and $Recall(c') = \frac{TP}{TP+FN}$. In this study, a true positive (*TP*) occurs when the class of time series is predicted positive; otherwise, the model generates a false positive (*FP*). Furthermore, a false negative (*FN*) occurs when the model miss that the class of time series is positive.

In applicability evaluation, we regard the percentage of testing dataset which can be classified by *REACT* as $Applicability(\%) = \frac{|\{mt|mt\in D_{testing}\cdot SDT(mt)\neq\emptyset\}|}{|D_{testing}|}$. On the other hand, we regard the average percentage of time points used for classification as the earliness evaluation $Earliness(\%) = \frac{1}{|D_{testing}|}\sum_{mt\in D_{testing}}\frac{EMT(mt)}{len(mt)}$.

Table 1. Characteristic on different datasets

Dataset	High Dimension	Missing Value	Imbalance	Multiple Classes	Large size	Long time series
MS70	O	O				
Robot				O		
ECG					O	O
Wafer			O		O	O
Asthma	O	O	O		O	

Table 2. Performances on different datasets (%)

		MS70	Robot (Avg.)	ECG	Wafer	Asthma
	REACT	**72.9**	**72.7**	76.7	91.9	**74.9**
	REACT-Full	68.5	70.4	76	**92.1**	69.3
Avg. f-score	MSD	60.6	39.6	58.8	---	---
	MSD-Full	60.2	41.1	47.5	---	---
	1NN-Full	44.6	71.9	**78.7**	87.2	---
	REACT	93.8	94.7	100	100	99.4
	REACT-Full	93.8	94.7	100	100	99.4
Applicability	MSD	97.8	96.3	100	---	---
	MSD-Full	97.8	96.3	100	---	---
	1NN-Full	100	100	100	100	---
	REACT	22.9	40.7	**10.5**	32.8	73.7
	REACT-Full	100	100	100	100	100
Earliness	MSD	27.4	**27.4**	12.8	---	---
	MSD-Full	100	100	100	---	---
	1NN-Full	100	100	100	100	---

Table 3. Ccomparison of computation overhead

Dataset	Max. Length	#Instances	MSD (sec)	*REACT* (sec)	*REACT*-GPU(sec)
MS70	5	53	62.8 sec	335.2	7.8
Robot(Avg.)	15	92.6	173.94	255.16	5.64
ECG	152	200	13168.4	15335.8	104.6
Wafer	198	1194	>2 weeks	150834.1	68228.2
Asthma	5	1805	NA	1488.1	370.7

Table 2 lists the results on all datasets where the similarity measurement of numeric time series is Euclidean distance. The wafer dataset cannot be handled by MSD as a result of enormous computation cost. In general, *REACT* outperforms MSD and achieves comparable accuracy to that of 1NN-Full because our algorithm can discover more potential information of multivariate time series. Although MSD makes the earliest classification since the criteria, namely weighted information gain, used in MSD prefers earliness and frequency rather than discrimination, the result demonstrates that it is too early to be accurate. Due to characteristic of being serial, the difference of Avg. f-score between *REACT* and *REACT*-Full is small, which shows that *REACT* can capture the key features with suitable lengths of prefixes and make confident classification at appropriate timestamp.

Table 3 compares the training time of *REACT*, *REACT* on GPUs and MSD using the caching technique described in section 3. The result shows that *REACT* is slower than MSD on small datasets since our approach requires feature extraction and feature-based sequential pattern discovery for each variable and class. However, on the datasets of long time series or large amount of instances, *REACT* is faster than MSD in execution time. The reason is that MSD have to generate a huge number of shapelet candidates and pick out a small rule set from them. In addition, *REACT* on GPUs runs faster than *REACT* over 40 to 150 orders of magnitude on MS70, Robot and ECG Datasets, and over 2 to 4 orders of magnitude on Wafer and Asthma Datasets. We observed that Wafer and Asthma Datasets are large size and Wafer is also a long time series, and they needed lots of distance calculations on subsequences of time series.

5   Conclusion and Future Work

In this paper, we have proposed a novel methodology named *REACT* (*Reliable EArly ClassificaTion*) for constructing a reliable (i.e., *serial*) early classifier on *MTS-NC*. In addition, we adopt equivalence classes with generators mining to efficiently extract numerical and categorical features. Our experimental results clearly show that *REACT* outperforms the state-of-the-art methods in terms of accuracy and earliness. In addition, the GPU implementation significantly runs faster than the baseline approach of *REACT* model by several orders of magnitudes.

Although this is the first work that addresses this issue, it still leaves ample room for exploration in the future work. For example, we aim to find the significant features in different time series and use these features to build the classifier. However, the combination of non-significant features in different time series may be identifiability for early classification. In addition, the signal transform analysis techniques, e.g. wavelet or Fourier transform, may be employed to transform *MTS-NC* to find the significant combination features.

References

1. Bache, K., Lichman, M.: UCI machine learning repository. University of California, Irvine (2013)
2. Batal, I., Hauskrecht, M.: Constructing classification features using minimal predictive patterns. In: 10th CIKM, New York, pp. 869–878 (2010)
3. Baranzini, S.E., Mousavi, P., Rio, J., Caillier, S.J., Stillman, A., Villoslada, P., Wyatt, M.M., Comabella, M., Greller, L.D., Somogyi, R., Oksenberg, J.R.: Transcription-based prediction of response to IFNβ using supervised computational methods. PLos Biology 3(1), 166–176 (2005)
4. Chang, K.W., Deka, B., Hwu, W.M.H., Roth, D.: Efficient Pattern-Based Time Series Classification on GPU. In: ICDM, Belgium, pp. 131–140 (2012)
5. Ding, H., Trajcevski, G., Scheuermann, P., Keogh, E.: Querying and mining of time series data: experimental comparison of representations and distance measures. PVLDB 1(2) (2008)
6. Gao, C., Wang, J.: Efficient itemset generator discovery over a stream sliding window. In: 9th CIKM, Hong Kong, pp. 355–364 (2009)
7. Ghalwash, M.F., Radosavljevic, V., Obradovic, Z.: Extraction of Interpretable Multivariate Patterns for Early Diagnostics. In: 13th ICDM, Dallas, pp. 201–210 (2013)
8. Ghalwash, M.F., Obradovic, Z.: Early classification of multivariate temporal observations by extraction of interpretable shapelets. BMC Bioinformatics 13(195) (2012)
9. Griffin, M.P., Moorman, J.R.: Toward the early diagnosis of neonatal sepsis and sepsis-like illness using novel heart rate analysis. PEDIATRICS 107(1), 97–104 (2001)
10. He, G., Duan, Y., Qian, T.Y., Chen, X.: Early prediction on imbalanced multivariate time series. In: 22th CIKM, Burlingame, pp. 1889–1892 (2013)
11. Lee, C., Chen, J.C., Tseng, V.S.: A novel data mining mechanism considering bio-signal and environmental data with application on asthma monitoring. Computer Methods and Program in Biomedicine 101(1), 44–61 (2011)

12. Li, J., Li, H., Wong, L., Pei, J., Dong, G.: Minimum description length principle: Generators are preferable to closed patterns. In: 21th AAAI, Boston , pp. 409–414 (2006)
13. Li, J., Liu, G., Wong, L.: Mining statistically important equivalence classes and delta-discriminative emerging patterns. In: 13th KDD, New York, pp. 430–439 (2007)
14. Lines, J., Davis, L.M., Hills, J., Bagnall, A.: A shapelet transform for time series classification. In: 18th KDD, New York, pp. 289–297 (2012)
15. Lo, D., Khoo, S., Li, J.: Mining and ranking generators of sequential patterns. In: SDM, Atlanta, pp. 553–564 (2008)
16. Olszewski, R.T.: Generalized feature extraction for structural pattern recognition in time-series data. PhD Thesis, School of Computer Science, Carnegie Mellon University, Pittsburgh (2011)
17. Pasquier, N., Bastide, Y., Taouil, R., Lakhal, L.: Discovering Frequent Closed Itemsets for Association Rules. In: Beeri, C., Bruneman, P. (eds.) ICDT 1999. LNCS, vol. 1540, pp. 398–416. Springer, Heidelberg (1998)
18. Xing, Z., Pei, J., Dong, G., Yu, P. S.: Mining sequence classifiers for early prediction. In: SDM, Atlanta, pp. 644–655 (2008)
19. Xing, Z., Pei, J., Yu, P.S.: Early classification on time series: A nearest neighbor approach. In: 21th IJCAI, Pasadena, pp. 1297–1302 (2009)
20. Ye, L., Keogh, E.: Time series shapelet: A new primitive for data mining. In: 15th KDD, Paris, pp. 947–956 (2009)

Distributed Document Representation
for Document Classification

Rumeng Li[1]([✉]) and Hiroyuki Shindo[2]

[1] Peking University, Beijing 100871, China
alicerumeng@foxmail.com
[2] Nara Institute of Science and Technology, Nara 630-0192, Japan
shindo@is.naist.jp

Abstract. The distributed vector representations learned from the deep learning framework have shown its great power in capturing the semantic meaning of words, phrases and sentences, from which multiple NLP applications have benefited. As words combine to form the meaning of sentences, so do sentences combine to form the meaning of documents, the idea of representing each document with a dense distributed representation holds promise. In this paper, we propose a supervised framework (Compound RNN) for document classification based on document-level distributed representations learned from deep learning architecture. Our framework first obtains the distributed representation at sentence-level by operating on the parse tree structure from recursive neural network, and then obtains the document presentation-level by convoluting the sentence vectors from a recurrent neural network. Our framework (Compound RNN) outperforms existing document representations such as bag-of-words, LDA in multiple text classification/regression tasks.

1 Introduction

For text classification or regression task, documents need to be represented by a fix-length feature vector, on which machine learning algorithms can operate. The most naive but commonly used approach is bag-of-words representations, which represent each document with a feature vector by counting its containing unigrams, bigrams or trigrams and then leveraging the vectors based on Tf-idf scheme [34]. Tf-idf presentations have some appealing features notably its basic identification of sets of words that are discriminative for documents, but also come with severe shortcomings: the simple word-counting statistics are incapable of grasping in-depth information such as word semantics, and usually end up with extremely high dimensional vector representations.

Distributional semantic models(DSM) [44] offer a way of document representation by approximating the meaning of words with vectors that keep track of the patterns of co-occurrence of the words in a corpus. It assumes that semantically related words should occur in similar contexts [10]. Hybrid DSM methods based on traditional topic models are also developed for various tasks [14,45].

© Springer International Publishing Switzerland 2015
T. Cao et al. (Eds.): PAKDD 2015, Part I, LNAI 9077, pp. 212–225, 2015.
DOI: 10.1007/978-3-319-18038-0_17

However, DSM are psychological models of how we humans acquire and use semantic knowledge. However we can rely not only on linguistic context, but also on our rich perceptual experience [21].

Another prevailed alternative representation is obtained from Latent Dirichlet Allocation (LDA) [3], which represents each document as a probability over a set of components characterized as different distributions over vocabularies. LDA serves as a dimensionality reduction technique. As it first does the word-clustering based on document-level word co-occurrence, it can represent each document with a vector based on more condensed but more meaningful components, called "topics". LDA suffers from the shortcomings like ignorance of word orders, features, and word-similarity. While a bunch of approaches (e.g., [31,50]) have been proposed to address the aforementioned shortcomings, the improvements are limited and these approaches often end up with very complicated learning procedure.

Recently, deep architectures, such as recurrent and recursive neural networks, have been successfully applied to various natural language processing tasks. Such deep architectures learn a dense, low-dimensional representation of their problem in a hierarchical way that is capable of capturing both semantic and syntactic aspects of tokens (e.g., [1]), entities, N-grams [47], or phrases [42]. A significant advantage of the deep learning framework is that it frees researchers from feature engineering, since its representations are emergent. Furthermore, recent research has begun looking at higher level distributed representations that transcend the token level such as discourse-level [12].

Inspired by the idea that words combine to form the meaning of sentences, so do sentences combine to form the meaning of paragraphs and then documents, in this paper, we propose a supervised framework that learns continuous distributed condensed vector representations at document level for text classification and regression task. Our approach is hierarchical and is founded on two basic learning structures in deep learning: recursive neural network and recurrent neural network.

We first obtain the distributed representations at sentence level. The sentential compositionality operation is performed relying on sentence parse trees from recursive neural network. The distributed representation for each node in the sentence parse tree is computed in a bottom-up fashion as in [42] until the root is reached.

Next, we introduce the compositionality strategy for paragraph and document which is based on a recurrent neural network architecture that is useful for sequences [25,43]. The distributed vectors for paragraphs are obtained by subsequently convoluting its containing sentences with the input from the previous step. Document-level representations are obtained by adding up the vectors of its containing paragraphs in the similar way based on recurrent neural network.

Given the satisfying results obtained from distributed word and sentence representation in previous work, it is natural to extend it to distributed paragraph and document representation to represent the meaning of texts and benefit other NLP tasks dealing with documents. Recursive neural network requires structured inputs and for text processing, it relies heavily on the parsing results. Document parsing is uncompetitive compared with sentence parsing. So we employ the recursive neural network for sentence level representation but recurrent neural network for paragraph and document level.

Our approach is a task-specific framework where parameters involved in intra- and inter- sentence compositionality are optimized through the task-specific tuning procedure, which can also be treated as a feature selection procedure. We experiment our approach (Compound RNN) on different document-classification tasks, i.e., binary classification, multi-class classification and regression. Experimental results illustrate the effectiveness of our model over existing baselines.

2   Related Work

2.1   Recursive Neural Networks

Recursive neural networks (RNN), as one kind of deep learning frameworks, was first proposed in [9]. Recursive framework relies and operates on structured inputs (e.g., parse tree) and computes the representation of each parent based on its children iteratively in a bottom-up fashion. To tailor different task-specific requirements, some variations of RNN have been proposed such as Recursive Neural Tensor Network [42] that allows the model to have greater interactions between the input vectors and Matrix-Vector RNN [39] which represents every word as both a vector and a matrix. There are also some work addressing the feature weight tuning for recursive neural networks [17] to make the model emphasize more on important information. Tasks have benefited from recursive framework including parsing [19,40], sentiment analysis [42], machine translation [20], textual entailment [4] and paraphrase detection [38].

2.2   Recurrent Neural Networks

Recurrent neural networks, as another learning structure of deep learning [35,43], takes a collection of tokens, phrases or sentences as a sequence and incorporates information from the past (i.e. preceding tokens) to get the current output. Specifically, at each step, recurrent network takes both the output of previous step and the current token as input, convolutes the inputs, and forwards it to the next step. It has been successfully applied to tasks such as language modeling [25] or spoken language understanding [22]. Recurrent network does not need external deeper structure (e.g., parse tree) and is able to preserve the embedding dimension when convoluting different number of components. However, in recurrent framework, long unit dependencies might be difficult to capture and the framework suffers from the vanishing gradient problem.

2.3   Distributed Representations

Both recurrent and recursive neural networks require vector representations for input tokens. Distributed representations for words were first proposed in [33] and have been used for statistical language modeling [8]. Various deep learning architectures have been explored to learn these embeddings in an unsupervised manner from a large corpus [1,6,23,27]. They might have different generalization capabilities and able to capture the semantic meanings for specific tasks. These vector representations capture interesting semantic relationships (or to some extent) such as $King-man \approx Queue-woman$ and have benefited multiple NLP applications such as name entity recognition, tagging or machine translation (e.g., [6,51]).

Recent researchers have also begun looking at distributed representations of phrases and sentences [12,18,38,39,42]. But their methods are not extended beyond sentences. [39] applied the matrix-vector RNN model to learn compositional vector representations for phrases and sentences. But their experiments mainly focused on phrase and sentence level classification like predicting sentiment distributions of adverb-adjective pairs etc. [12] extends the representation to sentence-level and discourse level by convoluting representations in a recurrent neural network. But their improvements are trivial.

[15] proposed a straightforward method to calculate the distributed representations for paragraphs. The basic idea is that the distributed representation of a piece of text should be able to predict the following word. Experiments on various tasks verify the effectiveness of such methods. But they concatenate sentence vectors, even paragraph vectors with word vectors. In this paper, we regard word vectors, sentence vectors and paragraph vectors as different layers in the model.

2.4   Approaches to Classification and Regression

As this paper mainly focuses on applying deep architectures to document classification tasks, we just give a brief review of other approaches to classification and regression. Many models can be used for the classification and regression task [36], such as k-NN, SVM, Naive Bayes, Neural Networks, ect., as long as documents can be represented in appropriate forms. To present documents, popular methods use topic models. Notable ones are pLSI [11], LDA [3] and some variations of LDA, such as sLDA [2], L-LDA [32] etc.

3   Sentence Model

In this section, we describe how we compute the distributed representation for a given sentence based on its parse tree structure and containing words. As the details can be found in a bunch of early work (e.g., [42]), we try to make this section brief and skip the details for brevity.

Let s denotes any given sentence, which is comprised of a sequence of tokens $s = \{w_1, w_2, ..., w_{n_s}\}$, where n_s denotes the number of tokens within sentence s. Each token w within the sentence is associated with a specific vector embedding $\mathbf{e_w} = \{e_w^1, e_w^2, ..., e_w^K\}$, where K denotes the dimension of the word embedding. Here the word embeddings are initialized by using word representations taken from RNNLM [24, 26]. The dimension of embeddings is 80. We wish to compute the vector representation h_s for current sentence, where $h_s = \{h_s^1, h_s^2, ..., h_s^K\}$.

Parse trees are obtained from Stanford Parser[1]. For a given parent p in the tree and its two children c_1 (associated with vector representation h_{c_1}) and c_2 (associated with vector representation h_{c_2}), standard recursive network calculates the distributed vector for parent p as follows:

$$h_p = f(W \cdot [h_{c_1}, h_{c_2}] + b) \tag{1}$$

where $[h_{c_1}, h_{c_2}]$ denotes the concatenating vector for children representations h_{c_1} and h_{c_2}. W is a $K \times 2K$ matrix and b is the $1 \times K$ bias vector. $f(\cdot)$ is tanh function.

Standard recurrent framework uses the same (tied) weights W at all nodes to compute the vector. This requires the compositionality function to be extremely powerful as it has to combine phrases with different syntactic roles, which is usually unrealistic. Several approaches have been proposed to address such weakness including Matrix-Vector RNN [39] or Recursive Neural Tensor Network [42]. In this work, we adopt a simple alternative where instead of using a single compositionality matrix W, we associate each of the sentence roles (i.e., VP, NP or NN) with a specific compositionality matrix (i.e, W_{VP}, W_{NP} or W_{NN}). Let W_{c_1} and W_{c_2} denote the matrices associated with children c_1 and c_2 based on their roles. Then the convolution is given by:

$$h_p = f([W_{c_1}, W_{c_2}] \cdot [h_{c_1}, h_{c_2}] + b) \tag{2}$$

where $[W_{c_1}, W_{c_2}]$ denotes the K*2K dimensional concatenating matrix for $W_{tag(c_1)}$ and $W_{tag(c_2)}$.

4   Document Model

In this section, we illustrate how we get the distributed vector representation for a given document based on its contained sentences, of which the distributed representations have already been obtained in the Sentence Model Section. Document d is comprised of a sequence of paragraphs $D = \{L_1, L_2, ..., L_{N_d}\}$ where N_d denotes the number of paragraphs within the document. Each paragraph is comprised of a sequence of sentences $L = \{s_1, s_2, ..., s_{N_L}\}$, where N_L denotes the number of sentences within the paragraph. To obtain the vector representation h_L for paragraph L, we turn to recurrent neural network, which successively takes in sentence s_t at step i, combines its vector representation h_{s_t} with former input h_L^{t-1} from step $i-1$, calculates the resulting current embedding h_L^t, and passes it to the next step. The convolution can be summarized as follows:

$$h_L^t = f(V_L \cdot h_L^{t-1} + W_L \cdot h_{s_t} + b_L) \tag{3}$$

---

[1] http://nlp.stanford.edu/software/lex-parser.shtml

where W_L and V_L are $K \times K$ matrixes. b_L denotes $K \times 1$ bias vector and $f = tanh$ is a standard element-wise nonlinearity. To note, the calculation for representation at time $t = 1$ is given by:

$$h_L^1 = f(V_L \cdot h_0 + W_L \cdot h_{s_t} + b_L) \tag{4}$$

where h_0 denotes the global sentence starting vector for paragraphs. To note, for documents with only one paragraph, their corresponding distributed vectors are obtained by using the strategy just described and no convolution between paragraphs is needed.

Similar as the paragraph-level compositionality, we compute document-level distributed representations as follows: given the vector presentation for its containing paragraphs $\{h_{L_1}, h_{L_2}, ..., h_{L_{N_d}}\}$, at each step i, the recurrent framework takes as input the vector representation h_{L_t} for current paragraph L_t, combines it with former input h_d^{t-1} from step $i - 1$, calculates h_d^t and passes it to the next step.

$$h_d^t = f(V_d \cdot h_d^{t-1} + W_d \cdot h_{L_t} + b_d) \tag{5}$$

where W_d and V_d are $K \times K$ matrixes. b_d denotes $K \times 1$ bias vector.

5   Document Classification

We train our classifier regarding three types of document classification tasks: binary classification, multi-class classification and regression.

5.1   Binary Classification

For binary classification, each document d is associated with a 0/1 binary valued variable t_d. For classification purpose, given the document distributed vector h_d, we first generate a scalar using linear function $U_{\text{binary}}^T \cdot h_d + b$ and then projects it into $[0,1]$ possibility space using a sigmoid function, as given by:

$$P(t_d = 1) = g(U_{\text{binary}} \cdot h_d + b_{\text{binary}}) \tag{6}$$

where U_{binary} is a $K \times 1$ dimensional vector and b_{binary} denotes the bias vector. $g(\cdot)$ denotes the sigmoid function.

For a given set of training data D, the cost function with regularization on the training set is given by:

$$J = \frac{1}{|D|} \sum_{d \in D} J_{\text{binary}}(d) + \frac{Q}{|D|} \sum_{\theta \in \Theta} \theta^2 \tag{7}$$

$$J_{\text{binary}}(d) = -t_d \log p(t_d = 1) - (1 - t_d) \log[1 - p(t_d = 1)] \tag{8}$$

The regularization part is parameterized by Q that pushes the weights from $\Theta = [\{W_{tag}\}, W_L, W_d, U_{binary}]$ to zero. For any parameter θ to optimize, the derivative of $J_{\text{binary}}(d)$ with respect to θ is given by:

$$\frac{\partial J_{\text{binary}}(d)}{\partial \theta} = [p(t_d = 1) - t_d] \frac{\partial p(t_d = 1)}{\partial \theta} \tag{9}$$

where $\frac{\partial p(t_d=1)}{\partial \theta}$ can be further obtained from the standard back-propagation.

5.2   Multi-Class Classification

For multi-class classification task, each document d is associated with a class tag t_d, which takes value from $[1,2,3,...T]$, where T denotes the number of potential classes. We associate each document d with a T dimensional binary vector R_d, which is the ground truth vector with a 1 at the correct label t_d and all other entries 0. The prediction task is done through a softmax classifier.

$$S_d = U_{\text{multi}} \cdot h_d \tag{10}$$

$$P_d(i) = \frac{\exp(S_d(i))}{\sum_j \exp(S_d(j))} \tag{11}$$

U_{multi} is a $T \times K$ matrix. S_d is the intermediate result and P_i is the probability of assigning class i^{th} to the current document.

For a given set of training document D, the cost function for multi-class classification is given by:

$$J = \frac{1}{|D|} \sum_{d \in D} J_{\text{multi}}(d) + \frac{Q}{|D|} \sum_{\theta \in \Theta} \theta^2 \qquad J_{\text{multi}}(d) = -\log p(t_d) \tag{12}$$

Similar to binary classification, Q is the regularization parameter that pushes elements in $\Theta = [\{W_{tag}\}, W_L, W_d, U_{multi}]$ to zero. For any parameter θ we wish to optimize, the derivative of $J_{\text{multi}}(d)$ with respect to θ is given by:

$$\frac{\partial J_{\text{multi}}(d)}{\partial \theta} = [P_d - R_d] \otimes \frac{\partial S_{(d)}}{\partial \theta} \tag{13}$$

where $\otimes$ denotes the Hadamard product between the two vectors.

5.3   Regression

For a given document $d \in D$ associated with regression tag t_d (e.g, review rating, website popularity), the deep learning framework makes prediction $\hat{t}_d$ for document d as follows:

$$\hat{t}_d = U_{\text{regression}} \cdot h_d \tag{14}$$

where $U_{\text{regression}}$ denotes the K dimensional vector. Parameters are estimated through minimizing the following cost function:

$$J = \frac{1}{|D|} \sum_{d \in D} ||\hat{t}_d - t_d||^2 + \frac{Q}{|D|} \sum_{\theta \in \Theta} \theta^2 \tag{15}$$

5.4   Optimization

The derivative for each parameter can be obtained from standard backpropagation [9,41]. For optimization, we turn to the diagonal variant of AdaGrad [7] with minibatches, which is widely applied in deep learning literature (e.g.,[30,38]).

The learning rate in AdaGrad is adapting differently for different parameters at different steps. Concretely, let g_τ^i denote the subgradient at time step t for parameter θ_i obtained from backpropagation, the parameter update at time step t is given by:

$$\theta_\tau = \theta_{\tau-1} - \frac{\alpha}{\sum_{t=0}^{\tau} \sqrt{g_\tau^{i2}}} g_\tau^i \tag{16}$$

where α denotes the learning rate.

5.5 Initialization

Many tricks have been reported regarding the initialization of neural networks [16]. We employed only two of them. The initialization of W were done according to the fan-in of the layer by randomly drawing from uniform distribution $[-\epsilon, \epsilon]$, where $\epsilon = \sqrt{\frac{6}{K+2*K}}$. For elements involved in recurrent network in paragraph and document compositionality, i.e., W_L, W_d, we initialize them by randomly drawing from uniform distribution $[-0.2, 0.2]$, preserving the same scale as W. All bias vectors are initialized as 0.

Previous work also discovered a huge performance boost by initializing word embeddings using vectors pre-trained from a large unlabeled data corpus instead of random initialization (e.g.,[48]). We therefore initialize word embeddings $\{e\}$ using word representations taken from RNNLM [24, 26]. The dimension of embeddings is 80.

6 Experiments

In this section, we show the experimental results regarding the three aforementioned document classification problems: multi-class classification, binary classification and regression.

6.1 Multi-class Classification

We perform multi-class classification task on the 20 Newsgroup dataset[2]. The data set has a balanced distribution over the 20 categories. The test set is comprised of 7,505 documents in total, with the smallest category containing 251 documents and the largest category containing 399 documents. The training set has a total number of 11,269 documents, the smallest and the largest categories of which contain 376 and 599 documents respectively. The naive baseline that predicts the most frequent category for all the test documents has the classification accuracy 0.0532. For comparison, we employ the following models as baselines:

- **tf-idf+SVM**: Each document is represented as vector of unigram based on tf-idf. A multi-class linear SVM classifier is trained using SVM[multi-class] package[3].

---

[2] http://qwone.com/~jason/20Newsgroups/
[3] www.cs.cornell.edu/people/tj/svm_light/svm_multiclass.html

- **LDA+SVM**: We run variational EM algorithm for LDA using package[4] on the 18,828 documents with topic number being set to 110, as suggested in [13,50]. Each document is therefore represented by a 110 dimensional vector. SVM$^{multi-class}$ takes as input these vectors as training data.
- **sLDA**: A supervised version of LDA for multi-class classification [46].

In addition to the aforementioned baselines, we also implement a simplified version of our proposed model **Simplified** which uses a unified convolution matrix when operating on parse tree structure. The prediction accuracy regarding different approaches is reported in Table 1. As bag-of-words models consider neither how each sentence is composed (e.g., word ordering) nor word semantics, it obtained the worst performance. LDA based models can, to some extent, take into consideration the latter (word semantics) by the word pre-clustering but fail to consider the former. The deep learning approaches (**Compound RNN** and **Simplified**) significantly outperform the other baselines. The original version takes into account different types of compositionality, performing better than the **Simplified** version, which uses one unified compositionality matrix when operating on sentence parse trees. As **Compound RNN** consistently outperforms **Simplified** version, the results for **Simplified** version are excluded in the later parts for brevity.

Table 1. Multi-class Classification Performance (Accuracy) for different approaches on 20 Newsgroup dataset

	tf-idf+SVM	LDA+SVM	sLDA	Simplified	Compound RNN
Acc	0.568	0.607	0.694	0.758	**0.782**

6.2 Binary Classification

News Group Classification: We first perform binary classification evaluation on distinguishing postings of newsgroup *alt.atheism* and *talk.religion.misc* from the 20 news groups, as Lacoste-Julien et [13] did. The training set contains 856 documents with a split of 480/376 over the two categories, and the test set contains 569 documents with a split of 318/251 regarding the two categories.

Table 2. Binary Classification Performance (Accuracy) for different approaches on *alt.atheism* and *talk.religion.misc* news group

	tf-idf+SVM	LDA+SVM	sLDA(regression)	sLDA(multi-class)	Compound RNN
Acc	0.628	0.668	0.724	0.758	**0.812**

The baselines we explore include tf-idf+SVM, LDA+SVM, and two versions of supervised LDA: the regression version of s-LDA [2] which uses the binary

---

[4] http://www.cs.princeton.edu/~blei/lda-c/

representation (0/1) of the classes, and uses a threshold 0.5 to make prediction, and multi-class sLDA as described in the previous section. For LDA based approaches, topic numbers are set to 30 and 35 as suggested in [13,50] and we report the better performance.

As we can see from Table 2, the deep learning approach outperforms all LDA based approaches and the native tf-idf approach. To note here, our deep learning approach does not yield significant performance boosting when compared with other sophisticatedly developed LDA-based baselines, such as MedLDA [49] (reported accuracy around 0.81) and DiscLDA [13] (reported accuracy around 0.80).

Truthful vs Deceptive Review Classification: We perform binary classification task on the hotel reviews for 20 Chicago hotels described in [28]. The dataset contains 400 deceptive fake reviews (positive) solicited from Amazon Mechanical Turk and 400 truthful reviews (negative). The algorithm makes prediction regarding whether a given review is deceptive or truthful. We perform 5-fold cross-validation experiments.

The baselines we implemented include SVM frameworks based on features suggested in [28]: LIWC+SVM, Unigram+SVM, Bigram+SVM, LIWC+Bigram+SVM and LDA+SVM. LIWC, short for the Linguistic Inquiry and Word, is an automatic analysis tool which counts and groups the number of instances of nearly 4,500 keywords into 80 psychologically meaningful dimensions. For LDA+SVM, we run gibbs sampling of LDA on the 800 documents with topic number ranging from 2 to 20 at interval of 2 and report the best performance. We report performance regarding each models in Table 3. Among

Table 3. Binary Classification Performance for different approaches on Myle et al [28]'s fake review dataset

Approach	Accuracy	Precision	Recall
LIWC+SVM	0.768	0.764	0.775
LDA+SVM	0.812	0.804	0.834
unigram+SVM	0.884	0.899	0.865
bigram+SVM	0.896	0.901	0.890
LIWC+Bigram	0.898	0.898	0.898
Compound RNN	**0.924**	**0.937**	**0.915**

the baselines, LIWC+Bigram setting achieve the best performance. The deep learning approach has an absolute improvement of 2.6% in terms of accuracy over the LIWC+Bigram setting.

6.3 Regression

Review Rating Prediction: For regression evaluation, we first evaluate Compound RNN on the movie review data set introduced in [29], which contains movie reviews paired with the number of stars given. We treat the rating prediction task as a regression problem and use the same settings as in [2,49]. The evaluation criterion is predictive R^2, which is defined as one minus the mean squared error divided by the data variance as defined in [2].

$$\text{pR}^2 = 1 - \frac{\sum_d (t_d - \hat{t}_d)^2}{\sum_d (t_d - \bar{t})^2}$$

where t_d and $\hat{t}_d$ are the true and estimated ratings of document d. $\bar{t}$ denotes mean of review ratings on the whole data set.

We run experiment with default settings as described in [2]. We employ the following approaches as baselines:

- **sLDA**: the supervised version of LDA introduced in [2].
- **LDA+SVR**: we train the Support Vector Regression (SVR) [37] on LDA topic representations using the LIBSVM toolkit [5].
- **L$_1$ regularized least-squares regression (Lasso)** as suggested in [2], which uses each document's empirical distribution over words as its lasso covariates.

Topic number is set to 30 for LDA-based approaches. And The pR^2 scores for LDA+SVR, sLDA, Lasso, and Compound RNN are 0.348, 0.502, 0.457 and **0.552** separately. As we can see, the Compound RNN again outperforms the standard baselines. To note, the performance of the Compound RNN is comparable to other derivations of LDA such as MedLDA [49] (reported pR^2 around 0.55).

7   Conclusion

In this paper, we propose a deep learning based framework for supervised document classification and regression. The classification task relies on document distributed representation obtained on the basis of recursive and recurrent neural networks. Our framework is task-specific as it does not aim to learn the general representations for sentences, paragraphs or documents but are based on the task-specific parameters in intra- and inter- sentence convolution which are optimized and guided by the optimization function, which can be viewed as a feature selection process specific for different tasks. Experiments on several text classification tasks such as binary classification, multi-class classification and regression demonstrate that the proposed algorithm is competitive and significantly outperforms prevailed existing baselines such as LDA and bag-of-words.

References

1. Bengio, Y., Schwenk, H., Senécal, J.S., Morin, F., Gauvain, J.L.: Neural probabilistic language models. In: Bengio, Y., et al. (eds.) Neural Probabilistic Language Models. STUDFUZZ, vol. 194, pp. 137–186. Springer, Heidelberg (2006)
2. Blei, D.M., McAuliffe, J.D.: Supervised topic models. Advances in Neural Information Processing Systems 7, 121–128 (2007)

3. Blei, D.M., Ng, A.Y., Jordan, M.I.: Latent dirichlet allocation. The Journal of Machine Learning Research **3**, 993–1022 (2003)
4. Bowman, S.R.: Can recursive neural tensor networks learn logical reasoning? arXiv preprint arXiv:1312.6192 (2013)
5. Chang, C.C., Lin, C.J.: Libsvm: a library for support vector machines. ACM Transactions on Intelligent Systems and Technology (TIST) **2**(3), 27 (2011)
6. Collobert, R., Weston, J.: A unified architecture for natural language processing: deep neural networks with multitask learning. In: Proceedings of the 25th International Conference on Machine Learning, pp. 160–167. ACM (2008)
7. Duchi, J., Hazan, E., Singer, Y.: Adaptive subgradient methods for online learning and stochastic optimization. The Journal of Machine Learning Research **12**, 2121–2159 (2011)
8. Elman, J.L.: Finding structure in time. Cognitive Science **14**(2), 179–211 (1990)
9. Goller, C., Kuchler, A.: Learning task-dependent distributed representations by backpropagation through structure. In: IEEE International Conference on Neural Networks, vol. 1, pp. 347–352. IEEE (1996)
10. Harris, Z.S.: Distributional structure. Word (1954)
11. Hofmann, T.: Probabilistic latent semantic indexing. In: Proceedings of the 22nd Annual International ACM SIGIR Conference on Research and Development in Information Retrieval, pp. 50–57. ACM (1999)
12. Kalchbrenner, N., Blunsom, P.: Recurrent convolutional neural networks for discourse compositionality. arXiv preprint arXiv:1306.3584 (2013)
13. Lacoste-Julien, S., Sha, F., Jordan, M.I.: Disclda: discriminative learning for dimensionality reduction and classification. In: Advances in Neural Information Processing Systems, pp. 897–904 (2009)
14. Larochelle, H., Lauly, S.: A neural autoregressive topic model. In: Advances in Neural Information Processing Systems, pp. 2708–2716 (2012)
15. Le, Q.V., Mikolov, T.: Distributed representations of sentences and documents. arXiv preprint arXiv:1405.4053 (2014)
16. LeCun, Y., Bottou, L., Bengio, Y., Haffner, P.: Gradient-based learning applied to document recognition. Proceedings of the IEEE **86**(11), 2278–2324 (1998)
17. Li, J.: Feature weight tuning for recursive neural networks. arXiv preprint arXiv:1412.3714 (2014)
18. Li, J., Hovy, E.: A model of coherence based on distributed sentence representation. In: Proceedings of Empirical Methods in Natural Language Processing (2014)
19. Li, J., Li, R., Hovy, E.: Recursive deep models for discourse parsing. In: Proceedings of Empirical Methods in Natural Language Processing (2014)
20. Liu, S., Yang, N., Li, M., Zhou, M.: A recursive recurrent neural network for statistical machine translation. In: Proceedings of Association for Computational Linguistics, pp. 1491–1500 (2014)
21. Louwerse, M.M.: Symbol interdependency in symbolic and embodied cognition. Topics in Cognitive Science **3**(2), 273–302 (2011)
22. Mesnil, G., He, X., Deng, L., Bengio, Y.: Investigation of recurrent-neural-network architectures and learning methods for spoken language understanding. In: Proceedings of Interspeech, pp. 3771–3775 (2013)
23. Mikolov, T., Chen, K., Corrado, G., Dean, J.: Efficient estimation of word representations in vector space. arXiv preprint arXiv:1301.3781 (2013)
24. Mikolov, T., Deoras, A., Povey, D., Burget, L., Cernocky, J.: Strategies for training large scale neural network language models. In: 2011 IEEE Workshop on Automatic Speech Recognition and Understanding (ASRU), pp. 196–201. IEEE (2011)

25. Mikolov, T., Karafiát, M., Burget, L., Cernocký, J., Khudanpur, S.: Recurrent neural network based language model. In: Proceedings of Interspeech, pp. 1045–1048 (2010)
26. Mikolov, T., Yih, W.T., Zweig, G.: Linguistic regularities in continuous space word representations. In: Proceedings of NAACL-HLT, pp. 746–751 (2013)
27. Mnih, A., Hinton, G.: Three new graphical models for statistical language modelling. In: Proceedings of the 24th International Conference on Machine Learning, pp. 641–648. ACM (2007)
28. Ott, M., Choi, Y., Cardie, C., Hancock, J.T.: Finding deceptive opinion spam by any stretch of the imagination. In: Proceedings of the 49th Annual Meeting of the Association for Computational Linguistics: Human Language Technologies, vol. 1, pp. 309–319. Association for Computational Linguistics (2011)
29. Pang, B., Lee, L.: Seeing stars: exploiting class relationships for sentiment categorization with respect to rating scales. In: Proceedings of the 43rd Annual Meeting on Association for Computational Linguistics, pp. 115–124. Association for Computational Linguistics (2005)
30. Pei, W., Ge, T., Baobao, C.: Max-margin tensor neural network for chinese word segmentation. In: Proceedings of the 52nd Annual Meeting on Association for Computational Linguistics (2014)
31. Petterson, J., Buntine, W., Narayanamurthy, S.M., Caetano, T.S., Smola, A.J.: Word features for latent dirichlet allocation. In: Advances in Neural Information Processing Systems, pp. 1921–1929 (2010)
32. Ramage, D., Hall, D., Nallapati, R., Manning, C.D.: Labeled lda: a supervised topic model for credit attribution in multi-labeled corpora. In: Proceedings of the 2009 EMNLP, vol. 1, pp. 248–256. Association for Computational Linguistics (2009)
33. Rumelhart, D.E., Hinton, G.E., Williams, R.J.: Learning representations by back-propagating errors. Cognitive Modeling 5 (1988)
34. Salton, G., McGill, M.J.: Introduction to modern information retrieval (1983)
35. Schuster, M., Paliwal, K.K.: Bidirectional recurrent neural networks. IEEE Transactions on Signal Processing 45(11), 2673–2681 (1997)
36. Sebastiani, F.: Machine learning in automated text categorization. ACM Computing Surveys (CSUR) 34(1), 1–47 (2002)
37. Smola, A.J., Schölkopf, B.: A tutorial on support vector regression. Statistics and Computing 14(3), 199–222 (2004)
38. Socher, R., Huang, E.H., Pennin, J., Manning, C.D., Ng, A.Y.: Dynamic pooling and unfolding recursive autoencoders for paraphrase detection. In: Advances in Neural Information Processing Systems, pp. 801–809 (2011)
39. Socher, R., Huval, B., Manning, C.D., Ng, A.Y.: Semantic compositionality through recursive matrix-vector spaces. In: Proceedings of the 2012 Joint Conference on Empirical Methods in Natural Language Processing and Computational Natural Language Learning, pp. 1201–1211. Association for Computational Linguistics (2012)
40. Socher, R., Lin, C.C., Manning, C., Ng, A.Y.: Parsing natural scenes and natural language with recursive neural networks. In: Proceedings of the 28th International Conference on Machine Learning, ICML 2011, pp. 129–136 (2011)
41. Socher, R., Manning, C.D., Ng, A.Y.: Learning continuous phrase representations and syntactic parsing with recursive neural networks. In: Proceedings of the NIPS-2010 Deep Learning and Unsupervised Feature Learning Workshop, pp. 1–9 (2010)

42. Socher, R., Perelygin, A., Wu, J.Y., Chuang, J., Manning, C.D., Ng, A.Y., Potts, C.: Recursive deep models for semantic compositionality over a sentiment treebank. In: Proceedings of the Conference on Empirical Methods in Natural Language Processing, EMNLP, vol. 1631, p. 1642. Citeseer (2013)
43. Sutskever, I., Martens, J., Hinton, G.E.: Generating text with recurrent neural networks. In: Proceedings of the 28th International Conference on Machine Learning, ICML 2011, pp. 1017–1024 (2011)
44. Turney, P.D., Pantel, P., et al.: From frequency to meaning: Vector space models of semantics. Journal of Artificial Intelligence Research **37**(1), 141–188 (2010)
45. Wan, L., Zhu, L., Fergus, R.: A hybrid neural network-latent topic model. In: Proceedings of International Conference on Artificial Intelligence and Statistics, pp. 1287–1294 (2012)
46. Wang, C., Blei, D., Li, F.F.: Simultaneous image classification and annotation. In: IEEE Conference on Computer Vision and Pattern Recognition, CVPR 2009, pp. 1903–1910. IEEE (2009)
47. Wang, S., Manning, C.D.: Baselines and bigrams: simple, good sentiment and topic classification. In: Proceedings of the 50th Annual Meeting of the Association for Computational Linguistics: Short Papers, vol. 2, pp. 90–94. Association for Computational Linguistics (2012)
48. Zheng, X., Chen, H., Xu, T.: Deep learning for chinese word segmentation and pos tagging. In: Proceedings of the 2013 Conference on Empirical Methods in Natural Language Processing, pp. 647–657 (2013)
49. Zhu, J., Ahmed, A., Xing, E.P.: Medlda: maximum margin supervised topic models for regression and classification. In: Proceedings of the 26th Annual International Conference on Machine Learning, pp. 1257–1264. ACM (2009)
50. Zhu, J., Xing, E.P.: Conditional topic random fields. In: Proceedings of the 27th International Conference on Machine Learning, ICML 2010, pp. 1239–1246 (2010)
51. Zou, W.Y., Socher, R., Cer, D.M., Manning, C.D.: Bilingual word embeddings for phrase-based machine translation. In: Proceedings of the 2013 Conference on Empirical Methods in Natural Language Processing, pp. 1393–1398 (2013)

Prediciton of Emergency Events: A Multi-Task Multi-Label Learning Approach

Budhaditya Saha$^{(\boxtimes)}$, Sunil K. Gupta, and Svetha Venkatesh

Centre for Pattern Recognition and Data Analytics School of Information
Technology, Deakin University, Geelong, Australia
budhaditya.saha@deakin.edu.au

Abstract. Prediction of patient outcomes is critical to plan resources in an hospital emergency department. We present a method to exploit longitudinal data from Electronic Medical Records (EMR), whilst exploiting multiple patient outcomes. We divide the EMR data into *segments* where each segment is a *task*, and all tasks are associated with multiple patient outcomes over a 3, 6 and 12 month period. We propose a model that learns a prediction function for each task-label pair, interacting through *two* subspaces: the *first* subspace is used to impose sharing across *all tasks for a given label*. The *second* subspace captures the task-specific variations and is shared across all the labels for a given task. The proposed model is formulated as an iterative optimization problems and solved using a scalable and efficient Block co-ordinate descent (BCD) method. We apply the proposed model on two hospital cohorts - Cancer and Acute Myocardial Infarction (AMI) patients collected over a two year period from a large hospital emergency department. We show that the predictive performance of our proposed models is significantly better than those of several state-of-the-art multi-task and multi-label learning methods.

1 Introduction

Resource allocation at hospital emergency departments is critical. To facilitate such allocation, accurate prediction of several patient outcomes is crucial - emergency presentation, readmission or length of stay are some examples. Routinely collected Electronic Medical Data (EMR) offers opportunity to make such prognosis. This data is longitudinal, containing information about evolving risk, capturing disease progression and health conditions amongst other factors. Multiple patient outcomes are related to underlying patient health and, therefore, their joint modeling can potentially help in building better prediction models.

Previous research efforts to build prediction models for longitudinal data have attempted to model the data as a time series [1] and use mixture distributions as data generative models. A set of latent states are learnt using mixture components and the change is modeled using the transitions over these states. Other works use random effect models for longitudinal measurements assuming that the risk over time remains nearly constant [2]. These approaches are insufficient to handle evolving risk.

© Springer International Publishing Switzerland 2015
T. Cao et al. (Eds.): PAKDD 2015, Part I, LNAI 9077, pp. 226–238, 2015.
DOI: 10.1007/978-3-319-18038-0_18

One way to tackle this problem is to divide the longitudinal data into segments, each of which can be considered as a task. We build one prediction model for each task in a joint framework of multi-task learning (MTL) [3,4]. The use of MTL ensures that the data from different tasks are appropriately combined to exploit their common relatedness whilst the distribution of risk for each segment can still differ. To assist joint modeling, MTL techniques employ various constraints on the task parameters e.g. sampling the task-parameters from a shared prior [5,6], modeling the task-parameters through a common low-dimensional subspace [4,7], or combining the tasks in proportion to their relatedness learnt using a task-to-task covariance matrix [8]. Focusing on our problem, we employ a MTL framework to jointly model different segments of the longitudinal data, and additionally exploit the relationship amongst multiple outcomes. Existing MTL models focus on multiple outcomes of a single task. *Therefore, the problem of developing a multi-task, multi-label prediction model remains open.*

Addressing this problem, we propose a framework for multiple outcomes or labels prediction in a MTL paradigm. For each task-label pair, we learn a prediction function. The prediction functions of the task-label pairs interact through *two* subspaces. The *first* subspace is used to impose sharing across *all tasks* for a given *label*. The *second* subspace, specific to a task, is used to allow task-specific variations and is shared across all the labels for the task. We term this model multi-task multi-label (MTML) learning. The proposed MTML is formulated as an iterative optimization problem and solved using a scalable and efficient block co-ordinate descent method. We empirically demonstrate both the scalability and convergence. We apply the proposed MTML models on two real-world cohorts - a Cancer Electronic Medical Records (EMR) with 3000 patients collected over two years involving 11 different cancer types, and an Acute Myocardial Infarction (AMI) cohort with 2652 patients collected over the same two year period. We predict multiple emergency related outcomes - future emergency attendances, admissions and length-of-stay- over a 3 month, 6 month and twelve month period. We show that the predictive performance of MTML models is better than those of several state-of-the-art baselines [9,4,10,11]. Our main contributions are:

- A *novel* multi-task-multi-label learning (MTML) model that extends the traditional MTL/MLL framework by jointly modeling multiple tasks with multiple labels simultaneously. The MTML model has two components (1) a subspace that spans across all the tasks for a given *label* and (2) a subspace that is "task-specific" spanning across all the labels of a task (section 3.1).
- Solution of the optimization problem using a scalable and efficient Block coordinate descent (BCD) method (section 3.2).
- Empirical validation of the model through experiments using two real world hospital datasets, Cancer and AMI cohorts containing 3000 and 2652 patients respectively. We show better performance of our proposed model in comparison with recent state-of-the-art MTL and MLL methods (section 4).

The significance of our approach is in providing solutions to classification problems in data that contain multiple tasks wherein examples of each task have multiple labels. Our solution helps in building accurate prediction system for better upfront planning of resources at hospital emergency departments.

2   Related Methods

We briefly review prediction models for longitudinal data analysis, multi-task and multi-label learning methods as following:

Prediction Models for Longitudinal Data: Longitudinal data has two crucial challenges: (1) uneven distributions of data points in temporal intervals and (2) evolving risk factors. Instead of modeling the temporal data using a time series method, MTL methods are used at multiple time-points of the data. For example, Zhang *et.al.* [12] proposed a MTL model to capture the disease progression pattern at multiple time points for Alzheimer's patients. Wang *et.al.* [13] used longitudinal phenotype markers for Alzheimer's disease (AD) progression prediction. A similar type of multi-task modeling of longitudinal data can be found at [1]. These models are not sutiable for evolving risk factors and multiple outcomes.

Multi-task Learning (MTL) Methods: Assume we have T supervised learning tasks and each task is associated with a predictive model f_t, where f_t is usually expressed as $f_t(\mathbf{x}) = \mathbf{u}_t^T \mathbf{x}$. Here, $\mathbf{u}_t$ is a task-parameter. The MTL framework aims to improve predictive performance of a task by learning multiple related tasks simultaneously. The predictive functions $\{f_t\}_{t=1}^{t=T}$ are learned jointly by minimizing the following regularized empirical risk

$$\{\mathbf{u}_t^\star\}_{t=1}^T = \min_{\mathbf{u}_t} \sum_{t=1}^T \frac{1}{N_t} \sum_{i=1}^{N_t} \left[\mathcal{L}_i(\mathbf{u}_t, \mathbf{x}_t^i, y_t^i) + \lambda \mathcal{R}(\mathbf{u}_t) \right] \tag{1}$$

where $\mathcal{L}_i$ is a loss function and $\mathcal{R}$ is a regularization function on $\mathbf{u}_t$ with a regularization parameter λ. Given the tasks are related, MTL techniques employ various constraints on the task parameters $\{\mathbf{u}_t\}_{t=1}^{t=T}$, e.g. sampling the task-parameters from a shared prior [6], modeling the task-parameters through a common low-dimensional subspace [4,6,7,14], or combining the tasks in proportion to their relatedness learnt using task-to-task covariance matrix [8].

Multi-label Learning (MLL) Methods: MLL models deal with examples with multiple labels. They express a task with multiple labels into multiple independent binary classification problem. Representative methods are as follows: Schapire *et.al.* [15] proposed a boosting technique for MLL problem, Chen *et.al.* [16] presented a semi-supervised MLL model, Zhang *et.al.* [17] extended k-nearest neighborhood method (kNN) to solve MLL problems. The major drawback of these methods is that they do not exploit the correlations amongst the labels. Representative methods which consider correlations amongst labels are as follows: Ghmrawi *et.al.* [18] exploits feature specific pairwise label correlations, Sun *et.al.* [19] presented a hypergraph spectral learning model and Hariharan *et.al.* [20] used a user specified prior matrix that encodes the correlation among the labels. A similar formulation as mentioned in equation (1) can be used for MLL problems. The parameter vector of label ℓ, i.e. $\mathbf{u}_\ell$ can be expressed as $\mathbf{u}_\ell = \mathbf{w}_\ell + \boldsymbol{\theta}^T \mathbf{p}$, where $\mathbf{w}_\ell$ encodes the information of the feature space and $\boldsymbol{\theta}$ is a low-diemensional subspace capturing the correlations amongst the labels. This model originally proposed by Ando $et.al.$ with an iterative solution [10] was later extended by Ji $et.al.$ [11] with a least-square loss function that admitted a closed form solution.

3   Proposed Framework

Assume we are given T supervised learning tasks wherein each task has N_t examples with M labels. Training data of the task t is expressed as $(\mathbf{X}_t, \mathbf{Y}_t) = \left[(\mathbf{x}_t^1, \mathbf{y}_t^1) \dots (\mathbf{x}_t^{N_t}, \mathbf{y}_t^{N_t}) \right]$, where the ith training example is denoted by $\mathbf{x}_t^i \in \mathcal{R}^D$ with labels $\mathbf{y}_t^i \in \{1, -1\}^M$. The ℓth element of label vector $\mathbf{y}_t^i$ i.e. $y_{t,\ell}^i$ is 1 if the ℓth label has been assigned to the example i and -1 otherwise. Overall, $\mathbf{Y}_t \in \mathcal{R}^{M \times N_t}$ has labels of task t. Our focus is on a learning linear predictor $f_{t,\ell}(\mathbf{x}) = \mathbf{u}_{t,\ell}^T \mathbf{x}$ where $\mathbf{u}_{t,\ell}$ is a prediction function of task t with label ℓ. We learn the prediction function $\mathbf{u}_{t,\ell}$ by minimizing following regularized empirical risk function

$$\min_{\mathbf{u}_{t,l}} \sum_{t=1}^{T} \sum_{\ell=1}^{M} \frac{1}{N_t} \sum_{i=1}^{N_t} \left[\mathcal{L}_i(\mathbf{u}_{t,\ell}, \mathbf{x}_t^i, y_{t,\ell}^i) + \lambda \mathcal{R}(\mathbf{u}_{t,\ell}) \right] \tag{2}$$

where $\mathcal{L}_i(\mathbf{u}_{t,l}, \mathbf{x}_{t,\ell}, y_{t,\ell}^i)$ is a loss function and $\mathcal{R}(\mathbf{u}_{t,\ell})$ is a regularization function of $\mathbf{u}_{t,\ell}$ with a penalizing parameter λ.

3.1   Multi-Task Multi-Label (MTML) Formulation

We propose a formulation for the multi-task multi-label problem inspired by the multi-label framework in [11]. We decompose the prediction function $\mathbf{u}_{t,\ell}$ in equation (2) into three components: The first component is derived from the original feature space, the second component learns a subspace shared across tasks and the third component is a shared subspace spanning across labels. We express the prediction function $\mathbf{u}_{t,\ell}$ as

$$\mathbf{u}_{t,\ell} = \mathbf{w}_{t,\ell} + \boldsymbol{\alpha}_\ell^T \mathbf{p}_{t,\ell} + \boldsymbol{\theta}_t^T \mathbf{v}_{t,\ell} \tag{3}$$

where $\mathbf{w}_{t,l} \in \mathcal{R}^D$ encodes the information of the original feature space, $\boldsymbol{\alpha}_\ell \in \mathcal{R}^{D_L \times D}$ with weight vector $\mathbf{p}_{t,\ell} \in \mathcal{R}^{D_L}$ parametrizes the subspace across all tasks of a given label ℓ. $\boldsymbol{\theta}_t \in \mathcal{R}^{D_T \times D}$ with weight vector $\mathbf{v}_{t,\ell} \in \mathcal{R}^{D_T}$ parametrizes the subspace across all labels of a given task t. The dimension of the shared subspaces i.e D_L and D_T are estimated by solving a generalized eigenvalue problem (detailed in Appendix 5). The prediction function $\mathbf{u}_{t,\ell}$ and other task and label related parameters can be obtained from following formulation:

$$\min \sum_{t=1}^{T} \sum_{\ell=1}^{M} \Big[\frac{1}{N_t} \sum_{i=1}^{N_t} \mathcal{L}_i(\mathbf{u}_{t,\ell}, \mathbf{x}_t^i, y_{t,\ell}^i) + \lambda_1 \|\mathbf{w}_{t,\ell}\|^2 + \lambda_2 \|\mathbf{w}_{t,\ell} + \boldsymbol{\alpha}_\ell^T \mathbf{p}_{t,\ell}\|^2 + \lambda_3 \|\mathbf{w}_{t,\ell} + \boldsymbol{\theta}_t^T \mathbf{v}_{t,\ell}\|^2 \Big]$$

$$s.t. \boldsymbol{\alpha}_\ell \boldsymbol{\alpha}_\ell^T = \mathbf{I}, \boldsymbol{\theta}_t \boldsymbol{\theta}_t^T = \mathbf{I}, t, \ell = 1, \dots, T, M \tag{4}$$

where the first regularization component on $\mathbf{w}_{t,\ell}$ with regularization coefficient λ_1 controls the amount of information of task-label pairs (t,ℓ), the second regularization component $\sum_{\ell=1}^{M} \sum_{t=1}^{T} \|\mathbf{w}_{t,\ell} + \boldsymbol{\alpha}_\ell^T \mathbf{p}_{t,\ell}\|^2 = \sum_{\ell=1}^{M} \|\mathbf{W}_\ell - \boldsymbol{\alpha}_\ell^T \mathbf{P}_\ell\|^2$, where $\mathbf{W}_\ell = [\mathbf{w}_{1,\ell}, \dots, \mathbf{w}_{T,\ell}]$ and $\mathbf{P}_\ell = [\mathbf{p}_{1,\ell}, \dots, \mathbf{p}_{T,\ell}]$ controls the amount of information in the shared subspace of tasks of a given label ℓ. Similarly, the third regularization component controls the amount of information in the shared subspace of labels of a given task t. The $\boldsymbol{\alpha}_\ell$ and $\boldsymbol{\theta}_t$ are assumed to be orthogonal to reduce label and task specific redundant information.

3.2 BCD Solution of MTML Formulation

By combining equations (3) and (4), we have

$$\min_{\mathbf{u}_{t,\ell}, \boldsymbol{\alpha}, \mathbf{p}_{t,\ell}, \boldsymbol{\theta}_t, \mathbf{v}_{t,\ell}} \sum_{t=1}^{T} \sum_{\ell=1}^{M} \frac{1}{N_t} \sum_{i=1}^{N_t} \Big[\mathcal{L}_i(\mathbf{u}_{t,\ell}, \mathbf{x}_t^i, y_{t,\ell}^i, c_{t,\ell}) + \mathcal{R}(\mathbf{u}_{t,\ell}, \boldsymbol{\alpha}_\ell, \mathbf{p}_{t,\ell}, \boldsymbol{\theta}_t, \mathbf{v}_{t,\ell}) \Big]$$

$$\tag{5}$$

where the regularization function $\mathcal{R}$ is defined as $\mathcal{R}(\mathbf{u}_{t,\ell}, \boldsymbol{\alpha}_\ell, \mathbf{p}_{t,\ell}, \boldsymbol{\theta}_t, \mathbf{v}_{t,\ell}) = \lambda_1 \|\mathbf{u}_{t,\ell} - \boldsymbol{\alpha}_\ell^T \mathbf{p}_{t,\ell} - \boldsymbol{\theta}_t \mathbf{v}_{t,\ell}\|^2 + \lambda_2 \|\mathbf{u}_{t,\ell} - \boldsymbol{\alpha}_\ell^T \mathbf{p}_{t,\ell}\|^2 + \lambda_3 \|\mathbf{u}_{t,\ell} - \boldsymbol{\theta}_t^T \mathbf{v}_{t,\ell}\|^2$. Considering $\mathcal{L}_i$ to be a hinge loss function, by change of variables, equation (5) can be decomposed into two separate equations as follows

$$\min_{\mathbf{a}_{t,\ell}, \mathbf{v}_{t,\ell}, \boldsymbol{\theta}_t} \sum_{t=1}^{T} \Big[\sum_{\ell=1}^{M} \frac{1}{N_t} \sum_{i=1}^{N_t} \tilde{\epsilon}_{t,i} + \lambda_1 \|\mathbf{a}_{t,\ell} - \boldsymbol{\theta}_t^T \mathbf{v}_{t,\ell}\|^2 + \lambda_2 \|\mathbf{a}_{t,l}\|^2 \Big]$$

$$s.\ t.\ \tilde{\epsilon}_{t,i} \geq 0,\ \tilde{\epsilon}_{t,i} \geq \tilde{r}_i - y_{t,\ell}^i(\mathbf{a}_{t,\ell}^T \mathbf{x}_t^i + c_{t,\ell}),\ \boldsymbol{\theta}_t \boldsymbol{\theta}_t^T = \mathbf{I}, \forall t \tag{6}$$

where $\mathbf{a}_{t,\ell} = \mathbf{u}_{t,\ell} - \boldsymbol{\alpha}_\ell^T \mathbf{p}_{t,\ell}$, $\tilde{r}_i = 1 - y_{t,\ell}^i \mathbf{p}_{t,\ell}^T \boldsymbol{\alpha}_\ell \mathbf{x}_t^i$. The other equation is

$$\min_{\mathbf{b}, \mathbf{p}, \boldsymbol{\alpha}} \sum_{\ell=1}^{M} \Big[\sum_{t=1}^{T} \frac{1}{N_t} \sum_{i=1}^{N_t} \bar{\epsilon}_{t,i} + \lambda_1 \|\mathbf{b}_{t,\ell} - \boldsymbol{\alpha}_\ell^T \mathbf{p}_{t,\ell}\|^2 + \lambda_2 \|\mathbf{b}_{t,\ell}\|^2 \Big]$$

$$\text{subject to } \bar{\epsilon}_{t,i} \geq 0,\ \bar{\epsilon}_{t,i} \geq \bar{r}_i - y_{t,\ell}^i(\mathbf{b}_{t,\ell}^T \mathbf{x}_t^i + c_{t,\ell}),\ \boldsymbol{\alpha}_\ell \boldsymbol{\alpha}_\ell^T = \mathbf{I}. \tag{7}$$

where, $\mathbf{b}_{t,\ell} = \mathbf{u}_{t,\ell} - \boldsymbol{\theta}_t^T \mathbf{v}_{t,\ell}$, and $\bar{r}_i = 1 - y_{t,\ell}^i \mathbf{v}_{t,\ell}^T \boldsymbol{\theta}_t \mathbf{x}_t^i$. As seen from equation (6), the formulation is decoupled over T tasks. where we can run T sub-problems in parallel. Each sub-problem is convex with respect to parameters $(\mathbf{a}_{t,\ell}, \boldsymbol{\theta}_t, \mathbf{v}_{t,\ell})$ respectively. We consider an iterative Block coordinate descent method (BCD) [21] to find an optimal solution for each sub-problem. A concise summary of MTML framework is provided in Algorithm 1 with an analysis of the optimization steps is in Appendix 5.

Algorithm 1. Multi-Task Multi-Label Method

Input:$\mathbf{X}_t$, $\mathbf{Y}_t$, λ_1, λ_2 and λ_3.
Output: $\mathbf{u}_{t,\ell}$, $\forall t, \ell$
do

- Compute ($\boldsymbol{\theta}_t$, $\mathbf{v}_{t,\ell}$, and $\mathbf{a}_{t,\ell}$) for fixed ($\boldsymbol{\alpha}_\ell$, $\mathbf{p}_{t,\ell}$) from equation (6) $\forall t \in \{1 \ldots T\}$.
- Compute $\boldsymbol{\alpha}_\ell$ and $\mathbf{p}_{t,\ell}$, for fixed ($\boldsymbol{\theta}_t$, $\mathbf{v}_{t,\ell}$) from equation (7) $\forall \ell \in \{1, \ldots, M\}$.

untill convergence

- Compute $\mathbf{u}_{t,\ell} = \mathbf{a}_{t,\ell} + \boldsymbol{\alpha}_\ell \mathbf{p}_{t,\ell}$, $\forall t, \ell$.

4 Experiments

We perform several experiments to evaluate the predictive performance of the proposed MTML models and compare them with state-of-the-art baseline methods. For evaluation, we use two real-world hospital cohorts: Cancer and AMI patients, and predict events at hospital emergency.

4.1 Healthcare Datasets

Cancer Electronic Medical Records: This dataset consists of electronic medical records (EMR) of the patients visiting the hospital from 2010-2012 [1]. There are 11 different types of cancers in the dataset, for example, Breast, Skin, Central Nervous System (CNS), Colorectal, Lung etc. The feature set has information about medical conditions relating to the previous visits of each patient including past diagnosis and procedure codes (ICD-10) [2], diagnosis related group codes (DRG), conditions relating to emergency admissions and cancer specific details. 3000 patients are reported in the dataset and the feature length is 531. The whole dataset is divided into non-overlapping segments where each segment is considered a *task*. It contains patient records from the past 3 months, past 3-6 months, past 6-12 months and past 12-24 months respectively. Our focus is on prediction of emergency events for patient where each patient is associated with 6 emergency labels: emergency attendances (E-ATND) and emergency admissions (E-ADM) in the future 3, 6, and 12 months. The dataset is detailed in [22].

Acute Myocardial Infarction (AMI) EMR: AMI is the medical term for heart attack. The AMI electronic medical records (EMR) is recorded over a period of two years. The data contains diverse information such as patient demographics, state of the emergency admissions, personal history of other diseases (*e.g.* nervous and musculoskeletal systems). 2652 patients are reported in the dataset and the feature length is 431. The pre-processing of the dataset and the outcome variables are similar to the Cancer EMR.

---

[1] Ethics approval obtained through University and the Hospital - 12/83.
[2] http://apps.who.int/classifications/icd10/browse/2010/en

4.2   Baselines

We compare the proposed MTML models against following baseline multi-task learning and multi-label learning methods:

Multi-Task Learning (MTL) baselines:

RMFL [9]: Robust multi-task feature learning method learns a task-parameter that has two components. The first component has the common feature set across related tasks, whereas the other component detects outlier tasks. The convex formulation is optimized by the accelerated gradient optimization method (AGM) [23].

MTFL [4]: MTFL model learns a common set of features across tasks by constraining the task-parameters with a mixed ℓ_2/ℓ_1 norm. The formulation is non-convex and translated to a convex optimization problem with an iterative solution.

DM [24]: The formulation of the Dirty model (DM) and RMFL is similar, however DM used a group-sparse ℓ_1/ℓ_∞ norm to learn the common features and a ℓ_1 norm for detecting outlier tasks. The proposed formulation is convex and is solved using AGM [23] method.

Multi-Label Learning (MLL) baselines:

ASOM [10]: The proposed multi-label learning (MLL) framework computes a common subspace that captures the correlations amongst labels. The formulation is non-convex and the optimization technique is iterative.

ML-LS [11]: The proposed framework is similar to the ASO method. However the formulation has a least-square loss function and provides a closed form solution.

M3L [20]: The proposed framework has a max-margin formulation for multi-label classification problem where the formulation has a user specified prior matrix that represents the correlation amongst the labels.

4.3   Experimental Analysis

Performance Evaluation Measures. For multi-task learning (MTL) methods, the performance of the learning system is evaluated by a measure that computed on each test example separately and computes the average value across the test dataset. Examples are *accuracy* (AC) [7] and area under ROC curve (AUC). The evaluation measures specific to multi-label learning (MLL) models are micro-F_1 and macro-F_1[25], the metrics are defined as

$$micro - F_1 = \frac{2 \times Microprecision \times Microrecall}{Microprecision + Microrecall}$$
$$macro - F_1 = \frac{2 \times Macroprecision \times Macrorecall}{M}$$

Table 1

(a) Comparison of MTML methods with baselines for Cancer EMR data. Baseline methods: DM[24], RMTL[9], MTFL [4], ML-LS[11], ASOM [10], M3L [20]. Numbers in parentheses are standard deviation. The best performances are highlighted.

Evaluation Metric	MTL Baselines			MLL Baselines			Proposed
	DM	RMFL	MTFL	ML-LS	ASOM	M3L	MTML
Accuracy	0.522 (0.021)	0.556 (0.028)	0.512 (0.022)	0.679 (0.023)	0.654 (0.023)	0.623 (0.025)	**0.742** (0.023)
AUC	0.609 (0.025)	0.690 (0.015)	0.629 (0.011)	0.766 (0.017)	0.740 (0.018)	0.700 (0.025)	**0.808** (0.015)
micro-F_1	0.659 (0.020)	0.695 (0.019)	0.656 (0.020)	0.781 (0.009)	0.762 (0.012)	0.712 (0.026)	**0.815** (0.012)
macro-F_1	0.643 (0.013)	0.685 (0.012)	0.607 (0.012)	0.691 (0.011)	0.664 (0.010)	0.612 (0.008)	**0.756** (0.009)

(b) Comparison of MTML methods with baselines for AMI EMR data. Baseline methods: DM[24], RMTL[9], MTFL [4], ML-LS[11], ASOM [10], M3L [20]. Numbers in parentheses are standard deviation. The best performances are highlighted.

Evaluation Metric	MTL Baselines			MLL Baselines			Proposed
	DM	RMFL	MTFL	ML-LS	ASOM	M3L	MTML
Accuracy	0.605 (0.010)	0.600 (0.010)	0.650 (0.014)	0.788 (0.012)	0.724 (0.018)	0.712 (0.012)	**0.837** (0.012)
AUC	0.713 (0.020)	0.732 (0.016)	0.712 (0.017)	0.864 (0.018)	0.766 (0.011)	0.788 (0.011)	**0.908** (0.011)
micro-F_1	0.710 (0.015)	0.732 (0.014)	0.714 (0.007)	0.856 (0.010)	0.759 (0.017)	0.786 (0.012)	**0.914** (0.017)
macro-F_1	0.654 (0.023)	0.665 (0.010)	0.650 (0.020)	0.815 (0.017)	0.734 (0.020)	0.776 (0.011)	**0.869** (0.020)

where Microprecision and Microrecall denotes the precision and recall averaged over all example/label pair. Macroprecison and Macrorecall are defined as $Macroprecision = \frac{1}{M}\sum_{\ell=1}^{M}\frac{Tp_\ell}{Tp_\ell+Fp_\ell}$ and $Macrorecall = \frac{1}{M}\sum_{\ell=1}^{M}\frac{Tp_\ell}{Tp_\ell+Fn_\ell}$, where Tp_ℓ, Fp_ℓ and Fn_ℓ are numbre of true positives, false positives and false negatives for label the ℓ.

We use these four measures to evaluate the predictive performance of the proposed MTML models and the baselines. Higher the value of AC, AUC, micro-F_1 and macro-F_1, the *better is* the classifier. As our dataset has T tasks and each task has M labels, we apply the MTL baseline models for each label separately and then average the performance across labels. Similarly, we use MLL baseline models for each task and compute the average performance over all tasks.

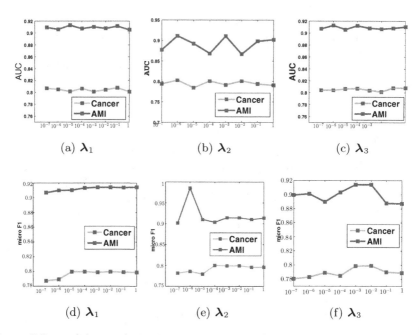

Fig. 1. Effects of the regularization parameters on the MTML model in terms of AUC and micro-F_1. To show the variation w.r.t. λ_1, we set λ_2, λ_3 to 0.0001 and vary λ_1 from 10^{-7} to 1. The resulting graphs for AUC and micro-F_1 are shown in the *first column*. We show similar graphs for λ_2 and λ_3 in the second and third columns respectively.

Experimental Setting and Results. We randomly select 500 examples from the cancer EMR dataset for training and use the rest of the examples for testing. We repeat this experiment over 10 randomly chosen training/test sets and the mean of the performances are reported in Table 1. The parameters of the MTML models are chosen by 5-fold cross-validation.The proposed MTML framework outperform all the baselines. Specifically, for the cancer EMR, the difference in performances of MTML and the closest contending MTL baseline model is 19% on accuracy, 10% on AUC, 10% on micro-F_1 and 7% on macro-F_1 respectively. Similarly, the difference in performance with the closest contending MLL model is 6% on accuracy, 4.2% on AUC, 2.4% on micro-F_1 and 6% on macro-F_1 respectively. Similar improvements exist for the AMI cohort.

4.4 Sensitivity Analysis

Effect of the Regularization Parameters: We randomly sample 20% examples from the Cancer EMR and the rest of the data is used as the test data. Fixing $\lambda_2, \lambda_3 = 10^{-4}$ and varying λ_1 from as small as 10^{-7} to 1 in steps 10^{-1}, we monitor the effect of λ_1 on the classification performance of the MTML model. We adopt a similar strategy to study λ_2 and λ_3. In Figure 1, we present classification

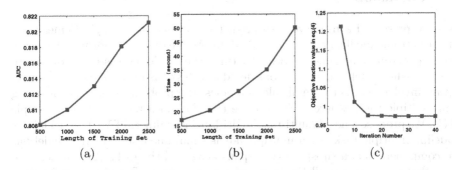

Fig. 2. (a) The effect of training set size: AUC vs length of the training set (b) Computational cost: Time (in second) vs. length of the training set (c) Convergence plot: objective function value in equation (4) vs. number of iterations.

performance of the MTML model with respect to the parameters λ_1, λ_2 and λ_3. The best AUC is obtained for $\lambda_1 = 10^{-4}$, $\lambda_2 = 10^{-2}$ and $\lambda_3 = 10^{-4}$ and the best micro-F_1 is obtained for $\lambda_1 = 10^{-4}$, $\lambda_2 = 10^{-2}$ and $\lambda_3 = 10^{-3}$ respectively. Similar plots are also provided for the AMI data.

Effect of Training Set Size: We vary the length of the training set by randomly sampling 500, 1000, 1500, 2000 and 2500 examples from the Cancer cohort, whereas the rest of the data is used for testing only. Figure 2a presents the classification performance of each training sets. The plot shows the length of the training sets vs. AUC. The performance of MTML improves with increasing size of the training set. This observation is according to our expectation that more training data leads to improved performance.

4.5   Computational and Convergence Analysis

We randomly select training sets with 500, 1000, 1500, 2000 and 2500 examples respectively and the rest of the data is used for testing. From each training dataset, we compute the mean computational time and the number of iterations for convergence required by the MTML model. Figure 2b presents the computational time taken by each training set averged over 10 repreated trials.We notice that the computation time is nearly linear in proportion to the length of the training set. Figure 2c present the convergence plot of the proposed MTML framework. The plot is between the cost value of equation (4) and number of iterations taken by the MTML model when length of the training set is fixed to 1000. The MTML framework (Algorithm 1) is considered to converge when the change in cost function (equation (4)) between two consecutive iterations is less than 10^{-4}. As we see, BCD method for MTML requires only 10 iterations for converging to an optimal point.

5   Conclusion

We have presented a framework to predict the events at a hospital emergency
using electronic medical records (EMR) data. As the data is longitudinal and the
risk of attending emergency varies over time, the whole data is partitioned into
segments where each *segment* is considered as a *task*. As hospital emergency prediction involves predicting multiple outcomes, the problem is posed as learning
from multiple tasks where each example (patient) in a task is having multiple
labels. We propose a novel multi-task-multi-label learning (MTML) model jointly
modeling multiple tasks with multiple labels simultaneously. MTML model has
two components (1) a subspace that spans across all the tasks (2) another subspace that spans across all the labels of a specific task. The objective function
of the proposed model is convex and guarantees an optimal solution. The optimal point is computed using an efficient and scalable Block Coordinate Descent
method. The proposed formulation has been applied on two hospital emergency
cohorts and performs better than many existing baselines. Although we apply
our framework for classification tasks, it can easily be adapted for regression.

  Appendix: **BCD for optimizing equation (6)**: We can write equation (6)
as

$$\min_{\mathbf{a}_\ell, \mathbf{v}_\ell, \boldsymbol{\theta}} \frac{1}{N} \sum_{\ell=1}^{M} \left[\sum_{i=1}^{N_t} \epsilon_i + \lambda_1 \|\mathbf{a}_\ell - \boldsymbol{\theta}^T \mathbf{v}_\ell\|^2 + \lambda_2 \|\mathbf{a}_\ell\|^2 \right],$$
$$\text{s.t. } \epsilon_i \geq 0, \ \epsilon_i \geq r_i - y_\ell^i(\mathbf{a}_\ell^T \mathbf{x}^i + c_\ell), \boldsymbol{\theta}\boldsymbol{\theta}^T = \mathbf{I} \qquad (8)$$

where we dropped subscript t for notational brevity. The three optimization
steps are as following: (1) **Minimization of $\mathbf{v}_\ell$**: fixing $\mathbf{a}_\ell$ and $\boldsymbol{\theta}$, $\mathbf{v}_\ell$ is obtained
by minimizing the formulation as $\min_{\{\mathbf{v}_\ell\}} \sum_{\ell=1}^{M} \|\mathbf{a}_\ell - \boldsymbol{\theta}^T \mathbf{v}_\ell\|^2$, for which, the
optimal solution is given by $\mathbf{v}_\ell^\star = \boldsymbol{\theta}\mathbf{a}_\ell, \ \forall \ell = [1, \ldots, M]$. (2) **Minimization of
$\mathbf{a}_\ell$**: By fixing $(\boldsymbol{\theta}, \mathbf{v}_\ell)$ and the problem in equation (8) becomes

$$\min_{\mathbf{a}_\ell} \sum_{i=1}^{N_t} \epsilon_i + \lambda_1 \|\mathbf{a}_\ell - \boldsymbol{\theta}^T \mathbf{v}_\ell\|^2 + \lambda_2 \|\mathbf{a}_\ell\|^2, \text{s.t. } \epsilon_i \geq 0, \ \epsilon_i \geq r_i - y_\ell^i(\mathbf{a}_\ell^T \mathbf{x}^i + c_\ell), \forall i$$

$$(9)$$

We solve the dual formulation of equation (9) by scalable SVM solving package
LIBSVM [26]. As we can write, $\lambda_1 \|\mathbf{a}_\ell - \boldsymbol{\theta}^T \mathbf{v}_\ell\|^2 + \lambda_2 \|\mathbf{a}_\ell\|^2 = \mathbf{a}_\ell^T [(\lambda_1 + \lambda_2)\mathbf{I} - \lambda_1 \boldsymbol{\theta}^T \boldsymbol{\theta}] \mathbf{a}_\ell$, the dual problem of this formulation becomes

$$\min_{\boldsymbol{\mu}} \boldsymbol{\mu}^T \mathbf{e} - \frac{1}{2} \boldsymbol{\mu}^T \text{Diag}(\mathbf{y}_\ell) \mathbf{Z} \text{Diag}(\mathbf{y}_\ell) \boldsymbol{\mu}, \text{s.t. } 0 \leq \boldsymbol{\mu} \leq r_i, \ \boldsymbol{\mu}^T \mathbf{y}_\ell = 0 \qquad (10)$$

where $\boldsymbol{\mu} \in \mathcal{R}^N$ is a dual variable and $\mathbf{Z} = \left[(\lambda_1 + \lambda_2)\mathbf{I} - \lambda_1 \boldsymbol{\theta}^T \boldsymbol{\theta} \right]$. From the dual
variable $\boldsymbol{\mu}$, we can compute the primal variable $\mathbf{a}_\ell$. For $r_i = 1$, the dual formu-
lation converges to a similar formulation of multi-task learning (MTL) method
as mentioned in [27]. (3) **Minimization of $\boldsymbol{\theta}$**: Fixing $\mathbf{A} = [\mathbf{a}_1, \ldots, \mathbf{a}_M]$ and $\mathbf{v}_\ell$,
the optimization formulation for $\boldsymbol{\theta}$ is as follows:

$$\boldsymbol{\theta} = \min_{\boldsymbol{\theta}} \sum_{\ell=1}^{M} \lambda_1 \|\mathbf{a}_\ell - \boldsymbol{\theta}^T \boldsymbol{\theta} \mathbf{a}_\ell\|_2^2 = \max_{\boldsymbol{\theta}} tr\left(\boldsymbol{\theta} \left[\mathbf{A}\mathbf{A}^T\right] \boldsymbol{\theta}^T\right) \text{ s.t. } \boldsymbol{\theta}\boldsymbol{\theta}^T = \mathbf{I} \quad (11)$$

where the solution of this problem is given by the eigenvalue decomposition of matrix $\mathbf{C}_1$ where $\mathbf{C}_1 = \mathbf{A}\mathbf{A}^T$ and the largest eigenvectors corresponding to the *largest D_T nonzero eigenvalues are optimal solution for $\boldsymbol{\theta}$.*

BCD for optimizing equation (7): We follow a similar procedure in steps (1) and (2) to compute $\mathbf{p}_{t,\ell}$ and $\mathbf{b}_{t,\ell}$ respectively. The transformation matrix $\boldsymbol{\alpha}_\ell$ is computed from the eigenvalue decomposition of matrix $\mathbf{C}_2 = \mathbf{B}_\ell \mathbf{B}_\ell^T$ where $\mathbf{B}_\ell = \left[\mathbf{b}_{1,\ell}, \ldots, \mathbf{b}_{T,\ell}\right]$. *The optimal solution $\boldsymbol{\alpha}_\ell^*$ is given by the largest eigenvectors of $\mathbf{C}_2$ corresponding to the largest D_C nonzero eigenvalues.*

References

1. Zhou, J., Yuan, L., Liu, J., Ye, J.: A multi-task learning formulation for predicting disease progression. In: Proceedings of the 17th ACM SIGKDD International Conference on Knowledge Discovery and Data Mining. ACM, pp. 814–822 (2011)
2. Albert, P.S.: A linear mixed model for predicting a binary event from longitudinal data under random effects misspecification. Statistics in Medicine **31**(2), 143–154 (2012)
3. Caruana, R.: Multitask learning. Machine Learning **28**(1), 41–75 (1997)
4. Argyriou, A., Evgeniou, T., Pontil, M.: Convex multi-task feature learning. Machine Learning **73**(3), 243–272 (2008)
5. Chelba, C., Acero, A.: Adaptation of maximum entropy capitalizer: Little data can help a lot. Computer Speech & Language **20**(4), 382–399 (2006)
6. Rai, P., Daume, H.: Infinite predictor subspace models for multitask learning. In: International Conference on Artificial Intelligence and Statistics, pp. 613–620 (2010)
7. Kang, Z., Grauman, K., Sha, F.: Learning with whom to share in multi-task feature learning. In: Proceedings of the 28th International Conference on Machine Learning, pp. 521–528 (2011)
8. Zhang, Y., Yeung, D.-Y.: A convex formulation for learning task relationships in multi-task learning. In: UAI 2010, pp. 733–442 (2010)
9. Gong, P., Ye, J., Zhang, C.: Robust multi-task feature learning. In: Proceedings of the 18th ACM SIGKDD, pp. 895–903. ACM (2012)
10. Ando, R.K., Zhang, T.: A framework for learning predictive structures from multiple tasks and unlabeled data. The Journal of Machine Learning Research **6**, 1817–1853 (2005)
11. Ji, S., Tang, L., Yu, S., Ye, J.: A shared-subspace learning framework for multi-label classification. ACM Transactions on Knowledge Discovery from Data (TKDD) **4**(2), 8 (2010)
12. Zhang, D., Liu, J., Shen, D.: Temporally-constrained group sparse learning for longitudinal data analysis. In: Ayache, N., Delingette, H., Golland, P., Mori, K. (eds.) MICCAI 2012, Part III. LNCS, vol. 7512, pp. 264–271. Springer, Heidelberg (2012)
13. Wang, H., Nie, F., Huang, H., Yan, J., Kim, S., Risacher, S., Saykin, A., Shen, L.: High-order multi-task feature learning to identify longitudinal phenotypic markers for alzheimer's disease progression prediction. In: Advances in Neural Information Processing Systems, pp. 1277–1285 (2012)

14. Gupta, S., Phung, D., Venkatesh, S.: Factorial multi-task learning: a bayesian non-parametric approach. In: International Conference on Machine Learning (2013)
15. Schapire, R.E., Singer, Y.: Boostexter: A boosting-based system for text categorization. Machine Learning **39**(2–3), 135–168 (2000)
16. Chen, G., Song, Y., Wang, F., Zhang, C.: Semi-supervised multi-label learning by solving a sylvester equation. In: SDM, pp. 410–419. SIAM (2008)
17. Zhang, M.-L., Zhou, Z.-H.: Ml-knn: A lazy learning approach to multi-label learning. Pattern Recognition **40**(7), 2038–2048 (2007)
18. Ghamrawi, N., McCallum, A.: Collective multi-label classification. In: Proceedings of the 14th ACM International Conference on Information and Knowledge Management. ACM, pp. 195–200 (2005)
19. Sun, L., Ji, S., Ye, J.: Hypergraph spectral learning for multi-label classification. In: Proceedings of the 14th ACM SIGKDD International Conference on Knowledge Discovery and Data Mining, pp. 668–676. ACM (2008)
20. Hariharan, B., Zelnik-Manor, L., Varma, M., Vishwanathan, S.: Large scale max-margin multi-label classification with priors. In: Proceedings of the 27th International Conference on Machine Learning (ICML 2010), pp. 423–430 (2010)
21. Xu, Y., Yin, W.: A block coordinate descent method for regularized multiconvex optimization with applications to nonnegative tensor factorization and completion. SIAM Journal on Imaging Sciences **6**(3), 1758–1789 (2013)
22. Gupta, S., et al.: Machine-learning prediction of cancer survival: a retrospective study using electronic administrative records and a cancer registry. BMJ Open 4(3) (2014)
23. Ji, S., Ye, J.: An accelerated gradient method for trace norm minimization. In: Proceedings of the 26th Annual International Conference on Machine Learning, pp. 457–464. ACM (2009)
24. Jalali, A., Sanghavi, S., Ruan, C., Ravikumar, P.K.: A dirty model for multi-task learning. In: Neural Information Processing Systems, pp. 964–972 (2010)
25. Zhang, M., Zhou, Z.: A review on multi-label learning algorithms. IEEE TKDE (2013)
26. Chang, C.-C., Lin, C.-J.: Libsvm: a library for support vector machines. ACM Transactions on Intelligent Systems and Technology (TIST) **2**(3), 27 (2011)
27. Chen, J., Tang, L., Liu, J., Ye, J.: A convex formulation for learning shared structures from multiple tasks. In: Proceedings of the 26th Annual International Conference on Machine Learning, pp. 137–144. ACM (2009)

Nearest Neighbor Method
Based on Local Distribution for Classification

Chengsheng Mao, Bin Hu$^{(\boxtimes)}$, Philip Moore, Yun Su, and Manman Wang

School of Information Science and Engineering, Lanzhou University, Lanzhou, China
{chshmao,ptmbcu}@gmail.com, {bh,suy13,mmwang12}@lzu.edu.cn

Abstract. The k-nearest-neighbor (kNN) algorithm is a simple but effective classification method which predicts the class label of a query sample based on information contained in its neighborhood. Previous versions of kNN usually consider the k nearest neighbors separately by the quantity or distance information. However, the quantity and the isolated distance information may be insufficient for effective classification decision. This paper investigates the kNN method from a perspective of local distribution based on which we propose an improved implementation of kNN. The proposed method performs the classification task by assigning the query sample to the class with the maximum posterior probability which is estimated from the local distribution based on the Bayesian rule. Experiments have been conducted using 15 benchmark datasets and the reported experimental results demonstrate excellent performance and robustness for the proposed method when compared to other state-of-the-art classifiers.

Keywords: Classification · Nearest neighbors · Local distribution · Posterior probability

1 Introduction

In classification problems, to classify an unknown sample, the k-nearest-neighbor (kNN) method searches the training set for the k closest samples, known as its k nearest neighbors, and then classifies the unknown sample based on its k nearest neighbors. One popular kNN method is the well known voting kNN (V-kNN) rule proposed by Cover & Hart [3]; in this method the unknown sample is assigned to the class represented by the majority of its k nearest neighbors in the training set.

Though the traditional V-kNN rule is popular and useful, a refinement to the kNN algorithm is to employ a weighted algorithm in which a weight is applied to each of the k neighbors based on their distance to the query point; a greater weight is given to a closer neighbor; the query sample x is assigned to the class in which the weights of the representatives of the k nearest neighbors sum to the greatest value. Dudani [5] has proposed a distance weighted kNN (DW-kNN) rule by assigning the ith nearest neighbor x_i a distanced-based weight

© Springer International Publishing Switzerland 2015
T. Cao et al. (Eds.): PAKDD 2015, Part I, LNAI 9077, pp. 239–250, 2015.
DOI: 10.1007/978-3-319-18038-0_19

w_i as Equation (1), where d_k and d_1 respectively represent the maximum and minimum distances of the k nearest neighbors to the test sample x.

$$w_i = \begin{cases} \frac{d_k - d_i}{d_k - d_1}, & d_k \neq d_1 \\ 1, & d_k = d_1 \end{cases} \tag{1}$$

Research has identified a number of advantages for kNN algorithm. (1) It is a non-parametric method and does not require a priori knowledge relating to probability distributions for the classification problem; (2) it has been demonstrated that the error rate for kNN approaches is the Bayes error (i.e., theoretically minimum error) when both the number of training samples and the value of k approximate to infinity [4]; (3) it can be implemented conveniently due to its simple algorithm.

Due to these advantages, the kNN method has been successfully applied to real-world applications and becomes one of the most popular algorithms for classification over several decades. It has been the subject of extensive development for use in Machine Learning (ML) and Data Mining (DM) [11,14]. Notwithstanding the inherent simplicity of the kNN rule, it has been shown to be one of the most useful and effective algorithms in DM where it has been considered to be one of the top 10 algorithms [20]. Moreover, the kNN algorithm provides support for classification problems and usually achieves very good performances in various research areas [9,17].

Notwithstanding the positive benefits discussed, the traditional V-kNN and DW-kNN rule is not guaranteed to be the optimal method for implementation using only quantity and distance information contained in the neighborhood; Organizing the information contained in the neighborhood to generate effective and efficient decision rules to improve the classification performance of the kNN method has remained an active research topic for several decades.

Research has investigated the decision rules generated from the k nearest neighbors resulting in a number of improvements to the kNN method. The Categorical Average Pattern (CAP) method proposed by Hotta and Kiyasu [13] uses the categorical k nearest neighbors of query sample to compute the local centers termed the categorical average pattern (CAP) for each class; the unseen query sample is classified to the class with the nearest CAP. The local mean-based nonparametric classifier proposed by Mitani and Hamamoto [18] shares the same idea with CAP, and it has excellent classification performance as compared to other state-of-the-art classifiers. Li [16] demonstrated improvements in the CAP classification method by introducing a notion of local probabilistic centers (LPC) to reduce the number of negative contributing samples. In LPC method, and the query sample is assigned to the class with the nearest LPC.

These methods control an equal neighborhood size for each class, and only take the distances to neighborhood into account. Though the distance to the neighborhood center can partly reflect the distribution of the corresponding class around the query sample, it would be arguably better if the distribution of the neighborhood were taken into account. Moreover, it would reduce the negative influences of noisy samples if the k nearest neighbors were considered integrally instead of individually as is the case for the voting kNN or the weighted kNN. In this paper we consider a query sample to be closely related to the distribution of its nearest neighbors and therefore analyze classification problems from the perspective of local distribution. Based on this approach, we propose a comprehensive kNN method based on the local distribution termed the Local Distribution based kNN (LD-kNN). The LD-kNN method estimates the local distribution of each class around the query sample to achieve the probability of the query sample belonging to each class and the query sample is assigned to the class with the greatest posterior probability.

2 LD-kNN Classification

As a kNN-type method, LD-kNN also performs the classification based on the neighborhood of the query sample. In LD-kNN, the local distribution is estimated from the neighborhood for each class and the classification tasks are performed by maximizing the posterior probability of each class based on the local distribution.

2.1 LD-kNN Formulation

Let X be the event that a data sample x is equal to the specified sample X described by measurements made on a set of attributes, i.e. $X : x = X$. Let C be the hypothesis that a data sample x belongs to the specified class C, i.e. $C : x \in C$. For classification problems, the purpose is to determine $P(x \in C | x = X)$ (abbreviated as $P(C|X)$), the probability that C holds given the event X. In other words, we are looking for the probability that sample X belongs to class C, given that we know the attribute description of X. $P(C|X)$ is the posterior probability of C conditioned on X, it should be maximized with respect to the class for the class label of the sample X (denoted by ω) as Equation (2).

$$\omega = \arg \max_C P(C|X) \tag{2}$$

LD-kNN method has been conceived to maximize the posterior probability of each class conditioned a query sample in local area. Let $\delta(X)$ denote the neighborhood of sample X and let $\delta(X)$ be the event that a sample x is in the neighborhood of X, i.e. $\delta(X) : x \in \delta(X)$. Since $\delta(X)$ is the neighborhood of sample X, we derive $X \in \delta(X)$, then $P(X, \delta(X)) = P(X)$. Through the theory of conditional probability, we can get

$$P(C|(X, \delta(X))) = P(C|X), \tag{3}$$

where we call $P(C|(X, \delta(X)))$ local posterior probability (LPP) of class C conditioned on X. That is to say, the probability of sample X belonging to class C is equal to the LPP of class C conditioned on X.

On the other hand, By Bayes' theorem [8], $P(C|X)$ can be computed as Equation (4), where $P(C)$ and $P(X)$ are the respective prior probabilities of C and X, and $P(X|C)$ is the posterior probability of X conditioned on C.

$$P(C|X) = \frac{P(X|C)P(C)}{P(X)} \qquad (4)$$

Equation (4) can be extended to local conditions, where each item should be estimated under the local condition $\delta(X)$. Then we get the Bayesian formula under local condition as Equation (5).

$$P(C|(X, \delta(X))) = \frac{P(X|(C, \delta(X)))P(C|\delta(X))}{P(X|\delta(X))} \qquad (5)$$

The Bayesian classifier [10] maximizes $P(C|X)$ according to formula (4) and estimates it in the whole dataset. While our method maximizes $P(C|X)$ according to formula (3) and (5), we estimate it in a local area around the query sample. Under the assumption that the near neighbors can represent the property of a query sample better than the more distant samples, estimating the LPP by formula (5) represents more reasonable than by formula (4).

To maximize $P(C|(X, \delta(X)))$ according to formula (5): as $P(X|\delta(X))$ is constant for all classes, only $P(X|(C, \delta(X)))P(C|\delta(X))$ needs to be maximized. Then, the optimization problem can be transformed to

$$\omega = \arg \max_{C} P(X|(C, \delta(X)))P(C|\delta(X)). \qquad (6)$$

2.2   Local Distribution Estimation

Given an arbitrary query sample X and a distance metric, its k nearest neighbors can be obtained from the training set. In this paper, we call the set of the k nearest neighbors k-neighborhood of sample X and denote it by $\delta_k(X)$. To solve the optimization problem (6), the two items $P(X|(C, \delta(X)))$ and $P(C|\delta(X))$ which are relevant to the local distribution of class C should be estimated based on $\delta_k(X)$ for each class.

$P(C|\delta(X))$ derives the probability of a sample belonging to class C given that the sample is in the neighborhood of X. If there are N_j samples from class C_j in the k-neighborhood of X, then $P(C|\delta(X))$ can be estimated by

$$P(C_j|\delta_k(X)) = N_j/k. \qquad (7)$$

$P(X|(C, \delta(X)))$ derives the probability of a sample being equal to X given that the sample is from class C and is in the neighborhood of X; this can be regarded as the local probability distribution density of class C at point X for continuous attributes.

To estimate $P(X|(C, \delta(X)))$ accurately we just consider the continuous attributes in our method; the estimation of $P(X|(C, \delta(X)))$ becomes a problem of probability density estimation in local area. In our method, we assume that the samples in the neighborhood follow a Gaussian distribution with a mean μ and covariance matrix Σ defined by Equation (8).

$$f(X; \mu, \Sigma) = \frac{1}{\sqrt{(2\pi)^d |\Sigma|}} e^{-0.5(X-\mu)^T \Sigma^{-1}(X-\mu)} \tag{8}$$

where d is the dimension of the data. So that, for $\delta_k(X)$ and a specified class C_j, we have

$$P(X|(C_j, \delta_k(X))) \propto f(X; \mu_{C_j}, \Sigma_{C_j}) \tag{9}$$

where μ_{C_j} and Σ_{C_j} respectively represent the mean and the covariance matrix of class C_j in $\delta_k(X)$.

Then we need to estimate the mean μ and the covariance matrix Σ from $\delta_k(X)$ for each class. In our approach, to ensure the covariance matrix is positive definite, we take the naive assumption of local class conditional independence that an attribute on each class does not correlate with the other attributes in local area; that is, the covariance matrix (Σ) would be a diagonal matrix. If there are N_j samples from class C_j in $\delta_k(X)$, denoted by $X_i^{C_j}(i = 1, \cdots, N_j)$, the two parameters the mean (μ_{C_j}) and the covariance matrix (Σ_{C_j}) can be estimated through maximum likelihood estimation by the following Formulae (10) and (11) [15].

$$\hat{\mu_{C_j}} = \frac{1}{N_j} \sum_{i=1}^{N_j} X_i^{C_j} \tag{10}$$

$$\hat{\Sigma_{C_j}} = diag(\frac{1}{N_j} \sum_{i=1}^{N_j} (X_i^{C_j} - \hat{\mu_{C_j}})(X_i^{C_j} - \hat{\mu_{C_j}})^T) \tag{11}$$

where $diag(\cdot)$ converts a square matrix to a diagonal matrix with the same diagonal elements.

Then, we plug the mean (μ) and covariance matrix (Σ) respectively estimated from Formulae (10) and (11) into Equation (8) to estimate $f(X; \mu_{C_j}, \Sigma_{C_j})$ and then estimate $P(X|(C_j, \delta_k(X)))$ from Formula (9).

2.3   Classification Rules

As k is constant for all classes, according to Formulae (7) and (9), the classification problem as defined in (6) can be transformed into an optimization problem finally formulated as shown in Formula (12).

$$\omega = \arg \max_{j=1, \cdots, N_C} \{N_j \cdot f(X; \mu_{C_j}, \Sigma_{C_j})\} \tag{12}$$

where N_C is the total number of classes, $f(\cdot)$, μ_{C_j} and Σ_{C_j} is denoted by Formulae (8), (10) and (11) respectively.

According to the aforementioned process, the LD-kNN approach classifies a query sample by the LPP estimated from local distribution. This is calculated according to the Bayesian Theorem in the local area. The query sample is then labeled with the class having a maximum LPP.

2.4   Related Methods

The traditional V-kNN classified the query sample only by the number of near-est neighbors for each class in the k-neighborhood (i.e. N_j for the jth class). Compared with the V-kNN rule, LD-kNN takes into account the local probabil-ity density around the query sample ($f(X; \mu_{C_j}, \Sigma_{C_j})$) besides the number ($N_j$). For different classes, the local probability densities are not always the same and may play a significant role for classification.

Another classification method related with LD-kNN is the Bayesian classifi-cation method. Bayesian classifier assigns the query sample to the class with the highest posterior probability, which is estimated through the global distribution. While LD-kNN estimates the posterior probability through the local distribu-tion around the query sample. Naive Bayesian Classification (NBC) method can be considered as a special case of LD-kNN with k approaching the size of the dataset. Thus, LD-kNN would be more effective and comprehensive for a special query sample.

In actuality, the LD-kNN method may be viewed as a compromise between the nearest neighbor rule and the Bayesian method. The parameter k denotes the locality in LD-kNN; when parameter k is close to 1, LD-kNN approaches the nearest neighbor rule. And when k is large and equal to the size of the dataset, the local area is extended to the whole dataset; in this case LD-kNN becomes a Bayesian classifier. Thus, LD-kNN may combine the advantages of the two classifiers and become a more effective and comprehensive classification method.

As for CAP and LPC, they consider an equal number of nearest neighbors for each class and the classification is based on the nearest center. As presented in Equation (12), CAP and LPC use a constant N_j for all classes and the other item ($f(X; \mu_{C_j}, \Sigma_{C_j})$) is estimated only from the center of the N_j samples in each class. Thus, CAP and LPC can be viewed as special cases of LD-kNN.

3   Experiments

3.1   The Datasets

In our experimentation we have selected 15 real datasets from the well-known UCI-Irvine repository of machine learning datasets [1]. The selected datasets include six two-class problems and nine multi-class problems, and vary in terms of their domain, size, and complexity. The estimation of probability density is only for continuous attributes and we only take into account continuous attributes in our experiments. Table 1 summarizes the relevant information for these datasets; for more information, please turn to http://archive.ics.uci.edu/ml.

3.2   Experimental Settings

Before classification, to prevent attributes with an initially large range from inducing bias by out-weighing attributes with initially smaller ranges, we use z-score normalization to linearly transform each of the numeric attributes of a

Table 1. Some Information about the datasets

datasets	#Instances	#Attributes	#Classes
Abalone	4177	7	3
Australian	690	6	2
Breast	106	9	6
Bupa Liver	345	6	2
Dermatology	366	33	6
Glass	214	9	6
ILPD	583	9	2
Iris	150	4	3
Letters	20000	16	26
Pageblock	5473	10	5
Sonar	208	60	2
Spambase	4601	57	2
spectf	267	44	2
Vehicle	846	18	4
Wine	178	13	3

dataset with mean value 0 and standard deviation 1 by $v' = \frac{v - \mu_A}{\sigma_A}$, where μ_A and σ_A are the mean and standard deviation, respectively, of attribute A.

In order to achieve an impartial evaluation, we have employed six competing classifiers to test the performance of alternative approaches and to provide a comparative analysis to evaluate the effectiveness of our LD-kNN algorithm. These competing classifiers include base classifiers (e.g. V-kNN, DW-kNN [5] and NBC), and the state-of-the-art classifiers (e.g. CAP [13], LPC [16] and SVM [2]).

For kNN-type classifiers, we use Euclidean distance to measure the distance between two samples in search of the nearest neighbors. In addition, the parameter k in kNN-type classifiers indicates the number of nearest neighbors, we use the average number of nearest neighbors per class (denoted by kpc) to indicate the neighborhood size, i.e. $kpc * N_C$ nearest neighbors are searched, where N_C is the number of classes.

To express the generalization capacity, i.e. the classification ability of a classifier classifying previously unseen samples, the training samples and the test samples should be independent. In our research we use stratified 5-fold cross validation to estimate the misclassification rate of a classifier on each dataset. The data are stratified into 5 folds. For the 5 folds, 4 folds constitute the training set with the remaining fold being used as the test set. The training and test sessions are performed 5 times with each session using a different test set and the corresponding training set. To avoid bias, the 5-fold cross validation process is applied to each dataset 10 times and the average misclassification rate (AMR) is calculated to evaluate the performance of the classifier.

Table 2. The AMR (%) of the seven methods with corresponding stds on the 15 UCI datasets (the best recognition performance is described in bold-face on each data set)

datasets	LD-kNN	V-kNN	DW-kNN	CAP	LPC	SVM	NBC
Abalone	35.26±0.23	34.92±0.20	34.95±0.28	35.52±0.26	35.54±0.43	**34.52±0.08**	41.55±0.09
Australian	24.49±0.83	24.57±0.47	24.55±0.59	25.00±0.80	25.12±0.65	**24.22±0.34**	27.87±0.46
Breast	30.19±2.63	33.77±1.45	31.89±1.62	30.38±0.57	**30.00±1.18**	41.13±2.00	35.38±1.21
Bupaliver	31.86±1.36	34.43±1.28	34.00±1.37	32.09±1.74	32.93±1.22	**29.83±1.02**	48.06±0.71
Dermatology	**1.75±0.44**	3.93±0.28	3.83±0.24	2.81±0.41	2.84±0.22	2.70±0.35	7.46±0.59
Glass	**26.73±1.47**	31.26±1.50	28.83±1.82	28.27±1.75	29.72±1.11	30.61±1.06	59.67±2.73
ILPD	29.97±1.34	**28.54±0.61**	28.99±0.83	30.67±1.36	30.81±1.25	29.11±0.54	44.80±0.49
Iris	**3.67±0.33**	3.87±0.65	3.80±0.60	3.73±0.68	3.73±0.68	4.07±0.31	4.33±0.58
Letter	5.08±0.09	8.02±0.08	5.29±0.07	**3.93±0.05**	4.14±0.10	5.54±0.08	35.75±0.07
Pageblock	3.27±0.10	3.30±0.06	3.19±0.06	**3.16±0.13**	3.17±0.11	3.99±0.09	13.12±1.39
Sonar	**11.30±1.40**	14.47±1.42	14.33±1.23	11.68±1.96	14.47±1.42	17.45±1.81	31.11±1.01
Spambase	7.77±0.17	8.60±0.10	7.85±0.20	7.97±0.21	8.13±0.23	**6.80± 0.12**	18.24±0.16
spectf	20.00±1.56	**19.40±0.71**	20.60±0.92	20.22±0.75	20.45±1.24	21.09±0.71	32.62±1.15
Vehicle	24.04±1.05	28.53±0.86	27.86±0.83	**23.96±1.20**	24.36±0.73	24.20±0.82	53.95±0.74
Wine	**0.84±0.38**	2.30±0.69	2.13±0.42	1.85±0.67	1.46±0.45	1.97±0.55	2.58±0.45
Average AMR	**17.08**	18.66	18.14	17.42	17.79	18.48	30.43
Average Rank	**2.13**	4.63	3.80	2.90	3.80	3.80	6.93

4   Results and Discussion

The parameter kpc is an important factor that can affect the performance of LD-kNN. If kpc is too small, the estimation of the local distribution may be unstable; however, if it is too large, there will be many distant neighbors that may have an adverse effect on the local distribution estimation. To investigate the influence of the parameter kpc on classification results for kNN-type classifiers, we tune the parameter kpc as an integer in the range 1 to 30 for each dataset, perform the classification tasks and achieve the corresponding AMR for each kpc value. This procedure will guide us in the selection of parameter kpc for classification. Fig. 1 shows the performance curves with respect to kpc of the five kNN-type methods on several real datasets. Because different real datasets usually have different distributions, the curves of AMR with respect to the kpc for LD-kNN are usually different. These performance curves show that, on average the LD-kNN method can be quite effective for these real problems, and validate that a modest kpc for LD-kNN can usually achieve a more effective performance.

We use the lowest AMR with the corresponding kpc ranging from 1 to 30 to evaluate the performance of a kNN-type classifier. Then, following experimental testing we obtained a comparative performance for our posited approach when compared with the alternative approaches. The classification results on each dataset for all the classifiers are shown in Table 2 in terms of AMR with the corresponding standard deviations (stds).

From the results in Table 2 we can see that LD-kNN offers the best performance on 5 datasets, more than all other classifiers; this is an improvement over the alternative classifiers. The overall average AMR and rank of LD-kNN on these datasets are 17.08% and 2.13 respectively, lower than all other classi-

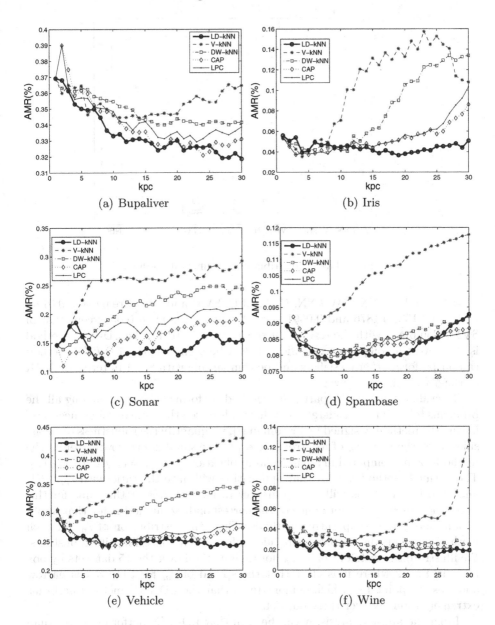

Fig. 1. The performance curves with respect to *kpc* on different real datasets

fiers, which means that the proposed LD-kNN may be more effective than other classifiers for these datasets.

To evaluate the statistical significance of the difference between LD-kNN and each other classifiers, we have performed a Wilcoxon signed rank test [12] between LD-kNN and each other classifiers. The p-values of the tests between

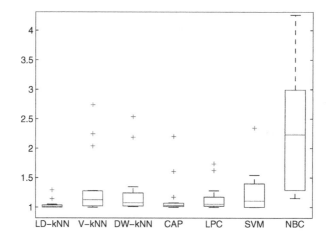

Fig. 2. The r_m distributions of different methods

LD-kNN and V-kNN, DW-kNN, CAP, LPC, SVM and NBC are 0.0103, 0.0125, 0.0181, 0.0103, 0.1876 and 0.0001 respectively, all less than 0.05 except that of SVM. Combined with the result that the average AMR for the LD-kNN method is the lowest among these classifiers, it can be seen that the LD-kNN method can outperform other classifiers and be comparable with SVM in terms of AMR at the 5% significance level.

To evaluate how well a particular method performs on average among all the problems taken into consideration we have addressed the issue of robustness. Following the method designed by Friedman [7], we quantify the robustness of a classifier m by the ratio r_m of its error rate e_m to the smallest error rate over all the methods being compared in a particular application (i.e. $r_m = e_m/\min_{1 \leq k \leq 7} e_k$). The optimal method m^* for that application will have the ratio with $r_{m*} = 1$, and all other methods will have a greater ratio. The greater the value for this ratio, the worse the performance of the corresponding method is for that application among the comparative methods. Thus, the distribution of r_m for each method, over all the datasets, provides information concerning its robustness. We illustrate the distribution of r_m for each method over the 15 datasets by box plots in Fig. 2 where it is clear that the spread of r_m for LD-kNN is narrow and close to point 1.0, which demonstrates that the LD-kNN method performs extremely robustly over these datasets.

From the above analysis, it can be seen that LD-kNN performs better than other classifiers in respect of the overall AMR. In considering the kNN-type classifiers, the DW-kNN improves the performance over the traditional V-kNN by weighting; the CAP and the LPC has improved the kNN method by local centering. The LD-kNN is a more comprehensive method and considers the nearest neighbor set integrally by local distribution; thus it is reasonable to conclude that among the kNN-type classifiers the LD-kNN performs best followed by CAP, LPC, DW-kNN and V-kNN.

The SVM, as an advanced and highly respected algorithm, can also achieve a comparable performance with LD-kNN for certain classification problems; however the performance of the LD-kNN is more robust to application than SVM; that is, SVM may perform effectively on certain datasets however it also performs badly on other datasets and is not as stable as the LD-kNN on the experimental datasets. NBC performs badly in the experimental classification tasks principally due to the fact that the class conditional independence assumption is too severe in practical problems.

The LD-kNN can be viewed in terms of a Bayesian classification method as it is predicated on the Bayes theorem. Since the classification is based on maximum posterior probability, the LD-kNN classifier can in theory achieve the Bayes error rate. Additionally, As a kNN-type classifier, LD-kNN can inherit the advantages of kNN method. Thus, it may be intuitively anticipated that LD-kNN can perform much more effectively than NBC and other kNN-type classifiers in most cases.

5 Conclusion

We have introduced the concept of local distribution to the kNN methods for classification. The proposed LD-kNN method essentially considers the k nearest neighbors of the query sample as several integral sets by the class labels and then estimates the local distribution of these integral sets to achieve the LPP for each class; then the query sample is classified based on the maximum LPP. This approach provides a simple mechanism for quantifying the probability of the query sample attached to each class and has been shown to present several advantages. The experimental results demonstrate the effectiveness and robustness for LD-kNN and show its potential superiority.

In the proposed method, a significant step is the estimation of local distribution. In our experiments, we assume that the local probability distributions of the instances for each class can be modeled as a Gaussian distribution. However, the Gaussian distribution assumption may not be always appropriate for all practical problems; there are other probability distribution estimation methods available, such as Gaussian mixture model [19] and kernel density estimation [6]. Different local distribution estimation methods for LD-kNN may produce different results. For a particular classification problem in a specific domain of interest various methods may be tested to achieve good results; this represents a future direction for our research.

References

1. Bache, K., Lichman, M.: UCI machine learning repository (2013). http://archive.ics.uci.edu/ml
2. Chang, C.C., Lin, C.J.: LIBSVM: A library for support vector machines. ACM Transactions on Intelligent Systems and Technology **2**, 27:1–27:27 (2011). http://www.csie.ntu.edu.tw/cjlin/libsvm

3. Cover, T., Hart, P.: Nearest neighbor pattern classification. IEEE Transactions on Information Theory **13**(1), 21–27 (1967)
4. Duda, R.O., Hart, P.E., Stork, D.G.: Pattern classification. John Wiley & Sons (2012)
5. Dudani, S.: The distance-weighted k-nearest-neighbor rule. IEEE Transactions on Systems, Man and Cybernetics **4**, 325–327 (1976)
6. Duong, T.: ks: Kernel density estimation and kernel discriminant analysis for multivariate data in r. Journal of Statistical Software **21**(7), 1–16 (2007)
7. Friedman, J., et al.: Flexible metric nearest neighbor classification. Unpublished manuscript available by anonymous FTP from playfair. stanford. edu (see pub/friedman/README) (1994)
8. Gelman, A., Carlin, J.B., Stern, H.S., Dunson, D.B., Vehtari, A., Rubin, D.B.: Bayesian data analysis. CRC Press (2013)
9. Govindarajan, M., Chandrasekaran, R.: Evaluation of k-nearest neighbor classifier performance for direct marketing. Expert Systems with Applications **37**(1), 253–258 (2010)
10. Han, J., Kamber, M., Pei, J.: Data mining: concepts and techniques. Morgan kaufmann (2006)
11. Hand, D., Mannila, H., Smyth, P.: Principles of data mining. MIT Press (2001)
12. Hollander, M., Wolfe, D.A.: Nonparametric statistical methods. John Wiley & Sons, NY (1999)
13. Hotta, S., Kiyasu, S., Miyahara, S.: Pattern recognition using average patterns of categorical k-nearest neighbors. In: Proceedings of the 17th International Conference on Pattern Recognition, ICPR 2004, vol. 4, pp. 412–415. IEEE (2004)
14. Kononenko, I., Kukar, M.: Machine learning and data mining. Elsevier (2007)
15. Lehmann, E.L., Casella, G.: Theory of point estimation, vol. 31. Springer (1998)
16. Li, B., Chen, Y., Chen, Y.: The nearest neighbor algorithm of local probability centers. IEEE Transactions on Systems, Man, and Cybernetics, Part B: Cybernetics **38**(1), 141–154 (2008)
17. Magnussen, S., McRoberts, R.E., Tomppo, E.O.: Model-based mean square error estimators for k-nearest neighbour predictions and applications using remotely sensed data for forest inventories. Remote Sensing of Environment **113**(3), 476–488 (2009)
18. Mitani, Y., Hamamoto, Y.: A local mean-based nonparametric classifier. Pattern Recognition Letters **27**(10), 1151–1159 (2006)
19. Reynolds, D.: Gaussian mixture models. In: Encyclopedia of Biometrics, pp. 659–663 (2009)
20. Wu, X., Kumar, V., Quinlan, J.R., Ghosh, J., Yang, Q., Motoda, H., McLachlan, G.J., Ng, A., Liu, B., Philip, S.Y., et al.: Top 10 algorithms in data mining. Knowledge and Information Systems **14**(1), 1–37 (2008)

Immune Centroids Over-Sampling Method for Multi-Class Classification

Xusheng Ai[1], Jian Wu[1](✉), Victor S. Sheng[2],
Pengpeng Zhao[1], Yufeng Yao[1], and Zhiming Cui[1]

[1] The Institute of Intelligent Information Processing and Application,
Soochow University, Suzhou 215006, China
jianwu@suda.edu.cn
[2] Department of Computer Science,
University of Central Arkansas, Conway 72035, USA

Abstract. To improve the classification performance of imbalanced learning, a novel over-sampling method, Global Immune Centroids Over-Sampling (Global-IC) based on an immune network, is proposed. Global-IC generates a set of representative immune centroids to broaden the decision regions of small class spaces. The representative immune centroids are regarded as synthetic examples in order to resolve the imbalance problem. We utilize an artificial immune network to generate synthetic examples on clusters with high data densities. This approach addresses the problem of synthetic minority oversampling techniques, which lacks of the reflection on groups of training examples. Our comprehensive experimental results show that Global-IC can achieve better performance than renowned multi-class resampling methods.

Keywords: Resampling · Immune network · Over-sampling · Imbalanced learning · Synthetic examples

1 Introduction

The class imbalance problem typically occurs when there are many more instances belonging to some classes than others in multi-class classification. Recently, reports from both academy and industry indicate that the imbalanced class distribution of a data set has posed a serious difficulty to most classification algorithms which assume a relatively balanced distribution. Furthermore, identifying rare objects is of crucial importance. In many real-world applications, the classification performances on the small classes are the major concerns in determining the property of a classification model.

In the research community of imbalanced learning, almost all reported solutions are designed for binary classification. However, multi-class imbalanced learning problems appear frequently. Identifying the concept for each class in these problems is usually equally important. When multiple classes are present in an application domain, solutions proposed for binary classification problems may

T. Cao et al. (Eds.): PAKDD 2015, Part I, LNAI 9077, pp. 251–263, 2015.
DOI: 10.1007/978-3-319-18038-0_20

not be directly applicable, or may achieve a lower performance than expected. For example, solutions at the data level suffer from the increased search space, and solutions at the algorithm level become more complicated, since they must consider small classes and it is difficult to learn the corresponding concepts for these small classes. Additionally, learning from multiple classes itself implies a difficulty, since the boundaries among the classes may overlap. The overlap would downgrade the learning performance.

There exist many researches on multi-class imbalance learning. However, most ex-isting researches transfer multi-class imbalance learning into binary using different class decomposition schemes and apply existing binary imbalance learning solutions. These decomposition approaches help reuse the existing binary imbalance learning solutions. However, they have their own shortcomings, which will be discussed in the next section related work. To overcome these shortcomings, in this paper we present a novel global multi-class imbalance learning approach, which does not need to transfer multi-class into binary. This novel approach is based on immune network theory, and utilizes an aiNet model [3] to generate immune centroids for the clusters of each small class, which have high data density, called global immune centroids over-sampling (denoted as Global-IC). Specifically, our novel approach Global-IC resamples each small class by introducing immune centroids of the clusters of the examples belonging to the small class. Our experimental results show that Global-IC achieves better performance, comparing with existing methods.

The rest of this paper is organized as follows. We review related work in Section 2. Section 3 presents our proposed over-sampling method Global-IC. Our experimental results and comparisons are shown in Section 4. Finally, we conclude this paper in Section 5.

2   Related Work

As we said before, most existing solutions for multi-class imbalance classification problems use different class decomposition schemes to convert a multi-class classifi-cation problem into multiple binary classification problems, and then apply binary imbalance techniques on each binary classification problem. For example, Tan et al. [4] used both one-vs-all (OVA) [2] and one-vs-one (OVO) [1] schemes to break down a multi-class problem to binary problems, and then built rule-based learners to im-prove the coverage of minority class examples. Zhao [20] used OVA to convert a multi-class problem into multiple binary problems, and then used under-sampling and SMOTE [5] techniques to overcome the imbalance issues. Liao [6] investigated a variety of over-sampling and under-sampling techniques with OVA for a weld flaw classification problem. Chen et al. [7] proposed an approach that used OVA to convert a multi-class classifica-tion problem to binary problems and then applied some advanced resampling methods to rebalance the data of each binary problem. All these methods are based on multi-class decomposition. Multi-class decomposition oversimplifies the original multi-class problem. It is obvious that each individual classifier learned for a binary sub-problem couldnt be trained with the full information of the original data. This can cause classification ambiguity or uncovered data regions with respect to each type of decomposition.

Different from the previous discussion, a global cost-sensitive algorithm (called Global-GS) was proposed [8], which re-weights the instances from each class accord-ing to their ratio without using class decomposition. In order to equi-librate the signi-ficance of the examples for different classes in an imbalanced framework, it resam-ples each class in a consistent manner by considering a fac-tor of N_i/N_{max}, where N_i the number of the examples of the ith class and N_{max} is the number of the examples for the majority class of the problem. Navarro et al. [9] presented a preprocessing mechanism based on SMOTE, which iteratively generates new synthetic samples from the least represented class at each step, known as Static-SMOTE. The synthetic examples are obtained by applying the SMOTE algorithm [5] only over the instances of the minority classes. Wang et al. [10] developed a study regarding the extension of boosting techniques for imbalance problems with "multi-minority" and "multi-majority" classes, called AdaBoost.NC. Their approach is based on AdaBoost [22], combining with neg-ative correlation learning. The initial weights of the examples in this boosting approach are assigned in inverse proportion to the number of instances in the corresponding class. Our novel approach Global-IC is closely related to these methods. However, it introduces a complete new approach, immune centroid gen-eration, to oversample the examples of the small classes for multi-class imbalance learning.

3   Global Immune Centroids Over-Sampling

Before we introduce our solution Global-IC, we first briefly introduce the basic con-cepts and knowledge of immune systems. After that, we present the details of Global-IC.

3.1   Immune Systems

Before discussing our method, we sketch a few aspects of the human adaptive im-mune system. The immune systems guard our bodies against infections due to the attacks of antigens. The surface receptors on B-cells (one kind of lym-phocyte) are able to recognize to specific antigens. The response of a receptor to an antigen can activate its hosting B-cell. Activated B-cell then proliferates and differentiates into memory cells. Memory cells secret antibodies to neutralize the pathogens through complementary pattern matching. During the proliferation of the activated B-cells, a mutation mechanism is employed to create diverse antibodies by altering the gene segments. Some of the mutants may be a better match for the corresponding antigen. In order to be protective, the immune sys-tem must learn to distinguish between our own (self) cells and malefic external (nonself) invaders. This process is called self/nonself discrimination: those cells recognized as self dont promote an immune response. The system is said to be tolerant to them, while those that are not provoke a reaction resulting in their elimination.

Immune network theory, originally proposed in [11], hypothesizes a novel viewpoint of lymphocyte activities, natural antibody production, pre-immune repertoire selection, tolerance and self/nonself discrimination, memory and the evolution of an immune system. It was suggested that the immune system is composed of a regulated network of cells and molecules that recognize one another. The immune cells can respond either positively or negatively to the recognition signal (antigen or other immune cell or molecule). A positive response would result into cell proliferation, cell activation and antibody secretion, while a negative response would lead to tolerance and suppression.

Learning in the immune system involves raising the population size and affinity of those lymphocytes that have proven themselves valuable by having recognized any antigen. Burnet [12] introduced clonal selection theory by modifying N.K. Jerne's theory. The theory states that in a pre-existing group of lymphocytes (specifically B cells), a specific antigen only activates (i.e. selection) its counter-specific cell so that a particular cell is induced to multiply (producing its clones) for antibody production. With repeated exposures to the same antigen, the immune system produces antibodies of successively greater affinities. A secondary response elicits antibodies with greater affinity than in a primary response. Based on the clonal selection principle, Castro proposed a computational implementation of the clonal selection principle that explicitly takes into account the affinity maturation of the immune response. He also defined aiNet (an artificial immune network model) for data analysis [3]. The aiNet is an edge-weighted graph, not necessarily fully connected, composed of a set of nodes, called antibodies, and sets of node pairs called edges with an assigned number called weight or connection strength, associated with each connected edge. The aiNet clusters serve as internal images (mirrors) responsible for mapping existing clusters in the data set into network clusters. These clusters map those of the original data set. The shape of the spatial distribution of antibodies follows that of the antigenic spatial distribution.

3.2   Immune Centroids Resampling

In order to directly handle the imbalance problem of multi-class classification, our Global-IC takes two separate major steps. Its first step resamples the examples of each class. Each class is resampled in a consistent manner by considering a factor of $(N_{max} - N_i)$, where N_i is the number of the examples of the ith class and N_{max} is the number of the examples for the majority class of the problem. The second step of Global-IC generates the synthetic examples for the small classes. The synthetic ex-amples are generated based on our proposed immune centroids over-sampling tech-nique (ICOTE), which is the core of our Global-IC. Briefly, the synthetic examples are derived from the immune network and represent the internal images of original small class examples. The details of ICOTE will be discussed later. The pseudo code of our Global-IC algorithm is shown in Algorithm 1. Note that the size of the majority class will not be increased.

Algorithm 1. Global-IC

Input: A training set D and The size of the training set N
Output: D with synthetic examples.
  for each example x in D **do**
    Obtain the class index i of x
    class_size[i]++   ▷ the size of the ith class
  end for
  obtain max_class_size   ▷ the size of the majority class
  for each class i **do**
    find training examples with the class index i Dc[]
    $R = max\_class\_size - class\_size[i]$
    while $R > 0$ **do**
      $Ic = ICOTE(Dc, class\_size[i], 10, 10)$ ▷ Ic is the synthetic examples
generated by ICOTE. Details of ICOTE in Algorithm 2
      if $length(Ic) > R$ **then**
        append the first R rows of Ic to D
      else
        append Ic to D
      end if
      $R = R - length(Ic)$
    end while
  end for

Global-IC samples the small class examples to generate memory antibodies (im-mune centroids). The shape of the spatial distribution of the immune centroids follows that of the original examples. As illustrated in Fig.1 and Fig. 2, Global-IC introduces the immune centroids and follows the shape of the neighboring minority class examples. Global-IC thus not only creates larger and less specific decision regions, but also overcomes small disjunct problem introduced by over-sampling [13].

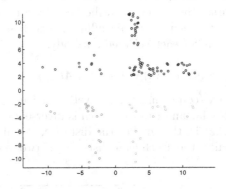

Fig. 1. A data set with four clusters with high density

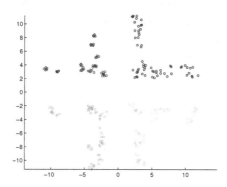

Fig. 2. The extended data set with original examples and the immune centroids

After we present the framework (Global-IC) of our solution, here we will discuss the details of the core of Global-IC, ICOTE. ICOTE uses the aiNet model [3] to generate antibody-derived synthetic examples from the original examples of a class. It includes five major steps as follows:

Step 1: Attribute selection In order to reduce computational cost, we first remove the attributes whose values are constant.

Step 2: Unit-based normalization Then we adjust the values of attributes on different scales to a notionally common scale $[0, 1]$.

$$norm\,(x) = \frac{x - x_{min}}{x_{max} - x_{min}} \tag{1}$$

Step 3: Immune centroids generation There are three sub-steps to generate immune centroids. First, the selected antibodies $\boldsymbol{Ab}$ are going to proliferate (clone) proportionally to their antigenic affinity. The higher the affinity, the larger the clone size nc is for each selected antibody.

$$clone\,(\boldsymbol{Ab}) = \{\boldsymbol{Ab_1} \ldots \boldsymbol{Ab_{nc}}\} \tag{2}$$

Note that the affinity (complementarity level) of the antigen-antibody match is measured by their Euclidean distance, which is inversely proportional to their Euclidean distance. That is, the smaller the distance, the higher the affinity is, and vice-versa. Formally, the Euclidean distance of two vectors is defined as follows.

$$dist(\boldsymbol{X_i}, \boldsymbol{X_j}) = \sqrt{\sum \left(X_i^m - X_j^m\right)^2} \tag{3}$$

where m is the dimension of each vector.

Next, each antibody $\boldsymbol{Ab}$ from the clone set will suffer a mutation with a rate α_k, which is inversely proportional to the antigenic affinity of its parent antibody $\boldsymbol{Ag}$.

$$mutate\,(\boldsymbol{Ab}) = \boldsymbol{Ab} + \alpha_k\,(\boldsymbol{Ab} - \boldsymbol{Ag})\,, \alpha_k = \frac{1}{d} * dist(\boldsymbol{Ab}, \boldsymbol{Ag}) \qquad (4)$$

And then we eliminate the memory antibodies (denoted as M) with a low antigen-antibody affinity (clonal suppression) f_{ij} and a high antibody-antibody affinity (network suppression) f'_{ij}.

$$suppress\,(M) = M - M_{f_{ij}>T_1} - M_{f'_{ij}>T_2}, T_1 \in R, T_2 \in R \qquad (5)$$

Step 4: De-normalization Next, we de-normalize memory antibodies M and make synthetic examples identical to sample distribution.

$$de\text{-}norm\,(x) = x_{min} + (x_{max} - x_{min}) * x \qquad (6)$$

Step 5: Attribute recovery The last step, we put back constant-value attributes that are removed in Step 1.

Correspondingly, the pseudo code of our ICOTE algorithm is shown in Algorithm 2:

4 Experiments

In this section, we will investigate the performance of our proposed method Global-IC, and compare it with existing well-known resampling methods.

4.1 Experimental Settings

Our experiments are conducted using three base classifiers: kNN [15], C4.5 [16] and SVM [17], respectively. We use these algorithms, since they are available within the KEEL software tool [14]. In the experiments, the parameter values are set based on the recommendations from the corresponding authors. The specific settings are as follows:

1. Instance based learning kNN [15]: In this algorithm, we set k=3 and use the Euclidean distance metric.
2. C4.5 Decision tree [16]: For C4.5, we set a confidence level as 0.25, the minimum number of item-sets per leaf as 2, and use pruning.
3. Support vector machines (SVM) [17]: We choose Polykernel reference functions, setting an internal parameter 1.0 for the exponent of each kernel function and a penalty parameter of the error term as 1.0.

Algorithm 2. ICOTE

Input: Original examples O, The number of original examples n, The number of initial
    antibodies K, the maximum number of generations R
Output: S with synthetic examples.
  for each example x in O **do**
    remove constant attributes of x
  end for
  for $i = 1$ to n **do**   ▷ generate antigens $\boldsymbol{Ag}$
    $Ag[i] = norm(x_i)$ ▷ Eq.(1)
  end for
  Generate K random antibodies $Ab[]$
  while $N < R$ **do**
    Initialize memory antibodies $M[]$
    for $i = 1$ to n **do**
      for $j = 1$ to K **do**
        $Dist[] = dist(Ag[i], Ab[j])$ ▷ Eq.(3)
      end for
      Clone K antibodies in proportion to antigen-antibody affinities   ▷ Eq.(2)
      Select a portion of antibodies to perform mutation   ▷ Eq.(4)
      Dispose antibodies with antigen-antibody affinity less than 0.05   ▷ Eq.(5)
      Append antibodies to $M[]$
    end for
    for $i = 1$ to the number of memory antibodies $M[]$ **do**
      $Dist[] = dist(M[i], M[i])$ ▷ Eq.(3)
    end for
    Dispose memory antibodies with antibody-antibody affinity less than 0.05   ▷
  Eq.(5)
    Fill $Ab[]$ with $M[]$ and new K random antibodies
    $N = N + 1$
  end while
  for $i = 1$ to the number of memory antibodies $M[]$ **do**
    $S[i] = de\text{-}norm(M[i])$ ▷ Eq.(6)
  end for
  for $i = 1$ to the number of memory antibodies $M[]$ **do**
    recovery the removed attributes for $S[i]$
  end for

We conduct experiments on 12 datasets from the KEEL dataset repository[1],
whose characteristics are summarized in Table 1, namely the number of examples
(#Ex.), number of attributes (#Atts.), and the number of examples in each
class(separated by comma). The experiments are evaluated in terms of one of the
popular metrics, the Area Under the ROC Curve (AUC) [18]. Our experimental
results are obtained based on 10-fold cross-validation.

---

[1] http://www.keel.es/dataset.php

Table 1. The characteristics of imbalanced datasets

Data	#Ex.	#Atts.	Examples in each class
bal	625	4	(49, 288, 288)
con	1473	9	(629, 333, 511)
der	358	33	(60, 111, 71, 48, 48, 20)
eco	336	7	(143, 77, 2, 2, 35, 20, 5, 42)
gla	214	9	(70, 76, 17, 13, 9, 29)
hay	160	4	(65, 64, 31)
lym	148	18	(61, 81, 4, 2)
new	215	5	(150, 35, 30)
pag	5472	10	(4913, 329, 87, 115, 28)
shu	57999	9	(8903, 45586, 3267, 49, 173, 13)
win	178	13	(59, 71, 48)
yea	1484	8	(244, 429, 463, 44, 35, 51, 161, 30, 20, 5)

4.2 Experimental Results

In this section, we investigate the performance of different methods on the imbalanced datasets listed in Table 1.

As shown in the previous work [19] on the keel datasets, the "OVO+over-sampling" and "OVO+cost-sensitive learning" outperforms both the direct multi-class decompo-sition schemes OVO and OVA in almost all the cases. So we will investigate our proposed method Global-IC with OVO+over-sampling and OVO+cost-sensitive learning. In order to study the combination of pre-processing and cost-sensitive ap-proaches for multi-class imbalance learning, we combine OVO with four representa-tive methods, namely ROS [21], SMOTE-ENN [21], SMOTE [5], and CS [20]. In addition, since our Global-IC is closely related to Global-GS[8], Static-SMOTE [9], and AdaBoost.NC [10]. All of them are the directed multi-class imbalance learning methods. It is nature for us to make comparisons among these methods. The average experimental results for each method are shown in Table 2-4 respectively, in term of three different base learners, i.e., KNN, C4.5, and SVM.

From Table 2-4, we can see that our method Global-IC performs much better than other seven resampling methods, on all the three base learners. When we used KNN as the base learner, our Global-IC performs the best on nine out of 12 datasets. Its average AUC over the 12 datasets is 85.08, which is much higher than the second highest (76.70) achieved by OVO+SMOTE. When we used C4.5 as the base learner, our Global-IC performs the best on 10 out of 12 datasets. Its average AUC over the 12 datasets is 88.38, which is much higher than the second highest (78.81) achieved by Ada-Boost.NC. When we used SVM as the base learner, our Global-IC also performs the best on 10 out of 12 datasets. Its average AUC over the 12 datasets is 84.57, which is much higher than the second highest (78.88) achieved by Global-CS.

Except the excellent performance of our Global-IC, we couldn't find the second best method among the eight approaches on all the three base learners. From

Table 2. AUCs of different resampling methods with kNN as the base learner

Data	Global-CS	Static-SMOTE	Ada-Boost.NC	OVO-ROS	OVO-SMOTE ENN	OVO-SMOTE	OVO-CS	Global-IC
bal	56.29	55.67	49.46	54.14	61.40	56.15	53.70	**81.37**
con	42.58	42.58	45.53	44.24	**47.52**	44.44	44.32	47.27
der	94.86	95.13	95.22	96.82	97.13	96.49	95.93	**97.76**
eco	71.79	70.53	70.43	73.54	72.75	74.38	72.70	**95.02**
gla	71.73	74.16	69.19	73.87	70.60	71.52	74.23	**87.74**
hay	48.06	49.40	61.83	73.29	44.80	72.82	**77.82**	39.17
lym	77.88	83.99	81.21	73.02	72.81	74.68	72.81	**92.56**
new	95.17	96.50	91.83	94.28	94.00	96.00	95.39	**98.67**
pag	83.93	84.97	84.63	85.38	92.65	92.51	86.20	**99.15**
shu	91.02	92.71	96.13	89.73	91.58	92.67	89.73	**99.88**
win	**98.10**	97.14	96.06	96.25	95.30	96.25	95.30	96.71
yea	50.45	51.22	54.45	50.17	50.15	52.45	51.91	**85.66**
Average	73.49	74.50	74.66	75.39	74.22	76.70	75.84	**85.08**

Table 3. AUCs of different resampling methods with C4.5 as the base learner

Data	Global-CS	Static-SMOTE	Ada-Boost.NC	OVO-ROS	OVO-SMOTE ENN	OVO-SMOTE	OVO-CS	Global-IC
bal	55.93	55.30	60.84	55.57	52.35	54.29	54.20	**81.82**
con	49.83	47.19	50.14	48.05	52.31	50.09	49.74	**55.01**
der	93.56	94.83	94.79	95.71	95.65	95.61	96.33	**96.72**
eco	66.28	65.15	74.79	72.89	70.97	70.99	73.65	**91.44**
gla	70.95	63.71	73.97	68.81	70.58	70.84	65.44	**87.29**
hay	83.49	**86.03**	88.17	82.86	70.08	83.49	83.49	83.00
lym	69.27	67.81	66.49	72.51	61.95	60.91	70.77	**91.35**
new	91.67	90.56	95.11	90.11	90.33	92.50	91.44	**97.33**
pag	88.28	85.55	90.59	91.52	89.88	90.24	90.69	**98.98**
shu	98.55	95.05	98.47	96.69	94.70	96.84	96.79	**99.89**
win	94.32	94.24	**96.46**	91.24	87.75	92.02	91.35	94.37
yea	47.66	51.77	55.94	50.72	50.70	52.10	52.30	**83.33**
Average	75.82	74.76	78.81	76.39	73.94	75.83	76.35	**88.38**

the average results shown in Table 2-4, we can see that the ranks of the performance of all other methods are varied with the base learner. Besides the average results shown in Table 2-4, we also rank these methods on each dataset with each base learner for further comparison analysis. The average rank of each method with each base learner is shown in Fig.3. From Fig.3, we can see that the average rank of Global-IC is the best under the three base learners. The ranks of the other methods depend on the base learner. The other three directed multi-class imbalance learning methods Global-CS, Static-SMOTE and AdaBoost.NC do not consistently rank higher than the OVO combination methods (i.e., OVO+ROS,

Table 4. AUCs of different resampling methods with SVM as the base learner

Data	Global-CS	Static-SMOTE	Ada-Boost.NC	OVO-ROS	OVO-SMOTE-ENN	OVO-SMOTE	OVO-CS	Global-IC
bal	91.63	91.63	90.64	91.63	91.63	91.63	91.63	**91.67**
con	51.66	49.01	**53.18**	50.95	50.95	51.72	50.48	51.46
der	95.78	95.60	97.08	95.93	95.93	95.78	95.44	**98.96**
eco	67.95	70.03	65.17	69.37	70.59	68.21	68.19	**85.23**
gla	64.72	58.31	55.62	62.42	61.69	63.95	67.91	**75.00**
hay	57.78	64.29	57.22	58.41	54.05	55.00	56.83	**68.67**
lym	82.60	82.74	82.04	82.81	70.33	70.79	82.39	**93.52**
new	96.89	95.78	92.67	94.67	95.56	97.11	96.89	**98.89**
pag	91.67	69.04	88.29	89.09	87.93	88.47	89.32	**96.95**
shu	92.68	63.70	83.87	84.25	84.17	84.39	84.14	**99.34**
win	**97.77**	97.22	95.98	97.22	97.68	97.22	**97.77**	95.78
yea	55.49	54.45	55.39	55.69	56.22	56.74	55.91	**59.40**
Average	78.88	74.31	76.43	77.70	76.39	76.75	78.08	**84.57**

OVO+SMOTE-ENN, OVO+SMOTE, and OVO+CS). OVO+CS has a relatively robust rank with the three base learners.

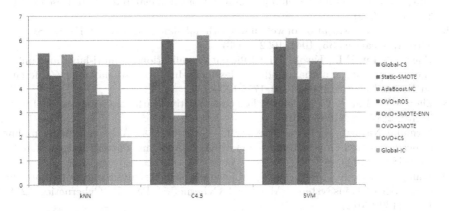

Fig. 3. The average ranks of the resampling methods under three base learners respectively

5 Conclusions

In this paper we present a novel global multi-class imbalance learning approach Global-IC, which does not need to transfer multi-class into binary. This novel approach is based on immune network theory, and generates immune centroids for the clusters of each small class, which have high data density. It is completely different from renowned resampling methods. Our experimental results showed that Global-IC achieves better performance, comparing with existing methods.

Acknowledgments. This research was partially supported by the Natural Science Foundation of China under grant No.61170020, 61402311 and 61440053, Jiangsu Province Colleges and Universities Natural Science Research Project under grant No. 13KJB520021, Jiangsu Province Technology Innovation Fund Project for Science and Technology Enterprises under grant No.BC2013124, 2013 Suzhou Municipal Special Fund Project for Speeding up the Information Construction, and the U.S. National Science Foundation (IIS-1115417), and the Jiangsu Province Postgraduate Cultivation and Innovation Project under grant No. ZY32001814.

References

1. Hastie, T., Tibshirani, R.: Classification by pairwise coupling. Annals of Statistics **26**(2), 451–471 (1998)
2. Rifkin, R., Klautau, A.: In defense of one-vs-all classification. Journal of Machine Learning Research **5**, 101–141 (2004)
3. Castro, L.N.D., Zuben, F.J.V.: aiNet: An artificial immune network for data analysis. In: Abbass, H.A., Sarker, R.A., Newton, C.S., (eds.) Data Mining: A Heuristic Approach. Idea Group Publishing, ch XII, pp. 231–259, USA (2001)
4. Tan, A.C., Gilbert, D., Deville, Y.: Multi-class protein fold classification using a new ensemble machine learning approach. Genome Informatics **14**, 206–217 (2003)
5. Chawla, N.V., Bowyer, K.W., Hall, L.O., Kegelmeyer, W.P.: Smote: synthetic minority over-sampling technique. Journal of Artificial Intelligent Research **16**, 321–357 (2002)
6. Liao, T.W.: Classification of weld flaws with imbalanced class data. Expert Systems with Applications **35**, 1041–1052 (2008)
7. Chen, K., Lu, B.L., Kwok, J.T.: Efficient classification of multi-label and imbalanced data using min-max modular classifiers. In: International Joint Conference on Neural Net-works, pp. 1770–1775 (2006)
8. Zhou, Z.H., Liu, X.Y.: On multi-class cost-sensitive learning. Computational Intelligence **26**(3), 232–257 (2010)
9. Fernndez-Navarro, F., Hervs-Martnez, C., Gutirrez, P.A.: A dynamic oversampling pro-cedure based on sensitivity for multi-class problems. Pattern Recognition **44**, 1821–1833 (2011)
10. Wang, S., Yao, X.: Multi-class imbalance problems: analysis and potential solutions. IEEE TransSystems, Man, and Cybernetics, Part B: Cybernetics **42**(4), 1119–1130 (2012)
11. Jerne, N.K.: Towards a Network Theory of the Immune System. Annales d'immunologie **125C**(1–2), 373–389 (1974)
12. Burnet, F.M.: A modification of Jerne's theory of antibody production using the concept of clonal selection. A Cancer Journal for Clinicians **26**(2), 119–121 (1976)
13. Jo, T., Japkowicz, N.: Class imbalances versus small disjuncts. ACM SIGKDD Explorations Newsletter **6**(1), 40–49 (2004)
14. Alcala-Fdez, J., Snchez, L., Garca, S., del Jesus, M.J., Ventura, S., Garrell, J.M., Otero, J., Romero, C., Bacardit, J., Rivas, V.M., Fernndez, J.C., Herrera, F.: KEEL: a software tool to assess evolutionary algorithms for data mining problems. Soft Computing **13**, 307–318 (2009)
15. McLachlan, G.J.: Discriminant Analysis and Statistical Pattern Recognition. John Wiley and Sons (2004)
16. Quinlan, J.R.: C4.5: Programs for Machine Learning. Morgan Kauffman (1993)

17. Cortes, C., Vapnik, V.: Support vector networks. Machine Learning **20**, 273–297 (1995)
18. Bradley, A.P.: The use of the area under the roc curve in the evaluation of machine learning algorithms. Pattern Recognition **30**(7), 1145–1159 (1997)
19. Fernndez, A., Lpez, V., Galar, M., Jesus, M.J.D., Herrera, F.: Analysing the classification of imbalanced data-sets with multiple classes: Binarization techniques and ad-hoc approach-es. Knowledge-Based Systems **42**, 97–110 (2013)
20. Zhao, H.: Instance weighting versus threshold adjusting for cost-sensitive classification. Knowledge and Information Systems **15**(3), 321–334 (2008)
21. Batista, G.E.A.P.A., Prati, R.C., Monard, M.C.: A study of the behaviour of several methods for balancing machine learning training data. SIGKDD Explorations **6**(1), 20–29 (2004)
22. Freund, Y., Schapire, R.E.: Experiments with a new boosting algorithm. In: Proceedings of the 13th International Conference, Machine Learning, pp. 148–156 (1996)

Optimizing Classifiers for Hypothetical Scenarios

Reid A. Johnson[1], Troy Raeder[2], and Nitesh V. Chawla[1]([✉])

[1] University of Notre Dame, Notre Dame, IN 46556, USA
{rjohns15,nchawla}@nd.edu
[2] Dstillery, 470 Park Ave. S., 6th Floor, New York, NY 10016, USA
troy@dstillery.com

Abstract. The deployment of classification models is an integral component of many modern data mining and machine learning applications. A typical classification model is built with the tacit assumption that the deployment scenario by which it is evaluated is fixed and fully characterized. Yet, in the practical deployment of classification methods, important aspects of the application environment, such as the misclassification costs, may be uncertain during model building. Moreover, a single classification model may be applied in several different deployment scenarios. In this work, we propose a method to optimize a model for uncertain deployment scenarios. We begin by deriving a relationship between two evaluation measures, H measure and cost curves, that may be used to address uncertainty in classifier performance. We show that when uncertainty in classifier performance is modeled as a probabilistic belief that is a function of this underlying relationship, a natural definition of *risk* emerges for both classifiers and instances. We then leverage this notion of risk to develop a boosting-based algorithm—which we call *RiskBoost*—that directly mitigates classifier risk, and we demonstrate that it outperforms AdaBoost on a diverse selection of datasets.

1 Introduction

Many real-world problems necessitate the use of a classification model to assign items in a collection to target categories or classes. The chief objective of a classification model is to accurately predict the target class for each case in the data. Accordingly, when evaluating a classification model, one desires an accurate assessment of its performance on unseen data. Accurate model assessments are important because they permit candidate models to be meaningfully compared and allow one to determine whether a model will perform at an "acceptable" level. The notion of acceptable performance may be defined solely by internal concerns (e.g., the benefit of a model must outweigh its implementation cost) or by external factors (e.g., regulators may hesitate to approve a diagnostic test with a high false negative rate). No matter how it is applied, however, sound model assessment is a critical element of any classification task [10].

There are many ways to quantifiably assess the performance of a classifier. In this work, we quantify classifier performance via a *simple linear cost model*:

$$\ell = c_0 \pi_0 e_0 + c_1 \pi_1 e_1, \tag{1}$$

© Springer International Publishing Switzerland 2015
T. Cao et al. (Eds.): PAKDD 2015, Part I, LNAI 9077, pp. 264–276, 2015.
DOI: 10.1007/978-3-319-18038-0_21

where c_i is the cost of misclassifying a class i instance, π_i is the proportion of class i instances in the data, and e_i is the error rate on class i instances. This cost model is convenient and is commonly used in cost-sensitive learning. However, cost-sensitive methods generally assume that the parameters π_i and c_i are known and constant (e.g., [5,12,18,21]), an assumption that is often not borne out in practice [14]. Zadrozny and Elkan [20] provide a framework for estimating costs and probabilities when sample data are available, but for the purpose of scenario analysis (i.e., the process of evaluating possible future events for which such information is not readily available) [11].

In this work, we focus on developing classification models for hypothetical future deployment scenarios engendered by uncertain operating environments. We begin in Section 2 by connecting the current techniques for dealing with uncertain operating environments with a notion of cost. We then demonstrate in Section 3 that, as a result of this connection, there exists an underlying theoretical relationship between several of these methods that leads to a natural definition of the *risk* of an individual classifier and instance. Further, we find that this risk can be substantially mitigated via a boosting-based algorithm we call *RiskBoost*. In Section 4, we demonstrate that RiskBoost outperforms AdaBoost [8] over a diverse collection of datasets. Finally, we present our conclusions in Section 5.

2   Addressing Uncertain Cost in Classifier Performance

Consider a binary classification task where we have several cases or instances, each of which may be assigned to one of two categories or classes that are labeled 1 (positive) and 0 (negative). Further, assume that any classifier learned on the training data is capable of producing, for each input vector $\mathbf{x}$, a real-valued score $s(\mathbf{x})$ that is a monotonic function of $p(1|x)$, which is the probability that $\mathbf{x}$ belongs to class 1. These scores are mapped to binary classifications by choosing a threshold t such that an instance $\mathbf{x}$ is classified as class 0 (negative) if $s(x) < t$ and class 1 (positive) if $s(x) \geq t$.

Each classification threshold produces a unique classifier, the performance of which can be characterized by a confusion matrix of particular true positive (tp), false positive (fp), true negative (tn), and false negative (fn) values. Presently, the *de facto* standard method for evaluating classification models from a confusion matrix is the receiver operating characteristic, though alternatives such as the H Measure and cost curves also exist. We first elaborate upon each of these below, after which we define a clear relationship between all three.

For the reader's convenience, a summary of the notations used in the this work is given as Table 1. For the remainder of the work, we use the term *classifier* to refer to a specific confusion matrix, whereas *classification algorithm* or *learning algorithm* is used to refer to a trained model for which a decision threshold has not been defined.

Table 1. The notation used in this work

Symbol	Description
ℓ	The total classification loss.
π_i	The proportion of class i instance in test data.
c_i	The cost of misclassifying a class i instance.
c	A normalized cost ratio, i.e., $c = c_0/(c_0 + c_1)$.
$u(c)$	The likelihood distribution over cost ratios.
e_i	The error rate on class i instances.
n_i	The number of class i test instances.
$\mathscr{L}_i$	The marginal cost of class i instances.
t	A classification threshold.
(r_{1_i}, r_{0_i})	The ith point on the ROC convex hull.
$f_i(x)$	The ith line segment on the lower envelope in cost space.
tp, fp	A true and false positive classification, respectively.
tn, fn	A true and false negative classification, respectively.
tpr, fpr	The true and false positive rate, respectively.
tnr, fnr	The true and false negative rate, respectively.

2.1 Addressing Cost with ROC Curves

The Receiver Operating Characteristic (ROC) curve [7,13] forms the basis for many of the techniques that we will discuss in the remainder of this work. An ROC curve is formed by varying the classification threshold t across all possible values. In a binary classification problem, each threshold produces a distinct confusion matrix that corresponds to a two-dimensional point (r_1, r_0) in ROC space, where $r_1 = fpr$ and $r_0 = tpr$.

A point p_1 in ROC space is said to "dominate" a point p_2 in ROC space if p_1 is both above and to the left of p_2. It follows, then, that only classifiers on the convex hull of the ROC curve are potentially optimal for some value of c_i and π_i, as a point not on the convex hull will be dominated by a point that is on it [14]. As each point on the ROC convex hull represents classification performance at some threshold t, different thresholds will be optimal under different operating conditions c and π_i. For example, classifiers with lower false negative rates will be optimal at lower values of c, while classifiers with lower false positive rates will be optimal at higher values of c.

Now, let $p_i = (r_{1i}, r_{0i})$ and $p_{i+1} = (r_{1(i+1)}, r_{0(i+1)})$ be successive points on the ROC convex hull. Then p_{i+1} will produce superior classification performance to p_i if and only if the change in the false positive rate is offset by a corresponding change in the true positive rate. That is, if we set $\Delta x_i = r_{1(i+1)} - r_{1i}$ and $\Delta y_i = r_{0(i+1)} - r_{0i}$, then p_{i+1} is optimal if

$$c < \frac{\pi_1 \Delta y}{\pi_0 \Delta x + \pi_1 \Delta y}. \qquad (2)$$

Similarly, given a fixed value for c, we can determine the optimal classifier at a given value of π_0. Then for p_{i+1} to outperform p_i, we require that

$$\pi_0 < \frac{(1-c)\Delta y}{c\Delta x + (1-c)\Delta y}. \tag{3}$$

Thus, the ROC convex hull can be used to select the optimal classification threshold (and classifier) under a variety of different operating conditions, a notion first articulated by Provost and Fawcett [14].

Relationship Between ROC Curves and Cost. Each point in ROC space corresponds to a misclassification cost that can be specified via our simple linear cost model as

$$\ell = c_0\pi_0 r_1 + c_1\pi_1(1 - r_0). \tag{4}$$

Note that only the ordinality (i.e., relative magnitude) of the cost is needed for ranking classifiers. Accordingly, if we assume that the cardinality (i.e, absolute magnitude) of the cost can be ignored, then, as $c = c_0/(c_0 + c_1)$, we find that

$$\ell = c\pi_0 r_1 + (1 - c)\pi_1(1 - r_0). \tag{5}$$

This formulation will be used frequently throughout the remainder of this work.

2.2   Addressing Uncertain Cost with the H Measure

An alternative to the ROC is the H Measure, proposed by Hand [9] to address shortcomings of the ROC. Unlike the ROC, the H Measure incorporates uncertainty in the cost ratio c by integrating directly over a hypothetical probability distribution of cost ratios. As the points on the ROC convex hull correspond to optimal misclassification cost over a contiguous set of cost ratios (see Equation 2), then, given known prior probabilities π_i, the average loss over all cost ratios can be calculated by integrating Equation 4 piecewise over the cost regions defined by the convex hull.

Relationship Between the H Measure and Uncertain Cost. To incorporate a hypothetical cost ratio distribution, we set $c = c_0/(c_0 + c_1)$ and weight the integral by the cost distribution, denoted as $u(c)$. The final loss measure is then defined as:

$$\ell_H = \sum_{i=0}^{m} \int_{c^{(i)}}^{c^{(i+1)}} \big(c\pi_0 r_{1i} + (1 - c)\pi_1(1 - r_{0i})\big)u(c)dc. \tag{6}$$

The H Measure is represented as a normalized scalar value between 0 and 1, whereby higher values correspond to better model performance.

2.3   Addressing Uncertain Cost with Cost Curves

Cost curves [6] provide another alternative to ROC curves for visualizing classifier performance. Instead of visualizing performance as a trade-off between false positives and true positives, they depict classification cost in the simple linear cost model against the unknowns π_i and c_i.

The marginal misclassification cost of class i can be written as $\mathscr{L}_i = \pi_i c_i$. This means that if the misclassification rate of class i instances increases by some amount Δe_i, then the total misclassification cost increases by $\mathscr{L}_i \Delta e_i$. The maximum possible cost of any classifier is $\ell_{max} = \mathscr{L}_0 + \mathscr{L}_1$, when both error rates are 1. Accordingly, we can define the normalized marginal cost (termed the probability cost by Drummond and Holte [6]) as $pc_i = \mathscr{L}_i / (\mathscr{L}_0 + \mathscr{L}_1)$, and the normalized total misclassification cost as $\ell_{norm} = \ell / \ell_{max}$. Intuitively, the quantity pc_i can be thought of as the proportion of the total risk arising from class i instances, since we have $pc_0 + pc_1 = 1$, while ℓ_{norm} is the proportion of the maximum possible cost that the given classifier actually incurs.

Each ROC point (r_{1i}, r_{0i}) corresponds to a range of possible misclassification costs that depend on the marginal costs $\mathscr{L}_i$, as shown in Equation 4. We can rewrite Equation 4 as a function of pc_1 as follows:

$$\ell_{norm} = (1 - pc_1)r_{1i} + pc_1(1 - r_{0i})$$
$$= pc_1(1 - r_{0i} - r_{1i}) + r_{1i}.$$

Thus any point in ROC space translates (i.e., can be transformed) into a line in cost space. Of particular interest are the lines corresponding to the ROC convex hull, as these lines represent classifiers with optimal misclassification cost. These lines enclose a convex region of cost space known as the lower envelope. The values of pc_1 for which a classifier is on the lower envelope provide scenarios under which the classifier is the optimal choice.

One can compute the area under the lower envelope to obtain a scalar estimate of misclassification cost. Here, we denote points on the convex hull by $(r_{1i}, r_{0i}), r_{00} < r_{01} < \ldots < r_{0m}$ in increasing order of x-coordinate, and we denote the corresponding cost lines as $f_i(x) = m_i x + b_i$, where m_i is the slope and b_i is the y-intercept of the ith cost line. The lower envelope is then composed of the intersection points of successive lines $f_i(x)$ and $f_{i+1}(x)$. We denote these points $p_i = (x_i, y_i)$, which can be calculated as

$$x_i = \frac{r_{1(i+1)} - r_{1i}}{(r_{0(i+1)} - r_{0i}) + (r_{1(i+1)} - r_{1i})}$$

$$y_i = \frac{r_{1i} - r_{1(i+1)}}{1 - r_{0(i+1)} - r_{1(i+1)}} + r_{1i}.$$

The area under the lower envelope can be calculated geometrically as the area of a convex polygon or analytically as a sum of integrals (the areas under the constituent line segments). For our purposes, it is convenient to express it as follows:

$$A(f_1 \ldots f_m) = \sum_{i=0}^{m} \int_{x_i}^{x_{i+1}} f_i(x)dx. \tag{7}$$

The function $A(\cdot)$ represents a loss measure, where higher values of A correspond to worse performance. This area represents the expected misclassification cost of the classifier, where all values of pc_1 are considered equally likely. In the next section, we discuss the implications of this loss measure.

3   Deriving and Optimizing on Risk from Uncertain Cost

In the previous section, we related several measures of classifier performance to a notion of cost. In this section, we elaborate on the consequences of these connections, from which we derive definitions of "risk" for classifiers and instances.

3.1   Relationship Between Cost Curves and H Measure

An interesting result emerges if we assume an accurate estimate of π_i, either from the training data or from some other source of background knowledge and replace the pair (c_0, c_1) with $(c, 1 - c)$. In this case, a hypothetical cost curve represents $\ell_c = c\pi_0 r_1 + (1-c)\pi_1(1-r_0)$ on the y-axis and c on the x-axis. We can rewrite this expression into the standard form of an equation for a line, which gives us $\ell_c = c(\pi_0 r_1 - \pi_1(1 - r_0)) + (1 - r_0)$.

The intersection points of successive lines, which would form the lower envelope, can similarly be derived as

$$x_i = \frac{\pi_1(r_{0i} - r_{0(i+1)})}{\pi_1(r_{0i} - r_{0(i+1)}) + \pi_0(r_{1i} - r_{1(i+1)})}. \tag{8}$$

Consequently, the area under the lower envelope can be expressed as:

$$A(f_1 \ldots f_m) = \sum_{i=0}^{m} \int_{x_i}^{x_{i+1}} \left(c\pi_0 r_1 + (1 - c)\pi_1(1 - r_0) \right) dc. \tag{9}$$

As the endpoints x_i are the same as those used in the computation of the H Measure (see Equation 2), it follows that the H Measure is equivalent to the area under the lower envelope of the cost curve with uniform $u(c)$ and prior probabilities π_i known. Further, Hand has demonstrated that, for a particular choice of $u(c)$, the area under the ROC curve is equivalent to the H Measure [9].

Thus, these three different techniques—ROC curves, H Measure, and cost curves—are simply specific instances of the simple linear cost model. Rather than debating the relative merits of these specific measures, which is beyond the scope of this work (cf. [3,9] for such discussions), we instead focus on the powerful consequences of adhering to the more general model.

Intuitively, since the simple linear model underlies several measures of classifier performance, it also provides an avenue for interpreting model performance. In fact, we find that it provides an insight into model performance under hypothetical scenarios—that is, a notion of risk—that cannot be explicitly captured by these other measures. We elaborate on this below.

3.2   Interpreting Performance Under Hypothetical Scenarios

As a consequence of the relationship between the H Measure and cost curves, we can actually represent the H Measure loss function in cost space. By representing different loss functions on a single set of axes, we form a series of scenario curves, each of which corresponds to a loss function.

Figure 1 depicts scenario curves for several different likelihood functions alongside a standard cost curve. Each curve quantifies the vulnerability of the classification algorithm over the set of all possible scenarios pc_1 for different probabilistic beliefs about the likelihood of different cost ratios. The likelihood distributions include: (1) the $Beta(2,2)$ distribution $u(c) = \frac{1}{6}c(1-c)$, as suggested by [9]; (2) a Beta distribution shifted so that the most likely cost ratio is proportional to the proportion of minority class instances (i.e., $c \propto \pi_0$); (3) a truncated Beta distribution where the probability of minority class instances is greater than the probability of majority class instances (i.e., $p(c_0 > c_1) = 0$), motivated by the observation that the minority class typically has the highest misclassification cost; (4) a truncated exponential distribution where the parameter λ is set to ensure that the expectation of class i is inversely proportional to the proportion of that class in the data (i.e., $c_i \propto 1/\pi_i$); and (5) the cost curve, which assumes uniform distributions over probabilities and costs.

From the figure, it is clear that the choice of likelihood distribution can have a significant effect on both the absolute assessment of classifier performance (i.e., the area under the curve) and on which scenarios we believe will produce the greatest loss for the classifier. These curves also have intuitive meanings that may be useful when analyzing classifier performance. First, as the cost curve makes no *a priori* assumptions about the likelihood of different scenarios, it can present the performance of an algorithm over any given scenario. Second, if and when information about the likelihood of different scenarios becomes known, the cost curve presents the set of classifiers the pose the greatest risk (i.e., the components of the convex hull).

Both interpretations are important. On the one hand, an unweighted cost curve can be used to identify the set of scenarios over which a classifier performs acceptably for any domain-specific definition of reasonable performance. On the other hand, a weighted scenario curve can be used to identify where an algorithm should be improved in order to achieve the maximum benefit given the available information. From the second observation arises a natural notion of risk.

3.3   Defining Risk

Given a likelihood distribution over the cost ratio c, each classifier on the convex hull is optimal over some range of cost ratios (see Equation 2). From this, we can derive two intuitive definitions: one for the risk associated with individual classifiers and one for the risk associated with individual instances.

Definition 1. *Assume that classifier $\mathscr{C}$ is optimal over the range of cost ratios* $[c_1, c_2]$. *Then the risk of classifier $\mathscr{C}$ is the expected cost of the classifier over the range for which it is optimal:*

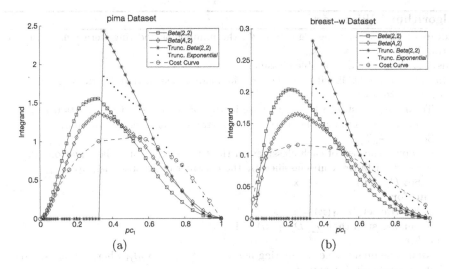

Fig. 1. Scenario curves for several different cost distributions $u(c)$ generated by a boosted decision tree model on the (a) `pima` and (b) `breast-w` datasets. The curves have been normalized such that (1) the area under each curve represents the value of the respective loss measure and (2) the maximum loss for the cost curve is 1.

$$risk(\mathscr{C}) = \int_{c_1}^{c_2} \ell_H(c)dc \qquad (10)$$

Definition 2. *The risk of instance* **x** *is the aggregate risk over all classifiers that misclassify* **x**.

We discuss how these definitions may be applied to improve to classifier performance below.

3.4   RiskBoost: Optimizing Classification by Minimizing Risk

Since we can quantify the degree to which instances pose the greatest risk to our classification algorithm, it is natural to strengthen the algorithm by assigning greater importance to these "risky" instances.

Standard boosting algorithms such as AdaBoost combine functions based on the "hardness" of correctly classifying a particular instance [8]. Instead, we propose a novel boosting algorithm that reweights instances according to their relative risk, which we call *RiskBoost*. RiskBoost uses the expected misclassification loss ℓ to reweight instances that are misclassified by the most vulnerable classifier according to both classifier performance and the hypothetical cost ratio distribution. Pseudocode for RiskBoost is provided as Algorithm 1.

Algorithm 1. RiskBoost

Require: A base learning algorithm $\mathcal{W}$, the number of boosting iterations n, and m training instances $\mathbf{x}_1 \ldots \mathbf{x}_m$.
Ensure: A weighted ensemble classifier.
  Initialize a weight distribution D over the instances such that $D_1(\mathbf{x}_i) = 1/m$.
  for $j = 1$ to n **do**
    Train a new instance $\mathcal{W}_j$ of the base learner $\mathcal{W}$ with weight distribution D_j.
    Compute the loss ℓ of the learner on the training data via Equation 6.
    Set $\beta_j = \frac{1-0.5*\ell}{0.5*\ell}$.
    Compute the risk of each classifier on the ROC convex hull via Equation 10.
    for each instance $\mathbf{x}$ misclassified by the classifier of greatest risk **do**
      Set $D_{j+1}(\mathbf{x}) = \beta_j \cdot D_j(\mathbf{x})$.
    end for
    Otherwise set $D_{j+1}(\mathbf{x}) = D_j(\mathbf{x})$.
    Normalize such that $\sum_i D_{j+1}(\mathbf{x}_i) = 1$.
  end for
  return The final learner predicting $p(1|\mathbf{x}) = z \sum_j p_j(1|\mathbf{x})\beta_j$, where z is chosen such that the probabilities sum to 1.

---

4 Experiments

To evaluate the performance of RiskBoost, we compare it with AdaBoost on 19 classification datasets from the UCI Machine Learning Repository [1]. We employ RiskBoost by setting its risk calculation (i.e., Equation 10) as $u(c) = Beta(2,2)$, as suggested by [9]. AdaBoost is employed with the AdaBoost.M1 variant [8]. For both algorithms, we use 100 boosting iterations of unpruned the C4.5 decision trees, which previous work has shown benefit substantially from AdaBoost [15].

In order to compare the classifiers, we use 10-fold cross-validation. In 10-fold cross-validation, each dataset is partitioned into 10 disjoint subsets or folds such that each fold has (roughly) the same number of instances. A single fold is retained as the validation data for evaluating the model, while the remaining 9 folds are used for model building. This process is then repeated 10 times, with each of the 10 folds used exactly once as the validation data. As the cross-validation process can exhibit a significant degree of variability [16], we average the performance results from 100 repetitions of 10-fold cross-validation to generate reliable estimates of classifier performance. Performance is reported as AUROC (area under the Receiver Operating Characteristic).

4.1 Statistical Tests

Previous literature has suggested the comparison of classifier performance across multiple datasets based on ranks. Following the strategy outlined in [4], we first rank the performance of each classifier by its average AUROC. The Friedman test is then used to determine if there is a statistically significant difference between the rankings of the classifiers (i.e., that the rankings are not merely

Table 2. AUROC performance of AdaBoost and RiskBoost on several classification datasets. Bold values indicate the best performance for a dataset. Checkmarks indicate the model performs statistically significantly better at the confidence level $1 - \alpha$.

Dataset	AdaBoost.M1	RiskBoost
breast-w	0.9829	**0.9899**
bupa	0.7218	0.7218
credit-a	0.8973	**0.9187**
crx	0.8970	**0.9191**
heart-c	0.8643	**0.8919**
heart-h	0.8531	**0.8723**
horse-colic	**0.8501**	0.8295
ion	**0.9753**	0.9744
krkp	0.9985	**0.9996**
ncaaf	0.8658	**0.9144**
pima	0.7803	**0.7872**
promoters	**0.9611**	0.8863
ringnorm	0.9793	**0.9849**
sonar	0.9281	**0.9344**
threenorm	0.9094	**0.9210**
tictactoe	**0.9994**	0.9986
twonorm	0.9834	**0.9885**
vote	0.9733	**0.9856**
vote1	0.9338	**0.9543**
Average Rank	1.79	1.21
$\alpha = 0.05$		✓

randomly distributed), after which the Bonferroni-Dunn post-hoc test is applied to control for multiple comparisons.

4.2 Results

From Table 2, we observe that RiskBoost performs better than AdaBoost in 14 of the 19 datasets evaluated, with 1 tie. Further, we find that RiskBoost performs statistically significantly better than AdaBoost at a 95% confidence level over the collection of evaluated datasets. The 95% critical distance of the Bonferroni-Dunn procedure for 19 datasets and 2 classifiers is 0.45; consequently, an average rank lower than 1.275 is statistically significant, which RiskBoost achieves with an average rank of 1.21. Similar results were achieved for 10 repetitions of 10-fold cross-validation (where RiskBoost's average rank was 1.11), 50 repetitions (1.26), and 500 repetitions (1.21).

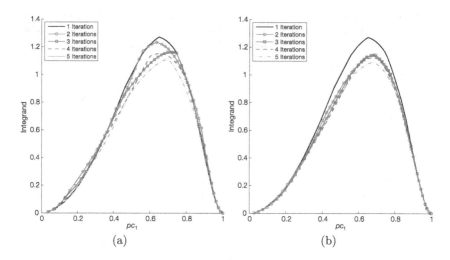

Fig. 2. Scenario curves for successive iterations of (a) AdaBoost and (b) RiskBoost ensembles on the `ncaaf` dataset

4.3 Discussion

For a better understanding of the general intuition behind RiskBoost, Figure 2 shows the progression for AdaBoost and RiskBoost when optimizing the H Measure with the $Beta(2,2)$ cost distribution. At each iteration, the RiskBoost ensemble directly boosts the classifier of greatest risk, which is represented by the global maximum in the figure. Successive iterations of RiskBoost lead to direct cost reductions for this classifier, resulting in a gradual but consistent reduction from peak risk. By contrast, AdaBoost establishes an arbitrary threshold for "incorrect" instances. As a result, AdaBoost does not always focus on the instances that contribute greatest to the overall misclassification cost, which ultimately results in the erratic behavior demonstrated by AdaBoost's scenario curves.

Though RiskBoost offers promising performance over a diverse array of classification datasets, we note that there is an expansive literature on cost-sensitive boosting (e.g., [12,18,19]) and boosting with imbalanced data (e.g., [2,17,18]) that can be used to tackle similar problems. A critical feature that sets our work apart from prior efforts, however, is that previous work tacitly assumes that misclassification costs are known, whereas RiskBoost can expressly optimize misclassification costs that are unknown and uncertain. Further, we demonstrate that this strategy for risk mitigation actually arises naturally from the framework of scenario analysis. We leave further empirical evaluation of RiskBoost with cost-sensitive boosting algorithms as future work.

5   Conclusion

Classification models are an integral tool for modern data mining and machine learning applications. When developing a classification model, one desires a model that will perform well on unseen data, often according to some hypothetical future deployment scenario. In doing so, two critical questions arise: First, how does one estimate performance so that the best-performing model can be selected? Second, how can one build a classifier that is optimized for these hypothetical scenarios?

Our work focuses on addressing these questions. By examining the current approaches for evaluating classifier performance in uncertain deployment scenarios, we derived a relationship between H Measure and cost curves, two well-known techniques. As a consequence of this relationship, we found that ROC curves, H Measure, and cost curves can be represented as specific instances of a simple linear cost model. We found that by defining scenarios as probabilistic expressions of belief in this simple linear cost model, intuitive definitions emerge for the risk of an individual classifier and the risk of an individual instance. These observations suggest a new boosting-based algorithm—RiskBoost—that directly mitigates the greatest component of classification risk, and which we find to outperform AdaBoost on a diverse selection of classification datasets.

Acknowledgments. This work is supported by the National Science Foundation (NSF) Grant OCI-1029584.

References

1. Bache, K., Lichman, M.: UCI machine learning repository (2013). http://archive.ics.uci.edu/ml
2. Chawla, N.V., Lazarevic, A., Hall, L.O., Bowyer, K.W.: SMOTEBoost: improving prediction of the minority class in boosting. In: Lavrač, N., Gamberger, D., Todorovski, L., Blockeel, H. (eds.) PKDD 2003. LNCS (LNAI), vol. 2838, pp. 107–119. Springer, Heidelberg (2003)
3. Davis, J., Goadrich, M.: The relationship between precision-recall and ROC curves. In: Proceedings of the 23rd International Conference on Machine Learning (ICML), pp. 233–240. ACM (2006)
4. Demšar, J.: Statistical comparisons of classifiers over multiple data sets. Journal of Machine Learning Research (JMLR) **7**, 1–30 (2006)
5. Domingos, P.: MetaCost: a general method for making classifiers cost-sensitive. In: Proceedings of the 5th ACM SIGKDD International Conference on Knowledge Discovery and Data Mining (KDD), pp. 155–164. ACM (1999)
6. Drummond, C., Holte, R.C.: Cost curves: An improved method for visualizing classifier performance. Machine Learning **65**(1), 95–130 (2006)
7. Fawcett, T.: An introduction to ROC analysis. Pattern Recognition Letters **27**(8), 861–874 (2006)
8. Freund, Y., Schapire, R.E.: Experiments with a new boosting algorithm. In: Proceedings of the 13th International Conference on Machine Learning (ICML), pp. 148–156 (1996)

9. Hand, D.J.: Measuring classifier performance: A coherent alternative to the area under the ROC curve. Machine Learning **77**(1), 103–123 (2009)

10. Hastie, T., Tibshirani, R., Friedman, J.: The Elements of Statistical Learning, vol. 2 (2009)

11. Lempert, R.J., Popper, S.W., Bankes, S.C.: Shaping the Next One Hundred Years: New Methods for Quantitative, Long-Term Policy Analysis, Rand Corp (2003)

12. Masnadi-Shirazi, H., Vasconcelos, N.: Cost-sensitive boosting. IEEE Transactions on Pattern Analysis and Machine Intelligence (TPAMI) **33**(2), 294–309 (2011)

13. Provost, F., Fawcett, T.: Analysis and visualization of classifier performance: comparison under imprecise class and cost distributions. In: Proceedings of the 3rd ACM SIGKDD International Conference on Knowledge Discovery and Data Mining (KDD), pp. 43–48. AAAI (1997)

14. Provost, F., Fawcett, T.: Robust classification for imprecise environments. Machine Learning **42**(3), 203–231 (2001)

15. Quinlan, J.R.: Bagging, boosting, and C4.5. In: Proceedings of the 13th National Conference on Artificial Intelligence (AAAI), pp. 725–730 (1996)

16. Raeder, T., Hoens, T.R., Chawla, N.V.: Consequences of variability in classifier performance estimates. In: Proceedings of the 10th IEEE International Conference on Data Mining (ICDM), pp. 421–430. IEEE (2010)

17. Seiffert, C., Khoshgoftaar, T.M., Hulse, J.V., Napolitano, A.: RUSBoost: Improving classification performance when training data is skewed. In: Proceedings of the 19th International Conference on Pattern Recognition (ICPR), pp. 1–4. IEEE (2009)

18. Sun, Y., Kamel, M.S., Wong, A.K.C., Wang, Y.: Cost-sensitive boosting for classification of imbalanced data. Pattern Recognition **40**(12), 3358–3378 (2007)

19. Ting, K.M.: A comparative study of cost-sensitive boosting algorithms. In: Proceedings of the 17th International Conference on Machine Learning (ICML), pp. 983–990

20. Zadrozny, B., Elkan, C.: Learning and making decisions when costs and probabilities are both unknown. In: Proceedings of the 7th ACM SIGKDD International Conference on Knowledge Discovery and Data Mining (KDD), pp. 204–213. ACM (2001)

21. Zadrozny, B., Langford, J., Abe, N.: Cost-sensitive learning by cost-proportionate example weighting. In: Proceedings of the 3rd IEEE International Conference on Data Mining (ICDM), pp. 435–442. IEEE (2003)

Repulsive-SVDD Classification

Phuoc Nguyen and Dat Tran[(⊠)]

Faculty of Education, Science, Technology and Mathematics,
University of Canberra, Canberra, ACT 2601, Australia
dat.tran@canberra.edu.au

Abstract. Support vector data description (SVDD) is a well-known kernel method that constructs a minimal hypersphere regarded as a data description for a given data set. However SVDD does not take into account any statistical distribution of the data set in constructing that optimal hypersphere, and SVDD is applied to solving one-class classification problems only. This paper proposes a new approach to SVDD to address those limitations. We formulate an optimisation problem for binary classification in which we construct two hyperspheres, one enclosing positive samples and the other enclosing negative samples, and during the optimisation process we move the two hyperspheres apart to maximise the margin between them while the data samples of each class are still inside their own hyperspheres. Experimental results show good performance for the proposed method.

Keywords: Repulsive SVDD · Support vector data description · Support vector machine · Classification

1 Introduction

Support vector data description (SVDD) [1] was proposed by Tax and Duin to train a hyperspherically shaped boundary around a normal dataset while keeping all abnormal data samples outside the hypersphere. This SVDD has been a successful approach to solving one-class problems such as outlier detection since the volume of this data description is kept minimal. One-class support vector machine (OC-SVM) [2] is a similar approach proposed earlier to estimate the support of a high-dimensional distribution. Although this method uses a maximal-margin hyperplane instead of a hypersphere to separate the normal data from the abnormal data, it has the same optimisation problem as SVDD. In both OC-SVM and SVDD, the boundary in the feature space when mapped back to the input space can produce a complex and tight description of the data distribution.

There are various extensions to SVDD. A small hypersphere and large margin approach was proposed in [3] for novelty detection problems where a minimal hypersphere was trained to include most of normal examples while the margin between the hypersphere and outliers is as large as possible. A further extension using two large margins instead of one was proposed in [4], where an interior margin between the hypersphere and the normal data and an exterior margin

© Springer International Publishing Switzerland 2015
T. Cao et al. (Eds.): PAKDD 2015, Part I, LNAI 9077, pp. 277–288, 2015.
DOI: 10.1007/978-3-319-18038-0_22

between the hypersphere and the abnormal data both are maximised. In [5], the authors define an optimisation problem as maximising the separation ratio $(R+d)/(R-d)$, where R is the hypersphere's radius and d is the hypersphere's margin. It is shown to be equivalent to minimising $(R^2 - kd^2)$ where k is a parameter to adjust between minimising R and maximising d. Hao et al. [6] also used a similar formulation in which several similarity functions were used to compute the distance to centres. Another extension of SVDD is [7] in which the use of two SVDDs for the description of data with two classes was proposed.

However all of those models are for one-class problems in which the task is to provide a tight data description or to detect outliers. When applying to a two-class problem where the numbers of data samples of two classes are not much different, the boundary of one-class methods is inappropriate. To overcome this problem, the first straight forward approach is to train two SVDDs, one for each class and define the decision boundary as the bisector between two surfaces of the hyperspheres. Although this approach improves the performance of one-class methods for two-class problems, they are limited by the small-sphere constraint of the data description.

In this paper, we propose a method using two SVDDs, one enclosing positive samples and the other enclosing negative samples, for binary classification tasks. The minimum bounding hypersphere constraint is relaxed to allow the hyperspheres to acquire larger regions. This is achieved by imposing a criterion that maximises the distance between two hyperspheres while still keeping the data inside the spheres. A margin variable is added to the optimisation to further improve the classification boundary. Since the proposed method trains two SVDDs that repel each other, we call it repulsive-SVDD classification (RSVC). RSVC decision boundary can be considered as a compromise between the boundary of a SVM boundary and a bisector boundary of two SVDDs' surfaces, this is controlled by a trade off parameter to adjust the balance between describing the data and maximising the distance between the two sphere centres.

The rest of the paper is organized as follows. The theory of the proposed RSVC will be presented in Section 2. Comparison of RSVC with Two SVDDs will be discussed in Section 3. Experimental results are presented to show the performance of the proposed method in Section 4. Finally, Section 5 presents our conclusions.

2   Proposed Approach: Repulsive-SVDD Classification (RSVC)

To apply SVDD for binary classification problems, we construct a hypersphere for each class to describe its data distribution with additional properties to discriminate the two classes. First, the hypersphere constraint in SVDD is relaxed to allow this hypersphere to acquire a larger area that is far from the other class. This is achieved by imposing a criterion that maximises the distance between two hyperspheres while still keeping all data samples of a class inside its hypersphere. Second, the margin (i.e., the distance between surfaces of the two hypersphere) is maximised, similar to the maximal margin philosophy of a support vector machine.

A visualisation of RSVC is demonstrated in Fig. 1. In the left figure, SVM determines a maximum margin hyperplane without considering data distributions of positive and negative classes. Whereas in the middle figure, SVDDs determine two minimal hyperspheres without considering the margin between the two classes, and the decision boundary is the perpendicular hyperplane of the line segment connecting the two hypersphere centres.

By contrast, our RSVC can provide an intermediate solution between SVM and SVDDs. Given the problem in Fig. 1, the RSVC optimisation problem attempts to keep the radii minimum while maximising the distance between the two hyperspheres. As a result, the hyperspheres will expand in the direction that increases the distance between the two hyperspheres. Moreover, the weights of these two directions can be controlled by a parameter.

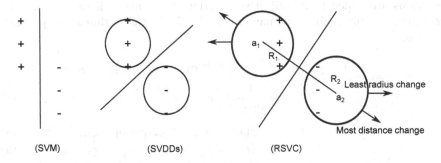

Fig. 1. SVM (left figure) determines a maximum margin hyperplane without considering data distributions of positive and negative classes. SVDDs (middle figure) determine minimum hyperspheres without considering the margin between two classes. RSVC (right figure) determines two minimal hyperspheres, one enclosing positive samples and the other enclosing negative samples, while maximising the distance between two centres to a degree controlled by a parameter.

2.1 Problem Formulation

Consider a dataset $\{x_i\}, i = 1, \ldots, n$ with two classes, positive class with n_1 data samples and negative class with n_2 data samples, $n_1 + n_2 = n$. The problem of RSVC is to determine two optimal hyperspheres (a_1, R_1) and (a_2, R_2), one encloses data samples of the positive class and the other encloses data samples of the negative class, and at the same time maximise the distance between the two centres. In addition, all positive and negative data samples are forced to stay outside the margin ρ_1 and ρ_2 of the positive hypersphere and the margin of the negative hypersphere respectively. The optimisation problem is formulated as follows:

$$\min_{R_1,R_2,a_1,a_2,\rho_1,\rho_2} \quad R_1^2 + R_2^2 - k||a_1 - a_2||^2 - \mu_1\rho_1 - \mu_2\rho_2 \tag{1}$$

$$s.t. \quad ||\phi(x_i) - a_1||^2 \leq R_1^2 - \rho_1, \qquad \forall i, y_i = +1 \tag{2}$$

$$||\phi(x_i) - a_1||^2 \geq R_1^2 + \rho_1, \qquad \forall i, y_i = -1 \tag{3}$$

$$||\phi(x_i) - a_2||^2 \leq R_2^2 - \rho_2, \qquad \forall i, y_i = -1 \tag{4}$$

$$||\phi(x_i) - a_2||^2 \geq R_2^2 + \rho_2, \qquad \forall i, y_i = +1 \tag{5}$$

$$\rho_1 \geq 0, \rho_2 \geq 0 \tag{6}$$

where k is a parameter which represents the repulsive degree between two centres, μ_1 and μ_2 are two parameters controlling the support vectors, and ϕ is the mapping to transform the vector x_i to a feature space.

The above problem is for separable datasets. In practice, to allow errors, the constraints are relaxed by introducing slack variables ξ_{1i} and ξ_{2i}, and penalized terms are added to its objective function. In addition, if we combine the constraints in this problem to have a simpler form, the optimisation problem becomes:

$$\min_{R_1,R_2,a_1,a_2,\rho_1,\rho_2,\xi_{1i},\xi_{2i}} \quad R_1^2 + R_2^2 - k||a_1 - a_2||^2 - \mu_1\rho_1 - \mu_2\rho_2$$

$$+\frac{1}{\nu_1 n_1}\sum_i \xi_{1i} + \frac{1}{\nu_2 n_2}\sum_i \xi_{2i} \tag{7}$$

$$s.t. \quad y_i||\phi(x_i) - a_1||^2 \leq y_i R_1^2 - \rho_1 + \xi_{1i}, \qquad \forall i \tag{8}$$

$$y_i||\phi(x_i) - a_2||^2 \geq y_i R_2^2 + \rho_2 - \xi_{2i}, \qquad \forall i \tag{9}$$

$$\rho_1 \geq 0, \rho_2 \geq 0 \tag{10}$$

$$\xi_{1i} \geq 0, \xi_{2i} \geq 0 \qquad \forall i \tag{11}$$

where ν_1 and ν_2 are parameters controlling the number of support vectors, together with μ_1 and μ_2. They will be explained in Proposition 1 below.

2.2 Convex Formulation of RSVC

Although the optimisation in (7) has a non-convex objective function, it can be reformulated to have a convex form as follows:

$$\min_{R_1,R_2,a_1,a_2,\rho_1,\rho_2,\xi_{1i},\xi_{2i}} \quad R_1^2 - a_1^2 + R_2^2 - a_2^2 + a_1^2 + a_2^2 - k||a_1 - a_2||^2$$

$$-\mu_1\rho_1 - \mu_2\rho_2 + \frac{1}{\nu_1 n_1}\sum_i \xi_{1i} + \frac{1}{\nu_2 n_2}\sum_i \xi_{2i} \tag{12}$$

$$s.t. \quad y_i\phi(x_i)^2 - 2y_i\phi(x_i)a_1 \leq y_i(R_1^2 - a_1^2) - \rho_1 + \xi_{1i}, \forall i \tag{13}$$

$$y_i\phi(x_i)^2 - 2y_i\phi(x_i)a_2 \geq y_i(R_2^2 - a_2^2) + \rho_2 - \xi_{2i}, \forall i \tag{14}$$

$$\rho_1 \geq 0, \rho_2 \geq 0 \tag{15}$$

$$\xi_{1i} \geq 0, \xi_{2i} \geq 0 \qquad \forall i \tag{16}$$

Let $\delta_1 = a_1^2 - R_1^2, \delta_2 = a_2^2 - R_2^2$ and $0 \leq \delta_0 \leq ||a_1 - a_2||^2$, (12) becomes

$$\min_{\delta_1,\delta_2,\delta_0,a_1,a_2,\rho_1,\rho_2,\xi_{1i},\xi_{2i}} \quad -\delta_1 - \delta_2 + a_1^2 + a_2^2 - k\delta_0 - \mu_1\rho_1 - \mu_2\rho_2$$

$$+\frac{1}{\nu_1 n_1}\sum_i \xi_{1i} + \frac{1}{\nu_2 n_2}\sum_i \xi_{2i} \tag{17}$$

$$s.t. \quad 2y_i\phi(x_i)a_1 - y_i\phi(x_i)^2 \geq y_i\delta_1 + \rho_1 - \xi_{1i}, \qquad \forall i \tag{18}$$

$$2y_i\phi(x_i)a_2 - y_i\phi(x_i)^2 \leq y_i\delta_2 - \rho_2 + \xi_{2i}, \qquad \forall i \tag{19}$$

$$\rho_1 \geq 0, \rho_2 \geq 0 \tag{20}$$

$$\xi_{1i} \geq 0, \xi_{2i} \geq 0 \qquad \forall i \tag{21}$$

$$||a_1 - a_2||^2 \geq \delta_0 \tag{22}$$

$$\delta_0 \geq 0 \tag{23}$$

We can construct the Lagrange function below using these following Lagrange multipliers $\alpha_{1i}, \alpha_{2i}, \gamma_{1i}, \gamma_{2i}, \theta_1, \theta_2, \beta, \lambda$:

$$L(\delta_1, \delta_2, \delta_0, a_1, a_2, \rho_1, \rho_2, \xi_{1i}, \xi_{2i}, \alpha_{1i}, \alpha_{2i}, \gamma_{1i}, \gamma_{2i}, \theta_1, \theta_2, \beta, \lambda) = -\delta_1 - \delta_2$$

$$+a_1^2 + a_2^2 - k\delta_0 - \mu_1\rho_1 - \mu_2\rho_2 + \frac{1}{\nu_1 n_1}\sum_i \xi_{1i} + \frac{1}{\nu_2 n_2}\sum_i \xi_{2i}$$

$$-\sum_i \alpha_{1i}(2y_i\phi(x_i)a_1 - y_i\phi(x_i)^2 - y_i\delta_1 - \rho_1 + \xi_{1i}) - \sum_i \gamma_{1i}\xi_{1i} - \theta_1\rho_1 \tag{24}$$

$$+\sum_i \alpha_{2i}(2y_i\phi(x_i)a_2 - y_i\phi(x_i)^2 - y_i\delta_2 + \rho_2 - \xi_{2i}) - \sum_i \gamma_{2i}\xi_{2i} - \theta_2\rho_2$$

$$-\beta(||a_1 - a_2||^2 - \delta_0) - \lambda\delta_0$$

Using KKT conditions, we have:

$$\frac{\partial L}{\partial \delta_1} = 0 \Rightarrow -1 + \sum_i \alpha_{1i}y_i = 0 \Rightarrow \sum_i \alpha_{1i}y_i = 1 \tag{25}$$

$$\frac{\partial L}{\partial \delta_2} = 0 \Rightarrow -1 + \sum_i \alpha_{2i}y_i = 0 \Rightarrow \sum_i \alpha_{2i}y_i = 1 \tag{26}$$

$$\frac{\partial L}{\partial \delta_0} = 0 \Rightarrow -k + \beta - \lambda = 0 \Rightarrow \beta - \lambda = k \tag{27}$$

$$\frac{\partial L}{\partial a_1} = 0 \Rightarrow (1 - \beta)a_1 + \beta a_2 = \sum_i \alpha_{1i}y_i\phi(x_i) = A \tag{28}$$

$$\frac{\partial L}{\partial a_2} = 0 \Rightarrow (1 - \beta)a_2 + \beta a_1 = -\sum_i \alpha_{2i}y_i\phi(x_i) = -B \tag{29}$$

$$\frac{\partial L}{\partial \xi_{1i}} = 0 \Rightarrow \frac{1}{\nu_1 n_1} - \alpha_{1i} - \gamma_{1i} = 0 \Rightarrow \alpha_{1i} + \gamma_{1i} = \frac{1}{\nu_1 n_1} \quad \forall i \tag{30}$$

$$\frac{\partial L}{\partial \xi_{2i}} = 0 \Rightarrow \frac{1}{\nu_2 n_2} - \alpha_{2i} - \gamma_{2i} = 0 \Rightarrow \alpha_{2i} + \gamma_{2i} = \frac{1}{\nu_2 n_2} \quad \forall i \tag{31}$$

$$\frac{\partial L}{\partial \rho_1} = 0 \;\Rightarrow\; -\mu_1 + \sum_i \alpha_{1i} - \theta_1 = 0 \;\Rightarrow\; \sum_i \alpha_{1i} - \theta_1 = \mu_1 \tag{32}$$

$$\frac{\partial L}{\partial \rho_1} = 0 \;\Rightarrow\; -\mu_2 + \sum_i \alpha_{2i} - \theta_2 = 0 \;\Rightarrow\; \sum_i \alpha_{2i} - \theta_2 = \mu_2 \tag{33}$$

Equations (28) and (29) leads to

$$\begin{cases} a_1 + a_2 = A - B \\ a_1 - a_2 = \frac{A+B}{1-2\beta} \end{cases} \Rightarrow \begin{cases} a_1 = \frac{(1-\beta)A+\beta B}{1-2\beta} \\ a_2 = \frac{-\beta A+(\beta-1)B}{1-2\beta} \end{cases} \tag{34}$$

By substituting the KKT conditions into the Lagrangian function we obtain the dual form of the optimisation:

$$\min \frac{1}{1-2k}\Big[(1-k)\sum_{i,j}\alpha_{1i}\alpha_{1j}y_iy_jK(x_i,x_j) + (1-k)\sum_{i,j}\alpha_{2i}\alpha_{2j}y_iy_jK(x_i,x_j)$$

$$+2k\sum_{i,j}\alpha_{1i}\alpha_{2j}y_iy_jK(x_i,x_j)\Big] + \sum_i \alpha_{1i}y_iK(x_i,x_i) - \sum_i \alpha_{2i}y_iK(x_i,x_i) \tag{35}$$

$$s.t. \quad \sum_i \alpha_{1i}y_i = 1 \;, \; \sum_i \alpha_{2i}y_i = -1 \tag{36}$$

$$\sum_i \alpha_{1i} = \mu_1 \;, \; \sum_i \alpha_{2i} = \mu_2 \tag{37}$$

$$0 \le \alpha_{1i} \le \frac{1}{\nu_1 n_1} \;, \; 0 \le \alpha_{2i} \le \frac{1}{\nu_2 n_2} \quad \forall i \tag{38}$$

where the inner product between vectors has been replaced by the kernel K, and the Lagrange multipliers $\gamma_{1i} \ge 0, \gamma_{2i} \ge 0, \theta_1 \ge 0, \theta_2 \ge 0, \lambda \ge 0$ have been removed from Equations (30), (31), (32), (33) and (27) respectively. Similarly to ν-SVC, $\sum_i \alpha_{1i}$ is set to μ_1, $\sum_i \alpha_{2i}$ is set to μ_2 and β is set to k, where k is a parameter chosen in the range $k \in [0, \frac{1}{2})$.

It can be seen that if k is set to 0 in the above optimisation problem then the RSVC optimisation problem (35) can be broken into two independent optimisation problems similar to SVDDs except for the extra constraints $\sum_i \alpha_{1i} = \mu_1$ and $\sum_i \alpha_{2i} = \mu_2$ resulting from the margin requirements in the original RSVC problem (1).

Solving the problem (35) gives a set of α_{1i}, α_{2i}. Then the centres a_1, a_2 can be determined from Equations (34).

To determine the radius R_1, the support vector x_t that lies on the surface of the hypersphere (a_1, R_1) and corresponds to the smallest $\alpha_{1t} \in (0, \frac{1}{\nu_1 n_1})$ is selected. Then the radius R_1 is calculated as $R_1 = d_1(x_t)$, where $d_1(x_t)$ is the distance from x_t to the centre a_1 and is determined as follows:

$$d_1^2(x_t) = \|\phi(x_t) - a_1\|^2 = K(x_t,x_t) - \frac{2}{1-k}\Big[(1-k)\sum_i \alpha_{1i}y_iK(x_t,x_i)$$

$$+ k\sum_i \alpha_{2i}y_iK(x_t,x_i)\Big] + a_1^2 \tag{39}$$

The radius R_2 is calculated similarly:

$$d_2^2(x_t) = \|\phi(x_t) - a_2\|^2 = K(x_t, x_t) - \frac{2}{1-k}\Big[- k\sum_i \alpha_{2i} y_i K(x_t, x_i)$$
$$+ (k-1)\sum_i \alpha_{2i} y_i K(x_t, x_i)\Big] + a_2^2 \qquad (40)$$

In the test phase, a sample x can be determined whether it belongs to the hypersphere (a_1, R_1) or (a_2, R_2), i.e. class +1 or class -1, by the following decision function:

$$sign(d_2^2(x) - d_1^2(x)) \qquad (41)$$

2.3 ν-Property

Following [8], a data sample x_i is called a *support vector* if it has Lagrange multiplier $\alpha_i > 0$; a data sample is called a *margin error* if it has positive slack variable $\xi_i > 0$.

Similarly to the property of the ν parameter in ν-SVC [8], we derive the property for the ν_1, ν_2, μ_1 and μ_2 parameters and use it for parameter selection to train the RSVC.

Proposition 1. *Let m_1 and m_2 denote the number of margin errors of the positive sphere and negative sphere respectively, and let s_1 and s_2 denote their numbers of support vectors. Then for parameters ν_1, ν_2, μ_1 and μ_2 we have:*

1. *$\mu_1\nu_1$ and $\mu_2\nu_2$ are upper bounds on the fraction of margin errors, and a lower bound on the fraction of support vectors for the positive sphere and negative sphere respectively:*

$$\frac{m_1}{n_1} \leq \mu_1\nu_1 \leq \frac{s_1}{n_1} \quad and \quad \frac{m_2}{n_2} \leq \mu_2\nu_2 \leq \frac{s_2}{n_2} \qquad (42)$$

2. *The feasible ranges of ν_1, ν_2, μ_1 and μ_2 are:*

$$0 < \nu_1 \leq 1 \,,\ 1 \leq \mu_1 \leq \frac{1}{\nu_1} \quad and \quad 0 < \nu_2 \leq 1 \,,\ 1 \leq \mu_2 \leq \frac{1}{\nu_2} \qquad (43)$$

Proof. We first prove for the positive hypersphere.

1. By the KKT conditions, all data points with $\xi_{1i} > 0$ imply $\gamma_{1i} = 0$. From (30) we have the equation $\alpha_{1i} = 1/(\nu_1 n_1)$ holds for every margin error. Summing up α_{1i} and using $\sum_i \alpha_{1i} = \mu_1$ from (37) we have:

$$\frac{m_1}{\nu_1 n_1} \leq \sum_i \alpha_{1i} = \mu_1 \qquad (44)$$

On the other hand, (38) indicates that each support vector of the positive hypersphere can get at most $1/(\nu_1 n_1)$. Therefore summing up α_{1i} for support vectors of positive hypersphere, plus $\alpha_{1i} = 0$ for non-support vectors, and from (37) we have:

$$\frac{s_1}{\nu n_1} \geq \sum_i \alpha_{1i} = \mu_1 \qquad (45)$$

Combining (44) and (45) we have the inequalities (42) for the positive hypersphere.

2. From (42) we have $0 < \mu_1 \nu_1 \leq 1$. In addition, from (36) we have $\sum_i \alpha_{1i} y_i = 1$,

or $\sum_{\{i:y_i=+1\}} \alpha_{1i} = 1 + \sum_{\{i:y_i=-1\}} \alpha_{1i}.$

Since $\alpha_{1i} \geq 0 \; \forall i$, this leads to $\mu_1 = \sum_i \alpha_{1i} \geq \sum_{\{i:y_i=+1\}} \alpha_{1i} \geq 1.$

Combining these results we have the proof of (43).

The proof of inequalities (42) and (43) for the negative hypersphere is similar.

The proposed RSVC is for binary classification problems. It can be extended for multi-class classification problems by using "one-against-the rest" approach or "one-against-one" approach. Following [9], we use the one-against-one approach in this paper where data of every pair of classes are used to train a binary classifier that separates the two classes, resulting in $M(M-1)/2$ classifiers in a M-class classification problem. In the test phase, a voting strategy is used: each binary classification of a test sample generates a vote, and the class with the maximum number of votes for this test data sample is output as the overall classification result. In case that two classes have identical votes, one can simply choose the class appearing first in the array of storing class names as in [9].

3   Comparison of RSVC with Two SVDDs

SVDD can be extended to two SVDDs to describe a data set of two classes. Consider a data set $\{x_i\}, i = 1, \ldots, n$ of two classes, positive class with n_1 data samples and negative class with n_2 data samples, $n_1 + n_2 = n$. The optimisation problem is formulated as follows [7]:

$$\min_{R_1, R_2, a_1, a_2, \xi_{1i}, \xi_{2i}} \quad R_1^2 + R_2^2 + \frac{1}{\nu_1 n_1} \sum_i \xi_{1i} + \frac{1}{\nu_2 n_2} \sum_i \xi_{2i}$$

$$\begin{aligned}
s.t. \quad &||x_i - a_1||^2 \leq R_1^2 + \xi_{1i}, &&\forall i, y_i = +1 \\
&||x_i - a_1||^2 \geq R_1^2 - \xi_{1i}, &&\forall i, y_i = -1 \\
&||x_i - a_2||^2 \leq R_2^2 + \xi_{2i}, &&\forall i, y_i = -1 \\
&||x_i - a_2||^2 \geq R_2^2 - \xi_{2i}, &&\forall i, y_i = +1 \\
&\xi_{1i} \geq 0, \xi_{2i} \geq 0 &&\forall i
\end{aligned} \qquad (46)$$

where (a_1, R_1) and (a_2, R_2) are two hyperspheres, ν_1, ν_2 are parameters.

This optimisation can produce a description of two minimal hyperspheres enclosing two classes. The decision boundary can be defined as the bisector between their surfaces. However this model is for one-class problems in which the task is to provide a tight data description or to detect outliers. When applying to a two-class problem where the data samples of two classes are balance the boundary of one-class methods is inappropriate. The RSVC can overcome this problem by allowing hyperspheres to acquire a larger area by minimising $-k||a_1 - a_2||^2$ and creating a larger margin by minimising $-\mu_1\rho_1 - \mu_2\rho_2$ while still trying to provide data description for two classes.

4 Experiments

4.1 2-D Demonstration of RSVC

Figure 2 shows visual results for experiments performed on a simple 2-D datasets using RSVC. When parameter $k = 0$, the RSVC optimisation function becomes the optimisation function for two SVDDs, hence two SVDDs is a special case of RSVC. It can be seen that when k increases, two hyperspheres repulsed each other, resulting in a larger margin in between. Those data samples outside the hyperspheres but inside this margin are penalised by a cost proportional to $1/(\nu_1 n_1)$ or $1/(\nu_2 n_2)$. The decision boundary is the bisector between the hyperspheres' surfaces. The first row in Figure 2 shows that when parameter k increases, the hypersphere enclosing positive samples is moving away from negative samples while keeping all the positive samples inside it. The second row in Figure 2 shows that when $\mu_1\nu_1$ and $\mu_2\nu_2$ increase, more positive samples are outside the hyperspheres.

Classification experiments were conducted on 9 UCI datasets[1]. Details of these datasets are listed in Table 1. The datasets were divided in to 2 subsets, the subset contained 50% of the data is for training and the other 50% for testing. The training process was done using 5-fold cross validation. The parameters for the methods are as follows. Gaussian mixture models (GMM) [10] use 64 mixture components. OC-SVM parameters are searched in $\gamma \in \{2^{-13}, 2^{-11}, \ldots, 2^1\}$ and $\nu \in \{2^{-5}, 2^{-4}, \ldots, 2^{-2}\}$. Parameters of SVDD and SVDD with negative examples (Two SVDDs) are searched in $\gamma \in \{2^{-13}, 2^{-11}, \ldots, 2^1\}$ and $\nu \in \{2^{-5}, 2^{-4}, \ldots, 2^{-2}\}$. SVM parameters are search in $\gamma \in \{2^{-13}, 2^{-11}, \ldots, 2^1\}$ and $C \in \{2^{-1}, 2^3, \ldots, 2^{15}\}$; and RSVC parameters are searched in $\gamma \in \{2^{-7}, 2^{-5}, \ldots, 2^{-1}\}$, $\nu_1 = \nu_2 \in \{0.001, 0.01\}$, $\mu_1 = \mu_2 \in \{10, 30, \ldots, 90\}$, and $k \in \{0.5, 0.7, 0.9\}$.

Note that the parameter γ in RSVC is searched in a narrower range than that in SVM, while $\nu_1 n_1$ and $\nu_2 n_2$ are searched in a roughly similar number of options as of parameter C. This is to produce a sparse number of support vectors and avoid over fitting of the two SVDDs. Parameter $k \in \{0.5, 0.7, 0.9\}$ is to favour classification more than tight description. After the best parameters are selected in the cross validation step, the models are trained again with them on the whole training set and are tested on the 50% unseen test set. Experiments

---

[1] Available online at http://www.csie.ntu.edu.tw/~cjlin/libsvmtools/datasets/

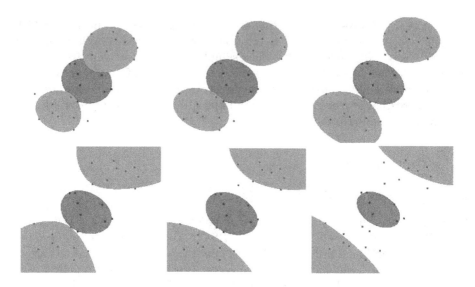

Fig. 2. The first row contains screenshots for RSVC when $k = 0, 0.3$ and 0.6, and $\mu_1 \nu_1 = \mu_2 \nu_2 = 0.2$. The second row contains screenshots for RSVC when $\mu_1 \nu_1 = \mu_2 \nu_2 = 0.1, 0.2$ and 0.5, and $k = 0.9$. A Gaussian RBF kernel was used, with $\gamma = 5$. Red points are positive samples and blue points are negative samples.

were repeated 10 times and the results were averaged with standard deviations given.

Table 2 shows the prediction rates in cross validation training. Table 3 shows the prediction rates on unseen test sets with best parameters selected.

It can be seen that the GMM, OCSVM and SVDD have undesirable performance in the classification task.

The two SVDDs have much higher performance than these one-class methods since they describe two minimal hyperspheres enclosing two classes and the

Table 1. Dataset information: number of classes, dataset size and number of features

Data set	#class	size	#feature
Fourclass	2	862	2
Liver disorders	2	345	6
Heart	2	270	13
Wine	3	178	13
Breast Cancer	2	683	10
Diabetes	2	768	8
Australian	2	690	14
Ionosphere	2	351	34
German numer	2	1000	24

decision boundary is the bisector between their surfaces. It can be seen that SVM has higher performance than two SVDDs, it trains a maximal-margin separating hyperplane rather than two minimal hyperspheres. RSVC show highest performance in most datasets. RSVC can overcome the limitation of two SVDDs for the classification task by training two SVDDs that repel each other, allowing spheres to acquire a larger area and creating a larger margin while still trying to provide data description for two classes.

Table 2. Prediction rates in cross validation training of classification methods

Dataset	GMM	OCSVM	SVDD	Two SVDDs	SVM	RSVC
Fourclass	67.24 ±5.73	62.15 ±3.3	54.01 ±4.97	71.44 ±4.15	75.08 ±4.4	77.98 ±4.52
Liver disorders	40.86 ±5.63	50.41 ±5.69	55.41 ±5.87	55.15 ±5.49	59.90 ±3.53	60.14 ±5.56
Heart	46.33 ±4.26	60.24 ±5.89	46.41 ±4.24	61.11 ±5.98	72.56 ±4.24	76.44 ±4.45
Wine	33.43 ±5.03	55.57 ±4.07	46.43 ±5.8	59.89 ±3.7	75.24 ±5.56	83.15 ±4.59
Breast cancer	56.52 ±3.34	73.85 ±4.11	62.91 ±4.15	77.16 ±4.08	81.29 ±4.44	81.49 ±4.46
Diabetes	55.24 ±5.06	51.84 ±3.99	40.24 ±5.16	50.95 ±5.63	63.47 ±5.93	66.87 ±3.28
Australian	54.15 ±5.84	58.36 ±5.52	48.23 ±5.16	61.03 ±5.75	70.96 ±3.53	71.90 ±3.66
Inosphere	57.12 ±3.61	65.48 ±3.85	34.26 ±3.48	68.92 ±4.59	73.69 ±4.51	75.86 ±4.29
German numer	40.09 ±5.49	58.96 ±5.39	58.14 ±5.31	59.65 ±5.51	64.04 ±5.99	65.75 ±3.35

Table 3. Prediction rates on unseen test sets; classification methods on 9 datasets

Dataset	GMM	OCSVM	SVDD	Two SVDDs	SVM	RSVC
Fourclass	67.24 ±5.73	59.08 ±3.24	54.44 ±5.09	72.24 ±5.04	70.72 ±5.64	75.65 ±5.92
Liver disorders	40.86 ±5.63	43.48 ±4.88	47.68 ±4.4	50.25 ±5.15	52.03 ±3.03	54.12 ±5.53
Heart	46.33 ±4.26	57.49 ±4.83	46.41 ±4.24	61.08 ±3.02	71.51 ±4.47	72.12 ±4.36
Wine	33.43 ±5.03	42.09 ±6.99	21.41 ±2.35	46.46 ±5.84	75.66 ±4.86	76.99 ±4.69
Breast cancer	56.52 ±3.34	73.08 ±4.01	48.34 ±7.87	75.03 ±4.33	79.92 ±4.34	79.79 ±5.07
Diabetes	55.24 ±5.06	55.68 ±5.34	39.30 ±4.98	54.10 ±5.71	60.21 ±3.16	59.00 ±3.81
Australian	54.15 ±5.84	56.44 ±5.53	48.38 ±5.03	55.75 ±3.83	69.71 ±3.42	68.95 ±3.53
Inosphere	57.12 ±3.61	62.55 ±4.71	38.41 ±2.7	65.79 ±5.72	69.07 ±4.3	70.74 ±4.63
German numer	40.09 ±5.49	58.07 ±5.47	58.40 ±5.34	57.46 ±5.32	62.30 ±5.7	63.90 ±5.67

5 Conclusion

We have proposed the repulsive-SVDD classification to extend SVDD for binary classification problems. Two hyperspheres are trained in an optimisation problem to describe the distribution of two classes. Additional requirements are added to the optimisation problem to help with the discrimination task. First, the distance between two hypersphere centres is maximised to allow hyperspheres to expand. Second, margins between the hypersphere surfaces and data are maximised. The resulting method can create a decision boundary that takes information not only from distributions of the classes but also the boundary's margins. Experimental results on 9 datasets validate the good performance of the proposed method.

References

1. Tax, D.M.J., Duin, R.P.W.: Support vector data description. Machine Learning **54**(1), 45–66 (2004)
2. Schlkopf, B., Platt, J.C., Shawe-Taylor, J., Smola, A.J., Williamson, R.C.: Estimating the support of a high-dimensional distribution. Neural Computation **13**(7), 1443–1471 (2001)
3. Wu, M., Ye, J.: A small sphere and large margin approach for novelty detection using training data with outliers. IEEE Transactions on Pattern Analysis and Machine Intelligence **31**(11), 2088–2092 (2009)
4. Le, T., Tran, D., Ma, W., Sharma, D.: An optimal sphere and two large margins approach for novelty detection. In: The 2010 International Joint Conference on Neural Networks (IJCNN), pp. 1–6. IEEE (2010)
5. Wang, J., Neskovic, P., Cooper, L.N.: Pattern classification via single spheres. In: Hoffmann, A., Motoda, H., Scheffer, T. (eds.) DS 2005. LNCS (LNAI), vol. 3735, pp. 241–252. Springer, Heidelberg (2005)
6. Hao, P.-Y., Chiang, J.-H., Lin, Y.-H.: A new maximal-margin spherical-structured multi-class support vector machine. Applied Intelligence **30**(2), 98–111 (2009)
7. Huang, G., Chen, H., Zhou, Z., Yin, F., Guo, K.: Two-class support vector data description. Pattern Recognition **44**(2), 320–329 (2011)
8. Schlkopf, B., Smola, A.J., Williamson, R.C., Bartlett, P.L.: New support vector algorithms. Neural Computation **12**(5), 1207–1245 (2000)
9. Chang, C.C., Lin, C.J.: LIBSVM: a library for support vector machines. ACM Transactions on Intelligent Systems and Technology (TIST) **2**(3), 27 (2011)
10. Bilmes, J.A., et al.: A gentle tutorial of the em algorithm and its application to parameter estimation for gaussian mixture and hidden markov models. International Computer Science Institute **4**(510), 126 (1998)

Centroid-Means-Embedding:
An Approach to Infusing Word Embeddings into Features for Text Classification

Mohammad Golam Sohrab[✉], Makoto Miwa, and Yutaka Sasaki

Faculty of Engineering, Toyota Technological Institute, 2-12-1 Hisakata,
Tempaku-ku, Nagoya 468-8511, Japan
{sohrab,makoto-miwa,yutaka.sasaki}@toyota-ti.ac.jp

Abstract. This paper presents word embedding-based approach to text classification. In this study, we introduce a new vector space model called Semantically-Augmented Statistical Vector Space Model (SAS-VSM) that is a statistical VSM with a semantic VSM for information access systems, especially for automatic text classification. In the SAS-VSM, we first implement a primary approach to concatenate continuous-valued semantic features with an existing statistical VSM. We, then, introduce the Centroid-Means-Embedding (CME) method that updates existing statistical feature vectors with semantic knowledge. Experimental results show that the proposed CME-based SAS-VSM approaches are promising over the different weighting approaches on the 20 Newsgroups and RCV1-v2/LYRL2004 datasets using Support Vector Machine (SVM) classifiers to enhance the classification tasks. Our approach outperformed other approaches in both micro-F_1 and categorical performance.

Keywords: Text classification · Word embedding · Machine learning · Term weighting · Semantic indexing

1 Introduction

Due to the growing availability of digital textual documents, automatic text classification (ATC) has been actively studied to organize a vast amount of unstructured documents into a set of categories, based on the textual contents of the document. Most automatic classification systems analyze documents statistically and linguistically, determine important terms from the documents, and generate vector representations from these important terms. A good text-to-vector representation is necessary in order to enhance ATC and accomplish effective document retrieval [1], [6], [14], [15].

In recent years, in addition to supervised statistical learning approaches, many studies have been carried out with showing success in adopting unsupervised methods for learning continuous word embedding [3], [5], [7] from unlabeled texts. Word embedding features have actively studied on word analogies, word similarity, chunking, and named entity recognition (NER). Word embeddings

© Springer International Publishing Switzerland 2015
T. Cao et al. (Eds.): PAKDD 2015, Part I, LNAI 9077, pp. 289–300, 2015.
DOI: 10.1007/978-3-319-18038-0_23

are also used in ATC [22], but there remains the task of investigating how word embedding features can be infused into existing statistical features.

In ATC, words or terms in a certain document are Zipf distributed, that is, most of the words in some documents appear a few times or completely absent in other documents or in some categories. These infrequent words usually cannot be fully trained by term frequency-based approaches. Weighting approaches like the TF.IDF thus give positive discrimination to infrequent terms by biasing them against frequent terms. Furthermore, the training set may not have enough discriminative features to obtain a good vector space model (VSM). Some documents in the training or test set may not share enough information to classify the test set properly. Therefore, word embedding can be useful as input to classification models or as additional features to enhance existing systems. In this paper embedding vectors are generated using the global vectors (GloVe) model [5].

The motivation for exploiting word embedding features for ATC can be attributed to two main properties. First, in generating a more information-rich VSM, it is interesting to understand how continuous embedding features may assist to enhance ATC. Second, there is a demand for document representation to integrate semantic VSM into statistical VSM.

In this paper, we propose a new Centroid-Means-Embedding (CME)-based Semantically-Augmented Statistical-VSM (SAS-VSM) approach that exploits infusing embedding features, where the degree of semantic similarity is estimated using word co-occurrence information from unlabeled texts into features for ATC. This study makes the following major contributions with introducing the CME based SAS-VSM approach to address ATC.

- The word embedding vectors help to enrich categorical performances, and the augmented approaches outperformed all baseline approaches.
- The CME approach enriches the existing statistical VSMs using semantic knowledge.
- The CME enriches every category performance on the 20 Newsgroups and RCV1-v2 datasets over the statistical VSM-based system.
- The proposed CME-based SAS-VSM is a prominent approach in ATC.

2   SAS-VSM: Semantically-Augmented Statistical-VSM

In the ATC, the construction of VSM has always been considered as the most important step. Most ATC systems analyze documents statistically, determine important terms from document space $D = \{d_1, d_2, ...d_n\}$ and generate a text-to-vector representation from these important terms in order to reduce the complexity of the documents and make them easier to handle. To generate text-to-vector representation, two properties are main concern to determine the important terms from documents: the widely used statistical-VSM and recently-focused semantic-based VSM.

Of greater interest with both of these VSMs to enhance ATC, we introduce SAS-VSM that merges together statistical VSM and word-co-occurrence-based continuous embedding vectors. The architecture of SAS-VSM for a document space can be represented as:

$$\text{SAS-VSM} = \text{Statistical-VSM} \parallel \text{Semantic-VSM},$$

where $\parallel$ denotes the concatenation of two different VSMs. An SAS-VSM feature vector $\boldsymbol{x}(d)$ for a document d is:

$$\boldsymbol{x}(d) = \left(\boldsymbol{x}^{Stat}(d), \boldsymbol{x}^{Sem}(d)\right),$$

where $\boldsymbol{x}^{Stat}(d)$ is a statistical feature vector and $\boldsymbol{x}^{Sem}(d)$ is a semantic feature vector.

In this work, the Statistical-VSM is formulated based on different weighting approaches for a given corpus. In contrast, to represent the Semantic-VSM, we consider a context prediction GloVe model for learning word embedding. Word embedding is useful to inject additional semantic features to the existing VSM. It is an open question how continuous word embedding features should be infused into discrete weights of term vectors. We considered two approaches to representing the SAS-VSM: (1) the primary approach, which shows the motivation to incorporate two different VSM and (2) the CME approach.

2.1   Primary Approach

In the above formulation of SAS-VSM, where the Statistical-VSM denotes term weightings based on discrete weights of terms for a corresponding document d. A Statistical-VSM vector is an $\boldsymbol{x}^{Stat}(d) = \left(x_1^{Stat}(d), \dots, x_M^{Stat}(d)\right)$. For term t_i, $x_i^{Stat}(d)$ is defined as:

$$x_i^{Stat}(d) = \begin{cases} f(t_i), & \text{if } t_i \in d \\ 0, & \text{otherwise} \end{cases}, \tag{1}$$

where $f(t_i)$ is a term weighting function representing any weighting approach for term t_i which will be later discussed on Section 3. In contrast, a Semantic-SVM vector is an $\boldsymbol{x}^{Sem}(d) = (x_1^{Sem}(d), \dots, x_N^{Sem}(d))$ for document d. Using a word embedding matrix $\boldsymbol{V}$, $\boldsymbol{x}^{Sem}(d)$ is defined as:

$$\boldsymbol{x}^{Sem}(d) = \boldsymbol{x}^{Stat}(d)\boldsymbol{V}. \tag{2}$$

In the ATC, document d consists of a sequence of terms or words $t_i = \{t_1, t_2, \dots t_n\}$. Let Σ be the vocabulary set of a given corpus. Word embedding matrix $\boldsymbol{V}$ is an $M \times N$ matrix with the vocabulary size M and the dimensionality N of word embedding vector. That is, $\boldsymbol{V}$ is defined as follows:

$$V = \begin{bmatrix} \boldsymbol{v}_1 \\ \boldsymbol{v}_2 \\ \vdots \\ \boldsymbol{v}_M \end{bmatrix} = \begin{bmatrix} v_{11} & v_{12} & \cdots & v_{1N} \\ v_{21} & v_{22} & \cdots & v_{2N} \\ \vdots & \vdots & \ddots & \vdots \\ v_{M1} & v_{M2} & \cdots & v_{MN} \end{bmatrix}.$$

Each row v_i represents the embedding vector for term t_i. This approach is our primary approach which leads to generate centroid-means-embedding vector. From the Eqn. 2, we can see that new updated augmented features for document d are incorporated with discrete and continuous weights. However the existing supervised discrete weights based on different term weighting schemes are remained free from getting continuous weight.

2.2   CME with SAS-VSM

In this proposed approach, we will introduce how to infuse continuous word embedding vectors into existing discrete weights by rewriting Eqn. 1 and 2. We first compute a sum centroid embedding (SCE) for a candidate document d. The SCE is the sum of all continuous embedding weights for the column vectors of V that correspond to word embedding of terms in a certain document d. Therefore, the SCE weight of a certain term $t_i \in \Sigma$ for a given document d can be represented as:

$$SCE(d) = A(d)V = (SCE_1(d), ..., SCE_N(d))$$ (3)

$$A_i(d) = \begin{cases} 1, & \text{if } t_i \in d \\ 0, & \text{otherwise} \end{cases},$$ (4)

where A is an M-dimensional row vector. In the next computational step, we compute the mean of the SCE which we call centroid-means-embedding (CME) for a certain term t_i in document d.

$$\overline{SCE(d)} = \frac{1}{N} \sum_{i=1}^{N} SCE_i(d).$$ (5)

We then rewrite Eqn.1 as:

$$x^{Stat'}(d) = x^{Stat}(d) \times \overline{SCE(d)}.$$ (6)

From the Eqn. 2, the new generated weight gets a larger weight than existing vector which may turn training ovefit. We therefore scale the embedding vectors by setting a hyper parameter. The goal of using hyper parameter is to scale large weights that overfit the training data. We introduce Gaussian or normal distribution based scaling function for a certain document d to scale each new generated weight for SAS-VSM. The hyper parameter $\lambda = (\lambda_1, ..., \lambda_N)$ for document d can be denoted as:

$$\lambda_i(d) = \frac{1}{\sqrt{2\pi\sigma_d^2}} \exp\left(-\frac{\left(x_i^{Sem}(d) - \mu_d\right)^2}{2\sigma_d^2}\right),$$ (7)

where the mean μ_d and standard deviation σ_d for a document d are calculated from candidate documents as:

$$\mu_d = \frac{1}{M} \sum_{i=1}^{M} f(t_i) \tag{8}$$

$$\sigma_d = \sqrt{\frac{1}{M} \sum_{i=1}^{M} (f(t_i) - \mu)^2}. \tag{9}$$

We rewrite Eqn. 2 as:

$$\boldsymbol{x}^{Sem'}(d) = \boldsymbol{\lambda}(d) \circ \left(\boldsymbol{x}^{Stat'}(d) \boldsymbol{V} \right), \tag{10}$$

where $\circ$ denotes the element-wise multiplication of two row vectors.

3   Term Weighting Schemes

Recently, several studies have been conducted using different term weighting approaches [2], [4], [10], [11], [18], [21] to address the ATC. TF.IDF is the most widely-used, conventional, document-indexing-based [1], [8], [9], [19], [21] term weighting approach for ATC. The common TF.IDF [12], [14], is defined as:

$$W_{TF.IDF}(t_i, d) = tf_{(t_i, d)} \times \left(1 + \log \frac{D}{\#(t_i)} \right), \tag{11}$$

where D denotes the total number of documents in the training corpus, $tf(t_i, d)$ is the number of occurrences of term t_i in document d, $\#(t_i)$ is the number of documents in the training corpus in which term t_i occurs at least once, $\#(t_i)/D$ is referred to as the documents frequency (DF), and $D/\#(t_i)$ is the inverse document frequency (IDF) of term t_i.

In terms of class-oriented indexing [1], [13], Ren and Sohrab [1] discussed two different weighting approaches, TF.IDF.ICF and TF.IDF.ICS$_\delta$F, where the global document-indexing-based IDF and class-indexing-based inverse class frequency (ICF) and inverse class space density frequency (ICS$_\delta$F) are incorporated with local weights term frequency (TF). We can also define two class-indexing-based weighting approaches TF.ICF and TF.ICS$_\delta$F. These two representations of class-indexing-based category mapping are represented as:

$$W_{TF.ICF}(t_i, d, c_k) = tf_{(t_i, d)} \times \left(1 + \log \frac{C}{c(t_i)} \right), \tag{12}$$

$$W_{TF.ICS_\delta F}(t_i, d, c_k) = tf_{(t_i, d)} \times \left(1 + \log \frac{C}{CS_\delta(t_i)} \right), \tag{13}$$

where C denotes the total number of predefined categories in the training corpus, $c(t_i)$ is the number of categories in the training corpus in which term t_i occurs at least once, $\frac{c(t_i)}{c}$ is referred to as the class frequency (CF), and $\frac{C}{c(t_i)}$ is the ICF

M.G. Sohrab et al.

of the term t_i. $\frac{CS_\delta(t_i)}{C}$ is referred to as the class space density frequency ($CS_\delta F$) and $\frac{C}{CS_\delta(t_i)}$ is the $ICS_\delta F$ of term t_i.

TF.IDF.ICF and TF.IDF.ICS$_\delta$F for a certain term t_i in document d with respect to category c_k, are defined in as:

$$W_{TF.IDF.ICF}(t_i, d, c_k) = tf_{(t_i,d)} \times \left(1 + \log \frac{D}{\#(t_i)}\right) \times \left(1 + \log \frac{C}{c(t_i)}\right), \quad (14)$$

$$W_{TF.IDF.ICS_\delta F}(t_i, d, c_k) = tf_{(t_i,d)} \times \left(1 + \log \frac{D}{\#(t_i)}\right) \times \left(1 + \log \frac{C}{CS_\delta(t_i)}\right). \quad (15)$$

4 Evaluation

In this section, we provide empirical evidence for the effectiveness of the proposed approaches. In this evaluation, we employ two commonly-used ATC datasets: 20 Newsgroups[1] and RCV1-v2/LYRL2004 [16]. We employ a 10-fold cross validation scheme for the 20 Newsgroups dataset in which the dataset is randomly divided into 10 subsets. For each fold, one subset is used for testing and the remaining subsets are used for training. For the RCV1-v2/LYRL2004 dataset, we split the corpus into training and test data, which is discussed in Ren and Sohrab [1]. We have kept the same splits and experiment setup that are used in Ren and Sohrab [1]. The standard evaluation metrics like precision, recall, F$_1$-measure and the micro-average of precision, recall, and F$_1$-measure are used to judge the system performances. Please refer to Ren and Sohrab [1] for more details.

4.1 Experimental Datasets

To evaluate the performance of the proposed model with existing different baseline weighting approaches, we conducted our experiments using the 20 Newsgroups and RCV1-v2/LYRL2004, which are widely used benchmark collections in the ATC task.

20 Newsgroups Dataset. The first dataset that we used in this experiment is the 20 Newsgroups, which is a popular dataset to use against machine learning techniques such as ATC and text clustering. It contains approximately 18,828 news articles across 20 different newsgroups. For convenience, we call the 20 categories: Atheism (Ath), CompGraphics (CGra), CompOsMsWindows-Misc (CMWM), CompSysIbmPcHardware (CSIPH), CompSysMacHardware (CSMH), CompWindowsx (CWin), MiscForsale (MFor), RecAutos (RAuto), RecMotorcycles (RMot), RecSportBaseBall (RSB), RecSportHockey (RSH), SciCrypt (SCry), SciElectronics (SEle), SciMed (SMed), SciSpace (SSpa), SocReligionChristian (SRChr), TalkPoliticsGuns (TPG), TalkPoliticsMideast (TPMid), TalkPoliticsMisc (TPMisc), and TalkReligionMisc (TRMi).

---

[1] Available at http://people.csail.mit.edu/jrennie/20Newsgroups/

RCV1 Dataset. The RCV1 dataset, RCV1-v2/LYRL2004 is adopted, which contains a total of 804,414 documents with 103 categories from four parent topics. As single-label classification is in concern in this study, we extract all the documents which are labeled with at least once. We found that only approximate 23,000 documents out of 804,414 are labeled with at least once. To create larger dataset for the single-label classification problem, we extracted all the documents which are labeled with two categories, a parent and a child category. Then we removed the parent category from the document label and child category is assigned in order to produce the single-label classification problem. From RCV1-v2/LYRL2004, a single category is assigned to a total of 219,667 documents and there are 54 different categories in total. We have kept the same split, the first 23,149 documents as for training and the remainder 196,518 documents are for testing according to RCV1-v2/LYRL2004.

4.2 Word Embedding Training with GloVe Model

In this paper, the word embedding matrix $V_{M \times N}$ is generated using GloVe model. We consider the GloVe model for learning word representation from unlabeled data to generate word embedding vectors, since it is outperformed other methods on word similarity and NER tasks. The GloVe model is an weighted least squares regression model that performs global matrix factorization with a local context window models. In this work, the word embedding vectors are generated from the available source code[2]. All parameters were left at default values in this toolbox.

4.3 Support Vector Machine Classifier

In the machine learning workbench, support vector machine (SVM) has been achieved great success in ATC and considered as one of the most robust and accurate methods among all well-known algorithms [15]. Therefore, as a learning classifier, SVM-based classification toolbox SVM-multiclass[3] is used in this experiment. All parameters were left at default values. The regularization parameter c was set to 1.0.

4.4 Results with the 20 Newsgroups and RCV1 dataset

In this paper, we compare our primary and CME approaches for SAS-VSM with baseline weighting schemes including TF, TF.IDF, TF.ICF, TF.ICS$_\delta$F, TF.IDF.ICF, and TF.IDF.ICS$_\delta$F approaches. In Tables 1, 2, 3, and 4, EV=4, EV=10, EV=20, and EV=40 indicate the vector sizes of word embedding that are injected in SAS-VSM with respect to different weighting approaches.

---

[2] Available at http://nlp.stanford.edu/projects/glove/

[3] Available at http://svmlight.joachims.org/svm_multiclass.html

Table 1. Primary approach performances on the 20 Newsgroups dataset

Term Weighting	Baseline (%)	Word Embedding with Primary Approach			
		EV=4 (%)	EV=10 (%)	EV=20 (%)	EV=40 (%)
TF	69.60	69.66(+0.06)	69.14(−0.47)	69.75(+0.14)	69.19(−0.41)
TF.IDF	85.01	85.52(+0.51)	85.46(+0.46)	85.44(+0.44)	85.40(+0.40)
TF.ICF	85.82	85.93(+0.11)	86.17(+0.36)	85.92(+0.10)	85.90(+0.08)
TF.ICS$_\sigma$F	85.45	85.57(+0.12)	85.40(−0.05)	85.35(−0.11)	85.43(−0.02)
TF.IDF.ICF	85.34	85.75(+0.41)	85.85(+0.51)	85.88(+0.54)	85.75(+0.41)
TF.IDF.ICS$_\sigma$F	92.78	92.93(+0.15)	92.94(+0.16)	92.94(+0.16)	**92.96**(+0.18)

Note: Results in parentheses indicating the performance in/decrease from baseline

Table 2. Primary approach performances on the RCV1 dataset

Term Weighting	Baseline (%)	Word Embedding with Primary Approach			
		EV=4 (%)	EV=10 (%)	EV=20 (%)	EV=40 (%)
TF	71.23	71.38(+0.15)	71.41(+0.18)	71.46(+0.24)	71.55(+0.32)
TF.IDF	76.86	76.91(+0.05)	77.16(+0.30)	77.13(+0.27)	77.05(+0.19)
TF.ICF	73.46	74.01(+0.56)	74.07(+0.61)	74.00(+0.54)	73.90(+0.44)
TF.ICS$_\sigma$F	76.27	76.40(+0.13)	76.44(+0.17)	76.54(+0.27)	76.46(+0.19)
TF.IDF.ICF	80.01	79.90(−0.12)	79.65(−0.37)	79.75(−0.28)	79.70(−0.31)
TF.IDF.ICS$_\sigma$F	84.79	84.80(+0.01)	84.80(+0.01)	84.80(+0.01)	**84.81**(+0.01)

Note: Results in parentheses indicating the performance in/decrease from baseline

Results with the Primary Approach. Tables 1 and 2 show the performance comparison with micro-F_1 on six different term weighting approaches over the 20 Newsgroups and RCV1 datasets using the SVM classifier. In Table 1, it is noticeable that the primary approach shows a marginal improvement over the baseline weighting approaches. In some cases when feeding with a bit larger augmented vectors including EV=10, EV=20, and EV=40, the performance shows a minimal drop from the baseline TF and TF.ICS$_\delta$F approaches. In Table 2, it is also noticeable that the results on RCV1 show marginal improvements over all baseline weighting approaches except for a minimal drop on TF.IDF.ICF.

Results with the CME Approach. Tables 3 and 4 show the performance comparison with micro-F_1 on six different term weighting approaches over the 20 Newsgroups and RCV1 datasets using the SVM classifier. Table 3 shows that by applying CME to SAS-VSM over the different weighting approaches, CME enriches system performance not only from the baseline approaches but also from our proposed primary approach using SAS-VSM. In Table 4, it is also noticeable that the CME-based SAS-VSM approach outperforms all the baselines and primary proposed approaches.

Table 3. Word embedding with CME performances on the 20 Newsgroups dataset

Term Weighting	Baseline (%)	Word Embedding with CME Approach			
		EV=4 (%)	EV=10 (%)	EV=20 (%)	EV=40 (%)
TF	69.60	76.82(+7.21)	76.22(+6.62)	77.35(+7.74)	77.92(+8.32)
TF.IDF	85.01	87.78(+2.77)	87.70(+2.70)	88.07(+3.07)	88.50(+3.50)
TF.ICF	85.82	87.04(+1.22)	86.82(+1.01)	86.98(+1.16)	87.35(+1.60)
TF.ICS$_\sigma$F	85.45	87.74(+2.29)	87.68(+2.25)	87.95(+2.50)	88.50(+3.05)
TF.IDF.ICF	85.34	89.17(+3.84)	89.06(+3.73)	89.35(+4.01)	89.69(+4.35)
TF.IDF.ICS$_\sigma$F	92.78	94.34(+1.60)	94.29(+1.51)	94.40(+1.62)	**94.41**(+1.62)

Note: Results in parentheses indicating the performance in/decrease from baseline

Table 4. Word embedding with CME performnaces on the RCV1 dataset

Term Weighting	Baseline (%)	Word Embedding with CME Approach			
		EV=4 (%)	EV=10 (%)	EV=20 (%)	EV=40 (%)
TF	71.23	73.06(+1.83)	72.99(+1.76)	73.13(+1.90)	74.07(+2.84)
TF.IDF	76.86	83.48(+6.63)	83.45(+6.59)	83.87(+7.01)	84.42(+7.57)
TF.ICF	73.46	77.51(+4.05)	77.27(+3.81)	77.59(+4.13)	78.49(+5.04)
TF.ICS$_\sigma$F	76.27	83.64(+7.37)	83.53(+7.27)	83.95(+7.68)	84.51(+8.24)
TF.IDF.ICF	80.01	85.05(+5.03)	84.93(+4.91)	85.08(+5.07)	85.28(+5.26)
TF.IDF.ICS$_\sigma$F	84.79	85.99(+1.20)	85.90(+1.11)	85.90(+1.10)	**85.91**(+1.12)

Note: Results in parentheses indicating the performance in/decrease from baseline

Categorical Performance Comparison. In ATC, besides overall performance, it is also important to judge the categorical performance for a certain dataset. Because of space limitation we only provide the TF.IDF categorical performance for 20 Newsgroups dataset. Fig. 1 shows the categorical performance based on F_1-measure, where our primary approach is performing lower than the baseline on some categories. In contrast, the CME-based SAS-VSM approach shows its superiority over the baseline classifiers on 19 out of 20 categories.

4.5 Discussions

The results of the above experiments show that the CME-based SAS-VSM consistently outperforms over the baseline approaches, including TF, TF.IDF, TF.ICF, TF.ICS$_\delta$F, TF.IDF.ICF, and TF.IDF.ICS$_\delta$F, which are used to create statistical VSMs. From the results with our primary approach, it is important to note that the combination of statistical and semantic VSM can marginally improve the system performance. In the primary approach, it is noticeable that continuous-valued embedding matrix is updated with both statistical and semantic knowledge but the discrete weights are remaining unchanged in the primary approach. In contrast, we introduce CME approach in the SAS-VSM

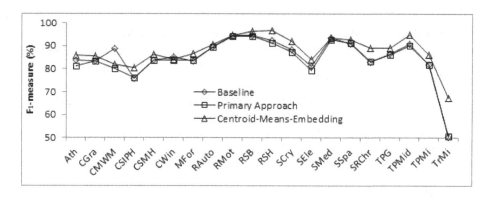

Fig. 1. Categorical performance based F_1-measure in the 20 Newsgroups dataset. Embedding vector EV=40 was employed for the CME.

to update the discrete weight and provide semantic knowledge into statistical VSMs. These results indicate that our CME-based SAS-VSM approach can significantly improve the system performance.

Our experiments also show that the CME-based SAS-VSM is a novel VSM that produces a consistently higher performance over different term weighting approaches. Thus, the word embedding vectors are useful to enhance ATC.

5   Related Works

Ren and Sohrab [1] performed their experiments with eight different weighing approaches: local weight TF incorporated with global weights including coefficient correlation (TF.CC), mutual information (TF.MI), odds ratio (TF.OR), probability based (TF.PB), relevance frequency (TF.RF), IDF (TF.IDF), IDF.ICF (TF.IDF.ICF), and IDF.ICS$_\delta$F (TF.IDF.ICS$_\delta$F). The results showed that the class-indexing-based TF.IDF.ICS$_\delta$F is useful with an SVM classifier. The TF.IDF-.ICS$_\delta$F approach showed its superiority in all the categories of the 20 Newsgroups and a majority of the Reuters-21578 datasets using SVM. This work emphasizes on statistical supervised approach with semantic information for a certain document, which is neither discussed nor empirically evaluated.

Jeffrey et al. [5] introduced global vectors for word representation where the work proposed specific weighted least square model that trains global word-word co-occurrence counts and produce a word vector space. The results demonstrate that the GloVe model outperforms existing models over word analogy, word similarity, and NER tasks. This work left a key note that word embedding vectors can be used as features in ATC.

Jiang et al. [3] introduced a distributional prototype approach for utilizing the embedding features applied on NER. The basic idea of the distributional prototype features is that similar words are supposed to be tagged with the same label. The experiment result shows that continuous embedding features improve the system performance for NER.

Luo et al. [18] proposed a semantic term weighting by exploiting the semantics of categories using WordNet and replaced the IDF function with a semantic weight (SW). The TF.SW approach that outperformed TF.IDF in overall system performance but it was unable to outperform TF.IDF on the categorical performance.

6 Conclusions

In this study, we investigated the effectiveness of exploiting word embedding in the ATC and proposed a novel CME-based SAS-VSM for ATC.

After analyzing the result, four conclusions seem warranted. First, from the experiment results, it is noticeable that the proposed CME-based SAS-VSM can significantly improve the performances of different weighting approaches, including TF, TF.IDF, TF.ICF, TF.ICS$_\delta$F, TF.IDF.ICF, and TF.IDF.ICS$_\delta$F. Therefore, this approach can apply to any existing weighting approaches to improve the existing system. Second, a properly feeding method for augmented features can be useful as input to a VSM, especially SAS-VSM to enhance ATC. Third, the results of this study indicate that the proposed CME-based SAS-VSM can significantly improve the categorical performance for different weighting approaches in two different datasets. The proposed approach is very effective to enhance ATC. Forth, SVM is considered one of the most robust and accurate classification methods in machine learning workbench, and here our results show that the CME-based SAS-VSM is effective with SVM method in two different datasets to address classification task.

Possible ideas for future work would be to conduct experiments on very large scale multi-label hierarchical text classification for Wikipedia medium and large datasets[4]. It might be interesting to investigate the behavior of SAS-VSM for the large scale datasets which have thousands of categories and one or more categories are assigned for a certain document in order to address multi-label hierarchical classification.

Acknowledgement. This work has been partially supported by JSPS KAKENHI Grant Number 25330271.

References

1. Ren, F., Sohrab, M.G.: Class-indexing-based term weighting for automatic text classification. Information Sciences **236**, 109–125 (2013)
2. Debole, F., Sebastiani, F.: Supervised term weighting for automated text categorization. In: 18th ACM Symposium on Applied Computing, pp. 784–788. Florida (2003)

---

[4] Available at http://lshtc.iit.demokritos.gr

3. Jiang, G., Wanxiang, C., Haifeng, W., Ting, K.: Revisiting embedding features for simple semi-supervised learning. In: 2014 Conference on Empirical Methods in Natural Language Processing, pp. 110–120. Qatar (2014)

4. Flora, S., Agus, T.: Experiments in Term Weighting for Novelty mining. Expert Systems with Applications **38**, 14094–14101 (2011)

5. Jeffrey, P., Richard, S., Christopher, D. M.: Glove: global vectors for word representation. In: 2014 Conference on Empirical Methods in Natural Language Processing, pp. 1532–1543. Qatar (2014)

6. Guo, Y., Shao, Z., Hua, N.: Automatic text categorization based on content analysis with cognitive situation models. Information Sciences **180**, 613–630 (2010)

7. Huang, E.H., Socher, R., Christopher, D.M., Andrew, Y.N.: Improving word representations via global context and multiple word prototypes. In: 50th Annual meeting of the Association for Computational Linguistics, pp. 873–882. Korea (2012)

8. Salton, G.: A theory of indexing. Bristol, UK (1975)

9. Kang, B., Lee, S.: Document indexing: A concept-based approach to term weight estimation. Information Processing and Management **41**(5), 1065–1080 (2005)

10. Kansheng, S., Jie, H., Hai-tao, L., Nai-tong, Z., Wen-tao, S.: Efficient text classification method based on improved term reduction and term weighting. The Journal of China Universities of Posts and Telecommunications **18**, 131–135 (2011)

11. Ko, Y., Seo, J.: Text Classification From Unlabeled documents with bootstrapping and feature projection techniques. Information Processing and management **45**, 70–83 (2009)

12. Xia, R., Zong, C., Li, S.: Ensemble of feature sets and classification algorithms for sentiment classification. Information Sciences **181**, 1138–1152 (2011)

13. Sohrab, M.G., Ren, F.: Class-indexing: the effectiveness of class-space-density in high and low-dimensional vector space for text classification, In: 2nd International Conference of Cloud Computing and Intelligence Systems, pp. 2034–2042. China (2012)

14. Sparck, K.J.: A statistical interpretation of term specificity and its application in retrieval. Journal of Documentation **28**(1), 11–21 (1972)

15. Wu, X., Kumar, V., et al.: Top 10 algorithms in data mining, Knowledge. Information Systems **14**, 1–37 (2008)

16. Lewis, D.D., Yang, Y., Rose, T., Li, F.: RCV1: A New Benchmark Collection for Text Categorization Research. Journal of Machine Learning Research **5**, 361–397 (2004)

17. Liu, Y., Loh, H., Sun, A.: Imbalanced text classification: A term weighting approach. Expert Systems with Applications **36**(1), 690–701 (2009)

18. Luo, Q., Chen, E., Xiong, H.: A semantic term weighting scheme for text classification. Expert Systems with Applications **38**(10), 12708–12716 (2011)

19. Salton, G., Yang, C.S., Yu, C.T.: Contribution to the theory of indexing. In: IFIP Congress 74, Stockholm. American Elsevier, New York (1973)

20. Sebastiani, F.: Machine learning in automated text categorization. ACM Computing Surveys **34**(1), 1–47 (2002)

21. Salton, G., McGill, M.J.: Introduction to modern information retrieval. New York (1983)

22. Quoc, L., Mikolov, T.: Distributed representation of sentences and documents. In: 31th International Conference on Machine Learning, pp. 1188–1196 (2014)

Machine Learning

Collaborating Differently on Different Topics: A Multi-Relational Approach to Multi-Task Learning

Sunil Kumar Gupta(✉), Santu Rana, Dinh Phung, and Svetha Venkatesh

Center for Pattern Recognition and Data Analytics,
Deakin University, Geelong 3216, Australia
{sunil.gupta,santu.rana,dinh.phung,svetha.venkatesh}@deakin.edu.au

Abstract. Multi-task learning offers a way to benefit from synergy of multiple related prediction tasks via their joint modeling. Current multi-task techniques model related tasks jointly, assuming that the tasks share the same relationship across features uniformly. This assumption is seldom true as tasks may be related across some features but not others. Addressing this problem, we propose a new multi-task learning model that learns separate task relationships along different features. This added flexibility allows our model to have a finer and differential level of control in joint modeling of tasks along different features. We formulate the model as an optimization problem and provide an efficient, iterative solution. We illustrate the behavior of the proposed model using a synthetic dataset where we induce varied feature-dependent task relationships: positive relationship, negative relationship, no relationship. Using four real datasets, we evaluate the effectiveness of the proposed model for many multi-task regression and classification problems, and demonstrate its superiority over other state-of-the-art multi-task learning models.

1 Introduction

In machine learning, one often encounters multiple prediction tasks that are related to each other. Multi-task learning (MTL) offers principled frameworks to benefit from synergy of these related tasks via their joint modeling. MTL has been used in diverse applications - digit recognition [1], face recognition [2], landmine detection [3], disease progression modeling [4], cancer mortality prediction [5] are some examples.

Multi-task learning techniques introduce an inductive bias in the common hypothesis space of all the tasks. Typically, it is done via using some commonality on task parameters e.g. the use of a common subspace [1,6,7], induction of a common prior in a probabilistic setting [3,8], structural regularization [9,10]. One of the major challenges in the MTL framework is to find "related" tasks and quantify task-to-task relatedness. Initial works in this area [9,11] assumed all the tasks to be well-related and naïvely combined them for joint learning.

© Springer International Publishing Switzerland 2015
T. Cao et al. (Eds.): PAKDD 2015, Part I, LNAI 9077, pp. 303–316, 2015.
DOI: 10.1007/978-3-319-18038-0_24

However, when tasks are unrelated or even negatively related, such combination may lead to poor performance. Addressing this, later works estimate some form of task relatedness and combine the tasks accordingly. For example, Jacob et al. propose a model [12] that clusters similar tasks into groups and joint modeling is achieved by maximizing the pairwise-similarity between task parameters of all tasks in a group. A similar model via alternating structure optimization is proposed by Zhou et al. in [13]. Taking a subspace learning approach, Argyriou et al. [6] develop a model that combines the knowledge from tasks sharing the same basis vectors. To encourage sparsity in the subspace representation, Kumar et al. [7] propose a model that alleviates noisy task relations. When there are many outlier tasks or unrelated tasks, the performance of these methods suffers as they include all the tasks in joint modeling without any grouping. To overcome this problem, Kang et al. [1] extend the model in [6] adding the flexibility to learn multiple task groups and then confining the joint modeling within a group. More advanced models along these lines using nonparametric Bayesian frameworks are done in [14,15]. Although all these models are able to separate unrelated tasks from joint modeling, they are unable to exploit negatively related tasks. This problem is addressed by Zhang et al. [16], who propose a model that uses a task covariance matrix to learn all types of task relationships. *A common problem of all the aforementioned techniques is that they assume same relationship between two tasks along all the features. This assumption is seldom true in reality and causes a problem when task-to-task relationship varies from feature-to-feature.*

Consider an example of predicting user ratings of desktop computers for a set of users based on features such as CPU speed, RAM size, screen size, price etc. Learning the rating function for each user can be considered as a task. Users may be related as they may have similar preferences over certain features e.g. most might like lower prices. However, users may differ on other features e.g. some prefer higher screen size, whilst others might prefer high performance (faster CPU and larger RAM). Consider two users, say A and B, where both give similar importance to price but user A may give more importance to screen size whilst user B give more importance to CPU speed. Conventional MTL algorithms would compute a single relatedness score based on all the features, which assumes both users to have similar level of agreement on both screen size and CPU speed. Clearly, this is not the case. Moreover, due to using a single relatedness score, which is averaged considering all the features, their agreement on price is underestimated. Therefore, a multi-task learning method that learns *feature-specific task relationships* is required. Learning task-to-task relationships for every single feature may be unnecessary as tasks may have similar relationships along many semantically related features. For example, high CPU speed and large RAM are associated to high performance. Therefore, we hypothesize that it may be sufficient to learn a task relationship for each semantically related feature group.

We propose a new multi-task learning model that (a) extracts groups of related features, (b) computes task relatedness based on each feature group and (c) uses these relationships for joint modeling of tasks. To extract the groups of related features we learn a low-dimensional subspace from the set of task parameters. Each subspace basis captures the set of semantically related features. Next, we compute a separate task relationship along each subspace basis. To capture all form of task relationships (low to high, positive and negative) we use a covariance matrix that is computed from the projection of the task parameters on each basis. Joint modeling of tasks is achieved via an optimization formulation that combines the standard least-squares loss with an appropriate regularization term involving the task covariance matrices. We derive an efficient iterative solution to this optimization problem. Due to the use of multiple relationships, our model is called *Multi-Relational Multi-Task Learning* (MR-MTL). We illustrate the behavior of our MR-MTL model using a synthetic dataset in scenarios where tasks relationships vary based on different feature groups. We evaluate the effectiveness of MR-MTL on two regression and two classification real-world datasets and demonstrate its superiority over other state-of-the-art multi-task learning methods.

Our contributions are:

- Proposal of a new multi-task learning model, capable of learning different task relationships between two tasks with respect to different feature subsets. This has implications in modeling partial relatedness and avoiding negative knowledge transfer.
- Formulation of the model as an optimization problem, providing an efficient iterative solution.
- Illustration of the behavior of the proposed model using a synthetic dataset that demonstrates algorithmic performance for varied feature-dependent task relationships: positive relationship, negative relationship, no relationship.
- Evaluation of the proposed MR-MTL model on four real datasets validating its effectiveness over a variety of regression and classification problems, and demonstrating its superiority over several state-of-the-art multi-task learning models.

The significance of our approach is that it is capable of exploiting knowledge across tasks from multiple heterogeneous sources that might differ in their features. For example, in a multi-hospital scenario, patient records extracted from different hospitals may contain hospital-specific features apart from the usual phenotypical features. Hospital-specific features may include additional information such as genomic data, which may not be widespread across all hospitals, or differential features may result from different interventions practices across hospitals. Our model can separate the hospital-specific features from the phenotypical features, and confine the joint modeling only along the common phenotypical features. This capability offers accurate modeling of real data and is absent in conventional models.

2 The Proposed Model

We propose a new multi-task learning model that can capture finer relationships between tasks by modeling feature-specific task relatedness. Let us assume we have T_0 learning tasks, indexed as $t = 1, \ldots, T_0$. For the t-th task, the training set is denoted as $\{(\mathbf{x}_{ti}, y_{ti}) \mid i = 1, \ldots, N_t\}$ where $\mathbf{x}_{ti} \in \mathbb{R}^M$ is a M-dimensional feature vector and y_{ti} is the target, usually real-valued for regression and binary-valued for binary classification problems. Let β_t denote the weight vector for the task t, we also refer to this as *task parameter*. Collectively, we denote the data of t-th task by $\mathbf{X}_t = (\mathbf{x}_{t1}, \ldots, \mathbf{x}_{tN_t})^T$ and $\mathbf{y}_t = (y_{t1}, \ldots, y_{tN_t})^T$ and all the task parameters as $\beta = (\beta_1, \ldots, \beta_{T_0})$. When tasks differ in some of the features, a common feature list can be obtained via their union.

2.1 Formulation

Since our goal is to develop a MTL model that allows multiple task relationships (one for each correlated feature subset) between any two tasks, we simultaneously learn several correlated feature subsets and use a task covariance matrix to capture task relationships with respect to each feature subset. In learning these correlated feature subsets, our idea is that relatedness of tasks along the features of a subset are similar. These feature subsets can be thought of the latent semantic bases of a low dimensional subspace. We represent this low dimensional subspace using a matrix $\mathbf{U}$ where each column is a basis vector of the subspace. The task parameter β_t is represented in this subspace using θ_t as $\beta_t = \mathbf{U}\theta_t$. Collectively, we denote these representations as $\Theta = (\theta_1, \ldots, \theta_{T_0})$. The k-th row of this matrix is denoted as $\theta_{(k)}$. We unify the subspace learning with the regularized multi-task learning to construct a model that allows joint modeling between tasks at a finer level using multiple task relatedness instead of a single aggregated relatedness. The proposed model is learnt by minimizing the following cost function

$$\min_{\mathbf{U}, \Omega_{1:K}, \theta_{1:T_0}} \sum_t ||\mathbf{X}_t \mathbf{U}\theta_t - \mathbf{y}_t||^2 + \eta||\mathbf{U}||_F^2 + \sum_{k=1}^{K} \left[\lambda_1 \theta_{(k)}^T \theta_{(k)} + \lambda_2 \theta_{(k)}^T \Omega_k^{-1} \theta_{(k)} \right] \tag{1}$$

$$\text{s.t. } \Omega_k \succeq 0, \ \text{tr}(\Omega_k) = 1 \ \forall k,$$

where the multi-task learning is achieved due to the last two terms that regularize the least-square loss using parameters λ_1, λ_2, and Ω_k is a task-to-task covariance matrix specific to k-th feature subset. We refer to this model as **Multi-Relational Multi-Task Learning (MR-MTL)**.

2.2 Optimization

The optimization of the cost function in Eq (1) involves minimization with respect to $\mathbf{U}$, $\Omega_{1:K}$ and $\theta_{1:T_0}$. Given $\mathbf{U}$, the cost function is jointly convex in $\Omega_{1:K}$ and $\theta_{1:T_0}$, and separable for each k. Similarly, given $\Omega_{1:K}$ and $\theta_{1:T_0}$, the cost function is convex in $\mathbf{U}$ and the optimal solution has a closed form expression. This property of the cost function suggests an iterative algorithm for optimization.

Optimizing U given Θ, $\Omega_{1:K}$: For a fixed Θ, $\Omega_{1:K}$, the cost in (1) becomes a regularized least square function in $\mathbf{U}$ and has a closed form solution. The optimal solution can be obtained by equating the gradient of Eq (1) to zero as below

$$\sum_t \mathbf{X}_t^T \left(\mathbf{X}_t \mathbf{U} \theta_t - \mathbf{y}_t \right) \theta_t^T + \eta \mathbf{U} = \mathbf{0}.$$

To solve the above equation, we apply 'vec' operator. This operator when applied to a matrix concatenates all the columns one-by-one below the previous columns to form a long vector. Applying 'vec' operator, the above linear equation in $\mathbf{U}$ can be written as

$$\mathrm{vec} \left(\sum_t \mathbf{X}_t^T \mathbf{X}_t \mathbf{U} \theta_t \theta_t^T \right) + \mathrm{vec} \left(\eta \mathbf{U} \right) = \mathrm{vec} \left(\mathbf{X}_t^T \mathbf{y}_t \theta_t^T \right),$$

which can be simplified to obtain the following linear equation for $\mathbf{U}$

$$\left[\sum_t \left(\theta_t \theta_t^T \right) \otimes \left(\mathbf{X}_t^T \mathbf{X}_t \right) + \eta \mathbf{I} \right] \mathrm{vec} \left(\mathbf{U} \right) = \mathrm{vec} \left(\mathbf{X}_t^T \mathbf{y}_t \theta_t^T \right) \tag{2}$$

where we use the following property of vec operator: $\mathrm{vec} \left(\mathbf{A} \mathbf{X} \mathbf{B} \right) = \left(\mathbf{B}^T \otimes \mathbf{A} \right) \mathrm{vec} \left(\mathbf{A} \right)$. The above equation can be solved using LU or QR factorizations, which are more efficient and offer better numerical stability than a matrix inverse based solution.

Optimizing Θ given U, $\Omega_{1:K}$: Given $\mathbf{U}$, $\Omega_{1:K}$, the optimization problem in (1) becomes

$$\min_{\theta_{1:T_0}} \sum_t \|\mathbf{X}_t \mathbf{U} \theta_t - \mathbf{y}_t\|^2 + \sum_{k=1}^K \left[\theta_{(k)}^T \left(\lambda_2 \Omega_k^{-1} + \lambda_1 \mathbf{I} \right) \theta_{(k)} \right],$$

where the first term involves the *columns* of matrix Θ and the second term involves the *rows* of matrix Θ. Although at first instance, it seems like a difficult problem to solve, we can optimize the above cost function in terms of rows of Θ, i.e. $\theta_{(k)}$ and obtain a closed form solution. For this, we take its derivative w.r.t. $\theta_{(k)}$ and set it to zero to obtain the following relation

$$\left[\theta_1^T \mathbf{C}_{(k)}^1, \ldots, \theta_{T_0}^T \mathbf{C}_{(k)}^{T_0} \right]^T + \left(\lambda_2 \Omega_k^{-1} + \lambda_1 \mathbf{I} \right) \theta_{(k)} = \left[d_k^1, \ldots, d_k^{T_0} \right]^T$$

where we define $\mathbf{C}^t \triangleq \mathbf{U}^T \mathbf{X}_t^T \mathbf{X}_t \mathbf{U}$ and $\mathbf{d}^t \triangleq \mathbf{U}^T \mathbf{X}_t^T \mathbf{y}_t$ for task t. We note that $\mathbf{C}_{(k)}^t$ and d_k^t denotes k-th row of matrix $\mathbf{C}^t$ and k-th element of vector $\mathbf{d}^t$ respectively. The above equation can be further simplified to a system of linear equations as below

$$\left[\lambda_2 \Omega_k^{-1} + \lambda_1 \mathbf{I} + \mathrm{diag} \left(\mathbf{c}_k \right) \right] \theta_{(k)} = \left[d_k^1, \ldots, d_k^{T_0} \right]^T - \mathbf{z}_{(k)}, \tag{3}$$

where we define $\mathbf{z}_{(k)} \triangleq \left[\sum_{k' \neq k} \theta_{k'1} \mathbf{C}_{kk'}^1, \ldots, \sum_{k' \neq k} \theta_{k'T_0} \mathbf{C}_{kk'}^{T_0} \right]^T$ and $\mathbf{c}_k \triangleq$ $\left[\mathbf{C}_{kk}^1, \ldots, \mathbf{C}_{kk}^{T_0} \right]$. We note that Eq (3) can be efficiently solved using Cholesky decomposition.

Optimizing $\Omega_{1:K}$ given Θ, U: Given Θ, U, the optimization problem in (1) becomes

$$\min_{\Omega_{1:K}} \sum_{k=1}^{K} \left[\theta_{(k)}^T \Omega_k^{-1} \theta_{(k)} \right] \text{ s.t. } \Omega_k \succeq 0, \text{ tr}(\Omega_k) = 1 \ \forall k, \tag{4}$$

which can be independently optimized for each Ω_k. To get the solution of above problem, define $\mathbf{S}_k = \theta_{(k)} \theta_{(k)}^T$ and consider

$$\theta_{(k)}^T \Omega_k^{-1} \theta_{(k)} = \text{tr} \left(\Omega_k^{-1} \mathbf{S}_k \right) \text{tr} \left(\Omega_k \right) = \text{tr} \left(\left(\Omega_k^{-\frac{1}{2}} \mathbf{S}_k^{\frac{1}{2}} \mathbf{S}_k^{\frac{1}{2}} \Omega_k^{-\frac{1}{2}} \right) \text{tr} \left(\Omega_k^{\frac{1}{2}} \Omega_k^{\frac{1}{2}} \right) \right).$$

In the above we have used $\text{tr}(\Omega_k) = 1$. Further defining $A = \Omega_k^{-\frac{1}{2}} \mathbf{S}_k^{\frac{1}{2}}$, $B = \Omega_k^{\frac{1}{2}}$ and noting the positive semi-definite property of these matrices, we can apply a *Cauchy-Schwarz Inequality* on the inner product of trace, i.e. $\text{tr}(A^2) \text{tr}(B^2) \geq (\text{tr}(AB))^2$ to get

$$\text{tr} \left(\left(\Omega_k^{-\frac{1}{2}} \mathbf{S}_k^{\frac{1}{2}} \mathbf{S}_k^{\frac{1}{2}} \Omega_k^{-\frac{1}{2}} \right) \text{tr} \left(\Omega_k^{\frac{1}{2}} \Omega_k^{\frac{1}{2}} \right) \right) \geq \left(\text{tr} \left(\Omega_k^{-\frac{1}{2}} \mathbf{S}_k^{\frac{1}{2}} \Omega_k^{\frac{1}{2}} \right) \right)^2 = \left(\text{tr} \left(\mathbf{S}_k^{\frac{1}{2}} \right) \right)^2.$$

In the above expression, an Ω_k that leads to the *equality*, corresponds to the optimal solution of (4). The equality is satisfied when $\Omega_k = \frac{1}{\alpha} \mathbf{S}_k^{\frac{1}{2}}$ where α is a scalar. Since the optimal Ω_k has to satisfy the constraint $\text{tr}(\Omega_k) = 1$, we get $\alpha = \text{tr} \left(\mathbf{S}_k^{\frac{1}{2}} \right)$. Therefore, $\Omega_k = \mathbf{S}_k^{\frac{1}{2}} / \text{tr} \left(\mathbf{S}_k^{\frac{1}{2}} \right)$. Algorithm 1 outlines step-by-step procedure for MR-MTL.

Computational Complexity: The order of complexity for updating $\theta_{(k)}$ in Eq (3) is $O(T_0^3)$. Similarly, the order of complexity to update U is $O(M^3 \times K^3)$. Finally, the order of complexity to update Ω_k for each k is $O(T_0^2)$. Therefore, overall complexity for the proposed MR-MTL per iteration is of the order $O(M^3 K^3 + T_0^3 + K T_0^2)$.

3   Experiments

We present our experimental results on synthetic and real datasets. Synthetic data is used to create a niche scenario where the proposed MR-MTL model is expected to work better than other models. To evaluate the effectiveness of our model for real world applications, we use two classification and two regression

---

Algorithm 1. The proposed MR-MTL

1: **Input**: Multi-task data $\{\mathbf{X}_t, \mathbf{y}_t\}_{t=1}^{T_0}$, parameters λ_1, λ_2, η and subspace dimension K.

2: **Output**: Task parameters $\beta_{1:T_0}$, matrix $\mathbf{U}$, matrix Θ and matrices $\Omega_{1:K}$.

3: **Initialization**: Initialize $\beta_{1:T_0}$ using single task learning and matrix $\mathbf{U}$ randomly. Initialize matrix Θ as $\Theta = \mathbf{U}^\dagger \beta$.

4: **repeat**

5:    update $\mathbf{U}$ using Eq. (2).

6:    **for** $k = 1 : K$ **do**

7:        update $\theta_{(k)}$ using Eq. (3).

8:        update Ω_k as $\Omega_k = \mathbf{S}_k^{\frac{1}{2}}/\mathrm{tr}\left(\mathbf{S}_k^{\frac{1}{2}}\right)$ where $\mathbf{S}_k = \theta_{(k)}\theta_{(k)}^T$.

9:    **end for**

10: **until** convergence

---

datasets and compare MR-MTL with single-task learning (**STL**) and three state-of-the-art multi-task learning baselines: **MTFL** [6], **GMTL** [1] and **MTRL** [16]. All these models are based on optimization frameworks. Similar to the proposed MR-MTL, the first two baselines (MTFL and GMTL) learn a low dimensional subspace for task parameters. In the optimization, MTFL uses a L_2/L_1 mixed norm penalty term for joint modeling of all tasks while GMTL first learns groups of related tasks and uses a L_2/L_1 mixed norm penalty for each task group. MTRL, on the other hand, uses a covariance matrix for learning task relationship allowing to exploit the knowledge from negatively related tasks. All these models use regularization parameters, which are learnt using a grid search over $\{10^{-3}, 10^{-2}, 10^{-1}, 10^0\}$ via cross-validation. For performance evaluation, we use the following metrics: *Explained variance* (R^2) and *root-mean-square-error* (RMSE) for regression; *area under ROC curve* (AUC) and *F1-measure* for classification.

Table 1. Performance evaluation for Synthetic data in terms of *Explained Variance* (R^2) and *root-mean-square-error* (RMSE). The performance is averaged over 40 randomly generated datasets. The numbers in parenthesis are the corresponding standard errors.

Performance Metric	STL	MTRL	MR-MTL
Explained Variance (↑)	0.882 (0.006)	0.798 (0.007)	**0.993 (0.000)**
RMSE (↓)	0.532 (0.012)	0.727 (0.015)	**0.132 (0.003)**

3.1 Experiments with Synthetic Data

Our synthetic data is generated by creating 30 related tasks where each task is to learn a linear regression model in a 9-dimensional feature space given 15 supervised training instances. We create three groups of tasks: group-1 (tasks 1-10), group-2 (tasks 11-20) and group-3 (tasks 21-30). We ensure that tasks in each

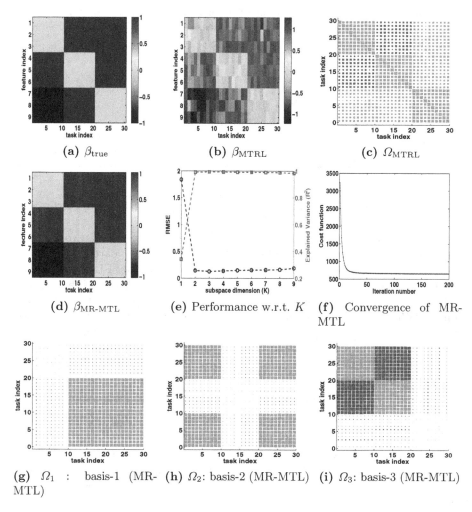

(a) β_{true} **(b)** β_{MTRL} **(c)** Ω_{MTRL}

(d) $\beta_{\text{MR-MTL}}$ **(e)** Performance w.r.t. K **(f)** Convergence of MR-MTL

(g) Ω_1 : basis-1 (MR-MTL) **(h)** Ω_2: basis-2 (MR-MTL) **(i)** Ω_3: basis-3 (MR-MTL)

Fig. 1. Experimental results for Synthetic data. (a) *True* task parameters (b) task parameters *estimated* by MTRL [16] (c) task relatedness *estimated* by MTRL, shown as Hinton plot ('green' denotes positive values and 'red' denotes negative values); (d) task parameters *estimated* by the proposed MR-MTL (e) Performance variations of MR-MTL w.r.t. subspace dimension (f) Convergence plot for MR-MTL algorithm (g)-(i) task relatedness *estimated* by MR-MTL for the features in the *first, second and third* basis respectively, shown as Hinton plot. The first basis is about features '1-3', the second basis is about features '4-6' and the third basis is about '7-9'.

group have the same parameters and are thus strongly correlated. Given these tasks, our idea is to create multiple relationships across task groups by using feature-dependent task relationships in various forms: positive relationship, negative relationship and no relationship. Figure 1 (a) depicts the simulated task parameters (i.e. β) for all the tasks along 9 features. Along the first three features, task group-2 and task group-3 are *positively* related but both are unrelated to task group-1. Similarly, along the next three features, task group-1 and task group-3 are *positively* related but both are unrelated to task group-2. Finally, along the last three features, task group-1 and task group-2 are *negatively* related but both are unrelated to task group-3. Given these task parameters, feature vectors are randomly drawn from a 9-dimensional multi-variate Gaussian distribution as $\mathbf{x}_{ti} \sim \mathcal{N}(\mathbf{0}, \mathbf{I})$. The corresponding target y_{ti} is randomly drawn as $y_{ti} \sim \mathcal{N}\left(\beta_t^T \mathbf{x}_{ti}, 0.1\right)$.

We randomly split the synthetic dataset in two parts using 70% instances for training and the remainder for test. We run our proposed MR-MTL algorithm and compare its performance to one of the related baseline, MTRL for illustration purposes. Figure 1 (b) and (d) show the task parameters estimated by MTRL and the proposed MR-MTL respectively. Clearly the task parameter estimates of MR-MTL are much closer to the true task parameters (Figure 1 (a)). The better estimates by MR-MTL can be explained by looking at the task relationships learnt by both methods, which are shown in Figure 1 (c) for MTRL and Figure (g)-(i) for MR-MTL using Hinton plots. The task relationship learnt by MTRL is averaged across all 9 features, causing *overestimation* of the unrelatedness while *underestimation* of the strong relatedness. In contrast, the proposed MR-MTL accurately estiamtes task relationships by using three separate feature groups (one feature group represented by each basis of the subspace as we use $K = 3$) and thus learning one task relationship matrix for each feature group. This added flexibility allows MR-MTL to have a finer and differential level of control in joint modeling of tasks along different features. We use the held out test set to evaluate the performance of MR-MTL and compare it with MTRL in Table 1. The reported results are averaged over 40 randomly generated datasets along with corresponding standard errors. As seen from the Table, MR-MTL clearly outperforms both STL and MTRL with respect to two evaluation metrics - Explained variance (R^2) and root mean square error (RMSE). Due to presence of different task relationships in data, MTRL is unable to estimate the task relationships and thus performs worse than STL. The performance variations of MR-MTL with respect to subspace dimension (K) is shown in Figure 1 (e), wherein the best performance is achieved at $K = 3$, however, the performance degrades very slowly with increasing values of K. An example of the convergence behavior of proposed MR-MTL is shown in Figure 1 (f) - the algorithm quickly converges within 50 iterations.

3.2 Experiments with Real Data

We use the following *classification* and *regression* datasets.

Landmine Data (Classification):   This dataset is created from radar images collected from 19 landmine fields. This is a benchmark dataset and used widely for multi-task learning. Each data instance is a 9-dimensional representation of each image formed by concatenating different image based features. The task is to detect images with landmines. Treating each landmine field as a task, we jointly model them via multi-task learning. For each task we randomly split the data in two parts: 30% instances for training and the remainder for testing. The results are averaged over 40 training-test splits.

Acute Myocardial Infarction (AMI) Data (Classification):   This dataset is collected from a hospital in Australia (Ethics approval #12/83). It contains records of patients who visited the hospital during 2007-2011 with AMI as the primary reason for admission. The cohort is first divided into two main AMI types: STEMI and Non-STEMI, each of which is further divided into 4 subcohorts based on the major interventions administered (coronary artery bypass surgery, coronary artery stenting, other intervention or no intervention at all), resulting in a total of 8 subcohorts. The task is to predict readmission within the first 30-days of discharge due to any heart related medical emergency. Out of the original 8 subcohorts only 5 are chosen as they have at least 2 positive examples per year. In the selected subcohorts, total number of patients varied from 50-182 per year. The features used are patients demography (gender, age, occupation) and health status in terms of Elixhauser comorbidities [17], aggregated over 3 time scales: 1 month, 3 months and 1 year prior to their AMI admission. Evaluation is performed progressively with patients from 2009, 2010 and 2011 for test whilst using all past patients data before the test year for training.

Computer Survey Data (Regression):   This dataset [6] contains ratings of 20 computers by 190 students based on 13 binary features (cf. Figure 2). Each rating value lies between 0-10 indicating likelihood of buying a computer. We treat ratings by each student as a task, thus having a total of 190 tasks. As these tasks are related, we jointly model them under the setting of multi-task learning. Following [15], we use the first 15 computer ratings for training and test using ratings of the last 5 computers.

SARCOS Data (Regression):   The data relates to an inverse dynamics problem for a seven degrees-of-freedom SARCOS anthropomorphic robot arm. The task is to map from a 21-dimensional input space (7 joint positions, 7 joint velocities, 7 joint accelerations) to the corresponding 7 joint torques, giving rise to 7 mapping tasks. For this dataset 100 random examples are sampled for training and another 400 are sampled randomly for test. This is to demonstrate the efficacy of multi-task learning algorithm for small data. We average the performance over 40 random training-test datasets.

Experimental Results. Table 2 presents a comparison of our proposed MR-MTL algorithm with STL, and other baseline MTL algorithms on Landmine dataset in terms of both prediction AUC and F1. Predictive performance of MR-MTL for different numbers of feature subsets (K=1, 2, and 3) are also reported. Clearly, MR-MTL with K=2 (AUC 0.775, F1 0.880) outperforms all other methods by a good margin. The closest performer is MTRL (AUC 0.760, F1 0.872), whilst other methods are further lower. The landmine dataset contains tasks which can be broadly divided into two groups based on whether a task is a landmine detection problem at a foliated region or in a desert region. Interestingly, MR-MTL also found K=2 to be the best for this dataset.

Table 3 presents a similar comparison of performance on the AMI dataset. Predictive performance at three different training-test scenarios are presented. For all those settings K=2 is found to give the best performance for MR-MTL. For the test year 2009, MR-MTL closely follows MTRL in terms of AUC and GMTL in terms of F1. For the two other test years, MR-MTL convincingly outperforms all other methods in terms of both AUC and F1. For both the scenarios, the AUC is above 0.6 and F1 is above 0.75, whilst the same for other methods are much lower. There is also gradual improvement of performance by MR-MTL as more and more training data is available when tested on later years, whilst all other methods behaved erratically.

Table 2. Comparative AUC of MR-MTL against baseline methods on Landmine dataset. Training and test splits are generated randomly with 30% for training and the rest for test. Average over 40 such splits are reported. Corresponding standard errors are reported in brackets.

	STL	MR-MTL			MTRL	MTFL	GMTL
		K=1	K=2	K=3			
AUC (std err)	0.734 (0.002)	0.664 (0.007)	**0.775** (0.003)	0.757 (0.002)	0.760 (0.001)	0.733 (0.002)	0.720 (0.002)
F1 (std err)	0.853 (0.013)	0.795 (0.012)	**0.880** (0.008)	0.873 (0.008)	0.872 (0.008)	0.847 (0.012)	0.839 (0.014)

Table 3. Comparative AUC and F1 of MR-MTL on AMI dataset against baseline methods. Test is performed progressively at 2009, 2010, and 2011 with corresponding past years data being used for training.

Training years	Test year	Measure	STL	MR-MTL K=2	MTRL	MTFL	GMTL
2007-08	2009	AUC	0.507	0.584	**0.588**	0.570	0.487
		F1	0.517	0.568	0.518	0.452	**0.613**
2007-09	2010	AUC	0.558	**0.606**	0.521	0.539	0.552
		F1	0.676	**0.781**	0.492	0.576	0.669
2007-10	2011	AUC	0.545	**0.614**	0.588	0.554	0.535
		F1	0.683	**0.826**	0.502	0.723	0.599

Table 4 & 5 presents results on two regression dataset namely, Computer and SARCOS datasets. For computer dataset, MR-MTL with K=3 performs (RMSE

Table 4. Comparative RMSE and explained variance (R^2) of MR-MTL on Computer dataset against the baselines. Rating data from the first 15 computers are used for training and the remaining 5 for test. MR-MTL is evaluated at four different numbers of latent basis (K=2, 3 and 4).

	STL	MR-MTL K=2	K=3	K=4	MTRL	MTFL	GMTL
RMSE	2.085	1.711	**1.664**	1.673	1.766	2.056	2.638
Explained Variance (R^2)	0.238	0.309	**0.318**	0.317	0.291	0.220	0.160

Table 5. Comparative RMSE and explained variance ($R^{2)}$ of MR-MTL on SARCOS dataset with respect to the baselines methods. Randomly selected 100 data points are used for training and 1400 for test. Average performance over 40 such random experiments are reported. Respective standard errors are reported in brackets.

	STL	MR-MTL K=5	K=6	K=7	MTRL	MTFL	GMTL
RMSE (std err)	3.449 (0.025)	3.257 (0.019)	3.248 (0.017)	**3.218** (0.018)	6.945 (0.032)	4.722 (0.030)	3.496 (0.025)
Explained Variance (R^2) (std err)	0.823 (0.001)	0.798 (0.003)	0.818 (0.003)	**0.829** (0.002)	0.379 (0.003)	0.640 (0.003)	0.821 (0.002)

1.664, R^2 0.318) the best followed by MTRL (RMSE 1.766, R^2 0.291). All other baselines have higher RMSE values. To illustrate the behavior of MR-MTL further, we present the basis vectors corresponding to K=3 in Fig 2 (a). The three basis vectors captures 3 different grouping of features. The first basis (U1) captures positive preference for high performance (CPU speed, RAM size) along with positive preference for having CD-ROM. The second basis (U2) captures positive preference for CD-ROM, whilst non-preference for higher CPU speed with larger cache. The third basis (U3) captures price of the unit as a major factor. Fig 2(b) shows the histogram of task relatedness along different basis. It is interesting to note that task-relatedness along U3, whose major factor is price shows higher prevalence of positive relatedness (the histogram for U3 is skewed on the positive side), which implies that many raters give importance to price similarly. This is intuitive since price is always a major factor in consumer spending. We see that histogram on U1 have high peak around zero, implying that preference for high performance and CD-ROM is more independent in nature. Conversely, highest disagreement among the raters is observed along U2. For SARCOS dataset, MR-MTL with K=7 performs (RMSE 3.198, R^2 0.832) the best, followed by GMTL (RMSE 3.349, R^2 0.821). Other baseline methods have considerable higher RMSE and lower R^2 values. For this dataset, the tasks are low-related, therefore, other MTL methods which tries to regularize strongly

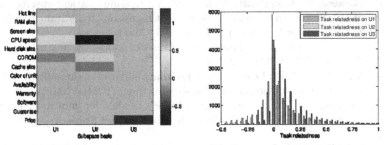

(a) Subspace basis matrix.   (b) Histogram of task relatedness.

Fig. 2. Illustration of results of MR-MTL with K=3 on Computer dataset. (a) Subspace basis matrix with basis U1, U2 and U3. Only weights with absolute value more than 0.1 are shown., and (b) Histogram of task relatedness with respect to each basis.

performed lower, whereas, MR-MTL with K=7 is able to offer the right balance between the flexibility and regularization leading to better performance.

4 Conclusion

We have presented a novel multi-task learning framework that allows joint modeling of tasks based on multiple relationship between them, where each relation is independently defined on a set of semantically related features. This helps in modeling scenarios where task-to-task relationships differ based on feature sets or where tasks have slightly different features sets. To model multiple task relatedness, we learn several feature subsets using a low dimensional subspace and use a task covariance matrix to capture task relationships (both positive and negative) along each feature subset. We formulate the model as an optimization problem and derive an efficient solution. Using both synthetic and real datasets, we demonstrate that the performance of proposed model is better than several state-of-the-art multi-task learning algorithms.

References

1. Kang, Z., Grauman, K., Sha, F.: Learning with whom to share in multi-task feature learning. In: International Conference on Machine Learning, pp. 521–528 (2011)
2. Saha, B., Gupta, S., Phung, D., Venkatesh, S.: Multiple task transfer learning with small sample sizes. Knowledge and Information Systems (2014). doi:10.1007/s10115-015-0821-z
3. Xue, Y., Liao, X., Carin, L., Krishnapuram, B.: Multi-task learning for classification with dirichlet process priors. The Journal of Machine Learning Research **8**, 35–63 (2007)
4. Zhou, J., Liu, J., Narayan, V.A., Ye, J.: Modeling disease progression via multi-task learning. NeuroImage **78**, 233–248 (2013)

5. Lin, H., Baracos, V., Greiner, R., Chun-nam, Y.: Learning patient-specific cancer survival distributions as a sequence of dependent regressors. In: Advances in Neural Information Processing Systems, pp. 1845–1853 (2011)
6. Argyriou, A., Evgeniou, T., Pontil, M.: Convex multi-task feature learning. Machine Learning **73**(3), 243–272 (2008)
7. Kumar, A., Daumé III, H.: Learning task grouping and overlap in multi-task learning. In: International Conference on Machine Learning (ICML) (2012)
8. Rai, P., Daume, H.: Infinite predictor subspace models for multitask learning. In: International Conference on Artificial Intelligence and Statistics, pp. 613–620 (2010)
9. Evgeniou, T., Pontil, M.: Regularized multi-task learning. In: ACM SIGKDD International Conference on Knowledge Discovery and Data Mining, pp. 109–117. ACM (2004)
10. Evgeniou, T., Micchelli, C.A., Pontil, M.: Learning multiple tasks with kernel methods. In: Journal of Machine Learning Research, pp. 615–637 (2005)
11. Caruana, R.: Multitask learning. Machine learning **28**(1), 41–75 (1997)
12. Jacob, L., Vert, J.-P., Bach, F.R.: Clustered multi-task learning: a convex formulation. In: Advances in neural information processing systems, pp. 745–752 (2009)
13. Zhou, J., Chen, J., Ye, J.: Clustered multi-task learning via alternating structure optimization. In: Advances in Neural Information Processing Systems, pp. 702–710 (2011)
14. Passos, A., Rai, P., Wainer, J., Daume, H.: Flexible modeling of latent task structures in multitask learning. In: Int'l Conference on Machine Learning, pp. 1103–1110 (2012)
15. Gupta, S., Phung, D., Venkatesh, S.: Factorial multi-task learning: a Bayesian nonparametric approach. In: International Conference on Machine Learning, pp. 657–665 (2013)
16. Zhang, Y., Yeung, D.-Y.: A convex formulation for learning task relationships in multi-task learning. In: Uncertainty in Artificial Intelligence, pp. 733–442 (2010)
17. Elixhauser, A., Steiner, C., Harris, D.R., Coffey, R.M.: Comorbidity measures for use with administrative data. Medical Care **36**(1), 8–27 (1998)

Multi-Task Metric Learning on Network Data

Chen Fang$^{(\boxtimes)}$ and Daniel N. Rockmore

Computer Science Department, Dartmouth College, Hanover, NH 03755, USA
{chenfang,rockmore}@cs.dartmouth.edu

Abstract. Multi-task learning (MTL) has been shown to improve prediction performance in a number of different contexts by learning models jointly on multiple different, but related tasks. In this paper, we propose to do MTL on general network data, which provide an important context for MTL. We first show that MTL on network data is a common problem that has many concrete and valuable applications. Then, we propose a metric learning approach that can effectively exploit correlation across multiple tasks and networks. The proposed approach builds on structural metric learning and intermediate parameterization, and has efficient an implementation via stochastic gradient descent. In experiments, we challenge it with two common real-world applications: citation prediction for Wikipedia articles and social circle prediction in Google+. The proposed method achieves promising results and exhibits good convergence behavior.

Keywords: Multi-task learning · Metric learning · Social network · Link prediction

1 Introduction

Multi-task learning (MTL) [2,3,6,7,21] considers the problem of learning models jointly and simultaneously over multiple, different but related tasks. Compared to single-task learning (STL), which learns a model for each task independently using only task specific data, MTL leverages all available data and shares knowledge among tasks, thereby resulting in better model generalization and prediction performance. The underlying principle of MTL is that highly correlated tasks can benefit from each other via joint training, but additional care should be taken to respect the distinct nature of each task, i.e., it is usually inappropriate to pool all available data and learn a single model for all tasks.

Despite the popularity and value of MTL, most MTL methods are developed for tasks on i.i.d. data. Standard examples include phoneme recognition [14] and image recognition [19]. Explicitly correlated data, often represented in the form of a network, is widely available, such as social network, citation network and influence network. It provides a rich source of new application contexts to MTL. Due to the diversity and variation in networks (e.g., multi-relational links or multi-category entities/nodes), various tasks can be performed and often a rich correlation exists between them. In the following, we give two common scenarios where there is abundant correlation between tasks and it is beneficial to apply

© Springer International Publishing Switzerland 2015
T. Cao et al. (Eds.): PAKDD 2015, Part I, LNAI 9077, pp. 317–329, 2015.
DOI: 10.1007/978-3-319-18038-0_25

MTL to exploit it. (These scenarios are also the settings for the experiments using real-world data that we present in Section 4).

Scenario 1: Article Citation Prediction

The citation prediction problem has been studied extensively [1,8–10,18]. People either build a predictive model for a unified network [10] (i.e., a citation network that contains papers across all subject areas) or build predictive models for each area independently [16]. Since article content and citation pattern varies across different areas, the former methodology ignores the difference between areas. However, some areas, while labeled as different are still related, in the sense of both content and citation pattern. Thus the latter methodology fails to exploit the correlation among subject areas. For example, computer science and electrical engineering articles may be classified or tagged as different areas, but in many cases they may still have much in common, or at least have significant similarity or overlap. In this case, to build predictive models for citations, a learning algorithm that is capable of utilizing these overlaps and explicit commonalities has advantages over traditional methods.

Scenario 2: Social Circle Prediction

Members of online social networks tend to categorize their links to followers/followees. For example, many social networking platforms enable coarse-scale categorizations such as "family members," or "friends and colleagues." Finer gradations allow for categorizations such as colleagues at particular companies or classmates at specific schools. A person's *social circle*, studied in [11], is the ego network of a social network user (or "ego"). This is the (star-shaped) subgraph on "ego" and all of ego's followers comprising all the links joining ego to ego's followers that belong to the same category. Given a friend or stranger, the goal of social circle prediction is to assign him/her to appropriate social circles. Because some social circles are related to each other (e.g., family members and childhood friends may share some common informative features such as geographical proximity), advantages may very well accrue if the relatedness of the entities is used for the various predictions, instead of building a predictive assignment model for each social circle independently.

As these scenarios suggest, correlations commonly exist among tasks on network data and there should be significant advantages to developing methods that can leverage it. Different from i.i.d. data, network data not only has attributes (metadata) associated with each entity (node), but also rich structural information, mainly encoded in the links. Therefore, we employ structural learning to exploit both attributes and structure of networks. Specifically, we adopt structure preserving metric learning (SPML) [16], which was originally developed for single-task learning on networks. Our proposed method, MT-SPML, empowers SPML with the ability of doing MTL over multiple tasks and networks. SPML learns a single Mahalanobis distance metric on node attributes for a single task by using network structure as supervision, so that the learned distance function encodes the structure. Our method learns Mahalanobis distance metrics jointly over all tasks. More precisely, it learns a common metric for all tasks and one metric for each individual task. The common metric construction follows the methodology of shared intermediate parameterization [7,12], which allows sharing knowledge between tasks. While a task specific metric alone captures task specific information, when combined they work together to preserve the connectivity structure of the corresponding network. The learned metrics of SPML and MT-SPML are useful to many tasks on network, one of which is predicting future link pattern. We further show that as in the case of SPML, MT-SPML can be optimized with efficient online methods similar to OASIS [4] and PEGASOS [15] via stochastic gradient descent. Finally, MT-SPML is designed for general networks, thus can be applied extensively in a wide variety of problems. In experiments, in order to demonstrate the advantages of MTL on network data, we apply MT-SPML to two common real-world prediction problems (citation prediction and social circle prediction), and achieve promising results for link prediction.

2 Related Work

MTL is a popular research topic and has been studied extensively and systematically for i.i.d. data. To name a few, Yu et al. [21] applied hierarchical Bayesian modeling for text categorization. Evgeniou et al. [7] extended Support Vector Machines (SVMs) to MTL via parameter sharing. Following [7], Parameswaran et al. [12] proposed the multi-task version of large margin nearest-neighbor metric learning [20]. However, there have been only few works focusing on MTL on relational data [5,17,22]. Of greatest relevance for our work is [13] wherein Qi et al. carefully designed a mechanism to sample across networks to predict missing links in a target network. Our paper differs from it in several ways. First, we aim at improving prediction performance of all networks, while [13] targets at a specific network and uses other networks as additional sources. Second, MT-SPML learns a joint embedding of both attribute features and network topological structure. Thus, the learned metrics can predict link patterns solely from node attributes while [13] tries to combine linearly attribute features with hand-constructed local structure information such as the number of shared neighbors between nodes. This suffers from the well-known "cold start" problem when structure information is limited (e.g. new nodes).

3 Our Approach

In this section, we first cover the technical details of SPML and then those of MT-SPML.

3.1 Notations and Preliminaries

Given a network on n nodes we represent it as a pair $\mathbf{G} = (\mathbf{X}, \mathbf{A})$, where $\mathbf{X} \in \mathbb{R}^{d \times n}$ represents the node attributes and $\mathbf{A} \in \mathbb{R}^{n \times n}$ is the binary adjacency matrix, whose entry $\mathbf{A}_{ij}$ indicates the linkage information between node

i and node j. Recall that a *Mahalanobis distance* is parameterized by a positive semidefinite (PSD) matrix $\mathbf{M} \in \mathbb{R}^{d \times d}$, where $\mathbf{M} \succeq 0$. The corresponding distance function is defined as $d_{\mathbf{M}}(x_i, x_j) = (x_i - x_j)^{\top} \mathbf{M} (x_i - x_j)$. This is equivalent to the existence of a linear transformation matrix $\mathbf{L}$ on the feature space such that $\mathbf{M} = \mathbf{L}^{\top} \mathbf{L}$. Given a metric $\mathbf{M}$, to predict the structure pattern of $\mathbf{X}$ we adopt a simple k-nearest neighbor algorithm, which is denoted as $\mathcal{C}$, meaning each node is connected with its top-k nearest neighbors under the defined metric. Mathematically, we say $\mathbf{M}$ is *structure preserving* or that *it preserves* $\mathbf{A}$, if $\mathcal{C}(\mathbf{X}, \mathbf{M})$ closely approximates $\mathbf{A}$.

Let $\mathcal{G} = \{\mathbf{G}_1, \mathbf{G}_2, \ldots, \mathbf{G}_Q\}$ denote a set of networks. Each individual network $\mathbf{G}_q$ has its own $\mathbf{X}_q$ and $\mathbf{A}_q$. We use q to index the network so that $\mathbf{A}_{qij}$ stands for element (i, j) in $\mathbf{A}_q$. Similarly, x_{qi} represents the feature of node i in $\mathbf{X}_q$. In algorithms, we will use a superscript to index over iteration, e.g., $\mathbf{M}^k$ refers to the k-th iteration of $\mathbf{M}$ under the relevant iterative process.

3.2   SPML

The goal of SPML is to learn $\mathbf{M}$ from a network $\mathbf{G} = (\mathbf{X}, \mathbf{A})$, such that $\mathbf{M}$ preserves $\mathbf{A}$. This problem has a semidefinite max margin learning formulation,

$$\min_{\mathbf{M} \succeq 0} \frac{\lambda}{2} \|\mathbf{M}\|_F^2 + \xi \tag{1}$$

subject to the following constraints:

$$\forall_{i,j}, \quad d_{\mathbf{M}}(x_i, x_j) \geq (1 - \mathbf{A}_{ij}) \max_l (\mathbf{A}_{il} d_{\mathbf{M}}(x_i, x_l)) + 1 - \xi. \tag{2}$$

In Eq.(1) $\|\cdot\|_F$ denotes the Frobenius norm and it takes on the role as a regularizer on $\mathbf{M}$ with λ representing the corresponding weight parameter. The key piece for achieving structure preserving is the set of linear constraints in Eq.(2). This essentially enforces that from node i, the distances to all disconnected nodes must be larger than the distance to the furthest connected node. Thus, when the constraints in Eq.(2) are all satisfied, $\mathcal{C}(\mathbf{X}, \mathbf{M})$ will exactly reproduce $\mathbf{A}$. Furthermore, to allow for violation (with penalty), the slack variable ξ is introduced.

With the many constraints in Eq.(2), optimizing Eq.(1) becomes unfeasible when the network has even a few hundred nodes. But a rewriting of the problem as follows makes possible the use of stochastic subgradient descent (see Algorithm 1):

$$f(\mathbf{M}) = \frac{\lambda}{2} \|\mathbf{M}\|_F^2 + \frac{1}{|S|} \sum_{(i,j,l) \in S} \max(\Delta_{\mathbf{M}}(x_i, x_j, x_l) + 1, 0) \tag{3}$$

where $\Delta_{\mathbf{M}}(x_i, x_j, x_l) = d_{\mathbf{M}}(x_i, x_l) - d_{\mathbf{M}}(x_i, x_j)$ and $S = \{(i, j, l) | \mathbf{A}_{i,l} = 1 \wedge \mathbf{A}_{i,j} = 0\}$. Thus, inclusion of the triplet (i, j, l) means that there is a link between node i and node l, but not between i and j. The subgradient of Eq.(3) can be calculated as

$$\nabla f = \lambda \mathbf{M} + \frac{1}{|S|} \sum_{(i,j,l) \in S_+} \left((x_i - x_l)(x_i - x_l)^\top - (x_i - x_j)(x_i - x_j)^\top \right) \quad (4)$$

where S_+ is the set of triplets whose hinge losses are positive. At every iteration t of Algorithm 1, B triplets are randomly sampled and the corresponding stochastic subgradient is calculated with regard to the current metric $\mathbf{M}^t$ and these triplets. Since Algorithm 1 is a variant of PEGASOS [15], its complexity does not depend on the training set size n, but on the feature dimensionality d. For the number of iterations T needed to reach convergence, as proved by [15,16] it depends on the parameter λ and the optimization error, which measures how close the final objective value is to the global optimal objective value. Notice that after updating $\mathbf{M}$, it is optional to project the current $\mathbf{M}$ to be positive semidefinite (PSD). Experiments in [16] show that delaying this operation to the end of the algorithm works well in practice and reduces computational complexity.

3.3 MT-SPML

In this section, we explain how MT-SPML extends SPML to the multi-task setting. The input is a set of networks $\mathcal{G} = \{\mathbf{G}_1, \mathbf{G}_2, \dots, \mathbf{G}_Q\}$. Once again, each network is $\mathbf{G}_q = (\mathbf{X}_q, \mathbf{A}_q)$. Our approach is a general method. It works in settings for which there either are or are not nodes overlapping between networks. Note that the nodes of all networks are assumed to have a common feature space. MT-SPML treats each network as a task. It follows the idea of *shared intermediate parametrization* [12] to enable knowledge transfer between tasks. The goal is to learn jointly over $\mathcal{G}$ a task specific metric $\mathbf{M}_q$ for each task and a common metric $\mathbf{M}_0$, through which knowledge transfers among tasks, so that the combined metric $(\mathbf{M}_0 + \mathbf{M}_q)$ respects the structure of $\mathbf{G}_q$, for all $\mathbf{G}_q \in \mathcal{G}$. The distance between two nodes $x_{qi}, x_{qj} \in \mathbf{G}_q$ is defined as $d_q(x_{qi}, x_{qj}) = (x_{qi} - x_{qj})^\top (\mathbf{M}_0 + \mathbf{M}_q)(x_{qi} - x_{qj})$. And MT-SPML is formulated as the solution to the regularized learning problem

$$\min_{\mathbf{M}_0, \mathbf{M}_1, \dots, \mathbf{M}_Q} \frac{\gamma_0}{2} ||\mathbf{M}_0 - \mathbf{I}||_F^2 + \sum_{q=1}^{Q} \frac{\gamma_q}{2} ||\mathbf{M}_q||_F^2 + \sum_{q=1}^{Q} \xi_q \quad (5)$$

subject to the following constraints:

$$\forall q, i, j, : d_q(x_{qi}, x_{qj}) \geq (1 - \mathbf{A}_{qij}) \max_l (\mathbf{A}_{qil} d_q(x_{qi}, x_{ql})) + 1 - \xi_q. \quad (6)$$

In order to solve this we rewrite it by incorporating the constraints

$$f(\mathbf{M}_0, \mathbf{M}_1, \dots, \mathbf{M}_Q) = \frac{\gamma_0}{2} ||\mathbf{M}_0 - \mathbf{I}||_F^2 + \sum_{q=1}^{Q} \frac{\gamma_q}{2} ||\mathbf{M}_q||_F^2$$

$$+ \sum_{q=1}^{Q} \frac{1}{|S_q|} \sum_{(i,j,l) \in S_q} \max(\Delta_q(x_{qi}, x_{qj}, x_{ql}) + 1, 0) \quad (7)$$

where $\Delta_q(x_{qi}, x_{qj}, x_{ql}) = d_q(x_{qi}, x_{ql}) - d_q(x_{qi}, x_{qj})$. Although Eq.(7) has more unknown variables than Eq.(3), with respect to each unknown, it is in the same form as Eq.(3). Therefore, Eq.(7) can be solved with the same stochastic subgradient descent method using partial subgradient. The partial subgradients of Eq.(7) with respect to $\mathbf{M}_0$ and $\mathbf{M}_q$ are

$$\nabla_{\mathbf{M}_0} f = \gamma_0 (\mathbf{M}_0 - \mathbf{I}) + \sum_{q=1}^{Q} \frac{1}{|S_q|} \sum_{(i,j,l) \in S_{q+}} \left((x_{qi} - x_{ql})(x_{qi} - x_{ql})^\top \right.$$
$$\left. - (x_{qi} - x_{qj})(x_{qi} - x_{qj})^\top \right) \quad (8)$$

and

$$\nabla_{\mathbf{M}_q} f = \gamma_q \mathbf{M}_q + \frac{1}{|S_q|} \sum_{(i,j,l) \in S_{q+}} \left((x_{qi} - x_{ql})(x_{qi} - x_{ql})^\top - (x_{qi} - x_{qj})(x_{qi} - x_{qj})^\top \right) \quad (9)$$

The optimization algorithm outlined in Algorithm 2 runs for T iterations. Within each iteration, it does two things: (1) Randomly samples B triplets for each task, then calculates the partial subgradient and updates the corresponding unknowns; (2) Calculates the partial subgradient of the common metric $\mathbf{M}_0$ and updates it using the $Q \times B$ triplets already sampled. Optionally, the metric matrices can be projected to be PSD. The analysis of Algorithm 1 still holds for Algorithm 2. Thus it scales with regard to feature dimensionality, optimization error and the parameters γ_q, but not the training set size.

Algorithm 1. Optimization of SPML

Input: $\mathbf{G} = (\mathbf{X}, \mathbf{A}), \lambda, T, B$
Output: $\mathbf{M} \succeq 0$
1: $\mathbf{M}^0 \leftarrow \mathbf{I}^{d \times d}$
2: **for** $t = 1, 2, \ldots, T$ **do**
3:     $\eta^t \leftarrow \frac{1}{\lambda \times t}$
4:     $s \leftarrow \emptyset$
5:     **for** $b = 1, 2, \ldots, B$ **do**
6:         Random sample (i, j, l) from S
7:         $s \leftarrow s \cup (i, j, l)$
8:     **end for**
9:     $\mathbf{M}^t \leftarrow \mathbf{M}^{t-1} - \eta^t \nabla f(\mathbf{M}^{t-1}, s)$
10:    $\mathbf{M}^t \leftarrow [\mathbf{M}^t]_+$
11: **end for**
12: return $\mathbf{M}^T$

Algorithm 2. Optimization of MT-SPML

Input: $\mathcal{G} = \{\mathbf{G}_1, \mathbf{G}_2, \ldots, \mathbf{G}_Q\}$, where $\mathbf{G}_q = (\mathbf{X}_q, \mathbf{A}_q)$,
$\gamma_0, \gamma_1, \ldots, \gamma_Q, T, B$
Output: $\mathbf{M}_0, \mathbf{M}_1, \ldots, \mathbf{M}_Q \succeq 0$
1: **for** $q = 0, 1, \ldots, Q$ **do**
2:     $\mathbf{M}_q^0 \leftarrow \mathbf{I}^{d \times d}$
3: **end for**
4: **for** $t = 1, 2, \ldots, T$ **do**
5:     **for** $q = 1, 2, \ldots, Q$ **do**
6:         $\eta_q^t \leftarrow \frac{1}{\lambda \times t}$
7:         $s_q \leftarrow \emptyset$
8:         **for** $b = 1, 2, \ldots, B$ **do**
9:             Random sample (i, j, l) from S_q
10:            $s_q \leftarrow s_q \cup (i, j, l)$
11:        **end for**
12:        $\mathbf{M}_q^t \leftarrow \mathbf{M}_q^{t-1} - \eta_q^t \nabla_{\mathbf{M}_q} f(\mathbf{M}_q^{t-1}, s_q)$
13:        $\mathbf{M}_q^t \leftarrow [\mathbf{M}_q^t]_+$
14:    **end for**
15:    $\mathbf{M}_0^t \leftarrow \mathbf{M}_0^{t-1} - \eta_0^t \nabla_{\mathbf{M}_0} f(\mathbf{M}_0^{t-1}, \{s_1, s_2, \ldots, s_Q\})$
16:    $\mathbf{M}_0^t \leftarrow [\mathbf{M}_0^t]_+$
17: **end for**
18: return $\mathbf{M}_0^T, \mathbf{M}_1^T, \ldots, \mathbf{M}_Q^T \succeq 0$

4 Experiments

In this section, we present experimental results on real-world data and we adopt link pattern prediction as the performance measurement. We apply MT-SPML to the two scenarios mentioned in Section 1: article citation prediction and social circle prediction. We show that in both cases, MT-SPML significantly improves performance and has various advantages.[1]

4.1 Citation Prediction on Wikipedia

The data is obtained from [16]. The articles of the following three areas were crawled from Wikipedia: search engine, graph theory and philosophy, each of which has 269, 223 and 303 articles respectively. The citations between articles within each area are also crawled. The number of citations within each area are 332, 917 and 921. The goal is, given an article, to predict the referencing of other articles within its area solely from its content. Therefore, at test time, no reference information from the test article is made available at all. The challenge of this problem is the fact that: (1) there is little node overlap between networks (i.e., an article belongs to only one area), thus the marginal distribution of node attributes $P(\mathbf{X}_q)$ may vary dramatically from area to area, which poses difficulty for knowledge transfer; (2) the conditional probability of structure on attributes $P(\mathbf{A}|\mathbf{X})$ may also vary, because some words are informative and indicative for some areas, but not for others. Bag-of-words (i.e., word frequency) is used to capture article content and the dimensionality is 6695. The high dimensionality reduces the need to learn full matrices. Therefore, we choose to learn diagonal metric matrices. This further reduces computational complexity. We split the dataset 80%/20% as training and testing respectively, then fix the testing part and vary the size of the training set by sampling from the training part. We end up sampling $20\%, 40\%, 60\%, 80\%$, and finally 100% of the training part. Model selection is carried out on the sampled training set via 5-fold cross-validation. At test time, the goal is to predict links between testing nodes from attributes. For every test example our algorithm ranks other articles for citation according to their distances. We build the receiver operator characteristic (ROC) curve for every test article, and use the average area under the curve (AUC) of the entire test set as performance measurement. We compare our results with two families of methods:

SVM Methods. We apply SVM-based methods as part of our baselines. Since SVM-based methods do not model network structure, we need to construct features to encode this information. The training examples are constructed by taking the pairwise difference of the attributes between two nodes. The training labels are binary, with 1 representing the existence of a link between two nodes and 0 the absence. For a given edge, we measure its distance/length using the output of the classification score, which represents the confidence of having a link. Although the classification score is inversely proportional to the notion of

---

[1] Code and data are available at author's website.

distance, a simple conversion can make the two variables proportional. Thus ROC and AUC can be calculated. The following specific methods are included:

(1) ST-SVM: This is the normal single-task SVM. An SVM is trained for each network independently. It does not explore the correlation between tasks.

(2) U-SVM: We train one SVM for all networks by pooling all data together. We use the capital letter "U" to denote the naive strategy of data pooling. This ignores the fact that training examples are from different tasks and treats it as single task learning.

(3) MT-SVM: This is the multi-task SVM in [7]. It jointly learns a common decision boundary for all and a specific boundary for each task. At test time, the common and task specific decision boundary together form the final model for each task. This method exploits task correlations via intermediate parameter sharing, but does not use network structure at the model level.

SPML methods: We apply three methods that are based on SPML. Compared to SVM-based methods, these methods explicitly model the network structure information. Therefore, the feature used here is simply the node attributes and links become linear constraints. Given an edge, its distance is just the Mahalanobis distance defined by learned metrics. The following methods are included:

(1) ST-SPML: This is the single-task SPML [16]. A metric is learned for each network independently. It does not model task correlations.

(2) U-SPML: "U" means data pooling. Training examples from all tasks are pooled together and the learning procedure is simply ST-SPML. This is a naive way of sharing knowledge between tasks, but it does not respect the differences between and distinctiveness of tasks. *Thus we expect inferior results, particularly for less related tasks.*

(3) MT-SPML: This is our proposed method. Comparison with these other methods demonstrates the fact that MT-SPML exploits relatedness missed by these other methods while respecting the distinctive nature of the individual tasks.

Finally, we also compare to the direct use of the original feature vector, i.e., using Euclidean distance. Other methods in link prediction literature, such as Adamic-Adar [10], typically heavily rely on local structure of test nodes, thus will suffer from "cold start" prescribed in our experimental framework.

The results are reported in Fig.1. The first thing we see is that SVM-based methods perform the worst when there are fewer training examples while the SPML family achieves good results in all settings, due to its ability to model structure information. We also find that among the SPML methods, MT-SPML consistently outperforms the others, which implies that MT-SPML is better at exploiting task correlations. The least amount of improvement from MT-SPML is found for philosophy articles. This is in line with intuition as papers related to search engines and graph (network) theory should have more in common with each other than either has with philosophy papers.

We also show the convergence behavior of MT-SPML by plotting the value of $|S_{q+}|$, the number of violated constraints among those randomly sampled triplets, for every task in each iteration. The fewer the number of violated constraints, the better the new metric respects the network structure. In experiments we set B, the time of random sampling, to be 10. In order to make a clearer demonstration, in Fig.2 we set B to be 100. As Fig.2 shows, the numbers of violated constraints of all tasks drop quickly within the first 1000 iterations and stabilizes after 4000 iterations. This is in accordance with the previous analysis and experiments of convergence and efficiency of SPML [16].

4.2   Social Circle Prediction on Google+

Every member of an online social network (e.g., Google+) is the ego of his/her (sub-)network and tends – or may be forced – to categorize his/her relationships (e.g. family members, college friends or childhood friends). For each type of relationship, there is a sub-network associate with it, the *social circle* (SC), which is directly formalized in the online structures of Google+ (see [11]). In this section, given a Google+ social network user (the ego) and his/her friends, we want to predict his/her SC, namely the type of relationships between ego and ego's friends based on profile information. We are only interested in the ego network, meaning that we do not predict the links between friends. A similar topic is studied by McAuley et al. [11], where the setup is very different from ours. They assume the observation of an entire ego network, including node attributes and structure, but not any SC labels, and the goal is to assign SC labels to links in an unsupervised manner. Our problem uses a supervised learning setting, where we observe only parts of the network and the corresponding SC labels. For the prediction of each social circle, we treat it as link prediction. However, as mentioned in Section 1, the correlation between social circles should be exploited. Thus, we treat the prediction of each social circle as a task, and MT-SPML is applied to learn metrics jointly over the underlying ego networks of all social circles. Note that, as reported in [11], SCs largely overlap with each other, which implies strong correlations and MTL is thus likely to achieve a more significant performance gain. We obtain data from [11], which was from Google+ users and information is anonymous. We randomly pick one user and his/her social circles for our experiment. The entire ego network has 4402 nodes and 5 social circles. The profile of all nodes is also preserved. There are 6 types of feature: gender, institution, job title, last name, place,

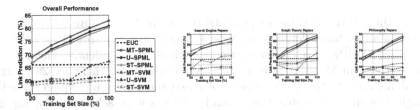

Fig. 1. Link prediction performance on Wikipedia article data. Training set size is varied. The larger figure on left is the average AUC performances over all three areas. Smaller figures on the right separate out the individual performance for each area.

and university. We build a bag-of-words feature for all feature types and concatenate them all, resulting in a feature vector of 2969 dimensions. The data are split
80%/20% as non-overlapping training and testing.

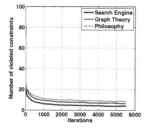

Fig. 2. Number of violated constraints within first 5500 iterations

Table 1. Node overlap between SCs. Overlapping ratios are presented in percentage
(%).

SC	1	2	3	4	5
1	100	1.1	81.9	89.6	84.1
2	1.1	100	0.9	1.1	1.1
3	81.9	0.9	100	73.5	68.9
4	89.6	1.1	73.5	100	93.7
5	84.1	1.1	68.9	93.7	100

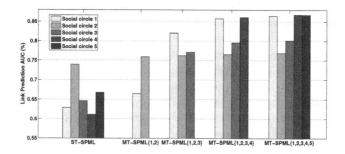

Fig. 3. Link prediction performance on Google+. SCs are color coded. The comparison is between ST-SPML and MT-SPML. The first group contains the prediction
performance of ST-SPML on all SCs, while the others show the performance of MT-
SPML that learned and tested on multiple combinations of SCs, for example, MT-
SPML{1,2,3} means learning and testing on SC 1, 2, 3.

In this experiment, we adopt a slightly different procedure for clear demonstration and a more detailed analysis. We index the SCs from 1 to 5. We start
by using ST-SPML to learn a metric for each SC independently. Then, for comparison we run MT-SPML on various numbers of SCs. There are 26 nontrivial
combinations of SCs, so for reasons of space and clarity we explore in detail a
single sequence of combinations. Similar results were achieved with the other
combinations. We begin by running on {1,2} and add one more SC at a time
in order, resulting in the following four combinations: {1,2}, {1,2,3}, {1,2,3,4},
{1,2,3,4,5}. We use these in later experiments as well. In this way, we can compare the behavior of the algorithms as more tasks join the process. In Fig.3, we

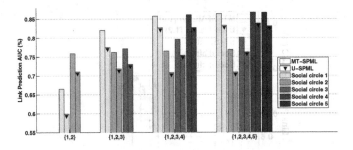

Fig. 4. Link prediction performance on Google+ data. The comparison is between MT-SPML and U-SPML. SCs are color coded. Different methods for the same task are compared side by side. U-SMPL is marked by a down pointing triangle. Each group is trained and tested on a set of SCs. For example, {1,2,3} means learning and testing on SC 1, 2, 3.

compare ST-SPML to MT-SPML on the four combinations of SCs. Note that, because of the inferior performance of SVM based methods on Wikipedia article data, we omit them in this experiment. There two clear observations from Fig.3: (1) All SCs benefit from MTL and the improvement is significant; (2) Performance continues to improve as more tasks are involved, which demonstrates the superior ability of joint learning. One exception is SC 2, where the performance gain is small. We speculate that SC 2 is not closely related to other circles (e.g., in terms of the number of overlapping nodes). We will discuss the case of SC 2 later.

Now we compare MT-SPML to U-SPML, which simply pools all data together and estimates a model for all tasks. Both MT-SPML and U-SPML are applied to the four combinations of SCs. As shown by Fig.4, MT-SPML consistently and significantly outperforms U-SPML at all locations.

Now we would like to further investigate SC 2. We first show some statistics in Table 1, where we show the percentage of node overlapping between SCs. The overlap is defined as the intersection of nodes over the union. As we can see, some circles largely overlap (e.g., SC 1 and 3 have 81.9% nodes in common), while SC 2 barely overlaps with the others. Although overlapping is not the only quantitative measurement of correlations between social circles, a substantial set of common nodes suggests that there are some shared semantics between two relationships. Thus Table 1 supports our earlier speculation as to why SC 2 does not benefit from joint learning as much as the others.

Furthermore, we would like to again show the advantage of MT-SPML by showing the results on a pair of tasks that are less correlated to each other. We choose SCs 1 and 2, since they have only 1.1% nodes in common. In Fig.5, MT-SPML is jointly learned on {1,2}, U-SPML is learned via data pooling, and ST-SPML is trained on 1 and 2 independently. The prediction performances of two tasks are reported in the two groups of bars respectively. As shown in Fig.5, MT-SPML still gets 2%-5% performance improvement over ST-SPML

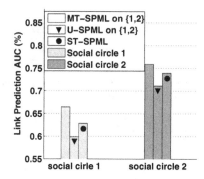

Fig. 5. U-SPML hurts performance when training on {1,2}, two less relevant tasks. MT-SPML is able to improve performance compared to ST-SPML by exploiting useful correlations.

(bars with circles on top). However, the naive data pooling strategy of U-SPML (bars with down pointing triangles) produces results even worse than ST-SPML. This observation suggests that on difficult cases where tasks are not as related, MTL is still able to utilize useful correlations, while respecting the boundaries between tasks.

5   Conclusions

In this paper, we deal with MTL on general network data. We first show that correlation widely exist between tasks on network data by giving two common scenarios where it is beneficial to employ MTL. Then we proposed MT-SPML, a multi-task structural metric learning method. It learns task specific metrics as well as a common distance metric. By combining them, the final metric preserves the structure of the networks. We applied MT-SPML to citation network and social network, and measure the performance in link prediction. Improvements were achieved and detailed analysis was provided. Moreover, its SGD implementation is easy and efficient with good convergence behaviour, thus the proposed method can scale up to larger problems and be a strong baseline approach to future research works.

References

1. Adafre, S.F., de Rijke, M.: Discovering missing links in wikipedia. In: Proceedings of the 3rd International Workshop on Link Discovery (2005)
2. Agarwal, A., III, H.D., Gerber, S.: Learning multiple tasks using manifold regularization. In: Neural Information Processing Systems (2010)
3. Caruana, R.: Multitask learning. Machine Learning **28**(1), 41–75 (1997)

4. Chechik, G., Sharma, V., Shalit, U., Bengio, S.: Large scale online learning of image similarity through ranking. Journal of Machine Learning Research **11**, 1109–1135 (2010)
5. Comar, P.M., Tan, P.N., Jain, A.K.: Multi-task learning on multiple related networks. In: ACM International Conference on Information and Knowledge Management (2010)
6. Daumé, III, H.: Bayesian multitask learning with latent hierarchies. In: Conference on Uncertainty in Artificial Intelligence (2009)
7. Evgeniou, T., Pontil, M.: Regularized multi-task learning. In: ACM International Conference on Knowledge Discovery and Data Mining (2004)
8. Hasan, M.A., Chaoji, V., Salem, S., Zaki, M.: Link prediction using supervised learning. In: SDM workshop on Link Analysis, Counterterrorism and Security (2006)
9. Kashima, H., Abe, N.: A parameterized probabilistic model of network evolution for supervised link prediction. In: ICDM (2006)
10. Liben-Nowell, D., Kleinberg, J.: The link prediction problem for social networks. In: ACM International Conference on Information and Knowledge Management (2003)
11. McAuley, J.J., Leskovec, J.: Learning to discover social circles in ego networks. In: Neural Information Processing Systems (2012)
12. Parameswaran, S., Weinberger, K.: Large margin multi-task metric learning. In: Neural Information Processing Systems (2010)
13. Qi, G.J., Aggarwal, C.C., Huang, T.S.: Link prediction across networks by biased cross-network sampling. In: ICDE (2013)
14. Seltzer, M., Droppo, J.: Multi-task learning in deep neural networks for improved phoneme recognition. In: IEEE International Conference on Acoustics, Speech, and Signal Processing (2013)
15. Shalev-Shwartz, S., Singer, Y., Srebro, N.: Pegasos: primal estimated sub-GrAdient SOlver for SVM. In: International Conference on Machine Learning (2007)
16. Shaw, B., Huang, B., Jebara, T.: Learning a distance metric from a network. In: Neural Information Processing Systems (2011)
17. Tang, W., Lu, Z., Dhillon, I.S.: Clustering with multiple graphs. In: IEEE International Conference on Data Mining (2009)
18. Taskar, B., fai Wong, M., Abbeel, P., Koller, D.: Link prediction in relational data. In: Neural Information Processing Systems (2003)
19. Wang, X., Zhang, C., Zhang, Z.: Boosted multi-task learning for face verification with applications to web image and video search. In: IEEE International Conference on Computer Vision and Pattern Recognition (2009)
20. Weinberger, K.Q., Saul, L.K.: Distance metric learning for large margin nearest neighbor classification. Journal of Machine Learning Research **10**, 207–244 (2009)
21. Yu, K., Tresp, V., Schwaighofer, A.: Learning gaussian processes from multiple tasks. In: International Conference on Machine Learning (2005)
22. Zhou, D., Zhu, S., Yu, K., Song, X., Tseng, B.L., Zha, H., Giles, C.L.: Learning multiple graphs for document recommendations. In: WWW (2008)

A Bayesian Nonparametric Approach
to Multilevel Regression

Vu Nguyen[1]([⊠]), Dinh Phung[1], Svetha Venkatesh[1], and Hung H. Bui[2]

[1] Deakin University, Geelong, Australia
{tvnguye,dinh.phung,svetha.venkatesh}@deakin.edu.au
[2] Adobe Research, San Francisco, USA
bui.h.hung@gmail.com

Abstract. Regression is at the cornerstone of statistical analysis. Multilevel regression, on the other hand, receives little research attention, though it is prevalent in economics, biostatistics and healthcare to name a few. We present a Bayesian nonparametric framework for multilevel regression where *individuals* including observations and outcomes are organized into *groups*. Furthermore, our approach exploits additional group-specific context observations, we use Dirichlet Process with product-space base measure in a nested structure to model group-level context distribution and the regression distribution to accommodate the multilevel structure of the data. The proposed model simultaneously partitions groups into cluster and perform regression. We provide collapsed Gibbs sampler for posterior inference. We perform extensive experiments on econometric panel data and healthcare longitudinal data to demonstrate the effectiveness of the proposed model.

1 Introduction

Real data is complex. They hardly conform to simple flat structure or a well-defined regular pattern. Multilevel, or hierarchical and nested, data structure persists in almost every day analysis tasks. Patients organized in different cohorts in multiple hospitals; economic activities of a city nested within a state, which is in turn influenced by national economic status and so on. Multilevel analysis [11,14,23] is an approach to analyze group contexts as well as the individual outcomes. In multilevel analysis, multilevel regression are commonly used in econometrics (panel data), biostatistics and sociology (longitudinal data) for regression estimation. Examples include panel data measures GDP observations over a period of time tracking in multiple states of the USA or longitudinal studies on a collection of patients' admissions to a hospital. To the best of our knowledge, almost no work of multilevel regression has attempted to model group context information to form 'optimal' cluster of groups to be regressed together. The main challenge is how to model the optimal or 'correct' clustering to leverage shared statistical strengths across groups.

In this paper, we consider the multilevel regression problem in multilevel analysis where *individuals* including observations and outcomes are organized into

© Springer International Publishing Switzerland 2015
T. Cao et al. (Eds.): PAKDD 2015, Part I, LNAI 9077, pp. 330–342, 2015.
DOI: 10.1007/978-3-319-18038-0_26

groups. Our modelling assumption is that individuals exhibit similar regression behaviours should be grouped and perform regression task together to leverage on their shared statistical strengths. For example, children with the same parents tend to be more alike in their physical and mental characteristics than individuals chosen at random from large population. Particularly, we focus on the multilevel regression problem for predicting individuals in *unseen groups*, the groups do not appear in the training set. For example, in health research - relied on patient's history of electronic medical record (EMR) - patient history records can be empty for patients have not admitted to a hospital before. Predicting individuals in unseen groups using multilevel regression presents another contribution of our work.

Traditional *single* regression method often treats hierarchical data as flat independent observations. Hence, it tends to mis-specify the regression coefficients, leading to poor fitting in overall populations. The well-known approach to multilevel regression is the Linear Mixed Effect model [16,20]. However, it is not well applicable for predicting individuals from unseen groups because the random effect is fixed to the given training groups.

Another way to multilevel regression is via multitask learning where each data group is treated as a *task* and individual seen as *examples*. Multi-task regression aims to improve generalization performance of related tasks by joint learning [4]. A few works have attempted to partition related tasks into task-groups [12]. Bayesian nonparametric approach is used to overcome the difficulty in defining the degree of relatedness among tasks [10]. For testing and evaluation, previous works use a proportion of examples in each task for training and the rest is further used for testing. Given a testing example, the task which the example belonged to, is identified from the hierarchical structure of the data. Nevertheless, given a testing example from unseen task, there is no proper way to perform prediction.

Addressing this gap, we present a Bayesian Nonparametric Multilevel Regression (BNMR) model. The model uses a Dirichlet Process as a product base-measure of group-context distribution and regression distribution to discover the unknown number of group clusters and do regression jointly. The group cluster is estimated based on the group-context observation and regression outcome of individuals. The goal is making the related groups strengthen each other in regression while unrelated groups do not affect themselves. In addition, simultaneously clustering groups and performing regression can prevent from overfitting to each training group. By using group-context information, the proposed model can assign the unseen group into an existing group-cluster for regression.

2 Multilevel Regression

Regression is a large research field. Within the scope of the paper, we focus on the model which can perform *multilevel regression* where the data presented in groups.

Observations in the same group are generally not independent, they tend to be more similar than observations from different groups. Standard single level regression models are not robust against violation of the independence assumption. That is why we need special multilevel treatment.

Dealing with grouped data, a popular setting known as multilevel analysis [11,23] has a board applications from multilevel regression [8] to multilevel document modelling and clustering [18].

We consider a pair of outcome and observation in hierarchical structure ($y_{ji} \in R, x_{ji} \in R^d$) where y_{ji} is an outcome (or response) and x_{ji} is an observation for trial i in group j. The multilevel models are the appropriate choice that can be used to estimate the intraclass correlation and regression in the multilevel data. Specifically, we consider Linear Mixed Effects models which are extensions of linear regression models for data that are organized in groups.

Linear Mixed Effects Model. The LME model [16] describes the relationship between a response variable and independent variables in multilevel structure, with coefficients that can vary with respect to one or more grouping variables. A mixed-effects model consists of two parts, fixed effects and random effects. Fixed-effects terms are usually the conventional linear regression part, and the random effects are associated with individual experimental units drawn randomly from population. The random effects have prior distributions whereas fixed effects do not. Linear Mixed Effects model can represent the covariance structure related to the grouping of data by associating the common random effects to observations in the same group. The standard form of a linear mixed-effects model is following:

$$y_{ji} = \beta_{j0} + x_{ji}^T \beta_{j1} + \epsilon_{ji} \qquad\qquad \epsilon_{ji} \sim \mathcal{N}\left(0, \sigma_\epsilon^2\right)$$

where the regression coefficients for group j: β_{j0} and β_{j1} are computed:

$$\beta_{j0} = \gamma_{00} + \gamma_{01} c_j + u_{j0} \qquad\qquad u_{j0} \sim \mathcal{N}\left(0, \sigma_{u0}^2\right)$$

$$\beta_{j1} = \gamma_{01} + \gamma_{11} c_j + u_{j1} \qquad\qquad u_{j1} \sim \mathcal{N}\left(0, \sigma_{u1}^2\right)$$

Therefore, the final form to predict the individual outcome variable y_{ji} using individual explanatory variables x_{ji} and group explanatory variable c_j is followed:

$$y_{ji} = \underbrace{\gamma_{00} + \gamma_{01} c_j + \gamma_{01} x_{ji} + \gamma_{11} c_j x_{ji}}_{\text{fixed effects}} + \underbrace{u_{j0} + u_{j1} x_{ji} + \epsilon_{ji}}_{\text{random effects}}$$

Fixed effects have levels that are of primary interest and would be used again if the experiment were repeated. Random effects have levels that are not of primary interest, but rather are thought of as a random selection from a much larger set of levels. We present the graphical representation of LME model in Fig. 1a. The common parameter estimation methods for linear mixed effect include Iterative Generalized Least Squares [9] and Expectation Maximization algorithm.

3 Preliminary

3.1 Linear Regression

Regression is an approach for modelling the relationship between a scalar *outcome* variable y and one or more *explanatory* variables denoted x. In linear regression, data are modelled using linear predictor functions, and unknown model parameters are estimated from the data. Given a data collection $\{y_i \in \mathcal{R}, x_i \in \mathcal{R}^d\}_{i=1}^N$ of N units, linear regression model assumes the relationship between the outcome variable y_i and the d-dimension vector of observation x_i is linear. Hence, the model takes the form: $y_i = x_i^T \beta + \epsilon_i$ where ϵ_i is a residual or error term, β is a regression coefficient, including *intercept* and *slope* parameters. The solution for β is: $\hat{\beta} = \left(X^T X\right)^{-1} X^T Y$ where $X = \{x_i\}_{i=1}^N$ and $Y = \{y_i\}_{i=1}^N$.

3.2 Bayesian Linear Regression

Bayesian linear regression is an approach to linear regression in which the statistical analysis is undertaken within the context of Bayesian inference with a prior distribution for parameter β. In this setting, the regression errors (or residual) is assumed to follow a normal distribution $\epsilon_i \sim \mathcal{N}\left(0, \sigma^2\right)$. Given a data point $x \in R^d$ and its respond variable y, the likelihood of Bayesian linear regression model with parameter β is defined as:

$$p\left(y \mid x, \beta, \sigma^2\right) = \frac{1}{\sqrt{2\pi}\sigma} \exp\left\{-\frac{1}{2}\|y - x^T \beta\|^2\right\}$$

Posterior probability distributions of the model's parameter under conjugate prior distribution $\beta \sim \mathcal{N}\left(0, \Sigma_0\right)$ is estimated following:

$$p\left(\beta \mid x_{1:N}, y_{1:N}, \Sigma_0, \sigma\right) \propto \mathcal{N}\left(\mu_n, \Sigma_n\right) \tag{1}$$

where the posterior mean $\mu_n = \Sigma_n \left\{X\sigma^{-1/2}Y\right\}$, and posterior covariance $\Sigma_n = \left(\Sigma_0^{-1} + X\sigma^{-1/2}X^T\right)^{-1}$ [3]. The likelihood for predicting new explanatory x_{new} with new response y_{new} is computed:

$$p\left(y_{\text{new}} \mid x_{\text{new}}, \mu_n, \Sigma_n\right) = \int_\beta p\left(y_{\text{new}} \mid x_{\text{new}}, \beta, \sigma^2\right) p\left(\beta \mid .\right) d\beta$$
$$= \mathcal{N}\left(x_{\text{new}}^T \mu_n, \sigma_n^2(x_{\text{new}})\right) \tag{2}$$

where $\sigma_n^2(x_{\text{new}}) = \sigma^2 + x_{\text{new}}^T \Sigma_n x_{\text{new}}$.

3.3 Bayesian Nonparametric

We provide a brief account of the Dirichlet Process Mixture and the Nested Dirichlet Process [21] which related to our work.

A *Dirichlet Process* [7] DP (γ, H) is a distribution over discrete random probability measure G on $(\Theta, \mathcal{B})$. Sethuraman [22] provides an alternative constructive definition which makes the discreteness property of a draw from a Dirichlet process explicit via the stick-breaking representation: $G = \sum_{k=1}^{\infty} \beta_k \delta_{\phi_k}$ where $\phi_k \overset{iid}{\sim} H, k = 1, \ldots, \infty$ and $\beta = (\beta_k)_{k=1}^{\infty}$ are the weights constructed through a 'stick-breaking' process. As a convention, we hereafter write $\beta \sim$ GEM (γ). Dirichlet Process has been widely used in Bayesian mixture models as the prior distribution on the mixing measures, resulting in a model known as the *Dirichlet Process Mixture model* (DPM) [1].

Dirichlet Process can also be constructed hierarchically to provide prior distributions over multiple exchangeable groups. One particular attractive approach is the *Hierarchical Dirichlet Processes* (HDP) [24] which posits the dependency among the group-level DPM by another Dirichlet process.

Another way of using DP to model multiple groups is to construct random measure in a nested structure in which the DP base measure is itself another DP. This formalism is the *Nested Dirichlet Process* [21], specifically $G_j \overset{iid}{\sim} U$ where $U \sim$ DP $(\alpha \times$ DP $(\gamma H))$. modelling G_j (s) hierarchically as in HDP and nestedly as in nDP yields different effects. HDP focuses on exploiting statistical strength across groups via sharing atoms ϕ_k (s), but it does not partition groups into clusters. Whereas, nDP emphasizes on inducing clusters on both observations and distributions, hence it partitions groups into clusters. Finally we note that this original definition of nDP in [21] does not force the atoms to be shared across clusters of groups, but this can be achieved by introducing a DP prior for the nDP base measure [18,19].

4 Bayesian Nonparametric Multilevel Regression

In this section, we describe our framework of Bayesian Nonparametric Multilevel Regression (BNMR). Our goal is to simultaneously clustering the *groups* and estimating regression for *individuals*. The fundamental assumption is that when the groups are related, the group-level explanatory variable (or group-context observation) is induced in the same distribution component (e.g., Gaussian distribution). Firstly, we aim to use the related groups to strengthen regression estimation for improving regression performance (prevent from overfitting to each group) while unrelated groups do not influence themselves. Second, the induced group-context distribution can be used to identify cluster for new groups (based on group-context observations in new groups).

Iteratively modelling and clustering group context and individual regression would gain benefit and mutually promote each other. First, good groups clustering will produce good regression estimation (e.g. we assume individuals in the same group-cluster have similar regression behavior). Second, the good regression estimation in return provides important information for the group-clustering process previously.

4.1 Model Representation

We consider data presented in a two-level structure. Denote by J the number of groups, we assume that the groups are exchangeable. Each group j contains N_j exchangeable explanatory variable and response variable, represented by $\left\{x_{ji} \in \mathcal{R}^d, y_{ji} \in \mathcal{R}\right\}_{i=1}^{N_j}$. The collection of $\{c_j\}_{j=1}^J$ represents group-level explanatory or group-level context (e.g., age of the patient, population of the state).

We now describe the generative process of BNMR (c.f Fig. 1b). Denote H is a base measure for generating group-context distribution and S is a base measure for generating regression coefficients. We use a product base measure of $H \times S$ to drawn a DP mixture for jointly clustering groups and regression individuals. Particularly, we have:

$$G \sim \mathrm{DP}\left(\alpha, H \times S\right) \qquad\qquad \left(\theta_j^c, \theta_j^y\right) = \theta_j \overset{\text{iid}}{\sim} G$$

Each realization θ_j includes a pair $\left(\theta_j^c, \theta_j^y\right)$ that θ_j^c is then used to generate the group-level explanatory observation c_j and θ_j^y is further used to drawn the individual response variables y_{ji} following:

$$c_j \sim F\left(\theta_j^c\right) \qquad\qquad y_{ji} \sim \mathcal{N}\left(x_{ji} \times \theta_j^y, \sigma_\epsilon^2\right)$$

where σ_ϵ is a standard deviation of residual error.

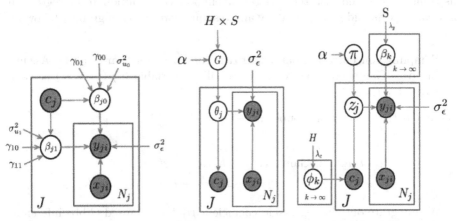

(a) Linear Mixed Effects model. **(b)** Bayesian Nonparametric Multilevel Regression.

Fig. 1. Graphical representation. Left: LME. Middle: BNMR stochastic process view. Right: BNMR stick-breaking view. There are J groups with group-level explanatory variable c_j, each group has N_j individuals including explanatory variable x_{ji} and response variable y_{ji}.

Stick-breaking representation. We further derive the stick-breaking representation for BNMR (c.f Right Fig. 1b) where all of the random discrete measures are characterized by a distribution over integers and a countable set of atoms.

The random measure G has the form: $G = \sum_{k=1}^{K} \pi_k \delta_{(\phi_k, \beta_k)}$ where $\pi \sim$ GEM (α), $\phi_k \stackrel{iid}{\sim} H(\lambda_c)$, and $\beta_k \stackrel{iid}{\sim} S(\lambda_y)$. Next, we draw an indicator cluster for each group $z_j \stackrel{iid}{\sim} \pi$ and generate group-context explanatory variable $c_j \sim F(\phi_{z_j})$. Accordingly, the response variables in group j given the cluster $z_j = k$ is drawn $y_{ji} \sim \mathcal{N}(x_{ji}^T \beta_k, \sigma_\epsilon^2)$.

4.2   Inference

We derive collapsed Gibbs sampling for BNMR. Due to the conjugacy property, we would integrate out ϕ_k, β_k, and π. The remaining latent variable z and hyperparameter α will be sampled.

- Sampling z_j. The conditional distribution for sampling z is:

$$p\left(z_j = k \mid c_j, \{y_{ji}, x_{ji}\}_{i=1}^{N_j}\right) \propto p(z_j = k \mid z_{-j}, \alpha)$$
$$\times p(c_j \mid z_j = k, c_{-j}, z_{-j}, H) \times p(y_{ji} \mid x_{ji}, z_j = k, S)$$

The first expression $p(z_j = k \mid z_{-j}, \alpha)$ is the Chinese Restaurant Process (CPR) with concentration parameter α. The second term is the predictive likelihood of group-context observation under component (or topic) k. This can be analytically computed due to conjugacy of likelihood distribution and prior distribution H. The last term is the likelihood contribution from regression observations (including explanatory and response variables) in group j following Eq. 2.

- Sampling concentration parameter α is similar to Escobar et al [6]. Assuming $\alpha \sim$ Gamma (α_1, α_2) with the auxiliary variable t: $p(t \mid \alpha, K) \propto$ Beta $(\alpha + 1, J)$

where J is the number of groups and $\frac{\pi_t}{1 - \pi_t} = \frac{\alpha_1 + K - 1}{J(\alpha_2 - \log t)}$.

$$p(\alpha \mid t, K) \sim \pi_t \text{Gamma}(\alpha_1 + K, \alpha_2 - \log(t))$$
$$+ (1 - \pi_t)\text{Gamma}(\alpha_1 + K - 1, \alpha_2 - \log(t))$$

We integrate out the regression coefficient β_k for collapsed Gibbs inference. However, for visualization and analysis of the regression coefficient β_k can be re-computed as $p(\beta_k \mid x_i, y_i, z_i = k, \Sigma_0)$ following Eq 1.

Given unseen groups of data include $\{x_{ji}^{\text{Test}}, c_j^{\text{Test}}\}$, we wish to estimate $\{y_{ji}^{\text{Test}}\}$. We observe that if β_k and σ_ϵ^2 are known, then y_{ji}^{Test} will be distributed by $\mathcal{N}\left(\beta_k^T x_{ji}^{\text{Test}}, \sigma^2 \mathbf{I}\right)$.

$$\hat{y}_{ji}^{\text{Test}} \propto \sum_{z_j^{\text{Test}}=1}^{K} \left[\boldsymbol{\beta}_{z_j}^{T} \boldsymbol{x}_{ji}^{\text{Test}} \right] \times p\left(z_j^{\text{Test}} \mid c_j^{\text{Test}} \right)$$

where $p\left(z_j^{\text{Test}} \mid c_j^{\text{Test}} \right) \propto p\left(z_j^{\text{Test}} \mid \boldsymbol{\pi} \right) p\left(c_j^{\text{Test}} \mid \phi_{z_j^{\text{Test}}} \right)$.

5 Experiment

We demonstrate the proposed framework on multilevel regression task, especially for regression individuals in unseen groups of data. Throughout this section, unless explicitly stated, the training and testing sets are randomly split, and repeated 10 times. The variables $\boldsymbol{x}_{ji}$ and y_{ji} is centralized to have the mean of 0 as recommended in regression tasks [11]. Our implementation is using Matlab. For synthetic and Econometric panel data, each iteration takes about 1-2 seconds and it takes 30-35 seconds for Heathcare dataset. All experiments are converged quickly within 30 iterations of collapsed Gibbs sampling. Initialization for concentration parameter $\alpha = 1$, $\alpha \sim \text{Gamma}(1,2)$. The conjugate distribution for group-level context is NormalGamma. We use four baseline methods for comparing the regression performance on individuals of unseen groups followings:

1. Naive Estimation: using the overall average of individuals outcome in training groups $\hat{y}_{\text{new}}^{\text{Test}} = \frac{1}{J} \sum_{j=1}^{J} \frac{1}{N_j} \sum_{i=1}^{N_j} y_{ji}^{\text{Train}}$ as the predicted value.
2. No-Group MultiTask Learning (NG-MTL) [2]: where all tasks are considered in a single group.
3. No-Group MultiTask Learning With Context (NG-MTL-Context): where all tasks are considered in a single group, and context is treated as another explanatory variable.
4. LME: $y_{ji} = \gamma_{00} + \gamma_{01} c_j + \gamma_{01} \boldsymbol{x}_{ji} + \gamma_{11} c_j \boldsymbol{x}_{ji} + \epsilon_{ji}$, we ignore random variables u_{j0} and u_{j1} from original LME for predicting unseen groups because we do not have $u_j(\text{s})$ for unseen groups. (u_j is representing for group j given in training set).

The regression performance is evaluated using two metrics: Root Mean Square Error (RMSE), and Mean Absolute Error (MAE). Since the errors are squared before they are averaged, the RMSE gives a relatively high weight to large errors. The MAE measures the average magnitude of the errors in a set of forecasts, without considering their direction. The regression algorithm is the ideal when it has lower error in both RMSE and MAE.

5.1 Synthetic Experiment

Our goal is to investigate BNMR's ability to recover the true group clusters and number of regression atoms. We first create three univariate Normal distributions $\phi_k(\text{s})$ with different variances (Fig. 2) for generating group-context

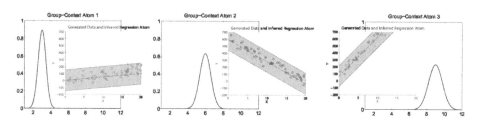

Fig. 2. Synthetic Experiment for Bayesian Nonparametric Multilevel Regression

observations. Conditional on these context distribution, we initialize three linear regression atoms β_k(s) with standard deviation for residual error $\sigma^2 = 50$. Then, we randomly sample $J = 200$ groups, each group comprises a group-context c_j and $N_j = 20$ pairs of observation (x_{ji}, y_{ji}).

Table 1. Regression performances on synthetic experiment. The lower is the better. Standard deviation is in a parenthesis.

Metrics\Methods	Naive Estimation	NG-MTL	NG-MTL-Context	LME	BNMR
RMSE	343.3 (11.3)	332.6 (6.9)	230.9 (8.7)	190.1 (8.5)	**118.0 (34.0)**
MAE	278.9 (8.1)	284.0 (4.1)	180.1 (9.2)	152.9 (5.4)	**56.0 (9.7)**

The model recovers correctly the ground truth atoms. Visualizations of the group-context distribution and generated data are plotted in Fig. 2. For evaluation, we split data into 70% number of groups for training and the rest (30% groups) for testing. The performance comparison is displayed in Table 1 so that our model gains great improvement in regression than the baseline methods.

5.2 Econometric Panel Data: GDP Prediction

The Panel Data [17] includes 48 states (ignoring Alaska and Hawaii) and 17 years of GDP collection from 1970 to 1986. There are nine divisions in the United States, e.g., New England, Mid-Atlantic, Pacific, and so on (Fig. 3a). Each division contains from 3 to 8 states.

The explanatory variable x_{ji} for each year i in a state j includes 11 dimensions, such as public capital stock, highways and streets capital stock, water and sewer facilities capital stock, employees on non-agricultural payrolls, unemployment rate, and so on. The response variable y_{ji} is a GDP.

We consider the state population (Wyoming has the lowest population of 0.57 millions and the highest population of 38 millions belongs to California, as of 2012) is an explanatory variable for group level. Population is one of the key factor determining the GDP [13,15]. Hence, states which alike number of population tend to have similar GDP outcome than other states in different number of population. We model the context distribution using univariate Gaussian distribution. The mean and precision for group context distribution are $(\mu, \tau) \sim \text{NormalGamma}(4, 0.25, 0.01, 1)$ and the standard deviation for regression residual error is set as $\sigma_\epsilon = 7000$.

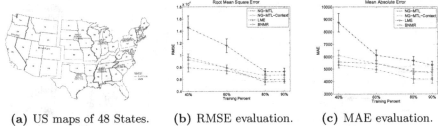

(a) US maps of 48 States.   (b) RMSE evaluation.   (c) MAE evaluation.

Fig. 3. Panel Data. Left: Maps have been modified from the Census Regions and Divisions www.census.gov/geo/maps-data/maps/pdfs/reference/us_regdiv.pdf. Middle and Right: Regression performance comparison.

We split the data into training set and testing set such that the states in the testing set do not appear in the training. We vary the proportion of training states from 40% to 90% and perform prediction on the rest. The number of state clusters are identified as $K = 3$ (indicating low, mid, and high population). The regression performance of BNMR versus NG-MTL, NG-MTL-Context and LME are plotted in Fig. 3. We do not include the scores of Naive Estimation into the figure because of its poor performance in this dataset. This poor performance of Naive Estimation can be explained by the high variance in the outcome (e.g., the GDPs of California and Texas are 10-20 times higher than GDPs of Vermont and Delaware). The proposed method achieves the best regression performance in term of RMSE (Fig. 3b) and MAE (Fig. 3c) scores. The more state we observe, the more accuracy in prediction we achieve.

5.3   Healthcare Longitudinal Data: Prediction Patient's Readmission Interval

Meaningful use, improved patient care and competition among providers are a few of the reasons electronic medical records are succeeding at hospitals. Readmission interval prediction could be used to help the delivery of hospital resource-intensive and care interventions to the patients. Ideally, models designed for this purpose would provide close estimation of the admission interval for the next admission. Very often, patients come to a hospital without any existed electronic medical records because they may have not been admitted before. This fact causes problem for existing multilevel regression approaches. We aim to use the proposed framework to improve performance for predicting readmission interval on new patients.

Our data collected from regional hospital (ethics approval 12/83.). Our main interest is in the chronic Polyvascular Disease (PolyVD) cohort. The collected data includes 209 patients with 3207 admissions in total. We consider the readmission interval within less than 90 days between two consecutive admissions. We treat a patient as a *group* consisting of multiple admissions as *individuals*.

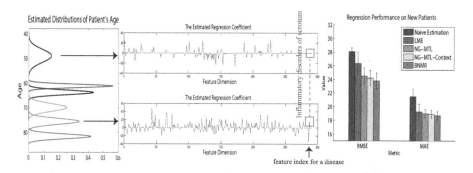

Fig. 4. Regression on HealthData with BNMR. Left: The estimated patient's age distributions. Middle: Two examples of the learned regression coefficient $(\boldsymbol{\beta}_k)$, discovering the correlation of disease code versus patient age (e.g., *Inflammatory disorders of scrotum* affects elder group of 78, not the group of 50). Right: Regression performance comparison on new patients.

The feature for each admission $\boldsymbol{x}_{ji}$ (in patient j) includes *External Factor Code*, and *Diagnosis Code* in 289 dimension.

The readmission interval outcome y_{ji} indicates how many days between this admission to the next admission. We use patient's age as a group-context c_j. We assume that patients within the same 'age region' would have the similar effects on diseases and readmission gap. For example, under the same diseases, patients in the age of 40-50 would be readmitted to a hospital differently from patients in the age of 70-80 because the prevalence of most chronic diseases increases with age [5].

The mean and precision for context distribution are $(\mu, \tau) \sim NG(40, 0.25, 0.2, 1.1)$ and the standard deviation for regression residual error is specified as $\sigma_\epsilon = 24$.

The data is split with 147 patients (70%) for training and the rest of 62 patients are used for testing (as unseen patients). The posterior inference results in $K = 6$ patient clusters. The univariate Normal distribution of age is plotted in Left Fig. 4 where we discover the patient's age distribution. In addition, we visualize the two conditional regression coefficients $(\boldsymbol{\beta}_k)$ on two patient's group of age 50 and 78 respectively. The estimated $\boldsymbol{\beta}_k$(s) also reveal the correlation among disease codes to patient age clusters (Middle Fig. 4). There are several disease codes, such as *Inflammatory disorders of scrotum* (feature dimension 287), affecting on the elder of 78 rather than the younger of 50 (resulting zero value in vector regression coefficient).

Our model uses group-level explanatory variable to identify patient's clusters, then do regression using the regression coefficients produced by the patients in the same cluster. Thus, we prevent from overfitting on each training patient and obtain better prediction on testing patients than the three baseline methods (Right Fig. 4).

6   Conclusion and Discussion

We have presented a *novel* approach for multilevel regression where prediction target is for individuals in new groups. The need of multilevel regression for individuals in unseen groups are commonly encountered in many data domains from econometrics panel data and healthcare longitudinal data domains. Our BNMR provides a join model for clustering groups and do regression for individuals. The unknown number of group cluster and regression coefficients are identified using Bayesian nonparametric setting. By clustering group, the estimated regression coefficients are more generalized and do not overfit to each training group.

References

1. Antoniak, C.E.: Mixtures of Dirichlet processes with applications to Bayesian nonparametric problems. The Annals of Statistics **2**(6), 1152–1174 (1974)
2. Argyriou, A., Evgeniou, T., Pontil, M.: Convex multi-task feature learning. Machine Learning **73**(3), 243–272 (2008)
3. Bishop, C.M.: Pattern recognition and machine learning, vol. 1. Springer, New York (2006)
4. Caruana, R.: Multitask learning. Machine Learning **28**(1), 41–75 (1997)
5. Denton, F.T., Spencer, B.G.: Chronic health conditions: changing prevalence in an aging population and some implications for the delivery of health care services. Canadian Journal on Aging **29**(1), 11 (2010)
6. Escobar, M.D., West, M.: Bayesian density estimation and inference using mixtures. Journal of the american statistical association **90**(430), 577–588 (1995)
7. Ferguson, T.S.: A Bayesian analysis of some nonparametric problems. The Annals of Statistics **1**(2), 209–230 (1973)
8. Gelman, A., Carlin, J.B., Stern, H.S., Rubin, D.B.: Bayesian Data Analysis. Chapman & Hall/CRC (2003)
9. Goldstein, H.: Multilevel mixed linear model analysis using iterative generalized least squares. Biometrika **73**(1), 43–56 (1986)
10. Gupta, S., Phung, D., Venkatesh, S.: Factorial multi-task learning: a bayesian nonparametric approach. In: Proceedings of the 30th International Conference on Machine Learning (ICML-13), pp. 657–665 (2013)
11. Hox, J.: Multilevel analysis: Techniques and applications. Routledge (2010)
12. Kang, Z., Grauman, K., Sha, F.: Learning with whom to share in multi-task feature learning. pp. 521–528 (2011)
13. Kitov, I.O.: Gdp growth rate and population (2008). arXiv preprint arXiv:0811.2125
14. Leyland, A.H., Goldstein, H.: Multilevel modelling of health statistics. Wiley (2001)
15. Maddison, A.: Statistics on world population, gdp and per capita gdp, 1–2008 ad. Historical Statistics (2010)
16. McLean, R.A., Sanders, W.L., Stroup, W.W.: A unified approach to mixed linear models. The American Statistician **45**(1), 54–64 (1991)
17. Munnell, A.H., Cook, L.M.: How does public infrastructure affect regional economic performance? In: Is There a Shortfall in Public Capital Investment? Proceedings of a Conference (1990)
18. Nguyen, V., Phung, D., Venkatesh, S., Nguyen, X., Bui, H.: Bayesian nonparametric multilevel clustering with group-level contexts. In: Proc. of International Conference on Machine Learning (ICML), pp. 288–296, Beijing, China (2014)

19. Phung, D., Nguyen, X., Bui, H., Nguyen, T.V., Venkatesh, S.: Conditionally dependent Dirichlet processes for modelling naturally correlated data sources. Technical report, Pattern Recognition and Data Analytics, Deakin University (2012)
20. Pinheiro, J.C., Bates, D.M.: Mixed-effects models in S and S-PLUS. Springer (2000)
21. Rodriguez, A., Dunson, D.B., Gelfand, A.E.: The nested Dirichlet process. Journal of the American Statistical Association **103**(483), 1131–1154 (2008)
22. Sethuraman, J.: A constructive definition of Dirichlet priors. Statistica Sinica **4**(2), 639–650 (1994)
23. Snijders, T.A.B.: Multilevel analysis. Springer (2011)
24. Teh, Y.W., Jordan, M.I., Beal, M.J., Blei, D.M.: Hierarchical Dirichlet processes. Journal of the American Statistical Association **101**(476), 1566–1581 (2006)

Learning Conditional Latent Structures from Multiple Data Sources

Viet Huynh[1]($\boxtimes$), Dinh Phung[1], Long Nguyen[2],
Svetha Venkatesh[1], and Hung H. Bui[3]

[1] Pattern Recognition and Data Analytics Centre, Deakin University,
Geelong, Australia
hvhuynh@deakin.edu.au
[2] Department of Statistics, University of Michigan, Ann Arbor, USA
[3] Adobe Research, San Francisco Bay Area, San Francisco, USA

Abstract. Data usually present in heterogeneous sources. When dealing with multiple data sources, existing models often treat them independently and thus can not explicitly model the correlation structures among data sources. To address this problem, we propose a full Bayesian nonparametric approach to model correlation structures among multiple and heterogeneous datasets. The proposed framework, first, induces mixture distribution over primary data source using hierarchical Dirichlet processes (HDP). Once conditioned on each atom (group) discovered in previous step, *context* data sources are mutually independent and each is generated from hierarchical Dirichlet processes. In each specific application, which covariates constitute content or context(s) is determined by the nature of data. We also derive the efficient inference and exploit the conditional independence structure to propose (conditional) parallel Gibbs sampling scheme. We demonstrate our model to address the problem of latent activities discovery in pervasive computing using mobile data. We show the advantage of utilizing multiple data sources in terms of exploratory analysis as well as quantitative clustering performance.

1 Introduction

We are entering the age of big data. The challenges are that these data not only present in massive amount but also co-exist in heterogeneous forms including texts, hypertexts, images, graphics, videos, speeches and so forth. For example, in dealing with social network analysis, data present in network connection accompanying with users' profiles, their comments, activities. In medical data understanding, the patients' information usually co-exists with medical information such as diagnosis codes, demographics, laboratory tests. This deluge of data requires advanced algorithms for analyzing and making sense out of data. Machine learning provides a set of methods that can automatically discover low-dimensional structures in data which can be used for reasoning, making decision and predicting. Bayesian methods are increasingly popular in machine learning due to their resilience to over-fitting. Parametric models assume a finite number of parameters and this number needs to be fixed in advance, hence hinders its practicality. Bayesian nonparametrics, on the other hand, relax the assumption

© Springer International Publishing Switzerland 2015
T. Cao et al. (Eds.): PAKDD 2015, Part I, LNAI 9077, pp. 343–354, 2015.
DOI: 10.1007/978-3-319-18038-0_27

of parameter space to be infinite-dimensional, thus the model complexity, e.g., the number of mixture components, can grow with the data[1].

Two fundamental building blocks in Bayesian nonparametric models are the (hierarchical) Dirichlet processes [14] and Beta processes [15]. The former is usually used in clustering models, whereas the later is used in matrix factorization problems. Many extensions of them are developed to accommodate richer types of data [12,16]. However, when dealing with multiple covariates, these models often treat them independently, hence fail to explicitly model the correlation among data sources. The presence of rich and naturally correlated covariates calls for the need to model their correlation with nonparametric models.

In this paper, we aim to develop a full Bayesian nonparametric approach to the problem of multi-level and contextually related data sources and modelling their correlation. We use a stochastic process, being DP, to conditionally "index" other stochastic processes. The model can be viewed as a generalization of the hierarchical Dirichlet process (HDP) [14] and the nested Dirichlet process (nDP) [12]. In fact, it provides an interesting interpretation whereas, under a suitable parameterization, integrating out the topic components results in a nested DP, whereas integrating out the context components results in a hierarchical DP. For simplicity, correlated data channels are referred as two categories: *content* and *context(s)*. In each application, which the covariates constitute *content* or *context(s)* is determined by the nature of data. For instance, in pervasive computing application, we choose the *bluetooth co-location* of user as content while contexts are *time and location*.

Our main contributions in this paper include: (1) a Bayesian nonparametric approach to model multiple naturally correlated data channels in different areas of real-world applications such as pervasive computing, medical data mining, etc.; (2) a derivation of efficient parallel inference with Gibbs sampling for multiple contexts; (3) a novel application on understanding latent activities contextually dependent on time and place from mobile data in pervasive applications.

2   Background

A notable strand in both recent machine learning and statistics literature focuses on Bayesian nonparametric models of which Dirichlet process is the crux. Dirichlet process and its existence was established by Ferguson in a seminal paper in 1973 [4]. A Dirichlet process DP (α, H) is a distribution of a random probability measure G over the measurable space $(\Theta, \mathcal{B})$ where H is a *base* probability measure and $\alpha > 0$ is the *concentration* parameter. It is defined such that, for any finite measurable partition $(A_k : k = 1, \ldots, K)$ of Θ, the resultant random vector $(G(A_1), \ldots, G(A_k))$ is distributed according to a Dirichlet distribution with parameters $(H(A_1), \ldots, H(A_k))$. In 1994, Sethuraman [13] provided an alternative constructive definition which makes the discreteness property of a Dirichlet process explicitly via a stick breaking construction. This is useful while dealing with infinite parametric space and defined as

---

[1] This characteristic is usually called "let the data speak for itself".

$$G = \sum_{k=1}^{\infty} \beta_k \delta_{\phi_k} \text{ where } \phi_k \stackrel{\text{iid}}{\sim} H, k = 1, \ldots, \infty \text{ and } \boldsymbol{\beta} = (\beta_k)_{k=1}^{\infty}, \tag{1}$$

$$\beta_k = v_k \prod_{s<k} (1 - v_s) \text{ with } v_k \stackrel{\text{iid}}{\sim} \text{Beta}(1, \alpha), \quad k = 1, \ldots, \infty.$$

It can be shown that $\sum_{k=1}^{\infty} \beta_k = 1$ with probability one, and as a convention in [11], we hereafter write $\boldsymbol{\beta} \sim \text{GEM}(\alpha)$. Due to its discreteness, Dirichlet processes is used as a prior for mixing proportion in Bayesian mixture models. Dirichlet processes mixture models (DPM) [1,7] which are nonparametric counterpart of well-known Gaussian mixture models (GMM)[2] with the relaxation of the number of components to be infinite were first introduced by Antoniak [1] and elaborated efficiently computational aspect by Neal [7].

However, in practice, data usually appear into collections which can be modelled together. From statistical perspective, it is interesting to extend the DP to accommodate these collections with dependent models. MacEachern [6] introduced framework that induces dependencies over these collections by using a stochastic process to couple them together. Following this framework, Nested Dirichlet process [12] induces dependency by using base measure as another Dirichlet process shared by collections which are modeled by Dirichlet process mixtures. Another widely used model driven by idea of MacEachern is hierarchical Dirichlet process [14] in which dependency is induced by sharing stick breaking representation of a Dirichlet process. All of these models are supposed to model single variable in data. In topic modeling, for instance, HDP is used as a nonparametric counterpart of Latent Dirichlet Allocation (LDA) to model word distributions over latent topics. In this application, the model ignores other co-existing variables such as time, authors.

When dealing with multiple covariates, one can treat the covariates as independent factors. With such independent assumption, he can not leverage the correlated nature of data. There are several works dealing with these situations. Recently, the work by Nguyen et. al. [8] tried to model secondary data channel (called *context*) attached with primary channel (*content*). In this model, secondary data channel is collected in group-level, e.g time or author for each document (consisting of words) or tags in each image. In the case of other data sets, observations are not at group-level but data point-level. For instance, in pervasive computing, each bluetooth co-location of each user includes several observations such as co-location, time stamp, location, etc. There is a motivation for modelling in these kind of applications. Dubey et. al. [2] tried to model topics over time where time are treated as context. The models can only handle one context while modelling but can not leverage the multiple correlated data channels. Another work by Wulsin et. al. [16] proposed the multi-level clustering hierarchical Dirichlet process (MLC-HDP) for clustering human seizures. In this model, authors assumed that data channels are clustered into multi-level which may not suitable for aforementioned data sets. In

---

[2] Indeed, DPM models are more general than (infinite) GMM since we can not only use Gaussian distribution but different kinds of distribution, e.g. Multinomial, Bernoulli, etc., to model each component.

consequence, there is the need for nonparametric models to handle naturally correlated data channels with certain dependent assumptions. In this paper, we propose a model that can model jointly the topic and the context distribution. Our method assumes a conditional dependence between two sets of stochastic process (content-context) which are coupled in a fashion similar to nested DP. The content models the primary observation with HDP and the dependent co-observations are modeled as nested DP with group index provided by the stochastic process from the content side. The set of DPs from the context side is further linked hierarchically in the similar fashion to HDP. Since our inference derivations rely on hierarchical Dirichlet processes, we briefly review hierarchical Dirichlet processes and some useful properties for inference. The justification for these properties can be found in [1,14, Proposition 3].

Let consider the case when we have a corpus with J documents. With the assumption that each document is related to several topics, we can model each document as a mixture of latent topics using Dirichlet process mixture. Though different documents may be generated from different topics, they usually share some of topics each others. Hierarchical Dirichlet process (HDP) models this topic sharing phenomenon. In HDP, the topics among documents are coupled using another Dirichlet process mixture G_0. For each document, a Dirichlet process G_j, $j = 1, \ldots, J$, is used to model its topic distribution. Formally, generative representation is as below:

$$G_0 \mid \gamma, H \sim DP(\gamma, H) \qquad G_j \mid \alpha, G_0 \sim DP(\alpha, G_0) \qquad (2)$$
$$\theta_{ji} \mid G_j \sim G_j \qquad x_{ji} \mid \theta_{ji} \sim F(\theta_{ji}).$$

Similar to DPs, stick breaking representation of HDP is described as follows

$$\beta = \beta_{1:\infty} \sim GEM(\gamma) \qquad G_0 = \sum_{k=1}^{\infty} \beta_k \delta_{\phi_k} \qquad \pi_j = \pi_{j1:j\infty} \sim DP(\alpha, \beta)$$

$$G_j = \sum_{k=1}^{\infty} \pi_{jk} \delta_{\phi_k} \qquad z_{ij} \sim \pi_j \qquad \phi_k \sim H(\lambda) \quad x_{ji} \sim F(\phi_{z_{ji}}). \quad (3)$$

Given the HDP model as described in Equation (3) and $\theta_{j1}, \ldots \theta_{jN_j}$ be i.d.d samples from G_j for all $j = 1, \ldots, J$. All of these samples of each group G_j are grouped into M^j factors $\psi_{j1}, \ldots, \psi_{jM^j}$. These factors from all groups can be grouped into K sharing atoms $\phi_1, \ldots, \phi_K$. Then the posterior distributions stick breaking of G_0 (denoted as $\beta = (\beta_1, \ldots, \beta_K, \beta_{new})$ is

$$(\beta_1, \ldots, \beta_K, \beta_{new}) \sim \text{Dir} (m_1, \ldots, m_K, \gamma), \qquad (4)$$

where $m_k = \sum_{j=1}^{J} \sum_{i=1}^{M^j} \mathbf{1} (\psi_{ji} = \phi_k)$.

Another useful property for posterior of number of cluster K of a Dirichlet process is that if $G \sim DP(\alpha, H)$ and $\theta_1, \ldots, \theta_N$ be N i.i.d samples from G. These θ's values can be grouped into K clusters where $1 \leq K \leq N$. The conditional probability of K given α and N is

$$p(K = k \mid \alpha, N) = \alpha^k \frac{\Gamma(\alpha)}{\Gamma(\alpha + N)} s(N, k), \qquad (5)$$

where $s(N, k)$ is the unsigned Stirling number of the first kind.

3   Framework

3.1   Context Sensitive Dirichlet Processes

Model Description: Suppose we have J documents in our corpus, and each has N_j words of which observed values are x_{ji}'s. From topic modeling perspective, there are a (specified or unspecified) number of topics among documents in corpus where each document may relate to several topics. We have an assumption that each of these topics is correlated with a number of realizations of context(s)[3] (e.g. time). To link the context with topic models we view context as distributions over some index spaces, governed by the topics discovered from the primary data source (content), and model both content and contexts jointly. We impose a conditional structure in which contents provide the topics, upon which contexts are conditionally distributed. Loosely speaking, we use a stochastic process to model content, being DP, and to conditionally "index" other stochastic processes which models contexts.

In details, we model the content side with a HDP, where x_{ji}'s are given in J groups. Each of group is modeled by a random probability distribution G_j, which shares a global random G_0 probability distribution. G_0 is draw from a DP with a base distribution H and concentration parameter γ. The distribution G_0 plays as a base distribution in a DP with concentration parameter α to construct G_j's for groups. The specification for this HDP is similar to Equation (2) in which the θ_{ji}'s are grouped into global atoms $\phi_k (k = 1, 2, \ldots)$.

For each observation x_{ji}, there is an associated context observation s_{ji} which is assumed to depend on the topic atom θ_{ji} of x_{ji}. Furthermore, the context observations of a given topic $S_k = \{ s_{ji} \mid \theta_{ji} = \phi_k \}$ are assumed to be distributed a mixture Q_k. Given the number of topics K, there are the same number of context groups. Now to link these context groups, we again use the hierarchical structure that have the similar manner with HDP [14] where Q_k's share the global random probability distribution Q_0. Formally, generative specification for conditional independent context is as follows

$$Q_0 \sim \mathrm{DP}\,(\eta, R) \qquad Q_k \sim \mathrm{DP}\,(\nu, Q_0) \qquad\qquad (6)$$
$$\varphi_{ji} \sim Q_k, \text{ s.t } \theta_{ji} = \phi_k \qquad s_{ji} \sim Y\,(\cdot \mid \varphi_{ji}).$$

The stick breaking construction for content side is similar to the HDP, however, for the context size we have to take into account of the partition as induced by the content atoms. The stick breaking construction for context is

$$\epsilon \sim \mathrm{GEM}\,(\eta) \qquad \tau_k \sim \mathrm{DP}\,(\nu, \epsilon) \qquad\qquad \psi \sim R$$
$$Q_0 = \sum_{m=1}^{\infty} \epsilon_m \delta_{\psi_m} \qquad Q_k = \sum_{m=1}^{\infty} \tau_{km} \delta_{\psi_m} \qquad l_{ji} \sim \tau_{z_{ji}} \qquad s_{ji} \sim Y\,(\psi_{l_{ji}}) \quad (7)$$

The graphical model for generative representation is depicted in Figure (1a).

Inference: we illustrate the auxiliary conditional approach using stick breaking scheme for inference. We briefly describe inference result of model. We also

---

[3] For simplicity, we will consider one context and generalize to multiple contexts.

assume conjugacy between F and H for content distributions as well as Y and R for context distributions since the conjugacy allows us to integrate out the atoms ϕ_k and τ_m. The sampling state space now consists of $\{z, \beta, l, \epsilon\}$. Furthermore, we endow Gamma distributions as priors for hyperparameters $\{\gamma, \alpha, \eta, \nu\}$ and sample through each Gibbs iteration. During sampling iterations, we maintain the following counting variables: n_{jk} - the number of content observations in document j belong to content topic k, the marginal counts are denoted as $n_{j\cdot} = \sum_k n_{jk}$, and $n_{\cdot k} = \sum_j n_{jk}$; w_{km} - the number of context observations given the topic k belong to context m. The marginal counts are denoted similarly to n_{jk}. Sampling equations for *content side* are described below.

Sampling z: the sampling of z_{ji} have to take into account of influence from the context apart from cluster assignment probability and likelihood.

$$p(z_{ji} = k \mid \boldsymbol{z}_{-ji}, \boldsymbol{l}, \boldsymbol{x}, \mathbf{s}) \propto p(z_{ji} = k \mid \boldsymbol{z}_{-ji}).$$

$$p(x_{ji} = k \mid z_{ji} = k, \boldsymbol{z}_{-ji}, \boldsymbol{x}_{-ji}) p(l_{ji} \mid z_{ji} = k, \boldsymbol{l}_{-ji}). \qquad (8)$$

The first term of above equation in the RHS is the predictive likelihood of prior at the content side similar to HDP in [14] while the second term indicates the predictive likelihood of the observation for content topic k (except x_{ji}), denoted as $f_k^{-x_{ji}}(x_{ji})$. The last term is the context predictive likelihood given the content topic k. As a result, conditional sampling for z_{ji} is

$$p(z_{ji} = k \mid \boldsymbol{z}_{-ji}, \boldsymbol{l}, \boldsymbol{x}, \mathbf{s}) = \begin{cases} \left(n_{\cdot k}^{-ji} + \alpha\beta_k\right) \frac{w_{km} + \nu\epsilon_m}{w_{k\cdot} + \nu} f_k^{-x_{ji}}(x_{ji}) & \text{if } k \text{ previously used} \\ \alpha\beta_{new}\epsilon_m f_{new}^{-x_{ji}}(x_{ji}) & \text{if } k = k_{new} \end{cases}$$

Sampling β: we use the posterior stick breaking of HDP in Equation (4).

In order to sample m, we use the result from Equation (5), i.e. $m_{jk} \propto (\alpha\beta_k)^m s(n_{jk}, m)$ for $m = 1 \ldots n_{jk}$ where $s(n_{jk}, m)$ is the unsigned Stirling number of the first kind and compute $m_k = \sum_{j1}^{J} m_{jk}$.

Next, we present sampling derivations for *context variables*.

Sampling l: given the cluster assignment of content observations $(\boldsymbol{z})$, context observations are grouped into K groups of context. Let $\boldsymbol{s}_k$ be the set of context observations indexed by the same content cluster k. i.e. $\boldsymbol{s}_k \triangleq \{s_{ji} : z_{ji} = k, \forall j, i\}$, while $\boldsymbol{s}_k^{-ji}$ is the same set as $\boldsymbol{s}_k$ but excluding s_{ji}. The posterior probability of l_{ji} is computed as follows

$$p(l_{ji} = m \mid \boldsymbol{l}_{-ji}, \boldsymbol{z}, \boldsymbol{s}, \boldsymbol{\nu}, \boldsymbol{\epsilon}) \propto p(l_{ji} = m \mid \boldsymbol{l}_{-ji}, z_{ji} = k, \boldsymbol{\epsilon}).$$

$$p(s_{ji} \mid l_{ji} = m, \boldsymbol{l}_{-ji}, z_{ji} = k, \boldsymbol{s}_{-ji}). \qquad (9)$$

The first term is the conditional Chinese restaurant process given content cluster k while the second term, denoted as $y_{k,m}^{-s_{ji}}(s_{ji})$, is recognized to be a form of predictive likelihood in a standard Bayesian setting of which likelihood function is Y, conjugate prior S and a set of observation $s_k^{-ji}(m) \triangleq \left\{ s_{j'i'} : l_{j'i'} = m, z_{j'i'} = k, j' \neq j, i' \neq i \right\}$. The sampling equation for l_{ji} is

$$p(l_{ji} = m \mid \boldsymbol{l}_{-ji}, \boldsymbol{z}, \boldsymbol{s}, \boldsymbol{\nu}, \boldsymbol{\epsilon}) = \begin{cases} (w_{km} + \nu\epsilon_m) y_{k,m}^{-s_{ji}}(s_{ji}) & \text{if } m \text{ previously used} \\ \epsilon_{new} y_{k,m_{new}}^{-s_{ji}}(s_{ji}) & \text{if } m = m_{new} \end{cases}$$

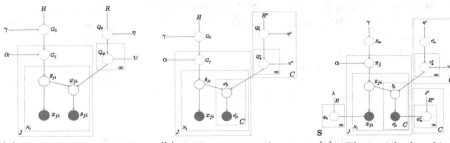

(a) Generative view for proposed model with single context

(b) Generative view for proposed model with C contexts

(c) The stick breaking view for the proposed model C contexts

Fig. 1. Graphical representation for the proposed model. (a) & (b) Generative view for single and multiple contexts which conditional independent given content topic. (c) Stick breaking view with C contexts, for single context, one can set $C = 1$.

Sampling ϵ: different from HDP, sampling ϵ requires more works as it is dependent on both z and l. Let isolate context variables l_{ji}^k's generated by the same topic $z_{ji} = k$ into one group $l^k \triangleq \{l_{ji} : z_{ji} = k, \forall j, i\}$, context observations are also isolated in the similar way $s^k \triangleq \{s_{ji} : z_{ji} = k, \forall j, i\}$. Now the context side is modeled with the structure similar to HDP in which the observations related Q_k are s^k. We can sample ϵ as follows $(\epsilon_1, \ldots, \epsilon_M, \epsilon_{new}) \sim$ $\text{Dir}(h_{.1}, \ldots, h_{.M}, \eta)$ where $h_{.m}$, $m = 1 \ldots M$ are auxiliary variables which represent number of active context factors associated with atom m. Similar to sampling m, the value of each $h_{.m}$ will be computed using samples $h_{km} \propto (\nu \epsilon_m)^h s(w_{km}, h)$ for $h = 1 \ldots w_{km}$ and summed up as $h_{.m} = \sum_{k=1}^K h_{km}$.

Moreover, there are four hyper-parameters in our model: α, γ, ν, η. Sampling α and γ is identical to HDP and therefore we refer to [14] for details. Sampling other hyperparameters is also doable, one can refer to [10] for details.

3.2 Context Sensitive Dirichlet Processes with Multiple Contexts

Model Description. When multiple contexts exist for a topic, the model can easily be extended to accommodate this. The generative and stick breaking specifications for content side remain the same as in Equation (2) and (3). The specification for multiple contexts will be duplicated from one context in Equation (6). Figure (1) depicts the graphical model for context sensitive Dirichlet process with multiple contexts. The generative model is

$$Q_0^c \sim \text{DP}(\eta^c, R^c) \qquad Q_k^c \sim \text{DP}(\nu^c, Q_0^c) \qquad \varphi_{ji}^c \sim Q_k^c, \text{ where } \theta_{ji} = \phi_k$$

$$x_{ji} \sim F(\cdot \mid \theta_{ji}) \qquad s_{ji}^c \sim Y^c(\cdot \mid \varphi_{ji}^c) \text{ for all } c = 1, \ldots, C.$$

The stick breaking construction for the context side is duplicated the specifications of context side in Equation (7) for C contexts which is provided below for all $c = 1, \ldots, C$:

Algorithm 1. Multiple Context CSDP Gibbs Sampler

1: **procedure** MCSDPGIBBSSAMPLER($\mathcal{D}$)   ▷ $\mathcal{D}$: input including x_{ij} and s_{ij}^c
2:   **repeat**   ▷ J: the number of groups
3:     **for** $j \leftarrow 1, J; i \leftarrow 1, N_j$ **do**   ▷ N_j: the number of data in j-th group
4:       Sample z_{ji} using Equation (10)   ▷ Sampling content side
5:       **for** $c \leftarrow 1, C$ **do**   ▷ Sampling context side (can be parallised)
6:         Sample l_{ji}^c using Equation (9)
7:       **end for**
8:     **end for**
9:     Sample β and ϵ using Equation (4) and hyperparameters
10:   **until** Convergence
11:   **return** $z, l^{1:C}, \beta, \epsilon$ ▷ return learned parameters of model
12: **end procedure**

$$\epsilon^c \sim \text{GEM}\,(\eta) \qquad \tau_k^c \sim \text{DP}\,(\nu, \epsilon) \qquad \psi^c \sim R^c$$

$$Q_0^c = \sum_{m=1}^{\infty} \epsilon_m^c \delta_{\psi_m^c} \qquad Q_k = \sum_{m=1}^{\infty} \tau_{km}^c \delta_{\psi_m^c} \qquad l_{ji}^c \sim \tau_{z_{ji}}^c \qquad s_{ji}^c \sim Y^c \left(\psi_{l_{ji}^c}^c \right).$$

Inference: using the same routing and assumptions on conjugacy of H and F, R^c and Y^c, we derive the sampling equations for variables as follows

Sampling z: in multiple context setting, the sampling equation of z_{ji} involves the influence from multiple context rather than one:

$$p(z_{ji} = k \mid \boldsymbol{z}_{-ji}, \boldsymbol{l}, \boldsymbol{x}, \boldsymbol{s}) \propto p(z_{ji} = k \mid \boldsymbol{z}_{-ji}). \tag{10}$$

$$p(x_{ji} = k \mid z_{ji} = k, \boldsymbol{z}_{-ji}, \boldsymbol{x}_{-ji}) \prod_{c=1}^{C} p(l_{ji}^c = m^c \mid z_{ji} = k, \boldsymbol{l}_{-ji}^c).$$

It is straightforward to apply the result for one context case. The final sampling equation for z_{ji} is

$$p(z_{ji} = k \mid \boldsymbol{z}_{-ji}, \boldsymbol{l}, \boldsymbol{x}, \boldsymbol{s}) = \begin{cases} \left(n_{\cdot k}^{-ji} + \alpha \beta_k \right) f_k^{-x_{ji}}(x_{ji}) \prod_{c=1}^{C} \frac{w_{kmc}^c + \nu^c \epsilon_{mc}^c}{w_{k\cdot}^c + \nu^c} & \text{if } k \text{ used} \\ \alpha \beta_{new} f_{k_{new}}^{-x_{ji}}(x_{ji}) \prod_{c=1}^{C} \epsilon_{mc}^c & \text{if } k = k_{new}. \end{cases}$$

Sampling derivation of β is unchanged compared with one context.

Sampling equations of $l^{1 \dots C}, \epsilon^{1 \dots C}$ are similar to one context case where each set of context variables $\{l^c, \epsilon^c\}$ is dependent given sampled values of z. We can perform sampling for each context in parallel thus the computation complexity in this case should remain the same as in the single context case given enough number of core processors to execute in parallel. We summarize sampling procedure for the model in Algorithm 1.

4   Experiments

In this section we demonstrate the application of our model to discover latent activities from social signals which is a challenging task in pervasive computing. We implemented model using C# and ran on Intel i7-3.4GHz machine with installed Windows 7. We then used Reality Mining, a well-known data set collected at MIT Media Lab [3] to discover latent group activities. The model not only improves grouping performance but also reveals when and where these activities happened. In the following sections, we briefly describe data set, data preparation, parameter settings for the model and exploratory results as well as clustering performance using our proposed model.

4.1   Reality Mining Data Set

Reality Mining [3] is a well-known mobile data set collected by MIT Media Lab on 100 users over 9 months (approximately 450.000 hours). The collected information includes proximity using Bluetooth devices, cell tower IDs, call logs, application usage, and phone status. To illustrate the capability of proposed model, we extract proximity data recorded by Bluetooth devices and users' location via cell tower IDs. In order to compare with the results from [9], we preprocessed to filter users whose affiliations are missing or who do not share affiliation with others and then sampled proximity data for every 10 minutes. In the end we had 69 users. For each user, at every 10 minutes, we obtained a data point of 69-dimension which represents co-location information with other users. Each data point is an indicator binary vector of which i-th element set to 1 if the i-th user is co-located and 0 otherwise (self-presence set to 1). In addition, we also obtain the time stamp and cell ID data vectors. As a consequence, we have 69-user data groups. Each data point in group includes three observations: co-location vector, time stamp, cell tower ID.

4.2   Experimental Settings and Results

In proposed model, one data source will be chosen as content, the rest will be considered as contexts. We use two different settings in our experiment.

In the first setting, co-location data source is modelled as content which is (69-dimension) *Multinomial* distribution (corresponding to F distribution in model), time and cell tower IDs are modelled as *Gaussian* and *Multinomial* distributions respectively (corresponding to Y^1 and Y^2 distribution in model). We use the conjugate prior H as *Dirichlet* distribution, while R^1 and R^2 are *GaussianGamma* and *Dirichlet* distributions, respectively. We run the data set with 4 different settings for comparison: *HDP* - standard use of HDP on co-location observations (similar to [9]); *CSDP-50% time* - co-location and 50% time stamp data (supposing 50% missing) used for CSDP; *CSDP-time* - similar to *CSDP-50% time*, except that whole time stamp data are used; *CSDP-celltower* - resembling to *CSDP-time* but additional cell tower ID observations are used.

When modelling with *HDP* as in [9], the model merely discovered hidden activities of users. It fails to answer more refined questions such as *when and where these activities happened?* Our proposed model can naturally be used to model the additional data sources to address these questions. In Figure (2a), the topic 1 (*sloan* students) usually happened at specific time on Monday, Tuesday and Thursday while topic 5 (*master frosh* students) mainly gathered on Monday and Friday (less often on the other days). Similarly, when we modelled cell tower IDs data, the results

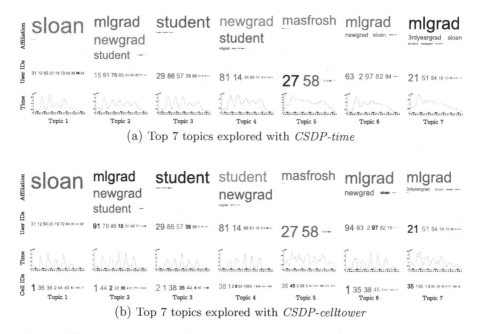

(a) Top 7 topics explored with *CSDP-time*

(b) Top 7 topics explored with *CSDP-celltower*

Fig. 2. Corresponding top 7 topics discovered by proposed model

revealed a deeper understanding on latent activities. In Figure (2b), we can observe the places (cell phone tower IDs)[4] where the activities took place. For topic 1 - *sloan* student group activities, apart from Sloan School building (*cell no.1* or *40*), they sometimes gathered at the restaurants (*cell no. 44*).

When using more contextual information, it does not only provide more exploratory information but also help the classification to be more discriminated. When using only time as context in Figure (2a), the user *no. 94* is (confusingly) recognized in both topic 2 and 6. But when location data is incorporated into our proposed model, the user *no. 94* is now dominantly classified into topic 6. To quantitatively evaluate proposed model when using more context data, we use the same setting with the work in [9]. First, we ran the data model to discover the latent activities among users. We then used the Affinity Propagation (AP) algorithm [5] to perform clustering among users with similar activities. We evaluated clustering performance using popular metrics: F-measure, cluster purity, rand index (RI) and normalized mutual information (NMI). As it can be clearly seen in Table 1, with more contexts we observed, CSDP achieves better clustering results. *Purity* and *NMI* are significantly improved when more contextual data are observed while other metrics slightly improved when modelling with contextual data.

---

[4] Since Reality Mining does not provide exact information about these cell towers however we can infer information about some of them by using users' descriptions. For example, cell no.1 and 40 are MIT Lab and Sloan School of Management which are two adjacent buildings. While cell no. 35 is located near Student Center and cell no. 44 is around some restaurants outside MIT campus.

Table 1. Clustering performance improved when more contextual data used in the proposed model

	Purity	NMI	RI	F-measure
HDP	0.7101	0.6467	0.9109	0.7429
CSDP-*50% time*	0.7391	0.6749	**0.9186**	**0.7651**
CSDP-*100% time*	0.7536	0.6798	0.9169	0.7503
CSDP-*celltower*	**0.7826**	**0.6953**	**0.9186**	0.7567

In the second setting, we model time as content and the rest (co-locations, cell towers) as contexts. The conjugate pairs are remained the same in previous setting. In Figure (3), we demonstrate top 4 time topics including Friday, Thursday (upper row), Tuesday, and Monday (lower row) which are Gaussian forms. The groups of users who gathered in that time stamp are depicted under each Gaussian. It is easy to notice that the group with user *27, 58* usually gathered on Friday and Monday whereas other groups met on all four time slots.

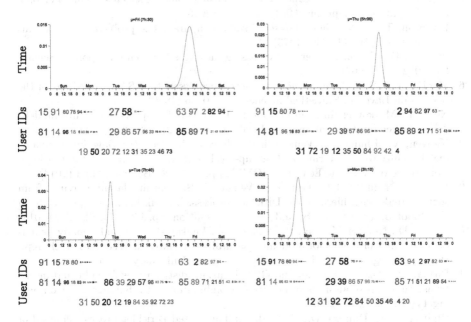

Fig. 3. Top 4 time topics and their corresponding conditional user-IDs groups discovered by proposed model

5 Conclusions

We propose a full Bayesian nonparametric approach to model explicit correlation structures in heterogeneous data sources. Our key contribution is the development of a context sensitive Dirichlet processes, its Gibbs inference and its parallelability. We have further demonstrated the proposed model to discover latent activities from mobile data to answer who (co-location), when (time) and where (cell-tower ID) – a central problem in context-aware computing applications. With its expressiveness, our model not only discovers latent activities (topics) of users but also reveals time and place information. Qualitatively, it was shown that better clustering performance than without them. Finally, although the building block of our proposed model is the Dirichlet process, based on HDP, it is straightforward to apply other stochastic processes such as nested Dirichlet processes or hierarchical Beta processes to provide alternative representation expressiveness for data modelling tasks.

References

1. Antoniak, C.: Mixtures of Dirichlet processes with applications to Bayesian nonparametric problems. The Annals of Statistics **2**(6), 1152–1174 (1974)
2. Dubey, A., Hefny, A., Williamson, S., Xing, E.P.: A non-parametric mixture model for topic modeling over time (2012). arXiv preprint arXiv:1208.4411
3. Eagle, N., Pentland, A.: Reality mining: Sensing complex social systems. Personal and Ubiquitous Computing **10**(4), 255–268 (2006)
4. Ferguson, T.: A Bayesian analysis of some nonparametric problems. The Annals of Statistics **1**(2), 209–230 (1973)
5. Frey, B., Dueck, D.: Clustering by passing messages between data points. Science **315**, 972–976 (2007)
6. MacEachern, S.: Dependent nonparametric processes. In: ASA Proceedings of the Section on Bayesian Statistical Science. pp. 50–55 (1999)
7. Neal, R.: Markov chain sampling methods for Dirichlet process mixture models. Journal of computational and graphical statistics **9**(2), 249–265 (2000)
8. Nguyen, V., Phung, D., Venkatesh, S. Nguyen, X., Bui, H.: Bayesian nonparametric multilevel clustering with group-level contexts. In: Proc. of International Conference on Machine Learning (ICML), pp. 288–296. Beijing, China (2014)
9. Phung, D., Nguyen, T.C., Gupta, S., Venkatesh, S.: Learning latent activities from social signals with hierarchical Dirichlet process. In: Sukthankar, G., et al. (ed.) Handbook on Plan, Activity, and Intent Recognition, pp. 149–174. Elsevier (2014)
10. Phung, D., Nguyen, X., Bui, H., Nguyen, T., Venkatesh, S.: Conditionally dependent Dirichlet processes for modelling naturally correlated data sources. Tech. rep., Pattern Recognition and Data Analytics, Deakin University(2012)
11. Pitman, J.: Poisson-Dirichlet and GEM invariant distributions for split-and-merge transformations of an interval partition. Combinatorics, Probability and Computing **11**(05), 501–514 (2002)
12. Rodriguez, A., Dunson, D., Gelfand, A.: The nested Dirichlet process. Journal of the American Statistical Association **103**(483), 1131–1154 (2008)
13. Sethuraman, J.: A constructive definition of Dirichlet priors. Statistica Sinica **4**(2), 639–650 (1994)
14. Teh, Y., Jordan, M., Beal, M., Blei, D.: Hierarchical Dirichlet processes. Journal of the American Statistical Association **101**(476), 1566–1581 (2006)
15. Thibaux, R., Jordan, M.: Hierarchical Beta processes and the Indian buffet process. In: Proc. of Int. Conf. on Artificial Intelligence and Statistics (AISTAT), vol. 11, pp. 564–571 (2007)
16. Wulsin, D., Jensen, S., Litt, B.: A hierarchical dirichlet process model with multiple levels of clustering for human eeg seizure modeling. In: Proc. of International Conference on Machine Learning (ICML) (2012)

Collaborative Multi-view Learning with Active Discriminative Prior for Recommendation

Qing Zhang and Houfeng Wang[(✉)]

Key Laboratory of Computational Linguistics (Peking University)
Ministry of Education, Beijing, China
{zqicl,wanghf}@pku.edu.cn

Abstract. Learning from multi-view data is important in many applications. However, traditional multi-view learning algorithms require the availability of the representation from multi-view data in advance, it is hard to apply these methods to recommendation task directly. In fact, the idea of multi-view learning is particularly suitable for alleviating the sparsity challenge faced in various recommender systems by adding additional view to augment traditional view of sparse rating matrix. In this paper, we propose a unified Collaborative Multi-view Learning (CML) framework for recommender systems, which can exploit task adaptive multi-view representation of data automatically. The main idea is to formulate a joint optimization framework, combining the merits of matrix factorization model and transfer learning technique in a multi-view framework. Experiments on real-life public datasets show that our model outperforms the compared state-of-the-art baselines.

Keywords: Collaborative filtering · Neural network · Representation learning

1 Introduction

With the explosive growth and variety of information available on the Web, the interest in recommender systems has dramatically increased from both research and industrial communities. In this filed, Collaborative Filtering (CF) approaches, especially matrix factorization models, have achieved significant success [11], based on users' previous interest encoded by the rating matrix reflecting the similarities of similar users or items. However, CF performs poorly when little collaborative information is available. This is referred to as the data sparsity problem [10,17], which is a common problem in many newly launched recommender systems.

There are two main directions for solving this challenging in various recommender systems. The first one lies in traditional single task setting, i.e., how to effectively use existing user-item[1] pairs combined with auxiliary information,

---

[1] Here item is a general term, e.g., book in Amazon and music in LastFM.

© Springer International Publishing Switzerland 2015
T. Cao et al. (Eds.): PAKDD 2015, Part I, LNAI 9077, pp. 355–368, 2015.
DOI: 10.1007/978-3-319-18038-0_28

such as content [1,14] or complex network structure [11] given in advance. The second one lies in recently developed transfer learning setting [9], i.e., how to exploit related task-beneficial auxiliary information, learned from source recommender containing dense interactive data, to strengthen target recommender performance. Specifically, 1) the first direction is extensively studied using side information. Although side information is beneficial for improving recommender performance, it is usually restricted by the availability of predictive data representations or restricted by relying on feature engineering. Thus, this motivates us to seek a general method that can automatically exploit side information for existing CF approaches. 2) The second direction becomes a recently hot research topic [17], but it needs more resources, i.e., cross-system information. For example, we need both Twitter and Facebook resources to improve the target system of Facebook, which is not always easy to acquire due to individual privacy or commercial issues. In addition, it usually makes strong assumptions for designing learning algorithms. For example, cross-system entity correspondence [10] is usually a crucial prerequisite. Thus, this motivates us to seek a novel cost-saving way to achieve knowledge transfer, i.e., attempting to employ the idea of transfer learning to the first direction.

In addition to the issues above for solving sparsity problem, we also consider how to incorporate discriminative power into existing CF algorithms, inspired by Supervised Matrix (or tensor) Factorization (SMF) approach [16]. The idea of SMF is to faithfully reconstruct the original matrix using discriminative priors (corresponding class labels), i.e., with additional discriminative constraints, such as max-margin criterion for classification [16] on the basis variables or factorized latent features. However, in our task, the supervised label information for unrated item, i.e., like or dislike for a user is uncertain. Thus, how to actively determine label certainty and use this predicted label information for improving discriminative power is a key challenge applying supervised setting to CF task. Due to this obstacle, most of CF algorithms in both mentioned directions act as unsupervised manners that failed to exploit the inherent discriminative priors of the data objects. In fact, this knowledge is useful in many real world applications, such as item with tags etc. Ideally, this obstacle and the above challenging issues for solving sparsity problem should be well considered jointly in a unified framework.

To solve all the above challenges, in this paper, we propose a unified multi-view framework for collaborative filtering. This framework can be seen as a compromised approach between the two directions of single task view and transfer learning setting. It combines the merits of these two directions with discriminative power to solve the rating sparsity problem. More specifically, our approach is from multi-view perspective [4,5], which can exploit side information automatically for CF task with the ability of knowledge transfer to construct discriminative prior from different views. In contrast with the two directions above, our method can work on a very extremely sparse rating matrix 1) without needing the multi-view data representation available in advance, 2) and only maintaining minimal external resources compared with transfer learning approach to achieve knowledge transfer. The following sections will discuss those in detail.

2 Preliminaries

In this paper, we address missing rating prediction problem from multi-view perspective. We employ the idea of transfer learning in a multi-view setting with side information[2], to complete a sparse rating matrix. For simplicity, we focus on the basic recommending case that the value of user rating has only binary states, which also can be extended easily for ordinal case in various applications.

Transfer Learning for CF. The idea of transfer learning [9,17] for addressing the data-sparsity problem in the target recommender system is to use the data from some related recommender systems. In this category, these approaches assume that the knowledge of a source CF model built with rich collaborative data can be extracted as a prior to assist the training of a more precise CF model for the target recommender systems. For example, many commercial Web sites often attract similar users (e.g., Twitter, Facebook, etc.), or provide similar product items (e.g., Amazon, eBay, etc.), thus we can bridge two related systems by cross-system entity correspondence [8,10] or using the group level similarity [7] to improve target system performance. However, all the algorithms of transfer learning rely on cross-system resources, which are usually not easy to acquire because of commercial competition or individual privacy. Thus, we extend this good idea to a more cost-saving resource setting, for CF task based on side information in a single system.

Multi-view Learning for CF. To achieve the goal of extending the idea of transfer learning to a more cost-saving resource setting mentioned above, we propose to use multi-view learning approach to incorporate the ability of knowledge transfer for CF. The basic idea of multi-view learning [4,5] is to leverage the redundancy and consistency among distinct views to strengthen the overall performance. We use this idea [4] originally for clustering problem to deal with data sparsity problem for recommendation. In traditional multi-view learning for classification problem, each view of objective function is assumed to be capable of correctly classifying labeled examples separately. Then, they are smoothed with respect to similarity structures in all views. Similarly, for the CF task in this paper, we also assume that our individual views of user-item rating matrix and side information are complementary with similar latent structure. The difference is that both views are bridged through a bi-directional prior with discriminative power, which extends the idea of transfer learning to multi-view framework.

3 The Overall Collaborative Multi-view Learning Framework

The key idea is that we exploit learning multi-view representations with multiple task oriented objectives (loss functions) in a unified optimization framework, to improve recommendation performance. Our framework is a general solution,

---

[2] We use item content as side information which also can be substituted by user content similarly.

which introduces different views of modeling for recommendation via item prior as a bridge.

The general framework[3] shows the high level generative process of the proposed basic model and the extension. The differences lie in that are modeled by different prior modeling approaches for each latent item representation, which will be discussed in the proposed basic model and the extension respectively.

- For each user i,
  - draw a user latent vector $\mathbf{u}_i \sim N(\mathbf{0}, \lambda_u^{-1} I_K)$, multivariate Gauss distribution with zero mean.
- For each item j,
  - draw a multi-view representation variable $\boldsymbol{\theta}_j$ via representation learning (that is different in the proposed basic model and the extension).
  - draw an item latent vector $\mathbf{v}_j \sim N(\boldsymbol{\theta}_j, \lambda_v^{-1} I_K)$, multivariate Gauss distribution.
- For each user-item pair (i, j),
  - draw the response $r_{ij} \sim N(\mathbf{u}_i^T \mathbf{v}_j, c_{ij}^{-1})$, univariate Gauss distribution, where c_{ij} is a confidence parameter [14] for rating r_{ij}, $a > b$. If r_{ij} is large, we trust r_{ij} more.

$$c_{ij} = \begin{cases} a, r_{ij} = 1 \\ b, r_{ij} = 0. \end{cases} \tag{1}$$

It is noted that the different ways to model prior as view specific representation lead to different recommendation models. The appropriate choice of data representation (or features) plays a key role in acquiring optimal performance of the state of arts machine learning methods. In particular, we use neural network approach to learn view specific prior automatically from data, instead of pre-defined fashion by hands in the framework.

4   Details of the Framework with Active Discriminative Prior

Our work shares similar intuition of a recent trend [14] which brings two well-established approaches together, i.e., probabilistic topic modeling and latent factor models. However, previous approach is not from multi-view perspective. These methods [3,11,14,15] cannot incorporate multi-view loss into a joint optimization framework and are not capable of transferring knowledge actively.

---

[3] For notations used in this paper, we use capital letters to represent matrices, use boldface lower-case letters to represent vectors, and use lowercase letters to represent scalars.

4.1 The Proposed Basic Model

Model Formulation: To model view specific prior, we incorporate Stacked Denoising Auto-encoder (SDA) [13] into our optimization framework as initial estimate for our neutral network. SDA is one of building blocks of (deep) representation learning as an extension of Auto-Encoder (AE). Given the input x representing document as a binary bag-of-words vector, Denoising Auto-encoder (DA) randomly masks 1 with 0 with a pre-defined probability. Since the missing components have to be recovered from partial input, DA has the chances of capturing general concepts and ignoring noise like function words. Then, as the standard AE, it performs encoding process $h(x)$ and decoding process $g(h(x))$, minimizing the reconstruction error $L(x, g(h(x)))$ to retain maximum information.

Specifically, in this basic model, we use one hidden layer DA to learn view specific representation of each item as prior in a joint optimization framework. The reconstruction criterion is given in *view 2* part of Eq. 2.

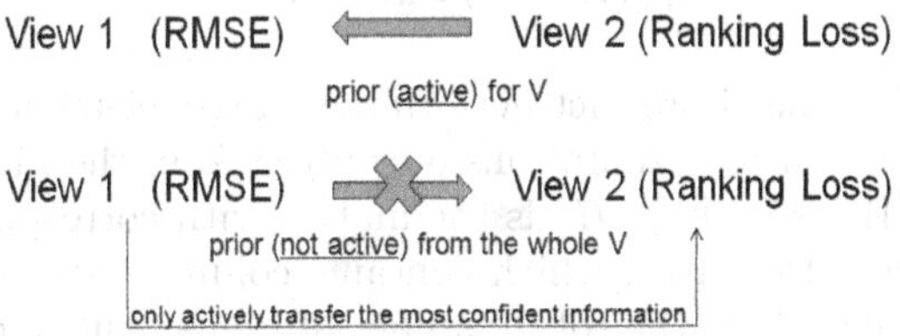

View 1 (RMSE) ⟸ View 2 (Ranking Loss)

prior (<u>active</u>) for V

View 1 (RMSE) ⟹✖⟹ View 2 (Ranking Loss)

prior (<u>not active</u>) from the whole V

only actively transfer the most confident information

Fig. 1. Active knowledge transfer with bi-directional prior mechanism in a multi-view framework

Model Learning: For learning the parameters, we develop a coordinate descent optimization algorithm to maximize a posteriori (MAP) estimate of U, V with column vectors $\mathbf{u}_i$ and $\mathbf{v}_j$ respectively. It is equivalent to minimizing the complete negative log-likelihood with respect to $W, U, V, \mathbf{b}, \mathbf{c}$:

$$\mathcal{L}_1 = \sum_{j=1} \| \underbrace{\sigma_2(W^T \sigma_1(W\mathbf{x}^{(j)} + \mathbf{b}) + \mathbf{c}) - \mathbf{x}^{(j)}}_{view2} \|_2^2 +$$

$$\frac{\lambda_v}{2} \sum_j \underbrace{(\mathbf{v}_j - \sigma_1(W\mathbf{x}^{(j)} + \mathbf{b}))^T (\mathbf{v}_j - \sigma_1(W\mathbf{x}^{(j)} + \mathbf{b}))}_{bridge} + \tag{2}$$

$$\frac{\lambda_u}{2} \sum_i \mathbf{u}_i^T \mathbf{u}_i + \sum_i \sum_j \frac{\lambda_{c_{ij}}}{2} \underbrace{(r_{ij} - \mathbf{u}_i^T \mathbf{v}_j)^2}_{view1} + \frac{\lambda_w}{2} \|W\|_F^2,$$

where $\sigma_1(x) = 1/1 + exp(-x)$ as nonlinear mapping and $\sigma_2(x) = x$ as linear reconstruction function in an element-wise way. W is the weights matrix of Neural Network (NN). Note that the prior $\boldsymbol{\theta}_j$ in general framework is modeled as

$$\boldsymbol{\theta}_j = \sigma_1(W\mathbf{x}^{(j)} + \mathbf{b}). \tag{3}$$

Specifically, we iteratively optimize the collaborative filtering variables U, V and the parameters of representation learning $W, \mathbf{b}, \mathbf{c}$. By setting the derivative of $\mathcal{L}_1$ with respect to $\mathbf{u}_i, \mathbf{v}_j$ to zero, we obtain the update rule

$$\mathbf{u}_i \leftarrow (VC_iV^T + \lambda_u I_K)^{-1} VC_i R_i, \tag{4}$$

$$\mathbf{v}_j \leftarrow (UC_jU^T + \lambda_v I_K)^{-1}(UC_j R_j + \lambda_v \sigma_1(W\mathbf{x}^{(j)} + \mathbf{b})), \tag{5}$$

where C_i is a diagonal matrix with elements c_{ij} for each j, R_i is a column vector with elements r_{ij} for each j. For item j, C_j and R_j are similarly defined.

Then, given U and V, we update the parameters $W, \mathbf{b}, \mathbf{c}$ via computing the corresponding gradient of $\mathcal{L}_1$, which is similar to back-propagation in NN but with additional regularization term and sharing weights constraint. Thus, to update W, b, c, we can only modify the existing optimization procedure of autoencoder (AE) by adding our regularization term. The additional adding gradient for W during each gradient descent iteration in our case is

$$\frac{\lambda_v}{M}(H - V) \circ dH \cdot X^T, \tag{6}$$

where $\circ$ denotes Hadamard product performing matrix element wise product, H is a matrix with column representations outputted from the hidden layer in NN for each item $\{\sigma_1(W\mathbf{x}^{(j)} + \mathbf{b})\}$. dH is the matrix with corresponding derivative value of H. X is a data matrix which contains column vectors as bag of word features for each item. M is the total number of items. The adding gradient for $\mathbf{b}$ to existing AE optimization is

$$\frac{\lambda_v}{M}(H - V) \circ dH \cdot \mathbf{1}, \tag{7}$$

where $\mathbf{1}$ is a column vector in which all elements are equal to one.

4.2   Extension with Active Discriminative Prior

Model Formulation: The proposed approach has three merits. First, we model item prior using a discriminative learning approach rather than a generative fashion. Thus, it allows us to flexibly incorporate any multiple task oriented objectives instead of a pre-defined generative process with lower bound of objective like CTR [14], for joint optimization. Second, this discriminative prior modeling naturally offers explicit weighs for mapping new samples out of training data, not needing a re-sampling procedure as in generative models, e.g., LDA [2]. Third, our method is a general framework which can be easily extended to exploiting other side information, such as social network data for modeling user prior.

a) **Discriminative Prior Modeling** We extend the basic model to multi-view multi-objective setting. Traditionally, collaborative filtering via matrix factorization is to solve an unsupervised matrix reconstruction problem under root mean squared error (RMSE) criterion. However, the ultimate goal of any recommender is to generate recommendation lists for users, which is a ranking problem in nature. Thus, we incorporate ranking based loss explicitly, into our joint optimization framework, with representation learning for prior modeling. Any measures for ranking can be incorporated into our framework. For simplicity, we define the pairwise ranking criterion as loss function appeared in *view 2* part of Eq.8, stacking on the output layer of SDA for joint optimization. Different from supervised matrix factorization case [16], in our CF task, accurately acquiring label information of unrated item is non-trivial.

   b) **Active Knowledge Transfer for Constructing Negative Samples**
To achieve discriminative prior modeling, the main obstacle is that the supervised information is not available. Here we cannot acquire the true label of unrated item for each user, i.e., like or dislike, because $r_{ij} = 0$ can be interpreted into two ways. One is that user i is not interested in item j; the other is that user i does not know about item j. Instead, inspired by the work [17], we actively compute the predicted (label) rating of each unrated item in each optimization iteration, and then use this predicted label as the supervised information for training discriminative prior. The selection rule for negative sample set is to choice the top-K unrated items for current user in the predicted rating list sorted in ascending order, according to the score of inner product of latent user $\mathbf{u}_i$ and item $\mathbf{v}_j$.

   c) **Bi-directional Prior Mechanism** In this mechanism, the prior as a bridge between two views (i.e., rating matrix and side information), is not identical for each direction while optimization as shown in Figure 1. The key idea behind this mechanism is that we assume side information for similarity learning is more reliable than for that using a extremely sparse rating matrix. More specifically, while we optimize the variables related to rating matrix view, the variables related to side information are active as a regularization for it. On the contrary, while we optimize the variables related to side information view, the regularization effect of variables related to rating matrix view is not allowed to be active explicitly, but with a way of using active transfer learning approach implicitly. Thus, the most confident knowledge encoded by actively constructed negative samples, learned from rating matrix view, can be utilized for correctly directing the optimization process with corresponding loss.

Model Learning: For the extended model, similarly, Maximizing A Posteriori (MAP) estimate of U, V is equivalent to minimizing the following complete negative log-likelihood with respect to $W_1, W_2, U, V, \mathbf{b}, \mathbf{c}$:

$$
\mathcal{L}_2 = \frac{\lambda_u}{2} \sum_i \mathbf{u}_i^T \mathbf{u}_i + \underbrace{\sum_i \sum_j \frac{\lambda_{c_{ij}}}{2} (r_{ij} - \mathbf{u}_i^T \mathbf{v}_j)^2}_{view1}
$$

$$
+ \underbrace{\frac{\lambda_v}{2} \sum_j (\mathbf{v}_j - \mathbf{h}_1^j)^T (\mathbf{v}_j - \mathbf{h}_1^j)}_{bridge} + \frac{\lambda_w}{2} \|W\|_F^2 \tag{8}
$$

$$
- \underbrace{\sum_i \sum_{j \in R_i} log \frac{exp(dot(\sigma(W_2^T \mathbf{h}_1^j + \mathbf{c}), \mathbf{u}_i))}{\sum_{k \in Cn_i} exp(dot(\sigma(W_2^T \mathbf{h}_1^k + \mathbf{c}), \mathbf{u}_i))}}_{view2},
$$

where $\mathbf{h}_1^{\cdot} = \sigma(W_1 x^{(\cdot)} + \mathbf{b})$, $W = [W_1, W_2]$ is the weights matrix in NN with two hidden layers, R_i denotes the rated item set of user i, Cn_i denotes the candidate set including the current item j and all other unrated items as negative samples constructed by active knowledge transfer for user i. σ is a nonlinear sigmoid function as shown in our basic model. Note that the prior $\boldsymbol{\theta}_j$ in general framework is modeled as

$$\boldsymbol{\theta}_j = \mathbf{h}_1^j = \sigma(W_1 x^{(j)} + \mathbf{b}). \tag{9}$$

Similarly, we follow the same strategy used for the basic model to derive the optimization procedure here. It is noted that we only consider two views of each item, ignoring the effect of user regularization through ranking based loss (view 2) when optimizing each latent user representation. Thus, for $\mathbf{u}_i$ and $\mathbf{v}_j$, we derive the similar update rule as shown in basic model to guarantee the closed optimal solution for updating $\mathbf{u}_i$ and $\mathbf{v}_j$ respectively, which can also reduce computational cost simultaneously for our extended model.

$$\mathbf{u}_i \leftarrow (VC_iV^T + \lambda_u I_K)^{-1} VC_i R_i, \tag{10}$$

$$\mathbf{v}_j \leftarrow (UC_jU^T + \lambda_v I_K)^{-1} (UC_j R_j + \lambda_v \mathbf{h}_1^j). \tag{11}$$

Then, we use the same way discussed in our basic model to modify standard AE optimization procedure by adding additional gradient for regularization of ranking based loss in view 2. Specifically, one modification refers to computing the desired partial derivatives for output layer of NN. We define the output value of NN for each item j, $\mathbf{h}_2^j = \sigma(W_2^T \mathbf{h}_1^j + \mathbf{c})$. The partial derivative of $\mathcal{L}_2$ with respect to $\mathbf{h}_2^j$ is

$$\frac{\partial \mathcal{L}_2}{\partial \mathbf{h}_2^j} = \sum_i \left(\frac{\mathbf{u}_i exp(dot(\mathbf{u}_i, \mathbf{h}_2^j))}{\sum_{k \in Cn_i} exp(dot(\mathbf{u}_i, \mathbf{h}_2^k))} - \mathbf{u}_i \right). \tag{12}$$

The other gradient modification of AE is the consideration for regularization term appeared in *bridge* part through view 2, which is similar to our basic model but with 2 hidden layers structure. Thus, the similar derivation can be obtained.

Speeding Up the Optimization: To reduce computational costs when updating u_i and v_j, we adopt the same strategy of matrix operation shown in [6]. Specifically, directly computing VC_iV^T and UC_jU^T requires time $O(K^2J)$ and $O(K^2I)$ for each user and item, where J and I are the total number of items and users respectively, K is the dimension of latent representation space. Instead, we rewrite

$$UC_jU^T = U(C_j - bI_K)U^T + bUU^T. \tag{13}$$

Then, bUU^T can be pre-computed and $C_j - bI_K$ has only I_r non-zeros elements, where I_r refers to the number of users who rated item j and empirically $I_r \ll I$. For VC_iV^T, it is similar. Thus, we can significantly speed up computation by this sparsity property.

Prediction: Using the learned parameters above, we can make in-matrix and out-of-matrix predictions defined in [14]. For in-matrix prediction, it refers to the case where those items that have been rated by at least one user in the system. To compute predicted rating, we use $r_{ij}^* \approx (\mathbf{u}_i^*)^T \mathbf{v}_j^*$. For out-of-matrix prediction, it refers to the case where those items that have never been rated by any user in the system. To compute predicted rating, we use $r_{ij}^* \approx (\mathbf{u}_i^*)^T \boldsymbol{\theta}_j^*$, where the corresponding $\boldsymbol{\theta}_j^*$ is defined in Equation 9.

5   Experiments

5.1   Data and Metric

Datasets 1) CiteULike Dataset: For a fair comparison, we use the same CiteULike dataset[4] as the benchmark, following the prior work in [14]. This dataset is challenging. Though it contains 204,986 pairs of observed ratings with 5551 users and 16,980 articles, the sparseness is quite low, i.e., merely 0.2175% , which is much lower than that of the well-known Movielens dataset with the sparseness 4.25% . On average, each user has 37 articles in the library, ranging from 10 to 403, and each article appears in 12 users libraries, ranging from 1 to 321. For each article, the title and abstract information are used as the bag-of-word representation. After the text processing by selecting informative words via tf-idf and removing stop words, 8,000 distinct words are remained in the corpus. **2) LastFM Dataset:** We further evaluate our proposed method on real life dataset [5] from LastFm[6]. This dataset is also challenging. Though it contains 92,834 pairs of observed ratings with 1892 users and 17,632 items, the sparseness is quite low, i.e., merely 0.2783% , which is also much lower than that of the well-known Movielens dataset with the sparseness 4.25%. On average, each user has 44.21 items in the play list, ranging from 0 to 50, and each item appears in 4.95 users libraries, ranging from 0 to 611. For each item, the tag information is used as bag-of-word representation. After text processing, 11,946 distinct words are remained in the corpus. In addition, we further remove noisy users which have no items.

Evaluation Metric. Two possible metrics are precision and recall. As discussed in [11,14,15], zero ratings are uncertain which may indicate that a user does not like an article or does not know about it. Thus, we use recall as our metric. $Recall@k = \frac{\sharp relevance@k}{\sharp total\ relevance}$, where $\sharp$ denotes the number of relevant items in top-k result and total relevant items respectively.

5.2   Baselines and Settings

Baselines

- **CML-ADP-Bi**: The proposed extended model with active discriminative prior, in which the bi-directional prior modeling mechanism is enabled. It is noted that the initialization of NN is from the values of CML-ADP.

---

[4] Data available at http://www.cs.cmu.edu/~chongw/citeulike/
[5] Data available at http://grouplens.org/datasets/hetrec-2011/
[6] http://www.last.fm/

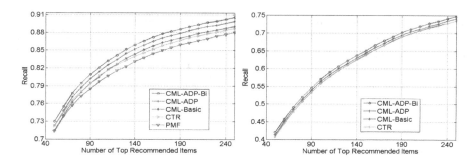

Fig. 2. Comparison of recall for our model and the state-of-the-arts. Left Plot: 'in matrix prediction' case and Right Plot: 'out-of-matrix prediction' case. It is noted that our method can achieve the significant improvements only using content information, in contrast with [3,11,15] which rely on both content and social information to achieve the comparable improvements.

- **CML-ADP**: The proposed extended model with active discriminative prior, in which the bi-directional prior modeling mechanism is not used.
- **CML-Basic**: The proposed basic model without active discriminative prior.
- **CTR**: The model described in [14], which is the most similar state-of-the-art approach combining the merits of traditional collaborative filtering and probabilistic topic modeling.
- **PMF**: The model described in [12], which is a state-of-the-art matrix factorization approach widely applied without using side information.

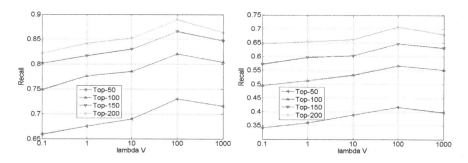

Fig. 3. Parameter sensibility analysis for lambda V on in-matrix (Left Plot) and out-of-matrix (Right Plot) prediction with the number of top recommended item at 50, 100, 150 and 200

Settings. We evaluate our models in three cases. **1) In Matrix and Out-of Matrix Cases (CiteULike Dataset):** We use 5-fold cross-validation scheme following [14] and we use grid search to find corresponding optimal parameters on a small heltout dataset. We found that the common parameters $v = 100$; $u = 0.01$; $a = 1$; $b = 0.01$; $K = 200$ gives good performance for PMF and CTR approach. For CTR, we set additional parameters $\lambda_u = 0.01$; For our model, we set additional parameters $\lambda_w = 10$; and vary parameter λ_v to study their effect on prediction accuracy. We also select our optimal parameter $\lambda_v = 10$. For DAE, the number of hidden variables $K = 300$ selected for the optimal performance. Particularly, we use a masking noise probability in 0.7 for the input layer and a Gaussian noise with standard deviation of 0.1 is used for higher output. For parameter analysis with different λ_v and ratio of unrated items for knowledge transfer, we perform this testing for our proposed models for different top items {50,100,150,200} with 300 factors. **2) Randomly Split Case (CiteULike Dataset):** We randomly split the dataset into two parts, training (90%) and test datasets (10%), with constraint that users in test dataset have more than half of the average number of rated items, i.e., 20. This expands the range of performance analysis for our evaluation compared with [11]. The optimal parameters are obtained on a small held-out dataset. For PMF, we set $\lambda_v = 100, \lambda_u = 0.01$. For all CTR, we set $a = 1, b = 0.01, \lambda_v = 0.1$. and set $\lambda_u = 0.01$. The remaining setting is the same as that described above.

5.3   Results and Analysis

1) CiteULike Dataset: For in matrix prediction, from Figure 2 (left), we can see that our models consistently outperform CTR model and PMF under recall and achieves considerable improvement. In addition, we study how the content parameter λ_v affect the overall performance of the recommendation system. From Figure 3 (left), we observe that the value of λ_v impacts the recommendation results significantly, which demonstrates that fusing representation learning with PMF improves recommendation accuracy considerably. For out of matrix

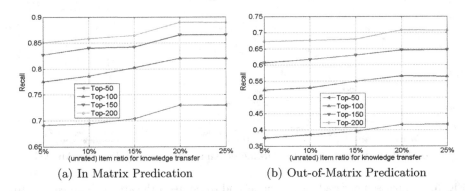

(a) In Matrix Predication   (b) Out-of-Matrix Predication

Fig. 4. Parameter sensibility analysis of the ratio of unrated items for knowledge transfer on in-matrix (Left Plot) and out-of-matrix (Right Plot) prediction

prediction task, PMF is useless in this problem. Thus, we only compare CTR with the proposed models in this paper. From Figure 2 (right), we can also see that our models consistently outperforms CTR model under recall metric. Similarly, we study how the content λ_v affects the overall performance in out of matrix prediction task setting. We also can find that λ_v impacts the recommendation results significantly from Figure 3 (right). It is noted that although the improvement compared with in matrix predication case is not considerable, it is also much better than that in original CTR which compares with LDA in its original paper [14]. This could be explained that the task-oriented optimization benefits from the discriminative learning approach compared with a generative fashion as in CTR, which makes a strong assumption in generative process. We further exploit how the ratio of unrated items for knowledge transfer can influence the recommendation performance in Figure 4. It is shown that the performance is increased with the ratio but the computation costs are also increased. Thus, we choose 20% as our optimal value. With a more larger one, it may introduce more some uncertain negative samples to undermine the performance. Moreover, we can see that the ranking based objective as additional optimization view in the extended model (CML-ADP and CML-ADP-Bi), to augment the RSME error criterion, is also a necessary, which is proven by our experiment results in both in-matrix and out-of-matrix tasks. This ability to incorporate multiple task related optimization objectives is a salient advantage in our proposed collaborative multi-view learning framework, which is not easily achieved in the generative approach, e.g., CTR.

2) LastFM Dataset: From Figure 5, we can find the similar results as shown in previous discussion on CiteULike dataset, which further demonstrates the effectiveness of the proposed method in randomly splitting case. Thus, all three cases in various real applications have proved the promising performance of the proposed method.

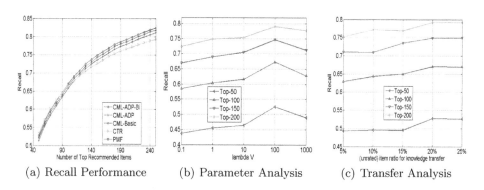

(a) Recall Performance   (b) Parameter Analysis   (c) Transfer Analysis

Fig. 5. Comparison of recall for our model and the state-of-the-arts. It is noted that to well illustrate, the result of PMF is omitted due to its poor performance in our experiment. Our method performs significantly better than CTR, using the same content information, without introducing additional side information as in [3,11,15] to achieve the comparable improvements.

6 Conclusions

In this paper, we propose a multi-view learning framework with the ability of knowledge transfer for recommendation. We can learn multi-view representation automatically from data, without needing multi-view data representation available in advance. Our method achieves significant improvements on all three cases compared with the state-of-the-arts. In particular, our models achieve such considerable improvements only using content information, in contrast with the models relying on more external resources, such as both content and social network information. Thus, the proposed method serves as a fundamental framework, which can be further improved by incorporating additional side information using the same fashion in [3,11,15].

Acknowledgments. This research was partly supported by National Natural Science Foundation of China (No.61370117,61433015) and Major National Social Science Fund of China (No.12&ZD227).

References

1. Agarwal, D., Chen, B.C.: flda: matrix factorization through latent dirichlet allocation. In: Proceedings of WSDM, pp. 91–100 (2010)
2. Blei, D.M., Ng, A.Y., Jordan, M.I.: Latent dirichlet allocation. The Journal of Machine Learning Research **3**, 993–1022 (2003)
3. Ding, X., Jin, X., Li, Y., Li, L.: Celebrity recommendation with collaborative social topic regression. In: IJCAI, pp. 2612–2618 (2013)
4. Gao, J., Han, J., Liu, J., Wang, C.: Multi-view clustering via joint nonnegative matrix factorization. In: Proceedings of SDM, pp. 252–260 (2013)
5. Guo, Y.: Convex subspace representation learning from multi-view data. In: Proceedings of AAAI, pp. 387–393 (2013)
6. Hu, Y., Koren, Y., Volinsky, C.: Collaborative filtering for implicit feedback datasets. In: Proceedings of ICDM, pp. 263–272 (2008)
7. Li, B., Yang, Q., Xue, X.: Can movies and books collaborate? cross-domain collaborative filtering for sparsity reduction. In: Proceedings of IJCAI, pp. 2052–2057 (2009)
8. Mehta, B., Hofmann, T.: Cross system personalization and collaborative filtering by learning manifold alignments. In: Freksa, C., Kohlhase, M., Schill, K. (eds.) KI 2006. LNCS (LNAI), vol. 4314, pp. 244–259. Springer, Heidelberg (2007)
9. Pan, S.J., Yang, Q.: A survey on transfer learning. IEEE Trans. Knowl. Data Eng. **22**(10), 1345–1359 (2010)
10. Pan, W., Xiang, E.W., Liu, N.N., Yang, Q.: Transfer learning in collaborative filtering for sparsity reduction. In: Proceedings of AAAI, pp. 230–235 (2010)
11. Purushotham, S., Liu, Y.: In: Proceedings of ICML, pp. 759–766
12. Salakhutdinov, R., Mnih, A.: Probabilistic matrix factorization. Proceedings of Advances in Neural Information Processing Systems **20**, 1–8 (2008)
13. Vincent, P., Larochelle, H., Lajoie, I., Bengio, Y., Manzagol, P.: Stacked denoising autoencoders: Learning useful representations in a deep network with a local denoising criterion. Journal of Machine Learning Research **11**, 3371–3408 (2010)

14. Wang, C., Blei, D.M.: Collaborative topic modeling for recommending scientific articles. In: Proceedings of KDD, 2011, pp. 448–456. ACM (2011)
15. Wang, H., Chen, B., Li, W.J.: Collaborative topic regression with social regularization for tag recommendation. In: Proceedings of IJCAI, pp. 2719–2725 (2013)
16. Wu, F., Tan, X., Yang, Y., Tao, D., Tang, S., Zhuang, Y.: Supervised nonnegative tensor factorization with maximum-margin constraint. In: Proceedings of AAAI, pp. 962–968 (2013)
17. Zhao, L., Pan, S.J., Xiang, E.W., Zhong, E., Lu, Z., Yang, Q.: Active transfer learning for cross-system recommendation. In: Proceedings of AAAI, pp. 1205–1211 (2013)

Online and Stochastic Universal Gradient Methods for Minimizing Regularized Hölder Continuous Finite Sums in Machine Learning

Ziqiang Shi$^{(\boxtimes)}$ and Rujie Liu

Fujitsu Research and Development Center, Beijing, China
{shiziqiang,rjliu}@cn.fujitsu.com

Abstract. Online and stochastic gradient methods have emerged as potent tools in large scale optimization with both smooth convex and nonsmooth convex problems from the classes $C^{1,1}(\mathbb{R}^p)$ and $C^{1,0}(\mathbb{R}^p)$ respectively. However, to our best knowledge, there is few paper using incremental gradient methods to optimization the intermediate classes of convex problems with Hölder continuous functions $C^{1,v}(\mathbb{R}^p)$. In order to fill the difference and the gap between the methods for smooth and nonsmooth problems, in this work, we propose several online and stochastic universal gradient methods, which we do not need to know the actual degree of the smoothness of the objective function in advance. We expanded the scope of the problems involved in machine learning to Hölder continuous functions and to propose a general family of first-order methods. Regret and convergent analysis shows that our methods enjoy strong theoretical guarantees. For the first time, we establish algorithms that enjoys a linear convergence rate for convex functions that have Hölder continuous gradients.

1 Introduction and Problem Statement

Online and stochastic gradient methods (or referred to as incremental gradient methods) are of the most promising approaches in large scale machine learning tasks in these days [5,9,10,13,15,16]. Important advances of incremental gradient methods have been made on sequential learning in the recent literature on similar and famous problems, including lasso, logistic regression, ridge regression, and support vector regression. Composite objective mirror descent (COMID) [2] generalizes mirror descent [1] to the online setting. Regularized dual averaging (RDA) [14] generalizes dual averaging [7] to online and composite optimization, and can be used for distributed optimization [3]. Online alternating direction multiplier method (ADMM) [12], RDA-ADMM [12] and online proximal gradient (OPG) ADMM [13] generalize classical ADMM [4] to online and stochastic settings. In stochastic gradient methods, more recent descent techniques like MISO [5], SAG [9] and SVRG [15] take update steps in the average gradient direction, and achieve linear convergence rate.

However, most current incremental gradient methods deal with smooth functions or non-smooth functions with Lipschitz-continues function values. In this

T. Cao et al. (Eds.): PAKDD 2015, Part I, LNAI 9077, pp. 369–379, 2015.
DOI: 10.1007/978-3-319-18038-0_29

paper, we consider incremental gradient methods with an objective function that has Hölder continuous gradients with degree v:

$$\|\nabla g(x) - \nabla g(y)\|_* \leq M_v \|x - y\|^v, \tag{1}$$

where $0 \leq v \leq 1$ and $\nabla g(x)$ means any subgradient if $g(x)$ is nonsmooth. It can be seen that $g(x)$ becomes smooth function with Lipschitz-continues gradients when $v = 1$ and becomes non-smooth Lipschitz-continues function when $v = 0$. M_v is mainly used to characterize the variability of the (sub)gradients, all of this kind of functions form the class $C^{1,v}(\mathbb{R}^p)$. In this paper, we consider the problems of the following form:

$$\min_{x \in \mathbb{R}^p} f(x) := \frac{1}{n} \sum_{i=1}^{n} g_i(x) + h(x), \tag{2}$$

where g_i is a convex loss function with Hölder continuous gradients associated with a sample in a training set, and h is a convex penalty function or regularizer. Let $g(x) = \frac{1}{n} \sum_{i=1}^{n} g_i(x)$.

If the Problem (2) is treated as minimizing of composite functions $g(x) + h(x)$, Nesterov has proposed the universal gradient methods (UGM) to solve it in [8]. However, UGM for Problem (2) is a learning procedure in batch mode, which cannot deal with training data appearing in succession, such as audio processing [11]. Furthermore, one can hardly ignore the fact that in reality the size of the data is rapidly increasing in various domain and thus training set for the data probably cannot be loaded into the memory simultaneously in batch mode methods. In such situation, sequential learning becomes powerful tools. In this paper, we generalize UGM to online and stochastic settings to deal with objective functions which have Hölder continuous gradients.

Assume x^* is a solution of Problem (2), and in this work, we introduce a novel kind of regret definition and seek bounds for this regret in the online learning setting with respect to x^*, defined as

$$R(T, x^*, \epsilon) := \sum_{t=0}^{T} f_{g_t}(x_t) - \sum_{t=0}^{T} f_{g_t}(x^*), \tag{3}$$

where ϵ if a pre-specified error limit. All of our algorithms need to first assume a fixed accuracy ϵ, and then the smaller the ϵ, the smaller the regret. For example, if we assume $\epsilon = 1/T$, then we will have a regret bound of $O(1)$ after T iterations. And if $\epsilon = 1/\sqrt{T}$, then we will have a regret bound of $O(\sqrt{T})$ after T iterations. Thus, we have the results that look too good to be true, since our algorithms are different from previous online algorithms, and we have an extra parameter describing the accuracy. The regret bound is not in a standard sense. Ours are in a sense that, for any fixed T, we can obtain an $O(1)$ bound after T iterations.

We now outline the rest of the study. In Section 2, we propose online **prime/ dual** universal gradient methods to solve the online optimization problem for the data that appear in succession and present the regret and convergence analysis.

Section 3 states the stochastic universal gradient (SUG) method for the data that cannot be loaded into the memory at the same time and show that the SUG achieves a linear convergence rate. We conclude in Section 4. Some applications of our theory will be presented in the appendix, but due to space limitation, the detail of these applications, the numerical experiments and further the proofs will be shown in another paper or a long version of this paper.

1.1 Notations and Lemmas

Before proceeding, we introduce the notations and some useful lemmas formally first. In this work, we most adopt the nomenclature used by Nesterov on universal gradient methods [8]. The functions encountered in this work are all convex if there are no other statements.

This inequality (1) ensures that

$$|g(x) - g(y) - \nabla g(y)^T (x - y)| \leq \frac{M_v}{1 + v} \|x - y\|^{1+v}. \tag{4}$$

Bregman distance is defined as

$$\xi(x, y) := d(y) - d(x) - \langle \nabla d(x), y - x \rangle, \tag{5}$$

where $d(x)$ is a *prox-function*, which is differentiable strongly convex with convexity parameter equal to one and its minimum is 0. Take derivative for y, we have

$$\nabla_y \xi(x, y) = \nabla d(y) - \nabla d(x).$$

Bregman mapping is defined as

$$\hat{x} = \arg \min_y \left[g(x) + \langle \nabla g(x), y - x \rangle + M\xi(y, x) + h(y) \right], \tag{6}$$

where $h(y)$ is the fixed regularizer.

The first-order optimality condition for Problem (6) is

$$\langle \nabla g(x) + M(\nabla d(\hat{x}) - \nabla d(x)) + \nabla h(\hat{x}), y - \hat{x} \rangle \geq 0. \tag{7}$$

Some useful lemmas and equations introduced by [8] are frequently employed in establishing the results and are stated below for the sake of completeness.

Lemma 1. *If $\epsilon > 0$ and $M > (\frac{1}{\epsilon})^{\frac{1-v}{1+v}} M_v^{\frac{2}{1+v}}$, then for any pair $t \geq 0$ we have*

$$\frac{M_v}{1 + v} t^{1+v} \leq \frac{1}{2} M t^2 + \frac{\epsilon}{2}. \tag{8}$$

This lemma play an important role in this paper, which is been used to transform the Hölder Continuous conditions to Lipschitz-continues conditions.

Lemma 2. *If g satisfy condition* (1), *assume* $\epsilon > 0$ *and* $M > (\frac{1}{\epsilon})^{\frac{1-v}{1+v}} M_v^{\frac{2}{1+v}}$, *then for any pair* x, y *we have*

$$g(y) \leq g(x) + \langle \nabla g(x), y - x \rangle + \frac{1}{2} M \|y - x\|^2 + \frac{\epsilon}{2}. \tag{9}$$

If $\hat{x}$ *is the Bregman mapping at* x *obtained by* (6), *then we have*

$$g(\hat{x}) + h(\hat{x}) \leq g(x) + \langle \nabla g(x), \hat{x} - x \rangle + M\xi(\hat{x}, x) + h(\hat{x}) + \frac{\epsilon}{2}. \tag{10}$$

Throughout this work, we denote $\gamma(M_v, \epsilon) := (\frac{1}{\epsilon})^{\frac{1-v}{1+v}} M_\infty^{\frac{2}{1+v}}$.

Lemma 3. *If* $\phi(x)$ *is convex and* $\phi(x) - Md(x)$ *is subdifferentiable, let* $\bar{x} = \arg\min_x \phi(x)$, *then we have*

$$\phi(y) \geq \phi(\bar{x}) + M\xi(\bar{x}, y). \tag{11}$$

These lemmas are proposed in [8], please refer there for proofs if interested.

2   Online Universal Gradient Method

In this section, we extend UGM to the online learning setting to deal with situation that the training data appearing in succession, such as multimedia information processing [11]. The modification of UGM that we proposed is simple: just change $f_T(x)$ to $f_{g_t}(x)$ in each iteration and output the average value in each iteration. Our online algorithms are almost the same as the UGM with an important difference: we only meet and process one sample (one function) at each iteration. This methodology mainly comes from [2] and [13]. In the sequel, we consider two types of methods according to the original work of [8], from whose proofs we also draw some ingredients in ours.

2.1   Online Universal Prime Gradient Method (O-UPGM)

Lemma 2 shows that the Bregman mapping can move the current point more close to the real solution, and this intuition form the core of the UGM and our online algorithms. In UGM, the Bregman mapping is employed to update the x_t in each iteration, and x_t is output as the solution after all the iterations. Here we offer the general online universal primal gradient method (O-UPGM) to solves Problem (2) in the following algorithm, where the same as UGM, Bregman mapping is also employed to update the x_t in each iteration seeing current sample, while unlike UGM that the average of these x_t is output as solutions after all the iterations.

Algorithm 1. A generic O-UPGM

Input: $L_0 > 0$ and $\epsilon > 0$.
1: **for** $t = 0, 1, \cdots, T$ **do**
2: Find the smallest $i_t \geq 0$ such that $g_t(\hat{x}) + h(\hat{x}) \leq g_t(x_t) + \langle \nabla g_t(x_t), \hat{x} - x_t \rangle + 2^{i_t} L_t \xi(\hat{x}, x_t) + h(\hat{x}) + \frac{\epsilon}{2}$.
3: Set $x_{t+1} = \hat{x}$ and $L_{t+1} = 2^{i_t - 1} L_t$.
4: $t = t + 1$.
5: **end for**
Output: $\bar{x} = \frac{1}{S_T} \sum_{t=1}^{T+1} \frac{1}{L_t} x_t$, where $S_T = \sum_{t=1}^{T+1} \frac{1}{L_t}$.

The above online UPGM is similar as batch UPGM except the x_t update in O-UPGM uses a time varying function f_{g_t}. The following establishes the regret bound and the convergence rate for UPGM for general convex function with Hölder continuous gradients.

Theorem 1. *Assume $M_v(g_t) < M_v$ and $h(x)$ is a simple convex function. Let the sequence $\{x_t\}$ be generated by the general O-UPGM in Algorithm 1. Then we have*

$$\sum_{t=0}^{T} \frac{1}{L_{t+1}} [f_{g_t}(x_{t+1}) - f_{g_t}(x^*)] \leq \frac{\epsilon}{2} S_T + 2r_0(x^*), \tag{12}$$

where $S_T = \sum_{t=1}^{T+1} \frac{1}{L_t}$.

The ideas of the proof is closed related to that of UPGM by Nesterov [8], but due to the space limitation, the proof will be given in a long version of this paper.

If we replace Step 2 and 3 in Algorithm 1 with $x_{t+1} = \mathfrak{B}_{2\gamma(M_v, \epsilon), g_t}(x_t)$, then $L_{t+1} = \gamma(M_v, \epsilon)$. Thus Theorem 1 becomes

Corollary 1. *Assume $M_v(g_t) < M_v$ and $h(x)$ is a simple convex function. Let the sequence $\{x_t\}$ be generated by O-UPGM with fixed steps $L_{t+1} = \gamma(M_v, \epsilon)$. Then we have the standard regret bound*

$$R(T, x^*, \epsilon) \leq \frac{\epsilon}{2}(T + 1) + 2r_0(x^*)\gamma(M_v, \epsilon). \tag{13}$$

Further, let $\epsilon = T^{-\frac{1+v}{2}}$, we have

$$R(T, x^*, \epsilon) = O(T^{\frac{1-v}{2}}). \tag{14}$$

We have the following remarks regarding the above result:

Remark 1. All of our online algorithms (O-UPGM and the following O-UDGM) need to first assume a fixed accuracy ϵ, and then the smaller the ϵ, the more accurate the solution. For example, if we assume $\epsilon = 1/T$, then we will have a regret bound of $O(1)$ after T iterations. And if $\epsilon = 1/\sqrt{(T)}$, then we will have a regret bound of $O(\sqrt{(T)})$ after T iterations. Thus, we have the results that look too good to be true, since our algorithms are different from previous online algorithms, and we have an extra parameter describe the accuracy. And the regret bound is not in a standard sense. Ours are in a sense that, for any fixed T, we can obtain an $O(1)$ bound after T iteration.

2.2   Online Universal Dual Gradient Method (O-UDGM)

The original batch UDGM is based on updating a simple model for objective function of Problem (2). We built a general online UDGM based on this principle for online or large scale problems.

Algorithm 2. A generic O-UDGM

Input: $L_0 > 0$, $\epsilon > 0$ and $\phi_0(x) = \xi(x_0, x)$.

1: **for** $t = 0, 1, \cdots, T$ **do**

2: Find the smallest $i_t \geq 0$ such that for point $x_{t,i_t} = \arg\min_x \phi_t(x) + \frac{1}{2^{i_t}L_t}[g_t(x_t) + \langle \nabla g_t(x_t), x - x_t \rangle + h(x)]$, we have $f_{g_t}(\mathcal{B}_{2^{i_t}L_t, g_t}(x_{t,i_t})) \leq \psi^*_{2^{i_t}L_t, g_t}(x_{t,i_t}) + \frac{1}{2}\epsilon$.

3: Set $x_{t+1} = x_{t,i_t}$, $L_{t+1} = 2^{i_t-1}L_t$ and $\phi_{t+1}(x) = \phi_t(x) + \frac{1}{2L_{t+1}}[g_t(x_t) + \langle \nabla g_t(x_t), x - x_t \rangle + h(x)]$.

4: $t = t + 1$.

5: **end for**

Output: $\bar{x} = \frac{1}{S_T}\sum_{t=1}^{T+1}\frac{1}{L_t}x_t$, where $S_T = \sum_{t=1}^{T+1}\frac{1}{L_t}$.

Theorem 2. *Assume $M_v(g_t) < M_v$ and $h(x)$ is a simple convex function. Let the sequence $\{x_t\}$ be generated by the general O-UDGM. Then we have*

$$\sum_{t=0}^{T}\frac{1}{2L_{t+1}}f_{g_t}(x_t) - \sum_{t=0}^{T}\frac{1}{2L_{t+1}}f_{g_t}(x^*) \leq S_T\frac{\epsilon}{4} + \xi(x_0, x^*) \qquad (15)$$

where $S_T = \sum_{t=1}^{T+1}\frac{1}{L_t}$.

We have the following remarks regarding the above result:

Remark 2. If we replace Step 2 and 3 in Algorithm 2 with

$$x_{t+1} = \arg\min_x\{\phi_t(x) + \frac{1}{2\gamma(M_v, \epsilon)}[g_t(x_t) + \langle \nabla g_t(x_t), x - x_t \rangle + h(x)]\} \qquad (16)$$

and

$$\phi_{t+1}(x) = \phi_t(x) + \frac{1}{2\gamma(M_v, \epsilon)}[g_t(x_t) + \langle \nabla g_t(x_t), x - x_t \rangle + h(x)] \qquad (17)$$

respectively, then $L_{t+1} = \gamma(M_v, \epsilon)$ and Theorem 2 becomes

Corollary 2. *Assume $M_v(g_t) < M_v$ and $h(x)$ is a simple convex function. Let the sequence $\{x_t\}$ be generated by O-UDGM with fixed steps $L_{t+1} = \gamma(M_v, \epsilon)$. Then we have the standard regret bound*

$$R(T, x^*, \epsilon) \leq \frac{\epsilon}{2}(T+1) + 2\xi(x_0, x^*)\gamma(M_v, \epsilon). \tag{18}$$

Further let $\epsilon = T^{-\frac{1+v}{2}}$, thus Corollary 2 becomes

Corollary 3. *Assume $M_v(g_t) < M_v$ and $h(x)$ is a simple convex function. Let the sequence $\{x_t\}$ be generated by the specific O-UDGM with x_t updated by (16) and (17). Then we have*

$$R(T, x^*, T^{-\frac{1+v}{2}}) = O(T^{\frac{1-v}{2}}). \tag{19}$$

3   Stochastic Universal Gradient Method

In this section, we propose the stochastic universal gradient (SUG) method to deal with situation that the data probably cannot be loaded into the memory at the same time in batch mode methods since the size of the data is rapidly increasing. We summarize the SUG method in Algorithm 3.

Algorithm 3. SUG: A generic stochastic universal gradient method

---

Input: start point $x^0 \in \text{dom } f$; for $i \in \{1, 2, .., n\}$, let $g_i^0(x) = g_i(x^0) + (x - x^0)^T \nabla g_i(x^0) + M_0^i \xi(x^0, x)$, and $G^0(x) = \frac{1}{n} \sum_{i=1}^n g_i^0(x)$.

1: **repeat**

2: Solve the subproblem for new approximation of the solution: $x^{k+1} \leftarrow \arg\min_x [G^k(x) + h(x)]$.

3: Sample j from $\{1, 2, .., n\}$, and update the surrogate functions:

$$g_j^{k+1}(x) = g_j(x^{k+1}) + (x - x^{k+1})^T \nabla g_j(x^{k+1}) + M_{k+1}^i \xi(x^{k+1}, x), \tag{20}$$

while leaving all other $g_i^{k+1}(x)$ unchanged: $g_i^{k+1}(x) \leftarrow g_i^k(x)$ $(i \neq j)$; and $G^{k+1}(x) = \frac{1}{n} \sum_{i=1}^n g_i^{k+1}(x)$.

4: **until** stopping conditions are satisfied.

Output: x^k.

3.1   Convergence Analysis of SUG

Theorem 3. *Suppose $g_i(x)$ satisfy condition (1) and $M \geq M_0^i > (\frac{2}{\epsilon})^{\frac{1-v}{1+v}} M_v^{\frac{2}{1+v}}$ for $i = 1, ..., n$, $d(x)$ satisfy $\|\nabla d(x) - \nabla d(y)\|_* \leq M_d \|x - y\|^d$, $h(x)$ is strongly convex with $\mu_h \geq 0$, then the SUG iterations satisfy for $k \geq 1$:*

$$\mathbb{E}[f(x^k)] - f^* \leq M\rho^{k-1}\|x^* - x^0\|^2 + \frac{3\epsilon}{4n\mu_h} \frac{1 - \rho^{k-1}}{1 - \rho} + \frac{3\epsilon}{4}, \tag{21}$$

where $\rho = \frac{1}{n} \frac{M}{\mu_h} + (1 - \frac{1}{n})$.

We have the following remarks regarding the above result:

- In order to satisfy $\mathbb{E}[f(x^k)] - f^* \leq \tilde{\epsilon}$, the number of iterations k needs to satisfy

$$k \geq (\log \rho)^{-1} \log \left[\left(\tilde{\epsilon} - \frac{3\epsilon}{4n\mu_h} \frac{1}{1-\rho} - \frac{3\epsilon}{4} \right) \frac{1}{M\|x^* - x^0\|^2} \right] + 1.$$

- Inequality (21) gives us a reliable stopping criterion for SUG method.

Since $\mathbb{E}[f(x^k)] - f^* \geq 0$, Markov's inequality and Theorem 3 imply that for any $\epsilon > 0$,

$$\text{Prob}\left(f(x^k) - f^* \geq \tilde{\epsilon} \right) \leq \frac{\mathbb{E}[f(x^k)] - f^*}{\tilde{\epsilon}} \leq \frac{M\rho^{k-1}\|x^* - x^0\|^2}{\tilde{\epsilon}} + \frac{3\epsilon}{4\tilde{\epsilon}n\mu_h} \frac{1}{1-\rho} + \frac{3\epsilon}{4\tilde{\epsilon}}.$$

Thus we have the following high-probability bound.

Corollary 4. *Suppose the assumptions in Theorem 3 hold. Then for any $\epsilon > 0$ and $\delta \in (0,1)$, we have*

$$\text{Prob}\left(f(x^k) - f(x^\star) \leq \tilde{\epsilon} \right) \geq 1 - \tilde{\delta}$$

provided that the number of iterations k satisfies

$$k \geq (\log \rho)^{-1} \log \left[\left(\tilde{\delta} - \frac{3\epsilon}{4\tilde{\epsilon}} - \frac{3\epsilon}{4\tilde{\epsilon}n\mu_h} \frac{1}{1-\rho} \right) \frac{\tilde{\epsilon}}{M\|x^* - x^0\|^2} \right] + 1.$$

4 Conclusions

In this paper, in order to fill the difference and gap between methods for smooth and nonsmooth problems, we propose efficient online and stochastic gradient algorithms to optimization the intermediate classes of convex problems with Hölder continuous functions $C^{1,v}(\mathbb{R}^p)$. We establish regret bounds for the objective and linear convergence rates for convex functions that have Hölder continuous gradients. There are some directions that the current study can be extended. In this paper, we have focused on the theory; it would be meaningful to also do the numerical evaluation and implementation details, and we give some simple applications in Section 4. Second, combine with randomized block coordinate method [6] for minimizing regularized convex functions with a huge number of varialbes/coordinates. Moreover, due to the trends and needs of big data, we are designing distributed/parallel SUG for real life applications. In a broader context, we believe that the current paper could serve as a basis for examining the method for the classes of convex problems with Hölder continuous functions $C^{1,v}(\mathbb{R}^p)$.

Appendix

In this appendix, we will present some applications of our methods.

1. Lasso Problem

The lasso problem is formulated as follows:

$$\min_{x \in \mathbb{R}^p} \frac{1}{n} \sum_{t=1}^{n} \|a_t^T x - b_t\|^2 + \mu \|x\|_1,$$

where $a_t, x \in \mathbb{R}^p$ and b_t is a scalar.

Throughout this subsection, let $g(x) = \frac{1}{n} \sum_{t=1}^{n} \|a_t^T x - b_t\|^2$ and $h(x) = \mu \|x\|_1$, $d(x) = \frac{1}{2}\|x\|^2$, then

The Bregman mapping associate with $g(x)$ and the component function $g_t(x) = \|a_t^T x - b_t\|^2$ are

$$\hat{x} = \arg\min_{y}\{\frac{1}{T}\sum_{t=1}^{T}\|a_t^T x - b_t\|^2 + \langle\frac{2}{T}\sum_{i=1}^{T}(a_t^T x - b_t)a_t, y - x\rangle + M\frac{1}{2}\|x - y\|^2 + \mu\|y\|_1\}$$

$$= \text{sign}(x - \frac{2}{MT}\sum_{i=1}^{T}(a_t^T x - b_t)a_t) \cdot \max\{\text{abs}\, x - \frac{2}{TM}\sum_{i=1}^{T}(a_t^T x - b_t)a_t - \frac{\mu}{M}, 0\}$$

and

$$\hat{x} = \arg\min_{y}\{\|a_t^T x - b_t\|^2 + \langle\, 2(a_t^T x - b_t)a_t, y - x\rangle + M\frac{1}{2}\|x - y\|^2 + \mu\|y\|_1\}$$

$$= \text{sign}(x - \frac{2}{M}(a_t^T x - b_t)a_t) \cdot \max\{\text{abs}\, x - \frac{2}{M}(a_t^T x - b_t)a_t - \frac{\mu}{M}, 0\}$$

respectively.

In online UDGM and SUG, we have

$$\begin{aligned}\phi_{t+1}(x) \quad &= \phi_t(x) + a_t[g_t(x_t) + \langle\nabla g_t(x_t), x - x_t\rangle + \mu\|x\|_1]\\ &= \xi(x_0, x) + \sum_{i=1}^{t} a_i[g_i(x_i) + \langle\nabla g_i(x_i), x - x_i\rangle + \mu\|x\|_1].\end{aligned}$$

Then we have

$$\begin{aligned}x_{t+1} &= \arg\min_{x}\phi_{t+1}(x) = \arg\min_{x}\{\frac{1}{2}\|x_0 - x\|^2 + \sum_{i=1}^{t} a_i[\langle\nabla g_i(x_i), x\rangle + \mu\|x\|_1]\}\\ &= \text{sign}(x_0 - \sum_{i=1}^{t} a_i\nabla g_i(x_i)) \cdot \max\{\text{abs}\, x_0 - \sum_{i=1}^{t} a_i\nabla g_i(x_i) - \mu\sum_{i=1}^{t} a_i, 0\}.\end{aligned}$$

2. Steiner Problem

In continuous Steiner problem we are given by centers $c_i \in \mathbb{R}^p$, $i = 1, ..., m$. It is necessary to find the optimal location of the service center x, which minimizes the total distance to all other centers. Thus, our problem is as follows:

$$\min_{x \in \mathbb{R}^p} g(x) := \frac{1}{m} \sum_{i=1}^{m} \|x - c_i\|,$$

where all norms in this problem are Euclidean. UGM solves that problem effectively. However, in real application, new locations will be added to the system, such as new shop opening or new warehouse establishing. Thus our online and stochastic gradient algorithms are needed.

Let $h(x) = 0$, $d(x) = \frac{1}{2}\|x\|^2$, then $\xi(x,y) = \frac{1}{2}\|x - y\|^2$. The subdifferential of the Euclidean norm $\|x\|$ is $\frac{x}{\|x\|}$ if $x \neq 0$ or $\{g\|\|x\| \leq 1\}$ if $x = 0$. In order to simplify the formula, we here denote $\nabla\|x\| = \frac{x}{\|x\|}$ instead distinguishing between $x = 0$ and $x \neq 0$.

The Bregman mapping associate with $\frac{1}{m}\sum_{i=1}^{m}\|x - c_i\|$ and the component function $\|x - c_i\|$ are

$$\hat{x} = \arg\min_y\{\frac{1}{m}\sum_{i=1}^{m}\|x - c_i\| + \langle\frac{1}{m}\sum_{i=1}^{m}\frac{x - c_i}{\|x - c_i\|}, y - x\rangle + M\frac{1}{2}\|x - y\|^2\}$$

$$= x - \frac{1}{mM}\sum_{i=1}^{m}\frac{x - c_i}{\|x - c_i\|}$$

and

$$\hat{x} = \arg\min_y\{\|x - c_i\| + \langle\frac{x - c_i}{\|x - c_i\|}, y - x\rangle + M\frac{1}{2}\|x - y\|^2\} = x - \frac{1}{M}\frac{x - c_i}{\|x - c_i\|}$$

respectively

In online UDGM and SUG for Steiner problem, we have

$$\phi_{t+1}(x) \quad = \phi_t(x) + a_t[g_t(x_t) + \langle\nabla g_t(x_t), x - x_t\rangle]$$
$$= \xi(x_0, x) + \sum_{i=1}^{t} a_i[g_i(x_i) + \langle\nabla g_i(x_i), x - x_i\rangle]$$

where $g_i(x_i) = \|x_i - c_i\|$ and $\nabla g_i(x_i) = \frac{x_i - c_i}{\|x_i - c_i\|}$. Thus we have

$$x_{t+1} \quad = \arg\min_x \phi_{t+1}(x) = \arg\min_x \frac{1}{2}\|x_0 - x\|^2 + \sum_{i=1}^{t} a_i\langle\frac{x_i - c_i}{\|x_i - c_i\|}, x\rangle$$
$$= \arg\min_x \frac{1}{2}\|x_0 - x\|^2 + \langle\sum_{i=1}^{t} a_i\frac{x_i - c_i}{\|x_i - c_i\|}, x\rangle = x_0 - \sum_{i=1}^{t} a_i\frac{x_i - c_i}{\|x_i - c_i\|}.$$

References

1. Beck, A., Teboulle, M.: Mirror descent and nonlinear projected subgradient methods for convex optimization. Operations Research Letters **31**(3), 167–175 (2003)
2. Duchi, J., Shalev-Shwartz, S., Singer, Y., Tewari, A.: Composite objective mirror descent (2010)
3. Duchi, J.C., Agarwal, A., Wainwright, M.J.: Dual averaging for distributed optimization. In: 2012 50th Annual Allerton Conference on Communication, Control, and Computing (Allerton), pp. 1564–1565. IEEE (2012)
4. Gabay, D., Mercier, B.: A dual algorithm for the solution of nonlinear variational problems via finite element approximation. Computers & Mathematics with Applications **2**(1), 17–40 (1976)
5. Mairal, J.: Optimization with first-order surrogate functions. arXiv preprint arXiv:1305.3120 (2013)
6. Nesterov, Y.: Efficiency of coordinate descent methods on huge-scale optimization problems. SIAM Journal on Optimization **22**(2), 341–362 (2012)

7. Nesterov, Y.: Primal-dual subgradient methods for convex problems. Mathematical Programming **120**(1), 221–259 (2009)
8. Nesterov, Y.: Universal gradient methods for convex optimization problems. CORE (2013)
9. Schmidt, M., Roux, N.L., Bach, F.: Minimizing finite sums with the stochastic average gradient. arXiv preprint arXiv:1309.2388 (2013)
10. Shalev-Shwartz, S., Zhang, T.: Proximal stochastic dual coordinate ascent. arXiv preprint arXiv:1211.2717 (2012)
11. Shi, Z., Han, J., Zheng, T., Deng, S.: Audio segment classification using online learning based tensor representation feature discrimination. IEEE transactions on audio, speech, and language processing **21**(1–2), 186–196 (2013)
12. Suzuki, T.: Dual averaging and proximal gradient descent for online alternating direction multiplier method. In: Proceedings of ICML 2013, pp. 392–400 (2013)
13. Wang, H., Banerjee, A.: Online alternating direction method. arXiv preprint arXiv:1206.6448 (2012)
14. Xiao, L.: Dual averaging methods for regularized stochastic learning and online optimization. The Journal of Machine Learning Research **11**, 2543–2596 (2010)
15. Xiao, L., Zhang, T.: A proximal stochastic gradient method with progressive variance reduction. arXiv preprint arXiv:1403.4699 (2014)
16. Zinkevich, M.: Online convex programming and generalized infinitesimal gradient ascent. In: Proceedings of ICML 2003 (2003)

Context-Aware Detection of Sneaky Vandalism on Wikipedia Across Multiple Languages

Khoi-Nguyen Tran[1]($\boxtimes$), Peter Christen[1], Scott Sanner[2], and Lexing Xie[1]

[1] Research School of Computer Science, The Australian National University,
Canberra, Australia
khoi-nguyen.tran@anu.edu.au
[2] Machine Learning Group, NICTA, Canberra, ACT 2601, Australia

Abstract. The malicious modification of articles, termed vandalism, is a serious problem for open access encyclopedias such as Wikipedia. Wikipedia's counter-vandalism bots and past vandalism detection research have greatly reduced the exposure and damage of common and obvious types of vandalism. However, there remains increasingly more sneaky types of vandalism that are clearly out of context of the sentence or article. In this paper, we propose a novel context-aware and cross-language vandalism detection technique that scales to the size of the full Wikipedia and extends the types of vandalism detectable beyond past feature-based approaches. Our technique uses word dependencies to identify vandal words in sentences by combining part-of-speech tagging with a conditional random fields classifier. We evaluate our technique on two Wikipedia data sets: the PAN data sets with over 62,000 edits, commonly used by related research; and our own vandalism repairs data sets with over 500 million edits of over 9 million articles from five languages. As a comparison, we implement a feature-based classifier to analyse the quality of each classification technique and the trade-offs of each type of classifier. Our results show how context-aware detection techniques can become a new counter-vandalism tool for Wikipedia that complements current feature-based techniques.

1 Introduction

Wikipedia is the largest free and open access online encyclopedia that attracts tens of thousands volunteer editors[1] and tens of millions of article views every day[2] [19,20]. The open nature of Wikipedia also facilitates many types of vandals that deliberately make malicious edits, such as changing facts, inserting obscenities, or deleting text. To combat vandalism, editors repair vandalised articles with an edit that removes the vandalised text or with a revert back to a previous revision, and commonly leave a comment indicating a repair. Wikipedia distinguishes many types of vandalism on its policy articles[3] and provides best practice guides to counter vandalism.

---

[1] http://stats.wikimedia.org/EN/TablesWikipediansEditsGt5.htm
[2] http://stats.wikimedia.org/EN/TablesPageViewsMonthly.htm
[3] http://en.wikipedia.org/wiki/Wikipedia:Vandalism

© Springer International Publishing Switzerland 2015
T. Cao et al. (Eds.): PAKDD 2015, Part I, LNAI 9077, pp. 380–391, 2015.
DOI: 10.1007/978-3-319-18038-0_30

The introduction and prevalence of counter-vandalism bots since 2006 [7] have reduced the exposure time of vandalism and the extra work needed by editors to repair vandalism [8,11]. Vandalism detection research has introduced new techniques that improve the detection rate. These techniques often focus on developing features as input to machine learning algorithms [10,22,23]. A variety of features based on the metadata, editor characteristics, article structure, and content of Wikipedia articles have shown to be effective in distinguishing normal revisions and revisions containing vandalism [19,20]. As new vandalism detection techniques are integrated into counter-vandalism bots on Wikipedia, vandalism of article content continues to become more sophisticated to avoid detection.

Wikipedia defines sneaky vandalism[3] as difficult to find, where the vandal may be using concealment techniques such as pretending to revert vandalism while introducing vandalism, or subtle changes in the article text that aim to deceive other editors to be legitimate changes. Subtle changes can be identified as vandalism because they may break the consistency of text used in other articles or past revisions, deviate from common or correct grammatical structure, introduce uncommon word patterns, or change the meaning of a sentence. Text features used in vandalism research do not inherently capture the context of the sentences being edited as they do not consider word dependencies [16].

In this paper, we propose a novel vandalism detection technique that is context-aware by considering word dependencies. Our technique focuses on a particular type of sneaky vandalism, where vandals make sophisticated modifications of text that change the meaning of a sentence without obvious markers of vandalism. We use a part-of-speech (POS) tagger [17] to tag types of words in sentences changed in each edit, and conditional random fields (CRF) [12,13] to model dependencies between tags to identify vandalised text.

We hypothesise that sneaky vandalism is out of context of sentences on Wikipedia, but seem normal with respect to the text features used in vandalism detection research. We evaluate our technique on the PAN data sets with over 62,000 edits, commonly used by related research; and the full vandalism repairs data sets with over 500 million edits of over 9 million articles from five languages: English, German, Spanish, French, Russian. As a comparison, we implement a feature engineering classifier, and analyse both classification results and the trade-offs of each type of classifier. Our results show how context-aware detection techniques can become a new state-of-the-art counter-vandalism tool for Wikipedia that complements current feature engineering based techniques.

Our contributions are (1) developing a novel context-aware vandalism detection technique; (2) demonstrating how our technique is scalable to the entire Wikipedia data set; (3) demonstrating the cross language application of classification models and the relationships between the languages considered; (4) replicating our experiments on the smaller PAN data sets often used in related work; and (5) demonstrating how our technique differs and contributes to traditional feature engineering approaches. These contributions backed by our results show how context-aware detection techniques can become a new counter-vandalism tool for Wikipedia that complements current feature-based techniques.

2   Related Work

The interpretation of vandalism differs amongst Wikipedia users, which can lead to incomplete or inconsistent labelling of vandalised revisions. [15] developed two corpora by crowd-sourcing votes on whether a Wikipedia revision contains vandalism using Amazon's Mechanical Turk. The PAN workshops in 2010 and 2011 held competitions to encourage development of machine learning based vandalism detection techniques.

For the PAN 2010 data set, Mola-Velasco [14] uses a set of 21 features to detect vandalism, which resulted in a first place ranking at the PAN 2010 competition. Adler et al. [2] improve on this winning entry by adding metadata, text, user reputation, and language features, totalling 37 features. Javanmardi et al. [10] further improve the classification results by introducing 66 features and applying feature reduction. For the PAN 2011 data sets, West et al. [23] develop 65 features that include many of the features from the entries from the PAN 2010 competition. The PAN data sets continue to be used to evaluate vandalism detection techniques after the workshops were held, with other types of features, such as syntactic and semantic features [21], statistical models of words and editor actions [5], or styles of words [9].

Other vandalism techniques used their own data sets constructed from sampled articles and revisions, or from a smaller Wikipedia [4,22].

Two vandalism detection techniques that are most similar to our work look at the relationship of words over time, and co-occurrence of pairs of words. Wu et al. [24] present a text-stability approach to find increasingly sophisticated vandalism. This technique builds on ideas presented in Adler et al. [1] on the longevity of words over time to determine the probability that parts of an article will be modified by a normal or a vandal edit. Ramaswamy et al. [16] propose two metrics that measure the likelihood of words contributed in an edit of a Wikipedia article belonging to that article with respect to the article's content and topic. The numerous words and word pairs resulting the data processing mean both techniques could only be evaluated using articles sampled from the PAN 2010 data set. Our work presents a feasible approach to context-aware vandalism detection with demonstrative evaluation on the full Wikipedia vandalism repairs data sets and all PAN data sets.

Overall, a variety of vandalism detection techniques has been developed and evaluated on different data sets, where many techniques are now evaluated on the PAN data sets. We show in our work that one of the many problems with using small data sets (the PAN data sets contain only around 2,000 vandalised edits) is that there are insufficient numbers of vandalism cases available for our classifiers – both context-aware and feature engineering – to effectively distinguish vandalism. Many features presented in related work show good classification performance on the PAN data sets, but they need to be evaluated on the full Wikipedia data set to truly gauge their effectiveness in distinguishing vandalism. Furthermore, while counter-vandalism bots have a strong presence on Wikipedia since 2006 [3,7] – especially in the English Wikipedia – they are not well represented in the PAN data sets.

Table 1. Number of edits and sentences in different Wikipedia languages, split by type. "all" means combining or union of all data sets.

Data Set		Edits		Sentences	
		Normal	Vandal Repairs	Normal	Vandal Repairs
Wiki	en	256,796,879 (98.4%)	4,909,181 (1.9%)	1,642,267,638 (96.6%)	58,183,825 (3.4%)
	de	52,895,509 (99.7%)	164,097 (0.3%)	370,010,973 (99.5%)	1,805,862 (0.5%)
	es	31,742,769 (99.0%)	330,135 (1.0%)	161,871,444 (98.9%)	1,879,431 (1.1%)
	fr	41,657,071 (99.5%)	189,849 (0.5%)	248,064,661 (99.3%)	1,671,695 (0.7%)
	ru	24,335,713 (99.8%)	39,234 (0.2%)	202,672,387 (99.6%)	747,854 (0.4%)
	all	407,427,941 (98.6%)	5,632,496 (1.4%)	2,624,887,103 (97.6%)	64,288,667 (2.4%)
Data Set		Normal	Vandal Cases	Normal	Vandal Cases
PAN	2010 en	23,025 (92.7%)	1,804 (7.3%)	236,721 (96.4%)	8,967 (3.6%)
	2011 en	6,876 (89.1%)	844 (10.9%)	82,256 (94.9%)	4,396 (5.1%)
	2011 de	7,359 (95.1%)	381 (4.9%)	80,308 (98.7%)	1,085 (1.3%)
	2011 es	6,922 (89.7%)	792 (10.3%)	42,998 (85.3%)	7,418 (14.7%)
	2011 all	21,157 (91.3%)	2,017 (8.7%)	205,562 (94.1%)	12,899 (5.9%)

3 Wikipedia Data Sets

We downloaded the first Wikipedia data dump available in 2013 and use all revisions of encyclopedic articles from 2001 to December 31st 2012 (our cut-off date) for the five languages English (en), German (de), French (fr), Spanish (es), and Russian (ru). When vandalism is discovered and repaired, the editor usually leaves a comment in the repaired revision with keywords indicating a repair of vandalism, such as "rvv" (revert due to vandalism), "vandalism", "...rv...vandal...", and analogues in the other languages.

As we are interested in sneaky vandalism introduced in edits, we can reduce the size of the revision content by using the Python unified diff[4] algorithm to obtain only the sentences (marked by a period) that were changed by an edit. We reason that changes within existing sentences are more difficult to find than additions or removals of text that are relatively easier types of vandalism to detect. For each sentence changed, we perform a sentence diff (subtracting common words) to obtain the words that were repaired in the vandalism case, and label each word with 'n' (normal) or 'v' (vandal).

Table 1 shows the number of edits and sentences obtained from our data processing (named 'Wiki') for the full Wikipedia, and the PAN data sets. We map these sentences to their edits to manually verify correctness, and compare classification results with a text-feature based detection technique. We find approximately 1.9% of all edits on the English encyclopedic articles are repairs of vandalism, which is consistent with results from Kittur et al. [11]. The PAN data sets show a higher percentage of vandalism because they estimate *all* vandal edits, whereas we are interested only in edits that repair vandalism.

To illustrate our data set, sneaky vandalism, and our detection technique, we present a running example in Fig. 1 that continues in Figs. 2 and 3.

---

[4] http://docs.python.org/2/library/difflib.html

We present a fictitious example sentence[a] with sneaky vandalism to illustrate our tagging and classification technique in the following sections:
- Repaired: Bread crust has been shown to **have more dietary fibers and** antioxidants.
- Vandalised (word label): Bread (n) crust (n) has (n) been (n) shown (n) to (n) **make** (v) **hair** (v) **curlier** (v) **because** (v) **of** (v) antioxidants (n).

The bolded words are changed words in the sentence diff that are identified as vandalised (v) or normal (n) from comparing the repaired and vandalised revisions. In the later examples, labels and tags are accumulated for each word are contained in the parentheses.

---

[a] Adapted from a vandalised revision of http://en.wikipedia.org/wiki/Bread.

Fig. 1. POS labelling example

4 Part-of-Speech Tagging

We process the labelled sentences further and tag each word with descriptive information that allows our context-aware classifier to exploit contextual information. We use part-of-speech (POS) tags provided by the TreeTagger[5] software, where the aim is to place words from a text corpus into text categories [17]. Tree-Tagger uses binary decision trees to estimate the transition probabilities of POS tags and select the most appropriate tag from the available training data. For each sentence in our data sets, a POS tagger analyses known words (trained from a large manually labelled corpus) and assigns each word the most probable tag that describes it. In sneaky vandalism cases on Wikipedia, small changes can alter the meaning of sentences while not disrupting the correctness of text patterns in words (spelling) or sentences (grammar).

Our example in Fig. 1 illustrates this sneaky vandalism case, where in Fig. 2, we show the output of the tagging by TreeTagger. We describe only the tags relevant to our example from the full English tag set documentation[5]: coordinating conjunction (CC), preposition or conjunction (IN), adjective (JJ), adjective - comparative (JJR), noun (NN), noun - plural (NNS), to (TO), verb - base form (VB), verb - past participle (VBN), verb - 3rd person (VBZ). We train the CRF classifier on these tag sequences to predict the sequence of labels.

Continuing our example from Fig. 1, we have tags generated by TreeTagger as:
- Repaired (tag, word label): Bread (NN, n) crust (NN, n) has (VBZ, n) been (VBN, n) shown (VBN, n) to (TO, n) **have** (VB, n) **more** (JJR, n) **dietary** (JJ, n) **fibers** (NNS, n) **and** (CC, n) antioxidants (NNS, n).
- Vandalised (tag, word label): Bread (NN, n) crust (NN, n) has (VBZ, n) been (VBN, n) shown (VBN, n) to (TO, n) **make** (VB, v) **hair** (NN, v) **curlier** (JJR, v) **because** (IN, v) **of** (IN, v) antioxidants (NNS, n).

The parentheses contain the accumulated labels and tags for each word that are to be used in the CRF classifier.

Fig. 2. TreeTagger tagging example

---

[5] http://www.cis.uni-muenchen.de/~schmid/tools/TreeTagger/

5   Context-Aware Vandalism Detection

Context-aware detection techniques are needed because some types of vandalism cannot be easily detected with feature engineering approaches [16]. Our running example illustrates a case of potential vandalism that would likely require a human editor to repair, because there are no clear markers of vandalism such as vulgarities, odd letter patterns in words, or radical changes to text.

Our vandalism detection technique uses conditional random fields (CRF) [13], a probabilistic undirected graphical model for segmenting and labelling sequence data. The full development and derivation of CRF are given by Lafferty et al. [13], and additional models and discussion by Sutton and McCallum [18].

From our processed data, we have for each sequence of words $\mathbf{s}$ (i.e. a sentence) and its word labels $\mathbf{l} = (l_1, l_2, ..., l_n)$ (i.e. n or v) and word tags $\mathbf{t} = (t_1, t_2, ..., t_n)$ (given by the POS tagger). To exploit the contextual information of the sequence of word tags, we define three binary feature functions f_j, g_j, and h_j – on the training data sets – for three separate experiments:

$$f_j(l_k, \mathbf{t}), \quad g_j(l_{k-1}, l_k, l_{k+1}, \mathbf{t}), \quad h_j(l_{k-2}, l_{k-1}, l_k, l_{k+1}, l_{k+2}, \mathbf{t}), \quad 1 \le k \le n \quad (1)$$

The feature functions f_j, g_j, and h_j return 1 when certain conditions – as learnt from the data set and explained below – are met, and 0 otherwise. This means for each tag, we define features that express some characteristics of the model only with its current label (f_j), with the labels of the two adjacent tags (g_j), or the four (two on each side) adjacent tags (h_j). We choose these number of adjacent tags to explore the benefits of context to detecting vandalised words.

For each feature function, such as f_j, we assign weights θ_j that are also learnt from the training data sets through maximum likelihood estimation. This creates a language model for each word from the surrounding words. Now, we can score a labelling $\mathbf{l}$ of tags $\mathbf{t}$ by summing the weighted features for each tag:

$$\text{sum}_k(\mathbf{l}|\mathbf{t}) = \sum_{j=1}^{m} \theta_j f_j(l_k, \mathbf{t}) \quad (2)$$

Note that feature function f_j can be interchanged with g_j or h_j, with the appropriate function parameters. Then we transform the scores into probabilities similar to the joint distribution of HMMs [18]:

$$p(\mathbf{l}, \mathbf{t}) = \frac{1}{Z} \prod_{k=1}^{K} exp\{\text{sum}_k(\mathbf{l}, \mathbf{t})\} \quad (3)$$

where Z is a normalisation constant to keep $p(\mathbf{l}, \mathbf{t})$ between 0 and 1, which is cancelled in the fraction of the next step below.

Finally, we have the conditional probability that models the conditional distribution as a linear-chain CRF [18]:

$$p(\mathbf{l}|\mathbf{t}) = \frac{p(\mathbf{l}, \mathbf{t})}{\sum_l p(\mathbf{l}, \mathbf{t})} \quad (4)$$

The training phase above gives us a model of the many sentences in each Wikipedia data set. To predict the labels (n or v) of a new input set of tags $\mathbf{t}$ (e.g. POS) extracted from an unseen sentence, we compute:

$$\mathbf{l}^* = \text{argmax}_{\mathbf{l}}\ p(\mathbf{l}|\mathbf{t}) \tag{5}$$

which gives us the predicted tags (e.g. POS), which are combined with the true labels, POS tags, and words of the sentence.

An advantage to using CRF in our application is the diversity of word labels that allow immediate identification of vandalised words for evidence or manual verification. A disadvantage of CRF is the potential slow convergence of training models when the feature functions are complex or have strong dependencies [18].

We use an open source implementation of CRF by Kudo [12], named CRF++, to evaluate our vandalism detection technique. We process our data further as required by CRF++ and recover classification results of test sentences for each edit for further evaluation. Our resulting testing data sets resemble our example below in Fig. 3, where we can now evaluate classification performance.

This final example continues from our example in Fig. 2. Assuming we have trained the CRF classifier on sentences, then we may have an optimal classification labelling of our vandalised sentence as:
- Vandalised (tag, word label, predicted label): Bread (NN, n, n) crust (NN, n, n) has (VBZ, n, n) been (VBN, n, n) shown (VBN, n, n) to (TO, n, n) make (VB, v, v) hair (NN, v, v) curlier (JJR, v, v) because (IN, v, n) of (IN, v, n) antioxidants (NNS, n, n).
The predicted labels are n and v, and the correct labelled vandal words are in bold text and coloured as green for a correct label and red for incorrect label.
The implications of these mislabellings are that they may be common phrases (as shown above), or incorrect patterns that need to be manually readjusted.

Fig. 3. CRF classification example

---

6   Results

We split each data set by the number of edits for 10-fold cross-validation. We perform sampling for the Wikipedia repairs data sets with different ratios of normal edits to vandal repair edits to investigate the effects of class imbalance and data sampling for context-aware classification techniques. For example, "2-to-1" means 2 normal edits for every 1 vandal repair edit.

We present our classification results compactly by plotting the area under the precision-recall (PR) curve (AUC-PR) against the area under the receiver-operator characteristic (ROC) curve (AUC-ROC) [6]. The AUC-PR score gives the probability that a classifier will correctly identify a randomly selected positive sample (e.g. vandalism) as being positive. The AUC-ROC score gives the probability that a classifier will correctly identify a randomly selected (positive or negative) sample. Both scores range from 0 to 1, where a score of 1 means 100% or complete correctness in labelling all samples considered by the measures.

6.1   CRF with POS Tags

The CRF classifier in our first set of results is trained and tested on the same source and target language, or named as "within" language classification. CRF

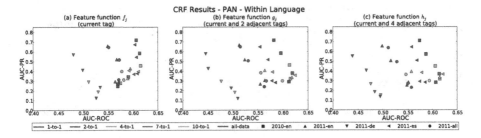

Fig. 4. CRF results for classification within the same language on the PAN data sets. Upper right is better.

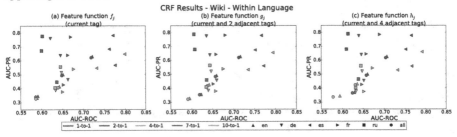

Fig. 5. CRF results for classification within the same language on the Wikipedia vandalism repairs data sets. Upper right is better.

classification results for the PAN data sets are presented in Fig. 4 and for the Wikipedia vandalism repairs data sets in Fig. 5.

The CRF classification results for the PAN data sets in Fig. 4 generally show consistent AUC-ROC scores for each data set. The 2010 English data set (2010-en) shows consistently high results for both AUC-PR and AUC-ROC scores compared to the 2011 data sets. Combining all 2011 data sets ("all") shows an average of the results for each 2011 data set.

The results for the Wikipedia data sets in Fig. 5 show significantly higher AUC-PR and AUC-ROC scores than the PAN data sets for each ratio of sampled data sets. Non-English Wikipedias have much higher scores than the English Wikipedia, suggesting vandalism in non-English Wikipedias more often break sentence structure detectable through changes in the sequence of POS tags. The different feature functions show minor improvements to AUC-PR and AUC-ROC classification scores, similar to the PAN data sets. Combining all data sets ("all") shows scores highly similar to the English (en) results because of the overwhelming number of English vandalism cases as seen from Table 1.

6.2   Reusing Models Across Languages

We investigate the cross-language performance of our context-aware technique, where Wikipedia vandalism detection models are trained on one language and reused to classify on other languages. The definition of CRF does not include a

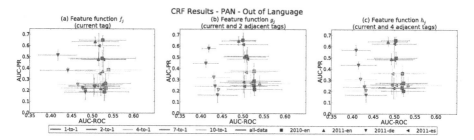

Fig. 6. CRF results with one standard deviation for out of language classification on the PAN data sets. Upper right is better.

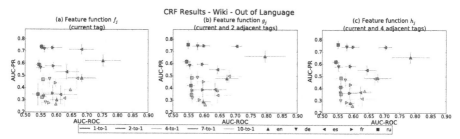

Fig. 7. CRF results with one standard deviation for out of language classification on the Wikipedia vandalism repairs data sets. Upper right is better.

model for the probability of tags $p(\mathbf{t})^6$, which makes CRF suitable for classifying unseen tags [18].

For a target language, we reuse the CRF models trained in other languages. For example, for the English (en) target language, we reuse the German (de), Spanish (es), French (fr), and Russian (ru) models, and report the average and one standard deviation of these classification scores. Our results are in Fig. 6 for the PAN data sets, and in Fig. 7 for the Wikipedia data sets.

The PAN data sets show lower classification scores compared to classification within the same language. The range of scores varies widely, especially for the AUC-ROC scores. Reusing CRF models trained on small data sets (e.g. German (de)) does not provide any significant benefits as observed by a lower convergence of average scores and clusters of results for the sampling ratios.

The Wikipedia data sets show higher classification scores compared to the PAN data sets, similar to within language classification. The feature functions with more adjacent tags also reduce the variance in the standard deviation, similarly to the PAN data sets, and especially for AUC-PR scores. This suggests the CRF classifier is more precise in classifying vandalism cases when it has contextual awareness of other tags. The non-English CRF models may be identifying sneaky vandalism that is lost within the English CRF model because of the large size difference in the training data sets.

---

[6] From the joint distribution of HMMs, which is often difficult to model because $p(\mathbf{t})$ may contain highly dependent features [18].

Table 2. Features for feature engineering vandalism detection. Features P01 to P12 are from winning entries from the PAN workshop competitions [2,10,14,23]. Features F01 to F12 are our contributions from previous work [20].

Feature	Description	Feature	Description
P01-PW	Pronoun words	F01-NWD	Number of unique words
P02-VW	Vulgar words	F02-TWD	Number of all words
P03-SW	Slang words	F03-UL	Highest ratio of upper to lower case letters
P04-CW	Capitalised words	F04-UA	Highest ratio of upper case to all letters
P05-UW	Uppercase words	F05-DA	Highest ratio of digit to all letters
P06-DW	Digit words	F06-NAN	Highest ratios of non-alphanumeric letters to all letters
P07-ABW	Alphabetic words	F07-CD	Lowest character diversity
P08-ANW	Alphanumeric words	F08-LRC	Length of longest repeated character
P09-SL	Single letters	F09-ZLIB	Lowest compression ratio, zlib compressor
P10-SD	Single digits	F10-BZ2	Lowest compression ratio, bz2 compressor
P11-SC	Single characters	F11-WL	Longest unique word
P12-LZW	Lowest compression ratio with lzw compressor	F12-WS	Sum of unique word lengths

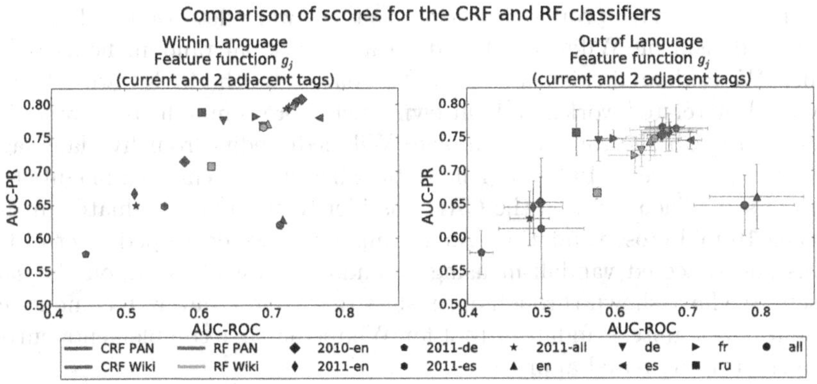

Fig. 8. Comparison of scores for the CRF and Random Forest (RF) classifiers

6.3 Comparing to Feature Classification

As a comparison to our context-aware technique, we implement a feature engineering based classifier with features in Table 2 following our previous work [20] and similar to related work [2,10,14,23]. We select a relevant subset of features from winning entries of the PAN workshop competitions (features P01-PW to P12-LZW), and contribute our own subset of features (features F01-NWD to F12-WS). We follow our previous work by extracting these features from the data sets in Sect. 3, and use 10-fold cross-validation with the same Random Forest (RF) classifier[7] that was shown to be the most robust and generally best performing classifier. We present our comparison plots for the 1-to-1 data sampling ratio in Fig. 8 for within language classification and for out of language classification.

For within language classification, the RF classifier has strong classification results for both PAN and Wikipedia data sets. For the PAN data sets, the RF

[7] http://scikit-learn.org

classifier performs consistently well, as expected from related work [2,10,14, 19,23]. The tight cluster of RF PAN results (Fig. 8) suggests the features are language independent and have strong performance. The RF classifier on the full Wikipedia data sets shows similar strong classification performance. The CRF and RF Wikipedia results show trade-offs in AUC-PR and AUC-ROC scores.

For out of language classification, we see a tight cluster of RF results for both the PAN and Wikipedia data sets (Fig. 8). This is expected as within language classification shows similar classification scores. Interestingly, the CRF and RF Wikipedia scores for the English (en) and "all" data set have almost opposite AUC-PR and AUC-ROC scores. This shows a trade-off in precision (P) and FPR when using each classifier. The CRF classifier has higher TPR and FPR scores instead of the higher precision (P) scores of the RF classifier.

7   Conclusion

In this paper, we have proposed a novel context-aware detection technique for sneaky vandalism on Wikipedia based on a conditional random fields (CRF) classifier. We evaluated this classifier on two data sets, the PAN data sets commonly used by related works, and our own much more comprehensive vandalism repairs data set built from the complete Wikipedia edits from five languages. We used part-of-speech (POS) tagging to tag all sentences changed in edits from both data sets. Then we used the CRF classifier to train and evaluate our data sets using 10-fold cross-validation. As a comparison, we developed a set of text features and detected vandalism using a random forest classifier on the same data sets. We have shown through our results that context-aware techniques can become a new counter-vandalism tool for Wikipedia that complements current feature engineering based approaches.

In future work, we aim to develop a language independent tag set that uses information from feature engineering approaches. Our working set of languages contains some shared POS tags, where we can unify these tags into higher level word tags that have direct mappings across languages, such as nouns, pronouns, verbs, adverbs, and adjectives. We plan to extend our linear-chain CRF to a general CRF that allows modelling of dependencies between articles, where vandals may also target adjacent internally linked articles. Our proposed novel context-aware vandalism detection technique is an exploratory step towards more complex detection techniques for progressively sneakier text vandalism on Wikipedia.

References

1. Adler, B.T., de Alfaro, L.: A content-driven reputation system for the wikipedia. In: WWW, pp. 261–270. Banff, Canada (2007)
2. Adler, B.T., de Alfaro, L., Mola-Velasco, S.M., Rosso, P., West, A.G.: Wikipedia vandalism detection: combining natural language, metadata, and reputation features. In: Gelbukh, A. (ed.) CICLing 2011, Part II. LNCS, vol. 6609, pp. 277–288. Springer, Heidelberg (2011)

3. Adler, B.T., de Alfaro, L., Pye, I., Raman, V.: Measuring author contributions to the wikipedia. In: WikiSym, pp. 15–24. Porto, Portugal (2008)
4. Chin, S.C., Street, W.N.: Divide and Transfer: an Exploration of Segmented Transfer to Detect Wikipedia Vandalism. JMLR **27**, 133–144 (2012)
5. Chin, S.C., Street, W.N., Srinivasan, P., Eichmann, D.: Detecting wikipedia vandalism with active learning and statistical language models. In: WICOW, pp. 3–10. Raleigh, NC (2010)
6. Davis, J., Goadrich, M.: The relationship between precision-recall and ROC curves. In: ICML, pp. 233–240. Pittsburgh, PA (2006)
7. Geiger, R.S.: The lives of bots. In: Critical Point of View: A Wikipedia Reader, pp. 78–93. Institute of Network Cultures, Amsterdam (2011)
8. Halfaker, A., Riedl, J.: Bots and Cyborgs: Wikipedia's Immune System. Computer **45**, 79–82 (2012)
9. Harpalani, M., Hart, M., Singh, S., Johnson, R., Choi, Y.: Language of vandalism: improving wikipedia vandalism detection via stylometric analysis. In: ACL: Short Papers, pp. 83–88. Portland, Oregon (2011)
10. Javanmardi, S., McDonald, D.W., Lopes, C.V.: Vandalism detection in wikipedia: a high-performing, feature-rich model and its reduction through lasso. In: WikiSym, pp. 82–90. Mountain View, California (2011)
11. Kittur, A., Suh, B., Pendleton, B.A., Chi, E.H.: He says, she says: conflict and coordination in wikipedia. In: CHI, Vancouver, BC, Canada, pp. 453–462 (2007)
12. Kudo, T.: CRF++: Yet Another CRF toolkit (2013)
13. Lafferty, J.D., McCallum, A., Pereira, F.C.N.: Conditional random fields: probabilistic models for segmenting and labeling sequence data. In: ICML, pp. 282–289. Williams College, MA (2001)
14. Mola-Velasco, S.M.: Wikipedia vandalism detection through machine learning: feature review and new proposals. In: CLEF. Padua, Italy (2010)
15. Potthast, M.: Crowdsourcing a wikipedia vandalism corpus. In: SIGIR, Geneva, Switzerland, pp. 789–790 (2010)
16. Ramaswamy, L., Tummalapenta, R.S., Li, K., Pu, C.: A content-context-centric approach for detecting vandalism in wikipedia. In: Collaboratecom, pp. 115–122. Austin, TX (2013)
17. Schmid, H.: Probabilistic part-of-speech tagging using decision trees. In: NeMLaP, Manchester, UK, pp. 44–49 (1994)
18. Sutton, C., McCallum, A.: An Introduction to Conditional Random Fields. Machine Learning **4**(4), 267–373 (2011)
19. Tran, K.-N., Christen, P.: Cross language prediction of vandalism on wikipedia using article views and revisions. In: Pei, J., Tseng, V.S., Cao, L., Motoda, H., Xu, G. (eds.) PAKDD 2013, Part II. LNCS, vol. 7819, pp. 268–279. Springer, Heidelberg (2013)
20. Tran, K.N., Christen, P.: Cross-Language Learning from Bots and Users to Detect Vandalism on Wikipedia. IEEE TKDE (2015)
21. Wang, W.Y., McKeown, K.R.: "Got You!": automatic vandalism detection in wikipedia with web-based shallow syntactic-semantic modeling. In: Coling, Beijing, China, pp. 1146–1154 (2010)
22. West, A.G., Kannan, S., Lee, I.: Detecting wikipedia vandalism via spatio-temporal analysis of revision metadata. In: EUROSEC, Paris, France, pp. 22–28 (2010)
23. West, A.G., Lee, I.: Multilingual vandalism detection using language-independent & ex post facto evidence. In: CLEF, Amsterdam, Netherlands (2011)
24. Wu, Q., Irani, D., Pu, C., Ramaswamy, L.: Elusive vandalism detection in wikipedia: a text stability-based approach. In: CIKM, Toronto, Canada, pp. 1797–1800 (2010)

Uncovering the Latent Structures
of Crowd Labeling

Tian Tian and Jun Zhu[✉]

State Key Lab of Intelligent Technology and Systems,
Tsinghua National Lab for Information Science and Technology,
Department of Computer Science and Technology, Tsinghua University,
Beijing 100084, China
`tiant13@mails.tsinghua.edu.cn, dcszj@mail.tsinghua.edu.cn`

Abstract. Crowdsourcing provides a new way to distribute enormous tasks to a crowd of annotators. The divergent knowledge background and personal preferences of crowd annotators lead to noisy (or even inconsistent) answers to a same question. However, diverse labels provide us information about the underlying structures of tasks and annotators. This paper proposes latent-class assumptions for learning-from-crowds models, that is, items can be separated into several latent classes and workers' annotating behaviors may differ among different classes. We propose a nonparametric model to uncover the latent classes, and also extend the state-of-the-art minimax entropy estimator to learn latent structures. Experimental results on both synthetic data and real data collected from Amazon Mechanical Turk demonstrate our methods can disclose interesting and meaningful latent structures, and incorporating latent class structures can also bring significant improvements on ground truth label recovery for difficult tasks.

1 Introduction

Researches and applications in the field of artificial intelligence are relying more and more on large-scale datasets as the age of Big-data comes. Conventionally, labels of tasks are collected from domain experts, which is expensive and time-consuming. Recently, online distributed working platforms, such as Amazon Mechanical Turk (MTurk) , provide a new way to distribute enormous tasks to a crowd of workers [1]. Each worker only needs to finish a small part of the entire task in this crowd labeling mode, so that the tasks can be done faster and cheaper. However, the labels given by the crowd annotators are less accurate than those given by experts. In order to well recover the true labels, multiple annotators are usually needed to evaluate every micro task. Furthermore, different annotators may have different backgrounds and personal preferences, and they may give inconsistent answers to a same question. This phenomenon brings us more difficulties to recover ground truth labels from noisy answers and raises a research topic in the crowdsourcing area.

On the other hand, the diverse labels can provide us with a lot of additional information for both data characteristics and people's behaviors [2]. For example, they may reflect some latent structures of the complicated data, such as the

T. Cao et al. (Eds.): PAKDD 2015, Part I, LNAI 9077, pp. 392–404, 2015.
DOI: 10.1007/978-3-319-18038-0_31

grouping structure of tasks according to their difficulty levels and/or the grouping structure of annotators according to their similar education background or preferences. In the perspective of psychology, users' labels actually show their understanding of the given tasks. For example, in a problem of classifying flowers in pictures, users' choices may be influenced by many different features, such as petal color, petal shape, background, size in the picture, etc; and personal choices of different users are influenced by users' tastes. These features are usually unknown. Some features are significantly related to the flower species and some features are not. So we think the observed user labels are generated from tasks' latent structures and annotators' abilities, but not directly from the truth category. By exploring these latent structures, we can have a better understanding of the data, and may also accomplish tasks like category recovery better.

Dawid and Skene's work [3] is a milestone in learning from crowds. They proposed an annotator-specific confusion matrix model, which is able to estimate the ground truth category well. Raykar et al. [4] extended Dawid and Skene's model by ways, such as taking item features into account or modifying the output model to fit regression or ranking tasks. Zhou et al. [5,6] proposed a minimax entropy estimator, which outperforms most previous models in category estimating accuracy, and later on they extended their model to handle ordinal labels. However, none of these models have taken latent structures into account. We extend some of them to learn latent structures from dataset. Welinder et al. [7] proposed a multidimensional annotation model, which was the earliest to consider latent structure in this field. But this model often suffers from overfitting and so performs averagely on many tasks [8]. Tian and Zhu [9] also proposed an idea on the latent structure for crowdsourcing but aimed at a different problem; our work draws some inspiration from their nonparametric ideas.

We propose two latent-class assumptions for learning from crowds: **(I)** each item belongs to one latent class, and annotators have a consistent view on items of the same class but maybe inconsistent views on items of different classes; and **(II)** several different latent classes consist in one label category. To recover the latent-class structures, we propose a latent class estimator using a nonparametric prior. We also extend the minimax entropy estimator to fine tune such latent class structures. Under the latent class assumptions, the estimators remain compact through parameter sharing. The experimental results on both synthetic and real MTurk datasets demonstrate our methods can disclose interesting and meaningful latent structures, and incorporating latent class structures can bring significant improvements on ground truth label recovery for difficult tasks. We summarize our contributions as: **(1)** We propose the latent-class assumptions for crowdsourcing tasks. **(2)** We develop appropriate nonparametric algorithms for learning latent-class structures, and extend previous minimax entropy principle. **(3)** We present an algorithm to recover category labels from latent classes, and empirically demonstrate its efficiency.

The rest paper of the is structured as follows. Sec. 2 describes related crowdsourcing models. Sec. 3 introduces latent-class assumptions and provides details of our latent class models. Sec. 4 presents category recovery methods. Sec. 5 shows empirical results for latent class and category recovery. Sec. 6 concludes.

2   Preliminaries

We introduce three major methods for label aggregation in learning from crowds. We focus on classification tasks in this paper. In a dataset consisting of M items (e.g., pictures or paragraphs), each item m has a specific label Y_m to denote its affiliated category. Y is the collection of these ground truth category labels, and all the possible label values form a set D. To obtain the unknown ground truth, we have N workers examine the dataset. W_{nm} is the label of item m given by worker n. W is the collection of these workers' labels. I is the collection of all worker-item index pairs corresponding to W. The goal of learning from crowds is to infer the values of Y from the observations of W.

2.1   Majority Voting (MV)

The simplest label aggregation model is the majority voting. This method assumes that: For every worker, the ground truth label is always the most common to be given, and the labels for each item are given independently. From this point of view, we just need to find the most frequently appeared label for each item. We use $q_{md} = P(Y_m = d)$ to denote the probability that the mth task has true label d, then

$$q_{md} = \frac{\sum_{(n,m)\in I} \delta_{W_{nm},d}}{\sum_{d,(n,m)\in I} \delta_{W_{nm},d}}, \forall m, \tag{1}$$

where $\delta_{.,.}$ is an indicator function: $\delta_{a,b}$ equals to 1 whenever $a = d$ is true, otherwise it equals to 0. The estimated label is represented by $Y_m = \max_d q_{md}, \forall m$.

2.2   Dawid-Skene Estimator (DS)

Dawid and Skene [3] proposed a probabilistic model, which is widely used in this area. They made an assumption that: The performance of a worker is consistent across different items, and his or her behavior can be measured by a confusion matrix. Diagonal entries of the confusion matrix indicate the probability that this worker gives correct labels; while off-diagonal entries indicate that this worker makes specific mistakes to label items in one category as another. Extensive analysis of this model's error bound has been presented [10,11].

More formally, we use p_n to denote the confusion matrix of worker n, with each element p_{ndl} being the probability that worker n gives label l to an item when the ground truth of this item is d. We use q_d to denote the probability that an item has the ground truth label d. Under these notations, parameters of workers can be estimated via a maximum likelihood estimator, $\{\hat{q}, \hat{p}\} =$ argmax $P(W|q,p)$, where the margined likelihood is

$$P(\boldsymbol{W}|\boldsymbol{q},\boldsymbol{p}) = \prod_m \left(\sum_d q_d \prod_{n,l} p_{ndl}{}^{\delta_{W_{nm},l}} \right), \tag{2}$$

by marginalizing out the hidden variables $\boldsymbol{Y}$. This problem is commonly solved using an EM algorithm.

2.3 Minimax Entropy Estimator (ME)

Minimax entropy estimator [5,6] is another well-performing method which combines the idea of majority voting and confusion matrix. This model assumes that: Labels are generated by a probability distribution over workers, items, and labels; and the form of the probability distributions can be estimated under the maximum entropy principle. For example, p_{nm} is a probability distribution on the label of item m given by worker n. To incorporate the idea of majority voting that ground truth labels are always the most common labels to be given, the count of empirical observations that workers give an item a certain label should match the sum of workers' probability corresponding to these observations within the model. So they come up with the first type of constraints:

$$\sum_n p_{nmd} = \sum_n \delta_{W_{nm},d}, \forall m, d. \tag{3}$$

To combine the confusion matrix idea that a worker is consistent across different items in the same category, the count of empirical observations that workers give items in the same category a certain label should match the sum of workers' probability corresponding to these observations within the model. So there is another type of constraints:

$$\sum_{\substack{m \\ s.t. Y_m = d}} p_{nmd} = \sum_{\substack{m \\ s.t. Y_m = d}} \delta_{W_{nm},d}, \forall n, d. \tag{4}$$

Under these constraints and the minimax entropy principle, we choose $\boldsymbol{Y}$ to minimize the entropy but choose $\boldsymbol{p}$ to maximize the entropy. This rationale leads to the learning problem:

$$\min_{\boldsymbol{Y}} \max_{\boldsymbol{p}} - \sum_{n,m,d} p_{nmd} \log p_{nmd}, \tag{5}$$

subject to constraints (3) and (4). In practice, hard constraints can be strict. Therefore, soft constraints with slack variables are introduced to the problem.

3 Extend to Latent Classes

Both DS and ME use specific probabilities to represent workers' behaviors. However, we can dig deeper into the structure of the items. For example, in a flower recognition task, we ask workers to decide whether the flower in a given picture is peach flower or not. When the standard DS estimator is used, the confusion matrix should contain 4 probabilities, that is, the probability that worker labels the picture correctly when it is peach flower; the probability that worker labels

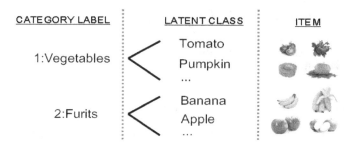

Fig. 1. Illustration for categories and latent classes of a vegetable vs. fruit classification

the picture incorrectly when it is peach flower; the probability that worker labels the picture correctly when it is not peach flower; and the probability that worker labels the picture incorrectly when it is not peach flower. If there are 2 breeds of peach flowers in the testing set, say mountain peach flowers and flowering peach flowers, then the probabilities that a worker recognizes them as peach flowers correctly might be different. For example, some workers who are very familiar with mountain peach may point out mountain peach flowers as peach flowers with an extraordinary high accuracy, but their accuracy of recognizing flowering peach might be close to random guess. Our experiments show that this phenomenon does exist. So we come to one conclusion that the latent structure of items can affect the workers' labeling results significantly, and we can take this influence into account in our label aggregation algorithm. *Latent class structure* is one of the simplest latent structures of items. The latent class here refers to a finer level structure of items than the category. In the flower example, the latent classes may correspond to the flower species such as flowering peach and mountain peach, while the categories can only recognize both these species as peach flower with no inner structure. If we restrict the number of latent classes to be the same as the number of categories, different classes will naturally correspond to the classification categories. Yet as a general rule, the number of latent classes should be larger than the category number.

A category of items might be divided into several latent classes, but a latent class belongs to one specific category only. Thus, we make two basic assumptions in the crowd labeling situations:

– **Assumption I.** Each item belongs to one specific latent class only.
– **Assumption II.** Items in a same latent class belong to a same category.

From another point of view, we believe that no label is spam. When the standards of solving our problems match the workers' own criterion, based on which they make their choices, the DS estimator works well. But if they do not, much information will be left unutilized by this estimator. In order to improve the aggregation performance and uncover more information hiding behind the noisy labels, we build up new models which take latent structures into account.

3.1 Nonparametric Latent Class Estimator (NDS)

For the DS estimator, a confusion matrix is used to measure workers' behavior, with each entry p_{ndl} representing the probability that worker n gives label l to an item when the ground truth of this item is d. Now we realise that the latent classes affect the output labels directly. We can replace the category dimension of the confusion matrix representation with the latent class dimension. Therefore, we have a latent class version confusion matrix $\boldsymbol{p}_n$ for each worker. An entry p_{nkl} denotes the probability that worker n gives label l to an item which belongs to latent class k. Similarly we use Z_m to represent the latent class that item m belongs to, and use $\boldsymbol{q}$ to denote the probability that each latent class appears, so that q_k denotes the probability that an item belongs to latent class k.

Probabilistic Model. Since it is hard to decide the number of latent classes K in advance, we put a nonparametric prior on the latent class assignment variable $\boldsymbol{Z}$, which can represent arbitrary number of classes. The *Chinese restaurant process* (CRP) is used here, it is a construction of Dirichlet process [12], and can be described using the metaphor of a restaurant with customers entering and sitting next to tables with different probabilities depending on the tables'jj relative sizes. α_c is the discount parameter of this process. We also put a Dirichlet prior Dirichlet(α_d) on every $\boldsymbol{p}_{nk}$, where α_d is the concentration parameter. So the probabilistic model is represented as follow,

$$\boldsymbol{Z}|\alpha_c \sim \text{CRP}(\alpha_c), \qquad \boldsymbol{p}_{nk}|\alpha_d \sim \text{Dirichlet}(\alpha_d), \ \forall n, k, \tag{6}$$

$$W_{nm}|\boldsymbol{Z}, \boldsymbol{p}_{n\cdot} \sim \text{Multinomial}(\boldsymbol{A}_{nm}), \ \forall n, m, \tag{7}$$

where $\boldsymbol{A}_{nm} = \{A_{nm1}, \cdots, A_{nmD}\}$, and $A_{nmd} = \prod_{k=1}^{K} p_{nkd}{}^{\delta_{Z_m,k}}$. Here $\boldsymbol{W}$ is the given labels, $\boldsymbol{p}$ is the parameters to learn, and $\boldsymbol{Z}$ is the hidden variable. If annotator n do not give item m a label, the probabilities of all W_{nm} values are set to be one.

Conditional Distribution. To infer their values, we build a Gibbs sampler to get samples from the joint posterior distribution. The conditional distribution for the confusion matrix parameter is

$$P(\boldsymbol{p}_{nk}|\boldsymbol{Z}, \boldsymbol{W}) \propto P(\boldsymbol{p}_{nk}) \prod_{m=1}^{M} P(W_{nm}|\boldsymbol{Z}, \boldsymbol{p}_{nk}) \tag{8}$$

$$\propto \Big(\prod_{d=1}^{D} p_{nkd}{}^{\alpha_d/D-1} \Big) \Big(\prod_{m=1}^{M} \prod_{d=1}^{D} p_{nkd}{}^{\delta_{W_{nm},d}\delta_{Z_m,k}} \Big).$$

So the conditional distribution $\boldsymbol{p}_{nk}|\boldsymbol{Z}, \boldsymbol{W} \sim \text{Dirichlet}(\boldsymbol{p}_{nk}|\boldsymbol{B}_{nk}), \forall n, k$, where $\boldsymbol{B}_{nk} = \{B_{nk1}, \cdots, B_{nkD}\}$, and $B_{nkd} = \sum_{m=1}^{M} \delta_{W_{nm},d}\delta_{Z_m,k} + \alpha_d/D$. As for the hidden variables, when $k \leq K$,

$$P(Z_m = k|\mathbf{Z}_{-m}, \mathbf{p}, \mathbf{W}) \propto P(Z_m = k) \prod_{n=1}^{N} P(W_{nm}|Z_m = k, \mathbf{p}_{nk}) \qquad (9)$$

$$\propto n_k \prod_{n=1}^{N} \prod_{d=1}^{D} p_{nkd}{}^{\delta_{W_{nm,d}}},$$

where n_k is the number of tasks that have latent class label k. When generating a new class,

$$P(Z_m = k_{new}|\mathbf{Z}_{-m}, \mathbf{p}, \mathbf{W}) \propto P(Z_m = k_{new}) \prod_{n=1}^{N} P(W_{nm}|Z_m = k_{new}) \qquad (10)$$

$$\propto P(Z_m = k_{new}) \prod_{n=1}^{N} \int P(W_{nm}|Z_m = k_{new}, \mathbf{p}_{nk_{new}}) P(\mathbf{p}_{nk_{new}}) d\mathbf{p}_{nk_{new}}$$

$$\propto \alpha_c \prod_{n=1}^{N} \frac{\prod_{d=1}^{D} \Gamma(\delta_{W_{nm,d}} + \alpha_d/D)}{\Gamma(1 + \alpha_d)}.$$

Then we can get samples from the posterior distribution of our model by iteratively updating hidden variables and parameters.

3.2 Latent Class Minimax Entropy Estimator (LC-ME)

Many existing estimators can be extended to learn latent class structures. The nonparametric latent class estimator can be regarded as an extension of DS estimator, we can also incorporate latent class structures into the minimax entropy estimator. Some constraints need to change for this extension, as detailed below.

We still assume that the ground truth label will always get more probability to be given by workers, so the first type constraints remain unchanged. As for the other constraints, now we apply the idea of latent class version DS estimator: When worker n deals with items in latent class k, he may label it as category d with a constant probability. So the constraints can be written as

$$\sum_{\substack{m \\ s.t.Z_m=k}} p_{nmd} = \sum_{\substack{m \\ s.t.Z_m=k}} \delta_{W_{nm,d}}, \forall n, k. \qquad (11)$$

To relax constraints, we introduce slack variables τ and σ and their regularization terms. Under these new constraints, the optimization problem is slightly changed comparing with the previous version:

$$\min_{\mathbf{Z}} \max_{\mathbf{p}, \tau, \sigma} - \sum_{n,m,d} p_{nmd} \log p_{nmd} - \sum_{m,d} \frac{\alpha_m \tau_{md}^2}{2} - \sum_{n,m,d} \frac{\beta_n \sigma_{ndk}^2}{2}$$

$$s.t. \sum_{n} (p_{nmd} - \delta_{W_{nm,d}}) = \tau_{md}, \forall m, d, \qquad (12)$$

$$\sum_{m} (p_{nmd} - \delta_{W_{nm,d}}) \delta_{Z_m,k} = \sigma_{ndk}, \forall n, k, \quad \sum_{d} p_{nmd} = 1, \forall n, m.$$

To solve this optimization problem, we update $\{\tau_{md}, \sigma_{ndk}\}$ and q_{mk} respectively. Since the inference procedure is similar to the original minimax entropy estimator in [5], we only express the final iterative formula here.

Step-1: we need to solve a simple sub-problem:

$$
\begin{aligned}
\{\tau_{md}^t, \sigma_{ndk}^t\} = \operatorname*{argmin}_{\tau,\sigma} \sum_{n,k,d} q_{mk}^{t-1} \Big[&\log \sum_d \exp(\tau_{md} + \sigma_{ndk}) \\
-\sum_d (\tau_{md} + \sigma_{ndk})\delta_{W_{nm},d} \Big] + &\sum_{m,d} \frac{1}{2}\alpha_m \tau_{md}^2 + \sum_{n,m,d} \frac{1}{2}\beta_n \sigma_{ndk}^2, \forall n, m, d, k,
\end{aligned}
$$
(13)

where $q_{mk}^t \propto P^t(Z_m = k)$ represents the probability that the item m is in latent class k. This optimization task can be solved by gradient descent and any other optimization methods.

Step-2: the probability distribution of each item's label is

$$
q_{mk}^t \propto q_{mk}^{t-1} \prod_n \frac{\exp\left(\sum_d (\tau_{md}^t + \sigma_{ndk}^t)\delta_{W_{nm},d}\right)}{\sum_d \exp(\tau_{md}^t + \sigma_{ndk}^t)}, \forall m, k.
$$
(14)

Iteratively updating $\{\tau_{md}, \sigma_{ndk}\}$ and q_{mk}, it will converge to a stationary point. Then we can get the latent class numbers Z by the peak positions of q. Since the algorithm is sensitive to the initial point, we use the result of NDS as the latent class number K and the initial point Z of the LC-ME.

4   Category Recovery

In order to obtain the ground truth labels, we need to recover the category information from latent classes. According to our second basic assumption that each latent class belongs to one specific category, we can recover the ground truth labels by associating latent classes to categories.

A re-estimating method can be used here to recover the categories. After we get the latent class information for items, we can regard items in a same class as one imaginary item, here we call it a *hyper-item*. Then there are totally K hyper-items, every hyper-item may have several different labels by each worker. This setting has been considered in the original Dawid-Skene estimator.

We use a generalized Dawid-Skene estimator with hyper-items to estimate the category assignments, which solves a maximum likelihood estimation problem. The margined likelihood of given labels is

$$
P(W|q, p) = \prod_k \left(\sum_d q_d \prod_{n,l} p_{ndl}^{n_{nkd}} \right),
$$
(15)

where $n_{nkd} = \sum_m \delta_{W_{nm},d}\delta_{Z_m,k}$ is the count of labels that worker n gives to hyper-item k. The EM algorithm for solving this problem also needs some modification. Specifically, we use C_k to represent the category of latent class k. Then in the **E-Step**, the probability distribution is

$$
P(C_k = d|W, q, p) \propto P(C_k = d)P(W|C_k = d) \propto q_d \prod_{n,l} p_{ndl}^{n_{nkd}},
$$
(16)

and the estimated category of each latent class is $C_k = \max_d P(C_k = d|\boldsymbol{W}), \forall k$. In the **M-Step**, we have the update equations:

$$q_d = \frac{1}{K}\sum_k \delta_{C_k,d}, \qquad p_{ndl} = \frac{\sum_k n_{nkl}\delta_{C_k,d}}{\sum_{k,l} n_{nkl}\delta_{C_k,d}}. \tag{17}$$

5 Experiment Results

We now present experimental results to evaluate the performance of the proposed models on both one synthetic dataset and real dataset collected from MTurk. We present both quantitative results on ground truth label recovery and quantitative results on latent structure discovery, with comparison to various competitors.

5.1 Synthetic Dataset

We designed a synthetic dataset to show the latent class recovery ability of each model. This dataset consists of 4 latent classes and 2 types of workers. We generated 40 items' parameters for each latent class and simulated 20 workers of each type. We set the confusion matrix for all simulating worker types and randomly sample labels. The probabilistic distribution values of different classes in the confusion matrices are dispersive, e.g. $[0.8, 0.2], [0.5, 0.5], [0.2, 0.8]$. So the effect of latent structure is more significant. The results on learning latent classes and category recovery are shown below.

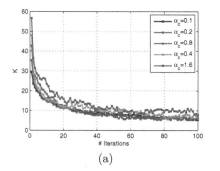 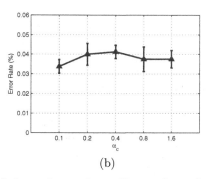

(a)   (b)

Fig. 2. Performance on synthetic dataset. (a) shows the numbers of latent classes found by NDS with different color. (b) shows the average category recovery error rates.

Sensitivity: We use the NDS model to recover the latent structure of this dataset. Fig. 2(a) shows the learnt latent class number K by models with different parameters. We set $\alpha_d = 2$ for all trials, and vary α_c from 0.1 to 1.60. We can see when parameter changes, the steady state value only changes a little, and all the values are close to the true latent class number. This result shows that our model is insensitive to the discount parameter. So when we use this model to learn latent structures for some purposes, we only need to find a rough range of the parameter with a validate dataset.

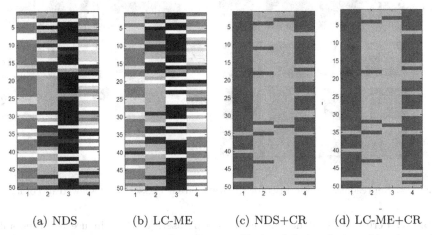

(a) NDS   (b) LC-ME   (c) NDS+CR   (d) LC-ME+CR

Fig. 3. Latent class and category visualization. Each subfigure shows a 50×4 matrix, with each entry corresponding to a flower image and each column corresponding to a unique flower species, which is flowering peach, sakura, apricot and mountain peach from left to right. For (a) and (b), each color denotes one latent class. For (c) and (d), each color denotes a classification category. (a) and (c) are learned by NDS, (b) and (d) are learned by LC-ME. (best viewed in color).

Category Recovery: To evaluate the ground truth category recovery accuracy, we compare the error rates of NDS with different α_c. We can see from Fig. 2(b) that the final accuracy is insensitive to the parameter α_c, and it is about 3.75% for all parameter settings. We also compare the NDS with other methods. Majority voting achieves error rate 9.38%, original Dawid-Skene estimator achieves error rate 12.50%, both of them are worse than NDS.

5.2 Flowers Dataset

To show the semantic meaning of the latent structure learned by our models, we designed a flower recognition task and collected crowd labeling data from MTurk annotators. Four flower species, mountain peach flower, flowering peach flower, apricot flower and sakura, make up the dataset of 200 images. Each species have 50 different pictures. Only mountain peach flower and flowering peach flower are peach flower while apricot flower and sakura are not. Workers were asked to choose whether the flower in picture is prunus persica (peach flower).

We collected labels on the Amazon Mechanical Turk (MTurk) platform. 36 of all the different participants completed more than 10 *Human Intelligence Tasks* (HIT) on each. And they provided 2366 HIT in total. During the annotating procedure, two hints are shown to make sure that workers can distinguish prunus persica and sakura or distinguish prunus persica and apricot. Each picture was labeled by 11.8 workers and each worker provided 65.7 labels on average.

To visualize the structures learned by our models, we draw colormaps to show the partitions of different latent classes and different categories in Fig. 3(b)-3(d). Each subfigure contains a 50×4 color matrix, with each entry representing a

Fig. 4. Representative pictures for different latent classes.(best viewed in color).

Table 1. Performance of models on flowers dataset. Workers in use are randomly selected for each trial, and the average error rate of 10 trials, together with standard deviation, are presented. $\alpha_c = 0.09$ and $\alpha_d = 1$ are used for latent class recovery.

#	20	25	30	35
MV	0.1998 ± 0.0506	0.2383 ± 0.0216	0.2153 ± 0.0189	0.2170 ± 0.0096
DS	0.1590 ± 0.0538	0.1555 ± 0.0315	0.1310 ± 0.0213	0.1300 ± 0.0041
NDS	0.1595 ± 0.0737	0.1605 ± 0.0434	0.1330 ± 0.0371	0.1475 ± 0.0354
ME	0.1535 ± 0.0695	0.1470 ± 0.0339	0.1315 ± 0.0200	0.1335 ± 0.0078
LC-ME	$\mathbf{0.1415 \pm 0.0382}$	$\mathbf{0.1430 \pm 0.0286}$	$\mathbf{0.1215 \pm 0.0133}$	$\mathbf{0.1190 \pm 0.0168}$

flower image in the dataset, and each column corresponding to a unique flower species. Specifically, the first column is flowering peach flower, second is sakura, third is apricot flower and forth is mountain peach flower.

In Fig. 3(a) and Fig. 3(b), each color denotes one latent class learned by the estimator. We can see that the first three columns almost have pure color boxes, which means these three latent classes are strongly related to the flower species. The fourth column is kind of miscellaneous, which means that lots of mountain peach flowers are misclassified into other species. This is because mountain peach flowers have no distinct features comparing with other flower species.

In Fig. 3(c) and Fig. 3(d), each color denotes a classification category, either peach flower or not. This result comes from putting blue and azure boxes into peach flower category and other two colors' boxes into another. Fig. 4 shows some representative flower pictures for different latent classes we learned. These results suggest that the structures we learned have explicit semantic meaning, and these latent class patterns could be used in many further applications.

Finally, we evaluate the category recovery performance. The average worker error rate in this flower recognition task is 30.00%, and majority voting gets an error rate of 22.00%. The latent class minimax entropy estimator (LC-ME) wins on this dataset with error rate 11.00%, and the nonparametric latent class estimator (NDS,$\alpha_c = 1.6, \alpha_d = 2$) achieves 11.50%. The original Dawid-Skene estimator (DS) achieves 13.00%. The minimax entropy estimator (ME) [1] also achieves 13.00%. We also generated some sub-datasets with different numbers of workers in order to make more comparisons. Results are shown in Table 1, which consistently show the improvements by exploring our latent class assumptions.

6 Conclusions and Future Work

We have carefully examined the effectiveness of latent class structures in crowdsourcing. Our methods characterize that items in one dataset can be separated into several latent classes and workers' annotating behaviors may differ among different classes. By incorporating such fine-grained structures, we can describe the generation mechanism of noisy labels more clearly. Our methods can disclose meaningful latent classes, as demonstrated in real data experiments. After we get the latent class assignments, a category label recovery algorithm is developed, which is empirically demonstrated to achieve higher accuracies on category recovery tasks. Our latent structure models can preserve the structure information of data. For the future work, we plan to investigate the effectiveness of such hidden structure information further in handling other interesting tasks, such as online task selection and user behavior analysis.

Acknowledgments. The work was supported by the National Basic Research Program (973 Program) of China (No. 2013CB329403), National Natural Science Foundation of China (Nos. 61322308, 61332007), and the Tsinghua National Laboratory for Information Science and Technology Big Data Initiative.

References

1. Snow, R., O'Connor, B., Jurafsky, D., Ng, A.Y.: Cheap and fast-but is it good?: evaluating non-expert annotations for natural language tasks. In: EMNLP (2008)
2. Zhu, J., Chen, N., Xing, E.P.: Bayesian inference with posterior regularization and applications to infinite latent svms. JMLR **15**, 1799–1847 (2014)
3. Dawid, A.P., Skene, A.M.: Maximum likelihood estimation of observer error-rates using the em algorithm. Applied Statistics, 20–28 (1979)
4. Raykar, V.C., Yu, S., Zhao, L.H., Valadez, G.H., Florin, C., Bogoni, L., Moy, L.: Learning from crowds. JMLR **11**, 1297–1322 (2010)
5. Zhou, D., Platt, J.C., Basu, S., Mao, Y.: Learning from the wisdom of crowds by minimax entropy. In: NIPS (2012)
6. Zhou, D., Liu, Q., Platt, J.C., Meek, C.: Aggregating ordinal labels from crowds by minimax conditional entropy. In: ICML (2014)
7. Welinder, P., Branson, S., Belongie, S., Perona, P.: The multidimensional wisdom of crowds. In: NIPS (2010)
8. Sheshadri, A., Lease, M.: Square: a benchmark for research on computing crowd consensus. In: First AAAI Conference on Human Computation and Crowdsourcing (2013)

[1] Implementation from http://research.microsoft.com/en-us/projects/crowd.

9. Tian, Y., Zhu, J.: Learning from crowds in the presence of schools of thought. In: ICDM (2012)
10. Li, H., Yu, B., Zhou, D.: Error rate analysis of labeling by crowdsourcing. In: ICML Workshop: Machine Learning Meets Crowdsourcing, Atalanta, Georgia, USA (2013)
11. Gao, C., Zhou, D.: Minimax optimal convergence rates for estimating ground truth from crowdsourced labels. arXiv preprint arXiv:1310.5764 (2013)
12. Neal, R.M.: Markov chain sampling methods for Dirichlet process mixture models. Journal of computational and graphical statistics **9**(2), 249–265 (2000)

Use Correlation Coefficients in Gaussian Process to Train Stable ELM Models

Yulin He[1]([✉]), Joshua Zhexue Huang[1], Xizhao Wang[1],
and Rana Aamir Raza[2]

[1] College of Computer Science and Software Engineering,
Shenzhen University, Shenzhen 518060, China
{yulinhe,zx.huang,xzwang}@szu.edu.cn
[2] College of Information and Communication Engineering,
Shenzhen University, Shenzhen 518060, China
aamir@szu.edu.cn

Abstract. This paper proposes a new method to train stable extreme learning machines (ELM). The new method, called StaELM, uses correlation coefficients in Gaussian process to measure the similarities between different hidden layer outputs. Different from kernel operations such as linear or RBF kernels to handle hidden layer outputs, using correlation coefficients can quantify the similarity of hidden layer outputs with real numbers in $(0, 1]$ and avoid covariance matrix in Gaussian process to become a singular matrix. Training through Gaussian process results in ELM models insensitive to random initialization and can avoid over-fitting. We analyse the rationality of StaELM and show that existing kernel-based ELMs are special cases of StaELM. We used real world datasets to train both regression and classification StaELM models. The experiment results have shown that StaELM models achieved higher accuracies in both regression and classification in comparison with traditional kernel-based ELMs. The StaELM models are more stable with respect to different random initializations and less over-fitting. The training process of StaELM models is also faster.

Keywords: Extreme learning machine · Correlation coefficient · Gaussian process · Neural network

1 Introduction

Extreme learning machine (ELM) is a special single-hidden layer feed-forward neural network (SLFN) [6]. Due to its lower computational complexity and better generalization performance, ELM has recently attracted a lot of interests in research and industry and is used in a wide range of applications [5]. ELM uses a random method to determine input weights/hidden layer biases and analytically computes the output weights. Therefore, it is extremely fast to train an ELM model. It has also been proved that ELM can guarantee the universal approximate capability of ELM [3].

© Springer International Publishing Switzerland 2015
T. Cao et al. (Eds.): PAKDD 2015, Part I, LNAI 9077, pp. 405–417, 2015.
DOI: 10.1007/978-3-319-18038-0_32

Currently, ELM has two main problems in practical applications. The first problem is that the trained ELM model is sensitive to the random initial settings [15]. Different initial settings often result in different performances, which implies that the training process produces instable ELM models from different initial settings. The second problem is over-fitting [3], which is usually caused by the numerous hidden layer nodes specified to best approximate the training data set. A number of improvements have been proposed to tackle these problems. One approach is to optimize the random weights with different evolutionary algorithms. Examples include E-ELM [15], SaE-ELM [1], and O-ELM [9]. Another approach is to select better architectures for ELM, for instance, I-ELMs [3,4], OP-ELM [10], and localized generalization error ELM [13]. Although the literatures reported the better performances of these improved ELM models, the higher computational complexity makes them impractical to deal with the regression and classification tasks with a large number of training instances.

A different direction to improve ELM without increase of computational complexity is to estimate the prior probability distribution of ELM models. Soria-Olivas *et al.* [12] designed a Bayesian ELM (BELM). Luo *et al.* [8,14] proposed sparse Bayesian ELM (SBELM).[1]. Chatzis *et al.* [2] proposed the one-hidden-layer nonparametric Bayesian kernel machine (1HNBKM). Because BELM and 1HNBKM used linear and RBF kernels to handle the hidden layer outputs, we call them kernel-based ELMs in this paper. The empirical analysis shows that kernel-based ELMs are still sensitive to random initialization. For example, there is an obvious difference between the predictive results of BELM and 1HNBKM on *Libras Movement* dataset[2] with random input weights in intervals $[0, 1]$ and $[-1, 1]$. In addition, the over-fitting still exists for 1HNBKM.

In this paper, we propose to use Gaussian process to train ELM models and present a stable extreme learning machine algorithm, StaELM. In this algorithm, we use correlation coefficients in Gaussian process to measure the similarity between different hidden layer outputs with real numbers in $(0, 1]$. The advantages of using Gaussian process in ELM model training over aforementioned training methods are that the training process is fast and the trained ELM models are insensitive to random initialization and can avoid over-fitting. In the training process, we use correlation coefficients to avoid the covariance matrix to become a singular matrix and make the inverse of covariance matrix solvable.

We have used 12 UCI and KEEL[3] datasets to conduct the experiments and compared the performances of accuracy and running time of StaELM, orginal ELM, BELM, and 1HNBKM. The experimental results show that StaELM models achieved higher accuracies and lower running time in both regression and

---

[1] The main difference between SBELM and BELM is that the independent regularization priors in SBELM are imposed on each weight instead of one shared prior for all weights in BELM. Because SBELM and BELM are homologous, we only discuss and analyse BELM in this paper due to its simplicity.

[2] http://archive.ics.uci.edu/ml/

[3] http://sci2s.ugr.es/keel/datasets.php

classification than other methods. The StaELM models are stable with respect to different random initializations and less over-fitting.

The rest of this paper is organized as follows: In Section 2, we briefly summarize kernel-based ELMs. Section 3 introduces our proposed StaELM. Experimental simulations are presented in Section 4. Finally, we conclude this paper in Section 5.

2 Kernel-based ELMs

In this section, we review three existing ELMs models, i.e., the original ELM, BELM, and 1HNBKM. Because the first two use linear kernels to handle the hidden layer outputs whereas the last one uses RBF kernel for that purpose. We call them kernel-based ELMs.

2.1 ELM

ELM [6] is a single-hidden layer feed-forward neural network (SLFN) and does not require any iterative optimization to input/output weights. Given the training dataset $(X_{N \times D}, Y_{N \times C})$ and testing dataset $(X'_{M \times D}, Y'_{M \times C})$:

$$X_{N \times D} = \begin{bmatrix} x_1 \\ x_2 \\ \vdots \\ x_N \end{bmatrix} = \begin{bmatrix} x_{11} & x_{12} & \cdots & x_{1D} \\ x_{21} & x_{22} & \cdots & x_{2D} \\ \vdots & \vdots & \ddots & \vdots \\ x_{N1} & x_{N2} & \cdots & x_{ND} \end{bmatrix}, Y_{N \times C} = \begin{bmatrix} y_1 \\ y_2 \\ \vdots \\ y_N \end{bmatrix} = \begin{bmatrix} y_{11} & y_{12} & \cdots & y_{1C} \\ y_{21} & y_{22} & \cdots & y_{2C} \\ \vdots & \vdots & \ddots & \vdots \\ y_{N1} & y_{N2} & \cdots & y_{NC} \end{bmatrix} \quad (1)$$

and

$$X'_{M \times D} = \begin{bmatrix} x'_1 \\ x'_2 \\ \vdots \\ x'_M \end{bmatrix} = \begin{bmatrix} x'_{11} & x'_{12} & \cdots & x'_{1D} \\ x'_{21} & x'_{22} & \cdots & x'_{2D} \\ \vdots & \vdots & \ddots & \vdots \\ x'_{M1} & x'_{M2} & \cdots & x'_{MD} \end{bmatrix}, Y'_{M \times C} = \begin{bmatrix} y'_1 \\ y'_2 \\ \vdots \\ y'_M \end{bmatrix} = \begin{bmatrix} y'_{11} & y'_{12} & \cdots & y'_{1C} \\ y'_{21} & y'_{22} & \cdots & y'_{2C} \\ \vdots & \vdots & \ddots & \vdots \\ y'_{M1} & y'_{M2} & \cdots & y'_{MC} \end{bmatrix}, \quad (2)$$

where N is the number of training instances, D is the number of input variables, M is the number of testing instances, and C is the number of output variables. Usually, $Y'_{M \times C}$ is unknown and needs to be predicted. ELM determines $Y'_{M \times C}$ as follows:

$$Y'_{M \times C} = H'_{M \times L} \beta_{L \times C} = \begin{cases} H' \left(H^T H \right)^{-1} H^T Y, \text{ if } N \geq L \\ H' H^T \left(H H^T \right)^{-1} Y, \text{ if } N < L \end{cases}, \quad (3)$$

where $\beta_{L \times C}$ is the output weights, L is the number of hidden layer nodes,

$$H_{N \times L} = \begin{bmatrix} h(x_1) \\ h(x_2) \\ \vdots \\ h(x_N) \end{bmatrix} = \begin{bmatrix} g(w_1 x_1 + b_1) & g(w_2 x_1 + b_2) & \cdots & g(w_L x_1 + b_L) \\ g(w_1 x_2 + b_1) & g(w_2 x_2 + b_2) & \cdots & g(w_L x_2 + b_L) \\ \vdots & \vdots & \ddots & \vdots \\ g(w_L x_N + b_L) & g(w_2 x_N + b_L) & \cdots & g(w_L x_N + b_L) \end{bmatrix} \quad (4)$$

is the hidden layer output matrix for training instances,

$$H'_{M \times L} = \begin{bmatrix} h\left(x'_1\right) \\ h\left(x'_2\right) \\ \vdots \\ h\left(x'_M\right) \end{bmatrix} = \begin{bmatrix} g\left(w_1 x'_1 + b_1\right) & g\left(w_2 x'_1 + b_2\right) & \cdots & g\left(w_L x'_1 + b_L\right) \\ g\left(w_1 x'_2 + b_1\right) & g\left(w_2 x'_2 + b_2\right) & \cdots & g\left(w_L x'_2 + b_L\right) \\ \vdots & \vdots & \ddots & \vdots \\ g\left(w_L x'_M + b_L\right) & g\left(w_2 x'_M + b_L\right) & \cdots & g\left(w_L x'_M + b_L\right) \end{bmatrix} \quad (5)$$

is the hidden layer output matrix for testing instances, $g\left(z\right) = \frac{1}{1+e^{-z}}$ is sigmoid activation function,

$$W_{D \times L} = \begin{bmatrix} w_1 \ w_2 \cdots w_L \end{bmatrix} = \begin{bmatrix} w_{11} & w_{21} & \cdots & w_{L1} \\ w_{12} & w_{22} & \cdots & w_{L2} \\ \vdots & \vdots & \ddots & \vdots \\ w_{1D} & w_{2D} & \cdots & w_{LD} \end{bmatrix} \ \text{and} \ b = \begin{bmatrix} b_1 \\ b_2 \\ \vdots \\ b_L \end{bmatrix} \quad (6)$$

are input weight and hidden layer biases which are randomly determined. ELM is sensitive to random initialization and has obvious over-fitting. In order to tackle these problems, two improvements, i.e., BELM and 1HNBKM, are discussed below.

2.2 BELM

BELM [12] optimizes the output weights β by using Bayesian linear regression as follows:

$$y = h\left(x\right)\beta + \varepsilon, \quad (7)$$

where $\varepsilon \sim N\left(0, \sigma_N^2\right)$ and $\beta \sim N\left(0, \alpha^{-1} I_{L \times L}\right)$. The posterior distribution over output weights β is expressed as

$$P\left(\beta \left| H, Y\right.\right) = N\left(\beta, S\right), \quad (8)$$

where $\beta = \sigma_N^{-2} S H^T Y$ and $S = \left(\alpha I + \sigma_N^{-2} H^T H\right)^{-1}$ are the mean and covariance matrix respectively. For a new instance $x' = \left(x'_1, x'_2, \cdots, x'_D\right)$, the output y' predicted with BELM obeys the Gaussian distribution $N\left(\mu, \sigma^2\right)$, where

$$\mu = h\left(x'\right)\beta, \quad (9)$$

$$\sigma^2 = \sigma_N^2 + h\left(x'\right) S h^T\left(x'\right). \quad (10)$$

In BELM, μ is deemed as the prediction of new instance x', i.e., $y' = \mu$, and σ^2 is the variance which is used to determine the confidence interval of prediction y'. There are two parameters that need to be determined in BELM: σ_N^2 and $\alpha > 0$. BELM effectively controls the over-fitting but still sensitive to randomly initial weights is still not solved.

2.3   1HNBKM

Given a training dataset $(X_{N \times D}, Y_{N \times C})$, 1HNBKM [2] predicts the output y' for new instance x' via the following joint probability distribution

$$\begin{bmatrix} Y \\ y' \end{bmatrix} \sim N \left(0, \begin{bmatrix} K(H,H) + \sigma_N^2 I & k^T(h(x'),H) \\ k(h(x'),H) & k(h(x'),h(x')) \end{bmatrix} \right), \tag{11}$$

where the meaning of σ_N^2 is same as in BELM,

$$K(H,H)_{N \times N} = \begin{bmatrix} k(h(x_1),h(x_1)) & k(h(x_1),h(x_2)) & \cdots & k(h(x_1),h(x_N)) \\ k(h(x_2),h(x_1)) & k(h(x_2),h(x_2)) & \cdots & k(h(x_2),h(x_N)) \\ \vdots & \vdots & \ddots & \vdots \\ k(h(x_N),h(x_1)) & k(h(x_N),h(x_2)) & \cdots & k(h(x_N),h(x_N)) \end{bmatrix} \tag{12}$$

is a kernel matrix,

$$k(h(x'),H) = \begin{bmatrix} k(h(x'),h(x_1)) & k(h(x'),h(x_2)) & \cdots & k(h(x'),h(x_N)) \end{bmatrix} \tag{13}$$

is a kernel vector, and

$$k(u,v) = \exp\left(-\frac{\|u-v\|^2}{2\lambda^2} \right). \tag{14}$$

is the RBF kernel function. We can find $k(u,v) = 1$, when $u=v$.

Then, the posterior distribution of predicted output y' is

$$P(y' | h(x'), H, Y) = N(\mu, \sigma^2), \tag{15}$$

where

$$\mu = k(h(x'),H)(K(H,H) + \sigma_N^2 I)^{-1} Y, \tag{16}$$

$$\sigma^2 = k(h(x'),h(x')) - k(h(x'),H)(K(H,H) + \sigma_N^2 I)^{-1} k^T(h(x'),H). \tag{17}$$

Similarly, μ is the prediction of x' and σ^2 is the variance of prediction. Parameters σ_N^2 and λ^2 are unknown and need to be determined. 1HNBKM suffers severe over-fitting due to the usage of the RBF kernel and is also sensitive to random initialization.

3   Gaussian Process-based Stable ELM

In this section, we describe our proposed StaELM which conducts the inference based on Gaussian process and uses the correlation coefficient to construct the covariance matrix. StaELM also predicts the output y' for new instance x' based on Eq. (15), where

$$\mu = q(h(x'),H)(Q(H,H) + \sigma_N^2 I)^{-1} Y, \tag{18}$$

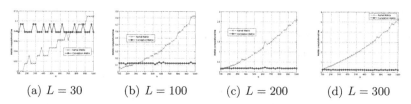

(a) $L = 30$ (b) $L = 100$ (c) $L = 200$ (d) $L = 300$

Fig. 1. Comparison on Matlab computational time between kernel and correlation matrices in HNBKM and StaELM respectively

$$\sigma^2 = \sigma_N^2 + q\left(h\left(x'\right), h\left(x'\right)\right) - q\left(h\left(x'\right), H\right)\left(Q\left(H, H\right) + \sigma_N^2 I\right)^{-1} q^T\left(h\left(x'\right), H\right), \quad (19)$$

$$Q\left(H, H\right)_{N \times N} = \begin{bmatrix} q\left(h\left(x_1\right), h\left(x_1\right)\right) & q\left(h\left(x_1\right), h\left(x_2\right)\right) & \cdots & q\left(h\left(x_1\right), h\left(x_N\right)\right) \\ q\left(h\left(x_2\right), h\left(x_1\right)\right) & q\left(h\left(x_2\right), h\left(x_2\right)\right) & \cdots & q\left(h\left(x_2\right), h\left(x_N\right)\right) \\ \vdots & \vdots & \ddots & \vdots \\ q\left(h\left(x_N\right), h\left(x_1\right)\right) & q\left(h\left(x_N\right), h\left(x_2\right)\right) & \cdots & q\left(h\left(x_N\right), h\left(x_N\right)\right) \end{bmatrix} \quad (20)$$

is a correlation matrix,

$$q\left(h\left(x'\right), H\right) = \left[q\left(h\left(x'\right), h\left(x_1\right)\right) \; q\left(h\left(x'\right), h\left(x_2\right)\right) \cdots q\left(h\left(x'\right), h\left(x_N\right)\right)\right] \quad (21)$$

is a correlation vector,

$$q\left(u, v\right) = \left(\frac{\rho_{uv} + 1}{2}\right)^2 \quad (22)$$

is correlation function, and

$$\rho_{uv} = \frac{Cov\left(u, v\right)}{\sqrt{D\left(u\right)}\sqrt{D\left(v\right)}} \quad (23)$$

is the correlation coefficient which measures the strength and direction of the linear relationship between two variables u and v. $Cov\left(u, v\right)$ is the covariance of variables u and v, and $D\left(u\right)$ and $D\left(v\right)$ are the standard deviations of u and v respectively. Note that Eq. (22) is to normalize the correlation coefficient into interval $(0, 1]$. Other normalization is also allowable. We can find that the inference process of StaELM is similar to 1HNBKM. The main difference between StaELM and 1HNBKM is that StaELM measures the similarity between two hidden layer outputs with correlation function in Eq. (22) instead of RBF kernel in 1HNBKM. The advantages of using correlation function are summarized as follows. (1) The correlation coefficient evaluates the relationship between two different hidden layer output $h\left(u\right)$ and $h\left(v\right)$ with probabilistic approach. This makes StaELM consider the inherent prior knowledge of training dataset more directly and comprehensively than kernel function based 1HNBKM. (2) The correlation coefficient reduce the chance of over-fitting of StaELM. For 1HNBKM with L hidden layer nodes, the prediction $\hat{Y}$ for training dataset is

$$\hat{Y} = K\left(H, H\right)\left(K\left(H, H\right) + \sigma_N^2 I\right)^{-1} Y. \quad (24)$$

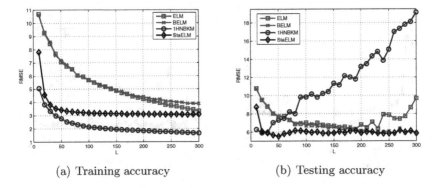

(a) Training accuracy   (b) Testing accuracy

Fig. 2. Curves for training and testing accuracies of different ELMs changing with L on *Concrete Compressive Strength* Dataset. StaELM is without over-fitting and obtains higher testing accuracies with less hidden nodes.

With the increase of L, $k(u, v) \rightarrow 0$ when $u \neq v$. This leads to $K(H, H) \rightarrow I$. Then, we can get $\hat{Y} \rightarrow Y$. This indicates that the RBF kernel easily results in the over-fitting of 1HNBKM. This is also confirmed by the following experimental validation. (3) Calculating the correlation matrix in Eq. (20) is more time-saving than kernel matrix in Eq. (12). We validate this fact via the following simulation on Matlab. For different L, we compare computational time of correlation matrix and kernel matrix. From Fig. 1, we can see that the computational time of kernel matrix grows exponentially with the increase of N.

StaELM is derived from Gaussian process regression (GPR) $y = h(x)\beta + \varepsilon$ with prior $\beta \sim N(0, \Sigma)$ and $\varepsilon \sim N(0, \sigma_N^2 I)$ [11]. The mean and covariance are

$$E[y] = h(x) E[\beta] + E[\varepsilon] = 0, \tag{25}$$

$$E[yy'] = h(x) E[\beta\beta^T] h^T(x') + E[\varepsilon\varepsilon^T] = h(x) \Sigma h^T(x') + \sigma_N^2. \tag{26}$$

The key of GPR is how to determine the term $h(x) \Sigma h^T(x')$ in Eq. (26). Because Σ is a symmetric positive definite matrix, Σ can be decomposed into AA^T, where A is a lower triangular matrix. Then, we can get

$$\begin{aligned} h(x) \Sigma h^T(x') &= h(x) AA^T h^T(x') = [h(x) A] [h(x') A]^T \\ &= \phi(h(x)) \phi^T(h(x')) = k(h(x), h(x')) \end{aligned} \tag{27}$$

where $k(u, v)$ is a kernel function which is used to measure the similarity between u and v. In StaELM, we replace $k(u, v)$ with $q(u, v)$ and use the correlation rather than distance to measure this similarity. In fact, we can find that the linear kernel $k(u, v) = uv^T$ is used in ELM and BELM. Then, ELM, BELM, and 1HNBKM are all the specials cases of StaELM which conducts the prediction based on GPR.

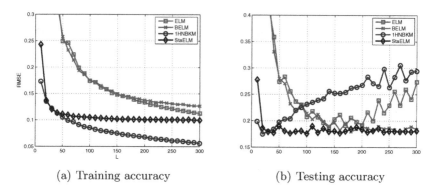

(a) Training accuracy   (b) Testing accuracy

Fig. 3. Curves for training and testing accuracies of different ELMs changing with L on *Treasury* Dataset

4   Experiments

In this section, we use 12 UCI and KEEL datasets to compare the performances of ELM, BELM, 1HNBKM, and StaELM, where 6 datasets are for regression problems and the other 6 datasets for classification problems. The basic descriptions to these datasets are listed in Tables 1 and 2 respectively. For the experimental procedure and parameter setting, we give the following descriptions.

- The input variables for regression datasets are normalized in $[-1, 1]$ and for classification datasets normalized in $[0, 1]$.
- We compare the training/testing accuracies and time for different learning algorithms. The accuracies for regression and classification problems are respectively measured with root mean square error (RMSE) and correct classification rate (CCR). The experimental results are the averages of 10 runs of 10-fold cross-validation.
- In our comparison, the parameters for different ELMs are set as $(\sigma_N^2, \alpha) = (0.001, 1)$ in BELM, $(\sigma_N^2, \lambda^2) = (0.001, 1)$ in 1HNBKM and $\sigma_N^2 = 0.001$ in StaELM respectively. The input weights and hidden biases are the random numbers in $[0, 1]$.

Tables 1 and 2 respectively give the comparative results on regression and classification datasets. According to the statistical analysis with Wilcoxon signed-ranks test at 95% significance level [7], we know that StaELM obtains the significantly better testing accuracies than other algorithms. Meanwhile, StaELM also has the better training accuracies than ELM and BELM. In addition, StaELM is also faster than 1HNBKM. On *Concrete Compressive Strength* and *Treasury* datasets, we give the curves for training and testing accuracies changing with the number of hidden layer nodes in Figs. 2 and 3. From these figures, we can clearly see that ELM and 1HNBKM have serious over-fitting problems. With the increase of L, the training RMSEs of ELM and 1HNBKM gradually decrease.

However, their testing RMSEs initially decrease with the increase of L, pass through a minimum, and then increase. Although the learning curves of BELM and StaELM all gradually decrease with the increase of L, we can find that StaELM has a faster convergence speed than BELM. This indicates that StaELM can obtain the lower RMSE with less hidden layer nodes.

The main reason that 1HNBKM has an obvious over-fitting is that the value of RBF kernel in Eq. (14) gradually approaches 0 with the increase of L. This leads to kernel matrix in Eq. (12) approximates an identity matrix. Assume the hidden layer outputs of instances u and v are

$$h_L(u) = \left[g(\hat{u}_1) \, g(\hat{u}_2) \cdots g(\hat{u}_L) \right], \tag{28}$$

$$h_L(v) = \left[g(\hat{v}_1) \, g(\hat{v}_2) \cdots g(\hat{v}_L) \right], \tag{29}$$

where $\hat{u}_l = w_l u + b_l$ and $\hat{v}_l = w_l v + b_l$, $l = 1, 2, \cdots, L$. With the increase of hidden layer nodes from L to L_1 $(L_1 > L)$, the hidden layer outputs of instances u and v are changed into

$$h_{L_1}(u) = \left[g(\hat{u}_1) \cdots g(\hat{u}_L) \, g(\hat{u}_{L+1}) \cdots g(\hat{u}_{L_1}) \right], \tag{30}$$

$$h_{L_1}(v) = \left[g(\hat{v}_1) \cdots g(\hat{v}_L) \, g(\hat{v}_{L+1}) \cdots g(\hat{v}_{L_1}) \right]. \tag{31}$$

Then, we can calaulate

$$k(h_L(u), h_L(v)) = \exp\left[-\frac{\sum_{l=1}^{L} \left[g(\hat{u}_l) - g(\hat{v}_l) \right]^2}{2\lambda^2} \right], \tag{32}$$

$$k(h_{L_1}(u), h_{L_1}(v)) = \exp\left[-\frac{\sum_{l=1}^{L} \left[g(\hat{u}_l) - g(\hat{v}_l) \right]^2 + \sum_{l_1=L}^{L_1} \left[g(\hat{u}_{l_1}) - g(\hat{v}_{l_1}) \right]^2}{2\lambda^2} \right]. \tag{33}$$

Because of $\sum_{l=1}^{L} \left[g(\hat{u}_l) - g(\hat{v}_l) \right]^2 < \sum_{l=1}^{L} \left[g(\hat{u}_l) - g(\hat{v}_l) \right]^2 + \sum_{l_1=L}^{L_1} \left[g(\hat{u}_{l_1}) - g(\hat{v}_{l_1}) \right]^2$,

$$k(h_L(u), h_L(v)) > k(h_{L_1}(u), h_{L_1}(v)) \tag{34}$$

can be derived. This indicates that the value of $k(h(u), h(v))$ gradually decreases with the increase of hidden layer nodes. $k(h(u), h(v))$ is non-negative, so $k(h(u), h(v)) \to 0$ with the increase of L. This leads to K $\to$ I and $\hat{Y} \to$ Y.

For the correlation function in Eq. (22), there is not an ordering relationship between $q(h_L(u), h_L(v))$ and $q(h_{L_1}(u), h_{L_1}(v))$, because the correlation coefficient in Eq. (23) measures the similarly between h(u) and h(v) with vectorial angle cosine rather than distance between them. The increase of vector dimension will not cause the vectorial angle cosine approaches 0. Then, we can know that Q $\nrightarrow$ I with the increase of L. This reduces the chance of over-fitting.

Table 1. Comparison of different ELMs on 6 REGRESSION datasets

Dataset	L	ELM TrainAcc	TrainTime	TestAcc	TestTime	BELM TrainAcc	TrainTime	TestAcc	TestTime	1HN3KM TrainAcc	TrainTime	TestAcc	TestTime	StaELM TrainAcc	TrainTime	TestAcc	TestTime
Airfoil Self-Noise (1503 instances, 5 inputs) UCI	20	4.507±0.349	0.00156	4.590±0.271	0.00000	4.507±0.349	0.01094	4.591±0.270	0.00000	2.549±0.055	15.10781	2.915±0.262	1.26875	2.832±0.066	1.53438	3.077±0.311	0.23125
	30	4.052±0.130	0.00781	4.113±0.342	0.00000	4.052±0.130	0.02187	4.113±0.342	0.00000	2.363±0.063	19.97969	2.763±0.252	1.56719	2.696±0.073	1.94687	2.967±0.238	0.26875
	100	3.055±0.099	0.04531	3.358±0.204	0.00313	3.149±0.078	0.04063	3.363±0.178	0.00000	1.905±0.042	23.82500	3.193±0.866	2.08125	2.531±0.045	1.98125	2.840±0.289	0.28906
	200	2.433±0.051	0.14219	3.141±0.438	0.00469	2.834±0.034	0.16875	3.074±0.241	0.00625	1.588±0.022	34.43125	5.282±1.859	3.20000	2.468±0.028	2.02031	2.767±0.261	0.37031
	300	2.337±0.045	0.22969	2.976±0.372	0.00937	2.738±0.034	0.29063	2.973±0.264	0.00000	1.384±0.021	36.05781	9.188±4.563	3.60312	2.481±0.028	1.82344	2.807±0.251	0.35625
Concrete Compressive Strength (1030 instances, 8 inputs) KEEL	20	9.477±0.531	0.00781	10.030±0.760	0.00000	9.477±0.531	0.00156	10.030±0.760	0.00000	3.365±0.120	9.82656	6.180±1.597	0.39844	4.686±0.210	1.85625	6.045±0.469	0.18281
	30	8.458±0.312	0.00469	8.679±0.896	0.00156	8.458±0.312	0.00781	8.679±0.896	0.00000	2.829±0.145	9.66719	7.184±3.123	0.39531	3.770±0.168	1.83281	5.411±0.642	0.19062
	100	5.743±0.138	0.04531	7.021±0.642	0.00156	5.745±0.137	0.05312	7.021±0.642	0.00156	1.965±0.071	10.56563	10.910±4.743	0.51719	3.224±0.091	1.88906	5.992±1.931	0.22812
	200	4.234±0.085	0.21875	6.158±0.834	0.00469	4.374±0.068	0.22344	6.041±0.652	0.00156	1.633±0.097	13.00156	20.997±8.170	0.75469	3.067±0.096	1.94531	5.722±1.300	0.30156
	300	3.338±0.108	0.56406	7.649±2.496	0.00156	3.838±0.093	0.63906	5.971±1.068	0.00156	1.420±0.108	15.64375	29.498±9.073	1.02969	3.075±0.054	1.99531	5.766±1.359	0.35938
Energy Efficiency (768 instances, 8 inputs) UCI	20	2.865±0.093	0.00625	2.957±0.160	0.00000	2.865±0.093	0.00469	2.957±0.160	0.00000	0.502±0.050	4.11250	0.826±0.119	0.18906	0.823±0.068	0.75313	1.144±0.144	0.09062
	30	2.645±0.079	0.00156	2.825±0.164	0.00000	2.645±0.079	0.01406	2.825±0.164	0.00000	0.320±0.030	4.15937	0.654±0.071	0.18750	0.526±0.058	0.76719	0.876±0.139	0.09062
	100	1.552±0.132	0.03750	1.929±0.196	0.00156	1.553±0.132	0.03906	1.916±0.195	0.00000	0.142±0.009	4.74062	0.965±0.179	0.24375	0.336±0.014	0.77969	0.639±0.083	0.11719
	200	0.652±0.043	0.17500	1.101±0.230	0.00156	0.729±0.040	0.18594	1.103±0.139	0.00156	0.044±0.005	5.89375	2.101±0.573	0.37812	0.306±0.012	0.83906	0.620±0.074	0.15781
	300	0.357±0.017	0.49844	0.795±0.089	0.00156	0.574±0.018	0.55312	0.912±0.121	0.00156	0.012±0.005	7.46094	2.087±0.473	0.52187	0.296±0.008	0.88594	0.622±0.074	0.19687
Mortgage data (1049 instances, 15 inputs) KEEL	20	0.576±0.137	0.00469	0.575±0.116	0.00000	0.576±0.137	0.00781	0.575±0.116	0.00000	0.041±0.002	9.54406	0.062±0.008	0.39531	0.051±0.001	1.82969	0.075±0.020	0.18281
	30	0.360±0.104	0.00937	0.376±0.105	0.00000	0.360±0.104	0.01250	0.376±0.105	0.00000	0.033±0.002	9.58594	0.063±0.008	0.40000	0.040±0.001	1.82656	0.067±0.013	0.17969
	100	0.082±0.012	0.04375	0.106±0.022	0.00156	0.082±0.012	0.05625	0.105±0.022	0.00156	0.019±0.001	10.60625	0.080±0.019	0.51875	0.031±0.001	1.89219	0.056±0.013	0.22969
	200	0.045±0.002	0.20000	0.088±0.032	0.00625	0.049±0.002	0.22187	0.078±0.022	0.00625	0.013±0.001	12.93750	0.100±0.028	0.77031	0.030±0.001	1.95781	0.057±0.013	0.29688
	300	0.033±0.001	0.57031	0.092±0.019	0.00781	0.041±0.001	0.64219	0.065±0.009	0.00313	0.011±0.000	16.06719	0.124±0.049	1.05469	0.030±0.001	2.08750	0.059±0.012	0.36406
Stock Prices data (950 instances, 9 inputs) KEEL	20	1.998±0.119	0.00625	2.014±0.192	0.00000	1.998±0.119	0.00937	2.014±0.192	0.00000	0.520±0.022	7.33594	0.757±0.048	0.31562	0.631±0.016	1.39063	0.835±0.060	0.13750
	30	1.425±0.071	0.00469	1.475±0.121	0.00000	1.425±0.071	0.01406	1.475±0.121	0.00000	0.432±0.015	7.36406	0.730±0.098	0.31719	0.505±0.010	1.41094	0.744±0.085	0.14531
	100	0.855±0.020	0.04688	0.971±0.056	0.00156	0.855±0.020	0.05312	0.969±0.057	0.00156	0.224±0.007	8.76719	0.895±0.136	0.42500	0.393±0.009	1.47656	0.685±0.078	0.19844
	200	0.611±0.016	0.20781	0.827±0.041	0.00156	0.636±0.014	0.22344	0.808±0.041	0.00313	0.124±0.007	10.91094	1.110±0.075	0.64531	0.379±0.007	1.56094	0.681±0.065	0.25625
	300	0.481±0.012	0.55781	0.860±0.085	0.00156	0.554±0.008	0.62187	0.778±0.053	0.00469	0.075±0.004	12.94531	1.251±0.291	0.86562	0.372±0.005	1.61094	0.678±0.059	0.31094
Treasury data (1049 instances, 15 inputs) KEEL	20	0.569±0.084	0.00781	0.608±0.098	0.00000	0.569±0.084	0.00625	0.608±0.098	0.00000	0.135±0.003	10.43281	0.175±0.034	0.42500	0.137±0.004	1.94844	0.192±0.049	0.19375
	30	0.408±0.086	0.00625	0.430±0.098	0.00000	0.408±0.086	0.01250	0.430±0.098	0.00000	0.121±0.004	10.65469	0.178±0.034	0.42656	0.120±0.005	1.94219	0.179±0.028	0.20313
	100	0.175±0.007	0.04688	0.217±0.028	0.00000	0.175±0.007	0.05156	0.216±0.029	0.00469	0.086±0.005	11.43906	0.225±0.058	0.54375	0.104±0.003	1.93906	0.180±0.033	0.23750
	200	0.133±0.005	0.21094	0.204±0.031	0.00000	0.138±0.004	0.22812	0.187±0.030	0.00469	0.067±0.004	13.58844	0.272±0.080	0.80625	0.101±0.003	1.99844	0.178±0.028	0.30938
	300	0.113±0.004	0.58750	0.291±0.189	0.00469	0.126±0.004	0.63594	0.194±0.041	0.00313	0.056±0.003	16.32500	0.293±0.060	1.06563	0.101±0.003	2.08906	0.177±0.034	0.38594

Note: • indicates that the testing accuracy of StaELM is significantly better than the corresponding algorithm based on Wilcoxon signed-ranks test at 95% significance level.

Table 2. Comparison of different ELMs on 6 CLASSIFICATION datasets

Dataset	L	ELM				BELM				1HNBKM				StaELM			
		TrainAcc	TrainTime	TestAcc	TestTime	TrainAcc	TrainTime	TestAcc	TestTime	TrainAcc	TrainTime	TestAcc	TestTime	TrainAcc	TrainTime	TestAcc	TestTime
Automobile (159 instances, 15 inputs, 6 classes) KEEL	20	0.701±0.023	0.00313	0.534±0.113	0.00000	0.699±0.021	0.00000	0.534±0.113	0.00000	0.843±0.018	0.07500	0.648±0.077	0.00625	0.995±0.006	0.01563	0.667±0.066	0.00156
	30	0.767±0.022	0.00156	0.559±0.092	0.00313	0.762±0.026	0.00937	0.572±0.104	0.00000	0.878±0.018	0.06875	0.679±0.114	0.00469	1.000±0.000	0.02031	0.717±0.118	0.00156
	100	0.993±0.007	0.02187	0.597±0.135	0.00000	0.936±0.016	0.02500	0.679±0.095	0.00000	0.976±0.007	0.08438	0.711±0.052	0.00781	1.000±0.000	0.01719	0.723±0.129	0.00156
	200	1.000±0.000	0.06563	0.573±0.152	0.00156	0.978±0.009	0.11875	0.666±0.139	0.00000	0.994±0.004	0.12188	0.710±0.123	0.01094	1.000±0.000	0.01719	0.735±0.164	0.00313
	300	1.000±0.000	0.06250	0.565±0.110	0.00000	0.987±0.005	0.37812	0.660±0.129	0.00156	0.998±0.003	0.15937	0.691±0.112	0.01563	1.000±0.000	0.02656	0.716±0.110	0.00625
Ecoli (336 instances, 5 inputs, 8 classes) UCI	20	0.881±0.010	0.00469	0.872±0.051	0.00000	0.881±0.009	0.00313	0.869±0.050	0.00313	0.897±0.007	0.40937	0.866±0.048	0.03125	0.899±0.009	0.07500	0.869±0.049	0.01094
	30	0.893±0.009	0.00781	0.851±0.071	0.00000	0.886±0.009	0.00313	0.857±0.069	0.00000	0.899±0.007	0.42969	0.857±0.069	0.03281	0.894±0.007	0.07031	0.860±0.065	0.01406
	100	0.918±0.005	0.02813	0.831±0.052	0.00000	0.893±0.005	0.02656	0.848±0.057	0.00000	0.909±0.004	0.51719	0.846±0.047	0.04531	0.896±0.004	0.07969	0.860±0.037	0.02031
	200	0.917±0.004	0.11563	0.839±0.049	0.00000	0.899±0.007	0.14688	0.860±0.056	0.00000	0.922±0.007	0.67500	0.851±0.054	0.06406	0.897±0.009	0.09219	0.863±0.056	0.02969
	300	0.917±0.007	0.34531	0.842±0.046	0.00156	0.898±0.005	0.45156	0.854±0.037	0.00156	0.928±0.009	0.88594	0.830±0.028	0.07969	0.895±0.005	0.10469	0.863±0.041	0.03750
Glass Identification (214 instances, 9 inputs, 7 classes) UCI	20	0.719±0.022	0.00313	0.651±0.086	0.00000	0.721±0.021	0.00313	0.655±0.081	0.00000	0.807±0.015	0.12188	0.673±0.078	0.01094	0.818±0.016	0.02187	0.678±0.062	0.00781
	30	0.766±0.010	0.00313	0.660±0.094	0.00000	0.763±0.013	0.00781	0.660±0.092	0.00000	0.822±0.014	0.12812	0.692±0.077	0.01406	0.821±0.019	0.01875	0.702±0.093	0.00469
	100	0.935±0.009	0.02813	0.616±0.082	0.00000	0.813±0.020	0.02344	0.644±0.107	0.00000	0.889±0.017	0.17031	0.639±0.113	0.01719	0.820±0.022	0.02969	0.650±0.139	0.00625
	200	0.949±0.010	0.08906	0.635±0.126	0.00000	0.827±0.010	0.12188	0.673±0.073	0.00469	0.938±0.012	0.22031	0.673±0.083	0.02187	0.828±0.017	0.03438	0.678±0.062	0.01250
	300	0.951±0.013	0.09531	0.642±0.127	0.00156	0.836±0.009	0.39531	0.682±0.066	0.00156	0.952±0.007	0.30156	0.655±0.103	0.02969	0.825±0.016	0.04219	0.682±0.054	0.01406
Phoneme (10%) (5404 instances, 5 inputs, 2 classes) KEEL	20	0.794±0.009	0.00156	0.766±0.056	0.00000	0.792±0.010	0.00313	0.768±0.060	0.00000	0.820±0.007	1.61406	0.781±0.040	0.07969	0.838±0.009	0.27969	0.815±0.032	0.03750
	30	0.811±0.011	0.00313	0.740±0.056	0.00000	0.804±0.010	0.00625	0.774±0.033	0.00000	0.839±0.011	1.65781	0.789±0.057	0.08594	0.840±0.011	0.28750	0.802±0.038	0.03750
	100	0.878±0.009	0.04375	0.829±0.028	0.00000	0.817±0.012	0.03750	0.781±0.053	0.00000	0.877±0.008	1.87344	0.824±0.038	0.11250	0.874±0.005	0.29375	0.831±0.039	0.05000
	200	0.875±0.005	0.12812	0.829±0.058	0.00156	0.825±0.009	0.16719	0.796±0.034	0.00000	0.891±0.006	2.37031	0.826±0.049	0.17656	0.876±0.004	0.31406	0.833±0.059	0.07656
	300	0.877±0.006	0.37188	0.824±0.044	0.00156	0.833±0.007	0.48594	0.783±0.048	0.00313	0.906±0.005	3.05000	0.822±0.036	0.24375	0.878±0.008	0.33594	0.829±0.041	0.09375
Vehicle Silhouettes (846 instances, 8 inputs, 4 classes) KEEL	20	0.668±0.016	0.00469	0.637±0.040	0.00156	0.668±0.016	0.00313	0.638±0.041	0.00156	0.828±0.008	5.44219	0.745±0.022	0.23281	0.875±0.009	1.02031	0.779±0.038	0.10312
	30	0.728±0.015	0.00313	0.694±0.082	0.00000	0.727±0.014	0.00313	0.694±0.077	0.00000	0.855±0.009	5.59375	0.779±0.049	0.24375	0.892±0.008	1.02813	0.786±0.054	0.11094
	100	0.853±0.009	0.04375	0.765±0.062	0.00000	0.822±0.009	0.04844	0.753±0.035	0.00000	0.901±0.008	6.15000	0.792±0.053	0.31875	0.906±0.007	1.05469	0.797±0.043	0.14219
	200	0.893±0.011	0.17813	0.783±0.043	0.00156	0.848±0.010	0.20469	0.764±0.046	0.00156	0.933±0.004	8.15781	0.786±0.030	0.49531	0.910±0.007	1.11250	0.800±0.032	0.18906
	300	0.893±0.005	0.46250	0.787±0.036	0.00313	0.860±0.010	0.55469	0.767±0.036	0.00313	0.956±0.003	9.70937	0.770±0.036	0.65625	0.910±0.009	1.13750	0.794±0.041	0.22969
Vowel Recognition-Deerding (528 instances, 10 inputs, 11 classes) UCI	20	0.657±0.023	0.00313	0.587±0.078	0.00000	0.657±0.023	0.00313	0.586±0.072	0.00000	0.858±0.020	1.48125	0.773±0.039	0.07813	0.991±0.006	0.26406	0.926±0.036	0.03438
	30	0.747±0.021	0.00313	0.651±0.066	0.00000	0.747±0.020	0.00781	0.653±0.067	0.00000	0.906±0.010	1.53125	0.829±0.043	0.08281	0.998±0.004	0.27500	0.953±0.030	0.03438
	100	0.967±0.009	0.03281	0.896±0.044	0.00000	0.933±0.009	0.03750	0.868±0.045	0.00000	0.981±0.005	1.81719	0.934±0.037	0.10781	1.000±0.000	0.29531	0.962±0.024	0.05000
	200	0.999±0.001	0.14688	0.958±0.019	0.00156	0.965±0.007	0.16250	0.905±0.033	0.00156	0.995±0.003	2.21875	0.951±0.025	0.16094	1.000±0.000	0.30156	0.975±0.018	0.06875
	300	1.000±0.000	0.42969	0.947±0.049	0.00000	0.977±0.004	0.47813	0.924±0.046	0.00000	0.999±0.001	3.00312	0.951±0.048	0.23906	1.000±0.000	0.33594	0.956±0.044	0.09062

Note: Only continues inputs are used for every dataset. For *Phoneme* dataset, 10% of 5404 instances are randomly selected.

5   Conclusion

In this paper, we have presented a stable ELM (StaELM) by using correlation coefficient to measure the similarity between different hidden layer outputs. We have further analysed the rationality of StaELM in the framework of Gaussian process regression. Compared with the kernel-based methods, StaELM obviously reduces the chance of singular covariance matrix and make ELM more stable to random initialization of input weights and hidden biases. In addition, our improvement does not cause the significant increase of computational complexity.

Acknowledgments. This work was supported by the National Natural Science Foundations of China under Grants 61170040, 71371063, and 61473194.

References

1. Cao, J., Lin, Z., Huang, G.B.: Self-Adaptive Evolutionary Extreme Learning Machine. Neural Process. Lett. **36**(3), 285–305 (2012)
2. Chatzis, S.P., Korkinof, D., Demiris, Y.: The one-hidden layer non-parametric bayesian kernel machine. In: 23rd IEEE International Conference on Tools with Artificial Intelligence, pp. 825–831. IEEE Press, New York (2011)
3. Huang, G.B., Chen, L., Siew, C.K.: Universal Approximation Using Incremental Constructive Feedforward Networks with Random Hidden Nodes. IEEE Trans. Neural Netw. **17**(4), 879–892 (2006)
4. Huang, G.B., Li, M.B., Chen, L., Siew, C.K.: Incremental Extreme Learning Machine with Fully Complex Hidden Nodes. Neurocomputing **71**(4–6), 576–583 (2008)
5. Huang, G.B., Wang, D.H., Lan, Y.: Extreme Learning Machines: A Survey. Int. J. Mach. Learn. & Cybern. **2**(2), 107–122 (2011)
6. Huang, G.B., Zhu, Q.Y., Siew, C.K.: Extreme Learning Machine: Theory and Applications. Neurocomputing **70**(1), 489–501 (2006)
7. Janez, D.: Statistical Comparisons of Classifiers over Multiple Data Sets. J. Mach. Learn. Res. **7**, 1–30 (2006)
8. Luo, J.H., Vong, C.M., Wong, P.K.: Sparse Bayesian Extreme Learning Machine for Multi-Classification. IEEE Trans. Neural Netw. Learn. Syst. **25**(4), 836–843 (2014)
9. Matias, T., Souza, F., Araújo, R., Antunes, C.H.: Learning of A Single-Hidden Layer Feedforward Neural Network Using An Optimized Extreme Learning Machine. Neurocomputing **129**, 428–436 (2014)
10. Miche, Y., Sorjamaa, A., Bas, P., Simula, O., Jutten, C., Lendasse, A.: OP-ELM: optimally pruned extreme learning machine. IEEE Trans. Neural Netw. **21**(1), 158–162 (2010)
11. Rasmussen, C.E., Williams, C.K.I.: Gaussian Processes for Machine Learning. The MIT Press, Cambridge (2006)
12. Soria-Olivas, E., Gomez-Sanchis, J., Jarman, I.H., Vila-Frances, J.: BELM: Bayesian Extreme Learning Machine. IEEE Trans. Neural Netw. **22**(3), 505–509 (2011)
13. Wang, X.Z., Shao, Q.Y., Miao, Q., Zhai, J.H.: Architecture Selection for Networks Trained with Extreme Learning Machine Using Localized Generalization Error Model. Neurocomputing **102**, 3–9 (2013)

14. Wong, K.I., Vong, C.M., Wong, P.K., Luo, J.H.: Sparse Bayesian extreme learning machine and its application to biofuel engine performance prediction. Neurocomputing **149**, 397–404 (2015)
15. Zhu, Q.Y., Qin, A., Suganthan, P., Huang, G.B.: Evolutionary Extreme Learning Machine. Pattern Recogn. **38**(10), 1759–1763 (2005)

Local Adaptive and Incremental Gaussian Mixture for Online Density Estimation

Tianyu Qiu, Furao Shen$^{(\boxtimes)}$, and Jinxi Zhao

National Key Laboratory for Novel Software Technology,
Nanjing University, Nanjing 210093, China
`tianyuqiu.nju@gmail.com`, {`frshen,jxzhao`}`@nju.edu.cn`

Abstract. In this paper, we propose an incremental and local adaptive gaussian mixture for online density estimation (LAIM). Using a similarity threshold based criterion, the method is able to allocate components incrementally to accommodate novel data points without affecting previously learned components. A local adaptive learning strategy is presented for estimating density with complex structure in an online way. We also adopt a denoising scheme to make the algorithm more robust to noise. We compared the LAIM to the state-of-art methods for density estimation in both artificial and real data sets, the results show that our method outperforms the compared online counterpart and produces comparable results to the compared batch algorithms.

Keywords: Online density estimation · Gaussian mixture · Local adaptive · Incremental learning

1 Introduction

Let $X = (x_1, x_2, ...x_n)^T$ be a sample of size n from an unknown distribution F with density function f, the problem of density estimation is to construct a estimator $\hat{f}$ from X to approximate f as good as possible. Traditionally, the methods for density estimation are generally divided into two categories, parametric methods and non-parametric ones. Finite mixture models [1] have been used for constructing parametric probabilistic models successfully, but they suffer from choosing the appropriate number of mixture components and are sensitive to initialization. The traditional nonparametric kernel density estimator (Parazen Window) [2] is guaranteed to converge to the underlying density under practical assumptions without worrying about the magic number k (every datapoint is itself a component) [3]. However, it is not easy to choose an appropriate bandwidth parameter to achieve a good performance. A lot of researches have been made to find a data-driven criterion to search for a good value of bandwidth parameter [4], [5], [6]. In addition, the non-parametric methods and the bandwidth selection algorithms generally need to store all the training data in memory which is unfeasible in many cases.

There have been several attempts to address the problems of parametric and nonparametric methods. RSDE [7] reduces the computational cost of full sample

© Springer International Publishing Switzerland 2015
T. Cao et al. (Eds.): PAKDD 2015, Part I, LNAI 9077, pp. 418–428, 2015.
DOI: 10.1007/978-3-319-18038-0_33

KDE through a sparsity induced optimization process. [8] presents an adaptive kernel density estimator based on linear diffusion process and achieves satisfactory performance. In [9], the author proposes a greedy algorithm for learning a gaussian mixture model, which starts with a single component and adds components sequentially until a maximum number k. Due to the global and local search procedure, this algorithm need to keep all training data around and is not suitable for online learning. [10] uses a user defined likelihood based threshold parameter to add new gaussian components for the purposed of incremental learning of gaussian mixture models. However, its learning strategy involves all the components in the current model for every input sample. That makes the model converges slowly and tends to over-smooth the density in the context of online learning. SOMN [11] adopts Self-Organizing Map as its structure and proposes a learning algorithm that minimizes the Kullback-Leibler distance between the estimator and the objective density function, the learning process is limited within a small number of nodes around the input data to accelerate the convergence of nodes. The problem of SOMN is the same as SOM, that is, it is difficult to specify a network topology in advance. oKDE [12] combines the mixture model and the KDE to realize online multivariate density estimation, it maintains and updates a mixture model of the observed data from which the KDE can be calculated, compression and revitalization procedures are executed regularly to balance the accuracy and model complexity. The final estimator is defined as a convolution of the sample distribution by a kernel. This convolution strategy makes oKDE easy to over-smooth the underlying density.

To realize online density estimation that sensitive to local density structure, we propose an incremental and local adaptive gaussian mixture which estimates object density function in an online way by maximizing the sample likelihood locally around each mixture component. Unlike the SOMN, LAIM need not to specify the network structure in advance. Using a similarity threshold based criterion, the method is able to allocate components incrementally to accommodate novel data-points without destroying previously learned components. We also adopt a density based denoising algorithm that make the model more robust to noise.

2 Proposed Method

For density estimation, the LAIM is the same as traditional gaussian mixture model. Every gaussian component of LAIM could be summarized by three parameters: the mean vector μ, covariance matrix Σ and n the effective number of data-points it possesses. We introduces n here for the purpose of extending the maximum likelihood estimation for single gaussian to a local adaptive learning strategy(see Section 2.2).

The final density estimator of LAIM is

$$\hat{f}(x) = \sum_{i=1}^{K} w_i \phi(x|\mu_i, \Sigma_i), \tag{1}$$

where $w_i = \frac{n_i}{\sum_j n_j}$ is the mixing proportion for component i, μ_i and Σ_i are mean and covariance matrix of the ith component. $\phi(x|\mu, \Sigma)$ is the gaussian density function with mean μ and covariance matrix Σ.

The LAIM has three key steps:

1. **Component Allocation.** Construct a neighborhood set for input data-point and decide whether it is necessary to insert a new gaussian component into the current model.
2. **Local Adaptive Learning.** Update the parameters of the components in the neighborhood set based on maximum likelihood principle.
3. **Denoising.** Eliminate the components induced by the noisy data.

We now give the details of each step.

2.1 Component Allocation

Suppose we have built a mixture model for a series of data $x_1, x_2, \ldots, x_{t-1} \in \mathbb{R}^n$

$$\hat{f}(x|\Theta) = \sum_{i=1}^{K} w_i \phi(x|\theta_i), \tag{2}$$

where $\phi(x|\theta_i)$ is the gaussian density function, w_i is the mixing proportion for component i and $\Theta = (\theta_1, \theta_2, \ldots, \theta_K)^T$ is the parameter matrix for the mixture model. When the new data point x_t is available, traditional EM-based algorithm would make a global adaption for all the components in the current model. This operation is guaranteed to increased the sample likelihood in the long run, however, it could also destroy the previously learned structures and trapped into local optimum if x_t and following inputs are somewhat novel to the current mixture model. It is also possible that x_t is just a noise that should not be learned.

Therefore, to fit the novel data without destroying the old model, we need to decide when to allocate a new component. If we could keep all the historical samples at hand, it is possible to make choice based on the sample likelihood or some model selection criterion. In the context of online learning, we must rely on the current learned model and make choices locally. To measure the novelty of the new coming data point x_t, we first evaluate its distance $D(x_t, i)$ to the components around it weighted by its covariance matrix

$$D(x_t, i) = \sqrt{(x_t - \mu_i)^T \Sigma_i^{-1}(x_t - \mu_i)}, \tag{3}$$

here $i = 1, 2, \ldots, K_t$, K_t is the number of gaussian components at time t. Then we construct the neighbourhood set S_t of x_t

$$S_t = \{i = 1, 2, \ldots, K_t | D(x_t, i) < T_i\}, \tag{4}$$

S_t contains the set of components in the current model that should most responsible for x_t, their parameters will be updated to fit x_t(see Section 2.2). T_i is the similarity threshold for component i, it controls the tendency of the corresponding component to absorb an input sample nearby, therefore the smaller the T_i, the more local the algorithm will be. We know that $D^2(x_t, i) \sim \chi_n^2$, if the current mixture model is indeed well fitted, we could just let T_i be the value such that

$$\Pr\left[(x_t - \mu_i)^T \Sigma_i^{-1}(x_t - \mu_i) < T_i^2\right] = q \,, \tag{5}$$

where q is the confidence level, in practice we just set it with 0.9. However, we can't make the assumption that the previously learned model is reliable due to the context of online learning, the initialized components need enough samples to converge to a state of well fitted. Therefore, we let T_i be some constant α times T_i, here α is a user defined parameter. To let the new components converge fast, we usually set it with value ranged 1.5 to 2.0 at insertion and decreases to 1 through the training procedure.

If the neighborhood set S_t is empty, we'll regard x_t as a novel data that deserves a new component to fit it. The initialization of the new component is as follows:

$$n_{new} = 1 \,, \mu_{new} = x_t \,, \Sigma_{new} = h^2 I \,. \tag{6}$$

where h is a user defined parameter, I is the identity matrix. h serves as a initial bandwidth for a new component, it should be relatively small compared to the actual standard deviation along each dimension in order to keep the locality sensitivity of LAIM.

2.2   Local Adaptive Learning

Once the neighborhood set S_t is determined for the input data at time t, we limit the learning process within S_t. The local learning strategy does not only accelerate the learning process, but also gives the LAIM the ability to fit new data without destroying the learned components far away from the current input. This property is essential for online learning due to the locality of the information, we could never have the global information about the whole training set, only the current model and current input are available. Doing things locally is a safe strategy so that we can handle the non-stationary input stream.

Starting from the incremental version of maximum likelihood estimation for a single gaussian density function [13]

$$\mu^{(n)} = \mu^{(n-1)} + \frac{1}{n}\left(x_n - \mu^{(n-1)}\right) \,, \tag{7}$$

$$\Sigma^{(n)} = \Sigma^{(n-1)} + \frac{n-1}{n^2}(x_n - \mu^{(n-1)})(x_n - \mu^{(n-1)})^T - \frac{1}{n}\Sigma^{(n-1)} \,. \tag{8}$$

It is easy to see the learning step here is $1/n$, where n is the current number of samples. We want the learning step of each component could be different according to the current model and the learning process within the neighborhood region could be accelerate further since the members of the set are supposed to generate the current sample with high probability. Therefore, for every component $i \in S_t$ we make the following updates:

$$
\begin{aligned}
r_i^{(t)} &= \frac{\phi\left(x_t|\theta_i^{(t-1)}\right)}{\sum_j \phi\left(x_t|\theta_j^{(t-1)}\right)} \\
n_i^{(t)} &= n_i^{(t-1)} + r_i^{(t)} \\
\mu_i^{(t)} &= \mu_i^{(t-1)} + r_i^{(t)}\frac{1}{n_i^{(t)}}\left(x_t - \mu_i^{(t-1)}\right) \\
\Sigma_i^{(t)} &= \Sigma_i^{(t-1)} + \frac{n_i^{(t-1)}}{\left(n_i^{(t)}\right)^2}\left(x_t - \mu_i^{(t-1)}\right)\left(x_t - \mu_i^{(t-1)}\right)^T - \frac{1}{n_i^{(t)}}\Sigma_i^{(t-1)}.
\end{aligned}
\tag{9}
$$

The quantity r_i evaluates the responsibility of component i to the current data x_t, it distributed the effective number of observed samples to each component in S_t respectively weighted by their responsibilies. We also use it to adapt the updating stepsize for each component $i \in S_t$. Notice that when multiple components exist in set S_t, this quantity would slow the process of the learning by shrinking the learning stepsize. In the case that S_t has only one element, the above updating rules degenerate to the original maximum likelihood estimation for single gaussian component naturally.

2.3   Denoising

For density estimation in practice, it is common that the training data are contaminated by noise. With the assumption that the noisy data are mostly distributed over the regions where the objective density function f has low probability density and their distribution is sparse enough so that the main structure of f could still be discovered, we adopt a denoising scheme based on the effective numbers of each components, which is used by some prototype based neural networks like [14], [15]. According to the insertion rule described by Section 2.1, those noisy data would lead to node insertion with high probability. However, the resulted components should not possess large effective numbers, i.e., their n is relatively small compared to the non-noise components. Let

$$
M = \sum_{i=1}^{K} \frac{n_i}{K}
\tag{10}
$$

be the mean effective number of the current mixture model, where K is the current number of components. We eliminates those components whose effective number is lower than some constant $\beta \in [0,1]$ times M after every λ input samples. Here β and λ are user defined parameters, large value of β and small values of λ should be set if the amount of noise is large.

2.4   Complete Algorithm

As a summary, we give the complete algorithm of LAIM here.

Algorithm 1.. Local Adaptive and Incremental Gaussian Mixture

---

1: Initialize a component for the first sample according to (6).
2: Input new sample $x \in \mathbb{R}^n$.
3: Determine the neighborhood set S for x according to (4).
4: If $S = \emptyset$, initialize a new component for x according to (6), then goto step 2.
5: If $S \neq \emptyset$, update the parameters of components in S according to (9).
6: If the number of input presented so far is a multiple of the parameter λ, make the denoising operation as Section 2.3.
7: Go to step 2 if there is new sample available, otherwise the algorithm terminates and return the trained mixture model.

---

3   Experiments

3.1   Artificial Data-Sets

The artificial density functions used here are the same to those in [8]. We first adopt the common used bimodal density to verify the effectiveness of our method

$$\frac{1}{2}N\left(0,(0.1)^2\right) + \frac{1}{2}N\left(5,1\right) . \tag{11}$$

We compared the proposed method with oKDE[12], a batch kernel density estimator(kernel density estimation via diffusion, KDE-d for simple)[8] and gaussian mixture models with 2 components (the optimal choice) trained by batch EM algorithm. The parameters are set as follows: oKDE($D_{th} = 0.1$), LAIM($h = 0.5, \alpha = 1.5, \beta = \lambda = \inf$). 3000 samples are drawn from (11) as training set, the resulted estimators are shown in fig. 1. We can see from fig. 1(a) and fig. 1(d) that our method and batch EM reconstruct the underlying density function almost perfectly. That is reasonable since the assumption of mixture of gaussian is perfect for (11), but LAIM doesn't need to specify the number of components due to the incremental nature of the algorithm. From fig. 1(b), we can see that oKDE fits the right hand gaussian pretty well but over-smoothes the left hand gaussian component, that's mainly because its estimation is based on a global convolution operation that lacks of locality sensitivity. fig. 1(c) also shows some under-smoothness on the right hand gaussian. This result shows that LAIM achieves the comparable performance to the batch EM algorithm in this simple bimodal situation.

Then we use the density function "claw", which has more complex structure. The parameters are set as follows: oKDE($D_{th} = 0.1$), LAIM($h = 0.5, \alpha = 1.5, \beta = \lambda = \inf$), the number of components for GMM is identical to LAIM, which is 10 in this case.

$$\frac{1}{2}N\left(0,(0.1)^2\right) + \sum_{k=0}^{4}\frac{1}{10}N\left(k/2 - 1,(0.1)^2\right) . \tag{12}$$

the results shown in fig. 2 suggest that our method could approximate the density function in each local region hence gives a reliable estimation. Batch EM

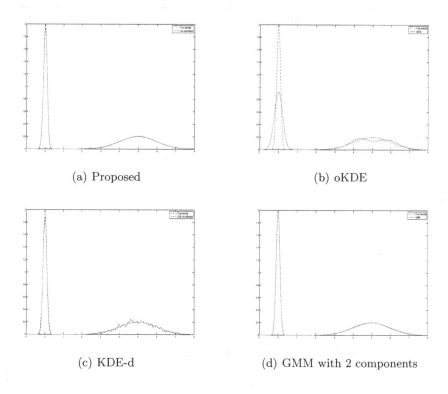

(a) Proposed

(b) oKDE

(c) KDE-d

(d) GMM with 2 components

Fig. 1. Blue dashed line is the density function (11), red solid line is the coresponding estimator. The estimator of proposed method and GMM with 2 components is almost perfect. oKDE over-smoothed the left hand gaussian while KDE-d under-smoothes the right hand gaussian.

algorithm fails to capture the whole structure of (12). Fig .2(b) is also a result of over-smoothing the sample distribution constructed by oKDE.

To quantify the approximation performance of our method, we did the numerical experiments on five artificial data sets (including (10), (11)). The criterion for the comparison is the numerical approximation to the following ratio,

$$Ratio = \frac{||\hat{f} - f||^2}{||\hat{g} - f||^2} \tag{13}$$

which was adopted by KDE-d [8]. Here $\hat{g}$ is the estimator with which we want to compare, $\hat{f}$ is our estimator and f is the underlying density function. (11) is the integrated squared error of the diffusion estimator to the integrated squared error of the alternative estimator. The results are shown in table .1. When compared to online method oKDE, our approach has lower integrated squared error, which means the corresponding estimator is more accurate. The proposed method outperformes the batch method KDE-d in case 1 and 3, that is reasonable since the density in those two cases are well-separated gaussians that are more suitable for the mixture models. In case 2 where the density function contains complex local structure, LAIM achieves comparable result to batch algorithm KDE-d.

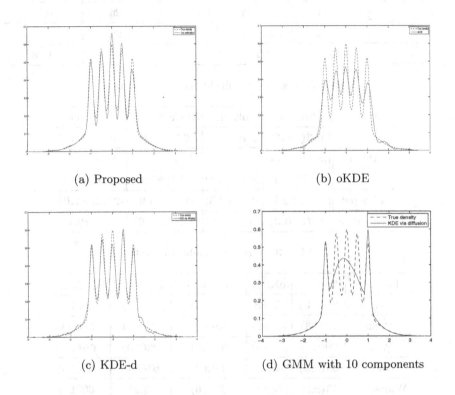

(a) Proposed

(b) oKDE

(c) KDE-d

(d) GMM with 10 components

Fig. 2. Blue dashed line is the density function (12), red solid line is the corresponding estimator. The proposed method and KDE via diffusion gave the satisfactory estimation results. oKDE over-smoothes the peaks of the density. Batch EM algorithm didn't capture the whole density structure.

3.2   Real Data-Sets

We compared our method with oKDE, RSDE and KDE with bandwidth selected by cross validation(CV) on the real datasets obtained from the UCI Machine learning Repository[16]. Five data-sets are used here for testing, Iris, Pima, Wine, WineRed and WineWhite. For the density estimation, we estimated the density for each class separately. The data were randomly reordered, 75% of the data in each class were used for training, and the rest for testing. We conduct the same experiment twelve times and recorded the mean and standard deviation

Table 1. Ratio of approximated integrated squared error

Case	Target density	Ratio(oKDE)	Ratio(KDE-d)
1	$\frac{1}{2}N\left(0,(1/10)^2\right) + \frac{1}{2}N\left(5,1\right)$	0.04	0.38
2	$\frac{1}{2}N\left(0,(1/10)^2\right) + \sum_{k=0}^{4}\frac{1}{10}N\left(k/2-1,(1/10)^2\right)$	0.67	1.05
3	$\frac{1}{2}N\left(-2,(1/4)^2\right) + \frac{1}{2}N\left(2,(1/4)^2\right)$	0.26	0.76
4	$\frac{2}{3}N\left(0,1\right) + \frac{1}{3}N\left(0,(1/10)^2\right)$	0.90	1.42
5	$\frac{3}{4}N\left(0,1\right) + \frac{1}{4}N\left(3/2,(1/3)^2\right)$	0.91	1.78

Table 2. Negative log likelihood on real datasets

Dataset	Proposed	oKDE	RSDE(batch)	CV(batch)
Iris	**0.3(±0.2)**	2.1(±0.5)	2.5(±0.9)	2.7(±0.9)
Pima	**28.8(±0.4)**	32.3(±0.3)	38.4(±11.3)	29.5(±0.5)
Wine	23.5(±3.1)	26.4(±3.4)	12.3(±1.9)	**11.6(±1.5)**
Winered	13.9(±2.7)	18.4(±3.5)	-12.3(±4.9)	**-27.2(±1.0)**
Winewhite	**8.7(±0.3)**	11.4(±0.3)	91.3(±44.6)	11.6(±0.4)

Table 3. Average classification rate

Dataset	Proposed	oKDE	RSDE(batch)	CV(batch)	SVM(batch)
Iris	**97(±3)**	**97(±3)**	96(±4)	96(±3)	96(±2)
Pima	73(±3)	72(±2)	65(±3)	72(±2)	**78(±3)**
Wine	89(±3)	94(±3)	91(±5)	92(±6)	**96(±3)**
Winered	**65(±2)**	64(±2)	44(±4)	64(±1)	63(±3)
Winewhite	**63(±2)**	55(±1)	25(±6)	62(±1)	60(±2)

as the result. The oKDE were initialized by the first 10 samples and parameter D_{th} was set to 0.1. Two parameters of LAIM are set as follows: $\alpha = 2.2$, $\beta = 0.05$, $\lambda = 0.1N$, where N is the total number of training samples. h is set according to the the scale of the data because it will affect the the complexity of the model: Iris(0.01), Pima(15), Wine(15), Winered(15), Winewhite(5). To measure the quality of estimation, we have computed the average negative likelihood(NLL) per test point, lower NLL generally suggests more accurate estimation.

The results of the experiments after observing all the data-points are summarized in Table 2. Compared to the online method oKDE, the proposed method achieves better results on all the data-sets. LAIM also ourperforms the batch methods on Iris, Pima and WineWhite.

We have also tested the classification ability of the proposed method on the previous five data-sets. Although density estimator is not generally the most accurate classifier, the classification results based on simple Bayesian criterion

$$\hat{y} = \arg\max_k p(x|c_k)p(c_k) \tag{14}$$

still reflects the quality of density estimation. We have chosen a multiclass SVM with RBF kernel[17] as the baseline classifer and compares our method with oKDE, RSDE, and KDE with cross validation. The results of classification are summarized in Table 3. From the table, we can see that the proposed method outperformes the online counterparts in most data-sets except for Wine and produces comparable results to the batch methods. Noticed that in the context of online learning, we don't store any historical data but the current input and the learned model, therefore, the time complexity and space complexity of LAIM are much smaller than the batch methods.

4 Conclusion

In this paper, the incremental and local adaptive gaussian mixture for online density estimation(LAIM) is proposed. With the similarity threshold, the method could allocate components incrementally while training without specifying the number of gaussian components in advance. We proposes a local learning algorithm for updating the parameters of mixture model based on maximum likelihood principle, this locality sensitivity enables the our model to discover the local density structure of the data samples in the context of online learning. A denoising scheme is used to eliminate the components initialized by noise. Experiments show that it outperforms the compared online density estimators and produces comparable results to the compared batch methods while keeping a lower model complexity.

Acknowledgments. The authors are very grateful to Youlu Xing for many useful discussions and programming assistance. We also acknowledge Tao Zhu and Haoran Xu for helpful comments. This work is supported in part by the National Science Foundation of China under Grant Nos. (61375064, 61373001) and Jiangsu NSF grant (BK20131279).

References

1. McLachlan, G., Peel, D.: Finite mixture models. Wiley-Interscience (2000)
2. Parzen, E.: On estimation of a probability density function and mode. Annals of Math. Statistics **33**, 1065–1076 (1962)
3. Wand, M.P., Jones, M.C.: Kernel Smoothing. Chapman Hall/CRC (1995)
4. Hall, P., Sheater, S.J., Jones, M.C., Marron, J.S.: On optimal data-based bandwidth selection in kernel density estimation. Biometrika **78**(2), 263–269 (1991)
5. Silverman, B.W.: Density Estimation. Chapman and Hall, London (1986)

6. Hall, P.: Large sample optimality of least squares cross-validation in density estimation. The Annals of Statistics, 1156–1174 (1983)
7. Girolami, M., He, C.: Probability density estimation from optimally condensed data samples. IEEE Trans. Pattern Anal. Mach. Intell. **25**(10), 1253–1264 (2003)
8. Botev, Z.I., Grotowski, J.F., Kroese, D.P.: Kernel density estimation via diffusion. The Annals of Statistics **38**(5), 2916–2957 (2010)
9. Vlassis, N., Likas, A.: A greedy EM algorithm for Gaussian mixture learning. Neural processing letters **15**(1), 77–87 (2002)
10. Engel, P.M., Heinen, M.R.: Incremental learning of multivariate gaussian mixture models. In: da Rocha Costa, A.C., Vicari, R.M., Tonidandel, F. (eds.) SBIA 2010. LNCS, vol. 6404, pp. 82–91. Springer, Heidelberg (2010)
11. Yin, H., Allinson, N.M.: Self-organizing mixture networks for probability density estimation. IEEE Transactions on Neural Networks **12**(2), 405–411 (2001)
12. Kristan, M., Leonardis, A., Skoaj, D.: Multivariate online kernel density estimation with Gaussian kernels. Pattern Recognition **44**(10), 2630–2642 (2011)
13. Bishop, C.M.: Pattern recognition and machine learning, vol. 1. Springer, New York (2006)
14. Ouyang, Q., Shen, F., Zhao, J.: A local distribution net for data clustering. In: Anthony, P., Ishizuka, M., Lukose, D. (eds.) PRICAI 2012. LNCS, vol. 7458, pp. 411–422. Springer, Heidelberg (2012)
15. Furao, S., Ogura, T., Hasegawa, O.: An enhanced self-organizing incremental neural network for online unsupervised learning. Neural Networks **20**(8), 893–903 (2007)
16. Asuncion, A., Newman, D.: UCI machine learning repository (2007)
17. Chang, C.-C., Lin, C.-J.: LIBSVM: a library for support vector machines. ACM Transactions on Intelligent Systems and Technology (TIST) **2**(3), 27 (2011)

Latent Space Tracking from Heterogeneous Data with an Application for Anomaly Detection

Jiaji Huang[1][(✉)] and Xia Ning[2]

[1] Department of Electrical Engineering, Duke University, Durham, NC 27708, USA
jiaji.huang@duke.edu
[2] Department of Computer and Information Science,
IUPUI, Indianapolis, IN 46202, USA
xning@cs.iupui.edu

Abstract. Streaming heterogeneous information is ubiquitous in the era of Big Data, which provides versatile perspectives for more comprehensive understanding of behaviors of an underlying system/process. Human analysis of these volumes is infeasible, leading to unprecedented demands for mathematical tools which effectively parse and distill such data. However, the complicated nature of streaming heterogeneous data prevents the conventional multivariate data analysis methods being applied immediately. In this paper, we propose a novel framework together with an online algorithm, denoted as LSTH, for latent space tracking from heterogeneous data. Our method leverages the advantages of dimension reduction, correlation analysis and sparse learning to better reveal the latent relations among heterogeneous information and adapt to slow variations in streaming data. We applied our method on both synthetic and real data, and it achieves results competitive with or superior to the state-of-the-art in detecting several different types of anomalies.

1 Introduction

In the era of Big Data, heterogeneity of various information generated from a same yet complex underlying system/process has become ubiquitous. Examples of such heterogeneous data include video and audio from a sensor network, acoustic and articulatory signals during a speech, etc. Such heterogeneous data provides complimentary or augmented depiction of the system from different perspectives, allowing more comprehensive understanding of the system than that from homogeneous data. Albeit the high dimensionality and heterogeneity, these data often exhibits low dimensional nature and can be characterized by a (low dimensional) latent space. Correctly identifying the latent space benefits classical machine learning tasks (e.g., classification [6]), as well as more novel applications (e.g., the anomaly detection). However, learning from heterogeneous data is highly nontrivial. The requirement of operating in real time imposes further challenges and prevents straightforward extensions of existing methods.

© Springer International Publishing Switzerland 2015
T. Cao et al. (Eds.): PAKDD 2015, Part I, LNAI 9077, pp. 429–441, 2015.
DOI: 10.1007/978-3-319-18038-0_34

Principal Component Analysis (PCA) [7] is arguably the most well-known method for extracting the low dimensional latent space. A common assumption in applying PCA is that most data is near the low dimensional space. The anomalies are assumed to be significantly deviated from the space such that using some simple statistics is sufficient to identify them. Inspired by this assumption, online PCA [12] techniques are developed to conduct anomaly detection on data streams. Representative online PCA algorithms include [4] as well as its extension [17] under union-of-subspace assumption. However, PCA based methods do not model the relations between the heterogeneous data sources. Therefore, PCA cannot identify anomalies corresponding to violation of the relations. In contrast, Canonical Correspondence Analysis (CCA) [5] is a classical method for analyzing the relation between multiple data sources. And online CCA through stochastic gradient on generalized Stiefel manifold has been applied to anomaly detection on time series [19]. However, it still does not fully consider the heterogeneous nature of the data.

Recently, learning from heterogeneous data has attracted much attention in machine learning community, particularly in transfer learning, multi-task learning and multi-view learning. Transfer learning utilizes an auxiliary source domain data to learn a better model in a target domain, where the two domains are often heterogeneous [13]. Multi-task learning leverages the relation between multiple tasks, each of which may work on a different/heterogeneous data domain [6]. Multi-view learning leverages multiple views of same instances for better models [18]. Many of these works assume a common low dimensional latent space, and learn a mapping from each data source/view to the latent space in a supervised fashion. However, adapting these methods to an online and unsupervised setting (e.g., anomaly detection task) is not straightforward.

In this paper, we tackle the problem of online learning of heterogeneous data via latent space tracking. In specific, we propose a framework to track the low-dimensional latent structures of heterogeneous data and learn their inherent relations. Our formulation incorporates the key insights underlying PCA, CCA, and sparse learning to enable dimension reduction together with feature selection for anomaly detection from heterogeneous data. We develop an efficient online algorithm that effectively conducts **L**atent **S**pace **T**racking from **H**eterogeneous data, denoted as LSTH. Based on the learned latent space, we further design an anomaly detection method that reports anomalies significantly outlying the latent space. We test LSTH on both synthetic and real datasets. Experimental results demonstrate that LSTH is effective in revealing relations among heterogeneous data for anomaly detection.

The paper is organized as follows. Section 2 formulates the latent space tracking problem. Section 3 presents the tracking algorithm. Section 4 further designs an anomaly detection method as an application of the learned latent space. Experimental results and conclusions are in Section 5 and 6 respectively.

2 Problem Formulation

Throughout this paper, vectors are represented by lower-case letters (e.g., x), and matrices are represented by upper-case letters (e.g., U). By default, all the vectors are column vectors, while row vectors are represented by having a transpose superscript$^\top$ (e.g., $x^\top$). We use subscript i and j to index an element in a matrix (e.g., $V_{i,j}$) and subscript t to index a data point at timestamp t (e.g., x_t) in a data steam. The estimate of a variable is represented by having a hat over the variable (e.g., $\widehat{U}$ represents the estimate of U).

We assume $x_t \in \mathbb{R}^{D_x}$ and $y_t \in \mathbb{R}^{D_y}$ are the high-dimensional heterogeneous data samples from a same system at timestamp t, where D_x and D_y are the number of features in x_t and y_t, respectively. The heterogeneity of particular interest in this paper is that x_t's features are correlated, whereas only very few features in y_t describe the states of the system. Heterogeneous data in many real-life applications exhibits such kind of property. For example, during a speech, y_t can be data recorded by articulatory sensors, which are highly correlated [3] due to connected muscles. In contrast, x_t can be Mel-frequency cepstrum coefficients (MFCC). Obtained by appending higher order derivatives of acoustic signal, it contains much redundancy and often need a feature selection [11] step before further processing. In a stock market, x_t could be the prices of multiple correlated stocks, and y_t is massive news about the market [14].

In order to learn the underlying structures and relations among x_t and y_t, we monitor the joint probability density $p(x_t, y_t)$ at each timestamp t:

$$p(x_t, y_t) = p(y_t|x_t)p(x_t). \tag{1}$$

However, since both x_t and y_t are of high dimensionality, online density estimation for $p(y_t|x_t)$ or $p(x_t)$ is prohibitively difficult. Therefore, we assume there is a d dimensional latent space ($d \ll D_x, D_y$) underlying the data, into which x_t and y_t can be transformed via two linear projectors $U \in \mathbb{R}^{D_x \times d}$ and $V \in \mathbb{R}^{D_y \times d}$. Their projections are denoted as $U^\top x_t$ and $V^\top y_t$, respectively, which can be considered as realizations of a common latent variable that determines the states of the underlying system. U and V will exhibit different structures. Specifically, while U may span a low-rank subspace as in PCA, V may model a latent space impacting only a subset of the features in y_t.

3 Proposed Approach

We constrain U to be orthonormal (i.e., $U^\top U = \mathbf{I}$, where $\mathbf{I}$ is the identity matrix) to preserve the magnitude of x_t. Thus, the reconstruction error of x_t is $\|x_t - UU^\top x_t\|^2$. In this case, we measure the probability distribution of x_t by the reconstruction error [17]:

$$p(x_t) \propto \exp\left(-\|x_t - UU^\top x_t\|^2/\sigma_x^2\right), \tag{2}$$

where σ_x^2 is the variance of reconstruction error in each dimension. Since the projections of x_t and y_t are considered as realizations of a common latent variable, they are expected to be close. Hence, we measure $p(y_t|x_t)$ by the distance of the projections in the latent space:

$$p(y_t|x_t) \propto \exp\left(-\|V^\top y_t - U^\top x_t\|^2/\sigma_y^2\right), \tag{3}$$

where σ_y^2 is the variance of the difference between x_t and y_t in the latent space.

By substituting Equation (2) and (3) into (1) and taking the logarithm, the log-likelihood can be represented as

$$\log p(x_t, y_t) \propto -\left[\frac{\|V^\top y_t - U^\top x_t\|^2}{\sigma_y^2} + \frac{\|(\mathbf{I} - UU^\top)x_t\|^2}{\sigma_x^2}\right].$$

In addition, we constrain V to exhibit "group sparse" structure so that applying V performs feature selection from y_t to identify the most informative features. We use the mixed norm $\|V\|_{1,2} \triangleq \sum_{i=1}^{D_y} \|v_i^\top\|_2$ to introduce sparsity into V, where $v_i^\top$ is the i-th row of V.

To enable tracking in a slowly evolving environment, we apply an exponentially decaying window to downweigh the historical samples. In addition, we define $\sigma = \sigma_y^2/\sigma_x^2$, and denote the estimates of U and V at timestamp t as $\widehat{U}_t$ and $\widehat{V}_t$, respectively. Then we formulate the following optimization problem to find the projectors U and V at timestamp t:

$$
\begin{aligned}
(\widehat{U}_t, \widehat{V}_t) &= \underset{U^\top U = \mathbf{I}, V}{\arg\min}\ F(U, V; t, \alpha, \sigma, \lambda) \\
&= \underset{U^\top U = \mathbf{I}, V}{\arg\min} \sum_{k=0}^{t-1} \frac{\alpha^k}{2}\left(\|U^\top x_{t-k} - V^\top y_{t-k}\|^2 + \sigma\|(\mathbf{I} - UU^\top)x_{t-k}\|^2\right) + \lambda\|V\|_{1,2},
\end{aligned}
\tag{4}
$$

where $\alpha \in (0, 1]$ is a forgetting factor over historical samples to implement the decaying window, σ balances between projection residual and discrepancy in the latent space, and λ is the regularization parameter for sparsity. Note that the data stream starts from $t = 1$.

In the above $F(U, V; t, \alpha, \sigma, \lambda)$, the first term measures the discrepancy of two data sources in the latent space. It has the flavor of CCA that maximizes the correlation of two projections. Same as PCA, the second term imposes low-dimensional structure in x_t. It is important to highlight the $\|V\|_{1,2}$ term here. $\|v_i^\top\|_2$ indicates the significance of the i-th feature in y_t. In addition, $\|V\|_{1,2}$ is invariant if multiplying an unitary matrix to the right of V. Therefore, the cost of (4) depends on the subspace spanned by $\widehat{U}_t$ and $\widehat{V}_t$ rather than the particular basis chosen.

3.1   A Batch Algorithm

We first present a batch algorithm, denoted as bLSTH, to solve U and V for simplicity. The bLSTH algorithm will be further modified into an online version in Section 3.2.

Algorithm 1. The Batch Algorithm bLSTH

Input: samples $X \in \mathbb{R}^{D_x \times L}$, $Y \in \mathbb{R}^{D_y \times L}$
        Parameters: λ, σ, latent dimension d
Output: $\widehat{U}$ and $\widehat{V}$
$\quad i \leftarrow 0, \quad U[0] \leftarrow$ the first d principal components of X
$\quad$ **repeat**
$\qquad i \leftarrow i + 1$
$\qquad Z \leftarrow U[i-1]^\top X$

$$V[i] \leftarrow \arg\min_V \frac{1}{2}\|V^\top Y - Z\|_F^2 + \lambda \|V\|_{1,2} \tag{6}$$

$\qquad W \leftarrow V[i]^\top Y$

$$U[i] \leftarrow \arg\min_{U^\top U = \mathbf{I}} \frac{1}{2}\left(\|U^\top X - W\|_F^2 + \sigma\|(\mathbf{I} - UU^\top)X\|_F^2\right) \tag{7}$$

$\quad$ **until** $U[i], V[i]$ converge or i is large enough
$\quad \widehat{U} \leftarrow U[i], \widehat{V} \leftarrow V[i]$

In bLSTH, L buffered samples $X = [x_{-L+1}, \cdots, x_0]$, $Y = [y_{-L+1}, \cdots, y_0]$ are used to solve the following optimization problem:

$$(\widehat{U}, \widehat{V}) = \arg\min_{U^\top U = \mathbf{I}, V} \frac{1}{2}\left(\|U^\top X - V^\top Y\|_F^2 + \sigma\|(\mathbf{I} - UU^\top)X\|_F^2\right) + \lambda\|V\|_{1,2}.$$

We use an alternating method to solve for $\widehat{U}$ and $\widehat{V}$, as presented in Algorithm 1. The optimization problem in Equation (6) of Algorithm 1 is a well-studied convex optimization problem. Now we focus on the optimization problem in Equation (7). The objective can be reformulated as:

$$\begin{aligned}
f(U; \sigma) &\triangleq \frac{1}{2}\left(\|U^\top X - W\|_F^2 + \sigma\|(\mathbf{I} - UU^\top)X\|_F^2\right) \\
&= \frac{1}{2}(1 - \sigma)\,\mathrm{tr}\left\{U^\top X X^\top U\right\} - \mathrm{tr}\left\{(XW^\top)U^\top\right\},
\end{aligned} \tag{5}$$

where $U^\top U = \mathbf{I}$ and $W = V^\top Y$. This orthonormality constrained problem is non-convex. However, we are able to find a local minimum within a few iterations and our experiments show that even local minimum is able to give good results. Following the idea in [8], we use a majorization minimization scheme. The basic idea is to construct a non-decreasing sequence $f(U[1]), \ldots, f(U[k]), \ldots$ that converges to a local minimum of $f(U)$. Specifically, suppose we are at $U[k]$, we construct a surrogate function $g_k(U)$ that satisfies

$$f(U) \leq g_k(U) \text{ and } f(U[k]) = g_k(U[k]). \tag{8}$$

That is, $g_k(U)$ is an upper bound of $f(U)$ and the equality holds when $U = U[k]$. Assign the global minimizer of $g_k(U)$ to $U[k+1]$, thus the sequence

$f(U[1]), \ldots, f(U[k]), \ldots$ is guaranteed to be non-increasing due to the properties of $g_k(U)$ as in Equation (8) and the notion of global minimizer. In practice, a surrogate function should be constructed such that its global minimizer is easily obtained. The following two lemmas suggest one form of such $g_k(U)$ and its global minimizer.

Lemma 1. *For any given orthonormal matrix* $U[k] \in \mathbb{R}^{D_x \times d}$, *the following* $g_k(U; a)$ *defined on the set of orthonormal matrices* $U \in \mathbb{R}^{D_x \times d}$

$$g_k(U; a) = \mathbf{tr} \left\{ \left[(1 - \sigma)(XX^\top - a\,\mathbf{I})U[k] - (XW^\top) \right]^\top U \right\} + c$$

is a surrogate function for the $f(U; \sigma)$ *in Equation (5), where c is some constant independent of U. And the scalar a chosen as*

$$a = \begin{cases} \lambda^* & \sigma < 1 \\ 0 & \sigma \geq 1 \end{cases},$$

where λ^* is the maximum eigenvalue of $XX^\top$.

Proof. The proof leverages Rayleigh quotient inequality and is omitted for conciseness.

Lemma 2. *[10] The global minimizer of*

$$\min_{U^\top U = \mathbf{I}} -\mathbf{tr}\{A^\top U\}$$

is $PQ^\top$, where $P\Sigma Q^\top = A$ is the Singular Value Decomposition (SVD) of A.

Using Lemma 2, the global minimizer of the surrogate function $g_k(U; a)$ has a closed form $\arg\min_{U^\top U = \mathbf{I}} g_k(U; a) = PQ^\top$, where $P\Sigma Q^\top$ is the SVD of $XW^\top - (1 - \sigma)(XX^\top - a\,\mathbf{I})U[k]$. Thus, by applying Lemma 1 and 2, the problem in Equation (7) can be solved via the iterative majorization minimization process as presented in Algorithm 2, where $G \triangleq XW^\top = XY^\top V$ and $C_x \triangleq XX^\top$. A special case is when $\sigma = 1$, in which the minimizer of $f(U; \sigma)$ is given by the closed-form solution directly by Lemma 2.

3.2   An Online Algorithm

Here we derive the online algorithm LSTH from bLSTH. We use the solution $(\widehat{U}, \widehat{V})$ by bLSTH on the samples $X = [x_{-L+1}, \cdots, x_0]$, $Y = [y_{-L+1}, \cdots, y_0]$ as the initialization $(\widehat{U}_0, \widehat{V}_0)$ for the online updates, assuming the online process starts from timestamp $t = 1$. We also use an alternating method to track (U_t, V_t) with the following definition of projections of x_t and y_t into the latent space:

$$z_t \triangleq U_t^\top x_t, \qquad w_t \triangleq V_t^\top y_t.$$

The online algorithm LSTH consists of an initialization via bLSTH and iterative online updates of U and V, as presented in Algorithm 3.

Algorithm 2. Updating U for bLSTH

Input: orthonormal U, scalar σ, cross-covariance matrix G
           auto-covariance matrix C_x
Output: U_{updated}
   $k \leftarrow 0,\ U[k] \leftarrow U$
   repeat
      $k \leftarrow k + 1$

      compute SVD: $P\Sigma Q^\top = G - (1-\sigma)(C_x - a\,\mathbf{I})U[k-1]$ (9)

      $U[k] \leftarrow PQ^\top$
   until $U[k]$ converged or k is large enough
   $U_{\text{updated}} \leftarrow U[k]$

Online Tracking of U_t. Upon arrival of new data (x_t, y_t) at t, we use $\widehat{V}_{t-1}$ to estimate the projection of y_t at t as follows:

$$\widehat{w}_t = \widehat{V}_{t-1}^\top y_t. \tag{10}$$

Substituting the $\widehat{w}_t$ into Equation (4), we will see that the objective function of U is of the same form as (5), except that the historical x_t are downweighed. Therefore it can be minimized via Algorithm 2 with the only modification that G in Equation (9) is replaced by $\sum_{k=0}^{t-1} \alpha^k x_{t-k}\widehat{w}_{t-k}^\top$, and C_x is replaced by $\sum_{k=0}^{t-1} \alpha^k x_{t-k}x_{t-k}^\top$. Both of these two summations can be incrementally updated.

Online Tracking of V_t. Given $\widehat{U}_t$ solved as in Section 3.2, we use $\widehat{U}_t$ to estimate z_t at current timestamp t as follows

$$\widehat{z}_t = \widehat{U}_t^\top x_t. \tag{11}$$

Substituting $\widehat{z}_t$ into Equation (4), we can get the following objective function w.r.t V,

$$F_V(V;t) = \sum_{k=0}^{t-1} \left[\frac{\alpha^k}{2} \left\| V^\top y_{t-k} - \widehat{z}_{t-k} \right\|_2^2 \right] + \lambda \|V\|_{1,2}. \tag{12}$$

For the above problem, we derive a Stochastic Coordinate Descent (SCD) method with a similar spirit as [9]. The SCD admits a row-wise updating of $\widehat{V}_t$, details can be found in Equation (13) in Algorithm 3.

3.3 Complexity Analysis

The complexity of LSTH is $O(c \cdot D_x^2 d + D_y^2 d)$, where $c \cdot D_x^2 d$ is due to the SVD step in Equation (9) and c is the number of iterations in majorization minimization for U ($c = 1$ suffices in practice). Efficient algorithms for computing the SVD of a sequentially updated matrix [2] can be applied to reduce the complexity. $D_y^2 \cdot d$ is due to the coordinate descent algorithm on V, for which further acceleration can be achieved via active set tricks. Our experiments show that LSTH is sufficiently fast for real applications, for example, 20 ms for the XRMB dataset (sampling interval: 25 ms/sample). The experimental details will be presented in Section 5. To reduce the complexity of LSTH is very important and it is left for future exploration for now.

4 Application: Anomaly Detection

The basic idea of our anomaly detection method is to monitor $\|U^\top x_t - V^\top y_t\|^2 + \sigma \|(\mathbf{I} - UU^\top)x_t\|^2$. We define the *a priori* error:

$$\xi_t \triangleq \|\widehat{U}_{t-1}^\top x_t - \widehat{V}_{t-1}^\top y_t\|^2 + \sigma\|x_t - \widehat{U}_{t-1}\widehat{U}_{t-1}^\top x_t\|^2, \qquad (14)$$

and use ξ_t as the detection statistic. An anomaly is claimed only when $p(x_t, y_t)$ appears to be significantly small, corresponding to ξ_t being significantly large. We maintain a sliding window over ξ_t with the mean μ_t and standard deviation ν_t within the window. When the new (x_{t+1}, y_{t+1}) arrives, we compare its ξ_{t+1}

Algorithm 3. The Online Algorithm LSTH

Parameters: d, α, λ, σ
Input: data stream: $\cdots, (x_0, y_0), \cdots, (x_t, y_t), \cdots$
Obtain $\widehat{U}_0$ and $\widehat{V}_0$ by Algorithm 1
for $t = 1, 2, \ldots$ **do**
    //update $\widehat{U}_t$
    $\widehat{w}_t \leftarrow \widehat{V}_{t-1}^\top y_t$
    $G_t \leftarrow \alpha G_{t-1} + x_t \widehat{w}_t^\top$ $/{*}G_0 = \sum_{\tau=-L+1}^0 x_\tau w_\tau^\top {*}/$
    $C_{x,t} \leftarrow \alpha C_{x,t-1} + x_t x_t^\top$ $/{*}C_{x,0} = \sum_{\tau=-L+1}^0 x_\tau x_\tau^\top {*}/$
    get $\widehat{U}_t$ via Algorithm 2 with $(\widehat{U}_{t-1}, \sigma, G_t, C_{x,t})$ as input
    //update $\widehat{V}_t$
    $\widehat{z}_t \leftarrow \widehat{U}_t^\top x_t$
    $H_t \leftarrow \alpha H_{t-1} + y_t \widehat{z}_t^\top$ $/{*}H_0 = \sum_{\tau=-L+1}^0 y_\tau z_\tau^\top {*}/$
    $C_{y,t} \leftarrow \alpha C_{y,t-1} + y_t y_t^\top$ $/{*}C_{y,0} = \sum_{\tau=-L+1}^0 y_\tau y_\tau^\top {*}/$
    for $i = 1, 2, \ldots, D_y$ **do**
        Calculate the i-th row of $\widehat{V}_t$:

$$\widehat{v}_{t,i}^\top = \frac{S(\|h_{t,i}^\top - \sum_{j \neq i} v_{t-1,j}^\top C_{y,t,i,j}\|, \lambda)}{C_{y,t,i,i}} \times \frac{h_{t,i}^\top - \sum_{j \neq i} v_{t-1,j}^\top C_{y,t,i,j}}{\|h_{t,i}^\top - \sum_{j \neq i} v_{t-1,j}^\top C_{y,t,i,j}\|}, \qquad (13)$$

        where $S(\cdot, \lambda)$ is the soft thresholding function with parameter λ.
    end for
end for

---

with a threshold $b_t = \mu_t + \gamma \nu_t$, where $\gamma > 0$ indicates the effect of variance. Once ξ_{t+1} exceeds the threshold, an anomaly is claimed.

Additional care need to be taken for the claimed anomalous data points. In specific, if the anomaly behaves as a sudden outlier after which the data stream goes back to normal state, then the anomalous data point should be excluded for model updating. The other case is that the anomaly is in fact the start of a different stage in the data stream, then the anomalous data point should be included in model updating. These two cases will be addressed in synthetic and real data experiments respectively.

5 Experiments

In this section, we conduct comparative experiments to demonstrate the performance of LSTH in tracking the latent space for anomaly detection. All types of tracking methods as well as their corresponding anomaly detection statistics are summarized in Table 1.

Table 1. Latent space tracking methods and corresponding detection statistics

method	detection statistics	semantics
LSTH	$\xi_t = \|\widehat{U}_{t-1}^{\top} x_t - \widehat{V}_{t-1}^{\top} y_t\|^2 + \sigma\|x_t - \widehat{U}_{t-1}\widehat{U}_{t-1}^{\top} x_t\|^2$	latent discrepancy and projection residual
(online) CCA	$\delta_t = \widehat{C}_{x,t-1} r_{x,t}/D_x + r_{y,t}^{\top}\widehat{C}_{y,t-1} r_{y,t}/D_y$ where $r_{x,t} = (\widehat{C}_{x,t-1}^{-1} - \widehat{U}_{t-1}\widehat{U}_{t-1}^{\top})x_t$ and $r_{y,t} = (\widehat{C}_{y,t-1}^{-1} - \widehat{V}_{t-1}\widehat{V}_{t-1}^{\top})y_t$	projection residual onto Generalized Stiefel manifold [19]
(online) PCAx PCAy PCAxy	$\epsilon_{x,t} = \|(\mathbf{I}-\widehat{U}_{t-1}\widehat{U}_{t-1}^{\top})x_t\|^2$ $\epsilon_{y,t} = \|(\mathbf{I}-\widehat{U}_{t-1}\widehat{U}_{t-1}^{\top})y_t\|^2$ $\epsilon_{xy,t} = \|(\mathbf{I}-\widehat{U}_{t-1}\widehat{U}_{t-1}^{\top})[x_t;y_t]\|^2$	projection residual onto individual or joint signal subspace [4]

5.1 Experiments on Synthetic Data

We generated a synthetic dataset with continuous data $x_t \in \mathbb{R}^{500}$ and sparse, discrete and non-negative data $y_t \in \mathbb{R}^{1000}$. The x_t's are generated via a linear model $x_t = A\theta_t + n_t$, $t = 1,\ldots,10500$. where $A \in \mathbb{R}^{500\times 10}$, $\theta_t \in \mathbb{R}^{10}$ and n_t is white Gaussian noise. The y_t's are generated as of dimension 1000. The first 50 features of y_t's are relevant to the underlying system, generated via $B\theta_t + m_t$, $t = 1,\ldots,10500$, where $B \in \mathbb{R}^{50\times 10}$ and m_t is white Gaussian noise. The rest 950 dimensions are padded as noise. We introduced sparsity into y_t by randomly setting half of its values to zero. In the end we round the y_t to non-negative integers. In this way, y_t is analogous to the real-world documents in bag-of-words representation. In this generated dataset, we introduced three types of anomalies, all of them are sudden outliers.

Table 2. Synthetic dataset: AUC and parameters

method	Type-1		Type-2		Type-3	
	AUC	parameters	AUC	parameters	AUC	parameters
LSTH	**0.863/0.860**	10,10,1,10	**0.995/0.993**	10,10,1,10	**0.984/0.979**	10,20,1,0
CCA	0.848/0.859	10	0.020/0.019	10	0.971/0.950	500
PCAx	0.500/0.525	10, 1	0.015/0.018	10, 1	0.013/0.016	10, 1
PCAxy	0.644/0.662	20, 1	0.744/0.730	20, 1	0.977/0.971	20, 1
PCAy	0.298/0.365	10, 1	0.015/0.015	10, 1	0.977/0.960	20, 1

The parameters for LSTH are d (dimension of the latent space), λ, α and σ, respectively, The parameter for CCA is d. The parameters for PCAx, PCAxy and PCAy are d and the forgetting factor, respectively. AUC of the precision-recall plot is used for evaluation; the larger the AUC value is, the better the performance is. The values under AUC column (i.e., x/y) are the performance on training and testing set, respectively. **Bold** numbers correspond to the best performance for each anomaly type among all the methods.

Type-1 anomaly: at $t = 500, 600, \ldots, 10400$, x_t is distorted to $\tilde{x}_t = \tilde{A}\theta_t + n_t$, where $\tilde{A}$ is identical to A except that one row of $\tilde{A}$ is randomly re-drawn from $\mathcal{N}(0, 1)$. At the same timestamps when A is distorted, B in generating y_t is also distorted to $\tilde{B}$ by randomly re-drawing 5 of its rows from $\mathcal{N}(1, 0.3^2)$. This corresponds to the scenario when both x_t and y_t behave anomalously at same time.

Type-2 anomaly: at $t = 500, 600, \ldots, 10400$, only x_t is distorted to $\tilde{x}_t = A\tilde{\theta}_t + n_t$ with $\tilde{\theta}_t \sim \mathcal{N}(3.5, 1)$, that is, the latent variable θ_t is distorted. In this way, a discrepancy is introduced between the latencies of $\tilde{x}_t$ and y_t. This corresponds to the scenario when x_t has anomalies but y_t behaves normally.

Type-3 anomaly: At $t = 500, 600, \ldots, 10400$, three relevant features and three among the rest 950 features of y_t are exchanged. This corresponds to the scenario when some relevant features in y_t are changed while x_t remains normal.

Experimental Results on Synthetic Data. We compare all methods in Table 1 for anomaly detection task. For all the methods, the first 100 samples are used for initialization. The γ in computing detection threshold is varied to produce a full precision-recall plot. The parameters are selected as the ones that maximize the Area Under Curve (AUC) of the precision-recall plot on a training set generated separately from the same data generation protocol. Results are presented in Table 2. For the three types of anomalies, LSTH consistently achieves the

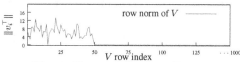

Fig. 1. Feature selection effects of LSTH

Table 3. XRMB results

method	AUC	parameters
LSTH	**0.342**	20,300,0.95,1000
CCA	0.045	30
PCAx	0.035	20, 1e-5
PCAxy	0.035	30, 1e-5
PCAy	0.033	30, 1e-5

The parameters for each method are same as those in Table 2 in paper.

best detection performance. CCA is competitive for Type-1 and Type-3 anomalies but completely fails for Type-2, due to the fact that its detection statistic cannot capture the changes in the signal/latent space. The failure of PCAx on Type-2 has a same reason as that of CCA on Type-2. On average, PCA based methods perform worst among all the methods except for Type-3. However, by joining two data sources properly, PCAxy is able to detect the change of the "joint" subspace so as to achieve better performance than PCAx and PCAy.

Figure 1 shows the norm of each row of the learned V, after all the updates of LSTH at $t = 10500$. For the relevant $(i = 1, \ldots, 50)$ features in y_t, $\|v_i^\top\|$ are non-zero. For the irrelevant features $(i > 50)$, $\|v_i^\top\|$ are zero or very small. This demonstrates that LSTH can successfully identify the relevant features via the mixed norm on V.

5.2   Experimental Results on Real Data: XRMB

XRMB [16] contains synchronous 273-dim MFCC and 112-dim articulatory information of length 51K. Each timestamp has a label indicating which word it corresponds to. Details on the data are available in [1]. Speech segmentation has attracted lots of attention for treating related diseases [15]. The task in our experiment is to detect the boundary of words from acoustic and articulatory features. During each segment, a tracking algorithm, e.g., LSTH, gradually learns the underlying latent subspace. Upon arrival of a new segment, the underlying latent space has a sudden change. This event may induce a drastic change of the detection statistics provided by the tracking algorithms, and therefore is considered as an anomaly. In this case, the claimed anomalous data point should be incorporated in learning the new latent space in the new segment.

When applying LSTH, we assign to x_t the articulatory features with highly correlated dimensions [3]. And y_t is designated as the MFCC, which is redundant and sparse filtering has been shown necessary for feature selection [11]. We randomly select 1000 frames for parameter tuning for all the methods, and use the tuned parameters for testing on the rest of the frames. Figure 2 shows the detection statistics of all methods on the parameter tuning dataset. Out of 25 words within the 1000 frames, LSTH is able to identify 15 words with clear and strong spikes in the detection statistics. After each alarm of anomaly (start of a new segment), it quickly adapts to the new latent space in the new segment. PCA based methods only show weak spikes. CCA fails in this case, as the conclusion in [1]. Based on their results, kernel CCA should be a better approach on this dataset than CCA. However, there is not a meaningful detection statistic for kernel CCA, so we leave this approach for later research.

We then applied all the methods on the rest of the data with their optimal parameters tuned on the training set. The parameters and the performance of different methods are presented in Table 3. LSTH has an AUC value 0.342 (note that a random guess would give an AUC of $412/51000 = 0.008$) and it is the only method that can detect the boundaries of the words from XRMB dataset. All the other methods fail with AUC values smaller than 0.05.

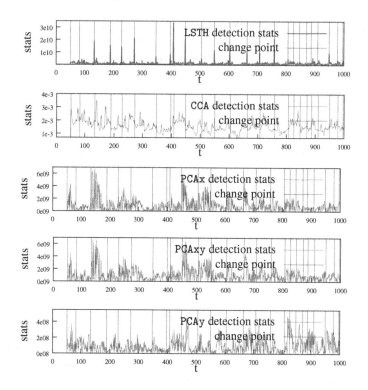

Fig. 2. Detection statistics on XRMB training data

6   Conclusions and Discussions

We developed LSTH, a latent space tracking method for heterogeneous streaming data. Under the assumption that anomalies significantly deviate from the latent space, we further designed an anomaly detection method based on LSTH. Experimental results demonstrate that LSTH's detection statistics outperform the other state-of-the-art in identifying anomalies. Therefore LSTH better characterizes the latent structure of heterogeneous data than does the other methods. Future work on LSTH includes non-linear mapping into the latent space via kernelization, online supervised learning in the latent space, and extending to cases with more than two views of a system.

References

1. Andrew, G., Arora, R., Bilmes, J., Livescu, K.: Deep canonical correlation analysis. In: Proceedings of the 30th International Conference on Machine Learning (ICML 2013), vol. 28, pp. 1247–1255 (2013)
2. Brand, M.: Fast low-rank modifications of the thin singular value decomposition. Linear algebra and its applications **415**(1), 20–30 (2006)

3. Cai, J., Laprie, Y., Busset, J., Hirsch, F.: Articulatory modeling based on semi-polar coordinates and guided pca technique. In: 10th Annual Conference of the International Speech Communication Association-INTERSPEECH (2009)
4. Chi, Y., Eldar, Y., Calderbank, R.: Petrels: Parallel subspace estimation and tracking by recursive least squares from partial observations. IEEE Transactions on Signal Processing **61**(23), 5947–5959 (2013)
5. Hotelling, H.: Canonical correlation analysis. Journal of Educational Psychology (1935)
6. Jin, X., Zhuang, F., Xiong, H., Du, C., Luo, P., He, Q.: Multi-task multi-view learning for heterogeneous tasks. In: Proceedings of the 23rd ACM International Conference on Conference on Information and Knowledge Management, pp. 441–450 (2014)
7. Jolliffe, I.T.: Principal Component Analysis. John Wiley & Sons, Ltd. (2005)
8. Kiers, H.A.L.: Majorization as a tool for optimizing a class of matrix functions. Psychometrika **55**(3), 417–428 (1990)
9. Mairal, J., Bach, F., Ponce, J., Sapiro, G.: Online dictionary learning for sparse coding. In: The 26th Annual International Conference on Machine Learning, pp. 689–696. ACM (2009)
10. Mardia, K.V., Kent, J.T., Bibby, J.M.: Multivariate Analysis. Academic Press, New York (1979)
11. Ngiam, J., Koh, P.W., Chen, Z., Bhaskar, S., Ng, A.: Sparse filtering. In: Advances in Neural Information Processing Systems, pp. 1125–1133 (2011)
12. dos Santos Teixeira, P.H., Milidiú, R.L.: Data stream anomaly detection through principal subspace tracking. In: Proceedings of the 2010 ACM Symposium on Applied Computing (2010)
13. Shi, X., Liu, Q., Fan, W., Yu, P.S., Zhu, R.: Transfer learning on heterogenous feature spaces via spectral transformation. In: 2010 IEEE 10th International Conference on Data Mining (ICDM), pp. 1049–1054 (2010)
14. Tang, X., Yang, C., Zhou, J.: Stock price forecasting by combining news mining and time series analysis. In: IEEE/WIC/ACM International Joint Conferences on Web Intelligence and Intelligent Agent Technologies, vol. 1, pp. 279–282 (2009)
15. Todder, D., Avissar, S., Schreiber, G.: Non-linear dynamic analysis of inter-word time intervals in psychotic speech. IEEE Journal of Translational Engineering in Health and Medicine **1**, 1–7 (2013)
16. Westbury, J.R.: X-ray microbeam speech production database user's handbook. Tech. rep., University of Wisconsin, Madison (1994)
17. Xie, Y., Huang, J., Willett, R.: Change-point detection for high-dimensional time series with missing data. IEEE Journal of Selected Topics in Signal Processing **7**(1), 12–27 (2013)
18. Xu, C., Tao, D., Xu, C.: A survey on multi-view learning. ArXiv e-prints (April 2013)
19. Yger, F., Berar, M., Gasso, G., Rakotomamonjy, A.: Adaptive canonical correlation analysis based on matrix manifolds. In: Proceedings of the 29th International Conference on Machine Learning, pp. 1071–1078 (2012)

A Learning-Rate Schedule for Stochastic Gradient Methods to Matrix Factorization

Wei-Sheng Chin, Yong Zhuang, Yu-Chin Juan, and Chih-Jen Lin[✉]

Department of Computer Science, National Taiwan University,
Taipei, Taiwan
{d01944006,r01922139,r01922136,cjlin}@csie.ntu.edu.tw

Abstract. Stochastic gradient methods are effective to solve matrix factorization problems. However, it is well known that the performance of stochastic gradient method highly depends on the learning rate schedule used; a good schedule can significantly boost the training process. In this paper, motivated from past works on convex optimization which assign a learning rate for each variable, we propose a new schedule for matrix factorization. The experiments demonstrate that the proposed schedule leads to faster convergence than existing ones. Our schedule uses the same parameter on all data sets included in our experiments; that is, the time spent on learning rate selection can be significantly reduced. By applying this schedule to a state-of-the-art matrix factorization package, the resulting implementation outperforms available parallel matrix factorization packages.

Keywords: Matrix factorization · Stochastic gradient method · Learning rate schedule

1 Introduction

Given an incomplete matrix $R \in \mathbb{R}^{m \times n}$, matrix factorization (MF) finds two matrices $P \in \mathbb{R}^{k \times m}$ and $Q \in \mathbb{R}^{k \times n}$ such that $r_{u,v} \simeq \boldsymbol{p}_u^T \boldsymbol{q}_v, \forall u,v \in \Omega$, where Ω denotes the indices of the existing elements in R, $r_{u,v}$ is the element at the uth row and the vth column in R, $\boldsymbol{p}_u \in \mathbb{R}^k$ is the uth column of P, $\boldsymbol{q}_v \in \mathbb{R}^k$ is the vth column of Q, and k is the pre-specified number of latent features. This task is achieved by solving the following non-convex problem

$$\min_{P,Q} \quad \sum_{(u,v) \in \Omega} (r_{u,v} - \boldsymbol{p}_u^T \boldsymbol{q}_v)^2 + \lambda(\|\boldsymbol{p}_u\|^2 + \|\boldsymbol{q}_v\|^2), \tag{1}$$

where λ is a regularization parameter. Note that the process to solve P and Q is referred to as the training process. To evaluate the quality of the used solver, we can treat some known elements as missing in the training process and collect them as the test set. Once P and Q are found, root-mean-square error (RMSE) on the test set is often used as an evaluation criterion. It is defined as

$$\sqrt{\frac{1}{|\Omega_{\text{test}}|} \sum_{(u,v) \in \Omega_{\text{test}}} e_{u,v}^2}, \quad e_{u,v} = r_{u,v} - \boldsymbol{p}_u^T \boldsymbol{q}_v, \tag{2}$$

© Springer International Publishing Switzerland 2015
T. Cao et al. (Eds.): PAKDD 2015, Part I, LNAI 9077, pp. 442–455, 2015.
DOI: 10.1007/978-3-319-18038-0_35

where Ω_{test} represents the indices of the elements belonging to test set.

Matrix factorization is widely used in recommender systems [11], natural language processing [16], and computer vision [9]. Stochastic gradient method[1](SG) is an iterative procedure widely used to solve (1), e.g., [2,7,14]. At each step, a single element $r_{u,v}$ is sampled to obtain the following sub-problem.

$$(r_{u,v} - \boldsymbol{p}_u^T \boldsymbol{q}_v)^2 + \lambda(\|\boldsymbol{p}_u\|^2 + \|\boldsymbol{q}_v\|^2). \tag{3}$$

The gradient of (3) is

$$\boldsymbol{g}_u = \frac{1}{2}(-e_{u,v}\boldsymbol{q}_v + \lambda\boldsymbol{p}_u), \quad \boldsymbol{h}_v = \frac{1}{2}(-e_{u,v}\boldsymbol{p}_u + \lambda\boldsymbol{q}_v). \tag{4}$$

Note that we drop the coefficient $1/2$ to simplify our equations. Then, the model is updated along the negative direction of the sampled gradient,

$$\boldsymbol{p}_u \leftarrow \boldsymbol{p}_u - \eta\boldsymbol{g}_u, \quad \boldsymbol{q}_v \leftarrow \boldsymbol{q}_v - \eta\boldsymbol{h}_v, \tag{5}$$

where η is the learning rate. In this paper, an update of (5) is referred to as an iteration, while $|\Omega|$ iterations are called an outer iteration to roughly indicate that all $r_{u,v}$ have been handled once. Algorithm 1 summarizes the SG method for matrix factorization. In SG, the learning rate can be fixed as a constant while some schedules dynamically adjust η in the training process for faster convergence [4]. The paper aims to design an efficient schedule to accelerate the training process for MF.

The rest sections are organized as follows. Section 2 investigates the existing schedules for matrix factorization and a per-coordinate schedule for online convex problems. Note that a per-coordinate schedule assigns each variable a distinct learning rate. We improve upon the per-coordinate schedule and propose a new schedule in Section 3. In Section 4, experimental comparisons among schedules and state-of-the-art packages are exhibited. Finally, Section 5 summarizes this paper and discusses potential future works. In summary, our contributions include:

Algorithm 1 . Stochastic gradient methods for matrix factorization.

Require: Z: user-specified outer iterations
1: **for** $z \leftarrow 1$ to Z **do**
2:     **for** $i \leftarrow 1$ to $|\Omega|$ **do**
3:         sample $r_{u,v}$ from R
4:         calculate sub-gradient by (4)
5:         update $\boldsymbol{p}_u$ and $\boldsymbol{q}_v$ by (5)
6:     **end for**
7: **end for**

---

1. We propose a new schedule that outperforms existing schedules.
2. We apply the proposed schedule to an existing package. The resulting implementation, which will be publicly available, outperforms state-of-the-art parallel matrix factorization packages.

---

[1] It is often called stochastic gradient descent method. However, it is actually not a "descent" method, so we use the term stochastic gradient method in this paper.

2   Existing Schedules

In Section 2.1, we investigate three schedules that are commonly used in matrix factorization. The per-coordinate schedule that inspired the proposed method is introduced in Section 2.2.

2.1   Existing Schedules for Matrix Factorization

Fixed Schedule (FS). The learning rate is fixed throughout the training process. That is, η equals to η_0, a pre-specified constant. This schedule is used in, for example, [8].

Monotonically Decreasing Schedule (MDS). This schedule decreases the learning rate over time. At the zth outer iteration, the learning rate is

$$\eta^z = \frac{\alpha}{1 + \beta \cdot z^{1.5}},$$

where α and β are pre-specified parameters. In [19], this schedule is used. For general optimization problems, two related schedules [6, 10, 12] are

$$\eta^z = \frac{\alpha}{z} \text{ and } \eta^z = \frac{\alpha}{z^{0.5}}, \tag{6}$$

but they are not included in some recent developments for matrix factorization such as [4, 19]. Note that [4] discusses the convergence property for the use of (6), but finally chooses another schedule, which is introduced in the next paragraph, for faster convergence.

Bold-Driver Schedule (BDS). Some early studies on neural networks found that the convergence can be dramatically accelerated if we adjust the learning rate according to the change of objective function values through iterations [1, 15]. For matrix factorization, [4] adapts this concept and considers the rule,

$$\eta^{z+1} = \begin{cases} \alpha\eta^z & \text{if } \Delta_z < 0 \\ \beta\eta^z & \text{otherwise,} \end{cases} \tag{7}$$

where $\alpha \in (1, \infty)$, $\beta \in (0, 1)$, and $\eta^0 \in (0, \infty)$ are pre-specified parameters, and Δ_z is the difference on the objective function in (1) between the beginning and the end of the zth outer iteration. Clearly, this schedule enlarges the rate when the objective value is successfully decreased, but reduces the rate otherwise.

2.2   Per-Coordinate Schedule (PCS)

Some recent developments discuss the possibility to assign the learning rate coordinate-wisely. For example, ADAGRAD [3] is proposed to coordinate-wisely control the learning rate in stochastic gradient methods for convex online optimization. For matrix factorization, if $r_{u,v}$ is sampled, ADAGRAD adjusts two matrices G_u and H_v using

$$G_u \leftarrow G_u + g_u g_u^T, \quad H_v \leftarrow H_v + h_v h_v^T,$$

and then updates the current model via

$$p_u \leftarrow p_u - \eta_0 G_u^{-1/2} g_u, \quad q_v \leftarrow q_v - \eta_0 H_v^{-1/2} h_v. \tag{8}$$

ADAGRAD also considers using only the diagonal elements because matrix inversion in (8) is expensive. That is, G_u and H_v are maintained by

$$G_u \leftarrow G_u + \begin{bmatrix} (g_u)_1^2 & & \\ & \ddots & \\ & & (g_u)_k^2 \end{bmatrix}, \quad H_v \leftarrow H_v + \begin{bmatrix} (h_v)_1^2 & & \\ & \ddots & \\ & & (h_v)_k^2 \end{bmatrix}. \tag{9}$$

We consider the setting of using diagonal matrices in this work, so the learning rate is related to the squared sum of past gradient elements.

While ADAGRAD has been shown to be effective for online convex classification, it has not been investigated for matrix factorization yet. Similar to ADAGRAD, other per-coordinate learning schedules such as [13,20] have been proposed. However, we focus on ADAGRAD in this study because the computational complexity per iteration is the lowest among them.

3   Our Approach

Inspired by PCS, a new schedule, *reduced per-coordinate schedule* (RPCS), is proposed in Section 3.1. RPCS can reduce the memory usage and computational complexity in comparison with PCS. Then, in Section 3.2 we introduce a technique called *twin learners* that can further boost the convergence speed of RPCS. Note that we provide some experimental results in this section to justify our argument. See Section 4 for the experimental settings such as parameter selection and the data sets used.

3.1   Reduced Per-Coordinate Schedule (RPCS)

The cost of implementing FS, MDS, or BDS schedules is almost zero. However, the overheads incurred by PCS can not be overlooked. First, each coordinate of p_u and q_v has its own learning rate. Maintaining G_u and H_v may need $\mathcal{O}((m+n)k)$ extra space. Second, at each iteration, $\mathcal{O}(k)$ additional operations are needed for calculating and using diagonal elements of G_u and H_v.

These overheads can be dramatically reduced if we apply the same learning rate for all elements in $\boldsymbol{p}_u$ (or $\boldsymbol{q}_v$). Specifically, at each iteration, G_u and H_v are reduced from matrices to scalars. Instead of (9), G_u and H_v are now updated by

$$G_u \leftarrow G_u + \frac{\boldsymbol{g}_u^T \boldsymbol{g}_u}{k}, \quad H_v \leftarrow H_v + \frac{\boldsymbol{h}_v^T \boldsymbol{h}_v}{k} \tag{10}$$

In other words, the learning rate of $\boldsymbol{p}_u$ or $\boldsymbol{q}_v$ is the average over its k coordinates. Because each $\boldsymbol{p}_u$ or $\boldsymbol{q}_v$ has one learning rate, only $(m+n)$ additional values must be maintained. This storage requirement is much smaller than $(m+n)k$ of PCS. Furthermore, the learning rates,

$$\eta_0(G_u)^{-\frac{1}{2}} \text{ and } \eta_0(H_v)^{-\frac{1}{2}},$$

become scalars rather than diagonal matrices. Then the update rule (8) is reduced to that in (5). However, the cost of each iteration is still higher than that of the standard stochastic gradient method because of the need to maintain G_u and H_v by (10). Note that the $\mathcal{O}(k)$ cost of (10) is comparable to that of (5). Further, because $\boldsymbol{g}_u$ and $\boldsymbol{h}_v$ are used in both (10) and (8), they may need to be stored. In contrast, a single **for** loop for (5) does not require the storage of them. We detailedly discuss the higher cost than (5) by considering two possible implementations.

1. Store $\boldsymbol{g}_u$ and $\boldsymbol{h}_v$.
   - A **for** loop to calculate $\boldsymbol{g}_u, \boldsymbol{h}_v$ and G_u, H_v. Then $\boldsymbol{g}_u$ and $\boldsymbol{h}_v$ vectors are stored.
   - A **for** loop to update $\boldsymbol{p}_u, \boldsymbol{q}_v$ by (8).
2. Calculate $\boldsymbol{g}_u$ and $\boldsymbol{h}_v$ twice.
   - A **for** loop to calculate $\boldsymbol{g}_u, \boldsymbol{h}_v$ and then G_u, H_v.
   - A **for** loop to calculate $\boldsymbol{g}_u, \boldsymbol{h}_v$ and update $\boldsymbol{p}_u, \boldsymbol{q}_v$ by (8).

Clearly, the first approach requires extra storage and memory access. For the second approach, its second loop is the same as (5), but the first loop causes that each SG iteration is twice expensive. To reduce the cost, we decide to use G_u and H_v of the previous iteration. Specifically, at each iteration, we can use a single **for** loop to calculate $\boldsymbol{g}_u$ and $\boldsymbol{h}_v$, update $\boldsymbol{p}_u$ and $\boldsymbol{q}_v$ using past G_u and H_v, and calculate $\boldsymbol{g}_u^T \boldsymbol{g}_u$ and $\boldsymbol{h}_v^T \boldsymbol{h}_v$ to obtain new G_u and H_v for the next iteration. Details are presented in Algorithm 2. In particular, we can see that in the **for** loop, we can finish the above tasks in an element-wise setting. In compared with the implementation for (5), Line 7 in Algorithm 2 is the only extra operation. Thus, the cost of Algorithm 2 is comparable to that of a standard stochastic gradient iteration.

Algorithm 2. One iteration of SG algorithm when RPCS is applied.

1: $e_{u,v} \leftarrow r_{u,v} - \boldsymbol{p}_u^T \boldsymbol{q}_v$
2: $\bar{G} \leftarrow 0, \quad \bar{H} \leftarrow 0$
3: $\eta_u \leftarrow \eta_0(G_u)^{-\frac{1}{2}}, \quad \eta_v \leftarrow \eta_0(H_v)^{-\frac{1}{2}}$
4: **for** $d \leftarrow 1$ to k **do**
5: $\quad (\boldsymbol{g}_u)_d \leftarrow -e_{u,v}(\boldsymbol{q}_v)_d + \lambda(\boldsymbol{p}_u)_d$
6: $\quad (\boldsymbol{h}_v)_d \leftarrow -e_{u,v}(\boldsymbol{p}_u)_d + \lambda(\boldsymbol{q}_v)_d$
7: $\quad \bar{G} \leftarrow \bar{G} + (\boldsymbol{g}_u)_d^2, \quad \bar{H} \leftarrow \bar{H} + (\boldsymbol{h}_v)_d^2$
8: $\quad (\boldsymbol{p}_u)_d \leftarrow (\boldsymbol{p}_u)_d - \eta_u(\boldsymbol{g}_u)_d$
9: $\quad (\boldsymbol{q}_v)_d \leftarrow (\boldsymbol{q}_v)_d - \eta_v(\boldsymbol{h}_v)_d$
10: **end for**
11: $G_u \leftarrow G_u + \bar{G}/k, \quad H_v \leftarrow H_v + \bar{H}/k$

In Figure 1, we check the convergence speed of PCS and RPCS by showing the relationship between RMSE and the number of outer iterations. The convergence speeds of PCS and RPCS are almost identical. Therefore, using the same rate for all elements in $\boldsymbol{p}_u$ (or $\boldsymbol{q}_v$) does not cause more iterations. However, because each iteration becomes cheaper, a comparison on the running time in Figure 2 shows that RPCS is faster than PCS.

We explain why using the same learning rate for all elements in $\boldsymbol{p}_u$ (or $\boldsymbol{q}_v$) is reasonable for RPCS. Assume $\boldsymbol{p}_u$'s elements are the same,

$$(\boldsymbol{p}_u)_1 = \cdots = (\boldsymbol{p}_u)_k,$$

and so are $(\boldsymbol{q}_v)$'s elements. Then (4) implies that all elements in each of $\boldsymbol{g}_v$ and $\boldsymbol{h}_v$ has the same value. From the calculation of G_u, H_v in (9) and the update rule (8), elements of the new $\boldsymbol{p}_u$ (or $\boldsymbol{q}_v$) are still the same. This result implies that learning rates of all coordinates are the same throughout all iterations. In our implementation of PCS, elements of $\boldsymbol{p}_u$ and $\boldsymbol{q}_v$ are initialized by the same random number generator. Thus, if each element is treated as a random variable, their expected values are the same. Consequently, $\boldsymbol{p}_u$'s (or $\boldsymbol{q}_v$'s) initial elements are identical in statistics and hence our explanation can be applied.

3.2   Twin Learners (TL)

Conceptually, in PCS and RPCS, the decrease of a learning rate should be conservative because it never increases. We observe that the learning rate may be too rapidly decreased at the first few updates. The reason may be that the random initialization of P and Q causes comparatively large errors at the beginning. From (4), the gradient is likely to be large if $e_{u,v}$ is large. The large gradient further results in a large sum of squared gradients, and a small learning rate $\eta_0(G_u)^{-\frac{1}{2}}$ or $\eta_0(H_v)^{-\frac{1}{2}}$.

To alleviate this problem, we introduce a strategy called *twin learners* which deliberately allows some elements to have a larger learning rate. To this end, we split the elements of $\boldsymbol{p}_u$ (or $\boldsymbol{q}_v$) to two groups $\{1, \ldots, k_s\}$ and $\{k_s + 1, \ldots, k\}$, where the learning rate is smaller for the first group, while larger for the second. The two groups respectively maintain their own factors, G_u^{slow} and G_u^{fast}, via

$$G_u^{\text{slow}} \leftarrow G_u^{\text{slow}} + \frac{(\boldsymbol{g}_u)_{1:k_s}^T (\boldsymbol{g}_u)_{1:k_s}}{k_s}, \quad G_u^{\text{fast}} \leftarrow G_u^{\text{fast}} + \frac{(\boldsymbol{g}_u)_{k_s+1:k}^T (\boldsymbol{g}_u)_{k_s+1:k}}{k - k_s}.$$

$$(11)$$

We refer to the first group as the "slow learner," while the second group as the "fast learner." To make G_u^{fast} smaller than G_u^{slow}, we do not apply the second rule in (11) to update G_u^{fast} at the first outer iteration. The purpose is to let the slow learner "absorb" the sharp decline of the learning rate brought by the large initial errors. Then the fast learner can maintain a larger learning rate for faster convergence. We follow the setting in Section 3.1 to use G_u^{slow}, H_v^{slow}, G_u^{fast}, and H_v^{fast} of the previous iteration. Therefore, at each iteration, we have

448   W.-S. Chin et al.

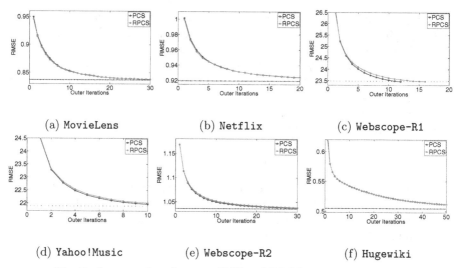

Fig. 1. A comparison between PCS and RPCS: convergence speed

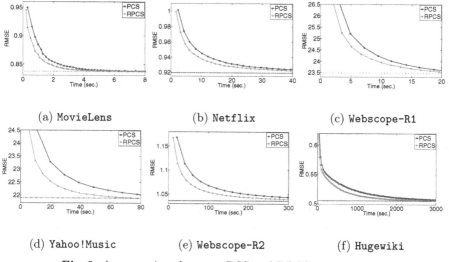

Fig. 2. A comparison between PCS and RPCS: running time

1. One **for** loop going through the first k_s elements to calculate $(\boldsymbol{g}_u)_{1:k_s}$, $(\boldsymbol{h}_v)_{1:k_s}$, update $(\boldsymbol{p}_u)_{1:k_s}$, $(\boldsymbol{q}_v)_{1:k_s}$, and obtain the next G_u^{slow}, H_v^{slow}.
2. One **for** loop going through the remaining $k - k_s$ elements to calculate $(\boldsymbol{g}_u)_{k_s+1:k}$, $(\boldsymbol{h}_v)_{k_s+1:k}$, update $(\boldsymbol{p}_u)_{k_s+1:k}$, $(\boldsymbol{q}_v)_{k_s+1:k}$, and obtain the next G_u^{fast}, H_v^{fast}.

Figure 3 shows the average learning rates of RPCS (TL is not applied), and slow and fast learners (TL is applied) at each outer iteration. For RPCS, the average learning rate is reduced by around half after the first outer iteration. When TL is applied, though the average learning rate of the slow learner drops even faster, the average learning rate of the fast learner can be kept high to ensure fast learning. A comparison between RPCS with and without TL is in Figure 4. Clearly, TL is very effective. In this paper, we fix k_s as 8% of k. We also tried $\{2, 4, 8, 16\}$%, but found that the performance is not sensitive to the choice of k_s.

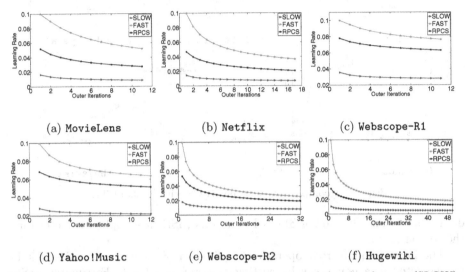

(a) MovieLens   (b) Netflix   (c) Webscope-R1

(d) Yahoo!Music   (e) Webscope-R2   (f) Hugewiki

Fig. 3. A comparison among the average learning rates of the slow learner (SLOW), the fast learner (FAST), and RPCS. Note that we use $\eta_0 = 0.1$ and initial $G_u = H_v = 1$ following the same settings in our experimental section. Hence the initial learning rate is 0.1.

4   Experiments

We conduct experiments to exhibit the effectiveness of our proposed schedule. Implementation details and experimental settings are respectively shown in Sections 4.1 and 4.2. A comparison among RPCS and existing schedules is in Section 4.3. Then, we compare RPCS with three state-of-the-art packages on both matrix factorization and non-negative matrix factorization (NMF) in Sections 4.4 and 4.5, respectively.

4.1   Implementation

For the comparison of various schedules, we implement them by modifying LIBMF,[2] which is a parallel SG-based matrix factorization package [21]. We choose it because of its efficiency and the ease of modification. Note that TL is applied to RPCS in all experiments. In LIBMF, single-precision floating points are used for

---

[2] http://www.csie.ntu.edu.tw/~cjlin/libmf

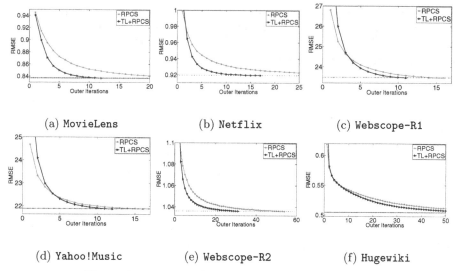

(a) MovieLens   (b) Netflix   (c) Webscope-R1

(d) Yahoo!Music   (e) Webscope-R2   (f) Hugewiki

Fig. 4. A comparison between RPCS with/without TL

data storage, and Streaming SIMD Extensions (SSE) are applied to accelerate the computation.

The inverse square root operation required in (8) is very expensive if it is implemented in a naive way by writing `1/sqrt(·)` in C++. Fortunately, SSE provides an instruction `_mm_rsqrt_ps(·)` to efficiently calculate the approximate inverse square roots for single-precision floating-point numbers.

4.2   Settings

Data Sets. Six data sets listed in Table 1 are used. We use the same training/test sets for MovieLens, Netflix, and Yahoo!Music following [21], and the official training/test sets for Webscope-R1 and Webscope-R2.[3] For Hugewiki,[4] the original data set is too large for our machine, so we sample first half of the original data. Within this sub-sampled data set, we randomly sample 1% as the test set, and using the remaining for training.

Platform and Parameters. We run the experiment on a machine with 12 cores on two Intel Xeon E5-2620 2.0GHz processors and 64 GB memory. We ensure that no other heavy tasks are running on the same computer.

A higher number of latent features often leads to a lower RMSE, but needs a longer training time. From our experience, 100 latent features is an acceptable balance between speed and RMSE, so we use it for all data sets. For the regularization parameter, we select the one that leads to the best test RMSE among {2, 1, 0.5, 0.1, 0.05, 0.01} and present it in Table 1. In addition, P and Q

---

[3] http://webscope.sandbox.yahoo.com/catalog.php?datatype=r
[4] http://graphlab.org/downloads/datasets/

Table 1. Data statistics, parameters used in experiments, and the near-best RMSE's (see Section 4.2 for explanation) on all data sets

Data Set	m	n	k	λ	#training	#test	RMSE MF	NMF
MovieLens	71,567	65,133	100	0.05	9,301,274	698,780	0.831	0.835
Netflix	2,649,429	17,770	100	0.05	99,072,112	1,408,395	0.914	0.916
Webscope-R1	1,948,883	1,101,750	100	1	104,215,016	11,364,422	23.36	23.75
Yahoo!Music	1,000,990	624,961	100	1	252,800,275	4,003,960	21.78	22.10
Webscope-R2	1,823,180	136,737	100	0.05	699,640,226	18,231,790	1.031	1.042
Hugewiki	39,706	25,000,000	100	0.05	1,703,429,136	17,202,478	0.502	0.504

are initialized so that every element is randomly chosen between 0 and 0.1. We normalize the data set by its standard deviation to avoid numerical difficulties. The regularization parameter and the initial values are scaled by the same factor as well. A similar normalization procedure has been used in [18].

The best parameters of each schedule are listed in Table 2. They are the fastest setting to reach 1.005 times the best RMSE obtained by all methods under all parameters. We consider such a "near-best" RMSE to avoid selecting a parameter that needs unnecessarily long running time. Without this mechanism, our comparison on running time can become misleading. Note that PCS

Table 2. The best parameters for each schedule used.

Data Set	FS η_0	MDS α	β	BDS η_0	PCS η_0
MovieLens	0.005	0.05	0.1	0.05	0.1
Netflix	0.005	0.05	0.1	0.05	0.1
Webscope-R1	0.005	0.05	0.1	0.01	0.1
Yahoo!Music	0.01	0.05	0.05	0.01	0.1
Webscope-R2	0.005	0.05	0.1	0.05	0.1
Hugewiki	0.01	0.05	0.01	0.01	0.1

and RPCS shares the same η_0. For BDS, we follow [4] to fix $\alpha = 1.05$ and $\beta = 0.5$, and tune only the parameter η_0. The reason is that it is hard to tune three parameters η_0, α, and β together.

4.3 Comparison Among Schedules

In Figure 5, we present results of comparing five schedules including FS, MDS, BDS, PCS, and RPCS. RPCS outperforms other schedules including the PCS schedule that it is based upon.

4.4 Comparison with State-of-the-art Packages on Matrix Factorization

We compare the proposed schedule (implemented based on LIBMF, and denoted as LIBMF++) with the following packages.
- The standard LIBMF that implements the FS strategy.
- An SG-based package NOMAD [19] that has claimed to outperform LIBMF.
- LIBPMF:[5] it implements a coordinate descent method CCD++ [17].

---

[5] http://www.cs.utexas.edu/~rofuyu/libpmf

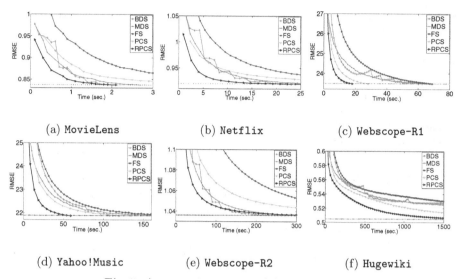

Fig. 5. A comparison among different schedules

For all packages, we use single-precision storage[6] and 12 threads. The comparison results are presented in Figure 6. For NOMAD, we use the same α and β parameters in [19] for Netflix and Yahoo!Music, and use parameters identical to MDS for MovieLens and Webscope-R1. We do not run NOMAD on Webscope-R2 and Hugewiki because of the memory limitation. Taking the advantage of the proposed schedule RPCS, LIBMF++ is significantly faster than LIBMF and LIBPMF. Our experimental results for NOMAD are worse than what [19] reports. In [19], NOMAD outperforms LIBMF and CCD++, but our experiments show an opposite result. We think the reason may be that in [19], 30 cores are used and NOMAD may have comparatively better performance if using more cores.

4.5 Comparison with State-of-the-art Methods for Non-negative Matrix Factorization (NMF)

Non-negative matrix factorization [9] requires that all elements in P and Q are non-negative. The optimization problem is

$$\min_{P,Q} \quad \sum_{(u,v)\in\Omega}(r_{u,v} - \boldsymbol{p}_u^T\boldsymbol{q}_v)^2 + \lambda(\|\boldsymbol{p}_u\|^2 + \|\boldsymbol{q}_v\|^2)$$

subject to $P_{du} \geq 0,\ Q_{dv} \geq 0,\ \forall d \in \{1,\ldots,k\},\ u \in \{1,\ldots,m\},\ v \in \{1,\ldots,n\}.$

SG can perform NMF by a simple projection [4], and the update rules used are

$$\boldsymbol{p}_u \leftarrow \max\left(\boldsymbol{0}, \boldsymbol{p}_u - \eta\boldsymbol{g}_u\right), \quad \boldsymbol{q}_v \leftarrow \max\left(\boldsymbol{0}, \boldsymbol{q}_v - \eta\boldsymbol{h}_v\right),$$

---

[6] LIBPMF is implemented using double precision, but we obtained a single-precision version from its authors.

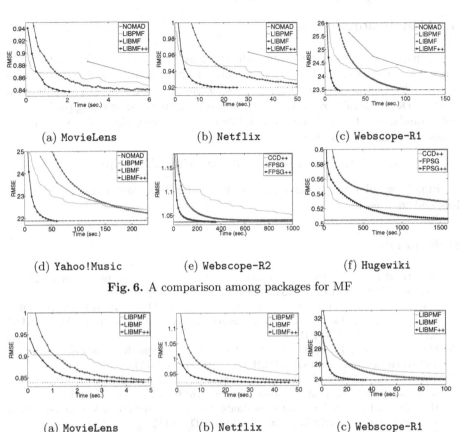

Fig. 6. A comparison among packages for MF

(a) MovieLens   (b) Netflix   (c) Webscope-R1

(d) Yahoo!Music   (e) Webscope-R2   (f) Hugewiki

Fig. 7. A comparison among packages for NMF

(a) MovieLens   (b) Netflix   (c) Webscope-R1

(d) Yahoo!Music   (e) Webscope-R2   (f) Hugewiki

where the max operator is element-wise. Similarly, the coordinate descent method in LIBPMF [5] solves NMF by projecting the negative value back to zero at each update. Therefore, except NOMAD, all packages used in the previous experiment can be applied to NMF. We compare them in Figure 7.

A comparison between Figure 6 and Figure 7 shows that all methods converge slower for NMF. This result seems to be reasonable because NMF is a more complicated optimization problem. Interestingly, we see the convergence degradation is more severe for CCD++ (LIBPMF) than SG (LIBMF and LIBMF++).

5 Conclusions

In this paper, we propose a new and effective learning-rate schedule for SG methods applied to matrix factorization. It outperforms existing schedules according to the rich experiments conducted. By using the proposed method, an extension of the package LIBMF is shown to be significantly faster than existing packages on both standard matrix factorization and its non-negative variant. The experiment codes are publicly available at

http://www.csie.ntu.edu.tw/~cjlin/libmf/exps

Finally, we plan to extend our schedule to other loss functions such as logistic loss and squared hinge loss.

References

1. Battiti, R.: Accelerated backpropagation learning: Two optimization methods. Complex Systems **3**(4), 331–342 (1989)
2. Chen, P.L., Tsai, C.T., Chen, Y.N., Chou, K.C., Li, C.L., et al.: A linear ensemble of individual and blended models for music rating prediction. In: ACM SIGKDD KDD-Cup WorkShop (2011)
3. Duchi, J., Hazan, E., Singer, Y.: Adaptive subgradient methods for online learning and stochastic optimization. JMLR **12**, 2121–2159 (2011)
4. Gemulla, R., Nijkamp, E., Haas, P.J., Sismanis, Y.: Large-scale matrix factorization with distributed stochastic gradient descent. In: KDD, pp. 69–77 (2011)
5. Hsieh, C.J., Dhillon, I.S.: Fast coordinate descent methods with variable selection for non-negative matrix factorization. In: KDD (2011)
6. Kiefer, J., Wolfowitz, J.: Stochastic estimation of the maximum of a regression function. The Annals of Mathematical Statistics **23**(3), 462–466 (1952)
7. Koren, Y., Bell, R.: Advances in collaborative filtering. In: Ricci, F., Rokach, L., Shapira, B., Kantor, P.B. (eds.) Recommender Systems Handbook, pp. 145–186. Springer US (2011)
8. Koren, Y., Bell, R.M., Volinsky, C.: Matrix factorization techniques for recommender systems. Computer **42**(8), 30–37 (2009)
9. Lee, D.D., Seung, H.S.: Learning the parts of objects by non-negative matrix factorization. Nature **401**, 788–791 (1999)
10. Polyak, B.T.: A new method of stochastic approximation type. Avtomat. i Telemekh. **7**, 98–107 (1990)
11. Ricci, F., Rokach, L., Shapira, B.: Introduction to recommender systems handbook. In: Ricci, F., Rokach, L., Shapira, B., Kantor, P.B. (eds.) Recommender Systems Handbook, pp. 1–35. Springer (2011)
12. Robbins, H., Monro, S.: A stochastic approximation method. The Annals of Mathematical Statistics **22**(3), 400–407 (1951)
13. Schaul, T., Zhang, S., LeCun, Y.: No more pesky learning rates. In: ICML, pp. 343–351 (2013)
14. Takács, G., Pilászy, I., Németh, B., Tikk, D.: Scalable collaborative filtering approaches for large recommender systems. JMLR **10**, 623–656 (2009)
15. Vogl, T., Mangis, J., Rigler, A., Zink, W., Alkon, D.: Accelerating the convergence of the back-propagation method. Biological Cybernetics **59**(4–5), 257–263 (1988)

16. Xu, W., Liu, X., Gong, Y.: Document clustering based on non-negative matrix factorization. In: SIGIR (2003)
17. Yu, H.F., Hsieh, C.J., Si, S., Dhillon, I.S.: Scalable coordinate descent approaches to parallel matrix factorization for recommender systems. In: ICDM (2012)
18. Yu, Z.Q., Shi, X.J., Yan, L., Li, W.J.: Distributed stochastic ADMM for matrix factorization. In: CIKM (2014)
19. Yun, H., Yu, H.F., Hsieh, C.J., Vishwanathan, S., Dhillon, I.S.: Nomad: non-locking, stochastic multi-machine algorithm for asynchronous and decentralized matrix completion. In: VLDB (2014)
20. Zeiler, M.D.: ADADELTA: An adaptive learning rate method. CoRR (2012)
21. Zhuang, Y., Chin, W.S., Juan, Y.C., Lin, C.J.: A fast parallel SGD for matrix factorization in shared memory systems. In: RecSys (2013)

Applications

On Damage Identification in Civil Structures Using Tensor Analysis

Nguyen Lu Dang Khoa[1]([✉]), Bang Zhang[1], Yang Wang[1], Wei Liu[2],
Fang Chen[1], Samir Mustapha[3], and Peter Runcie[1]

[1] National ICT Australia, Eveleigh, NSW 2015, Australia
{khoa.nguyen,bang.zhang,yang.wang,fang.chen,
peter.runcie}@nicta.com.au
[2] Advanced Analytics Institute, University of Technology, Sydney, Australia
wei.liu@uts.edu.au
[3] Department of Mechanical Engineering,
American University of Beirut, Beirut, Lebanon
sm154@aub.edu.lb

Abstract. Structural health monitoring is a condition-based technology to monitor infrastructure using sensing systems. In structural health monitoring, the data are usually highly redundant and correlated. The measured variables are not only correlated with each other at a certain time but also are autocorrelated themselves over time. Matrix-based two-way analysis, which is usually used in structural health monitoring, can not capture all these relationships and correlations together. Tensor analysis allows us to analyse the vibration data in temporal, spatial and feature modes at the same time. In our approach, we use tensor analysis and one-class support vector machine for damage detection, localization and estimation in an unsupervised manner. The method shows promising results using data from lab-based structures and also data collected from the Sydney Harbour Bridge, one of iconic structures in Australia. We can obtain a damage detection accuracy of 0.98 and higher for all the data. Locations of damage were captured correctly and different levels of damage severity were well estimated.

Keywords: Tensor analysis · Structural health monitoring · Damage identification · Unsupervised learning

1 Introduction

Most structural and mechanical system maintenance is time-based, which an inspection is carried out after a predefined amount of time. Structural health monitoring (SHM) is a condition-based technology to monitor infrastructure using sensing systems. The potential for life-safety and economic benefits has motivated the needs for SHM, facilitating the shift from time-based to condition-based maintenance [8].

© Springer International Publishing Switzerland 2015
T. Cao et al. (Eds.): PAKDD 2015, Part I, LNAI 9077, pp. 459–471, 2015.
DOI: 10.1007/978-3-319-18038-0_36

Damage identification is a key problem in SHM. It is classified by Rytter into four different levels of complexity [14]:

- Level 1 (Detection): to detect if damage is present in the structure.
- Level 2 (Localization): to locate the position of the damage.
- Level 3 (Assessment): to estimate the extent of the damage.
- Level 4 (Prediction): to give information about the safety of the structure, e.g. a remaining life estimation.

Among the four, level 4 requires an understanding of the physical characteristics of the damage progression in the structure. Machine learning methods can solve levels from 1 to 3, which level 1 can be solved using an unsupervised learning while levels 2 and 3 usually require a supervised learning approach [18]. Since we usually only have data associated with healthy states of structures, an unsupervised approach is more practical.

In SHM, the data are usually highly redundant and correlated. There are many sensors at different locations collecting similar vibration data over time. For instance, numerous sensors are installed at different locations on a long-span bridge to measure vibration signals due to traffic loading over long periods of time. One vehicle event at a specific time has multiple signals measured by different sensors. The measured variables are not only correlated with each other at a certain time but also autocorrelated themselves over time. Two-way analysis using matrix, which is usually used in SHM, can not capture all these relationships and correlations together. It is normally based on a matricization of a multiway array and then matrix-based techniques such as principal component analysis (PCA) or singular value decomposition (SVD) are used to analyse the data. However, unfolding the multiway data and analyse them using two-way methods may result in information loss and misinterpretation, especially when the data are noisy [1]. Tensor analysis allows us to analyse data in multiple modes at the same time [9].

This work is part of the efforts which have applied SHM to the Sydney Harbour Bridge. Unsupervised tensor analysis combined with one-class support vector machine (SVM) is used for damage identification including detection, localization and estimation of the damage. The contribution of the paper is as follows.

- SHM sensing data are formed as a tensor, from which tensor analysis is used to obtain the latent subspaces from the multiway data. Using tensor decomposition, data are mapped to a subspace with much lower dimension so that the learning can be done effectively and efficiently.
- Damage detection, localization and estimation are achieved in an unsupervised approach, which is more practical for a SHM problem.
- Experiments using data obtained from laboratory-based structures and the Sydney Harbour Bridge show the effectiveness of the approach in damage identification.

The remainder of the paper is organized as follows. Sections 2 and 3 summarize the related work and background for this work. Section 4 describes our damage identification approach using tensor analysis and one-class SVM. Experimental results are in Section 5. We conclude our work in Section 6.

2   Related Work

Unsupervised methods in SHM normally train the model using only healthy data. Events which significantly deviate from the normal behaviour of the trained model are considered as damage. Worden et al. used Mahalanobis distance to find anomalies in the data, which are likely to be damage [17]. Chan et al. [6] studied auto-associative neural networks for damage detection of the three cable-supported bridges in Hong Kong. However, due to the limitation of unsupervised learning techniques as noted in [18], these methods are only able to detect damage. Not much work available to discuss damage localization and estimation in an unsupervised approach.

Tensor analysis has been successfully applied in many application domains including chemistry, neuroscience, social network analysis and computer vision [1,10]. Prada [13] used three-way analysis of SHM data for damage detection and feature selection. However, this work was purely studied to detect damage, not to localize and estimate the extent of damage.

Sun et al. [16] proposed different methods on dynamically updating component matrices from a Tucker decomposition for online applications like computer network intrusion detection. Liu et al. [12] utilized the common substructures of graphs to accelerate the Tucker factorization for dynamic graphs. A difference with our work is they focus on Tucker analysis while we do that for CP, which has its simplicity in interpretation of the results.

3   Background

3.1   Tensor Analysis for SHM Data

In SHM, usually many sensors at different locations are used to measure the vibration signals over time. The data can be considered as a three-way tensor (*feature* × *location* × *time*) as described in Figure 1. Feature is the information extracted from the raw signals in time domain (e.g. features in frequency domain). Location represents sensors, and time is data snapshots at different timestamps. Each cell of the tensor is a feature value extracted from a particular sensor at a certain time. Each slice along the time axis shown in Figure 1 is a frontal slice representing all feature signals across all locations at a particular time. For simplicity, in this paper we represent a tensor as a three-way array, which is often a case in SHM. However, it is also possible to generalize all the theories for a n-way array.

Two typical approaches for tensor decomposition are CP decomposition (CANDECOMP/ PARAFAC decomposition) and Tucker decomposition [9]. After a decomposition from a three-way tensor, three component matrices can be obtained representing information in each mode. In the case of SHM data as in Figure 1,

Fig. 1. Tensor data in SHM

they are associated with feature (denoted matrix A), location (matrix B) and time modes (matrix C), respectively. In CP method, it is able to interpret the artifact in each mode separately using its corresponding component matrix. In Tucker method, any component can interact with other component in other mode quantified by the core tensor [2]. It makes the interpretation of a Tucker model more difficult than CP. Therefore, in this work we only use CP method for damage identification.

CP Decomposition. The CP decomposition factorizes a tensor as a sum of a finite number of rank-one tensors. In case of a three-way tensor $\mathcal{X} \in \mathbb{R}^{I \times J \times K}$, it is expressed as

$$\mathcal{X} = \sum_{r=1}^{R} \lambda_r A_{:r} \circ B_{:r} \circ C_{:r} + \mathcal{E}, \tag{1}$$

where R is the latent factor, $A_{:r}$, $B_{:r}$ and $C_{:r}$ are r-th columns of component matrices $A \in \mathbb{R}^{I \times R}$, $B \in \mathbb{R}^{J \times R}$ and $C \in \mathbb{R}^{K \times R}$, and λ is the weight vector so that the columns of A, B, C are normalized to length one. The symbol '$\circ$' represents a vector outer product. $\mathcal{E}$ is a three-way tensor containing the residuals. It can also be written in term of the k-th frontal slice of $\mathcal{X}$:

$$X_k = A D_k B^T + E_k, \tag{2}$$

where the diagonal matrix $D_k = diag(\lambda C_{k:})$ ($C_{k:}$ is the k-th row of matrix C). CP decomposition is typically solved using alternating least square (ALS) technique. The technique iteratively solves each component matrix using a least square method by fixing all the other components and repeats the procedure until it converges [9]. The results by CP are unique provided that we permute the rank-one components [10].

3.2 One-class Support Vector Machine

In this work, we use one-class SVM [15] as an anomaly detection method. SVM is well-known for its strong regularization property which is the ability to generalize the model to new data. One-class SVM finds a small region containing most of data points and the anomalies elsewhere. It is done by mapping data into a feature space using kernel and then separating them from the origin with maximum margin. This can be shown as an optimization problem:

$$\min_{w,\xi,\rho} \frac{1}{2} \| w \|^2 + \frac{1}{vn} \sum_{i=1}^{n} \xi_i - \rho \tag{3}$$

$$s.t. \quad w \cdot x_i \geq \rho - \xi_i, \quad \xi_i \geq 0, \quad i = 1, \dots, n,$$

where w and ρ are parameters of the model and can be learned from a training process. ξ_i is a slack variable for controlling how much training error is allowed. $\{x_i\}_{i=1}^n$ is a training event, '$\cdot$' is the dot product, and v controls the rate of anomalies in the data.

This optimization problem can be solved by Lagrangian multiplier and quadratic programming. Once a model is obtained from training data, it can generate a decision value for every new instance. A new instance with a negative decision value is an anomaly, indicating a damaged event [15].

4   Tensor Analysis for Damage Identification

This section describes an approach to identify damage using tensor analysis. Excitations to structures are measured over time by accelerometers or other kinds of sensors. Next, features are extracted from the raw data of all accelerometers, which form a three-way tensor data. Then the tensor is decomposed into matrices of different modes as described in previous section. Analysis of these factor matrices will help to identify the damage of the structure.

4.1   Damage Detection

Given a three-way tensor $\mathcal{X}$ (feature × location × time) which represents data in a healthy condition of a structure, we want to decide if a new event X_n (a frontal slice of size feature × location) is an anomaly with respect to all other healthy events in the training data. Therefore, subspace corresponding to the time mode after decomposition will be used to detect damage.

Building a Benchmark Model. $\mathcal{X}$ is decomposed into three component matrices A, B and C using CP decomposition. Each row of C represents an event in time mode. Using one-class SVM, we build a model using healthy training events which are represented by rows of the component matrix C.

Damage Detection. Due to an arrival of a new event (a new frontal slide in time mode), an additional row will be added to the component matrix C. As in Equation 2, $X_k = AD_kB^T$. When a new frontal slice X_n comes, we have:

$$X_n \approx AD_nB^T,$$

where $D_n = diag(\lambda C_{n:})$ which is a diagonal matrix based on the new row $C_{n:}$ of component matrix C caused by the new slice X_n. The new row $C_{n:}$ can be obtained via D_n [13]:

$$D_n = \arg\min \|X_n - AD_nB^T\|, \tag{4}$$

which can be solved using a least square method.

Algorithm 1. Damage detection

Input: Component matrices A, B, C, a trained SVM model, a new frontal slice X_n
Output: $+1/-1$ if X_n corresponds to a healthy/damaged event, respectively

Compute the new row $x = C_{n:}$ of C so that $D_n = \arg\min \|X_n - AD_nB^T\|$ and $D_n = diag(\lambda C_{n:})$.
Feed this event to the trained SVM model and estimate its decision value s.
If $s < 0$ return -1, otherwise return $+1$.

After having $C_{n:}$, this new row will be checked if it agrees with the benchmark model built in the training, answering the condition of the structure. In case of one-class SVM, a negative decision value indicates that the new event is likely a damaged event. The damage detection method is described in Algorithm 1.

4.2 Damage Localization and Estimation

In order to locate the position of the damage, components of the decomposed matrix in location mode are analysed to extract meaningful artifacts from different states of the structure. By analysing and comparing these components, it is able to find anomalies, which correspond to damaged locations.

To estimate the extent of the damage, we analyse decision values returned from the one-class SVM model. The rationality is that a structure with a more severe damage (e.g. a longer crack) will behave more differently from a normal behaviour. Different ranges of the decision values may present different severity levels of damage. These analyses will be shown in the experimental results.

5 Experimental Results

5.1 Case Studies

We conducted experiments on two case studies, representing two typical types of civil structures. One case study is an laboratory-based building structure obtained from Los Alamos National Laboratory (LANL) [11], and the other is the Sydney Harbour Bridge. For the Sydney Harbour Bridge data, it includes both laboratory testing and field trial.

Building Data. A dataset was obtained from LANL [11]. The data are from a three-story building structure constructed of Unistrut columns and aluminium floor plates. Plates and columns were connected by bolts and brackets. Dimensions of the structure and floor layout are presented in Figure 2. A shaker was used to generate excitation. As it appears in Figure 2, two accelerometers were attached to each joint, resulting in eight accelerometers within each floor.

There were 270 vibration events generated. Each event contained 8192 samples, which were sampled at 1600 Hz. Among those events, 150 healthy events

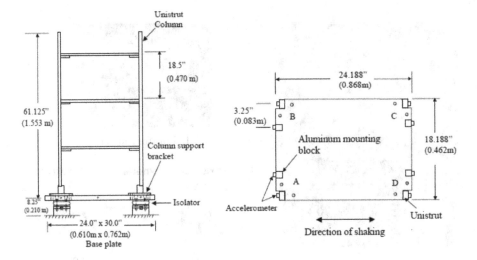

Fig. 2. Three-story building and floor layout [11]

were created using different shaker input levels and bandwidths to represent different environmental and operational conditions. In the remaining 120 events, there were 30 events with damage in location 1A (i.e. corner A at level 1), 60 events with damage in location 3C, and 30 events with damage in both locations (i.e. 1A and 3C). The damage was introduced by loosening the bolts and then hand tightening them, or by removing bolts and brackets at the joints, allowing the plate to move freely relative to the column.

The Sydney Harbour Bridge. The Sydney Harbour Bridge is one of major bridges in Australia, which was opened in 1932. There are 800 jack arches on the underside of the deck of the bus lane (lane seven) needed to be monitored, as shown in Figure 3a. Vibration data caused by passing vehicles were recorded by three-axis accelerometers installed under the deck of lane seven. Each joint was instrumented with a sensor node, which connected to three accelerometers mounted to the joint in left, middle and right positions as shown in Figure 3c.

There are two datasets used: a bridge specimen built from laboratory and real data collected from the bridge.

Specimen Data: A steel reinforced concrete beam was manufactured with a similar geometry to those on the Sydney Harbour Bridge (Figure 3b). The data were collected from two sets of sensor nodes placed on the base of the joint, one nodes is positioned at the tip while the other was mounted 750 mm away from the tip. The locations of three accelerometers from each node are similar to those on the joints of the bridge. The excitation was made using an impact hammer. Once the node was triggered by a hammer, it records data for 3 seconds at a sampling rate of 500 Hz, resulting in 1500 samples for each event.

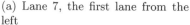

(a) Lane 7, the first lane from the left

(b) Laboratory specimen

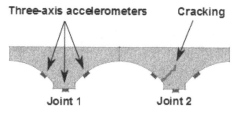

(c) Field trial with cracking

Fig. 3. The Sydney Harbour Bridge

After testing the benchmark in a healthy condition, a crack was gradually introduced into the specimen with four level of crack dimensions: (75×50)mm, (150×50)mm, (225×50)mm and (270×50)mm. The same test was conducted again in each damage severity. About 200 events were collected in healthy condition and in each level of damage severity.

Bridge Data: For this case study, only two instrumented joints on the bridge were considered (named joints 1 and 2 as in Figure 3c). A known crack existed at joint 2 at the time of measurement while joint 1 was in a good condition.

An event is defined as a time period during which a motor vehicle is driving across the joint. An event is normally triggered after the acceleration value is greater than a pre-set threshold. After the triggering occurs, the node records for a period of 1.5 seconds at a sampling rate of 400 Hz. Each event contains 100 samples before the event started and 500 samples are collected during and after the event. Denote A_i an instantaneous acceleration at i-th sample, A_r the rest vector which is the average of three readings (x, y, z) from the first 100 samples. One metric is extracted from three-axis readings: $V = |A_i| - |A_r|$, which is independent on the accelerometer orientations.

5.2   Feature Extraction

For all datasets, the features in the frequency domain were created as follows. For every vibration event, the data from each accelerometer were standardized to have zero mean and one standard deviation. Then the data were converted to the frequency domain using Fourier transform. Differences between vibrations of adjacent accelerometers in each location in the frequency domain were used as features. The rationality is that if a joint is healthy the accelerometers attached to it would move together. If the joint is damaged they would move differently. These features will be reflected in the differences of the signals.

Building Data. For every event, the difference between signals collected by two accelerometers at a joint in the frequency domain was taken (in total there are 12 joints in three stories). Then only frequency up to 150Hz was selected as features. So the data is a tensor of (768 features $\times$ 12 locations $\times$ 270 events).

Specimen Data. For each sensor node, differences between V feature mentioned above of accelerometers 1 and 2, 1 and 3, and 2 and 3 in the frequency domain were used as features. Only frequency up to the first 150Hz was selected. Finally we had a tensor of ($450 \times 6 \times 960$).

Bridge Data. Since only accelerometers in the same joints of the bridge are synchronized in time, only data from one joint was put in the tensor. It is all right since the vibration of each joint is quite independent to each other in this case. Since we have healthy data in joint 1 and damaged data in joint 2, to demonstrate the effectiveness of the method we combined the data from joints 1 and 2. Events from three accelerometers from joint 2 were used as damaged events while data from three accelerometers in joint 1 were used as the healthy events.

Then differences between V feature of accelerometers 1 and 2, 1 and 3, and 2 and 3 in the frequency domain were used as features. Only frequency up to the first 150Hz was selected. Finally we had a tensor of ($150 \times 3 \times 1341$).

5.3 Results

For building dataset, 100 healthy events were randomly selected as training data. The other were used for testing. There were 150 healthy events randomly selected as training data for specimen dataset and 500 random healthy events were used for training in bridge dataset. As described in Section 4.1, a benchmark model was built on the training tensor data and each new event (a tensor frontal slide) was test against the model to detect damage.

To increase the reliability of the results, multiple testing was used. In stead of computing the decision values for each event (single testing), we took a median value of a block of 10 sequential events in a chronological order. The reported accuracy was a block accuracy. The rationality for this is that there may be noisy events overtime but the health status of sequential events in a short time should be very similar. All the results shown were averaged over ten trials of experiment.

The tensor toolbox for Matlab [4] was used for tensor operations and LIBSVM for Matlab was used for one-class SVM [7]. In order to decide the number of rank-one tensors R in the CP method, core consistency diagnostic technique

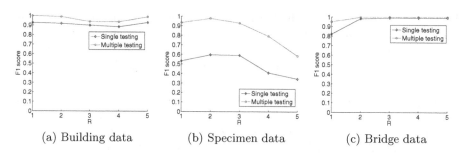

(a) Building data   (b) Specimen data   (c) Bridge data

Fig. 4. Damage detection accuracy using different values of R

(CORCONDIA) described in [5] was applied, in which a Tucker core is used for assessing the appropriateness of a CP model. This technique was implemented using the N-way Toolbox for Matlab [3].

Damage Detection. In this section, the damage detection using method described in Section 4.1 was investigated. For CP, we tried different values of R from one to five and used CORCONDIA method to decide the appropriate one. For one-class SVM model, the rate of anomalies $\nu = 5\%$ was selected.

We use $F_1 = 2 \frac{precision*recall}{precision+recall}$ as a damage detection accuracy. Figure 4 shows the F_1 scores for all new test instances in three datasets using both single testing and multiple testing. The best results almost agree with the parameter selection method (CORCONDIA). CORCONDIA selected $R = 2$ for the specimen and bridge data and $R = 1$ for the building data. Since $R = 2$ also gave similar results for the building data, we selected $R = 2$ for all datasets and it will be used for damage localization and estimation. Then we have F_1 scores of 0.99, 0.98 and 1 (using multiple testing) for the building, specimen and bridge datasets, respectively. The results also show that multiple testing can significantly improve the detection results, especially for the specimen data.

Damage Localization. Figure 5 shows two components of the location mode ($R = 2$) with color values for all sensor locations in the building data. The first component corresponded to a healthy state of the building when there was no damage while the second component presented a damaged state. In the damaged state, the colors with high values correctly associated with locations of known damage (1A and 3C). Therefore, this analysis is promising to localize damage in structures. For the bridge datasets, since all accelerometers were in the same joint, there was no need to localize the damage.

Damage Estimation. The decision values returned from the one-class SVM model were used to characterize the level of damage. The result in Figure 6 presents the decision values of every block of test events in the building and specimen data using one-class SVM (we did not do that for the bridge data

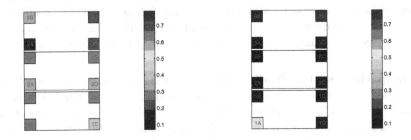

(a) Location mode - Healthy state   (b) Location mode - Damaged state

Fig. 5. Damage localization for building dataset

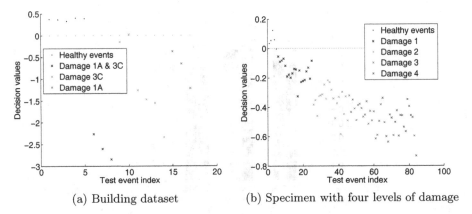

(a) Building dataset   (b) Specimen with four levels of damage

Fig. 6. Damage characterization using decision values obtained by CP and one-class SVM

since there are no ground truth for damage severity in this case). The dotted lines show the boundary between healthy and damaged events. Different ranges of the decision values matched with different severity levels of damage described in the datasets.

In Figure 6a, the first 50 events (i.e. 5 blocks) were decision values from healthy data. The next 30 events were damaged data when the damage occurred in both locations 1A and 3C (among them the first and the next 15 events had different levels of severity). The following 60 events corresponded to damage in location 3C with four levels of damage severity. And the last 30 events presented the decision values for damaged events in location 1A in two levels of severity. Moreover, for the same kind of damage, the decision values were lower when damage happened in both locations compared with those occurred in one location only. Figure 6b shows that the decision values successfully separated healthy events and four different levels of damaged events. In addition, events with more severe damage tended to have lower decision values. Therefore, it suggests that we can use the decision values obtained by CP and one-class SVM as structural health scores to characterize the damage severity in an unsupervised manner.

Comparison with a Traditional Approach. In this section, we will compare between the tensor approach with the approach without using tensor. For all datasets, individual one-class SVM model was built for each sensor location using the same train and test data with the same feature as in previous experiments. An average detection accuracy of all the location models was used to compared against the tensor approach. Figure 7 shows the F_1 score (multiple testing) of the two approaches for all three datasets. It shows that the tensor approach had better detection accuracies than the one without using tensor, especially for the specimen dataset. Moreover, it is impossible for the damage localization and estimation using the results obtained from each location model.

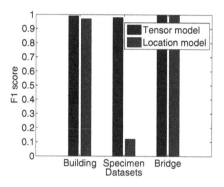

Fig. 7. Comparison between tensor and location models

6   Conclusion

This work presents a damage identification approach using tensor analysis for SHM applications where sensing data were converted to a three-way tensor form. A structural benchmark model was built using one-class SVM on a component matrix in time mode learned from the tensor decomposition. Then new events were updated using a simple least square approach and were tested against the benchmark model to detect damage. Moreover, in our approach damage detection, localization and assessment were achieved in an unsupervised manner. The approach was shown to work very well using data from lab-based structures and real data from the Sydney Harbour Bridge. We can obtain a damage detection F_1 scores of 0.98 and higher for all the datasets. Damage was localized correctly and different levels of damage severity were well estimated.

References

1. Acar, E., Yener, B.: Unsupervised multiway data analysis: A literature survey. IEEE Transactions on Knowledge and Data Engineering **21**(1), 6–20 (2009)
2. Acar, E., Aykut-Bingol, C., Bingol, H., Bro, R., Yener, B.: Multiway analysis of epilepsy tensors. Bioinformatics **23**(13), i10–i18 (2007)
3. Andersson, C.A., Bro, R.: The n-way toolbox for MATLAB. Chemometrics and Intelligent Laboratory Systems **52**(1), 1–4 (2000)
4. Bader, B.W., Kolda, T.G., et al.: Matlab tensor toolbox version 2.5 (January 2012). http://www.sandia.gov/tgkolda/TensorToolbox/
5. Bro, R., Kiers, H.A.L.: A new efficient method for determining the number of components in parafac models. Journal of Chemometrics **17**(5), 274–286 (2003)
6. Chan, T.H., Ni, Y.Q., Ko, J.M.: Neural network novelty filtering for anomaly detection. In: Cheng, F. (ed.) 2nd International Workshop on Structural Health Monitoring, pp. 133–137. Technomic Pub. Co., Standford (1999)
7. Chang, C.C., Lin, C.J.: LIBSVM: A library for support vector machines. ACM Transactions on Intelligent Systems and Technology **2**, 27:1–27:27 (2011)
8. Farrar, C.R., Worden, K.: An introduction to structural health monitoring. Philosophical Transactions of the Royal Society A: Mathematical, Physical and Engineering Sciences **365**(1851), 303–315 (2007)
9. Kolda, T.G., Bader, B.W.: Tensor decompositions and applications. SIAM Review **51**(3), 455–500 (2009)
10. Kolda, T.G., Sun, J.: Scalable tensor decompositions for multi-aspect data mining. In: ICDM 2008: Proceedings of the 8th IEEE International Conference on Data Mining, pp. 363–372 (December 2008)
11. LANL: Los alamos national laboratory website (2013). http://institute.lanl.gov/ei/software-and-data/ (last visited January 6, 2013)
12. Liu, W., Chan, J., Bailey, J., Leckie, C., Kotagiri, R.: Utilizing common substructures to speedup tensor factorization for mining dynamic graphs. In: Proceedings of the 21st ACM International Conference on Information and Knowledge Management, CIKM 2012, pp. 435–444. ACM, New York (2012)
13. Prada, M.A., Toivola, J., Kullaa, J., Hollmn, J.: Three-way analysis of structural health monitoring data. Neurocomputing **80**, 119–128 (2012). Special Issue on Machine Learning for Signal Processing 2010
14. Rytter, A.: Vibration-based inspection of civil engineering structures. Ph.D. thesis, University of Aalborg, Denmark (1993)
15. Schölkopf, B., Williamson, R.C., Smola, A.J., Shawe-Taylor, J., Platt, J.C.: Support vector method for novelty detection. In: NIPS, pp. 582–588 (1999)
16. Sun, J., Tao, D., Papadimitriou, S., Yu, P.S., Faloutsos, C.: Incremental tensor analysis: Theory and applications. ACM Trans. Knowl. Discov. Data **2**(3), 11:1–11:37 (2008)
17. Worden, K., Manson, G., Fieller, N.: Damage detection using outlier analysis. Journal of Sound and Vibration **229**(3), 647–667 (2000)
18. Worden, K., Manson, G.: The application of machine learning to structural health monitoring. Philosophical Transactions of the Royal Society A: Mathematical, Physical and Engineering Sciences **365**(1851), 515–537 (2007)

Predicting Smartphone Adoption
in Social Networks

Le Wu[1], Yin Zhu[2], Nicholas Jing Yuan[3], Enhong Chen[1(✉)],
Xing Xie[3], and Yong Rui[3]

[1] University of Science and Technology of China, Hefei, China
cheneh@ustc.edu.cn
[2] Hong Kong University of Science and Technology, Hong Kong, China
[3] Microsoft Research, Beijing, China

Abstract. The recent advancements in online social networks and mobile devices have provided valuable data sources to track users' smartphone adoption, i.e., the usage of smartphones over time. An incisive understanding of users' smartphone adoption can benefit many useful applications, ranging from user behavior understanding to targeted marketing. This paper studies smartphone adoption prediction in social networks by leveraging the wisdom of an online world. A critical challenge along this line is to identify the key factors that underline people's adoption behaviors and distinguish the relative contribution of each factor. Specifically, we model the final smartphone status of each user as a result of three influencing factors: the social influence factor, the homophily factor, and the personal factor. We further develop a supervised model that takes all three factors for smartphone adoption and at the same time learns the relative contribution of each factor from the data. Experimental results on a large real world dataset demonstrate the effectiveness of our proposed model.

Keywords: Smartphone adoption · Social network · Social influence · Homophily

1 Introduction

Smartphones (e.g., iPhone and Android based mobile phones) are now ubiquitous in our daily lives. There were 1.82 billion smartphones being used worldwide at the end of 2013. Furthermore, according to a forecast by International Data Corporation, the smartphone market is expected to increase to 70.5% in 2017 in terms of all smart devices, including desktop PCs, portable PCs, tablets and smartphones [1].

With the expanding opportunities in the smartphone market, an incisive understanding of smartphone adoption among users has significant applications ranging from user behavior understanding in scientific disciplines [4,19] to targeted advertising for marketing strategies [7,10]. Thus, acceptance or adoption of smartphones has long been studied in the past from a variety of angles, such as cultural factors [23], technology needs [22] and perceived usefulness [18]. Nearly

© Springer International Publishing Switzerland 2015
T. Cao et al. (Eds.): PAKDD 2015, Part I, LNAI 9077, pp. 472–485, 2015.
DOI: 10.1007/978-3-319-18038-0_37

all these studies were based on traditional survey based approaches, e.g., by surveying hundreds of people. With both the time and money costs of collecting data, few researchers have attempted to investigate the smartphone adoption within a large-scale social network.

Luckily, with the recent advancements of online social networks and smartphones, an increasing number of people are sharing their daily lives with friends on these platforms through smartphones. Due to the mobile nature of the login devices, these *mobilized social networks* record the smartphone footprints of users. To illustrate this, we provide the following example. *Weibo* (weibo.com) is the leading microblog service in China. When a user posts a message in *Weibo*, the platform forwards an enriched message to all of the user's followers as shown in Fig. 1, which includes the post message, the timestamp and the sending device (iPhone). This device information creates valuable data sources to track smartphone adoption within a large-scale social network.

Fig. 1. A sample post from Weibo

As a matter of fact, even with the mobilized social network data, accurately understanding a user's smartphone adoption is still technically challenging from at least two aspects. On the one hand, there are various factors that underline person's decision-making process. How can we leverage them in a unified framework? Researchers have long identified three key factors for this process: the social influence factor that argues users are influenced by their social neighbors to make decisions [11,25]; the homophily factor refers to linked users performing similar decisions [15]; and the personal factor states users have their own personalized preferences [29,30]. On the other hand, though all these key factors help predict users' adoption behaviors, they lead to significantly different results [2,14]. Accurately understanding and distinguishing the relative contribution of each factor is critically important to guiding the firm's marketing strategy. E.g., if the social influence is responsible for users' decisions, then it is effective for the firm to incentivize several seed customers to trigger a cascade of information diffusion [11]. If the homophily factor dominates, then the firm can identify new potential customers based on each user's neighbors' decisions. If the final adoption behavior is driven by the personal factor, a better idea is to select the targeted customers based on their historical preferences for marketing. Nevertheless, few previous methods have incorporated all these principal factors together for product adoption prediction. Therefore, how to leverage all these key factors and distinguish them at the same time for smartphone adoption prediction remains pretty much open.

In order to solve both the data barrier and technical challenges mentioned above, in this paper, we propose a supervised machine learning model for smartphone adoption prediction. As a preliminary, we leverage a mobilized online social network to discover the smartphone usage patterns of a large group of networked users. Then, by borrowing the traditional user segmentation concept, we identify two groups of users based on their current smartphone status, i.e., potential first-time smartphone adopters and potential brand changers, respectively. After that, we develop a *S*upervised *H*omophily-*I*nfluence-*P*ersonality (SHIP) model for smartphone adoption prediction, in which the key factors that underline people's adoption are explicitly integrated. In fact, the proposed model can easily be extended to other product adoption tasks. Finally, the experimental results on 200K active mobile users show the effectiveness of our proposed model. To the best of our knowledge, this is the first comprehensive attempt to predict smartphone adoption from a social perspective with large-scale real world data.

2   Data Description and Problem Definition

Given a snapshot of a social network as a directed graph $G = <U, F, \mathbf{T}>$, where the node set $U = \{1, 2, ..., N\}$ is the users and F represents the relationships of users. $\mathbf{T} = [t_{ji}]_{N*N}$ is an edge strength matrix, where t_{ji} represents the tie strength from user j to user i. Specifically, if user i follows user j, then $(i, j) \in F$ and $t_{ji} > 0$, otherwise $t_{ji} = 0$. Since we mainly focus on the smartphone adoption of users, for ease of later explanation, a *mobile post* is defined as a post that is sent from a smartphone, rather than a PC client or a tablet. If a user sends more than τ mobile posts in a time period, we regard him/her as a *mobile user*.

Data Collection and Description. During the data crawling process, we collected the post streams of nearly 235 thousand users from January 2013 to July 2013 from *Weibo*. For data cleaning, we only selected mobile users (i.e., $\tau = 10$ empirically) and their associated relations. After pruning, we still had nearly 200 thousand users, 15 million edges, and 45 million post streams. Now we introduce how to infer each user's smartphone status from the continuous post streams. Similar to many smartphone marketing research [1], we treat each quarter as a time slice and further split each user's device streams into two time slices, i.e., the first quarter (2013Q1) and the second quarter (2013Q2) of 2013. Then we take the most popular device brand as the smartphone status of the user at that time. Note that a user may use several smartphones during a time slot, however, it is reasonable to discard the infrequent uses of other smartphone brands since users prefer those phones that they use the most frequently. Table 1 shows an example of the inferred smartphone status of two typical users.

Problem Definition. Generally, our goal is to predict the smartphone adoption status of users in time t based on the available data in the previous time $t-1$. In marketing research, a common practice is to first divide a broad target market

Table 1. Examples of two typical users' post device streams in Weibo

	Alice				Bob		
Time slice	Timestamp	Sent device	Device status	Time slice	Timestamp	Sent device	Device status
2013Q1	20130211	Web Weibo		2013Q1	20130211	Nokia 5230	
	20130212	Web Weibo	⟹ Desktop		20130220	Nokia 5230	⟹ Nokia
	...	...			...	...	
	20130331	Web Weibo			20130331	Nokia 5230	
2013Q2	...	...		2013Q2	...	...	
	20130421	Web Weibo			20130419	Nokia 5230	
	20130425	Samsung Galaxy S2	⟹ Samsung		20130422	iPhone	⟹ iPhone
	20130428	Samsung Galaxy S2			20130423	iPhone	
	...	...				...	

into subsets of consumers and then design strategies to target each group of consumers. Following this approach, we divided users into two groups based on their smartphone status in $t-1$: potential first-time smartphone buyers and potential brand changers. The potential brand changers are those who have already used a smartphone in $t-1$ and their next action is deciding whether to change brands in t. E.g., as illustrated in Table 1, *Bob* is a potential brand changer as he had the *Nokia* smartphone in 2013Q1 (i.e., time $t-1$). We regard those who do not use any smartphone in $t-1$ as potential first-time smartphone buyers. This assumption may not be accurate when applied to each person, but the overall trend is well supported by the high penetration of mobilized social networks in our everyday life. *Alice* is a potential first-time buyer as shown in Table 1. After segmenting users into these two groups, we set the target for each group as follows:

Task 1: First-time Buying Prediction. If a user is a potential first-time buyer in $t-1$, we predict whether she/he will buy a particular brand b in time t or not.

Task 2: Brand Change Prediction. If a user is a potential brand changer in time $t-1$, we predict whether this user will change to another brand in the next time period t.

We next assigned a label to each user based on the group information and the smartphone adoption status in t. E.g., *Alice* is a member of Task 1 and buys a Samsung in t, so she is a positively labeled user if we focus on predicting whether she will buy a *Samsung*. *Bob* is a positively labeled user in Task 2 as he changed from *Samsung* to *iPhone* in t. In summary, after user segmentation and label assignment for each task, these two tasks can be summarized in a unified prediction problem: Given a snapshot of a directed social network $G = <U, F, \mathbf{T}>$ with a positively labeled user set UP and a negatively labeled user set UN in time t, our goal is to predict the labels of all unknown users at time t as accurate as possible. In the next section, we focus on the model.

3   The Proposed SHIP Model

Researchers have long converged on the idea that there are three principal factors that drive people's adoption decisions: the social influence factor and the homophily factor that lead to correlated user behaviors among linked users and the personal factor that states users' unique preferences [15,25,29]. Obviously, each of these three factors can exploit a specific part of users' decisions. In addition, as illustrated before, different factors result in significantly different marketing strategies [2,14]. Thus, simply aggregating all these factors for prediction will not be the best choice. A better idea is to distinguish the relative effect of each factor in the decision making process. In the following, we propose a *Supervised Homophily-Influence-Personlity* (SHIP) model that can automatically learn the contribution of each factor for users' smartphone adoption. Next we describe how to construct the SHIP model step by step.

Overview of Smartphone Adoption Function. For each user i, we explicitly model the smartphone adoption status p_i as a combination of the three key factors:

$$p_i = (1 - \alpha) \sum_{j \in F_i} t_{ji}[(1 - \beta)p_j + \beta u_{ji}] + \alpha b_i, \qquad (1)$$

where p_i is the predicted smartphone adoption probability that ranges from 0 to 1. F_i are the users that i follows in this network. t_{ji} represents the strength between i and j. If i follows j, then $t_{ji} = \frac{1}{|F_i|}$, otherwise $t_{ji} = 0$. We have two parts in this equation, the first part captures the social network effect (including social influence and homophily) and the second part (b_i) mimics the personal bias. Specifically, for user i and any user j that i links ($j \in F_i$), i's adoption probability p_i is balanced by the influence of j's adoption status p_j (social influence) and the homophily effect u_{ji}, where β controls the relative contribution of these two factors in social networks. α ($0 \leq \alpha \leq 1$) is a parameter that controls the relative effect of the social network and personal bias. The larger the α, the more personal preference plays a role in the task.

Since for each pair of linked users, we have a vector e_{ji} that captures the various features between them, we model the homophily, i.e., the similarity between each pair of linked users as:

$$u_{ji} = s(\boldsymbol{w} \cdot (\boldsymbol{e_{ji}})) = s(\sum_k w_k \times e_{jik}), \qquad (2)$$

where e_{jik} is the k-th element of $\boldsymbol{e_{ji}}$. Similarly, for each user i, we have $\boldsymbol{x_i}$, which captures her various characteristics. Then the personal bias can be defined as:

$$b_i = s(\boldsymbol{v} \cdot \boldsymbol{x_i}) = s(\sum_k v_k \times x_{ik}). \qquad (3)$$

In the above two equations, $s(l)$ can be set as any monotonically increasing function. As $\forall i \in U, 0 \leq p_i \leq 1$, for fair comparison of the different effects, these values are better ranges in $[0, 1]$. Thus, a natural idea is to set $s(l)$ as a logistic function $s(x) = \frac{1}{1+e(-x)}$.

Note that the proposed smartphone adoption probability function (Eq.(1)) has close relationship with the recent progress in supervised random walk based models. These models incorporated the node and edge features to supervise the random walk process for node classification tasks [3,28]. E.g., the works of

[3] utilized the social influence factor and [28] further extended this work by incorporating the personal bias. Nevertheless, the homophily factor is neglected by all these previous works in the modeling process, while we explicitly depict the homophily factor between each pair of linked users. In other words, the previous works for node classification can be seen as special cases of our models, e.g., our work is reduced to the models proposed by Zeng et al. when excluding the homophily factor [28].

Optimization Function Construction. Based on Eq. (1) , in order to get the final label preference p_i for each user i, we have to learn four parameters $\theta = [\alpha, \beta, \boldsymbol{w}, \boldsymbol{v}]$ in the training process. As we have a set of users' labels in the training data at time t, an intuitive idea is to train a supervised model that automatically learns the parameters θ such that all labeled positive users in the training data have larger probabilities than the labeled negative ones. Next, we model the objective learning function as:

$$\min_{\theta} L = \sum_{i \in UP} \sum_{j \in UN} h(p_j - p_i) + \lambda[\boldsymbol{w}'\boldsymbol{w} + \boldsymbol{v}'\boldsymbol{v}], \qquad (4)$$

where the first term models the goodness for fitting the data and the second term controls model complexity. Since j is a negatively labeled user and i a positively one, the larger the p_i the better and the smaller the p_j the better. Based on the above, we empirically set $h(x)$ as:

$$h(x) = \begin{cases} 0 & \text{if } x < 0 \\ \frac{1}{1+e^{-cx}} & \text{if } x \geq 0. \end{cases} \qquad (5)$$

Thus if $p_j - p_i > 0$, the loss value is about 1. Otherwise, it approximates to zero.

Model Learning. We apply the power iteration method to solve the optimization problem in Eq.(4) [17]. Specifically, we write the derivatives of each parameter of θ as:

$$\frac{\partial L}{\partial \alpha} = \sum_{i,j} \frac{\partial h(\delta_{ij})}{\partial \delta_{ij}} (\frac{\partial p_j}{\partial \alpha} - \frac{\partial p_i}{\partial \alpha}), \qquad \frac{\partial L}{\partial \boldsymbol{w}} = \sum_{i,j} \frac{\partial h(\delta_{ij})}{\partial \delta_{ij}} (\frac{\partial p_j}{\partial \boldsymbol{w}} - \frac{\partial p_i}{\partial \boldsymbol{w}}) + 2\lambda \boldsymbol{w},$$

$$\frac{\partial L}{\partial \beta} = \sum_{i,j} \frac{\partial h(\delta_{ij})}{\partial \delta_{ij}} (\frac{\partial p_j}{\partial \beta} - \frac{\partial p_i}{\partial \beta}), \qquad \frac{\partial L}{\partial \boldsymbol{v}} = \sum_{i,j} \frac{\partial h(\delta_{ij})}{\partial \delta_{ij}} (\frac{\partial p_j}{\partial \boldsymbol{v}} - \frac{\partial p_i}{\partial \boldsymbol{v}}) + 2\lambda \boldsymbol{v}. \qquad (6)$$

According to Eq. (1) of the predicted adoption rate, we have:

$$\frac{\partial p_i}{\partial \alpha} = -\sum_{j \in F_i} t_{ji}[(1 - \beta)p_j + \beta u_{ji}] + (1 - \alpha) \sum_{j \in F_i} t_{ji}(1 - \beta) \frac{\partial p_j}{\partial \alpha} + b_i,$$

$$\frac{\partial p_i}{\partial \beta} = (1 - \alpha)[\sum_{j \in F_i} t_{ji}[-p_j + (1 - \beta) \frac{\partial p_j}{\partial \beta} + u_{ji}]], \qquad \frac{\partial p_i}{\partial \boldsymbol{w}} = (1 - \alpha)[\sum_{j \in F_i} t_{ji}[(1 - \beta) \frac{\partial p_j}{\partial \boldsymbol{w}} + \beta \frac{\partial u_{ji}}{\partial \boldsymbol{w}}]],$$

$$\frac{\partial p_i}{\partial \boldsymbol{v}} = (1 - \alpha) \sum_{j \in F_i} t_{ji}[(1 - \beta) \frac{\partial p_j}{\partial \boldsymbol{v}}] + \alpha \frac{\partial u_i}{\partial \boldsymbol{v}}. \qquad (7)$$

Now it is easy to determine the remaining derivative of $\frac{\partial u_{ji}}{\partial \boldsymbol{w}}$ and $\frac{\partial b_i}{\partial \boldsymbol{v}}$:

Algorithm 1. Parameter Learning Process for the Proposed SHIP Model

---

Initialize $\alpha^0 = \beta^0 = 0.5$, $\boldsymbol{w}$, $\boldsymbol{v}$ and $\mathbf{p}^0$ with small positive values;
for $k = 1; k \leq K; k++$ **do**

$\quad$ //Part 1:Given $\theta^{(k-1)} = [\alpha^{(k-1)}, \beta^{(k-1)}, \boldsymbol{w}^{(k-1)}, \boldsymbol{v}^{(k-1)}]$, calculate $\mathbf{p}^k$;
$\quad$ **while** *not converged for* $\mathbf{p}^k$ **do**
$\quad\quad$ **for** *each user* $i \in U$ **do**
$\quad\quad\quad$ Calculate p_i^k based on Eq.(1);

$\quad$ //Part 2: Given $\mathbf{p}^k$, calculate the following equations;
$\quad$ **while** *not converged* **do**
$\quad\quad$ **for** *each user* $i \in U$ **do**
$\quad\quad\quad$ compute the equations in (7);

$\quad$ //Part 3: Given $\mathbf{p}^k$, calculate θ^k based on Eq.(6) ;
$\quad$ $\alpha^k = \alpha^{(k-1)} - step\_size * \frac{\partial^k L}{\partial \alpha}$, $\quad\quad \boldsymbol{w}^k = \boldsymbol{w}^{(k-1)} - step\_size * \frac{\partial^k L}{\partial \boldsymbol{w}}$;
$\quad$ $\beta^k = \beta^{(k-1)} - step\_size * \frac{\partial^k L}{\partial \beta}$, $\quad\quad \boldsymbol{v}^k = \boldsymbol{v}^{(k-1)} - step\_size * \frac{\partial^k L}{\partial \boldsymbol{v}}$;

Return $\mathbf{p}^K, \theta^K = [\alpha^K, \beta^K, \boldsymbol{w}^K, \boldsymbol{v}^K]$;

$$\frac{\partial u_{ji}}{\partial \boldsymbol{w}} = \frac{\partial s(\boldsymbol{w} \cdot e_{ji})}{\partial e_{ji}} e_{ji}, \quad\quad \frac{\partial b_i}{\partial \boldsymbol{v}} = \frac{\partial s(\boldsymbol{w} \cdot \mathbf{x_i})}{\partial (\boldsymbol{w} \cdot \mathbf{x_i})} \mathbf{x_i} \tag{8}$$

Convergence Analysis. Algorithm 1 shows the entire optimization process of our proposed model. There are two power iterations as shown in Part 1 (Eq. (1)) and Part 2 (Eq. (7)) of the algorithm. For all of these equations, they could be rewritten as a unified form as $\mathbf{z_i} = (1 - d)\sum_{k \in F_i} t_{ki}\mathbf{z_k} + d\mathbf{y}$. This unified representation defines a linear system problem and its closed form is $\mathbf{Z} = \alpha(\mathbf{I} - (1 - \alpha)\mathbf{T})^{-1}\mathbf{Y}$. This closed form satisfies the convergence condition of Gauss-Seidel iterative method [16]. In conclusion, all of the iterations can be solved in linear time with a convergence guarantee.

4 Experiments

4.1 Experimental Settings

We conduct experiments on the collected Weibo data as described in Section 2. We focus on predicting smartphone adoption in 2013Q2 (time t) based on the smartphone status in 2013Q1 (time $t-1$). Given a snapshot of the social network, for each task, we randomly split users into five equal parts and each time we select 80% of the users as labeled users for training and the remaining 20% users are used for prediction. We conduct five-fold cross validation and report the average results. In fact, we only choose the leading four brands (i.e., iPhone, Samsung, Nokia and Xiaomi) in Task 1 for prediction as the remaining brands take less than 1% market share. The detailed data statistics can be found in Table 2. As shown in this table, the data is very unbalanced, for most tasks, the number of negative records is much larger than that of the positive records.

Table 2. Dataset statistics of the two tasks. #P: the number of labeled positive users, #N: the number of labeled negative users (#P+#N=#users). $P\_ratio = \frac{\#P}{\#P+\#N}$.

Task	Users	Edges	Brand	#P	#N	P_ratio
			iPhone	5,152	7,154	41.9%
Task 1	12,306	125,876	Samsung	854	11,452	6.94%
			Xiaomi	458	11,848	3.72%
			Nokia	313	11,993	2.54%
Task 2	144,567	15,829,075	/	22,192	122,374	15.35%

For the evaluation, we first use the AUC (i.e., Area Under the ROC Curve) measure, which is especially useful for evaluating the performance of an unbalanced dataset [27]. A random guess would result in an AUC value around 0.5 and the larger the value the better the performance. In addition, as we focus on the most likely positive users of the test data, which can be used for marketing. We measure the relative gain of the precision as $Rel\_gain@N = \frac{Pre@N}{P\_ratio} - 1 = \frac{\#hits}{N*P\_ratio} - 1$. This measure evaluates how the proposed models improve the precision compared to random guess. A random guess will lead to a Rel_gain@N result of 0.0 and the larger the value the better the performance.

Table 3. Summarization of different kinds of features

Type	Feature Description	Type	Feature Description
Social	# of followers that have positive labels in $t-1$	Profile	gender, location
	# of followers that have negative labels in $t-1$		is this user a verified account
	# of friends that have positive labels in $t-1$		#followers, #followees, #friends
	# of friends that have negative labels in $t-1$		#posts that the user sent in $t-1$
			#posts that the user sent in $t-1$
Edge	whether they are friends	Brand	the brand the user used in $t-1$ (only available in Task 2)
	# of co-followers that have positive labels in $t-1$		
	# of co-friends that have positive labels in $t-1$		
	# of co-followers that have negative labels in $t-1$		
	# of co-friends that have negative labels in $t-1$		

Baselines. To the best of our knowledge, few researchers have tried to explore the smartphone adoption problem with real world collected data. However, we can borrow several classic models that are widely used for the binary class prediction task in a social network: the first category builds classifiers using the extracted graph information as features, and the second category directly propagates the existing labels via random walks in this graph [5]. In the first kind, we choose the logistic regression (LR) model. Specifically, we implemented the LRS baseline that purely relies on *S*ocial network features and the LRSP baseline which uses both *S*ocial network features and the user *P*rofiles. For the second kind, we choose label propagation (LP), which performs node class prediction based on partially labeled data in a graph [31]. Specifically, LP can be seen as an unsupervised version of our model that only utilizes the graph structure information, i.e., the social influence factor in our model.

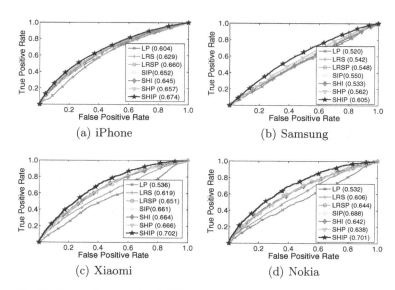

Fig. 2. Comparison of the AUC results of different models for Task 1

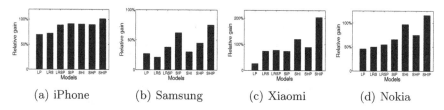

Fig. 3. Comparison of the Rel_gain@100 results of different models for Task 1

Also, to demonstrate the fitness of the three proposed factors in our model, we compare the proposed SHIP with three related models: SHI (Supervised Homophily-Influence), SHP (Supervised Homophily-Personlity) and SIP (Supervised Influence-Personality). Please note that the simplified SIP model can be seen as a superior version of the work proposed in [28], which can automatically learn the relative importance parameter α between influence and personality. In both LR and our proposed models, we have the regularization parameter λ. As the dataset is very large, choosing λ in a reasonable range (e.g., [0.01, 100]) has little impact on the final prediction results. For the remaining experiments, we empirically set $\lambda = 10$. We summarize the profile, edge, and social features we used in this paper in Table 3.

4.2 Experimental Results

Overall Performance. Task 1 focuses on predicting whether a user will buy a particular brand as first-time buying behavior. Fig. 2 reports the AUC results of different models, where each brand's prediction result is shown in the sub figure and the detailed AUC value is followed by each method in the legend. First, we observe that all models have better performance than a random guess.(i.e., an AUC value of 0.5) Among them, our proposed SHIP model is better than all baselines with regard to all brands, followed by the three related models (i.e., SIP, SHI and SHP), indicating the superiority of our proposed model and the importance of combining the three key factors together for predicting smartphone adoption. Although the overall trend is the same, the detailed AUC results vary. Among all brands, "whether to choose a Samsung for a first-time buying" is the hardest to predict and the best AUC result is only 0.605. One possible reason is that Samsung has too many device types, ranging from high-end smartphones that compete with the iPhone to entry-level smartphones. The reasons why people buy Samsung smartphones vary and are harder to predict. For the other brands, the AUC reaches about 0.7 for SHIP. The average improvement is 3% to 10% over LRSP and 15% to 30% for the remaining baselines. Similar trend can be found for the Rel_gain@100 comparison as shown in Fig. 3. Based on the above analysis, we conclude that the proposed SHIP can help better capture the decision process for first-time buying behavior, thus generating better results than other baselines and related models for Task 1.

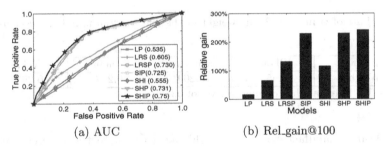

(a) AUC   (b) Rel_gain@100

Fig. 4. Overall comparison of Task 2: brand change prediction

In Task 2, we predict whether users will change brand in the next time period. Fig. 4 reports both the AUC and the Rel_gain@100 for different models in Task 2. The overall trends are the same as Task 1. For both metrics, SHIP performs the best, followed by our two related models (i.e., SIP and SHP) and LRSP baseline. However, the related SHI model and the LRS baseline, which do not consider the user preference factor, perform badly. In other words, after adding the user preference factor (i.e., the user profile features as shown in Table 3 and the brand feature), the performance improvement is very significant. E.g., the improvement of LRSP over LRS is 20.66% for AUC and 100% for Rel_gain@100, the improvement of SHIP over SHI is 35% for AUC and more than 100% for Rel_gain@100. Why is the improvement so significant after adding the user preference factor? We leave the explanations for the next section.

Table 4. The learned relative weight of each factor

Factor	Weight Representation	Task 1				Task 2
		iPhone	Samsung	Xiaomi	Nokia	
Influence	$(1-\alpha) \times (1-\beta)$	0.822	0.644	0.540	0.692	0.157
Homphily	$(1-\alpha) \times \beta$	0.081	0.138	0.181	0.219	0.42
Personality	α	0.097	0.218	0.278	0.089	0.423

Impacts of the Parameters. As shown in Eq. (1), α and β are two important parameters that control the relative effects of the three key factors for decision-making. We summarize the learned relative weight of each factor of the two tasks in Table 4. As shown in this table, in Task 1, the personality effect (α) and the homophily effect (i.e., $(1-\alpha) * \beta$) for all brands are very small while the relative contribution of social influence for all brands is larger than 50%. That is to say, users are easily influenced by social neighbors for first-time buying behavior. In contrast to this, the social influence effect is very small in Task 2 (i.e., 15.7%) while there is a high impact of the personality factor and the homophily effect for brand change behavior. In other words, users are not easily influenced by social neighbors for changing brand. Their brand change behavior is more liked caused by the homophily effect and their own preferences.

Table 5. Part of v in Task 2: the weight of the feature "the brand the user used in $t-1$"

Brand	iPhone	Samsung	Xiaomi	Nokia
Weight	-0.555*	-0.269*	-0.222*	0.266*

*Pass the T test at the confidence level of 0.005.

For brand manufactures, they would like to explore the inherent reasons that may prevent customer loss, i.e., the brand change behavior in Task 2. As explained before, after adding the user preference factor, the performance improvement is prominent in Task 2. Also, the user personality effect contributes more than 40% to brand change behavior. So we will focus on the user personality effect of Task 2 in this section. Specifically, each dimension of parameter v controls the importance of the corresponding user related feature for smartphone adoption (Eq.(3)). The larger the absolute value of this dimension, the greater the corresponding feature weights for smartphone adoption. In Task 2, we have two kinds of user personality features: user profile features and the brand feature, which describes the brand a user used in $t-1$. To our surprise, all profile features' weights are around 0 and the weight of the brand feature dominates. Next, we try to use this brand feature only in the logistic regression model and the AUC reaches 0.7229, while LRSP's result is 0.7300. The improvement is less than 1% when adding so many user profiles and social features, which also indicates social neighbors' smartphone adoption status does not have a large impact on users' choices of changing smartphones. Thus, we argue, the most prominent factor that determines whether a user will change brands later is the current brand she/he uses. A user's decision on whether to change brands follows the overall brand loyalty. If most people that uses a particular brand in the current time period are likely to change brand in the next time, then this user is also likely to change without a discussion. Table 5 shows the learned weights of the brand feature in Task 2. Among all the listed brands, iPhone users are most loyal. They do not like to change to another brand in the next time period, followed by Samsung and Xiaomi.

5   Related Work

Smartphone usage mining has attracted considerable attentions due to the rapid growth of the smartphone market in recent years. Some researchers revealed the correlation between mobile phone usage and user profiles [12,20,21]. Others attempted to consider the factors that affect people's choices when adopting a mobile device from various perspectives, such as culture [23], technology needs [22] and perceived usefulness [18]. Among them, Harsha et al. found compelling evidence of social influence in the purchase of mobile phones by sample surveys from Asian countries [8]. However, nearly all these works relied on small-scale questionnaires without considering the smartphone adoptions in a large-scale social network.

Our work is closely related to the problem of production adoption prediction in social networks. Generally, some models purely utilized user's profiles in social networks for product adoption prediction [29,30]. Others further incorporated the aggregated features extracted from social networks to boost product adoption performance [6,9]. However, the global product diffusion process among linked users is rarely analyzed, not to mention distinguishing the relative contributions of each factor. While the importance of distinguishing various factors underlining people's correlated decisions in social networks has been well recognized, the related work mainly focused on the homophily factor and the social influence factor that lead to correlated user behaviors [2,14]. The proposed solutions either estimated the upper bound of each factor or needed additional group information of users. On the contrary, our proposed model explicitly balances the correlated user behaviors and each user's own preference. Also, the relative performance of each factor can be learned automatically in the training process.

Our proposed model is also related to the node classification task, i.e., predict the classes of unlabeled nodes with partially labeled nodes in this graph [5,13,26]. A basic assumption of these models is the label correlations in the network, thus we can propagate the labels with respect to the intrinsic graph structure [24,31]. To leverage both the social network structure and the edge features, in recent years, [3] first proposed a supervised random walk algorithm that guides label propagation, where the social influence factor is explicitly modeled. Zeng et al. [28] further extended the supervised random walk model for user affiliation prediction by incorporating both the social influence and the user bias factors. Nevertheless, the homophily effect between linked users was neglected by all these works. Thus our model can be seen as generalizing the recent advances of these related methods in node classification tasks. Moreover, in contrast to these previous approaches, our proposed model can automatically learn the relative effect of each factor while others needed to tune the parameters manually.

6   Conclusion

In this paper, we have proposed a SHIP model for predicting smartphone adoption in a social network. Our model identified the three key factors in the decision-making process and can automatically distinguish the relative contributions of each factor. Experimental results on a large-scale dataset showed the strong prediction power of our model. An incisive conclusion is that the potential first-time smartphone buyers are largely influenced by social neighbors' choices while a user's decision on whether to change to another brand follows overall brand loyalty. In fact, the proposed model is also generally applicable to other node classification tasks. In the future, we plan to apply our model to other smartphone markets, and we will study the adoption in a finer granularity of time periods.

Acknowledgments. This research was partially supported by grants from the National Science Foundation for Distinguished Young Scholars of China (Grant No. 61325010) and the National High Technology Research and Development Program of China (Grant No. 2014AA015203).

References

1. Idc's forcast research. http://www.idc.com/getdoc.jsp?containerId=prUS24314413
2. Aral, S., Muchnik, L., Sundararajan, A.: Distinguishing influence-based contagion from homophily-driven diffusion in dynamic networks. PNAS **106**(51), 21544–21549 (2009)
3. Backstrom, L., Leskovec, J.: Supervised random walks: predicting and recommending links in social networks. In: WSDM, pp. 635–644. ACM (2011)
4. Baum, W.M.: Understanding behaviorism: Science, behavior, and culture. Harper-Collins College Publishers (1994)
5. Bhagat, S., Cormode, G., Muthukrishnan, S.: Node classification in social networks. In: Social Network Data Analytics, pp. 115–148. Springer (2011)
6. Bhatt, R., Chaoji, V., Parekh, R.: Predicting product adoption in large-scale social networks. In: CIKM, pp. 1039–1048. ACM (2010)
7. Boe, B.J., Hamrick, J.M., Aarant, M.L.: System and method for profiling customers for targeted marketing (2001). US Patent 6,236,975
8. De Silva, H., Ratnadiwakara, D., Zainudeen, A.: Social influence in mobile phone adoption: Evidence from the bottom of the pyramid in emerging asia. Information Technologies & International Development **7**(3) (2011)
9. Guo, S., Wang, M., Leskovec, J.: The role of social networks in online shopping: information passing, price of trust, and consumer choice. In: EC, pp. 157–166. ACM (2011)

10. Iyer, G., Soberman, D., Villas-Boas, J.M.: The targeting of advertising. Marketing Science **24**(3), 461–476 (2005)
11. Kleinberg, J.: Cascading behavior in networks: Algorithmic and economic issues. Algorithmic Game Theory **24**, 613–632 (2007)
12. Lins, L., Klosowski, J.T., Scheidegger, C.: Nanocubes for real-time exploration of spatiotemporal datasets. IEEE Trans. VCG **19**(12), 2456–2465 (2013)
13. London, B., Getoor, L.: Collective classification of network data. Data Classification: Algorithms and Applications, 399 (2014)
14. Ma, L., Krishnan, R., Montgomery, A.: Homophily or influence? an empirical analysis of purchase within a social network (2010)
15. McPherson, M., Smith-Lovin, L., Cook, J.M.: Birds of a feather: Homophily in social networks. Annual Review of Sociology, 415–444 (2001)
16. Meijerink, J.A., van der Vorst, H.A.: An iterative solution method for linear systems of which the coefficient matrix is a symmetric m-matrix. Mathematics of Computation **31**(137), 148–162 (1977)
17. Page, L., Brin, S., Motwani, R., Winograd, T.: The pagerank citation ranking: Bringing order to the web (1999)
18. Park, Y., Chen, J.V.: Acceptance and adoption of the innovative use of smartphone. Industrial Management & Data Systems **107**(9), 1349–1365 (2007)
19. Skinner, B.F.: Science and human behavior. Simon and Schuster (1953)
20. Smith, A.: 46% of american adults are smartphone owners. Pew Internet & American Life Project (2012)
21. Smith, A.: Smartphone ownership-2013 update. Pew Internet & American Life Project (2013)
22. Van Biljon, J., Kotzé, P.: Modelling the factors that influence mobile phone adoption. In: SAICSIT, pp. 152–161. ACM (2007)
23. Van Biljon, J., Kotzé, P.: Cultural factors in a mobile phone adoption and usage model. Journal of Universal Computer Science **14**(16), 2650–2679 (2008)
24. Wang, F., Zhang, C.: Label propagation through linear neighborhoods. IEEE Trans. KDE **20**(1), 55–67 (2008)
25. Xiang, B., Liu, Q., Chen, E., Xiong, H., Zheng, Y., Yang, Y.: Pagerank with priors: an influence propagation perspective. In: IJCAI, pp. 2740–2746. AAAI Press (2013)
26. Xu, H., Yang, Y., Wang, L., Liu, W.: Node classification in social network via a factor graph model. In: Pei, J., Tseng, V.S., Cao, L., Motoda, H., Xu, G. (eds.) PAKDD 2013, Part I. LNCS, vol. 7818, pp. 213–224. Springer, Heidelberg (2013)
27. Yan, L., Rober, D., Mozer, M.C., Wolniewicz, R.: Optimizing classifier performance via an approximation to the Wilcoxon-Mann-Whitney statistic. In: ICML, pp. 848–855 (2003)
28. Zeng, G., Luo, P., Chen, E., Wang, M.: From social user activities to people affiliation. In: ICDM, pp. 1277–1282. IEEE (2013)
29. Zhang, Y., Pennacchiotti, M.: Predicting purchase behaviors from social media. In: WWW, pp. 1521–1532 (2013)
30. Zhang, Y., Pennacchiotti, M.: Recommending branded products from social media. In: Recsys, pp. 77–84. ACM (2013)
31. Zhou, D., Bousquet, O., Lal, T.N., Weston, J., Schölkopf, B.: Learning with local and global consistency. NIPS **16**, 321–328 (2003)

Discovering the Impact of Urban Traffic Interventions Using Contrast Mining on Vehicle Trajectory Data

Xiaoting Wang[1,2(✉)], Christopher Leckie[1,2], Hairuo Xie[2],
and Tharshan Vaithianathan[3]

[1] NICTA Victoria, University of Melbourne, Melbourne, VIC, Australia
[2] Department of Computing and Information Systems,
University of Melbourne, Melbourne, VIC, Australia
Wangx5@student.unimelb.edu.au, caleckie@unimelb.edu.au
xieh@unimelb.edu.au
[3] Department of Electrical and Electronic Engineering,
University of Melbourne, Melbourne, VIC, Australia
tva@unimelb.edu.au

Abstract. There is growing interest in using trajectory data of moving vehicles to analyze urban traffic and improve city planning. This paper presents a framework to assess the impact of traffic intervention measures, such as road closures, on the traffic network. Connected road segments with significantly different traffic levels before and after the intervention are discovered by computing the growth rate. Frequent sub-networks of the overall traffic network are then discovered to reveal the region that is most affected. The effectiveness and robustness of this framework are shown by three experiments using real taxi trajectories and traffic simulations in two different cities.

1 Introduction

Urban traffic planning and control is a critical problem for many cities. One specific problem in urban traffic management is the analysis and assessment of the impact of traffic interventions, such as road closures, on the road traffic. For example, when a new subway line is to be built in a city, many roads may be closed for a considerable period. Consequently, the traffic flow in the city may be disturbed and traffic congestion may worsen. Therefore, it is essential that an evaluation of the impact of road closures on the traffic be conducted. However, this evaluation is challenging due to the complexity of the traffic network. Traditional technologies such as induction loop sensors installed at intersections may provide part of the solution, but they are extremely limited as the *trajectory* information of cars is lost and the change in the *flow* of traffic cannot be traced. Fortunately, vehicle GPS trajectories data, which has already been used to tackle many urban computing challenges [1], can assist the development of a solution. In this paper, we propose a framework that uses contrast mining on vehicle trajectories to analyze the change in traffic flow due to road closure events. Specifically, our **contributions** are:

© Springer International Publishing Switzerland 2015
T. Cao et al. (Eds.): PAKDD 2015, Part I, LNAI 9077, pp. 486–497, 2015.
DOI: 10.1007/978-3-319-18038-0_38

1. Traffic Modelling. We model the road traffic network of a city as a graph G, generate "n-Edgesets" and analyze vehicle trajectories to find traffic volume on connected subsets of edges in G.

2. Mining Emerging n-Edgesets. We propose the MineEmergingEdgeset algorithm to extract the n-Edgesets with significantly different traffic patterns before and after the traffic intervention (the Emerging n-Edgesets).

3. Mining Frequent Emerging Network. We propose the MineFreqNetwork algorithm to find frequently occurring emerging networks using Emerging n-Edgesets. This frequent network reveals the most severely affected sub-network of the city as a result of the intervention.

The rest of the paper is organized as follows. Section 2 reviews related work. Section 3 defines the problem and gives an overview of our framework. In Section 4, we discuss our methodology in detail. Section 5 presents the experiments and evaluation of our algorithms. In Section 6, we conclude and discuss future research directions.

2   Related Work

Urban computing using vehicle trajectory data has received a considerable amount of attention recently. Zheng et. al. [1] have given an extensive review on the current research progress and outlined some major challenges. Liu et. al. [2] have discovered spatio-temporal causal relationships between anomalous traffic links using a tree structure. Zheng et. al. [3] have used taxi GPS trajectories to find badly connected regions and urban design flaws in a city. Chawla et. al. [4] have mined traffic anomalies using Principle Component Analysis and investigated root cause of these anomalies. These works mainly focus on discovering urban traffic issues, whereas this paper addresses a further question: once some traffic intervention occurs, for instance, building a new road or closing a road segment, what will be its impact? To the best of our knowledge, this problem has not been given sufficient attention. Miller and Chetan [5] have studied the impact of short-term highway traffic incidents. In our framework, we are not restricted to highways and also consider long-term traffic interventions that can last for weeks or months. Salcedo-Sanz et. al. [6] performed simulations to reconfigure one-way streets in a town to find the shortest path in order to cross an area affected by a large event. However, real traffic data were not used, whereas we use both real vehicle GPS data and simulated vehicle trajectories to identify regions of impact.

Traditional transport systems research mainly uses loop sensors embedded under the road network to monitor traffic. Farnoush et. al. [7] have used loop sensor data in Los Angeles to cluster road segments and found a limited number of distinctive "signature" traffic patterns on all road segments. However, the connections between road segments are not discussed, which is a major limitation in using loop sensors. Our approach can be configured to study single road segment or a connected set of roads, which is more powerful than using loop sensors.

3   Overview

In this section, we introduce some definitions, formally define our problem of emerging sub-trajectory mining and provide an overview of our framework.

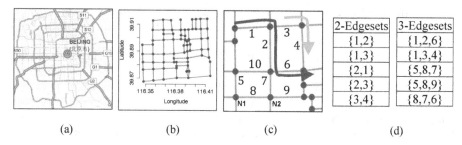

Fig. 1. An example illustrating the proposed framework. (a) Road network of Beijing. (b) Graph model extracted from the road network. (c) GPS trajectories showing traffic flow. (d) Edgesets representing directions of traffic flow.

3.1   Preliminaries

Definition 1. *Road Network Graph.* A Road Network Graph G (Figure 1 (b)) is an undirected graph where road intersections form its nodes and road segments form its edges. Each node in G is a triplet <NodeID, longitude, latitude> and each edge is a triplet <EdgeID, N_1, N_2> where N_1 and N_2 are the node pair connecting the edge.

Definition 2. *Trajectory.* A trajectory, *Traj*, is a sequence of time-stamped geographical locations of a moving object. A *Traj* can be expressed as a vector triplets <longitude, latitude, time>. Figure 1 (c) shows two trajectories.

Definition 3. *n-Edgeset.* An n-Edgeset *nES* is an itemset consisting of a sequence of n connected edges in a Road Network Graph. An *nES* represents the direction of traffic flow though the edges. Figure 1 (d) illustrates some 2-Edgesets and 3-Edgesets. Traffic flow from edge 1 to edge 2 is denoted as {1,2}. The trajectories in Figure 1 (c) traverse 3-Edgeset {1,2,6} and 2-Edgeset {3,4}.

Definition 4. *Emerging n-Edgesets.* Given two time periods T_1 and T_2, if the traffic profile of an n-Edgeset changes significantly during these times, it is called an Emerging n-Edgeset (EnES).

Definition 5. *Frequent Emerging Network.* A connected sub-network of a Road Network Graph is a Frequent Emerging Network (FEN) if it consists of frequently occurring Emerging n-Edgesets.

3.2   Problem Statement

In this paper, we propose a method to model the traffic network in a city and characterizing the difference in traffic conditions before and after an event. Specifically, given road map data R and vehicle trajectories *Traj*, we address the problem of constructing a Road Network Graph G, finding all Emerging n-Edgesets (EnES) using G and *Traj* and lastly finding the Frequent Emerging Network (FEN) using all EnES.

3.3   Framework

Our framework is illustrated in Figure 2. Using the GPS trajectories and road network, the first step is modeling the city traffic by matching trajectories to edges. Next, we generate the n-Edgesets, compute the Growth Rate and find Emerging n-Edgesets (Section 4.2). The final step is to detect a Frequent Emerging Network from the Emerging n-Edgesets (Section 4.3).

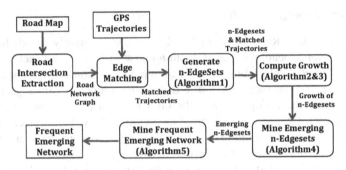

Fig. 2. Framework of our method

4   Methodology

In this section, we describe our method in three sub-sections: traffic modeling, detecting Emerging n-Edgesets (EnES) and detecting the Frequent Emerging Network (FEN). We focus on finding the Growth Rate of the traffic on n-Edgesets, which gives information about the change in traffic flow following an event.

4.1   Traffic Network Modelling

We model a road network as an undirected graph where intersections are the nodes and road segments are the edges. This approach is different with some popular modelling approaches such as modelling the map as regions enclosed by major roads [8]. Region-based approaches can be useful to provide higher-level semantic information, but the capability of inferring traffic conditions on individual roads is limited. On the other hand, modelling road segments can provide finer details about the traffic network. We extracted road segments and intersections manually from OpenStreetMap to ensure robustness. The road network is modelled as an undirected graph since direction information is embedded in the trajectory data, and n-Edgesets will be able to capture it.

The second step in our traffic network modeling is to map GPS trajectories to the edges of our Road Network Graph. This is performed by a nearest neighbor (NN) algorithm. Between any pair of connected nodes, a linear interpolation of "dummy nodes" is created. These dummy nodes carry the same edge ID as the edge connecting the real nodes. For each trajectory point, we find its nearest dummy node within a certain radius r and the edge ID of that dummy node is used as the edge ID of the point, thereby converting its representation <longitude, latitude, time> to <EdgeID, time> (see Definition 1 & 2). We use this nearest-neighbor-based approach mainly due to its computational efficiency ($O(n\log(n))$ time when implemented using a kd-tree).

4.2   Mining Emerging n-Edgesets

For each edge e in a Road Network Graph G, we exhaustively search for all connected sets of non-repeating edges of depth n that start from e, and each of these sets form an n-Edgeset. Table 1 illustrates an example of several 3-Edgesets in Figure 1 (c). Edgeset {1,2,6} represents the traffic flow in the following path: edge 1 to edge 2 to edge 6. Using the n-Edgesets and the GPS trajectories, we calculate the traffic volume on each n-Edgeset for different time periods. The vehicle trajectory database is divided into two parts, D_1 and D_2, where: D_1 = data collected before a road closure event; D_2 = data collected after the closure (Section 5.1 gives more details). Given the n-Edgesets, at each time step i, n new trajectory points are retrieved from the database. The id of these new trajectory points are compared to ensure that they belong to the same car. The edge labels are then matched to the list of n-Edgesets. If a match is detected, the traffic volume count of that n-Edgeset is incremented.

Table 1. An example of mining Emerging 3-Edgesets

3-Edgesets	TraffBe	SuppBe	TraffAf	SuppAf	Growth Rate
{1,2,6}	9	0.09	0	0.00	0.00
{1,3,4}	10	0.10	16	0.16	1.58
{5,8,7}	19	0.19	16	0.16	0.83
{5,8,9}	30	0.30	31	0.30	1.02
{8,7,6}	33	0.33	39	0.38	1.17

TraffBe: Number of Cars before event. SuppBe: Support of Edgeset before event. TraffAf: Number of Cars after event. SuppAf: Support of Edgeset after event.

Using the traffic volume on each n-Edgeset, we compute the support of all n-Edgesets in D_1 and D_2 where the support of an n-Edgeset X, $Supp(X)$, is the traffic volume on X divided by the sum of traffic volume on all n-Edgesets. Assuming that the routing behaviour of most drivers does not change significantly on a daily basis, the difference in the support is likely to carry information about the impact of the event. Therefore, we compute the Growth Rate (or simply Growth) of support of an n-Edgeset using the following definition:

Growth Rate: Given two datasets D_1 and D_2, the Growth of itemset X is:

$$\begin{cases} 0, & if\ supp_1(X) \neq 0\ and\ supp_2(X) = 0 \\ \infty, & if\ supp_1(X) = 0\ and\ supp_2(X) \neq 0 \\ \dfrac{supp_2(X)}{supp_1(X)}, & otherwise \end{cases}$$

This concept was initially proposed by Dong [9], but to the best of our knowledge, it has never been used in road traffic analysis. The original definition defines *Growth* = 0 if $Supp_1(X)$ = 0 and $Supp_2(X)$ = 0. However, we filter and remove those n-Edgesets with no trajectory data to reduce noise. Hence, we define Growth = 0 $Supp_1(X) \neq 0$ and $Supp_2(X)$ = 0. Table 1 shows the *Growth* of several 3-Edgesets. For example, Edgeset {1,3,4} has support before event = 0.10 and after event = 0.16. Thus its Growth = 0.16/0.10 = 1.58.

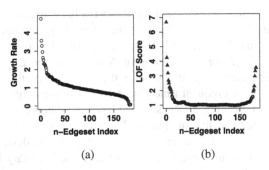

(a)   (b)

Fig. 3. Growth Rate and LOF scores. (a) Growth Rate for 2-Edgesets. (b) LOF scores with number of neighbors = 25, threshold = 1.6. The blue triangles are above the threshold.

After computing the Growth Rate of n-Edgesets, we sort Growth from high to low. Figure 3 (a) shows a plot of the sorted Growth for 2-Edgesets between consecutive days. It can be seen that the majority of 2-Edgesets have Growth of approximately 1.0. This is expected since the total traffic volume on most major roads is unlikely to change significantly between consecutive days. If viewed from a probability theory perspective, the middle region of the graph approximately corresponds to values drawn from a uniform distribution, whereas the side regions are from some form of tail distributions. Growth of the 2-Edgesets at the tails of the sorted edgesets deviates significantly from the majority. These 2-Edgesets are most likely under the effect of the road closure. Thus we apply the definition of *emerging patterns* [9] and define *Emerging n-Edgesets* as the n-Edgesets whose Growth deviate from "normal" by a certain threshold value. To find the upper and lower thresholds, we observe that the tail regions have much lower density than the middle region. Therefore, the Local Outlier Factors (LOF) [10] is used to select Emerging n-Edgesets as follows:

1. Find the LOF scores for each *n-Edgeset* in all n-Edgesets using sorted Growth
2. Find all n-Edgesets with LOF scores larger than a threshold *thres*. They are the Emerging n-Edgesets (*EnES*).
3. If an *EnES* has *growth* larger 1, add it to the list of *EnESes* with increased traffic. If its *growth* is less than 1, add to the list of *EnESes* with decreased traffic.

Although the LOF method is used, this is not anomaly detection. In anomaly detection, there is no prior knowledge about the time of occurrence of the anomaly, whereas we compare the traffic system before and after traffic intervention such as road closure. We evaluate the effect of the traffic intervention rather than detecting its presence. Figure 3 (b) shows the resulting LOF scores for 2-Edgesets where the tails with low density have been successfully discovered and are marked with blue triangles. The weakness of LOF is that it requires two parameters, k, the number of neighbours and *thres*, a threshold. In Section 5.1 we show that our method is robust against variations in these two parameters.

4.3   Mining Frequent Emerging Network

The Emerging n-Edgesets reveals the paths that have been affected before and after a traffic intervention. As shown in Figure 3 (b), many Edgesets can be affected. To identify the region that has suffered the most significant impact, we proposed the method of finding the ***Frequent Emerging Network*** in the Emerging n-Edgesets (Algorithm MineFreqNetwork). We require that an edge in the list of Emerging n-Edgeset is a part of the Frequent Emerging Network if and only if it has occurred at least m times in the Emerging n-Edgesets and has a neighbouring edge which is also frequent. The advantage of this method is that it can extract the "core" part of the network that is most affected. In a real trajectory dataset (more details in Section 5.1), noise in the trajectories can cause issues in defining the boundary between affected and unaffected areas. By applying frequent itemset mining, this problem can be mitigated since the same noise is unlikely to occur repeatedly.

Algorithm. MineFreqNetwork

Input: *EmerEdgesets:* Emerging n-Edgesets, *RNGraph:* Road Network Graph
Output: *FreqEmergNetwork:* Frequent Emerging Network

1:    *Uniq* ← unique edges in *EmerEg*
2:    *Network* ← empty list
3:    **forall** *edge* in *Uniq* **do**   // count frequency of each *edge*
4:        *count* ← frequency of occurance of *edge* in *EmerEg*
5:        **if** *count* > *freq_thres* **then** *Network*.add(*edge*) **end if**
6:    **end for**
7:    **forall** *freqEdge* in *Network* **do**   // find neighbours that are also in *Network*
8:        *Neighb* ← neighbours of *freqEdge* in *RNGraph*
9:        **if** there exists no *edge* in *Neighb* that is also in *Network* **then**
10:            *Network*.remove(*freqEdge*)
11:        **end if**
12:    **end for**
13:    **return** *Network*

---

5   Experiments and Evaluation

In this section, we present two case studies to evaluate the effectiveness and robustness of our algorithms. We first present a real-life case study using taxi GPS trajectories and two traffic simulations with a microscopic traffic simulator.

5.1   Real-life Case Study

We used a Beijing taxi GPS trajectory dataset, which is publicly available from Microsoft Research Asia [11, 12]. The dataset consists of trajectories generated by 10,357 taxis travelling in Beijing and its surroundings over a seven-day period (02/Feb/2008 – 08/Feb/2008). The road map of Beijing is obtained from openstreepmap.org. During the seven days, one road closure event is examined in this paper.

Road Closure Event: South Xinhua Street was closed daily between 8:00 – 18:00, from 07/Feb/2008 to 11/Feb/2008 due to Chinese New Year. Therefore, two days of the road closure event are captured in the taxi trajectory dataset.

We average the data of the closed days (07/Feb – 08/Feb) and mine emerging patterns from the average of two previous days without road closure (05/Feb – 06/Feb). A smoothing step is required since taxi trajectories can suffer from noise issues caused by insufficient data on smaller roads due to the GPS sampling rate. To further reduce the effect of noise, we filter and remove the Edgesets with fewer than 10 cars in total during the two days. Since our focus is on the major roads, the above steps will not affect our results.

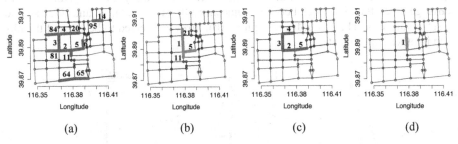

(a)   (b)   (c)   (d)

Fig. 4. Emerging 2-Edgesets (E2ES) and Frequent Emerging Network (FEN) extracted using these Edgesets. (a) E2ES with increased traffic. (b) E2ES with decreased traffic. (c) FEN with increased traffic. (d) FEN with decreased traffic.

Effectiveness
Figure 4 shows the emerging patterns we found due to the road closure. Figure 4 (a) depicts the 2-Edgesets with increased traffic after the road closure (red), while Figure 4 (b) illustrates the decreased traffic (green). Edge 1, shown in green in Figure 4 (b) was the road that was closed. As expected, the area surrounding the closed road has increased traffic levels since cars would have had to detour around the closed road. The paths that traverse the closed road, for example, path 5 to 1 and 1 to 21 have reduced traffic after the road closure, which is also expected as no cars can travel through the closed road. Figure 4 (c) and (d) depicts the Frequent Emerging Network extracted from the Emerging 2-Edgesets in Figure 4 (a) and (b). For roads with increased traffic (Figure 4 (c)), the edges adjacent to the road closure are part of the Frequent Emerging Network. This is expected since the impact should be inversely correlated to distance. For decreased traffic, only the closed road is reported. Therefore, the overall impact of the road closure on the traffic is mostly localized. Some isolated Emerging Edgesets in Figure 4 (a) can be observed, such as {64, 65} and {95,14}. Due to the unsupervised nature of the problem, whether these Edgesets are

true positives or noise is unknown. However, as they are infrequent and removed from FEN (Figure 4 (c)), they are unlikely to have a large impact on the traffic. Therefore, these Emerging Edgesets are likely to be noise, and the robustness of FEN is demonstrated.

Parameter Sensitivity

For the LOF algorithm, two parameters are essential, k, the number of neighbours; and *thres*, the threshold of LOF scores. We varied one parameter while fixing the other, and compare the number of Edgesets (NoE) selected (Figure 5). To choose k, one typically starts from a small positive integer and then increase k. This is because k defines the size of the neighbourhood that a data point can compare against, and a larger k provides more "smoothing" effect. Too much smoothing will not reflect the actual property of the system. In Figure 5 (a), NoE only varies slightly when k is between 15 and 30. Therefore, results are not sensitive to the selection of k in that range, and k is set to 25 for future experiments. To choose *thres*, one starts from a number slightly large than 1.0 and increase *thres*. This is because the majority of the Edgesets usually have LOF scores of near 1.0, which indicates that their Growth values are similar. When *thres* increases from 1.0, NoE decreases steadily. Finally, NoE tends to converge when the difference between adjacent LOF values is large, and *thres* should be set in this area. Therefore, from Figure 5 (b), *thres* is selected to be 1.6 and thereby used in other experiments.

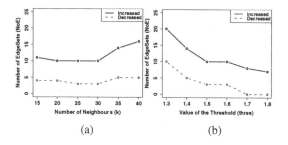

(a)   (b)

Fig. 5. Number of 2-Edgesets selected by LOF with varying parameters. (a) Fixing *thres* = 1.6 and vary k. (b) Fixing k = 25 and vary *thres*.

During our experiment, we can easily extract emerging patterns for 2 and 3-Edgesets. However, due to the low sampling rate of GPS devices (1 to 10 minutes), it is difficult to extract meaningful patterns for Edgesets longer than three edges. In the next section, we show the results of using a traffic simulator, which overcomes this limitation.

5.2   Traffic Simulation

We used a microscopic traffic simulator, which can perform large-scale and highly detailed traffic simulations, to validate our algorithms. Vehicles are individually modelled to simulate realistic car-following and lane-changing. Various traffic rules are implemented, such as how to give way to trams at tram stops. The simulator can also simulate traffic lights, whose timing can be controlled by static or dynamic strategies.

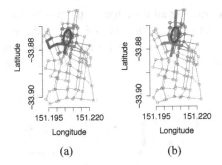

Fig. 6. Simulation results and Frequent Emerging Network for 4-Edgesets in Sydney with George Street closed. (a) Increased Traffic. (b) Decreased Traffic.

We simulated traffic in two cities, Beijing and Sydney. For Beijing, we replicated the road closure event on South Xinhua Street; for Sydney, we closed a few segments of George Street, which is one of the main streets in the CBD of Sydney. Since the sampling frequency of data can be set to a high value and no longer poses a limitation, we easily extracted Emerging 2,3,4-edgesets. The results of extracting the Frequent Emerging Network of 4-Edgesets for Sydney are shown in Figure 6. The simulation was initially performed with 900 cars and the closed segments of George Street are circled in blue. Figure 6 (a) shows that the edges near the closed street have increased traffic levels (red), while there is a decrease in the traffic on the closed road (Figure 6 (b)). The simulation results are consistent with our real-life case study using taxi GPS trajectories from Beijing.

Robustness

We evaluate the robustness of the LOF-based method using different numbers of cars (300, 600, 900, 1200, 1500) in the simulation. The common Edgesets are found and treated as ground truth. Using the common Edgesets, we evaluate the precision and recall of each single experiment for the case of increased traffic (Figure 7 (a) and (b)) and decreased traffic (Figure 7 (c) and (d)). The LOF-based method was compared with a baseline method, which sets arbitrary thresholds to the Growth value to select Emerging Edgesets. For precision, it can be seen that overall, the LOF-method outperforms the baseline. When traffic volume is high (1500 cars), the precision of the LOF method is 0.89 for both increased traffic and decreased traffic, whereas the baseline method only has a precision of 0.39 for increased traffic and 0.26 for decreased traffic. This implies that the variance in the value of Growth can be large for a large number of cars and the results of the baseline method can be sensitive to the threshold chosen. However, the LOF method is not affected since the variance in the data density of Growth can be much smaller than the variance in the value of Growth. When the traffic volume is very low (300 cars), the precision of the LOF method (0.61 for increased traffic, 0.64 for decreased traffic) is slightly below the baseline (0.69 for both increased and decreased traffic). This might be caused by the fact that a small number of cars are less likely to traverse all the edges in the whole road network. Consequently, traffic volume calculations can become noisy, and the performance of the LOF method can be slightly affected. For other experiments, the precision of the LOF method is at least comparable (600 cars) or better than the baseline (900 and 1200 cars).

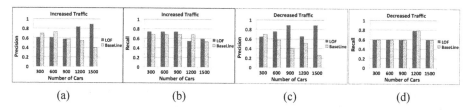

(a)   (b)   (c)   (d)

Fig. 7. Precision and recall of LOF and the baseline method under different settings of traffic load in the simulation. (a) Precision of increased traffic. (b) Recall of increased traffic. (c) Precision of decreased traffic. (d) Recall of decreased traffic.

For recall, the LOF-based method and the baseline show comparable results for both increased and decreased traffic (Figure 7 (b) and (d)). This is because the Edgesets being significantly affected by the road closure are usually adjacent to the closed road (Figure 4 and 6). Therefore, these Edgesets appear significantly more dominating than other Edgesets, and both methods are able to find them. The recall for decreased traffic of both methods is the same, which is also expected since the traffic reduction on the closed road is obvious (Growth dropping to 0). From the evaluation of both precision and recall, it can be seen that the LOF-based method is more robust than the baseline.

5.3   Computational Complexity

The most computationally intensive part of our framework is the traffic volume calculation for each n-Edgeset. Let s denote the total number of n-Edgesets, and d denote total number of trajectory points in our database. Note that the number of n-Edgesets is constrained by the connectivity between edges in the road network. This problem is equivalent to finding multiple matches of s strings in d, which is a well-studied problem and an average case of $O(s+d)$ time can be achieved using a hash table. For other parts of our framework, mining both Emerging n-Edgeset and Frequent Emerging Network take $O(s)$ time since each n-Edgeset is visited only once. Therefore, the overall complexity is $O(s+d)$.

6   Conclusions and Future Work

We have proposed a framework to assess the impact of traffic intervention on the road traffic system of cities. We have mined the Emerging n-Edgesets by computing the Growth Rate and Local Outlier Factor (LOF) score. We have also proposed the algorithm MineFreqNetwork to identify Frequent Emerging Networks, which is the sub-network of a road system that is significantly affected by the intervention. Our experiments using real taxi GPS data and simulated vehicle trajectories show that our approach using the LOF method outperforms a baseline algorithm. A possible future research direction is to integrate our approach with optimization techniques to improve decision making for applications in navigation and logistics.

Acknowledgement. This work was supported by NICTA.

References

1. Zheng, Y., Capra, L., Wolfson, O., Yang, H.: Urban Computing: concepts, methodologies, and applications. ACM Transaction on Intelligent Systems and Technology (ACM TIST) (2014)
2. Liu, W., Zheng, Y., Chawla, S., Yuan, J., Xing, X.: Discovering spatio-temporal causal interactions in traffic data streams. In: SIGKDD, pp. 1010–1018. ACM (2011)
3. Zheng, Y., Liu, Y., Yuan, J., Xie, X.: Urban computing with taxicabs. In: UbiComp, pp. 89–98. ACM, Beijing (2011)
4. Chawla, S., Zheng, Y., Hu, J.: Inferring the Root Cause in Road Traffic Anomalies. In: ICDM, pp. 141–150 (2012)
5. Miller, M., Gupta, C.: Mining traffic incidents to forecast impact. In: SIGKDD Workshop on Urban Computing, pp. 33–40. ACM (2012)
6. Salcedo-Sanz, S., Manjarrés, D., Pastor-Sánchez, Á., Del Ser, J., Portilla-Figueras, J.A., Gil-López, S.: One-way urban traffic reconfiguration using a multi-objective harmony search approach. Expert Systems with Applications **40**, 3341–3350 (2013)
7. Banaei-Kashani, F., Shahabi, C., Pan, B.: Discovering patterns in traffic sensor data. In: SIGSPATIAL Workshop on GeoStreaming, pp. 10–16. ACM, Chicago (2011)
8. Yuan, J., Zheng, Y., Xie, X.: Discovering regions of different functions in a city using human mobility and POIs. In: SIGKDD, pp. 186–194. ACM (2012)
9. Dong, G., Li, J.: Efficient mining of emerging patterns: discovering trends and differences. In: SIGKDD, pp. 43–52. ACM (1999)
10. Breunig, M.M., Kriegel, H.-P., Ng, R.T., Sander, J.: LOF: identifying density-based local outliers. In: SIGMOD, pp. 93–104. ACM (2000)
11. Yuan, J., Zheng, Y., Zhang, C., Xie, W., Xie, X., Sun, G., Huang, Y.: T-drive: driving directions based on taxi trajectories. In: SIGSPATIAL, pp. 99–108. ACM, San Jose (2010)
12. Yuan, J., Zheng, Y., Xie, X., Sun, G.: Driving with knowledge from the physical world. In: SIGKDD, pp. 316–324. ACM, San Diego (2011)

Locating Self-Collection Points for Last-Mile Logistics Using Public Transport Data

Huayu Wu[✉], Dongxu Shao, and Wee Siong Ng

Institute for Infocomm Research, A*STAR, Singapore, Singapore
{huwu,shaod,wsng}@i2r.a-star.edu.sg

Abstract. Delivery failure and re-scheduling cause the delay of services and increase the operation costs for logistics companies. Setting up self-collection points is an effective solution that is attracting attentions from many companies. One challenge for this model is how to choose the locations for self-collection points. In this work, we design a methodology for locating self-collection points. We consider both the distribution of a company's potential customers and the people's gathering pattern in the city. We leverage on citizens' public transport riding records to simulate how the crowds emerge for particular hours. We reasonably assume that a place near to a people crowd is more convenient for customers than a place far away for self parcel collection. Based on this, we propose a kernel transformation method to re-evaluate the pairwise positions of customers, and then do a clustering.

1 Introduction

The last-mile logistics is the final stage to deliver freight to urban customers from the port or consolidation centers in a city. The efficiency of the last-mile logistics directly affects the quality of delivery services. On the other hand, the last-mile logistics is a costly phase in a supply chain. Since the destinations are quite diverse within a city and in most cases delivering vehicles cannot be fully loaded, the last-mile logistics can take up to 28% of total cost for goods shipment and delivery [7]. For small parcel and package delivery services, whose major customers are the general public, the operating cost on last-mile could be even higher, due to the high chance of unsuccessful deliveries. According to the data from our partner company, 18.8% delivery jobs were failed in their first attempt due to the absence of customers, and the company has to re-schedule a second delivery. This significantly increases the cost.

One common way to reduce cost for delivery failures is to set up self-collection points in each neighborhood [12]. For example, Singapore's leading delivery service provider, SingPost uses their 59 post offices all over the island to temporarily store the parcels that were failed to deliver to the customers. Those customers will find a slip under the door to notify them that the delivery was failed and their parcels are ready for collection in the neighborhood post office. This method can effectively reduce the number of re-deliveries and thus reduce the additional cost brought by delivery failures. However, for most foreign logistics companies,

© Springer International Publishing Switzerland 2015
T. Cao et al. (Eds.): PAKDD 2015, Part I, LNAI 9077, pp. 498–510, 2015.
DOI: 10.1007/978-3-319-18038-0_39

the few number of physical offices becomes the obstacle for the adoption of self collection of parcels. To break the bottleneck, the companies need to look for suitable partner stores for setting up self-collection points.

In this work, we help our partner company to solve the problem that given a fixed budget, i.e., the rough number of self-collection points to be set up, how to effectively select the locations for partnership so that more customers can be served at their convenience. This problem is different from the traditional facility location problem in operations research, which will be discussed in Section 2. It is also different from classic data mining problems. Suppose we divide the customers into clusters based on their locations and consider the center of each cluster as a self-collection point. When we examine the result with background knowledge, we find it not good. First, the customers living far away from each center are anyway not convenient to visit it though we take them into account when we choose the centers. In other words, shifting a self-collection point towards a outlying area by considering the customers living there may still not benefit those people. Second, the center in each cluster may not be a convenient place for customers. The reason is that we did not consider, and it is also hard to define the "convenience" of a self-collection point. Based on the knowledge provided by our partner, most customers would like to collect their parcels on their ways back home from offices. This information can guide us on selecting convenient places for customers.

We adopt a heuristic approach to help the logistics company to find suitable locations for self-collection points. The key point we use to tackle the problem is to find the gathering pattern of people, and use it to estimate the "convenient" places for customers. For example, if we find an MRT station (MRT is the public transport system in Singapore) in suburban is dense of people during the evening peak hours, we can say the surrounding places near the MRT station are more convenient for people to collect their parcels than those places far away.

We first leverage on the public transport data to get insight of the temporal crowd pattern, which will be used later to guide us sensing the places that are convenient for customers. In particular, we fit a multivariate Gaussian mixture model (GMM) to describe the distribution of people for a certain time period. The model is supposed to reflect how people gather and where the crowd centers are. Intuitively, we should put self-collection points nearer to the crowd centers identified by the GMM model. On the other hand, we consider the distribution of potential customers. By combining their physical locations and their gathering patterns learnt from public transport data, we design a kernel function to redefine the positional relationship among customers and crowd centers, and finally cluster the customers and find locations for self-collection points.

Technical Novelty. Compared to the existing clustering methods, which are generally considered unsupervised, our approach uses a non-parametric clustering method that is supervised by the knowledge learnt from another data source by a parametric model. As more and more data are published and fused, this methodology to correlate multiple independent data sources and pipeline the learning result to get more useful insight can be widely adopted.

Social Impact. Most governments maintain rich data about the residents and the cities, and would like to promote analytics on their data for the social good. This work explores the opportunity to make use of government data to service customers from private sectors. It showcases the possibility and advantage to do this, and probably opens a new business model.

The rest of the paper is organized as follows. We define the problem in Section 2. We introduce the datasets in Section 3. Section 4 is the main section to present our approach. We show some analysis and experiment result in Section 5. Related work is reviewed in Section 6. Finally, we conclude this paper in Section 7.

2   Problem

2.1   Formulation

The customers are represented by a set of locations. The target is to set up a number K of self-collection points to serve utmost number of customers at their convenience. As suggested by domain experts, we assume that the places with dense crowd tend to be more convenient for customers. Then there are two objectives when we choose the locations for self-collection points: (1) to minimize the average distance between a customer and her nearest self-collection point, and (2) to maximize the neighborhood people flow of the self-collection points.

The problem can be formulated as a multi-objective optimization problem. Let $C = \{y_j | j = 1, \ldots, m\}$ be the set of customer locations. Let $D = \{(x_i, f_i) | i = 1, \ldots, n\}$ be the collection of people crowd where x_i stands for the center of a crowd with people flow of f_i. Let $S = \{s_1, \ldots, s_K\}$ be the K self-collection points, and ε be the distance threshold to define neighborhood of a self-collection point for the people flow estimation. Let z_j be the distance between customer y_j and her nearest self-collection point and $h_i \in \{0, 1\}$ indicate whether a people crowd x_i lies within the circle centered at some self-collection point with a radius of ε. Then the objective of the problem is: $\min(\sum_{j=1}^{m} z_j, -\sum_{i=1}^{n} h_i f_i)$.

To maximize the neighborhood people flow of the self-collection points, one naïve way is to set up self-collection points at the centers of people crowds. Then the problem can be modeled as a mixed integer linear programming problem with two objectives. The classic way to solve a multi-objective problem is to convert it into a single objective problem by defining a cost function of the objectives [8]. The new cost function can be a importance order of the objectives or a scalarization of the objectives. However, it can hardly be applies to our problem. First, since both of the two objectives are on continuous values, it is almost impossible to have two feasible solutions with the same objective value. Hence the importance order will trivially lead to optimizing the more important objective and ignoring the other one. Second, since the objectives are distance and number of people, it is meaningless to aggregate the two to obtain a cost.

3   Data Description

3.1   Delivery Data

We study the 3-month (April to June, 2014) parcel delivery data provided by our partner. The detailed data schema is omitted, as we are only interested in customer locations and delivery status. To find out what locations are suitable for self-collection points, we need to analyze the clusters of potential customers that will need this service. We have three choices to simulate potential customers' locations, i.e., the citizens' residences, the company's historical customer locations, and the locations of the company's historical customers who failed to receive parcels. We select the last choice and explain why.

(a) Delivered vs. failed locations   (b) Distribution comparison

Fig. 1. Comparison between customers and citizens

We visualize the successful and failed delivery locations in Fig. 1(a), in which successful deliveries are marked as blue points, and the failed attempts are marked as red points. We can see that the two sets of points are not following the same distribution. There are more dense blue points in the city center, and more red point clusters in suburban areas, e.g., center north and east. Thus, in our study, we should use failed locations for learning, which is more representative for future customers of self-collection points.

Also, we do not use the entire residents. For example, some residential regions have been existing for over 30 years, and accommodates more senior citizens than those newly developed regions. There may be more delivery business in new regions because young people are more active in online shopping and overseas delivery. Fig. 1(b) further validates our assumption, which visualizes the distribution comparison between local residents and the company's customers by different colors that indicate the inconsistency of the two distributions. Due to the space limit, we do not further explain it.

3.2   Public Transport Data

The other dataset we use is the public transport data. The public transport, including MRT (subway) and bus, is the major transportation tool used by most residents in Singapore. In 2012, there are 77.8% working people using public transport[1]. Thus it is reasonable to assume that the crowd pattern of working people can be estimated from the public transport data, by studying passengers' riding records. Furthermore, since the public transport commuters are the main target customers that self-collection points would like to benefit (the location is less important for private car drivers), the use of public transport data to estimate customers' gathering pattern is meaningful.

4   Approach

4.1   Overview

The basic idea of our work is to incorporate crowd gathering pattern into the distribution of customers in Singapore, to find better locations for self-collection point setup. Fig. 2 shows the general workflow. We first learn spatial temporal models to represent the people gathering pattern, based on the public transport data. Then according to the company's suggestion, we choose the model for the evening peak hour period, i.e., 5pm to 9m, to guide self-collection point locating.

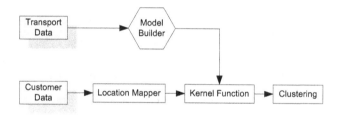

Fig. 2. Approach overview diagram

On the other hand, we extract the customer locations from the unsuccessful delivery data. By incorporating the model built from the public transport data, we design a kernel function to shift customers' locations to nearby crowd centers, and thus re-define the distance among customer locations. Finally, based on the new locations of customers, we divide the customers into a few clusters according to the budget.

4.2   Model Fitting

We are interested in the gathering pattern of residents on a time basis. Then we would try to shift the self-collection points toward the center of a crowd according to the common time period that customers tend to collect their parcels.

In a certain time period, there will be multiple crowds in Singapore. To model the crowds for the whole population across the Singapore island, we should exploit a mixture model because the crowd in different regions may distribute differently. For each sub-population, we assume a multivariate Guassian model to

---

[1] Estimated from LTA's annual report: http://www.lta.gov.sg/

represent the spatial distribution of people. For example, suppose a popular MRT station is located at (x_i, y_i), where (x_i, y_i) are latitude-longitude coordinates, then we assume the probabilities that a person appears in this neighborhood during morning peak hours follow a multivariate Guassian distribution, where the mean along the two dimensions of latitude and longitude takes place at x_i and y_i respectively. We explain why we use Guassian model to represent the distribution of people. In this example, MRT, as the main transport tool attracts working people from all directions in the neighborhood. As a result, there are more people appearing at the MRT stations. Furthermore, since the people centered at the MRT stations are converged from surrounding places, for a particular moment or taking the average over a short period, the density distribution should be in a bell shape. Note that other distribution models can also be used if they can reasonably capture the characteristics of people flow. In our assumption, if all crowds follow a multivariate Guassian distribution, for the whole population at a certain period, the probability that a person appears in each place in Singapore will follow a Gaussian Mixture Model (GMM).

We process the public transport data by finding the total number of alighting passengers in each MRT station and bus stop during the period, and consider it as the crowd size in each place. We use this data to fit a mixture model f of M Gaussian functions:

$$f = \sum_{i=1}^{M} w_i \mathcal{N}(\mu_i, \Sigma_i),$$

where w_i is the weight of the i-th Gaussian component (i.e., the peak passenger flow), μ_i is the mean of the corresponding Gaussian component (i.e., the center of a crowd) and Σ_i is the corresponding covariance matrix (i.e., how passenger flow spread from the center to the two dimensions).

Let Θ be the collection of all the parameters (w, μ, Σ). Let $D = \{(x_i, y_i, f_i)|i = 1, \ldots, n\}$ be the collection of each bus stop or MRT station's geographic location (x_i, y_i) with the passenger flow f_i. Then the likelihood of Θ given the observation D is defined as:

$$\mathcal{L}(\Theta|D) = \Pr[D|\Theta]$$

Suppose the observations are independent. Then

$$\Pr[D|\Theta] = \prod_{i=1}^{n} \Pr[f(x_i, y_i) = f_i|\Theta]$$

Θ can be estimated by the Expectation-Maximization (EM) algorithm.

4.3   Kernel Transformation

We propose a transformation function based on which the attractiveness between customers and crowds can be and the closeness between customers can be redefined.

Let x be a crowd center and y be a customer. If their distance square $||x - y||^2$ is large, we just move y very slightly. On the other hand, we should move y to a new location that is closer to x. By doing this, we can shift a customer's location towards a crowd center. Thus we define the kernel function between a customer and a crowd center as

$$k_\sigma(x, y) = e^{-\frac{||x-y||^2}{2\sigma^2}}$$

Note that the purpose here is to define a rule to shift a customer location towards a crowd center according to the above mentioned property. Other kernel functions may also be applied to achieve the same goal. Then the movement of y according to x is

$$v(x, y) = k_\sigma(x, y) \cdot (x - y)$$

It is normal that a customer will be attracted by more than one crowd centers. A natural way to aggregate the moving vectors to different crowd centers is to take the weighted sum of these vectors. However, this aggregation does not work well for our case. For example, if some customers are lying in between of two crowd centers with identical weight along the opposite direction, and the distances from the customer to the two crowd centers are also identical, then no matter how near the customer is to the crowd centers, she will not be moved by the function. In order to avoid such situations, we define the final moving vector of a customer to be the moving vector to its nearest crowd center. If there are more than one nearest centers, one will be randomly chosen.

Let x_y^* be the nearest crowd center to the customer y.

$$x_y^* = \arg\min_x ||x - y||$$

and the total movement of a client towards a set of crowd centers is:

$$v(y) = v(x_y^*, y)$$

The new location of a customer with a location y will be

$$T(y) = v(y) + y$$

After doing the transformation, the customers' locations are supposed to be closer to each crowd center. Then we can apply the K-means method to $\{T(y)|y \in C\}$, to find locations for self-collection point setup.

4.4 Optimizing Variance Parameter

The variance parameter σ in the kernel function controls how the movement decreases with the distance, and therefore determines the new locations and the final clustering performance. Note that our purpose is to minimize average distance from customers to self-collection points, and also to maximize the neighborhood people flow of the self-collection points. Thus we try to optimize σ according to these two criteria.

In the first scheme, we focus on the minimization of average distance. With different σ, we have different clustering results. We define the optimal variance to be the one returning minimal average distance after clustering.

$$\sigma_d = \arg\min_\sigma \sum_{j=1}^{m} z_j$$

In the second scheme, we focus on the maximization of neighborhood people flow. Let ε be the radius defining the neighborhood area, which is pre-set. Then we define the optimal variance to be the one returning maximal neighborhood people flow after clustering.

$$\sigma_f = \arg\max_\sigma \sum_{i=1}^{n} h_i f_i$$

By enabling user's input on choosing different optimization schemes for σ, the overall flow of the algorithm can be summarized in Algorithm 1.

Algorithm 1.. Kernel-based Clustering

Input: set of customers C, set of passenger records D, radius to define neighborhood ε, number of clusters K, set of candidate variance parameters $S = \{\sigma_l | l = 1, \ldots, L\}$ and choice of optimization schemes H

Output: Resulting locations for self-collection points

1: Fit the GMM with D and obtain the set of crowd centers A
2: Initiate result set O
3: **for** $l = 1$ to L **do**
4:    Compute the new customer locations $C^{(l)} = T(C)$ with respect to A and S
5:    Apply K-means clustering algorithm to $C^{(l)}$ with cluster number K and let $O(l)$ be the centers of resulting clusters
6:    **if** $H == 1$ **then**
7:       Compute $cost(l)$ the average distance from each customer to her nearest cluster center
8:    **else if** $H == 2$ **then**
9:       Compute $flow(l)$ the total people flow within the circle centered at some cluster centers with radius ε
10: **if** $H == 1$ **then**
11:    Let $opt = \arg_l \min cost(l)$
12: **else if** $H == 2$ **then**
13:    Let $opt = \arg_l \max flow(l)$
14: **return** $O(opt)$

---

5   Result Presentation and Discussion

5.1   GMM Fitting

We take the exit records of passengers in each MRT station and bus stop between 5pm and 9pm. We consider that 1) most customers prefer collecting parcels on

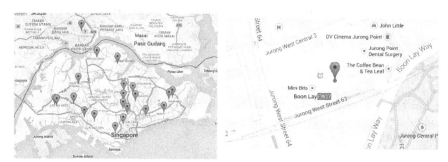

(a) Means of the Gaussian components   (b) Means of the Gaussian components

Fig. 3. GMM training result

their ways home, as assumed, thus we only use the exit records; 2) 5m to 9pm is a reasonable period of off hours for working people.

There are several criteria to balance the likelihood and the complexity of the model. We use the R package $mixmod$ [2], which considers Bayesian Information Criterion [14], Integrated Completed Likelihood [1], and Normalized Entropy Criterion [3]. The best model is chosen based on the lowest value among all.

The result shows that the best fitted model has 22 components. The means of the components are shown in Fig. 3(a). The result is reasonable, as most of the means are located near the major MRT stations or bus exchanges. For example, Fig. 3(b) shows a zoom-in view of a mean, which is quite close to the Boonlay MRT station, which is a main transportation hub in the west of Singapore.

5.2   Location Transformation

We set the range of the candidate variance parameters in the kernel function to be $[0.05, 0.1]$. Each variance parameter gives a different movement of customers. In this section, we only show an example in in Fig. 4. The figure on the left shows the original customer distribution in the northeast of Singapore. After applying the kernel transformation, as shown in the right figure, the customers are clearly shifted towards different centers.

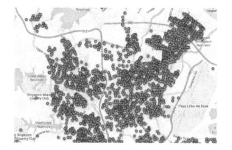

Fig. 4. Zoom-in distribution comparison before and after kernel transformation

5.3   Result

We try different values of K in [10, 50] and use the elbow method to choose $K = 21$ as the number of clusters. This number is also consistent with the company's budget. We present the clustering results using the two functions in Fig. 6, in which the center of each cluster is marked out on the map. For the kernel method, we use the second scheme, i.e., based on neighborhood people flow, to optimize the variance parameter, as it returns better result as discussed later.

Basically, both approaches can identify reasonable customer clusters for self-collection point setup. The one using the kernel function (Fig. 5(b)) chooses more locations in the center of the island and those places are more in line with the crowd centers returned by GMM (Fig. 3(a)), compared to the one using original distance function (Fig. 5(a)). Although these areas are not major residential areas, there are many people taking bus or MRT transfer there, as shown in Fig. 3(a). Thus, it is reasonable to put more collection points in these areas. Another highlight is the removal of the leftmost location from Fig. 5(a). Although there are customers over there, based on the crowd patterns, that place may not attract many people. As a result, in the new clustering result by the kernel transformation, that location is no longer there.

(a) Using original distance function   (b) Using metric function

Fig. 5. Clustering results

5.4   Quantitative Comparison

In this section, we try to quantify the results generated by using the Euclidean distance and the kernel function (under two optimization approaches) respectively, to compare the quality. In the first test, we compare the average distance from customers to the chosen self-collection points. As mentioned, the purpose of self-collection points is not for serving all the customers. In Fig. 6(a), we show the comparison of average distance under five clustering results, for different portion of customers to be served. The x-axis represent the percentage of nearest customers to each selected location, and the y-axis shows the corresponding average distance. The five results include the clustering result based on original distance, kernel function with variance parameter optimized by average distance, and kernel function with variance parameters optimized by people flow using radiuses of 1 to 3. From Fig. 6(a) we can see that there is not much difference between the five results, which means our proposed methods can guarantee a good average distance that is similar to the original approach which returns the optimal value for this metric.

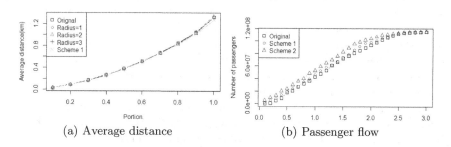

(a) Average distance   (b) Passenger flow

Fig. 6. Quantitive analysis

Next, we compare the number of people that can be potentially served within each size of neighborhood, as shown in Fig. 6(b). We can see that the method using the original customer locations performs worst. For each given value of the radius, after kernel transformation, the chosen self-collection points will be nearer to the crowd centers, and thus can potentially serve more people. Further, the second scheme performs better than the first. Combining with the result in Fig. 6(a), in which all the results give similar performances, we can conclude that the kernel method with the second optimization scheme is the best choice.

6   Related Work

6.1   Facility Location Problem

Facility location problem studies the optimal placement of facilities to minimize the cost of facility opening and the cost of servicing customers from the facilities. It is an important branch of operations research. There are many variants of the facility location problem, and most of them are proven NP-hard [6]. Different kinds of facility location problems have been extensively explored by researchers, in which many works focus on approximation. Since it is not quite related to this paper, we do not review the works here. Some surveys include [5].

As mentioned, the major reason that the existing works are not suitable for our problem is that our problem can be hardly reduced to a unique objective for optimization. Besides, we do not assume the self-collection points can serve all the customers. Our purpose is to serve a reasonable number of customers who can conveniently visit the collection points. As a result, although the problems sound similar, they should be approached differently.

6.2   Clustering

Clustering is a useful method for descriptive data analysis. The popular clustering methods can be divided into two categories, parametric model-based methods and non-parametric similarity (or distance) -based methods. In the first category, the dataset is assumed to follow a mixture model of finite number of components. Each component is a probability distribution. The most widely used mixture model is the one with multivariate Gaussian components [15]. There are also other types of mixture model. A detailed review can be found in [10]. The Expectation-Maximization (EM) algorithm [4] is commonly used to estimate the parameters of a mixture model. Since the EM algorithm is sensitive to initial values and the number of components needs to be pre-defined, there are a lot of research work focusing on the initialization problem of EM (e.g., for the recent works [11,16]). The second category of clustering methods are based on objective functions of similarity or distance measurement. K-means and its variants [9,13] are the most popular algorithms in this category. In this approach, similarity or distance between each pair of data points is defined, based on which the whole dataset will be partitioned into different clusters.

Basically, clustering is considered as an unsupervised learning process. In our approach, we do have some guidance for clustering, which is not from the labeled customer data, but from the residential data which is not directly linked to the customer data. Thus, in our work, we cannot apply existing methods directly.

7   Conclusion

In this paper, we jointly use the public transport data and the customer data to help our partner logistics company choose suitable locations for self-collection point setup. In particular, we use the public transport data to learn the people's gathering pattern, based on which, we design a kernel transformation to re-define the pairwise locations of customers. Finally, we cluster customers based on their new locations which are closer to people's crowds and considered more convenient for the customers. We also demonstrate the effectiveness of our approach.

Acknowledgment. This work was supported by the A*STAR SERC Grant No. 1224200004.

References

1. Biernacki, C., Celeux, G., Govaert, G.: Assessing a mixture model for clustering with the integrated completed likelihood. IEEE Transactions on Pattern Analysis and Machine Intelligence **22**(7), 719–725 (2000)
2. Biernacki, C., Celeux, G., Govaert, G., Langrognet, F., Noulin, G., Vernaz, Y.: Mixmod-statistical documentation (2008)
3. Celeux, G., Soromenho, G.: An entropy criterion for assessing the number of clusters in a mixture model. Journal of Classification **13**(2), 195–212 (1996)
4. Dempster, A.P., Laird, N.M., Rubin, D.B.: Maximum likelihood from incomplete data via the em algorithm. Journal of the Royal Statistical Society. Series B (Methodological), 1–38 (1977)
5. Farahani, R.Z., Asgari, N., Heidari, N., Hosseininia, M., Goh, M.: Covering problems in facility location: A review. Computers & Industrial Engineering **62**(1), 368–407 (2012)

6. Gonzalez, T.F.: Clustering to minimize the maximum intercluster distance. Theoretical Computer Science **38**, 293–306 (1985)
7. Goodman, R.W.: Whatever you calll it, just don't think of last-mile logistics, last. Glabal Logistics & Supply Chain Strategies (2005)
8. Hwang, C., Masud, A.: Multiple objective decision making, methods and applications: a state-of-the-art survey. Lecture notes in economics and mathematical systems. Springer-Verlag (1979)
9. MacQueen, J., et al.: Some methods for classification and analysis of multivariate observations. In: Proceedings of the fifth Berkeley symposium on mathematical statistics and probability, California, USA, vol. 1, pp. 281–297 (1967)
10. McLachlan, G., Peel, D.: Finite mixture models. John Wiley & Sons (2004)
11. Melnykov, V., Melnykov, I.: Initializing the em algorithm in gaussian mixture models with an unknown number of components. Computational Statistics & Data Analysis **56**(6), 1381–1395 (2012)
12. Morganti, E., Dablanc, L., Fortin, F.: Final deliveries for online shopping: The deployment of pickup point networks in urban and suburban areas. Research in Transportation Business & Management (2014)
13. Pollard, D.: Quantization and the method of k-means. IEEE Transactions on Information theory **28**(2), 199–204 (1982)
14. Schwarz, G., et al.: Estimating the dimension of a model. The Annals of Statistics **6**(2), 461–464 (1978)
15. Wolfe, J.H.: Normix: Computational methods for estimating the parameters of multivariate normal mixtures of distributions. Technical report, DTIC Document (1967)
16. Yang, M.-S., Lai, C.-Y., Lin, C.-Y.: A robust em clustering algorithm for gaussian mixture models. Pattern Recognition **45**(11), 3950–3961 (2012)

A Stochastic Framework for Solar Irradiance Forecasting Using Condition Random Field

Jin Xu[1]([✉]), Shinjae Yoo[2], Dantong Yu[2], Hao Huang[1], Dong Huang[2], John Heiser[2], and Paul Kalb[2]

[1] Stony Brook University, Stony Brook, NY, USA
{jin.xu,hao.huang}@stonybrook.edu
[2] Brookhaven National Laboratory, Upton, NY, USA
{sjyoo,dtyu,dhuang,heiser,kalb}@bnl.gov

Abstract. Solar irradiance volatility is a major concern in integrating solar energy micro-grids to the mainstream energy power grid. Accounting for such fluctuations is challenging even with supplier coordination and smart-grid structure implementation. Short-term solar irradiance forecasting is one of the crucial components for maintaining a constant and reliable power output. We propose a novel stochastic solar prediction framework using Conditional Random Fields. The proposed model utilizes features extracted from both cloud images taken by Total Sky Imagers and historical statistics to synergistically reduce the prediction error by 25-40% in terms of MAE in 1-5 minute forecast experiments over the baseline methods.

Keywords: Conditional random field · Stochastic model · Solar forecasting

1 Introduction

As the integration of photovoltaic (PV) power plants into the electricity network becomes increasingly prevalent, utilities and grid operators confront major challenges in maintenance and regulation stemming from the variability of solar irradiance largely due to atmospheric interference (cloud and aerosol contents). Within one minute, the ground solar irradiance can decrease more than 80%, causing drastic drops in power output (see example in Figure 1), and such rapid fluctuations can be constantly observed. Therefore, reliable short-term solar irradiance forecasting is the basis to control the usage of auxiliary systems such as batteries and gas generators [1].

Global Horizontal Irradiance (GHI), which consists of both a direct solar beam and a diffuse component, is an important indicator for PV power production. Current GHI forecasting methods can be categorized into two classes: statistical or physics based models. Statistical models use historical GHI data to train models such as ARMA [2] and ANN [3] to predict future irradiances. These tend to ignore physical atmospheric phenomenon, such as cloud microphysics and their interactions, and only showed a marginal improvement over

© Springer International Publishing Switzerland 2015
T. Cao et al. (Eds.): PAKDD 2015, Part I, LNAI 9077, pp. 511–524, 2015.
DOI: 10.1007/978-3-319-18038-0_40

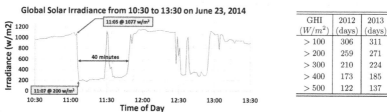

Fig. 1. The ground solar irradiance can decrease by more than 80% due to cloud interference. The table shows the number of days with GHI drops over $100\text{-}500 W/m^2$ within 5 minutes in 2012 and 2013, at the Long Island Solar Farm (LISF), New York.

the benchmark persistent model (PM), which directly uses the present irradiance as the prediction. On the other hand, physics based models, such as Numerical Weather Prediction (NWP), utilize meteorological observations and measurements with wind, temperature, and humidity as key variables [4]. While NWP is preferred for forecast horizons of six hours or beyond, cloud imagery based techniques (satellite or ground-based) produce more accurate short-term forecasts by propagating cloud movement into the future. Ground-based sensors with high spatial and temporal resolution, such as a Total Sky Imager (TSI) or Whole Sky Imager (WSI), are ideally suited to capture the local cloud variation for intra-hour predictions [5]. Deterministic methods were subsequently adopted as the standard approach, where the prediction is simply determined by the amount of cloud cover in the predicted sky image [6–8]. However, they overlook the correlated nature of time series data (temporal) and the strong dependence on environmental variables (spatial).

With the support of cloud imagery, a stochastic model, rooted in observational data and accounts for temporal dependency, is hence ideal for capturing the intrinsically non-deterministic nature of irradiance fluctuations. From the theoretical end, a Markov process was used to simulate the stochastic behavior of sunshine and cloud cover with respect to irradiance [9]. In practice, the Hidden Markov Model (HMM), which is widely used in natural language processing (NLP) tasks, made its first appearance in daily GHI prediction using temperature as observations and irradiance as hidden states [10].

In this paper, we are the first to propose a stochastic solar irradiance forecasting framework using Conditional Random Fields (CRFs). It accommodates a rich set of overlapping, interdependent and multi-granularity features, which boost performance over HMM. We estimate the cloud motion and extract features from the predicted TSI images, to provide a better context of the future than only historical statistics. Therefore, given the historical and predicted observation sequence, CRFs can output the most probable sequence of irradiance levels. Similar to making distinctions between appearances of identical phrases within different contexts in NLP, CRFs distinguish subtle variations of different combinations of features in our application. In this paper, we make the following contributions :

1. **Stochastic Modeling Framework.** We examine the stochastic nature of the solar irradiance and propose a GHI prediction model using *Linear-Chain* Conditional Random Field, which has not been previously applied to solar energy forecasting.
2. **Correlated-Feature Engineering.** We design a novel feature set that harnesses the advantage of multiple sources such as TSI images, meteorological measurements and historical statistics and utilize CRFs as a method to incorporate those overlapping features.
3. **Systematic Evaluation.** We implement a complete framework and evaluate the proposed models by cross-validation on 345,600 raw images (24 days of data). For 1-5 minute GHI forecasts, we observe an average of 36% MAE improvement over the benchmark persistent model, which is reported to be difficult to surpass [11].

The rest of the paper is organized as follows: in Section 2, we introduce TSI imagery, the features we extracted from images, historical statistics, and meteorological measurements. Section 3 contains a brief overview of the theory and application of the Linear-Chain CRF model, and other models for comparison. In Section 4, we explain our experiment setup and model specifications, report our experimental findings and compare the performance of both stochastic and non-stochastic models. Finally in Section 5 we summarize our work.

2   Background and Methodology

2.1   Problem Setting and Related Work

GHI consists of Direct Normal Irradiance (DNI), the solar beam component, and Diffuse Horizontal Irradiance (DHI), which emanates from the sky. The former corresponds particularly to the interference of clouds on the optical path of sunlight, and the latter is rather a result of more complex atmospheric factors. The challenge of sky imaging based prediction is to accurately correlate image pixel values to irradiance for intra and inter-hour prediction over a location of interest. Satellite images are not suitable for very short-term irradiance predictions since they are taken every 30 minutes over a large area. TSI images of the hemispherical sky (see example of Figure 2(a)) have a much higher spatial (sub-kilometer) and temporal (seconds) resolution to reflect the complexity of local meteorological conditions.

Various TSI cloud detection and tracking pipelines have been proposed recently: Fu et al. [12] solely made use of historical cloud features extracted from the sky imager and predicted GHI several minutes ahead via linear regression. Another class of methods calculates cloud velocity vectors allowing for the forward propagation of clouds [7]. Deterministic forecasting methods based on the

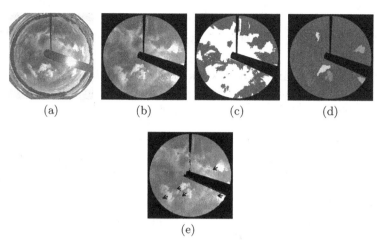

Fig. 2. TSI cloud tracking steps. (a) Original image. (b) Image undistortion and SB dispatch. (c) Cloud cover. (d) Multi-layer detection and cloud segment identification. (e) Cloud Motion prediction and SB filling.

predicted level of cloud cover in images often relies heavily on the robustness of the cloud motion estimation under the assumption that cloud cover is the only predictor of GHI.

In our framework, we carefully analyze and examine TSI images with regards to their correlations with GHI and propose the use of stochastic modeling to utilize features that interdependently reflect complex atmospheric conditions. The inherent flexibility of this model allows us to introduce new features such as historical statistics of irradiance and meteorological measures, which are the key inputs of the existing time series models, creating a synergistic improvement on solar irradiance predictions.

2.2   Feature Engineering

We process a TSI image dataset collected over a two-month period in the summer of 2014 from the Long Island Solar Farm (LISF). Corresponding irradiance and meteorological data were retrieved from the same location. We then carefully analyze and construct a combined set of features extracted from these sources. The full feature set is provided in Table 1. These features are categorized by two aspects according to the time and type.

Imagery-based Features. Figure 2 outlines our TSI image processing pipeline [13] that serves as a crucial step in our stochastic modeling framework. A narrow strip on the TSI mirror, termed shadowband, blocks the intense sunlight and casts a band from the image center to the perimeter (2(a)). The sun's position can be located on the image geometrically. We first sample the image streams to discover meaningful cloud movement and undistort each frame from the curved optical surface to a horizontal plane (2(b)). We then determine the cloud cover

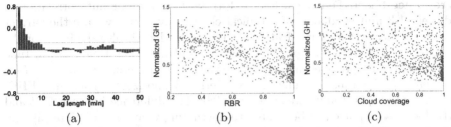

Fig. 3. TSI image feature extraction. (a) Partial autocorrelation of GHI on a cloudy day. (b) Correlation between GHI and circumsolar RBR. High GHIs usually correspond to low RBRs which indicate clear sky pixels. GHI decreases as RBR approaches one due to the presence of clouds around the sun. (c) Correlation between GHI and circumsolar cloud coverage. GHI is minimized when cloud coverage approaches one, but the negative correlation is weaker.

Table 1. Feature Set

Historical statistics			Propagation	
Irradiance-based	Meteorological-based		Imagery-based	
$rad(t-1)$	$RH(t-1)$	RBR_{mean}	$RBR(t+i)$	$mv_{sum}(t+i)$
$\Delta rad(t-1)$	$Temp_{panel}(t-1)$	RBR_{var}	$Cc(t+i)$	$mv_{count}(t+i)$
$Diff_{integ}$	$Temp_{air}(t-1)$	Cc_{mean}	$b_{max}(t+i)$	$mv_{mean}(t+i)$
$Diff_{derv}$		Cc_{var}	$b_{min}(t+i)$	
		$SBbr(t-1)$	$int(t+i)$	
(1)	(2)	(3)	(4)	$i=0,1,...$

(2(c)) and motion vectors for each cloud segment using cross correlation algorithms between consecutive images (2(d)2(e)). To produce a predicted sky image, the cloud movements are propagated to various future time points.

Among all of the properties of sky images, the pixel red-blue ratio (RBR) is one of the most effective features for cloud identification [5]. In 3(b), circumsolar RBR indicates the presences of clouds and is negatively correlated with GHI. The spread of the data points, however, indicates the variance of cloud color. Another suitable attribute, cloud cover (Cc), shows a similar negative trend, yet it is more of a global statistic 3(c). Cloud motion vectors at the sun's vicinity derived from the image processing pipeline suggest potential irradiance fluctuations. Therefore we incorporate the sum, number and mean quantity of motion vectors (mv_{sum}, mv_{count}, mv_{mean}) in the feature sets. The shadowband brightness (SBbr) directly corresponds to DNI, a component of GHI. In order to retain enough complexity for better representation of various types of weather, we compare and select additional features such as the maximum and minimum blue channel value of an image (b_{max}, b_{min}), and the image intensity (int).

Because the image streams are being propagated into the future, we are able to extract not only historical statistics but also from the predicted cloud images as well. Thus, imagery-based features in Table 1 are categorized into *Historical Statistics* such as mean and variance, and *Propagated*.

Irradiance-based Features. Figure 3(a) displays a strong autocorrelation of adjacent 1-3 min irradiance of a typical cloudy day. Thus, we use the current irradiance measurement as a feature ($\text{rad}(t-1)$), which is also the basis for the benchmark, PM. In addition, the change in irradiance from time $t-2$ to time $t-1$ ($\Delta\text{rad}(t-1)$) provides information on the trend, discretized to *rising, falling,* and *unchanged.* We also examine the extraterrestrial horizontal irradiance (EHI), which is atmospherically attenuated to GHI. The difference between EHI and GHI has been reported to be related to fluctuating weather conditions [3], and thus we adopted both the 5-min integral of the calculated difference ($\text{Diff}_{\text{integ}}$) and its third-order derivative ($\text{Diff}_{\text{derv}}$) as features.

Meteorological-based Features. Meteorological measurements provide general weather attributes, and we collected the relative humidity ($\text{RH}(t-1)$), solar panel temperature ($\text{Temp}_{\text{panel}}$), and surface air temperature (Temp_{air}) from the LISF.

3 Solar Irradiance Model

In this Section, we propose the stochastic models for the short-term solar irradiance prediction (minutes ahead). First we introduce two stochastic models, the *Linear Chain* Conditional Random Field (CRF) model [14] and Hidden Markov Model (HMM). For comparison, we explain non-stochastic baseline models. Section (4) will provide the detailed performance comparison among all models introduced here.

3.1 Stochastic Modeling

Linear-Chain Conditional Random Field. Conditional Random Field belongs to the discriminative probabilistic models [14,15]. It is encoded by a bipartite graph (Figure 4). As the name suggests, there are two types of nodes in the graph, one for the set of factors denoted as F (shaded boxes), the other for random variables denoted as V (circles). The graph is denoted by $G = (V, F, E)$. The edges E reflects the probabilistic dependency of factors F on elements of V.

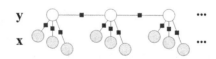

Fig. 4. Graph of Linear-Chain CRF model

CRF directly models the conditional probability distribution $P(\mathbf{y}|\mathbf{x})$, instead of the joint probability $P(\mathbf{x}, \mathbf{y})$. The later can lead to difficulties when overlapping, multi-granularity and non-independent features are involved. Both input $\mathbf{x} \in X$ (dark circles) and output $\mathbf{y} \in Y$ (light circles) can be treated as real valued random variables, while in (Figure 4) $V = X \cup Y$ includes both.

$P(\mathbf{y}|\mathbf{x})$ is the product of all *local* factors depending on a subset of all variables, labeled x_A and y_A for $A \subset V$.

$$P(\mathbf{y}|\mathbf{x}) = \frac{1}{Z} \prod_A \Psi_A(\mathbf{x_A}, \mathbf{y_A}),$$

$$\text{with} \quad Z = \sum_{\{\mathbf{x},\mathbf{y}\}} \prod_A \Psi_A(\mathbf{x_A}, \mathbf{y_A}). \tag{1}$$

The normalizing factor is the partition function Z that sums over all configurations of y_A. $\Psi_A(\mathbf{x_A}, \mathbf{y_A}) \in \mathbf{R}^+$ has the following form

$$\Psi_A(\mathbf{x_A}, \mathbf{y_A}) = \exp\left\{ \sum_k \theta_{Ak} f_{Ak}(\mathbf{x_A}, \mathbf{y_A}) \right\}, \tag{2}$$

where θ_{Ak} are real valued parameters and $f_{Ak}(\mathbf{y_A}, \mathbf{x_A})$ are termed as feature functions. In the experiments, the values of θ_{Ak} can be obtained by maximizing the log-likelihood using training data.

We chose to use the sequential Linear-Chain CRF with the following probability distribution

$$P(\mathbf{y}|\mathbf{x}) = \frac{1}{Z(\mathbf{x})} \exp\left\{ \sum_{t=1}^{T} \sum_{k=1}^{K} \theta_k f_k(y_t, y_{t-1}, \mathbf{x_t}) \right\}, \tag{3}$$

where $Z(\mathbf{x})$ is an instance specific normalization function

$$Z(\mathbf{x}) = \sum_{\mathbf{y}} \exp\left\{ \sum_{t=1}^{T} \sum_{k=1}^{K} \theta_k f_k(y_t, y_{t-1}, \mathbf{x_t}) \right\}. \tag{4}$$

Notice the feature functions depend only on the input sequence $\mathbf{x}$ between $\{1,t\}$ and the output y at $\{t-1,t\}$. This aspect of linear-chain CRF is similar to that of HMM.

To estimate the CRF parameters, the log-likelihood is calculated from the sample training data. If we use N sets of sample data, $(\mathbf{x}^{(i)}, \mathbf{y}^{(i)})$, $i = 1, 2, \ldots, N$, each set forming a time sequence over $t = 1, 2, \ldots, T$, the overall likelihood is

$$\ell(\theta) = \sum_{i=1}^{N} \sum_{t=1}^{T} \sum_{k=1}^{K} \theta_k f_k(y_t^{(i)}, y_{t-1}^{(i)}, \mathbf{x}_t^{(i)}) - \sum_{i=1}^{N} \log Z(\mathbf{x}_t^{(i)}). \tag{5}$$

As $P(\mathbf{y}|\mathbf{x}) \in [0,1]$ the function $\ell(\theta)$ has negative values. Given the exponential form in the probability distributions, $\ell(\theta)$ is a concave function, and each local optimum is also a global optimum.

Optimization involves computing the gradients of the objective function such as L-BFGS algorithms [16,17], which requires the marginal distributions $p(y_{t-1}, y_t|\mathbf{x_t})$. For Linear-chain CRF, distributions $p(y_{t-1}, y_t|\mathbf{x_t})$ can be obtained using the recursive backward-forward algorithms (for details see [18]). One finds

$$p(x_{t-1}, x_t|y_t) = \alpha_{t-1}(y_{t-1}) \Psi_t(x_{t-1}, x_t, y_t) \beta_t(y_t) \tag{6}$$

where α_t and β_t are the forward and backward recursions, namely

$$\alpha_t(y_t) = p(x_{\langle 1...t\rangle}|y_t), \quad \beta_t(y_t) = p(x_{\langle t+1...T\rangle}|y_t) \tag{7}$$

which are computed by

$$\alpha_t(y_t) = \sum_{x_{t-1}} \Psi(x_t, x_{t-1}|y_t)\alpha_{t-1}(x_{t-1})$$

$$\beta_t(y_t) = \sum_{x_{t+1}} \Psi(x_{t+1}, x_t|y_{t+1})\beta_{t+1}(x_{t+1}) \tag{8}$$

The combination of L-BFGS and backward-forward algorithms then furnishes our training algorithm package, allowing statistical inference of the sample data.

For forecasting, we use the dynamic programming algorithm to estimate the Viterbi path of a sequence of features $\mathbf{x}$ which maximizes the probability

$$\hat{\mathbf{y}}_{\text{vi}} = \arg\max_{\mathbf{x}} P(\mathbf{y}|\mathbf{x}) \tag{9}$$

Each recursion gives us a prediction to the event for next step. Multistep forecasting can be similarly achieved [19], leading to the predictions of 1-, 2-, $\cdots$ and 5-minutes ahead.

Hidden Markov Model. HMM is a canonical probabilistic model for sequential or time series data that considers not only considers the transitions of the value in the sequence, but also introduces a corresponding dependent sequence of events. It is well known that HMM can be recast in a CRF form shown as follows:

$$P(\mathbf{x}, \mathbf{y}) = \frac{1}{Z}\exp(\sum_{t=1}^{T}(\sum_{i,j\in S}\lambda_{ij}f_{ij}(y_t, y_{t-1}, x_t)+$$

$$\sum_{j\in S, o\in O}\theta_{jo}f_{jo}(y_t, y_{t-1}, x_t))), \tag{10}$$

where,
$f_{ij}(y_t, y_{t-1}, x_t) = \delta_{y_t=j}\delta_{y_{t-1}=i}$, $f_{io}(y_t, y_{t-1}, x_t) = \delta_{y_t=i}\delta_{x_t=o}$. δ is a Kronecker delta function, S stands for the space of all hidden states, λ_{ij} encodes the transition probability between two states and θ_{jo} for the emission probability. Note that we limit the feature functions to be binary indicator functions for both one state transition and one output emission instead of K real values with the state transitions. HMM models joint probability instead of conditional probability, which leads to these differences in the feature functions between the general CRF and HMM. The statistical inference algorithms for HMM are similar as the aforementioned Linear-Chain CRF. Historically, these inference algorithms were in fact first developed for HMM. In summary, HMM is a highly restricted Linear chain CRF model with a single feature function, which captures the RBR dependence, but no other useful features are used from TSI images.

3.2   Non-Stochastic Models

For comparison to the stochastic models, we also consider two non-stochastic models as our baselines. First, our baseline Persistent Model (PM) is a valid benchmark in the solar forecasting field [11]. Next, a straightforward extension of PM is to combine all terms with linear relationships, also known as Linear Regressions (LRs).

Persistent Model (PM). The baseline persistent model (PM) assumes that the irradiance at time t is best predicted with its value previously observed at time $t-1$. Therefore, PM contains only one feature $rad(t-1)$ from our feature set aforementioned in Section 2.2:

$$\hat{rad}(t) = rad(t-1) \tag{11}$$

PM is simple but highly effective especially for very short term prediction. Recent studies have shown that this baseline is very difficult to beat for forecasts of GHI and DNI within 15 minutes [7,8,11].

Linear Regression (LR). The Linear Regression (LR) model gives a linear relation between solar irradiance and the features introduced in the Linear-Chain CRF. We have

$$\mathbf{y} = \mathbf{w}X + b. \tag{12}$$

We then minimize the objective function

$$||\mathbf{y} - \mathbf{w}X - b||_2^2 + k||\mathbf{w}||_2^2, \tag{13}$$

where $\mathbf{w}$ is the weight coefficient vector, b is an intercept, X is an $N \cdot K$ matrix and k is a regularization parameter. With the $l2$-norm of the second term, the linear regression is also called Ridge Regression. k controls the degree of regularization and it helps the learned model from being overfitted.

   We considered two LR models. As discussed in Section 2.2, pixel RBR around the sun in the image is the most direct indicator of cloud conditions and highly correlated with irradiance. We subsequently combine the persistent model (PM) with the extracted RBR feature as LR_2 which uses two features, $rad(t-1)$ and $RBR(t)$. We then expand LR_2 to LR_all , which incorporates all candidate features. Note that LR does not consider a sequence of predictions together. Although we incorporated the previous time stamp irradiance values as a feature, it solely focus on the current state prediction and prior or later prediction could not affect the current status. So, non-stochastic model is myopic compared to CRFs.

4   Experiments

In this Section, we present strong experimental evidence that the stochastic multi-feature CRF model outperforms all the other models, including Linear Regression, Hidden Markov model and Persistent Model.

4.1   Experimental Setup and model specification

Our experimental data set consists of 345,600 raw TSI images and corresponding Pyranometer recordings for 24 days gathered at the LISF (Long Island Solar Farm), from April to August of 2013. The TSI images are received once per second, and sampled every 20 seconds to produce effective and accurate motion estimation. The length of our sample data sequence is $T = 5760$, obtained by uniformly sampling predicted images every minute from 10:00 am to 14:00 pm on each experiment day. Corresponding meteorological data from this location is retrieved from the LISF station. We included all cloud conditions which represent all weather types. Compared to average annual weather statistics, our dataset includes a smaller percentage of overcast and clear sky, and a larger fraction of the fluctuating weather conditions. It is therefore more difficult to predict, which is where the real challenge of solar irradiance forecasting lies. We evaluate the prediction performance using mean-absolute-error (MAE) and ran 24 fold cross validations to tune model parameters and evaluate performances.

For CRF, we included all feature types shown in Table 1 for every timestamp. Since we could not see the future data in real time forecasting, we separate the sequence T into short segments of length $n_\ell = 5$, which imitates the real-time forecasting of 1-5 minutes ahead respectively. Recall the features we introduced in Section 2:

$$
\begin{aligned}
\{\mathbf{x}_t\} = \{ &\mathrm{rad}(t-1), \Delta\mathrm{rad}(t-1), \mathrm{Diff}_{\mathrm{integ}}, \mathrm{Diff}_{\mathrm{derv}}, \mathrm{RH}(t-1), \mathrm{Temp}_{\mathrm{panel}}(t-1), \\
&\mathrm{Temp}_{\mathrm{air}}(t-1), \mathrm{RBR}_{\mathrm{mean}}, \mathrm{RBR}_{\mathrm{var}}, \mathrm{Cc}(t+i), \mathrm{Cc}_{\mathrm{mean}}, \mathrm{Cc}_{\mathrm{var}}, \\
&\mathrm{RBR}(t+i), \mathrm{b}_{\mathrm{max}}(t+i), \mathrm{b}_{\mathrm{min}}(t+i), \mathrm{mv}_{\mathrm{sum}}(t+i), \mathrm{mv}_{\mathrm{count}}(t+i), \\
&\mathrm{mv}_{\mathrm{mean}}(t+i), \mathrm{SBbr}(t-1), \mathrm{int}(t+i)\}.
\end{aligned}
$$

$$(14)$$

To be comparable to the HMM features, we discretize the numerical features into categorical data. Regarding the output variable, irradiance readings are also discretized, and each value is mapped to an integer if the value falls in the corresponding interval. We tested different discretization levels (number of states), including $n_S = 5, 10, 15, 20, 40$ states and the best result was obtained with 10 states. We also tested with different feature combinations from Table 1, which will be compared later.

4.2   Model Performance Comparisons

The results of the different models are shown in Figure 5. The performance of baseline PM is sensitive to cloud variability and decreases with respect to time span.Our stochastic models CRF and HMM show an average of 36% and 16% improvement, respectively, over the baseline PM and maintain a quite stable performance even with an increased prediction time span. Even though for a one minute prediction, HMM does not perform as well as PM or LR, but its performance is relatively constant throughout the 1-5 minute predictions. The LR model on the other hand, deteriorates quickly as the forecast horizon increases.

LR_all shows a similar trend as PM, while LR_2 particularly stands out for one minute predictions by taking advantage of the key features and the simplicity of the modeling. It merges with LR_all after 3 minutes. CRF performs the best, or second best, for 1-5 minute predictions, and is consistent as the forecasting horizon expands.

In terms of feature sets, we summarize the results in Table 3. The best performance is achieved using a combination of *Irradiance-based* and *Imagery-based* features $(1, 3, 4)$. *Imagery-based* features alone $(3, 4)$ cannot provide sufficient information about the previous trend of GHI, but *Historical statistics* features $(1, 2, 3)$, on the other hand, do not include information on future cloud movement, and thus both showed less accuracy. Note that *Meteorological-based* features are generally useful proxies for longer term predictions which involve more seasonal changes and tend to introduce noise to our short-term predictions.

We also provide comparisons on the performance of HMM and CRF with different number of states. CRF finds the best result at 10 states and HMM at 15 states. Both of them suffer from a small number of states which increases discretization error, and a large number of states where models require a larger amount of training datasets.

The performance of CRF is further verified by Figure 6 which gives the distribution of the difference ΔGHI between the forecasted and measured solar irradiance. The peaks at zero for the CRF are higher than other models, and it has a smaller dispersion over the non-stochastic models. Within $\pm 50 \text{W/m}^2$, the CRF model achieves an accumulated precision of roughly 80%.

As an example, in Figure 7, June 6th, 2013 shows a strong agreement between the 5-minute forecasted GHI with the ground measurement using CRF. Under such cloudy conditions, PM shows noticeable shifting effect for 5-minutes forecast, missing apparent spikes, and thus has a limited utility for grid operators. Both HMM and CRF cover the correct range of values for the measured irradiance from the pyranometer, while LR either over-predicts or under-predicts the solar irradiance. HMM captured all the large spikes but suffers from the lack of varying features to resolve finer changes. CRF is able to catch most of the fluctuations, and is more responsive to minor variations in irradiance compared to all the other models.

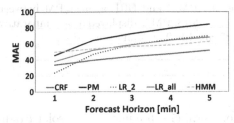

Table 2. MAE values for Figure 5. For comparison, the number of states of HMM and CRF presented here are both chosen to be 10.

Fig. 5. Comparison among different models for 1-5 min forecast

Method	1 min	2 min	3 min	4 min	5 min
CRF	33.71	39.53	44.20	47.48	51.79
PM	45.52	64.59	72.87	79.66	84.67
LR_2	23.75	46.95	58.83	65.52	70.19
LR_all	38.18	52.11	58.70	64.83	68.34
HMM	49.78	53.69	56.01	58.46	62.95

522   J. Xu et al.

Table 3. MAE for CRF with different feature combinations

Feature Set	1 min	2 min	3 min	4 min	5 min
1-4	41.39	45.02	48.62	50.81	53.93
1-3	38.24	44.66	48.06	53.36	57.08
1,3	35.22	41.71	43.18	48.13	53.27
3,4	49.47	50.30	51.99	53.84	58.27
1,3,4	33.71	39.53	44.20	47.48	51.80

Table 4. MAE for different numbers of states of CRF and HMM

States	CRF	HMM
5	53.98	61.15
10	43.34	56.18
15	45.42	54.97
20	50.33	59.44
40	60.15	66.37

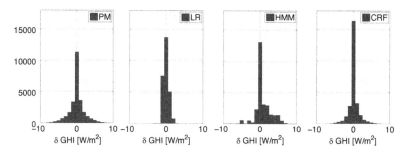

Fig. 6. Comparison of forecasting precision using histogram distribution of the discrepancy between forecasted and observed values of solar irradiance, with 28,800 points in each diagram (1-5 min).

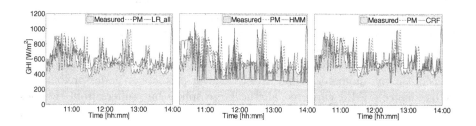

Fig. 7. Forecasting of 5-min ahead using LR, HMM and CRF against PM between 10:10 am and 14:00 pm on the day of June 6th, 2013. PM is displayed as a comparative baseline. We use LR_all here which outperforms LR_2 at 5-min prediction.

5 Conclusion

In this paper, a novel framework of short-term $(1-5$ minute scale) solar irradiance forecasting based on Linear-Chain Conditional Random Field was proposed and evaluated. Our stochastic model can integrate a rich array of correlated TSI features from the cloud patterns to cope with the stochastic nature of irradiance fluctuation. The combination of historical statistics and cloud images from the TSI synergistically improves prediction performance. The experimental results showed that the CRF model demonstrated significant improvements over the baseline models in terms of the MAE measurement metrics, reduced the forecasting discrepancy distribution, and generated a high-precision prediction. In particular, the averaged error rate for $1 - 5$ minute predictions of the CRF model, measured with the MAE score, is 36% on average less than the baseline persistent model (PM).

References

1. Lave, M., Kleissl, J., Arias-Castro, E.: High-frequency irradiance fluctuations and geographic smoothing. Solar Energy **86**(8), 2190–2199 (2012)
2. Reikard, G.: Predicting solar radiation at high resolutions: A comparison of time series forecasts. Solar Energy **83**(3), 342–349 (2009)
3. Wang, F., Mi, Z., Shi, S., Zhao, H.: Short-term solar irradiance forecasting model based on artificial neural network using statistical feature parameters. Energies **5**(5), 1355–1370 (2012)
4. Lorenz, E., Heinemann, D.: Prediction of solar irradiance and photovoltaic power. Comprehensive Renewable Energym, pp. 239–292. Elsevier, Oxford (2012)
5. Tapakis, R., Charalambides, A.G.: Equipment and methodologies for cloud detection and classification: A review. Solar Energy **95**, 392–430 (2013)
6. Wai Chow, C., Urquhart, B., Lave, M., Dominguez, A., Kleissl, J., Shields, J., Washom, B.: Intra-hour forecasting with a total sky imager at the uc san diego solar energy testbed. Solar Energy **85**(11), 2881–2893 (2011)
7. Quesada-Ruiz, S., Chu, Y., Tovar-Pescador, J., Pedro, H.T.C., Coimbra, C.F.M.: Cloud-tracking methodology for intra-hour DNI forecasting. Solar Energy **102**, 267–275 (2014)
8. Yang, H., Kurtz, B., Nguyen, D., Urquhart, B., Chow, C.W., Ghonima, M., Kleissl, J.: Solar irradiance forecasting using a ground-based sky imager developed at UC san diego. Solar Energy **103**, 502–524 (2014)
9. Morf, H.: Sunshine and cloud cover prediction based on markov processes. Solar Energy **110**, 615–626 (2014)
10. Hocaoğlu, F. O.: Stochastic approach for daily solar radiation modeling. Solar Energy **85**(2), 278–287 (2011)
11. Diagne, M., David, M., Lauret, P., Boland, J., Schmutz, N.: Review of solar irradiance forecasting methods and a proposition for small-scale insular grids. Renewable and Sustainable Energy Reviews **27**, 65–76 (2013)
12. Chia-Lin, F., Cheng, H.-Y.: Predicting solar irradiance with all-sky image features via regression. Solar Energy **97**, 537–550 (2013)
13. Xu, J., Yoo, S., Yu, D., Huang, D., Heiser, J., Kalb, P.: Solar irradiance forecasting using multi-layer cloud tracking and numerical weather prediction. In: Proceedings of the 30th Annual ACM Symposium on Applied Computing, SAC 2015. ACM, New York (2015)
14. Lafferty, J., McCallum, A., Pereira, F.C.N.: Probabilistic models for segmenting and labeling sequence data, Conditional random fields (2001)
15. Sutton, C., McCallum, A.: An introduction to conditional random fields for relational learning. Introduction to Statistical Relational Learning, pp. 93–128 (2006)
16. Bertsekas, D.P.: Nonlinear programming (1999)

17. Byrd, R.H., Nocedal, J., Schnabel, R.B.: Representations of quasi-newton matrices and their use in limited memory methods. Mathematical Programming **63**(1–3), 129–156 (1994)
18. Rabiner, L.: A tutorial on hidden markov models and selected applications in speech recognition. Proceedings of the IEEE **77**(2), 257–286 (1989)
19. Culotta, A., McCallum, A.: Confidence estimation for information extraction. In: Proceedings of HLT-NAACL 2004: Short Papers, pp. 109–112. Association for Computational Linguistics (2004)

Online Prediction of Chess Match Result

Mohammad M. Masud[1]([✉]), Ameera Al-Shehhi[1], Eiman Al-Shamsi[1],
Shamma Al-Hassani[1], Asmaa Al-Hamoudi[1], and Latifur Khan[2]

[1] College of IT, United Arab Emirates University, Al Ain, UAE
{m.masud,201002468,201003965,201004128,201006839}@uaeu.ac.ae
[2] Department of CS, University of Texas at Dallas, Richardson, TX, USA
lkhan@utdallas.edu

Abstract. In this work we propose a framework for predicting chess match outcome while the game is in progress. We make this prediction by examining the moves made by the players. For this purpose, we propose a novel ensemble based learning technique where a profile-based segmentation is done on the training dataset, and one classifier is trained from each such segment. Then the ensemble of classifiers is used to predict the outcome of new chess matches. When a new game is being played this ensemble model is used to *dynamically* predict the probabilities of *white winning*, *black winning*, and *drawing* after every move. We have evaluated our system with different base learning techniques as well as with different types of features and applied our technique on a large corpus of real chess matches, achieving higher prediction accuracies than traditional classification techniques. We have achieved prediction accuracies close to 66% and most of the correct predictions were made with nine or more moves before the game ended. We believe that this work will motivate the development of online prediction systems for other games, such as other board games and even some field games.

Keywords: Prediction · Classification · Chess · Data mining · Feature extraction

1 Introduction

Chess is a well structured game as the states and moves of the games are well defined. Unlike the games such as soccer, basketball and the like, where the position of the players, movement of the ball, etc. are not predefined, chess has predefined set of states (although very large) and moves. Our goal is to investigate that with the help of machine learning, how well the outcome of such a structured game can be predicted before the game ends just by examining the moves made by each player. We believe that the results from this research will *motivate* building prediction models for other structured or semi-structured games, such as other board games, cricket and baseball. This prediction system will be a useful *application* for spectators, players as well as trainers for various purposes. For example, spectators may use this system to understand the status

© Springer International Publishing Switzerland 2015
T. Cao et al. (Eds.): PAKDD 2015, Part I, LNAI 9077, pp. 525–537, 2015.
DOI: 10.1007/978-3-319-18038-0_41

of each opponent's hold in the game, players can use it to get an early warning before a move is made, and trainers can use it during the post analysis of the match to identify the turning points and bad moves in the match.

Computers have been programmed to play chess using AI based search techniques since the second half of the 20th century. IBM's Deep Blue was the first computer to defeat World Chess Champion Garry Kasparov in 1997 [6]. Since then, powerful chess engines have been evolving rapidly and now many commercial chess playing software are available. However, in this work, rather than proposing another smart chess playing agent, we focus on predicting the outcome of a chess match before the game ends.

There has been several attempts in the past to predict chess game outcomes [3,4]. These predictions are static, meaning, they don't consider the dynamism of the game, and predict the outcome before the game has started. On the contrary, our proposed system dynamically updates the winning (or drawing) probabilities after each move of the game. In our system, an ensemble training and classification approach is used for the prediction task. We use a profiling technique to separate chess games into different profiles based on the number of moves in the games. Then we extract move-based features from each chess game and develop feature vectors. These feature vectors are used to train a classifier for each profile. We use an ensemble of the classifiers to predict the outcome of a new game. We show both *analytically* and *empirically* that the ensemble based classifier is superior to a single classifiers trained with all the historical data.

To the best of our knowledge, this is the first approach to dynamically predict chess match outcome with the help of machine learning using only the information obtained form the moves of the game. Our contributions are as follows. First, we propose a learning technique that extracts move based features from the games. Second, we have proposed a novel technique for training classification models by carefully choosing and segmenting the training data with our profiling-segmentation technique and applying the trained models to predict new games. Third, we have proposed an online and adaptive ensemble classification technique for improving the prediction accuracy. Finally, we have applied our technique on a large corpus of real chess matches and obtained higher prediction accuracy than traditional classification techniques.

The rest of the paper is organized as follows. Section 2 discusses the work closely relevant to our work. Section 3 and 4 describe the proposed method in details and Section 5 reports the experiments, results, and analyzes the results. Finally, Section 6 concludes with direction to future works.

2 Related Work

Several works have been proposed for chess game prediction using machine learning or data mining techniques. A Chess Game Result Prediction System was proposed by Fan et. al. [3]. Their project was to train a classifier with the World Chess Federation (FIDE) rating system using a training dataset of a recent eleven-year period, which ranges from year 2000 to 2011, and then use their system to predict the outcome of chess games played by the same players in the following half year. Their success rate of the prediction was 55.64%.

Another similar work was reported by Ferreira [4], who also applied a prediction model trained on the outcomes of chess games in past few years and used to predict future games played by the same players. The model was trained iteratively, and it contained several parameters for tuning. However, it is not clear what was the prediction accuracy of the proposed method. In another interesting work, Ferreira [5] proposed a technique to determine the strength of a chess player based on his actual moves in a game.

However, our approach is different from the above in several ways. First, we do not use any information of the player (i.e. Elo rating) for training, and second, we use an ensemble based classification technique. Finally, our approach is dynamic, meaning, the prediction is updated as the game progresses. On the contrary, all previous approaches are static prediction, meaning, the predictions are fixed and computed before the game is played.

We now discuss some state-of-the-art ensemble classification techniques. Ensemble classification models have been effectively used in recent years for classifying large data evolving data streams [1,8]. Ensemble classification models consist of an ensemble (collection) of classifiers. When classifying an unknown instance, each individual classifier outputs its own prediction and a majority vote is taken to choose the winning class. This winning class is the predicted class for the ensemble. There are different majority voting techniques available, namely, simple majority and weighted majority. The ensemble approaches mentioned above are good in handling dynamic data stream, where the data are always evolving. However, as per our observation, the problem of predicting chess matches is rather noisy data than evolving data. So we propose our own noise reduction technique to achieve high prediction accuracy.

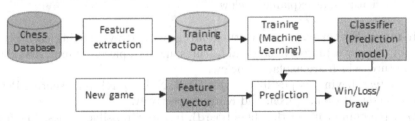

Fig. 1. Overview of the prediction system

3   Proposed Method

In this paper, we propose a general framework for the prediction system. The high level design of the system is shown in figure 1. The chess database is a collection of past chess games (Section 3.1). From this database, we extract and select features (Section 3.2) to generate the training data, which is used to train a classification model (Section 4.1). Finally, when a new game is played, the probabilities of win/loss/draw are predicted after every move using the trained classifier (Section 4.2). Following sections detail each component of the system.

3.1   Chess Database

There are many chess databases available online. However, we choose the database from ChessOK.com [2] because of several reasons. First, the database is free. Second, it contains the games of all major tournaments from the year 2011. Finally, the games in the database are well organized - month by month basis, which is useful for our purpose.

The games in the database are stored in PGN format. Each month's games are divided into smaller chunks of games, each chunk containing approximately 1,900-2,000 games. Each game in the database contains several information about the match, such as date, tournament name, players' names and ratings. Then the actual moves of the game are listed. An example of a PGN file is shown in figure 2.

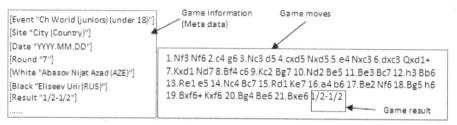

Fig. 2. Format of the PGN file

3.2   Feature Extraction and Selection

For each game we utilize the move-based information for the training and prediction. We extract two types of features from the moves, namely, split-move and n-ply features, as explained below.

Split-move Features: These features are generated by splitting each ply (a move by one player) into five nominal features, namely, piece, column, row, check and capture, which are explained below:

Piece type, having six possible values: {P, R, N, B, Q, K}, representing Pawn, Rook, Knight, Bishop, Queen, and King, respectively.
Column (column or file of the chess board), having 8 possible values: {a - h }
Row (row of the chess board), having 8 possible values: { 1 - 8 }
Capture (captured opponent's piece or not)= {0, 1}
Check (checked opponent's King or not) = {0, 1}

For each game, we convert each ply into these five nominal features and generate the feature vector. Therefore, if the game consists of 21 moves, meaning 42 plies, then there will be 42 * 5 = 210 nominal features, arranged in the same order as the moves appear in the game. For example, for the game in figure 2, the split-move features (comma separated) will be as follows:
{N,f,3,0,0,N,f,6,0,0,P,c,4,0,0,P,g,6,0,0,....}.

A game may have any number of moves. Therefore,this feature-based approach generates different number of features for different games. However, most machine learning techniques require fixed size feature vector (i.e., same number of features for all data). This issue is discussed in details in the next section (Section 4). The advantage of this approach is that that the order of moves are preserved in the feature vector.

n-ply Features: These features are generated by considering each ply (e.g. 'Nf6+') as a feature. There are two steps in generating the feature vector for these features. In the first step, we scan the whole database of games and collect all possible features from all games. Let S be the set of all such features obtained from all the games. The next step is to generate a feature vector for each game. The feature vector will be a vector of bits, where $bit_i=0$ of the i-th feature S_i is absent in the game and $bit_i=1$, if the i-th feature S_i is present. One problem in this approach is that the number of features, i.e., the size of S may be very large. In this case, we apply a feature selection technique based on information gain, and choose the best K features to be used in the feature vector.

Algorithm 1. PROSEGEN

Input: n: number of segments
    T: Training data
Output: H: the segment boundaries of the segmented training data, $H = \{H_1, ... H_m\}$
    E: the ensemble classifier, $E = \{E_1, ... E_m\}$
1: $N \leftarrow |T|$ //Training data size
2: $Max \leftarrow$ Maximum number of moves in any game in T
3: **for** $i \leftarrow 1$ to Max **do**
4:    $B_i \leftarrow \{x \in T | Moves(x) = i\}$ /* Separate games into bins of i moves */
5: **end for**
6: $i \leftarrow 0, j \leftarrow 0$
7: $G \leftarrow \phi$ //segmented data
8: $H \leftarrow \phi$ //segment boundaries
9: **while** $j < Max$ **do**
10:    $G_i \leftarrow G_i \cup B_j$
11:    **if** $|G_i| \geq N/n$ **then**
12:        $H_i \leftarrow j$ /* set the current boundary */
13:        $i \leftarrow i + 1$ /* segment is full, goto next one */
14:    **end if**
15:    $j \leftarrow j + 1$
16: **end while**
17: $m \leftarrow i$ //actual number of segments
18: **for** $i \leftarrow 0$ to $m - 1$ **do**
19:    $E_i \leftarrow$ TrainClassifier(G_i)
20: **end for**

---

4   Training and Prediction

The training process consists of two stages: i) profiling and segmentation (PRObreakSEG) and ii) building the ensemble (EN) of classifiers. The prediction involves using the ensemble classifier for predicting results of new games.

4.1   Profiling and Segmentation of Data

As mentioned earlier, variable number of moves in the games lead to variable number of split-move features, which is problematic for learning techniques that rely on vector-based features. However, we can manipulate the feature vectors to keep them equal. For example, suppose we choose vector size to be 200 moves. Therefore, a game having only 20 moves will have to be padded by 180 null or default values. This introduces noise in the training data, reducing the performance of the trained classifier. To solve this problem, we propose a profiling (PRO) and segmentation (SEG) technique, implemented with algorithm 1.

Description of Algorithm 1: It takes as input the training data T, and the number of segments n to created. Let N be the dataset size, i.e., total number of games in T. First, we create separate bins and keep all games having i moves into bin i (lines 3-5). Then we create each segment by joining the adjacent bins in a way such that each segment contains approximately the same number of games (lines 9-16). This is done to avoid dominance by any data segment. Besides, since we are going to build an ensemble of classifiers, we also need to ensure that each classifier gets equal amount of training data so that all classifiers achieve same quality (i.e., prediction power). Finally, we train one classifier from each data segment. In the experiments (Section 5), we show a detailed analysis of the bin sizes and performance of the ensemble on each bin.

Noise Reduction by Segmentation: Now we show that the segmentation reduces noise in the training data. Let m be the maximum number of moves considered in building the *unsegmented* training data, T. As mentioned earlier, all instances (i.e., games) in the feature vector must have equal length. Therefore, games having less than m moves must be padded with default feature values (e.g. zero) to make the feature vectors equal length. Therefore, the padded values can be considered as noise added to the features. For simplicity of representation, suppose we divide the training data into two equal segments, T_1 and T_2, i.e., $T = T_1 \cup T_2$ and $|T_1| = |T_2|$, such that T_1 contains all the games having m_1 or less moves and T_2 contains all the games having $m_1 + 1$ or more (upto m) moves. Also, let f_i = number of instances (i.e., games) having i moves. Therefore,

$$\sum_{i=1}^{m} f_i = |T|; \quad \sum_{i=1}^{m_1} f_i = |T_1|; \quad \sum_{i=m_1+1}^{m} f_i = |T_2| \qquad (1)$$

Let $\eta(T)$ = Added noise in T = Total padded moves in T. Note that a game having $i < m$ moves is padded with $m - i$ moves. Therefore,

$$\eta(T) = \sum_{i=1}^{m}(m-i)f_i = \sum_{i=1}^{m} mf_i - \sum_{i=1}^{m} if_i = m|T| - \sum_{i=1}^{m} if_i \text{ (using equation (1))}$$

$$= m|T| - \sum_{i=1}^{m_1} if_i - \sum_{i=m_1+1}^{m} if_i > m|T| - \sum_{i=1}^{m_1} if_i - \sum_{i=m_1+1}^{m} mf_i \text{ (since } m \geq i)$$

$$= m(|T_1| + |T_2|) - \sum_{i=1}^{m_1} i f_i - m(|T_2|) \text{ (using equation (1))}$$

$$= m(|T_1|) - \sum_{i=1}^{m_1} i f_i > m_1 |T_1| - \sum_{i=1}^{m_1} i f_i = \eta(T_1) \tag{2}$$

Where $\eta(T_1)$ is the added noise in the segment T_1. The same proof is applicable to segment T_2. Therefore, equation (2) concludes that the added noise in segmented training data is less than that of the combined training data. If we keep dividing the segments into smaller and equal segments, the noise reduction will continue upto a point where this advantage in noise reduction will be outweighed by insufficient amount of training data. This trade-off between noise reduction and diminished training data must be considered during training.

4.2 Ensemble Classification

The ensemble of classifiers is then used to classify new games. The classification process is shown in algorithm 2.

Algorithm 2. Classification

Input: x: The game to classify
    E: the ensemble classifier, $E = \{E_1, ... E_m\}$
    H: the segment boundaries of the segmented training data, $H = \{H_1, ... H_m\}$
Output: Y: the prediction vector for x, $Y = \{P_{white}, P_{black}, P_{draw}\}$
  1: $index \leftarrow$ SegId(H, Moves(x)) // Find the segment index for x
  2: $Y \leftarrow$ GetPrediction(E_{index}, x)

Description of Algorithm 2: The input to the algorithm are the ensemble (E) and the game to classify (x). The game x may be an ongoing game (i.e., not finished yet). First, based on the number of moves already made in the game, we find the appropriate segment where the game fits in (line 1). Then we predict the game using the corresponding classifier (line 2). This classification is analogous to an weighted ensemble classification, where the weight for the classifier corresponding to x's segment is 1, and the weight of all other classifiers in the ensemble is 0. This is done to minimize feature noise in the test data.

5 Experiments

In this section we describe the datasets, experimental environment, and discuss and analyze the results.

5.1 Data Sets and Experimental Setup

The dataset [2] contains a collection of more than 300,000 chess games dated from October 2011 upto August 2014. The data are divided into approximately equal sized batches of about 2000 games, organized in a single PGN file. For each month, there are 4-6 such batches. We use the games from 2011-2013 as training and the games from 2014 for testing, i.e., evaluating the trained prediction model.

Competing Approaches:

BC: The traditional batch learning classifier, which is trained with all the training data.

EE: An ensemble classifier with equal weight given to each individual classification model in the ensemble. The predicted class is the class that has majority vote. Each such individual model is trained from one segment of the training data, segmented using the proposed segmentation technique.

PE: This is the PROSEGEN ensemble model, which is an weighted ensemble with highest weight (1) given to the individual classification model that belongs to the same bin as the test instance. Other classifiers are given zero weight.

Base Classifiers: We have experimented with several classifiers. For each classifier, we used the WEKA machine learning API [9]. The classifiers are NaiveBayes (NB), Decision Tree (J48 in WEKA) Random Forest (RF), and Support Vector Machine (SVM).

Besides these classifiers, we have also experimented with sequence-based classifiers by considering the game as a sequence of moves. The classifiers are: Hidden Markov Model (HMM) classifier and Bounded Coordinate-Descent sequence classifier [7]. However, their prediction accuracies are far below the others and so, we do not report them.

Feature Set: We have used both the split-move and n-ply features in our experiments but due to space limitation, we report only the results for split-move features because they achieve higher prediction accuracy.

Parameter Settings: The only parameter in our approach is the number of segments (i.e., n), which also decides the ensemble size. In all experiments, we keep $12 < n < 16$ as we obtain the best results for these values. For the classifiers used in WEKA, we use the default parameter values provided by the API.

Hardware and Software: The experiments were done on a standalone workstation having Intel Core i5 2.4GHz processor with 8GB RAM and 750GB Hard Drive. The OS was Windows 7. All programs have been developed with Java with NetBeans IDE, and Weka API [9] has been used for the base classifiers.

5.2   Evaluation

Unless mentioned otherwise, we use prediction accuracy (Acc %) as the evaluation metric, which is the percentage of instance correctly classified. Here the classification problem is considered multi-class because there are three classes, namely, white winning, black winning, and draw. Also, unless mentioned otherwise, the accuracy are obtained by evaluating complete games (i.e., all moves are considered). Also, in all the evaluations, the training data are the games played in the years 2011 to 2013, and the evaluation (i.e., test) data are the games played in year 2014.

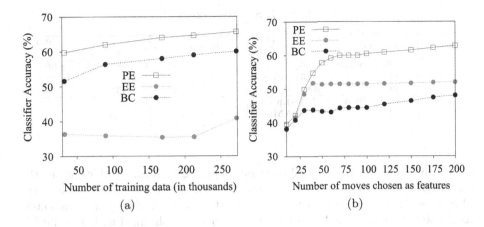

Fig. 3. (a) Number of training data vs accuracy, and (b) Feature set size vs accuracy

Effect of Varying Training Data: First, we evaluate the effectiveness of the learning techniques by varying the size of training data. Figure 3(a) shows the effect for decision tree (J48) classifier. Similar effect is obtained for other classifiers. We vary the training data from 33,000 to 270,000, where 33,000 dataset contains only the games from year 2011 and 270,000 dataset contains all the games from 2011 to 2013 obtained from [2]. We see that all competing approaches show the similar trend, i.e., increasing training data increases the classification accuracy. For example, for 33K training data, the accuracies of PE, EE, and BC are 59.8, 36.4, and 51.6, respectively, whereas for 270K training data, the values are 65.6, 40.8, and 59.8, respectively. In all cases, PE has the highest accuracy being at least 5% higher than that of BC, and 15% higher than that of EE.

Feature Set Size vs Accuracy: We varied the size of feature sets by capping the number of moves to a lower value and this effect is shown in figure 3(b) for Naive Bayes, using the 270,000 training dataset. The X axis shows the number of moves chosen as features and the Y axis shows the accuracy for each completing approach. For example, X=40 means we choose only the first 40 moves of each game as features and ignore the remaining moves. For this value of X, we obtain accuracies of PE, EE, and BC as 57.8%, 51.4%, and 43.3%, respectively. The general trend here is that with increasing size of feature set, the accuracy increases. This is because as more moves are considered for features, more information are obtained from the game, which facilitates achieving higher accuracy. Also, here in all cases, PE has the highest accuracy.

Summary on All Datasets, Classifiers, and Competitors: Table 1 shows the summary of the results for each competing methods using each classifier and datasets. The columns under "training data size" report the accuracy for different sizes of training data, ranging from 33K to 270K. Also, the rows corresponding to each classifier report the classification accuracy of each competing approach when trained with different sizes of training data. Note that in all cases, the test data

Table 1. Summary result

Classifier	Competitor	Training data size (x1000)					Classifier	Competitor	Training data size (x1000)				
		33	90	170	210	270			33	90	170	210	270
NB	PE	60.2	61.7	62.6	62.7	62.8	RF	PE	47.4	48.8	50.6	50.9	50.9
	EE	43.4	40.5	39.5	39.6	46.8		EE	40.6	41.1	39.9	41.6	39.1
	BC	47.5	47.8	48.1	47.8	48.0		BC	43.7	43.9	44.2	44.7	45.3
J48	PE	59.8	62	64	64.6	65.6	SVM	PE	58.4	62.0	—	—	—
	EE	36.4	35.9	35.4	35.5	40.8		EE	45	46.4	—	—	—
	BC	51.6	56.4	58.2	59.1	59.8		BC	51	—			

is the same, which is the set of 23,000 games played in 2014. As an example, the row headed by "J48", and "PE" shows the accuracy of PE using decision tree (J48) classifier when trained with training data having sizes ranging from 33,000 to 273,000. Note that the highest accuracy (65.6%) is obtained with J48 and PE for 273,000 training data. We are only able to train the SVM with 33,000 training data, and with the 90,000 training data for PE and EE, because for larger datasets, SVM crashed due to insufficient memory (we used upto 6GB memory for JVM). Therefore, we are unable to report the accuracies for SVM for larger training datasets. Also, note that in all settings, PE has the highest prediction accuracy. The main reason of PE having higher accuracy than BC is because of the noise reduction using profiling-segmentation and ensemble classification. The reason for PE having higher accuracy than EE is because PE assigns proper weight to the classifiers, whereas EE assigns equal weights, without judging the relative importance of the classifiers in the ensemble.

How Early Can We Predict? In this experiment, we answer the question, i.e., how early can we correctly predict the outcome of the game? We answer this in the number of moves, i.e., we say, we predict K move early, meaning, we correctly predict the outcome of the chess match K moves before the game ended. Figure 4(a) shows the summary on the same test dataset, with PE using Naive Bayes and 270,000 training data.

Figure 4(a) reports only the games that are correctly predicted. Note that out of 23,000 games in the test data, about 14,500 were correctly predicted. Each segment of the pie chart shows two values: the number of moves, and a percentage. For example, consider the segment: 1 move, 9%. This means 9% of the correctly predicted games (about 1,300 games) had the correct prediction 1 move before the game ended. Similarly, the segment 4-5 moves, 15% means 15% of the correctly predicted games (about 3,500 games) had the prediction 4 or 5 moves before the game ended. If we examine carefully, we would notice that about one third (31%) of the correctly predicted games (about 7,000 games) obtained the correct prediction at least 16 moves before the game ended, which is really promising. This means that with this data mining technique, we have a good chance to predict correctly long before the game ends. In order to demonstrate the effectiveness of this prediction system, we apply it on the so called "game of the century", played between Donald Byrne (white) and 13-year-old Bobby Fischer (black) in 1956. The probabilities of white winning, draw, and black

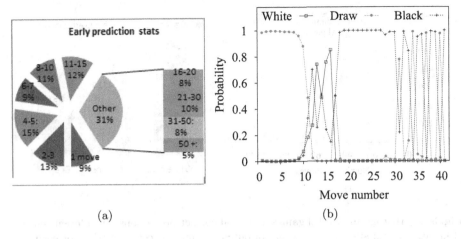

(a)   (b)

Fig. 4. (a) Early prediction stats: what % of games were correctly predicted how many moves early, (b) The move-by-move update of probability for the game of the century (Byrne-vs-Fischer:1956)

winning after each move is shown in the graphs in figure 4(b). At the beginning of the game ($<$ 10 moves), the chances of draw is high. Between 10 and 16, we observe a seesaw between white and black but white seems to have the upper hand. However, note that from move 17 and onward, winning probability of black is raised to near 100%. This reflects the brilliant counter made at the 17th move by Fischer, which is sometimes called the "counter of the century". We see that after this move, black is always on the top and black winning has highest probability (some fluctuations between draw and black is observed between 30-40 moves but the final outcome is correct). Therefore, our system could detect well the turning point in the game and also predict the final outcome about 22 moves earlier than the game ended.

Other Statistics: We also report several statistics about the games and predictions. In figure 5(a), we report the number of games having a particular number of moves and how many of these games are correctly predicted in the test data. The X-axis shows the number of moves. For a particular value of X, say 40, the Y values indicate how many games have exactly 40 moves, which is 550, according to the graph (the higher line); and how many games out of these 550 are correctly predicted by our approach (the lower line), which is 335 in this case. From these two histograms we can come to two conclusions. First, most of the games have 30-60 moves. We observe similar distributions in the training data. The correct predictions are also similarly distributed, i.e, the prediction accuracy does not depend on number of moves in the game.

Figure 5(b) shows the distribution of games having different outcomes (white winning, black winning, and draw) in the test data, and the distributions are similar to the training data. Note that these three outcomes have almost the same probabilities, therefore, the datasets have balanced class distribution. Also,

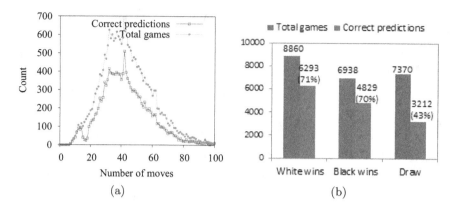

Fig. 5. (a) Histogram of total games and total correct predictions for different number of moves (b) Statistics of three types of outcomes and correct prediction for each

about 70% of the white winning (6293/8860) or black winning (4829/6938) games are correctly predicted. However, this accuracy is lower (43% = 3212/7370) for the games ending with draw. This happens because of the higher uncertainty involved in these games compared to the deciding games.

Table 2. Running time (in seconds) comparison

PE				EE				BC			
NB	J48	RF	SVM	NB	J48	RF	SVM	NB	J48	RF	SVM
23	40	24	8,478	205	204	189	55,107	147	512	137	102,400

Table 2 shows the running times (training + testing, in seconds) of different techniques when trained with 33,000 training data and tested with 23,000 test data. It is evident that PE is the fastest approach. PE is faster than BC because BC uses all training data together to build the classifier, whereas PE divides the data into n approximately equal segments and builds one classifier out of each segment. Therefore, training of PE is faster than that of BC. Also, PE has lower running time than EE because PE uses only one classifier for prediction, whereas EE uses all classifiers and takes the majority.

6   Conclusion

In this paper we proposed an online and dynamic prediction system for early prediction of chess match results. To the best of our knowledge, this is the first approach to use the move-based features, profiling-segmentation based ensemble training for building the prediction models, and dynamic prediction of chess game while the game is in progress. We have applied our technique on a large corpus of real chess matches and obtained higher prediction accuracy than benchmark contemporary techniques. In the future we would like to investigate more on retrieving deeper domain knowledge and experiment with other features, such as sequence of moves and sequence of states, and use relevant sequence or time series classification techniques.

References

1. Bifet, A., Holmes, G., Pfahringer, B., Kirkby, R., Gavald, R.: New ensemble methods for evolving data streams. In: Proc. SIGKDD, pp. 139–148 (2009)
2. ChessOK.com: Chessok. www.chessok.com
3. Fan, Z., Kuang, Y., Lin, X.: Chess game result prediction system. Tech. rep. Stanford University (2013)
4. Ferreira, D.R.: Predicting the outcome of chess games based on historical data. Tech. rep., IST - Technical University of Lisbon (November 2010)
5. Ferreira, D.R.: Determining the strength of chess players based on actual play. ICGA Journal 35(1), 3–19 (2012)
6. Hsu, F.H.: Behind Deep Blue: Building the Computer That Defeated the World Chess Champion. Princeton University Press, Princeton (2002)
7. Ifrim, G., Wiuf, C.: Bounded coordinate-descent for biological sequence classification in high dimensional predictor space. In: KDD 2011, pp. 708–716 (2011)
8. Katakis, I., Tsoumakas, G., Vlahavas, I.: Tracking recurring contexts using ensemble classifiers: an application to email filtering. Knowl. and Info. S. 22, 371–391 (2010)
9. U-of Waikato ML group: Weka. http://www.cs.waikato.ac.nz/ml/weka/

Learning of Performance Measures from Crowd-Sourced Data with Application to Ranking of Investments

Greg Harris[1]([✉]), Anand Panangadan[2], and Viktor K. Prasanna[2]

[1] Department of Computer Science, University of Southern California,
Los Angeles, CA, USA
gfharris@usc.edu
[2] Ming-Hsieh Department of Electrical Engineering,
University of Southern California, Los Angeles, CA, USA
{anandvp,prasanna}@usc.edu

Abstract. Interestingness measures stand as proxy for "real human interest," but their effectiveness is rarely studied empirically due to the difficulty of obtaining ground-truth data. We propose a method based on learning-to-rank algorithms that enables pairwise rankings collected from domain community members to be used to learn a domain-specific measure. We apply this method to study the interestingness measures in finance, specifically, investment performance evaluation measures. More than 100 such measures have been proposed with no way of knowing which most closely matches the preferences of domain users. We use crowd-sourcing to collect gold-standard truth from traders and quantitative analysts in the form of pairwise rankings of equity graphs. With these rankings, we evaluate the accuracy with which each measure predicts the user-preferred equity graph. We then learn a new investment performance measure which has higher test accuracy than the currently proposed measures, in particular the commonly used Sharpe ratio.

1 Introduction

The goal of data mining is to automatically identify "interesting" patterns in a dataset. Data mining algorithms therefore utilize an *interestingness measure*, a function that assigns a numerical score to a given pattern, to evaluate and rank patterns. Several interestingness measures have been proposed, surveyed, and evaluated for different domains [3,13,16,19,20,26,27]. The choice of interestingness measure depends on the specific domain since a pattern can exhibit multiple desirable attributes which must be traded-off against each other.

Designing an interestingness measure for a specific domain is challenging and typically requires a domain expert to create a new function and identify a set of features that can be calculated from the dataset attributes [22]. As an alternate approach, we propose a method to *learn* an interestingness measure

© Springer International Publishing Switzerland 2015
T. Cao et al. (Eds.): PAKDD 2015, Part I, LNAI 9077, pp. 538–549, 2015.
DOI: 10.1007/978-3-319-18038-0_42

from crowd-sourced data collected from end-users in the domain community. In our approach, domain users are presented with pairs of candidate patterns and are asked to rank one over the other. Pairwise ranking is a non-arduous way for domain users to share preference information. It also facilitates the combining of preference information from multiple users. The collected pairwise rankings are then provided as input to a learning-to-rank algorithm to learn a model of user preference which can be used as an interestingness measure. The features in the learning model are previously proposed interestingness measures for the domain. The result is a custom measure that represents "real human interest" [22] in the domain as expressed by its users.

We demonstrate the proposed approach and evaluate its effectiveness in the domain of finance, specifically the task of learning an investment performance measure that reflects the preferences of investment professionals. Investment preference rankings are collected from users of online discussion forums comprised of quantitative analysts and traders. The model features that are used in the learning-to-rank algorithm include currently used investment performance metrics and ratios. The learned model achieves an accuracy of 80% for predicting the domain users' preference, while the highest accuracy of any single existing performance measure is 77%.

We believe that learning such an interestingness measure can benefit this domain since there is a large number of investment choices. For instance, the United States has over 5,000 exchange-traded stocks and over 7,000 mutual fund choices. Our proposed approach can enable individuals to locate investments that match their specific interests. Moreover, the learned interestingness measure can also be used as an objective function for portfolio selection and optimization.

The contributions of this work are as follows:

1. We propose a novel approach based on learning-to-rank algorithms that enables a domain-specific performance measure to be learned from domain community contributions. The method requires only pairwise preferences from domain experts.
2. We evaluate this approach in the domain of investment ranking and show that the learned performance measure has higher accuracy than existing domain-specific measures. We also address issues of data quality that are critical in crowd-sourced datasets.
3. We provide all data collected as part of this study to encourage further research in this area[1].

2   Related Work

Ohsaki et al. [22] experimentally compared interestingness measures against *real human interest* in medical data mining. They generated prognosis-prediction rules from a clinical dataset on hepatitis. They then had a medical expert evaluate rules as *Especially-Interesting*, *Interesting*, *Not-Understandable*, and *Not-Interesting*. Carvalho et al. [5] build on [22] with evaluations on eight datasets.

---

[1] http://thames.usc.edu/rank.zip

They presented nine rules to each expert for each interestingness measure: the best three, the worst three, and three in the middle. Experts were asked to assign a subjective degree of interestingness to each rule. Tan et al. [26] studied ways to select the best interestingness measure for association rules – instead of using actual experts to rank contingency tables, they consider a held-out measure as the expert (and repeat over all measures). None of these works attempt to *learn* an interestingness measure from domain experts as we propose in this work.

To the best of our knowledge, no work has been published on comparing investment performance measure rankings against real human interest. For related work in finance, we summarize publications that describe the relative performance of different evaluation measures in this domain. Justification for these proposed measures is axiomatic, based on the properties of the measures [1,17]. Farinelli et al. [11] compare eleven performance ratios. Their work includes a limited empirical simulation, evaluating how well each ratio performed forecasting five stock indexes. They find that asymmetrical performance ratios work better and recommend that more than a single performance ratio be used. Cogneau and Hübner [7] survey over 100 investment performance measures. They provide a taxonomy and classification of measures based on their objectives, properties, and degree of generalization. Bacon [2] also provides a thorough survey of measures grouped into categories.

Some of the current research indicates that different performance metrics produce substantially the same rank orders. Hahn et al. [14] used 10 performance measures to rank data from two proprietary trading books and found high values of Spearman's rank correlation. Eling and Schuhmacher [10] find high rank correlation (0.96) between 13 performance measures that were used to rank the returns of 2,763 hedge funds. Eling [9] confirmed the high rank correlation between measures when applied to 38,954 mutual funds from 7 asset classes. On the other hand, Zakamouline [28] describe several less correlated measures and suggest the use of Kendall's tau instead of Spearman's rho for measuring rank correlation. None of these four studies considered the Pain, Ulcer, and Martin-related measures discussed in Section 3.3.

3   Finance Background

Investment performance measures are designed to weigh the risk as well as the reward, and are therefore called "risk-adjusted returns." Metrics are structured as ratios, with return on investment in the numerator, and risk in the denominator. In this way, a single metric can compare two investment options with different risk profiles.

While return on investment is a standard measure of reward, there are multiple measures of risk and hence consensus has not yet been reached as to which performance measure is best [11]. New performance metrics continue to be proposed [7,21], and investors have to choose from among them [2].

We first describe equity graphs which provide a visualization of asset performance, followed by a summary of performance measures that will be used as features in our learning model.

3.1   Equity Graphs

Historical performance is often presented as an *equity graph*, which shows the value of one's investment account over time. Equity graphs enable domain experts to rapidly evaluate historical performance. While there are different types of equity graphs, in our work we use the common variant where the graph presents a cumulative sum of daily returns. This is equivalent to assuming exactly one dollar was invested each day, with profits removed from the account. Such a graph is easy to examine, since the ideal is a straight line from the lower left corner to the upper right corner. Examples are shown in Figures 1, 2, and 3.

3.2   Distribution-Based Measures

Many performance measures calculate risk based on the distribution of *returns*. For a time series R, the return on investment for each period, R_t is:

$$R_t \equiv \frac{S_t - S_{t-1}}{S_{t-1}}$$

where S_t is the asset value at time t.

The baseline investment performance measure is the reward to variability ratio, the *Sharpe ratio* [23]. The Sharpe ratio is widely used [9], with surveys showing its use by up to 93% of money managers [2]. This performance measure is "optimal" if the return distribution is normal. The Sharpe ratio is closely related to the t-statistic for measuring the statistical significance of the mean differential return [24].

Using the same notation as Sharpe [24], let R_{Ft} be the return of the investment in period t, R_{Bt} the return of the benchmark security (commonly the risk-free interest rate) in period t, and D_t the differential return in period t:

$$D_t \equiv R_{Ft} - R_{Bt}$$

Let $\bar{D}$ be the average value of D_t from period $t = 1$ through T:

$$\bar{D} \equiv \frac{1}{T} \sum_{t=1}^{T} D_t$$

and σ_D be the standard deviation over the period:

$$\sigma_D \equiv \sqrt{\frac{\sum_{t=1}^{T} (D_t - \bar{D})^2}{T - 1}}$$

The Sharpe Ratio (S_h) is:

$$S_h \equiv \frac{\bar{D}}{\sigma_D}$$

Many performance evaluation measures are modifications of the Sharpe ratio. Given that asset returns are often non-normal, researchers have developed measures that incorporate higher moments of the distribution [17]. The Sortino

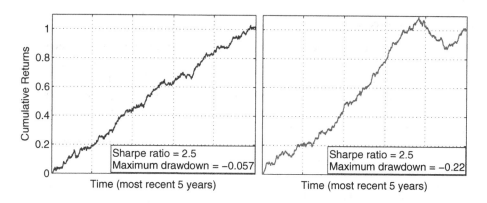

Fig. 1. The red chart on the right was generated by randomly permuting the daily returns from the blue chart on the left. Both have the same distribution of daily returns, and hence the same daily Sharpe ratio. This figure illustrates how distribution-based performance measures cannot capture some features preferred by traders, such as a small maximum drawdown.

ratio [25] is similar to the Sharpe ratio, except it uses the semi-standard deviation (downside risk) in the denominator. Other measures consider only the very worst returns in the tail of the return distribution [1,8].

3.3   Multi-Period-Based Measures

Shape-based measures focus on multi-period drawdowns instead of return distributions. The *Maximum Drawdown* is defined as the maximum peak-to-valley decline in the equity graph. Figure 1 shows how two orderings of returns can have very different maximum drawdowns while still having the same daily Sharpe ratio. The chart on the right has an unappealing drawdown of 22%, yet it has the exact same distribution of returns as the chart on the left (with a drawdown of only 6%).

Drawdown can also be defined as a string of consecutive negative returns. Many performance measures consider aspects of the distribution of such drawdowns instead of returns, including the mean, standard deviation, and selected number of worst drawdowns.

The *Martin ratio*, or "Ulcer performance index" has the same numerator as the Sharpe ratio, but has the *Ulcer index* as the denominator. Using the notation in Bacon [2], let D_i' be the drawdown since the previous peak in period i. The Ulcer index is then defined as:

$$\text{Ulcer index } UI = \sqrt{\sum_{i=1}^{n} \frac{D_i'^2}{n}}$$

Figure 2 shows an equity graph with each D_i' shown in black. The Ulcer index penalizes long drawdowns.

The *Pain ratio* also has the same numerator as the Sharpe ratio. The denominator is the *Pain index*, a modified form of the Ulcer index:

$$\text{Pain index } PI = \sum_{i=1}^{n} \frac{|D'_i|}{n}$$

The Pain index also penalizes long drawdowns but does not penalize deep drawdowns as severely as the Ulcer index.

Max Days Since First at This Level is an intuitive measure that we define as the longest horizontal line that can be drawn between two points on the graph, as shown in Figure 3. We introduce it here because it is not found in the literature, and we find it ranks highly in our experiments.

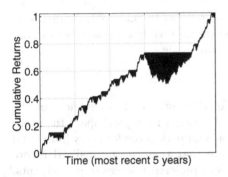

Fig. 2. The Pain index is the area colored black. The Ulcer index is the root mean squared height of each vertical black line.

Fig. 3. "Max Days Since First at This Level" is the longest horizontal line that can be drawn between two points on the graph

4   Approach

We now describe our approach to learn an investment performance measure with higher rank prediction accuracy than the current performance measures, using crowd-sourced domain user input. The steps of our approach are as follows:

1. Generate equity graphs simulating reasonable investment performance.
2. Collect preference data for the generated equity graphs from domain users in the form of pairwise rankings.
3. Use learning-to-rank algorithms with individual performance measures as features to create a new performance measure.

4.1   Generating Equity Graphs

Our approach uses equity graphs as a means for enabling domain experts to rapidly compare two strategies or investments. We generated (synthetic) equity graphs that follow a log-normal random walk. In this model, the asset price, S_t, follows the stochastic differential equation:

$$dS_t = \mu S_t dt + \sigma S_t dW_t$$

where μ is the constant drift, σ is the constant volatility, and dW_t is a Wiener process.

We generated discrete differential simple returns representing five years with 252 business days per year. The returns are normally distributed with a mean of 0.125 and a standard deviation of 1. These values were chosen to lead to a broad distribution of Sharpe ratios centered around 2. Of these, only graphs with Sharpe ratios between 1.5 and 2.5 are retained. This range corresponds to the range of Sharpe ratios typically encountered. Ratios below 1.5 are unattractive as an investment, and ratios greater than 2.5 are very rare in practice. In total, we generated 2,000 charts.

For each graph, we normalize the set of returns to sum to 1. Normalizing the cumulative return enables domain experts to directly compare risk metrics (such as the maximum drawdown) on the same scale.

4.2   Collection of Ranking Data

One of our innovations is the collection of domain expert preferences in the form of pairwise rankings. We believe that it is easier for a participant to choose between two equity graphs than to decide on a numeric score for every individual graph. In particular, numeric scores require that these be normalized before aggregating scores to account for the different preference scales of participants. This normalization would be difficult for cases where a participant only labeled a small number of charts. In contrast, our pairwise ranking-based method is fast for human users with median ranking time between 3 and 4 seconds.

We created a web page that described our research goal and presented two randomly chosen equity graphs side-by-side. A participant is asked which of these two investments is more attractive to invest in for the future. We requested participation from domain experts in two online forums. The first forum targets quantitative analysts and risk managers. The second forum targets individual traders, although some members run small hedge funds or are commodity trading advisors. 66 different anonymous people from these forums ranked a total of 1,004 chart pairs. We believe that the participation of many professionals is validation of community interest in improving investment performance measurement.

One author also ranked 1,659 equity graph pairs, including a re-ranking of every pair ranked by the community. In order to estimate self-consistency of rankings, the author later re-ranked each of the same 1,659 graph pairs. The estimate of self-consistency is 90%. In all rankings and re-rankings, the equity graph positions (i.e., left or right side) were chosen randomly.

4.3   Data Quality

Ensuring quality of crowd-sourced data is a recognized problem [18]. As expected, we found that some of the crowd-sourced data was of low quality. In this section, we describe the steps performed to derive a higher quality data subset from the crowd-sourced annotations.

One author tagged each of the pairs of equity graphs used for crowd-sourced ranking as either "close call" (81%) or "clear choice" (19%). A "clear choice" tag indicates that the author's preference was strong and this view was likely to reflect universal preferences. The author was 100% self-consistent when re-ranking "clear choice" equity graph pairs.

To identify low quality contributions, we evaluated each contribution according to the following characteristics:

- Small median time between clicks
- A high fraction of times the participant clicked the same button (i.e., left or right), rather than alternating approximately uniformly between the two
- A systematic preference for the chart with the lower Sharpe ratio
- A relatively high fraction of rankings that contradict the author's "clear choice" rankings

Overall, we filtered out 129 rankings, leaving 875 of the original 1,004. As such a data quality filter is subjective, we also ran all experiments on the unfiltered dataset in addition to making the data publicly available.

4.4   Learning-to-Rank

A *learning-to-rank* algorithm predicts the order of two objects given training data consisting of partial orders of objects (and their features). We use the learning-to-rank algorithm proposed by Herbrich et al. [15]. In this method, the ranking task is transformed into a supervised binary classification task by considering the *difference* between corresponding features. This transformation also enables the use of other learning algorithms in addition to support vector machines as originally proposed by Herbrich et al. [15].

The three classification algorithms we use in this work are:

1. Logistic regression, with L_1-norm regularization [12]
2. Random forests [4]
3. SVM with linear and RBF kernels [6]

Given two objects, A and B, the learning-to-rank task is to predict if $A > B$ based on their respective features. It is redundant to include both $A > B$ and $B > A$ (with negated feature differences) when training a model. In order to ensure balanced numbers of classes for the model to learn, we chose one of either $A > B$ or $B > A$ for each instance such that there were equal numbers of positive and negative instances in the training data. Balancing the training data also ensures that the intercept or bias term will be zero for logistic regression.

Features. The features we use as inputs to the machine learning models include relevant risk and performance metrics found in Bacon's comprehensive survey [2] which also provides descriptions of each measure using uniform notation. Note

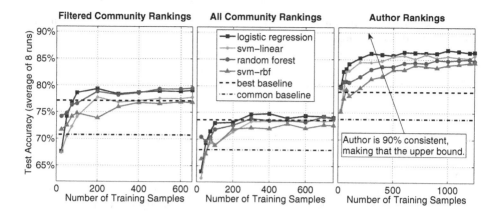

Fig. 4. Accuracy of learning-to-rank models trained and tested on crowd-sourced "real human interest" data in the form of pairwise rankings

that for our normalized charts, risk metrics produce identical rank orderings as their respective performance measures. We nevertheless include both, because models such as logistic regression use linear combinations of features, and we do not know *a priori* which feature will combine best with other features.

5   Experiments and Results

In our experiments, we consider the following three datasets:

1. The full set of all 1,004 community rankings (ACR)
2. The filtered set of 875 community rankings (FCR)
3. The set of 1,659 author rankings (AR)

Each experiment followed these steps for evaluation:

1. Randomly shuffle the data
2. Separate 25% of the data for testing
3. Choose optimal hyper-parameters using 5-fold cross-validation on the training data
4. Test the accuracy of the final model on the held-out test data

We performed each experiment 8 times and averaged the test accuracies. All models were trained and tested on the same random shuffle of the data to better compare their accuracies.

In order to estimate the impact of the number of pairwise rankings needed for training on the accuracy of the learned performance measure, we tested progressively increasing amounts of training data. The data was not reshuffled as training instances were added, i.e., for $n = 200$, the first 100 data points are the same ones used for $n = 100$. Figure 4 shows accuracies obtained for each of the three datasets, using each of the models, trained with an increasing number of pairwise ranking samples. Each point on the graphs represents the average of 8 runs. For reference, we show the most commonly used performance measure as a baseline, the monthly Sharpe ratio. In addition, we also show the performance of the *ex post facto* best measure for each dataset, although in practice which measure would perform the best on a given dataset would not be known.

From these experiments, we observed that none of the established performance measures in this domain is able to fully predict domain expert preferences. Our performance measure trained from domain expert preferences is able to achieve better prediction accuracy. For the filtered community ranking dataset, the random forests approach narrowly outperformed logistic regression, with 80% accuracy. The best baseline for this dataset is the monthly Pain index, with 77% accuracy. For the dataset containing all community rankings, logistic regression has the best performance, with 74% accuracy. The best baseline for this dataset is the daily Pain index, with 74% accuracy. For the dataset containing author rankings, logistic regression again has the best performance, with 86% accuracy. Note that for this dataset, the same author performed each pairwise ranking twice. As these two sets of rankings have an agreement rate of 90%, this forms an upper bound for any model's predictive accuracy. The best baseline for this dataset is the daily Martin ratio, with 79% accuracy.

Learning-to-rank accuracies are lower for the community datasets than the author dataset. This is because community members have idiosyncratic preferences, contributing inconsistency to the community training and test data.

The learning curves in Figure 4 are relatively flat. This indicates that ranking more equity graph pairs would not lead to higher accuracies, given the models and features we have chosen. A small number of rankings (approximately 300) is adequate to learn a trader's preferences. Given median ranking times between 3 and 4 seconds, a trader would likely spend 15 to 20 minutes ranking 300 chart pairs.

6 Conclusion

We presented a novel method using crowd-sourcing to learn a domain-specific performance measure. This method uses pairwise learning-to-rank algorithms with previously proposed performance measures as input features. We demonstrated and evaluated this approach for the case of learning a performance measure to rank investments. Our experimental results showed that machine learning algorithms can find linear combinations of performance measures that improve accuracy in this domain.

We provide all data[2] (equity graphs, measure calculations, and rankings) to encourage further study. With the data, we also include a table unable to fit in this paper, showing the accuracy of the individual baseline performance measures on each dataset.

---

[2] http://thames.usc.edu/rank.zip

Acknowledgments. We acknowledge the help of 66 different anonymous people from two online forums who provided the training data.

This work is supported by Chevron USA, Inc. under the joint project Center for Interactive Smart Oilfield Technologies (CiSoft), at the University of Southern California.

References

1. Alexander, G.J., Baptista, A.M.: Portfolio performance evaluation using value at risk. The Journal of Portfolio Management **29**(4), 93–102 (2003)
2. Bacon, C.R.: Practical Risk-adjusted Performance Measurement. John Wiley & Sons (2012)
3. Blanchard, J., Guillet, F., Gras, R., Briand, H.: Using information-theoretic measures to assess association rule interestingness. In: Fifth IEEE International Conference on Data Mining, p. 8. IEEE (2005)
4. Breiman, L.: Random forests. Machine Learning **45**(1), 5–32 (2001)
5. Carvalho, D.R., Freitas, A.A., Ebecken, N.F.F.: Evaluating the Correlation Between Objective Rule Interestingness Measures and Real Human Interest. In: Jorge, A.M., Torgo, L., Brazdil, P.B., Camacho, R., Gama, J. (eds.) PKDD 2005. LNCS (LNAI), vol. 3721, pp. 453–461. Springer, Heidelberg (2005)
6. Chang, C.-C., Lin, C.-J.: LIBSVM: a library for support vector machines. ACM Transactions on Intelligent Systems and Technology (TIST) **2**(3), 27 (2011)
7. Cogneau, P., Hübner, G.: The (more than) 100 ways to measure portfolio performance. part 1: standardized risk-adjusted measures. Journal of Performance Measurement 13 (Summer 2009)
8. Dowd, K.: Beyond Value at Risk: The New Science of Risk Management, vol. 3. Wiley, Chichester (1998)
9. Eling, M.: Does the measure matter in the mutual fund industry? Financial Analysts Journal, 54–66 (2008)
10. Eling, M., Schuhmacher, F.: Does the choice of performance measure influence the evaluation of hedge funds? Journal of Banking & Finance **31**(9), 2632–2647 (2007)
11. Farinelli, S., Ferreira, M., Rossello, D., Thoeny, M., Tibiletti, L.: Beyond Sharpe ratio: Optimal asset allocation using different performance ratios. Journal of Banking & Finance **32**(10), 2057–2063 (2008)
12. Friedman, J., Hastie, T., Tibshirani, R.: Regularization paths for generalized linear models via coordinate descent. Journal of Statistical Software **33**(1), 1 (2010)
13. Geng, L., Hamilton, H.J.: Interestingness measures for data mining: A survey. ACM Computing Surveys (CSUR) 38(3), 9 (2006)
14. Hahn, C., Peter Wagner, F., Pfingsten, A.: An empirical investigation of the rank correlation between different risk measures. In: EFA 2002 Berlin Meetings Presented Paper, pp. 02–01 (2002)
15. Herbrich, R., Graepel, T., Obermayer, K.: Large margin rank boundaries for ordinal regression. Advances in Neural Information Processing Systems, pp. 115–132 (1999)
16. Hilderman, R.J., Hamilton, H.J.: Evaluation of Interestingness Measures for Ranking Discovered Knowledge. In: Cheung, D., Williams, G.J., Li, Q. (eds.) PAKDD 2001. LNCS (LNAI), vol. 2035, p. 247. Springer, Heidelberg (2001)
17. Keating, C., Shadwick, W.F.: A universal performance measure. Journal of Performance Measurement **6**(3), 59–84 (2002)

18. Lease, M.: On quality control and machine learning in crowdsourcing. In: Human Computation (2011)
19. Lenca, P., Meyer, P., Vaillant, B., Lallich, S.: On selecting interestingness measures for association rules: User oriented description and multiple criteria decision aid. European Journal of Operational Research **184**(2), 610–626 (2008)
20. McGarry, K.: A survey of interestingness measures for knowledge discovery. Knowledge Eng. Review **20**(1), 39–61 (2005)
21. Mistry, J., Shah, J.: Dealing with the limitations of the Sharpe ratio for portfolio evaluation. Journal of Commerce and Accounting Research **2**(3), 10–18 (2013)
22. Ohsaki, M., Kitaguchi, S., Okamoto, K., Yokoi, H., Yamaguchi, T.: Evaluation of Rule Interestingness Measures with a Clinical Dataset on Hepatitis. In: Boulicaut, J.-F., Esposito, F., Giannotti, F., Pedreschi, D. (eds.) PKDD 2004. LNCS (LNAI), vol. 3202, pp. 362–373. Springer, Heidelberg (2004)
23. Sharpe, W.F.: Mutual fund performance. Journal of Business, 119–138 (1966)
24. Sharpe, W.F.: The Sharpe ratio. Journal of Portfolio Management **21**, 49–58 (1994)
25. Sortino, F.A., Van Der Meer, R.: Downside risk. The. Journal of Portfolio Management **17**(4), 27–31 (1991)
26. Tan, P.-N., Kumar, V., Srivastava, J.: Selecting the right interestingness measure for association patterns. In: Proceedings of the Eighth ACM SIGKDD International Conference on Knowledge Discovery and Data Mining, pp. 32–41. ACM (2002)
27. Vaillant, B., Lenca, P., Lallich, S.: A Clustering of Interestingness Measures. In: Suzuki, E., Arikawa, S. (eds.) DS 2004. LNCS (LNAI), vol. 3245, pp. 290–297. Springer, Heidelberg (2004)
28. Zakamouline, V.: The choice of performance measure does influence the evaluation of hedge funds. Available at SSRN **1403246** (2010)

Hierarchical Dirichlet Process for Tracking Complex Topical Structure Evolution and Its Application to Autism Research Literature

Adham Beykikhoshk$^{(\boxtimes)}$, Ognjen Arandjelović,
Svetha Venkatesh, and Dinh Phung

Pattern Recognition and Data Analytics Centre, Deakin University,
Geelong, Australia
{abeyki,ognjen.arandjelovic,svetha.venkatesh,dinh.phung}@deakin.edu.au

Abstract. In this paper we describe a novel framework for the discovery of the topical content of a data corpus, and the tracking of its complex structural changes across the temporal dimension. In contrast to previous work our model does not impose a prior on the rate at which documents are added to the corpus nor does it adopt the Markovian assumption which overly restricts the type of changes that the model can capture. Our key technical contribution is a framework based on (i) discretization of time into epochs, (ii) epoch-wise topic discovery using a hierarchical Dirichlet process-based model, and (iii) a temporal similarity graph which allows for the modelling of complex topic changes: emergence and disappearance, evolution, splitting and merging. The power of the proposed framework is demonstrated on the medical literature corpus concerned with the autism spectrum disorder (ASD) – an increasingly important research subject of significant social and healthcare importance. In addition to the collected ASD literature corpus which we made freely available, our contributions also include two free online tools we built as aids to ASD researchers. These can be used for semantically meaningful navigation and searching, as well as knowledge discovery from this large and rapidly growing corpus of literature.

1 Introduction

The Autism Spectrum Disorder (ASD) is a life-long neurodevelopmental disorder with poorly understood causes on the one hand, and a wide range of potential treatments supported by little evidence on the other. The disorder is characterized by severe impairments in social interaction, communication, and in some cases cognitive abilities. Considering the social and economic burden of ASD it is unsurprising that it has been attracting an increasing amount of research attention which has resulted in a rapid growth of the relevant corpus of literature. Navigating this vast amount of data by conventional, manual means is difficult and limiting. Consequently, the potential benefit of tools based on novel data-mining and machine learning techniques is immense [1]. More meaningful ways for visualising or searching for data could provide invaluable information in

© Springer International Publishing Switzerland 2015
T. Cao et al. (Eds.): PAKDD 2015, Part I, LNAI 9077, pp. 550–562, 2015.
DOI: 10.1007/978-3-319-18038-0_43

clinical and administrative decision making as well as aid research, while automatic knowledge discovery would in its own right advance the understanding of the underlying phenomena (e.g. epidemiological patterns). In the present paper we describe a novel method which contributes towards this goal.

More specifically, we describe a general framework for the analysis of medical literature capable of (i) discovering the underlying topical structure, (ii) inferring the relationships between different discovered topics, and (iii) tracking the evolution of topics over time. The proposed framework uses hierarchical Dirichlet process (HDP) to extract topics automatically, and then constructs a similarity graph over them using an inter-topic similarity measure; topic evolution over time can be inferred from this graph. The effectiveness of our approach is demonstrated on the specific example of a large longitudinal data corpus of medical literature on ASD which we collected. This corpus includes more than 18,000 articles published over the course of 42 years. Another contribution is this corpus which is made publicly available.

The results we report on the collected ASD literature corpus illustrate the usefulness of our method and its ability to extract and track over time abstract topical knowledge, inferring the point at which a certain topic comes into existence, how its evolves, splits into multiple new topics or merges with the existing ones, and lastly when it ceases to exist. This is demonstrated on examples of well-known research directions in the field. Our additional contributions come in the form of two free online tools which allow researchers to (i) navigate and search the literature in a semantically meaningful manner (see www.undersdtanfigutism.tk), and (ii) understand the development and relationships between different ideas which permeate research in the domain of ASD (see http://goo.gl/Ws7V64).

2 Previous Work

In this section we review the most relevant previous work on topic modelling. We focus our attention first on latent topic models which have dominated the field in the last decade, and then on biomedical text mining, given the application domain in which our framework in evaluated in Section 4.

2.1 Latent Topic Models

An important early approach is the latent semantic indexing (LSI) [2] which remains popular. Two notable limitations of LSI are its inability to deal effectively with polysemy and to produce an explicit description of the latent space. A probabilistic improvement of LSI [3] overcomes these by explicitly characterizing the latent space with semantic topics, and by employing a probabilistic generative model that addresses the polysemy problem. Nevertheless, probabilistic LSI is prone to parameter overfitting caused by an uncontrolled growth in the number of parameters as the document corpus is increased. In addition, the necessary assignment of probabilities to documents is a nontrivial task [4].

The recently proposed latent Dirichlet allocation (LDA) method [4] overcomes the overfitting problem by adopting a Bayesian framework and a generative process at document level. While LDA has quickly become a standard tool for topic modelling, it too experiences challenges when applied on real-world data. In particular, being a parametric model the number of desired output topics has to be specified in advance. The HDP model as the nonparametric counterpart of LDA was introduced by Teh *et al.* [5] and addressed this limitation by using a Dirichlet process (DP) (as opposed to a Dirichlet distribution) as the prior on topics. Therefore, each document is modelled using an infinite mixture model, allowing the data to inform the complexity of the model and infer the number of resulting topics automatically. We discuss this model in further detail in Section 3.

Temporal Topic Modelling: A notable limitation of most models described in the previous section lies in their assumption that the data corpus is static. However, in many practical applications documents are added to the corpus in a temporal manner. Therefore their ordering has significance and at best they might be exchangeable in short time slices. As a consequence, the topical structure of the corpus changes over time. Existing work can be divided into two groups.

First, the models that hold a Markov assumption over time by discretizing and dividing it into multiple *epochs*. Then a topic model is fit to each epoch where the parameters of adjacent models are tied together [6–9]. Whilst they capture how the comprising words of a topic evolve over time, they assume the data arrives in a uniform fashion whereas in our application documents may arrive at irregular time intervals. Indeed we adopt the time desensitization from this group. However our approach diverges from those in the current literature thereafter. We do not consider the Markov assumption to obtain a model with less complexity and easier inference. Second, the models that treat the document time-stamps as an observed continuous random variable [10,11]. These models are capable of modelling the life span of a topic, but not the capturing its evolution and trajectory (i.e. split and merge). The topic model used in both groups can be parametric [6,7,10] or nonparametric [8,9]. Parametric models will still suffer from the same problem as LDA in requiring the number of topics to be specified in advance.

2.2 Biomedical Text Mining

The idea that the medical literature could be mined for new knowledge is typically attributed to Swanson [12]. For example by manually examining medical literature databases he hypothesised that dietary fish oil could be beneficial for Raynaud's syndrome patients, which was later confirmed by experimental evidence. Work that followed sought to develop statistical methods which would make this process automatic. Previous work on biomedical text mining has rather focused on (i) the tagging of names of entities such as genes, proteins, and diseases [13], (ii) the discovery of relationships between different entities e.g. functional associations between genes [14], or (iii) the extraction of information pertaining to events such as gene expression or protein binding [15].

Most existing work on biomedical knowledge discovery is based on what may be described as traditional data mining techniques (neural networks, support vector machines etc); comprehensive surveys can be found in [15,16]. The application of state-of-the-art Bayesian methods in this domain is scarce. Amongst the notable exceptions is the work by Blei *et al.* who showed how latent Dirichlet allocation (LDA) can be used to facilitate the process of hypothesis generation in the context of genetics [17]. Arnold *et al.* used a similar approach to demonstrate that abstract topic space representation is effective in patient-specific case retrieval [18]. In their later work they introduced a temporal model which learns topic trends and showed that the inferred topics and their temporal patterns correlate with valid clinical events and their sequences [19]. Wu *et al.* used LDA for gene-drug relationship ranking [20].

3   Proposed Framework

We begin this section by reviewing the relevant theory underlying HDP mixture modelling which plays the central rule in the proposed framework. Then we turn our attention to the main technical contribution of our work and explain how the HDP is employed to discover the topical content of a literature corpus and track its structural changes over time.

3.1   Hierarchical Dirichlet Process Mixture Models

Dirichlet process as the building block of Bayesian non-parametric methods allows the document collection to accommodate potentially infinite number of topics. A Dirichlet process [21] $\mathrm{DP}(\gamma, H)$ is defined as a distribution of a random probability measure G over a measure space $(\Theta, \mathcal{B}, \mu)$, such that for any finite measurable partition $(A_1, A_2, \ldots, A_r)$ of Θ the random vector $(G(A_1), \ldots, G(A_r))$ is a Dirichlet distribution with parameters $(\gamma H(A_1), \ldots, \gamma H(A_r))$. An alternative view of the DP emerges from the so-called stick-breaking process which adopts a constructive approach using a sequence of discrete draws [22]. Specifically, if $G \sim \mathrm{DP}(\gamma, H)$ then $G = \sum_{k=1}^{\infty} \beta_k \delta_{\phi_k}$ where $\phi_k \overset{iid}{\sim} H$ and $\beta = (\beta_k)_{k=1}^{\infty}$ is the vector of weights obtained by the stick-breaking process that is $\beta_k = v_k \prod_{l=1}^{k-1}(1 - v_l)$ and $v_l \overset{iid}{\sim} \mathrm{Beta}(1, \gamma)$.

Owing to the discrete nature and infinite dimensionality of its draws, the DP is a highly useful prior for Bayesian mixture models. By associating different mixture components with atoms ϕ_k of the stick-breaking process, and assuming $x_i | \phi_k \overset{iid}{\sim} F(x_i | \phi_k)$ where $F(.)$ is the likelihood kernel of the mixing components, we can formulate the Dirichlet process mixture model (DPM). The DPM is suitable for nonparametric clustering of exchangeable data in a single group e.g. words in a document where the DPM models the underlying structure of the document with potentially an infinite number of topics. However, many real-world problems are more appropriately modelled as comprising multiple groups of exchangeable data (e.g. a collection of documents). In such cases it is usually desirable to model the observations of different groups jointly, allowing them to share their generative clusters to remain linked. This idea is known as the "sharing statistical strength" and it is naturally obtained by hierarchical architecture in Bayesian modelling.

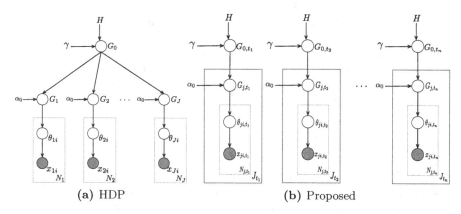

(a) HDP   **(b) Proposed**

Fig. 1. (a) Graphical model representation of HDP. Each box represents one document whose observed data (words) is shown shaded. Unshaded nodes represent latent variables. An observed datum x_{ji} is assigned to a latent mixture component parameterized by θ_{ji}. γ and α are the concentration parameters and H is the corpus-level base measure. (b) Graphical model representation of the proposed framework. The corpus is temporally divided into t_n epochs and each epoch modelled using an HDP (outer boxes). Different epochs' HDPs share their corpus-level DP and hyperparameters.

Amongst different ways of linking group-level DPMs, HDP [5] offers an interesting solution whereby base measures of group-level DPs are drawn from a corpus-level DP. In this way the atoms of the corpus-level DP (i.e. topics in our case) are shared across the documents. Formally, if $\mathbf{x} = \{\mathbf{x}_1, \ldots, \mathbf{x}_J\}$ is a document collection where $\mathbf{x}_j = \{x_{j1}, \ldots, x_{jN_j}\}$ is the j-th document comprising N_j words, each document is modelled with a DPM $G_j | \alpha_0, G_0 \overset{iid}{\sim} \mathrm{DP}\,(\alpha_0, G_0)$ where its DP prior is further endowed by another DP $G_0 | \gamma, H \sim \mathrm{DP}\,(\gamma, H)$. This is illustrated schematically in Figure 1a. Since the base measure of G_j is drawn from G_0, it takes the same support as G_0. Also the parameters of the group-level mixture components, θ_{ji}, share their values with the corpus-level DP support on $\{\phi_1, \phi_2, \ldots\}$. Therefore G_j can be equivalently expressed using the stick-breaking process as $G_j = \sum_{k=1}^{\infty} \pi_{jk} \delta_{\phi_k}$ where $\boldsymbol{\pi}_j | \alpha_0, \gamma \sim \mathrm{DP}\,(\alpha_0, \gamma)$[5]. The posterior for θ_{ji} has been shown to follow a Chinese restaurant franchise process which can be used to develop inference algorithms based on Gibbs sampling [5].

3.2 Modelling Topic Evolution Over Time

In this section we show how the described HDP-based model can be applied to the analysis of temporal topic changes in a longitudinal data corpus. We begin by dividing the literature corpus by time into multiple *epochs*. Each epoch is then modelled separately using an HDP. Different epochs' models share their hyperparameters and the corpus-level base measure. Hence if n is the number of epochs, we obtain n sets of topics $\boldsymbol{\theta} = \{\boldsymbol{\theta}_{t_1}, \ldots, \boldsymbol{\theta}_{t_n}\}$ where $\boldsymbol{\theta}_t = \{\theta_{1,t}, \ldots, \theta_{K_t,t}\}$ is the set of topics that describe epoch t, and K_t their number (which is inferred automatically, as described previously). This is illustrated in Figure 1b. In the next section we describe how given an inter-topic similarity measure the evolution of different topics across epochs can be tracked.

3.3 Measuring Topics Similarity

Our goal now is to track changes in the topical structure of a data corpus over time. The simplest changes of interest include the emergence of new topics, and the disappearance of others. More subtly, we are also interested in how a specific topic changes – how it evolves over time in terms of the contributions of different words it comprises, as well as how it splits into new topics or merges with the existing ones. Clearly this information can provide valuable insight into the refinement of ideas and findings in the scientific community, effected by new research and accumulating evidence.

The key idea behind our approach stems from the observation that while topics may change significantly over time, by their very nature their change between successive epochs is limited. Therefore we infer the continuity of a topic in one epoch by relating it to all topics in the immediately subsequent epoch which are sufficiently similar to it under some similarity measure. This can be seen to lead naturally to a similarity graph representation whose nodes correspond to topics and whose edges link those topics in two epochs which are related. Formally, the weight of the directed edge that links $\phi_{j,t}$, the j-th topic in epoch t, and $\phi_{k,t+1}$ is set equal to $\rho(\phi_{j,t}, \phi_{k,t+1})$ where ρ is an appropriate similarity measure. Given that in our HDP-based model each topic is represented by a probability distribution, suitable similarity metrics include the Jaccard similarity, the Jenson-Shannon divergence, and the L_2-norm.

A conceptual illustration of a similarity graph is shown in Figure 2a. It shows three consecutive time epochs $t-1, t$, and $t+1$ and a selection of topics in these epochs. Graph edge weight i.e. inter-topic similarity is encoded by varying the thickness of the corresponding line connecting two nodes – a thicker line signifies more similar topics. We use a threshold to eliminate automatically weak edges, retaining only the edges which correspond to sufficiently similar topics in adjacent epochs. It can be seen that this readily allows us to detect the disappearance of a particular topic, the emergence of new topics, as well as the splitting or merging of different topics:

Emergence If a node does not have any edges incident to it, the corresponding topic is taken as having emerged in the associated epoch (e.g. ϕ_{j+2} at time t in Figure 2a).

Disappearance If no edges originate from a node, the corresponding topic is taken to vanish in the associated epoch (e.g. ϕ_j at time t in Figure 2a).

Splitting If more than a single edge originates from a node, the corresponding topic is understood as being split into multiple topics in the next epoch (e.g. ϕ_i is split into ϕ_j and ϕ_{j+1} in Figure 2a).

Merging If more than a single edge is incident to a node, the topics of the nodes from which the edges originate are understood as having merged together to form a new topic (e.g. ϕ_i and ϕ_{i+1} merge to form ϕ_{j+1} in Figure 2a).

4 Experimental Evaluation

Having introduced the main technical contribution of our work we now illustrate its usefulness on the example of ASD literature analysis, and describe additional contributions in the form of two free online tools that we developed to aid ASD researchers.

4.1 Data Collection

fe To the best of our knowledge there are no publicly available corpora of ASD-related medical literature. Hence we collected a comprehensive dataset ourselves that we describe its collection methodology and the pre-processing of data we performed to extract standard features used for text analysis.

Raw Data Collection: We used the PubMed search engine that allows users to access the United States National Library of Medicine for abstracts and references of life science and biomedical scholarly articles. We assumed a paper is related to ASD if the term "autism" is present in its title or abstract, and collected only papers written in English. The earliest publication fitting our criteria is that by Kanner [23], and we collected all matching publications up to the final one indexed by PubMed on 24th July 2014, yielding a corpus of 20,138 publications. We discarded the 1,946 which do not have an abstract indexed, ending with the total of 18,192 papers in our dataset. We used the abstracts text to evaluate our method.

Data Pre-processing: Following the standard practice in text processing literature we applied soft lemmatization on the abstracts in our dataset, using the freely available WordNet tool [24]. No stemming was performed to avoid potential distortion of words which is sometimes effected by heuristic rules used by stemming algorithms. After lemmatization and the removal of so-called stop words, we obtained 1.9 million terms in the entire corpus when repetitions are counted, and 37,278 unique terms. We construct the vocabulary for our method by selecting the subset of the most frequent unique terms which explain 90% of the energy of the corpus, which resulted in a 3,738 term vocabulary.

4.2   Proposed Method Implementation

We divided the 42 year timespan of our data corpus into overlapping five year epochs, with a two year lag between consecutive epochs, resulting in 18 epochs in total. The topics of each epoch were then extracted as described in Section 3.2 and their dynamics inferred as per Section 3.3. The number of latent topics of different epoch is plotted in Figure 2b. Notice the exponential rise in the number of topics which mirrors the exponential increase in the number of publications over time in our dataset. This increasing interest in ASD can be illustrated by the observation that in 2013 there are five times as many publications as in 2000. For our inter-topic similarity described in Section 3.3 we adopted the use of the well-known Jaccard similarity; this similarity measure was used to obtain all results reported in this section. Lastly, Gibbs sampling was used for HDP inference, implemented in Python 2.7, with hyperparameter resampling as described by Teh *et al.* [5].

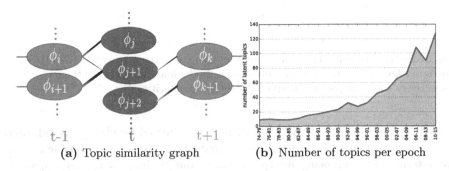

(a) Topic similarity graph   (b) Number of topics per epoch

Fig. 2. (a) Conceptual illustration of the proposed similarity graph that models topic dynamics over time. A node corresponds to a topic in a specific epoch; edge weights are equal to the corresponding topic similarities. (b) As the document corpus grows so does the number of topics needed to model its latent structure.

4.3   Case Study 1: ASD and Genetics

While the exact aetiology of the ASD is still poorly understood, the existence of a significant genetic component is beyond doubt [25]. Work on understanding complex genetic factors affecting the development of autism, which possibly involve multiple genes which interact with each other and the environment, is a major theme of research and as such a good case study on which the usefulness of the proposed method can be illustrated.

We started by identifying the topic of interest as that with the highest probability of the terms "gene" or "genetic" conditioned on the topic, and tracing it back in time to the epoch in which it originated. This led to the discovery of the relevant topic in the epoch spanning the period 1986–1991. Figure 4 shows the evolution of this topic from 1992 revealed by our method (due to space constraints only the most significant parts of the similarity graph are shown;

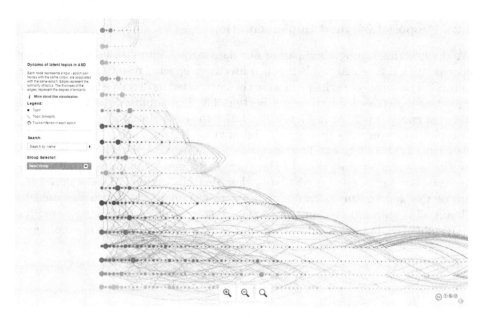

Fig. 3. Interactive similarity graph analysis tool (see http://goo.gl/Ws7V64). Word clouds of a few topics are shown for illustration. Nodes and links between them represent respectively topics in particular epochs and their similarities.

minor changes to the topic before 1992 are also omitted for clarity, as indicated by the dotted line in the figure). Each topic is labelled with its first few dominant terms. The following interpretation of our findings is readily apparent. Firstly, in the period 1992–1997, the topic is rather general in nature. Over time it evolves and splits into topics which concern more specific concepts (recall that such splitting of topics cannot be captured by any of the existing methods). For example by the epoch 2002–2007 the single original topic has evolved and split into four topics which concern:

- the relationship between mutations in the gene mecp2 (essential for normal functioning of neurone), and mental disorders and epilepsy (it is estimated that one third of ASD individuals also have epilepsy),

- gene alternations, for example the duplication of 15q11--13 and deletion of 16p11.2 both of which are associated with ASD,

- genetic linkage association analysis and heritability of autism, and

- observational work on autistic twins and probands with siblings on the spectrum.

Our framework also allows us to look 'back' in time. For example, by examining the topics that the 1992 genetics topic originate from we discovered that the topic evolved from the early concept of "infantile ASD" (originated by Kanner [23]).

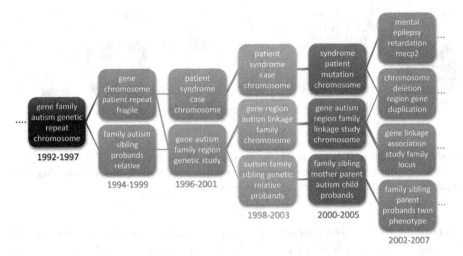

Fig. 4. Dynamics of the topic most closely associated with the concept of "genetics". A few dominant words are shown for each topic (shaded boxes).

4.4   Case Study 2: ASD and Vaccination

For our second case study we chose to examine research on the relationship between ASD development and vaccination. This subject has attracted much attention both in the research community, as well as in the media and the general public. The controversy was created with the publication of the work by Wakefield [26] which reported epidemiological findings linking MMR vaccination and the development of autism and colitis. Despite the full retraction of the article following the discovery that it was fraudulent, and numerous subsequent studies who failed to show the claimed link, a significant portion of the general public remains concerned with the issue.

As in the previous example, we begun by identifying the topic with the highest probability of the terms "vaccine" and "vaccination" conditioned on the topic, and tracing it back to the epoch in which it first emerged. Again, a single topic was readily identified, in the epoch spanning the period 1996–2001. Notice that this is consistent with the publication date of the first relevant publication by Wakefield [26]. The evolution of the topic is illustrated in Figure 5 in the same way as in the previous section. It can be seen that the original topic concerned the subjects initially brought to attention such as "measles", "vaccine", and "autism". In the subsequent epoch, when the original claim was still thought to have credibility, the topic evolves and splits into numerous others mirroring research directions taken by various researchers. Following this period and the revelations of its fraudulence, the topic assumes mainly single-threaded evolution, at times incorporating various originally separate ideas. For example observe the independent emergence of the term "mercury". Though initially unrelated to it this topic merges with the topic that concerns vaccination which can be explained by the widely publicized thiomersal (vaccine preservative)

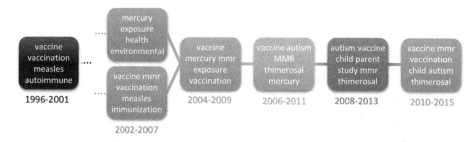

Fig. 5. Dynamics of the topic most closely associated with the concept of "vaccination". Notwithstanding the rejection of any link between vaccination and autism, this topic remains active albeit in a form which evolved over time.

controversy (again note that such merging of topics cannot be captured by the existing methods). Although rejected by the medical community due to a lack of evidence, this topic can be seen as persisting to date.

4.5   Topic Browser

A topic model can be seen as a dimensionality reduction framework that reduces documents into a topic space. This transformation of data can provide powerful insight and allow for the browsing of documents in a more subject-specific, semantic manner. For example by describing documents in the topic space, documents most related to a particular topic of interest can be readily identified and retrieved. To provide this functionality to the research community interested in ASD we used the framework described in this paper to model the entire literature corpus we collected, and built a website to facilitate free and ready use of our model and data. Researchers can use our online tool to browse topics, annotate them, and navigate through publications by topic. The website is available at http://www.understandingautism.tk.

5   Conclusions

We described a novel framework for temporal modelling of the topical structure of a longitudinal document corpus. Our approach consists of discretizing time into overlapping epochs, modelling the static topic structure within each epoch using an HDP, and tracking the evolution of topics over time using an inter-topic similarity measure. The resultant similarity graph captures relationships between topics in different epochs and allows for the automatic inference of the time of emergence and disappearance of topics, their evolution over time, merging and splitting. The power of the proposed general framework was demonstrated on the example of ASD-related medical literature. On two case studies which concern two important research issues in ASD literature we demonstrated that our method extracts meaningful topics and their temporal changes. A novel data corpus and free online tools are made freely available to researchers.

References

1. Beykikhoshk, A., Arandjelovic, O., Phung, D., Venkatesh, S., Caelli, T.: Data-mining twitter and the autism spectrum disorder: A pilot study (2014)
2. Deerwester, S.C., Dumais, S.T., Landauer, T.K., Furnas, G.W., Harshman, R.A.: Indexing by latent semantic analysis. JASIS **41**, 391–407 (1990)
3. Hofmann, T.: Probabilistic latent semantic indexing. SIGIR, 50–57 (1999)
4. Blei, D.M., Ng, A.Y., Jordan, M.I.: Latent Dirichlet allocation. JMLR **3**, 993–1022 (2003)
5. Teh, Y.W., Jordan, M.I., Beal, M.J., Blei, D.M.: Hierarchical Dirichlet processes. Journal of the American Statistical Association 101 (2006)
6. Blei, D.M., Lafferty, J.D.: Dynamic topic models. In: ICML, pp. 113–120 (2006)
7. Wang, C., Blei, D., Heckerman, D.: Continuous time dynamic topic models. In: UAI, pp. 579–586 (2008)
8. Ren, L., Dunson, D.B., Carin, L.: The dynamic hierarchical Dirichlet process. In: ICML, pp. 824–831 (2008)
9. Zhang, J., Song, Y., Zhang, C., Liu, S.: Evolutionary hierarchical Dirichlet processes for multiple correlated time-varying corpora. In: SIGKDD, pp. 1079–1088 (2010)
10. Wang, X., McCallum, A.: Topics over time: a non-Markov continuous-time model of topical trends. In: SIGKDD, pp. 424–433 (2006)
11. Dubey, A., Hefny, A., Williamson, S., Xing, E.P.: A nonparametric mixture model for topic modeling over time. In: SDM, pp. 530–538 (2013)
12. Swanson, D.R.: Undiscovered public knowledge. Library Quarterly **56**, 103–118 (1986)
13. Settles, B.: ABNER: an open Source tool for automatically tagging genes, proteins and other entity names in text. Bioinformatics **21**, 3191–3192 (2005)
14. Rhodes, D.R., Yu, J., Shanker, K., Deshpande, N., Varambally, R., Ghosh, D., Barrette, T., Pander, A., Chinnaiyan, A.M.: A cancer microarray database and integrated data-mining platform. Neoplasia **6**, 1–6 (2004)
15. Simpson, M.S., Demner-Fushman, D.: Biomedical text mining: a survey of recent progress. In: Mining Text Data, pp. 465–517 (2012)
16. Kumar, V.D., Tipney, H.J.: Biomedical Literature Mining. Springer (2014)
17. Blei, D.M., Franks, K., Jordan, M.I., Mian, I.S.: Statistical modeling of biomedical corpora: mining the Caenorhabditis genetic center bibliography for genes related to life span. BMC Bioinformatics **7**, 250 (2006)
18. Arnold, C.W., El-Saden, S.M., Bui, A.A., Taira, R.: Clinical case-based retrieval using latent topic analysis. AMIA **2010**, 26 (2010)
19. Arnold, C.W., Speier, W.: A topic model of clinical reports. SIGIR, pp. 1031–1032 (2012)
20. Wu, Y., Liu, M., Zheng, W., Zhao, Z., Xu, H.: Ranking gene-drug relationships in biomedical literature using latent Dirichlet allocation. In: Pacific Symposium on Biocomputing, pp. 422–433 (2012)
21. Ferguson, T.S.: A Bayesian analysis of some nonparametric problems. The Annals of Statistics, 209–230 (1973)
22. Sethuraman, J.: A constructive definition of Dirichlet priors. Technical report, DTIC Document (1991)
23. Kanner, L.: Irrelevant and metaphorical language in early infantile autism. American Journal of Psychiatry **103**, 242–246 (1946)

24. Miller, G.A., Beckwith, R., Fellbaum, C.D., Gross, D., Miller, K.: WordNet: An online lexical database. Int. J. Lexicograph **1**, 235–244 (1990)
25. Miles, J.H.: Autism spectrum disorders - a genetics review. Nature **13**, 278–294 (2011)
26. Wakefield, A.J., Murch, S.H., Anthony, A.: Ileal-lymphoid-nodular hyperplasia, non-specific colitis, and pervasive developmental disorder in children. The Lancet, 637–641 (1998) (retracted)

Automated Detection for Probable Homologous Foodborne Disease Outbreaks

Xiao Xiao[1], Yong Ge[2], Yunchang Guo[3], Danhuai Guo[1], Yi Shen[1],
Yuanchun Zhou[1], and Jianhui Li[1]($\boxtimes$)

[1] Computer Network Information Center, Chinese Academy of Sciences,
Bejing, China
{xiaoxiao,guodanhuai,shenyi,zyc,lijh}@cnic.cn
[2] University of North Carolina at Charlotte, Charlotte, USA
yong.ge@uncc.edu
[3] Division of Foodborne Disease Surveillance,
China National Center for Food Safety Risk Assessment, Beijing, China
gych@cfsa.net.cn

Abstract. Foodborne disease, a rapid-growing public health problem, has become the highest-priority topic for food safety. The threat of foodborne disease has stimulated interest in enhancing public health surveillance to detect outbreaks rapidly. To advance research on food risk assessment in China, China National Center for Food Safety Risk Assessment (CFSA) sponsored a project to construct an online correlation analysis system for foodborne disease surveillance beginning in October 2012. They collect foodborne disease clinical data from sentinel hospitals across the country. They want to analyze the foodborne disease outbreaks existed in the collected data and finally find the link between pathogen, incriminated food sources and infected persons. Rapid detection of outbreaks is a critical first step for the analysis. The purpose of this paper is to provide approaches that can be applied to an online system to rapidly find local and sporadic foodborne disease outbreaks out of the collected data. Specifically, we employ DBSCAN for local outbreaks detection and solve the parameter self-adaptive problem in DBSCAN. We also propose a new approach named K-CPS (K-Means Clustering with Pattern Similarity) to detect sporadic outbreaks. The experimental results show that our methods are effective for rapidly mining local and sporadic outbreaks from the dataset.

Keywords: Foodborne disease outbreak detection · Clustering · Parameters self-adaptive · Frequent patterns

This work is partly supported by Special Research Funding of National Health and Family Planning Commission of China under grant No.201302005, Natural Science Foundation of China under Grant No. 41371386, 91224006, the Strategic Priority Research Program of the Chinese Academy of Sciences under Grant No. XDA06010307, XDA05050601, 12th Five-Year Plan for Science & Technology Support under Grant No.2013BAD15B02.

T. Cao et al. (Eds.): PAKDD 2015, Part I, LNAI 9077, pp. 563–575, 2015.
DOI: 10.1007/978-3-319-18038-0_44

1   Introduction

The threat of foodborne disease has stimulated interest in public health surveillance [1][2]. Theoretically, for foodborne disease, there is a link between pathogens, incriminated food sources and each infected person. How to find out the link is crucial for foodborne disease surveillance. Analysis of foodborne outbreak data is one approach to find the link and it can estimate the proportion of human cases of specific enteric diseases attributable to a specific food item. [3] employed multiple correspondence analysis(MCA) to further explore the relationship between micro-organism, region and food vehicle. The analysis of foodborne outbreak data is perceived as food attribution and is an important tool in food safety risk analysis [4][5]. To advance research on food risk assessment in China, CFSA sponsored a project to construct an online correlation analysis system for foodborne disease surveillance beginning in October 2012. CFSA collects foodborne disease clinical data from sentinel hospitals. They want to analyze the foodborne disease outbreaks exist in the collected data and finally find out the link between pathogen, incriminated food source and infected persons. A primary purpose of the project is to detect problems in food and water production and delivery systems that might otherwise have gone unnoticed. Rapid detection of outbreaks is a critical first step to abate these active hazards and preventing their further recurrences. But how to rapidly find the outbreaks in the data is a problem for them. The purpose of this paper is to provide approaches that can be applied to the online system to rapidly find the local and sporadic foodborne disease outbreaks out of the collected data.

There are some researches focused on disease outbreak detection. Clearly, when an epidemic sweeps through a region or a foodborne outbreak emerges, there will be extreme perturbations in the number of hospital visits. So some anomaly detection approaches have been used to detect disease outbreaks based on the change of morbidity number. These methods require a baseline number which can be derived from historical data. [6] employs Fishers Exact Test to examine whether a rule occurs today is abnormal or not based on the historical occurrences. [7] develops a simple randomization-based framework to recognize significant increases in event counts. Besides, intense spatial aggregation was often observed in disease outbreaks. [8] presents a fast multi-resolution method to detect significant spatial disease clusters. Given a grid of squares, where each square has a count and an underlying population, the goal of the paper is to find the square region with the highest density, and to calculate its significance by randomization. [9] is the improvement of [8], which uses a novel overlap-kd tree data structure to reduce the time complexity to find the spatial disease clusters. [10] introduces a novel fast spatial scan algorithm, generalizing the 2D scan algorithm of [9] to arbitrary dimension. The work above on cluster detection is purely spatial in nature. But for most disease cluster problems, time is an essential component. Fortunately, there exist methods for the detection of emerging space-time disease clusters. [11] proposes a new class of spatio-temporal cluster detection methods designed for the rapid detection of emerging space-time clusters. It focuses on detecting space-time clusters of disease cases resulting from an emerging disease outbreak.

Although many approaches are proposed for disease outbreak detection, they are unsatisfactory for the problem we want to solve. In our scenario, the data collection started in October, 2012. So the data available is very limited and there are not enough historical data to predict a baseline. In fact, for foodborne disease outbreaks detection, the method employed depends on what data is available. Buckeridge et. give a practical classification for outbreak detection algorithms by considering the types of information encountered in surveillance analysis [12]. In our situation, we have a database of clinical cases from the 615 sentinel hospitals in 34 provinces, municipalities or autonomous regions across the country. Each record in this database contains information about the individual who has seen a doctor.

When many people infect a foodborne disease in a short time in a nearby location, we call that a local foodborne disease outbreak (LFDO), while if the locations are not limited in a small area, we call that a sporadic foodborne disease outbreak (SFDO). In this paper, we employ a density-based algorithm for discovering clusters in large spatial databases with noise (DBSCAN) [13] to detect LFDO and solve the parameter self-adaptive problem in DBSCAN. We propose a new approach to detect SFDO.

The rest of this paper is organized as follows. After a description of the dataset used in this paper in Section 2, we detail the detection approaches in Section 3. Then in the following section, we present our experiments. In Section 5, we give an analysis and a discussion of our experiment results. Finally, in Section 6 we conclude our work and provide an outlook of the future work.

2 Data Collection

Diarrhea is the commonest symptom of foodborne illness. Foodborne diarrhea provides one of the strongest signals for food safety. CFSA started to collect information from diarrheal patients who visit the sentinel hospitals in October 2012. The rules of data collection are: a) the information collected are all the diarrheal cases, but not all the diarrheal cases of sentinel hospitals are collected; b) only the cases with diarrhea 3 or more than 3 times per day and character of stool is abnormal are recorded; c) each sentinel hospital is required to collect at least 10 cases per week. The above data collection strategy is waiting to be improved. Currently, the detection has certain limitations in the way data is recorded. Clearly, under this record strategy, the number of the cases doesn't reflect a true disease occurrence. But the change of the number of cases is a significant signal for an outbreak. Thus, for this dataset, we can't make use of the methods that based on the number of cases to detect outbreaks.

Each record includes the information about the individuals. This information contains fields such as age, gender, career, symptoms exhibited, home location, diseased time and sampling or not (collected anal swab and stool). Parts of these records have incriminated food information. This information includes food name, food band, manufactures, place of purchase, place of eating, time of eating and sampling or not. In this paper, we mainly use home location, diseased time and symptoms exhibited to detect probable local homologous foodborne disease outbreaks. The incriminated food information is mainly used to preliminarily verify whether the outbreak clusters detected are homologous or not. For sporadic outbreaks, we use symptoms exhibited field combined with the food name to detect sporadic outbreaks. Note that the preliminary verification results made by our method are not completely reliable. It just provides a possible clue for researchers who will verify the results by professional analysis of the bacteria, such as Salmonella, Shigella and Sapovirus, examined in the patient samples (such as anal swab and stool), through a molecular typing system.

3 Approaches for Local and Sporadic Outbreaks Detection

An outbreak of foodborne disease was defined as when a group of people consume the same contaminated food and two or more of them come down with the same illness. According to the definition, we give a hypothesis that patients in an outbreak caused by the same contaminated source will exhibit similar or same symptoms. In reality, patients in LFDO are not distributed randomly, and the temporal and spatial clusters are obvious. So we use diseased time, home location and symptoms exhibited as features for LFDO clustering. For sporadic outbreaks, we hope to cluster the cases which have common symptoms and similar food information. So we use symptoms combined with the corresponding food information as features for sporadic outbreaks clustering.

3.1 LFDO Detection

Data Preprocessed. We use cases collected between 1 January 2013 and 16 January 2014 for local outbreak detection (We named it Dataset 1). The raw disease time is a time format and we convert it to a long type. The raw home location is a textual address, and we use the Google Geocoding API to capture the longitude and latitude of each address. Then we use gausskruger projection [14] to convert spherical coordinates to plane coordinates. Longitude and latitude correspond to the x and y axes, respectively. We separate the raw symptom text into symptom terms. In addition, we divided all the frequency of diarrhea, such as 5 times per day, into 4 grades: Low, Medium, High, Ultrahigh (Specifically, 0-3,4-6,7-9,10 or more than 10 corresponding to Low, Medium, High, Ultrahigh respectively). We also divided the temperature into the same 4 grades (Specifically, 37°C-37.9°C, 38°C-38.9°C, 39°C-39.9°C, 40°C or above corresponding to Low, Medium, High, Ultrahigh respectively). Besides, since all the cases contain diarrhea, we eliminate it as a stop word for each case. Then, we generate a 0-1 vector for each symptom description. Specifically, we use symptom terms included in all cases of a dataset as features, and the value of each feature is 0 or 1. If the case contains the symptom term, the value of the feature is 1, otherwise 0. Fig.1(a) is a simple example of the process of generating vectors. Finally, we combine the processed disease time, home location and symptom into a vector as the input for DBSCAN. Fig.1(b) is an example of a combined vector. It is also worth noting that the combined vectors need a normalization process and the weight of these three different features should be adjusted.

Method Description. DBSCAN is a density based algorithm which discovers clusters with arbitrary shape and with minimal number of input parameters. The input parameters required for this algorithm are the radius of the cluster (Eps) and minimum points required inside the cluster (Minpts). Based on combination of the feature of DBSCAN and LFDO, DBSCAN has the following advantages for LFDO: a) local disease clusters are arbitrary, b) the parameter Minpts enable users to flexibly detect clusters of different sizes as required, for example, users can set Minpts to 2 according to the definition of foodborne disease outbreak, c) the local outbreaks maybe have different density because of the different population density of regions, the parameter Eps enables users to detect clusters of different density.

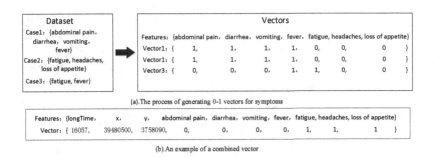

Fig. 1. The process of generating vectors for cases

The research on DBSCAN methods within the project tend to focus on the practical issues of applying existing algorithms for foodborne outbreak detection rather than on the development of new algorithms. There is one major problem for DBSCAN applied in foodborne disease clustering. How to automatically find a proper Eps for a specific dataset. Martin et. proposed a simple and effective heuristic to determine a desired Eps [13]. The heuristic defined a function k-dist from a dataset D to real numbers, mapping each point to the distance from its k-th nearest neighbor. We can get some hints about the density attribution of D when sorting the points of D in ascending order of their k-dist values. The threshold point is the first point in the first "valley" of the sorted k-dist graph (see Fig.2) and the corresponding k-dist value is the desired Eps. Also, the paper proposed an interactive approach for determining the threshold point. The interactive approach based on a realistic assumption that a user could easily see the valley in a graphical representation. We hope to reduce the user participation and provide an easy-to-use approach to be integrated into the online correlation analysis system for foodborne disease surveillance. This paper proposes an adaptive approach to determine the threshold point for a specific dataset without user participation.

In Fig.2, the ideal threshold point is the one that points before it are normal while after it are noises. Take 3-dist graph as an example, we can find the threshold point T as illustrated in Fig.3. We take the first point P1 and the last point P2 of the 3-dist graph to determine a line L. We take the point that has the maximum distance to L as the threshold point. Connect all the points to a curve and the threshold point T divides the curve into two parts. The slopes of left curves are smaller than the slope of L while that of the right curves are bigger. Intuitively, the 3-dist values in Part I increases slowly while the 3-dist values in Part II increase rapidly. We regard the points in part II as noises, because the 3-dist values of normal points are small and almost the same, while the 3-dist values of noise points are big and vary a lot. This method is very intuitive, simple and effective.

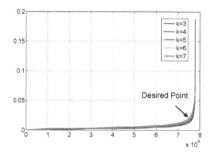

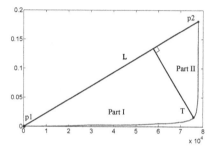

Fig. 2. The k-dist graph of Dataset 1   **Fig. 3.** How to find the desire point

3.2   SFDO Detection

For sporadic outbreak, the temporal and spatial clusters are not obvious. Symptoms exhibited and food information are the only and useful signals for sporadic outbreak detection. We describe a new approach for sporadic outbreak clustering based on pattern similarity which produces more easily interpretable and usable clusters. This approach is motivated by the following observation: an outbreak happen when two or more people get ill because of a same contaminated food source. So cases in a cluster may all contain a same symptom frequent pattern and consume a same food. Since clustering algorithms have no knowledge of these patterns, we propose and evaluate a new clustering algorithm for symptom clustering K-Means Clustering with Pattern Similarity (K-CPS). We use cases collected between 1 January 2013 and 9 June 2014 for sporadic outbreak detection (We named it Dataset 2). We separate the raw symptom text to symptom terms. In addition, we divided all the frequency of diarrhea and temperature into four grade as in Dataset 1. Besides, the term "diarrhea" is eliminated as a stop word for each case. Most part of the food name of each case contains only one term. There are a few cases contains two or more terms. In these cases, we take them as one term.

Basic Concept of Frequent Patterns. We quickly review some standard definitions of frequent patterns mining, which is a necessary and important step for mining association rules [15]. Let $I = i_1, i_1, ..., i_m$ be a set of items. Let T be a set of transactions, where each transaction t is a set of items, $t \subset I$. The support of an item-set $X, X \subset I$ is the fraction of transactions contain X. If the support is above a user-specified minimum, then we say that X is a frequent pattern.

K-CPS: K-Means Clustering with Pattern Similarity. In this subsection, we describe the details of K-CPS algorithm. First, Algorithm 1 shows the pseudo-code for K-CPS.

K-CPS consists of two phases. In the first phase, K-CPS computes the closed frequent patterns. In the second phase, the K-CPS algorithm computes the similarity between frequent patterns and objects in Dataset2. We define the similarity

---

Algorithm 1.. $K - CPS Algorithm$

Input: A dataset D; A minimum support threshold α; A denoising threshold β
Output: Clustering Result CR
    Phase I
1: $FP \leftarrow frequent\_pattern\_miner(\alpha, D)$ $\triangleright$ Mining the frequent patterns(FP) of D
2: $CFP \leftarrow closed\_frequent\_pattern(\text{FP})$ $\triangleright$ Screening out closed frequent patterns CFP
3: $SCFP \leftarrow screen\_closed\_frequent\_pattern(\text{CFP})$ $\triangleright$ Screening out the patterns which contains both symptoms and food
4: $\#MaxFP \leftarrow maximal\_frequent\_pattern(\text{SCFP})$ $\triangleright$ Screening out maximal frequent patterns $MaxFP$ from SCFP, $\#MaxFP$ is the number of the $MaxFP$
    Phase II
5: **for** $i = D_1 \rightarrow D_m$ **do**
6:    **for** $j = SCFP_1 \rightarrow SCFP_n$ **do**
7:       **if** D_i contains all the terms in $SCFP_j$ **then**
8:         $Similarity_{ij} \leftarrow d^{JAS}{}_{ij} = \frac{|D_i \cap SCFP_j|}{|D_i \cup SCFP_j|} = \frac{|SCFP_j|}{|D_i|}$
9:       **else**
10:         $Similarity_{ij} \leftarrow 0$
11:       **end if**
12:    **end for**
13: **end for**
14: $D \leftarrow Denoising(\beta, D)$ $\triangleright$ Removing the cases with every similarity is less than or equal to a specified threshold β
15: $CR \leftarrow Cluster(\#MaxFP, D)$ $\triangleright$ Running WEKA simple K-means on the processed dataset D
16: **return** CR

---

using Jaccard similarity [16]. It is defined as the quotient between the intersection and the union of the pairwise compared variables among two objects. Equation (1) illustrates the Jaccard similarity between object X and object Y.

$$d_{JAS}(X,Y) = \frac{X \bigcap Y}{X \bigcup Y} \tag{1}$$

After the computation, we get an n-dimension feature vector for each object. And n is the number of the frequent patterns. Then the processed data is clustered using simple K-Means. K-CPS assigns all the objects that have similarity symptom-food pattern to a same cluster.

4 Experimental Evaluation

In this section, we present an experimental evaluation of the parameter adaptive DBSCAN and K-CPS algorithms. We use Dataset1 and Dataset2 for local and sporadic outbreak detection respectively. Some characteristics of these two data sets are shown in Table 1. We use cases both with and without food information of Dataset1 for clustering and choose the cases with food information to evaluate the effectiveness of the local outbreak detection. We choose cases with food information of Dataset2 for sporadic outbreak detection. We don't have any training set or human-annotated data. So we can't use the common evaluation methods of clustering, such as purity, rand index (RI) and f measure [17] to evaluate our algorithms. For our experiments, we will associate the cluster results with food category. The local outbreaks will be statistically described by disease time, home location, symptoms and food category. The sporadic outbreaks will be statistically described by symptoms and food category. If the cases with a same cluster label all relate to a same food category, then the cluster is a probable foodborne disease outbreak. And these probable outbreaks we find are evidences to illustrate the usefulness of our algorithms.

Table 1. Some characteristics of experimental data sets

Data set	#cases	Time span	#province contained	#cases with food information
Dataset 1	77829	2013.01.01- 2014.01.16	31	26993
Dataset 2	91599	2013.01.01- 2014.06.09	31	33435

4.1 The Clustering Effect of Adaptive DBSCAN

In this experiment, we use the approach in section 3.1 to find an appropriate Eps for every dataset. There are several implementation details. Firstly, time, location and symptoms are three different types of data. We use normalization to unify these data from different sources into a same reference frame. Specifically, we normalize the data in every dimension to an interval $[0, 1]$. Secondly, since symptoms are 0-1 vectors, the difference caused by a different symptom term between two cases is much bigger than that of time and location. As a result, the time and location have no effect on clustering result. So we reduce the weight of symptoms. In practice, we set the weight of each dimension of symptoms to 1×10^{-7}, a heuristic weight derived from experiments and adjustment. Thirdly, these three different features are not equally important in local foodborne disease outbreak. And we use a rank-order weighting method [18]to derive a weight for each feature. By doing this, our responsibility is reduced to ranking the features based on their importances. It is easier and more reliable than specifying exact values. Specifically, we hold that the order of importance is time>location>symptoms in local outbreaks. Based on the importance order, we employ the rank-order centroid (ROC) method [18][19][20]to compute weights for these three features. Equation (2), where w_k is the weight of the $k\text{-}th$ dimension, generalizes weights for n features.

$$w_k(ROC) = \frac{1}{n} \cdot \sum_{i=k}^{n} \frac{1}{i}, k = 1, 2, ...n \tag{2}$$

According to Equation (2), the weight of time, location and symptoms is $\frac{11}{18}, \frac{5}{18}, \frac{2}{18}$ respectively. The location has two dimensions, the weight of each dimension is $\frac{1}{2} \times \frac{5}{18}$, and the same for symptoms. If the symptoms contain m dimensions, the weight of each dimension is $\frac{1}{m} \times \frac{2}{18}$. Last but not the least, the DBSCAN uses a global Eps for a dataset, but local outbreaks in different provinces may have different density because of the different population density and cases density of provinces. A global Eps can't satisfy all provinces. To solve this problem, we split Dataset 1 by provinces. And run each subset of Dataset 1 separately. We take the data of Anhui, Gansu, Guangxi, Henan, Hubei, Jiangsu, Jiangxi, Sichuan, Yunnan and Zhejiang for experiments. These 10 provinces have the most cases. Table 2 shows some statistical information of the experimental results. The probable outbreak is hand-marked by an expert who has experience in foodborne disease surveillance. The main basis of the hand-marking are the following four: a) whether the disease time and location are close to each other; b) whether the symptoms exhibited is similar or not; c) whether they are related to a same incriminated food or not; d) whether they are infected by a same bacteria (only very a few cases has the bacteria information). The experimental results show our method is promising. With the adaptive Eps, DBSCAN can effectively find all of the probable local outbreaks in the data. Rapid detection of outbreaks is the critical first step for foodborn disease surveillance.

4.2   The Clustering Effect of K-CPS

We compare clustering results of K-CPS and WEKA simple K-means on Dataset2 to show the effect of K-CPS on symptoms-food clustering. We use 33435 cases which the food information are not null. For further preprocessing, we

Table 2. Statistics of the local outbreak detection results

Statistics Provinces	#total cases	#total clusters	#probable outbreaks	#cases in outbreaks	incriminated food
Anhui	4107	138	5	40	kelp, roast, milk, sprouts
Gansu	3663	129	5	64	milk, noodle
Guangxi	3742	164	8	75	rice, pork, mushroom, beans, vinegar
Henan	4965	204	0	0	
Hubei	5645	256	3	52	wild mushroom, breast milk
Jiangsu	7024	516	12	107	soybean milk, pork, milk
Jiangxi	2427	107	7	96	rice soup, milk
Sichuan	5328	245	7	30	preserved egg, porridge, spiced crispy duck
Yunnan	1526	106	2	9	wild mushroom, grape
Zhejiang	16606	994	14	196	cake, fish, seafood, duck intestines, watermelon, banquet food

"#total cases" is all the clinic cases collected; "#total clusters" is the number of clusters; "#probable outbreaks" is the number of clusters that are probable outbreaks hand-marked by an expert in CFSA; "#cases in outbreaks" is the number of cases in probable outbreaks.

delete the cases with unclear food information, such as the terms "unknown". Finally, there left 21898 cases for experiments. And each case contains a symptom description and a kind of food. Note that there are some implementation details. First, as we know, the determination of parameter k is a hard algorithmic problem [21][22]. In K-CPS, we set parameter k to the number of the maximum frequent patterns based on the assumption that each frequent pattern represents a specific class. Second, in simple K-means, we only use the terms (symptoms and food) which the number of occurrences are greater than γ, $\gamma = \alpha \times 21898$ as features. And we generate a 0-1 vector for each case by using the same way that of illustrated in Fig. 1. The parameter k is set to the same value as in the K-CPS. Third, we use "contain" not "equal to" when we decide whether a case contains a specified symptom term or food or not. For example, a case C1 which consists of the following terms: abdominal pain, nausea and frozen watermelon. And there is a frequent pattern FP1 which consists of the following terms: abdominal pain and watermelon. We think the C1 contains the FP1. Accordingly, in simple K-means, we think C1 contains the term watermelon. Fig.4 shows the mean entropy of food at different support thresholds of K-CPS and simple K-means. As shown in Fig.4, the K-CPS has the outstanding performance on clustering the same food together while balancing the similarity of symptoms. Since the definition of foodborne outbreak is two or more people get ill after consuming the same contaminated food. We can make an obvious point that the K-CPS is more reasonable than simple K-means on the application of sporadic foodborne outbreak detection. It can find out the probable sporadic outbreaks in the data.

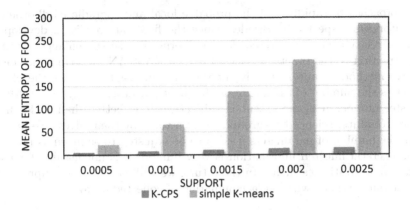

Fig. 4. The mean entropy of food at different support thresholds of K-CPS and simple K-means

5 Discussion

Based on the analyses mentioned above and the definition of foodborne diseases outbreak, we give a deep insight on the characteristics of LFDO and SFDO, which helps to find proper algorithms to detect outbreaks exist in data.

(1) In LFDO detection, the patient home location and disease time are the most useful signals for an outbreak. The cases of a LFDO show obvious spatio-temporal aggregation. Compared to time-space features, the symptoms exhibited is not so significant in clustering, so a weighting strategy is needed to reduce its weight.

(2) In SFDO detection, the time-space features are no longer the indicated information of an outbreak. As a result, we have to make the most use of symptoms exhibited. However, because the diversity among individuals, even if two people infected of same bacteria may have different symptoms exhibited. So we take food information into account simultaneously. Then the found outbreaks will have higher reliability. Through combining the experiment result to the infected bacteria(in present, very few cases have the bacteria information), we found that patients with very similar symptoms and at the same time consume a same type of food are very likely to be homological infection.

Note that our method focuses on probable outbreaks only. We hope to provide effective and rapidly screening of the raw collected data for experts worked on disease surveillance and food safety. Based on our experimental results, there are still a lot of work need to be done. And our work is the first critical step.

6 Conclusion

The detection of foodborne disease outbreak is important for food safety and is a complicated task at the same time. The contribution of this paper is to

574   X. Xiao et al.

find approaches to rapidly find the probable local and sporadic foodborne disease outbreaks respectively. Besides, since the final purpose is to develop an online correlation analysis system for foodborne disease surveillance, we paid great attention to the practical issues. First, in DASCAN, we give a parameter adaptive method to automatically set the Eps. In K-CPS, we set the cluster number as the number of the maximum frequent patterns. Though the selection of cluster number to some extent is subjective, it is better than to let users without computer science background to set a value by themselves. We hope to reduce the number of parameters needed to be adjusted to as much as possible. As the bacteria infected information is very limited, it is not feasible for us to do exact quantitative estimate. In the future, we will continuously improve and optimize our methods with the accumulation of collected data.

References

1. Schlundt, J.: New directions in foodborne disease prevention. International Journal of Food Microbiology **78**, 3–17 (2002)
2. Allos, B.M., Moore, M.R., Griffin, P.M., Tauxe, R.V.: Surveillance for sporadic foodborne disease in the 21st century: the foodnet perspective. Clinical Infectious Diseases **38**, S115–S120 (2004)
3. Greig, J., Ravel, A.: Analysis of foodborne outbreak data reported internationally for source attribution. International Journal of Food Microbiology, 77–87 (2009)
4. on Microbiological Specifications for FoodsICMSF, I.C.: Use of epidemiologic data to measure the impact of food safety control programs. Food Control 17, 825–837 (2006)
5. Havelaar, A., Bräunig, J., Christiansen, K., Cornu, M., Hald, T., Mangen, M.J., Mølbak, K., Pielaat, A., Snary, E., Van Pelt, W., et al.: Towards an integrated approach in supporting microbiological food safety decisions. Zoonoses and Public Health **54**, 103–117 (2007)
6. Wong, W.K., Moore, A., Cooper, G., Wagner, M.: Rule-based anomaly pattern detection for detecting disease outbreaks. In: AAAI/IAAI, pp. 217–223 (2002)
7. Heino, J., Toivonen, H.: Automated Detection of Epidemics from the Usage Logs of a Physicians' Reference Database. In: Lavrač, N., Gamberger, D., Todorovski, L., Blockeel, H. (eds.) PKDD 2003. LNCS (LNAI), vol. 2838, pp. 180–191. Springer, Heidelberg (2003)
8. Neill, D.B., Moore, A.W.: A fast multi-resolution method for detection of significant spatial disease clusters. Advances in Neural Information Processing Systems (2003)
9. Neill, D.B., Moore, A.W.: Rapid detection of significant spatial clusters. In: Proceedings of the tenth ACM SIGKDD International Conference on Knowledge Discovery and Data Mining, pp. 256–265. ACM (2004)
10. Neill, D.B., Moore, A.W., Pereira, F., Mitchell, T.M.: Detecting significant multi-dimensional spatial clusters. Advances in Neural Information Processing Systems, 969–976 (2004)
11. Neill, D.B., Moore, A.W., Sabhnani, M., Daniel, K.: Detection of emerging space-time clusters. In: Proceedings of the Eleventh ACM SIGKDD International Conference on Knowledge Discovery in Data Mining, pp. 218–227. ACM (2005)

12. Buckeridge, D.L., Burkom, H., Campbell, M., Hogan, W.R., Moore, A.W., et al.: Algorithms for rapid outbreak detection: a research synthesis. Journal of Biomedical Informatics **38**, 99–113 (2005)

13. Ester, M., Kriegel, H.P., Sander, J., Xu, X.: A density-based algorithm for discovering clusters in large spatial databases with noise. In: KDD, vol 96, pp. 226–231 (1996)

14. Deakin, R., Hunter, M., Karney, C.: The gauss-krüger projection. In: Victorian Regional Survey Conference. The Institution of Surveyors (2010)

15. Agrawal, R., Srikant, R., et al.: Fast algorithms for mining association rules. In: Proc. 20th Int. Conf. Very Large Data Bases, VLDB, vol. 1215, pp. 487–499 (1994)

16. Jaccard, P.: Gesetze der pflanzenverteilung in der alpinen region auf grund statistisch-floristischer untersuchungen. In: Flora, pp. 349–377 (1902)

17. Manning, C.D., Raghavan, P., Schütze, H.: Introduction to information retrieval. vol. 1. Cambridge university Press, Cambridge (2008)

18. Barron, F.H., Barrett, B.E.: Decision quality using ranked attribute weights. Management Science **42**, 1515–1523 (1996)

19. Barron, F.H.: Selecting a best multiattribute alternative with partial information about attribute weights. Acta Psychologica **80**, 91–103 (1992)

20. Jia, J., Fischer, G.W., Dyer, J.S.: Attribute weighting methods and decision quality in the presence of response error: a simulation study. Journal of Behavioral Decision Making, 85–105 (1998)

21. Greg Hamerly, C.E.: Learning the k in k-means (2002)

22. Pham, D.T., Dimov, S.S., Nguyen, C.: Selection of k in k-means clustering. Proceedings of the Institution of Mechanical Engineers, Part C: Journal of Mechanical Engineering Science **219**, 103–119 (2005)

Identifying Hesitant and Interested Customers for Targeted Social Marketing

Guowei Ma, Qi Liu[(✉)], Le Wu, and Enhong Chen

University of Science and Technology of China, Hefei, China
{gwma,wule}@mail.ustc.edu.cn, {qiliuql,cheneh}@ustc.edu.cn

Abstract. Social networks provide unparalleled opportunities for marketing products or services. Along this line, tremendous efforts have been devoted to the research of targeted social marketing, where the marketing efforts could be concentrated on a particular set of users with high utilities. Traditionally, these targeted users are identified based on their potential interests to the given company (product). However, social users are usually influenced simultaneously by multiple companies, and not only the user interest but also these social influences will contribute to the user consumption behaviors. To that end, in this paper, we propose a general approach to figure out the targeted users for social marketing, taking both user interests and multiple social influences into consideration. Specifically, we first formulate it as an Identifying Hesitant and Interested Customers (IHIC) problem, where we argue that these valuable users should have the best balanced influence entropy (being "Hesitant") and utility scores (being "Interested"). Then, we design a novel framework and propose specific algorithms to solve this problem. Finally, extensive experiments on two real-world datasets validate the effectiveness and the efficiency of our proposed approach.

1 Introduction

Recent years have witnessed the development of the social networking services and famous companies usually have their official accounts on many social network sites. For instance, Samsung, Huawei, HTC, and Xiaomi, all have their official accounts on Weibo, (weibo.com, the largest social platform in China). As users on social networks will follow the users (companies) they are interested in and receive messages and information posted by these followees [1], social network sites (e.g., Twitter, Facebook and Weibo) have become new resources and platforms to conduct marketing campaign [2].

Like traditional marketing strategies, it is also essential to figure out one or a few customer segments to target on for social marketing [3]. These targeted users should have high utilities, and then the marketing efforts (e.g., personalized recommendation [4], viral marketing [5]) could be concentrated on them. Generally speaking, both user profiles [6] and user's historical consumption records [7] are helpful for measuring their potential interests to the given company (product). On the other hand, since multiple companies simultaneously have their accounts

© Springer International Publishing Switzerland 2015
T. Cao et al. (Eds.): PAKDD 2015, Part I, LNAI 9077, pp. 576–590, 2015.
DOI: 10.1007/978-3-319-18038-0_45

in the social network, they will significantly influence social users' choice [8,9]. Thus, when conducting targeted social marketing, the given company should consider both user interests and the social influences as they will contribute to the users' final consumption behaviors.

For instance, during one targeted social marketing campaign for Samsung, we find three candidate targeted users u_1, u_2, u_3 may be interested in the products of Samsung (Interested Customers, i.e., mined based on their consumption records). Then, the problem becomes who is the most valuable user among them? Suppose there are two other competing companies of Samsung: Huawei and HTC, and suppose we could compute that the influence value distribution from these three companies to the three users are $[0.8, 0.1, 0.1]$ for u_1, $[0.1, 0.1, 0.8]$ for u_2, $[0.35, 0.34, 0.31]$ for u_3, respectively. Let's take u_3 as an example, it means that u_3 has the probability of 35%,34%,31% to be influenced by Samsung, Huawei, HTC, respectively. If these three users showed the similar interests to Samsung, then the most valuable user for targeted marketing should be u_3 rather than u_1 and u_2. Actually, u_1 is already a big fun of Samsung and thus we do not have to market on him, while u_2 is deeply influenced by our competitor (i.e., HTC) and thus he will have lower probability to choose our product[1]. Furthermore, it means we should spare our energy for other users, e.g., u_3, who has not been deeply influenced by any company and has no bias on any company (we call these users as the "Hesitant Customers"). If we pay attention on u_3, this user may choose our products. Thus, u_3 is actually the most valuable user that Samsung should market on. In summary, we argue that when the companies want to market their products, it's energy-efficient for them to target on the users, who not only have the interest to buy the specific product (e.g., one Smartphone) but also have no bias on any company and have not yet decided to choose which company's products (such as Samsung Galaxy or HTC one).

In this paper, we formulate the problem of figuring out these targeted users, like u_3, as an Identifying Hesitant and Interested Customers (IHIC) problem. As a matter of fact, there are several challenges along this line, e.g., how to compute the multiple companies' influences on users efficiently, how to measure the user hesitancy and the user interest. To address these challenges and to solve the IHIC problem effectively and efficiently, we design a novel framework. Specifically, we first propose an efficient algorithm(MIP) to compute the multiple companies' influences on users, and identify the hesitant customers by using hesitant functions. Then, we use the collaborative filtering approaches to measure the user's utilities (interests). Finally, the targeted users are those having the best balanced hesitancy scores and utility scores. Extensive experimental results demonstrate that the targeted users selected by our framework could bring in more benefit for the company than the users who are only interested or hesitant.

To the best of our knowledge, this is the first attempt on a comprehensive study of targeted marketing that considers both user interests and multiple social influences. Our solution could identify the most profitable potential targeted users to optimize the marketing performance. Meanwhile, the proposed targeted

---

[1] We will support this assumption by experimental analysis.

marketing approach is a general framework and each step could be open to some other algorithms.

2   Related Work

Marketers incline to conduct marketing campaign on social networks by employing various techniques and approaches [2], and some of them focus on (1) identifying the targeted customers who maybe interested in the specific product, or (2) exploiting the information diffusion effect to influence social customers' consumption decisions.

For identifying targeted customers [10,11], the techniques that are related to recommender systems could be easily adopted. Along this line, there are generally three types of techniques: content-based, collaborative filtering and social recommendations. The content-based methods leverage users' profile (e.g., age, job, and location) to predict whether the user's interest matches the product [12]. In contrast, collaborative filtering usually relies on users' past behaviors without requiring the explicit profiles[13]. Furthermore, the social recommendation takes the users' social ties into consideration and predicts a user's interest based on his neighbors' interests [10]. In real world scenarios, hybrid techniques are also widely used, for instance, Jamali et al. proposed to combine the social-based and the collaborative filtering approaches together to infer customer preference [7]. Unfortunately, few existing studies in this category pay attention to mining the influences coming from the companies to customers.

For exploiting the information diffusion effect, i.e., social influence, researchers first try to learn the information propagation probability between two social neighbors [14,15]. Then some related work proposed to model/simulate the entire process of information propagation, e.g., Independent Cascade (IC) model [16] and Linear Threshold (LT) model [17] are two widely used ones. However, both of them require Monto Carlo simulations to estimate the influence spread, which is very time-consuming; some efficient (or tractable) influence models are proposed, such as the stochastic information flow model [18] and the linear social influence (Linear) model [19]. Though it's convenient for these models to get the influences of a given node on others, the computation of the influences from multiple seed nodes on a given node is still inefficient. Actually, we will address this inefficiency problem by proposing a novel way of computing the influences from multiple companies on users. Meanwhile, note that social influence is often used to change customers' consumption decisions in viral marketing (e.g., via social influence maximization) [5,9], in this paper, we will show that it could also be helpful for targeted social marketing (i.e., identifying targeted customers).

3   Problem Statement and Formulation

The nodes in social networks could generally be classified into two categories, namely users' personal accounts and companies' official accounts. Usually, each company may maintain several official accounts at the same time. E.g., both "Xiaomi mobile" and "Xiaomi Company" are all official accounts of the *Xiaomi* Company on Weibo. Here, let the directed graph $G(V, E, T)$ represent a social network, where $V = C \cup U = \{1, 2, ..., c\} \cup \{c + 1, ..., n\}$ are the two types of nodes in this network, i.e., $C = \{C_1, C_2, ..., C_{|C|}\}$ is the set of an ensemble of $|C|$ companies' official accounts and U is the users' personal account set with $|U| = n - c$ users. Specifically, C_i denotes the set of associated accounts of the i-th company, thus we have $\sum_{i=1}^{|C|} C_i = c$. E represents the relationship/links between nodes and $T = [t_{ij}]_{n*n}$ is the influence propagation probability matrix. For each directed link $(i, j) \in E$, $t_{ij} \in (0, 1)$ denotes the influence propagation probability from node j to i [2]; for any link $(i, j) \notin E$, $t_{ij} = 0$.

In addition, users can consume or buy many different products[3] produced by the same company. We use a user-item matrix $R_{|U| \times |M|}$ to represent users' past consumption behaviors/records, where M is the item set. In R, the value of r_{uj} denotes user u's consumption for item j. In fact, the detailed value r_{uj} depends on the applications. E.g., it could be binary, indicating whether users bought this product before. Also, it could be a rating value, (e.g., 1 to 5 rating).

According to the illustrations in Introduction, when a company wants to market their products, it is energy efficient to target on the most valuable users that have not been deeply influenced by any company (being "Hesitant") and are also interested in the marketed products (being "Interested"). In this paper, we formulate this problem as an identifying hesitant and interested customers (IHIC) problem.

Problem Formulation. *Given a social network $G(V, E, T)$ and the user's past consumption behaviors $R_{|U| \times |M|}$, when a company wants to market a product t to K customers with energy efficient, our goal is to automatically identify these K targeted customers S who have both the hesitant quality and interested quality.*

4   The Proposed Framework

Fig. 1 shows the flowchart of the proposed framework for solving the IHIC problem. Given a social network $G(V, E, T)$ and the users' consumption behaviors $R_{|U| \times |M|}$, our proposed framework could identify the hesitant users and the interested users in parallel. On one hand, we first propose an efficient MIP(Multiple Influence Propagation) algorithm to construct an influence matrix $\mathbf{F}_{|V| \times |C|}$, where f_{ij} represents the j-th company's influence on node i. Then we define a function $H(u)$ to measure the hesitancy of user $u \in U$. On the other hand, based on $R_{|U| \times |M|}$, we infer the user preference by using collaborative filtering approaches. For each user u, we use a utility function $r(u, t)$ to measure the user u's interest. Finally, we combine the two functions $H(u)$ and $r(u, t)$ together and

---

[2] Learning the specific propagation probability between neighbors is outside the scope of this paper; in the experiments, we simply assign $t_{ij} = 1/indegree(i)$ as widely used [5,19].

[3] We will use terms customers and products as synonyms to the users and items, respectively.

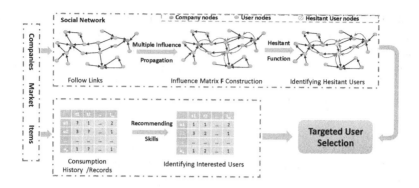

Fig. 1. Flowchart of the Framework

use a parameter η to balance the effect between user hesitancy and user interest. In the following subsections, each step is illustrated in detail.

4.1 Identifying Hesitant Users

In social networks, multiple companies influence each user simultaneously and the users also influence each other. Since we only identify the hesitant users, we focus on the influences of companies on users.

Multiple Influence Computation. Following the modeling of influence propagation [19], we propose to compute the influence of j-th company on each node as:

$$f_{i \leftarrow j} = \begin{cases} 1 & , i \in C_j \\ 0 & , i \in \{C - C_j\} \\ \sum_{k \in N(i)} t_{ik} f_{k \leftarrow j} & , i \in \{V - C\}, \end{cases} \qquad (1)$$

where $N(i)$ is the neighbors of node i. In this definition, if the node i belongs to the j-th company's official accounts, then $f_{i \leftarrow j} = 1$, which means a company always influence its official accounts; if i belongs to the other companies' official accounts, we assign $f_{i \leftarrow j} = 0$ for the reason that the company's official accounts is hard to be influenced by other companies; if i represents a user, we assign the $f_{i \leftarrow j}$ to the sum of the influence of j-th company on user i's neighbors.

Based on the definition above, we define an influence matrix $\mathbf{F}_{|V| \times |C|}$, where the first c rows represent the company nodes and the remaining $|U|$ rows represent user nodes; the j-th column represents the j-th company's influence on nodes. We propose the MIP (Multiple Influence Propagation) algorithm to construct $\mathbf{F} = [f_{ij}]_{|V| \times |C|}$, and Alg. 1 shows the details of the MIP. There are 4 steps: (1) Initialize the $\mathbf{F}^0$ (lines 1-7 in Alg. 1); (2) Multiple influences propagation (line 9 in Alg. 1); (3) Reset the companies' influences (lines 10-15 in Alg. 1); (4) Repeat step 2 and 3 until $\mathbf{F}$ converges (lines 8-16 in Alg. 1). We should note that step 3 is critical: Instead of letting the company nodes' influences "fade away", we reset their values to the entries in $\mathbf{F}^0$, so the influence probability mass is concentrated on these company nodes.

Algorithm 1. The MIP algorithm

Input: G $= (V, E, T)$, where $V = C \cup U$ 8 **repeat**
Output: $\mathbf{F}_{|V| \times |C|}$ 9 $\mathbf{F} = T\,\mathbf{F}$;
1 //initialize $\mathbf{F}^0$;   10   **foreach** $i \in C$ **do**
2 **foreach** $i \in V$ **do**   11   **for** $j = 1$ to $|C|$ **do**
3   **for** $j = 1$ to $|C|$ **do**   12   **if** $i \in C_j$ **then**
4   **if** $i \in C_j$ **then**   13 $f_{ij} = 1$;
5        $f_{ij} = 1$;   14   **else**
6   **else**   15 $f_{ij} = 0$;
7        $f_{ij} = 0$;

16 **until F** *converges*;
17 **Return F**;

Next, we analyze the convergence of the computation of influence matrix **F**. For better presentation, we split **F** after the c-th row into 2 sub-matrices and split T after the c-th row and the c-th column into 4 sub-matrices, namely,

$$\mathbf{F} = \begin{bmatrix} \mathbf{F}_c \\ \mathbf{F}_{|U|} \end{bmatrix}, \quad T = \begin{bmatrix} T_{cc} & T_{c|U|} \\ T_{|U|c} & T_{|U||U|} \end{bmatrix}.$$

Thus, step 2 (line 8 in Alg. 1) can be rewritten as follows:

$$\begin{bmatrix} \mathbf{F}_c \\ \mathbf{F}_{|U|} \end{bmatrix} = \begin{bmatrix} T_{cc} & T_{c|U|} \\ T_{|U|c} & T_{|U||U|} \end{bmatrix} \begin{bmatrix} \mathbf{F}_c \\ \mathbf{F}_{|U|} \end{bmatrix}.$$

Notice that $\mathbf{F}_c$ never really changes since it is reset after each iteration, and we are solely interested in $\mathbf{F}_{|U|}$. Obviously, $\mathbf{F}_{|U|} = T_{|U|c}\,\mathbf{F}_c + T_{|U||U|}\,\mathbf{F}_{|U|}$, which leads to $\mathbf{F}_{|U|} = \lim_{t \to \infty} T^t_{|U||U|}\,\mathbf{F}^0_{|U|} + [\sum_{i=1}^{t} T^{i-1}_{|U||U|}]\,T_{|U|c}\,\mathbf{F}^0_c$, where $\mathbf{F}^0_c$ and $\mathbf{F}^0_{|U|}$ are the top c rows and the remaining $|U|$ rows of the initial **F**. According to step 1, we know that $\mathbf{F}^0_c = \mathbf{0}$. It's obvious that $\mathbf{F}_{|U|} = (I - T_{|U||U|})^{-1}\,T_{|U|c}\,\mathbf{F}_c$ is the fixed point. Therefore, the iterative algorithm converges to the unique fixed point.

Hesitant Function $H(u)$. Based on $\mathbf{F}_{|U| \times |C|}$ (the remaining $|U|$ rows of $\mathbf{F}_{|V| \times |C|}$), we use a hesitant function $H(u)$ to measure the user hesitancy and identify the hesitant users (like u_3, not u_1 and u_2 in the example of Introduction). The higher of the value $H(u)$, the higher the "degree of hesitancy" of the user u. If the value $H(u)$ is very low, that means the user u is influenced deeply by some company. In this part, we introduce two different hesitant functions and compare their performance in the experiments. The first $H_E(u)$ is transferred from the information entropy [20] and the second $H_D(u)$ is transferred from the information diversity [21]; their formulations are as below:

$$H_E(u) = \sum_{j=1}^{|C|} (-f_{uj} \log_{|C|} f_{uj}), \quad H_D(u) = \sum_{j=1}^{|C|} \frac{f_{uj}}{1 + f_{uj}}.$$

Now we could recognize the hesitant nodes by using the hesitant functions. Please note that other rational functions $H(u)$ are also acceptable, such as the Gini index [20].

4.2   Identifying Interested Users

Besides measuring the user hesitancy, another key point for selecting targeted users is to measure the user interest, i.e., users' preference on a product and how likely users would consume the product. Many proposed recommendation methods estimate a utility function $r(u,t)$ to measure the user u's interest on item t and predict how the u will like t [13]. Since the focus of this paper is not to devise more sophisticated recommendation methods, we choose the two existing methods of *collaborative filtering*: item-based collaborative filtering (ICF) [22] and user-based collaborative filtering (UCF) [13]. The corresponding formulas are as below:

$$r(u,t)_{ICF} = \frac{\sum_{k \in M(u)} sim(k,t)\, r_{uk}}{\sum_{k \in M(u)} sim(k,t)}, \qquad r(u,t)_{UCF} = \frac{\sum_{v \in S(u)} sim(u,v)\, r_{vt}}{\sum_{v \in S(u)} sim(u,v)},$$

where $M(u)$ is the items that user u have consumed, $S(u)$ is the users who are most similar to u, $sim(k,t)$ is the similarity between items k and t, and $sim(u,v)$ is the similarity between users u and v; both of them are computed based on the user's past consumption behaviors $R_{|U| \times |M|}$. In this paper, we choose the Jaccard measure to calculate the similarities, and the formulas are as below:

$$sim(k,t) = \frac{\mathbb{U}_k \cap \mathbb{U}_t}{\mathbb{U}_k \cup \mathbb{U}_t}, \qquad sim(u,v) = \frac{\mathbb{I}_u \cap \mathbb{I}_v}{\mathbb{I}_u \cup \mathbb{I}_v},$$

where $\mathbb{U}_k$ and $\mathbb{U}_t$ represent the users who have consumed k and t, respectively; $\mathbb{I}_u$ and $\mathbb{I}_v$ represent the items which have been consumed by u and v, respectively.

Without loss of generality, we use ICF and UCF to select users with highest values as the interested users, and we will experimentally compare the performance of them.

4.3   Targeted User Selection

According to the illustrations above, we could compute the $H(u)$ and $r(u,t)$ to measure the user hesitancy and the user interest. Finally, we combine the two characters of user and propose the function $P(u,t)$ to measure the overall quality of each user. The final function $P(u,t)$ is as follows:

$$P(u,t) = \eta \frac{H(u)}{\bar{H}(u)} + (1-\eta) \frac{r(u,t)}{\bar{r}(u,t)}, \tag{2}$$

where η is used to balance the effect of the hesitancy and the interest, and the $\bar{H}(u)$ and $\bar{r}(u,t)$ are the maximum $H(u)$ and $r(u,t)$, respectively. The smaller the η, the more we pay on the interest measure. When η reduces to 0, the function only considers the user interests, thus the approach turns to traditional collaborative filterings.

Finally, a set S of K users with the highest value of $P(u,t)$ will be selected as the targeted customers for the given company when it wants to market the product t.

5 Experiments

5.1 Experimental Setup

The experiments are conducted on real-world social datasets: *Weibo* and *Epinions*.

(1) *Weibo*. We crawled from the social media weibo.com, where nodes represent the users or the companies' official accounts, and edges are nodes' followships. When a user posts a message in Weibo, the sending device (e.g., the mobile devices like Samsung Galaxy Note and iPhone5) are also recorded. That is, we can obtain the mobile purchasing behaviors of users by their posted messages. For instance, if a user send messages using Samsung Galaxy Note, we say he is a consumer of Samsung, and then he send another message by using "Xiaomi 2", we say the user is also a consumer of Xiaomi[4]. This data is collected in March 2013. For better illustration, we sample a small network which only contains the verified users and the official accounts of the five($|C| = 5$) famous mobile companies (namely, Samsung, Huawei, Xiaomi, HTC and ZTE) [5]; more specifically, each mobile company contains two or three official accounts(namely, $C_i = 2\ or\ 3$). In this way, we could obtain the consuming records R.

(2) *Epinions*. Epinions.com is a well known knowledge sharing and review site. In this site, registered users can submit their personal opinions on some topics such as products, movies or the reviews issued by other users, and assign products or reviews integer ratings from 1 to 5. These ratings and reviews will influence future customers when they are about to decide whether a product is worth buying or a movie is worth watching, and each rating could be regarded as a consumption from a user to an item (product or movie). Every member of Epinions maintains a "trust" list which presents a social network of trust relationships between users. To use this dataset, we select the $|C| = 5$ influential users (having the most followers) and treat them as the companies (i.e., $|C_i| = 1$ for this dataset as each selected user/company has only one account). Then, the items are the opinions/products shown by each user/company.

Table 1. Statistics of the Datasets

Data	#Nodes	#Items	#Social Edges	#Consumptions
Weibo	140,876	89	1,792,835	2,822,315
Epinions	49,290	139,738	487,183	664,824

Detailed information about the two datasets can be found in Table 1. For each dataset, we split it into a training set and a test set, by selecting the first 80 percentage of the consumptions for training and the remaining ones to be part of the test set. In this way, we could validate the performance of our methods on the test set.

---

[4] A user's consumptions/messages using the same mobile device will be integrated into a tuple record(user id, mobile id, number of records).

[5] As Apple Inc. has none official account in Weibo, our collections does not contain it.

5.2 The Correlation Analysis

In this subsection, we use *Weibo* as an example to show the strong correlations between social follow-links (between users and companies) and users' consumption behaviors.

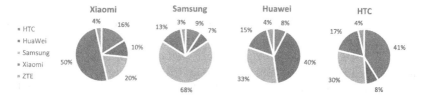

Fig. 2. Product Adoption Rate of the Company's Followers

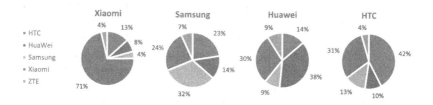

Fig. 3. Following Rate of the Companies' Consumers

First, we calculate the product (mobile) adoption rate of each company's followers and the results are shown in Fig. 2. From each subfigure (with respect to one company), we could observe that most of the followers adopt each company's products (Due to space limitation, we omit the result on ZTE). For instance, the first subfigure shows the product adoption rate of Xiaomi's followers, where 50% of these followers use the products (cellphones) produced by Xiaomi. Then, we calculate the following rate of the companies' consumers (the percentage of each company's consumers that follow this given company) and the results are shown in Fig. 3. We can see that most of each company's consumers incline to follow this company's accounts. For instance, the first subfigure of Fig. 3 shows that among the users who consumed Xiaomi's products, 71% of them followed Xiaomi's official accounts in Weibo, while only 13%, 8%, 4% of them followed the official accounts of HTC, Huawei and Samsung, respectively.

The above results demonstrate that there exists an obvious correlation between the social relations (influence) and the users' consumption behaviors. Thus, it is necessary to exploit the social influence for targeted marketing. More deeply understanding will be shown in the following subsection.

5.3 Evaluation of the MIP Algorithm

We further validate the assumption (the users who are deeply influenced by a company have much higher probability to choose the company's new product) by our MIP algorithm, and then demonstrate the performance of MIP.

Assumption Validation based on MIP. We first use MIP to compute the influence matrix $\mathbf{F}_{|V| \times |C|}$ on the training data. Then we select the top K users who have been deeply influenced by each company $j = 1, 2, ..., 5$, and compute the adoption (buy or rate the company's products) rate of these deeply influenced users on the test data, i.e., InfRate. We compare this InfRate with the average adoption rate of all users, i.e., AvgRate. The results are shown in Fig. 4, where we set $K = 30$ for *Weibo* and $K = 500$ for Epinion due to the different data sparsity. Specifically, the left and middle subfigures of Fig. 4 are the comparison results on *Weibo* and Epinion, and the right subfigure shows the Growth Rate of InfRate compared to AvgRate. For instance, On *Weibo*, the users who are deeply influenced by HTC (Company ID 1 in the right subfigure) have 60% probability to choose HTC's products (the left subfigure) and this rate is 16 times higher than AvgRate (the right subfigure). Once again, we could conclude that the users who are deeply influenced by a company have higher probability to choose the company's new product and lower probability to choose the competitors'. Hence, when conducting targeted marketing, the company should pay more attention to the hesitant customers, since the deeply influenced customers will choose the influencer's product.

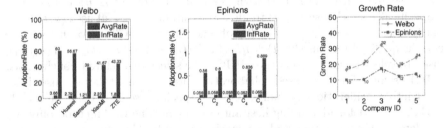

Fig. 4. Adoption Rate of the deeply influenced users

Efficiency and Effectiveness of MIP. Although several influence models (such as IC [5], LT [17] and Linear [19]) have been proposed to compute the influence of one node to another, none of them are efficient or suitable for measuring the influences from multiple companies on users. Without loss of generality, we use the recently proposed Linear model (which is both efficient and effective) as the baseline of our MIP algorithm, and the parameter settings (including the entries in T) are same to that in [19].

Specifically, we first use MIP and Linear, respectively, to get $\mathbf{F}$, and then compute the users' hesitant value $H(u)$. We run the process 100 times for different number of companies (i.e., $|C|$ equals to 1,2,..., or 5) and then compare the average runtime of the two methods. The results shown in Fig. 5 demonstrate that MIP is much more efficient and is also invariant to the number of companies. In addition, we compare the average hesitant values ($H(u)^{MIP}$ and $H(u)^{Linear}$) of the randomly selected 25 nodes, and then compute the difference rate $(H(u)^{MIP} - H(i)^{Linear})/H(u)^{Linear}$ under the $|C| = 3$. This result is shown in Fig. 6. The results illustrate that the value of $H(u)^{MIP}$ and $H(u)^{Linear}$ are very similar as the difference rate are almost less than 0.05. In other words, MIP has the similar ability to measure the social influence prorogation process as Linear, while MIP is much more efficient for computing the influences from multiple companies on users.

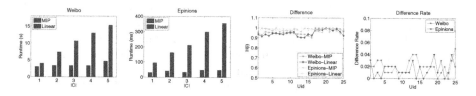

Fig. 5. Runtime Comparisons   **Fig. 6.** $H(u)$ Comparisons

5.4   Evaluation of Function $P(u, t)$

In this subsection, we first show the effects of the parameter η for final targeted user selection, and then compare the performance of $P(u, t)$ with two baselines. Finally, we generalize the $P(u, t)$ with different functions of $H(u)$ and $r(u, t)$.

Specifically, we choose $H_E(u)$ to identify hesitant users, and use $r(u, t)_{ICF}$ to identify interested users (the generality of other functions will be evaluated later). Then the quality function $P(u, t) = \eta H_E(u)/\bar{H}(u) + (1-\eta)r(u, t)_{ICF}/\bar{r}(u, t)_{ICF}$ is used to select targeted users (i.e., IHIC). We select two related benchmarks:

- **ICF** is short for item-based collaborative filtering, only takes the user interest into consideration. Actually, ICF is a special case of IHIC when $\eta = 0$.
- **IHC** is short for identifying hesitant customers. The IHC method only considers the user hesitancy and is a special case of IHIC when $\eta = 1$ for $P(u, t)$.

The Effects of η. We first show whether different values of η can help select different targeted customers. For comparison, we first use ICF to identify a targeted user set S_{ICF}. Then we change the values of η from 0 to 1 with a stepsize of 0.1, and each time we select another targeted use set S_{IHIC} with the η value and compute the Jaccard similarity between the two targeted user sets. We also further compute the Jaccard similarity of the consumed items of the selected users. In practice, the size of the targeted user set is set to 30 and the results are averaged over 50 randomly selected items. The final results are shown in Fig. 7; we can see that the larger the η, the bigger the difference between the user sets and the consumption behaviors of the selected targeted users. We conclude that IHIC($P(u, t)$) is able to select different targeted users with different η, and these users also have different item preferences.

Performance Comparisons. For a marketing item, we select the targeted users S on the training data by each method. Then, we compute the precision and recall of S on the test data to measure the performance of these methods, e.g., precision equals to the percentage of these targeted users that consumed the item.

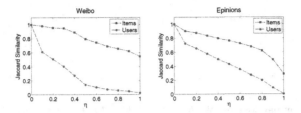

Fig. 7. The Jaccard Similarity of the user sets and their consumed items

Before detailed comparison with other models, we need to first select the best η for our proposed IHIC model. We randomly select 50 items and get the target users by IHIC with different η values under different target user size ($|S| = \{10, 20, 30\}$). Fig. 8 and Fig. 9 show the results on *Weibo* and *Epinions* respectively, which illustrate that the performance (both precision and recall) of IHIC changes with different η; additionally, IHIC could achieve the better performance with some value $\eta \in (0.1, 0.4)$. This implies that we could change the performance of selected targeted users by considering the user hesitancy with different weight(η), and choose the η leading to the best performance.

Then, we compare the performances of targeted users with different $|S|$ selected by the three methods (IHIC, ICF, IHC); we set $\eta = 0.2$ for better performance [6]. Fig. 10 and Fig. 11 show the results of the performance comparisons on *Weibo* and *Epinions* respectively. The results illustrate that the traditional recommendation method ICF not always achieve the best results. Nevertheless, IHIC, which considers both the user interest and hesitancy, usually obtains better performance. This implies that many higher interested users selected by ICF are deeply influenced by another company, and they incline to choose the similar products of the competitor's. Hence, it's necessary to consider the user hesitancy when conducting product marketing.

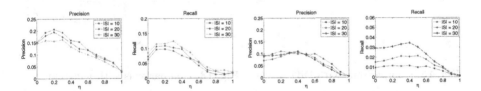

Fig. 8. Precision and Recall on *Weibo*   **Fig. 9.** Precision and Recall on *Epinions*

Comparing $P(u, t)$ with Different $H(u)$ and $r(u, t)$. We show the generality of the $P(u, i)$, that is, we will compare the performance of $P(u, t)$ with different $H(u)$ and $r(u, t)$. For better comparisons, we combine these functions reported in Section 4 and propose the following methods to calculate $P(u, t)$:

---

[6] In fact, the optimum η could be estimated by using a validation set.

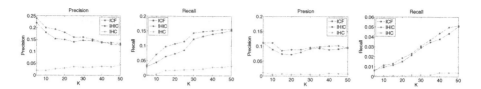

Fig. 10. Comparisons on *Weibo*   **Fig. 11.** Comparisons on *Epinions*

- HE-ICF. $P(u,t) = \eta H_E(u) / \bar{H}_E(u) + (1 - \eta)r(u,t)_{ICF} / \bar{r}(u,t)_{ICF}$.
- HE-UCF. $P(u,t) = \eta H_E(u) / \bar{H}_E(u) + (1 - \eta)r(u,t)_{UCF} / \bar{r}(u,t)_{UCF}$.
- HD-ICF. $P(u,t) = \eta H_D(u) / \bar{H}_D(u) + (1 - \eta)r(u,t)_{ICF} / \bar{r}(u,t)_{ICF}$.
- HD-UCF. $P(u,t) = \eta H_D(u) / \bar{H}_D(u) + (1 - \eta)r(u,t)_{UCF} / \bar{r}(u,t)_{UCF}$.

We randomly select 30 items for marketing. For each item, we select targeted users ($|S| = 30$) by the 4 methods under different $\eta \in [0,1]$, and then validate the precision and recall of the users on the test data. Fig. 12 shows the average precision and recall of these select targeted users on *Weibo*[7]; the results demonstrate that the performance of the 4 methods are affected by different η and all of them could achieve the best performance under some η value. For instance, HD-UCF achieves its best performance under $\eta = 0.1$. Then, we set $\eta = 0.1$ for the 4 methods and use them to select targeted users with different size $|S| = 5, 10, ..., 50$. Fig. 13 shows the average precision and recall of these targeted users; the results demonstrate that the different $H(u)$ or $r(u,t)$ also affect the performance of the selected targeted users. Hence it's necessary to devise more rational hesitant function $H(u)$ and propose more accurate $r(u,t)$.

In summary, when using recommending methods to conduct product marketing, we should take both the hesitant quality and the interest of users into consideration.

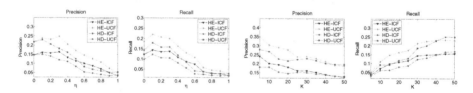

Fig. 12. Comparisons with Different η **Fig. 13.** Comparisons with Different $|S|$

6   Conclusion

In this paper, we proposed a novel framework to solve the problem of identifying hesitant and interested customers (IHIC) for targeted social marketing. Specifically, we first proposed an efficient MIP algorithm to calculate the multiple companies' influences on users, and defined two hesitant functions $H(u)$ to measure the user hesitancy. Then we measure the user interest on items by

---

[7] Similar results could be observed on *Epinions*, we omit it due to space limitation.

using two collaborative filtering approaches. Finally, we combined the two types of measures together and proposed the function $P(u, t)$ to identify the hesitant and interested customers. Extensive experiments validated the performance of our proposed approaches.

Acknowledgments. This research was partially supported by grants from the National Science Foundation for Distinguished Young Scholars of China (Grant No. 61325010), the Natural Science Foundation of China (Grant No. 61403358), the Fundamental Research Funds for the Central Universities of China (Grant No. WK0110000042) and the Anhui Provincial Natural Science Foundation (Grant No. 1408085QF110). Qi Liu acknowledges the support of the Youth Innovation Promotion Association, CAS.

References

1. Kwak, H., Lee, C., Park, H., Moon, S.: What is twitter, a social network or a news media? In: WWW, pp. 591–600. ACM (2010)
2. Hartline, J., Mirrokni, V., Sundararajan, M.: Optimal marketing strategies over social networks. In: WWW, pp. 189–198. ACM (2008)
3. Liu, L., Yang, Z., Benslimane, Y.: Conducting efficient and cost-effective targeted marketing using data mining techniques. In: GCIS, pp. 102–106. IEEE (2013)
4. Ricci, F., Rokach, L., Shapira, B., Kantor, P.B.: Recommender systems handbook. vol. 1. Springer (2011)
5. Kempe, D., Kleinberg, J., Tardos, É.: Maximizing the spread of influence through a social network. In: SIGKDD, pp. 137–146. ACM (2003)
6. Alowibdi, J.S., Buy, U.A., Yu, P.: Empirical evaluation of profile characteristics for gender classification on twitter. In: ICMLA. vol. 1, pp. 365–369. IEEE (2013)
7. Jamali, M., Ester, M.: Trustwalker: a random walk model for combining trust-based and item-based recommendation. In: SIGKDD, pp. 397–406. ACM (2009)
8. He, X., Song, G., Chen, W., Jiang, Q.: Influence blocking maximization in social networks under the competitive linear threshold model. In: SDM, pp. 463–474. SIAM (2012)
9. Chen, W., Lakshmanan, L.V., Castillo, C.: Information and influence propagation in social networks. Synthesis Lectures on Data Management 5(4), 1–177 (2013)
10. Tang, J., Hu, X., Liu, H.: Social recommendation: a review. Social Network Analysis and Mining 3(4), 1113–1133 (2013)
11. Koren, Y., Bell, R., Volinsky, C.: Matrix factorization techniques for recommender systems. Computer 42(8), 30–37 (2009)
12. Pazzani, M.J., Billsus, D.: Content-Based Recommendation Systems. In: Brusilovsky, P., Kobsa, A., Nejdl, W. (eds.) Adaptive Web 2007. LNCS, vol. 4321, pp. 325–341. Springer, Heidelberg (2007)
13. Su, X., Khoshgoftaar, T.M.: A survey of collaborative filtering techniques. Advances in Artificial Intelligence 2009, 4 (2009)
14. Goyal, A., Bonchi, F., Lakshmanan, L.V.: Learning influence probabilities in social networks. In: WSDM, pp. 241–250. ACM (2010)
15. Kutzkov, K., Bifet, A., Bonchi, F., Gionis, A.: Strip: stream learning of influence probabilities. In: KDD, pp. 275–283 (2013)
16. Goldenberg, J., Libai, B., Muller, E.: Talk of the network: A complex systems look at the underlying process of word-of-mouth. Marketing Letters 12(3), 211–223 (2001)

17. Granovetter, M.: Threshold models of collective behavior. American Journal of Sociology, 1420–1443 (1978)
18. Aggarwal, C., Khan, A., Yan, X.: On flow authority discovery in social networks. In: SDM, pp. 522–533 (2011)
19. Xiang, B., Liu, Q., Chen, E., Xiong, H., Zheng, Y., Yang, Y.: Pagerank with priors: An influence propagation perspective. In: IJCAI, pp. 2740–2746. AAAI Press (2013)
20. Tan, P.N., Steinbach, M., Kumar, V., et al.: Introduction to data mining. Pearson Addison Wesley, Boston (2006)
21. Tang, F., Liu, Q., Zhu, H., Chen, E., Zhu, F.: Diversified social influence maximization. In: ASONAM, pp. 455–459. IEEE (2014)
22. Deshpande, M., Karypis, G.: Item-based top-n recommendation algorithms. ACM Transactions on Information Systems (TOIS) **22**(1), 143–177 (2004)

Activity-Partner Recommendation

Wenting Tu[✉], David W. Cheung, Nikos Mamoulis,
Min Yang, and Ziyu Lu

Department of Computer Science, The University of Hong Kong,
Pokfulam, Hong Kong
{wttu,dcheung,nikos,myang,zylu}@cs.hku.hk

Abstract. In many activities, such as watching movies or having dinner, people prefer to find partners before participation. Therefore, when recommending activity items (e.g., movie tickets) to users, it makes sense to also recommend suitable activity partners. This way, (i) the users save time for finding activity partners, (ii) the effectiveness of the item recommendation is increased (users may prefer activity items more if they can find suitable activity partners), (iii) recommender systems become more interesting and enkindle users' social enthusiasm. In this paper, we identify the usefulness of suggesting activity partners together with items in recommender systems. In addition, we propose and compare several methods for activity-partner recommendation. Our study includes experiments that test the practical value of activity-partner recommendation and evaluate the effectiveness of all suggested methods as well as some alternative strategies.

1 Introduction

In real-world recommendation applications, many items are related to activities that people like to participate with their folks. For example, items such as online game invitations, movie tickets, dinner discounts are related to social activities (playing games, watching movies, and dining). We call such items (social) *activity items*. Activity items are commonly found in real-world e-commerce websites such as Groupon (www.groupon.com) and Meituan (www.meituan.com), as shown in the examples of Figure 1(a).

Previous work on recommending activity items typically focused on improving the precision, recall, or diversity of recommended items [1]. In this paper, we follow a totally new direction: as Figure 1 shows, instead of recommending only activity items to users, we combine the activity-item sale platform and social network platform to make the activity-item sales benefit from also recommending *activity partners* for the suggested items. Our rationale is that, for activities in which people like to participate with their folks, if a system recommends a related item alone, the user may give up attending the activity (i.e., reject the item) if s/he cannot immediately think of someone to invite to attend the activity together. The Figure 1(c) illustrates the effectiveness of recommending activity partners via an example. The recommended product "tickets of Bruno Mars' concert" is an activity item and the corresponding activity can be described as

© Springer International Publishing Switzerland 2015
T. Cao et al. (Eds.): PAKDD 2015, Part I, LNAI 9077, pp. 591–604, 2015.
DOI: 10.1007/978-3-319-18038-0_46

"watching Bruno Mars' concert". Imagine that you have some interest in Bruno Mars' show; however, when you see the recommendation message, it may be hard for you to think of suitable partners for watching the show together. This could be a good reason for you to give up attending this activity since you don't feel like going to a concert alone. On the other hand, if the recommendation also includes suggestions for possible partners, you can try inviting them and enjoy the show together. Based on this example, we designed a simple questionnaire to collect feedback from real web-users on the potential effectiveness of recommending activity partners. The results (shown in Section 4.1) demonstrate that the great majority of web users would favor such an approach as opposed to a simple activity item recommender. In summary, we assert that including partner recommendations not only improves the quality of recommender systems, but may also increase the positive response rate of users, improving therefore the revenue of the involved businesses.

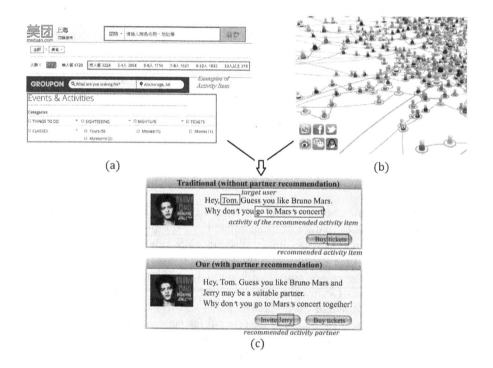

Fig. 1. Our recommendation service

As discussed above, recommending activity partners is likely to improve the success rate of activity item recommendations. On the other hand, to the best of our knowledge, there are no previous studies or applications of this idea in the research community or the industry, respectively. This motivates us to investigate methods for activity-partner recommendation. We firstly explore how attendance preference and social context can be used to recommend activity partners.

Then, we propose a method that analyzes the historical records of user preferences on activity partners to predict future activity partners. This is a reasonable methodology, since the past user preferences on activity partners would be available after the activity-partner recommendation system has been set up and used to collect training data.

In summary, the contributions of this paper are as follows:

- We bring in the idea of recommending suitable activity partners, in order to improve the effectiveness of activity item recommendation. A survey was conducted to confirm that real users favor the recommendation of activity partners together with the proposed items. We formulate the problem of activity-partner recommendation, accordingly.
- We study how to derive activity-partner recommendations using user-item preferences and the social context of users. Since such data are commonly tracked in current recommendation systems, our results can directly be embedded into an existing recommender system to turn it into an activity-partner recommender.
- We also propose a methodology for recommending activity partners based on past partner knowledge of users. This method extends conventional collaborative filtering techniques to make them more suitable for our problem.
- We conduct an experimental evaluation based on real data that tests the effectiveness of all proposed methods in recommending activity partners.

The remainder of this paper is organized as follows. Section 2 formulates our problem. Section 3 describes our methods for activity-partner recommendation. Section 4 includes our experiments. Section 5 discusses related work. Finally, Section 6 concludes with a discussion about future work.

2   Problem Formulation

As Figure 2 shows, typically there are two types of objects (i.e., *user* and *item*) in recommendation systems for activity items. Let $\mathcal{U} = \{u_1, u_2, \ldots, u_{n_u}\}$ be the

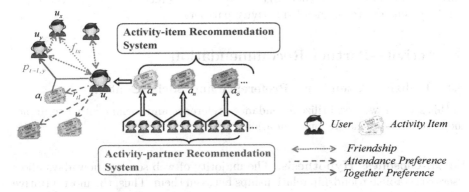

Fig. 2. Illustration of activity-partner recommendation

set of users and $\mathcal{A} = \{a_1, a_2, \ldots, a_{n_a}\}$ be the set of activity items. Two common types of relationships exist among these entities. First, users can be connected to each other in a social network; we use $f_{i,j}$ to represent the *friendship* status between users u_i and u_j, i.e., $f_{i,j} = 1$ if u_i and u_j are friends and $f_{i,j} = 0$ otherwise. Second, users may indicate their preference to activity items. Since, in our case, items are related to activities, we call the preference of users to items *attendance preference*. For each user u_t and activity item a_l, we use $pf(u_t \rightarrow a_l)$ to denote how much u_t prefers a_l. $pf(u_t \rightarrow a_l)$ can take value from a range (e.g., 1 to 5) or it can be binary number (i.e., $pf(u_t \rightarrow a_l) = 1$ means that u_t likes a_l). Besides the above two types of relationships (i.e., friendship and attendance preference), we bring in another relationship, called *together preference*, which indicates whether or how much a user prefers to attend a given activity item together with another user. For example, if Tom clicks the "Invite Jerry" button in the exemplary user interface in Figure 1(c), this indicates that Tom prefers to attend activity "Bruno Mars' show" together with Jerry. The together preference relates a user and an activity item $[u_t, a_l]$, i.e., a *ua-pair*, to another user u_x. For example, the fact that Tom prefers the tickets of Bruno Mars' concert creates pair [Tom, tickets of Bruno Mars' concert]; if Tom likes Jerry to join him to the concert, then there is a relationship between [Tom, tickets of Bruno Mars' concert] and Jerry. We use $pf([u_t, a_l] \rightarrow u_x)$ to indicate how much user u_t prefers to attend the activity of a_l together with u_x. $pf([u_t, a_l] \rightarrow u_x)$ can take numerical or binary values, similar to the attendance preference defined above. For example, we can set the binary value of $p([\text{Tom, tickets of Bruno Mars'}$ concert$] \rightarrow$ Jerry) to 1 if Tom clicks the "Invite Jerry" button or to 0 if Tom does not click the button.

As Figure 2 shows, the objective of our work is, for each activity item recommended by an activity-item recommendation system, to predict the users' together preference on the activity item. With the above notation, our problem can be stated as follows. Given a target user u_t and an activity item a_l (recommended by some activity-item recommender), use any known friendship, attendance preference, and together preference relationships to estimate $p([u_t, a_l] \rightarrow u_c)$, where u_c is any candidate activity partner. Then, rank the partner candidates u_c by their $pf([u_t, a_l] \rightarrow u_c)$ values and extract the top-k candidates as the recommended activity partners.

3   Activity-Partner Recommendation

3.1   Utilizing Attendance Preference and Social Context

In this section, we first utilize attendance preference and social context to implement activity-partner recommendation.

Social-Closeness Hypothesis. The majority of web services nowadays allow users to establish friendship relationships between them. Thus, the most intuitive relationship between users is their social closeness. Here we use the *neighborhood* *overlap* [2] (commonly used owing to its low computational complexity) to model the social closeness $SC(u_t, u_c)$ between two users: Thus, one user-user relationship that may help predict together preference $pf([u_t, a_l] \rightarrow u_x)$ is the social closeness $SC(u_t, u_c)$ between u_t and u_x. Here we assume that people prefer to attend activities with users who are socially close to them. Therefore, we can predict together preference as follows:

$$pf([u_t, a_l] \rightarrow u_c) \propto SC(u_t, u_c) = \frac{\mathcal{F}^t \bigcap \mathcal{F}^c}{\mathcal{F}^t \bigcup \mathcal{F}^c}, \tag{1}$$

where $\mathcal{F}^t$ ($\mathcal{F}^c$) is the friends set of u_t (u_c). In order to recommend activity partners to a target user u_t, we rank the activity-partner candidates according to their social closeness to u_t and return the top ones as the recommended partners. We call this method *Social-Closeness based Activity-Partner Recommendation (SCAPR)*.

Similar-Interest Hypothesis. The similarity between user interests (homophilly) is commonly used in previous recommender systems. For recommending activity partners based on user homophilly, we can rank the activity-partner candidates according to their similarity to the target user. This approach assumes that users prefer to participate in activities with people who have similar interests to them. For example, we can measure the cosine similarity between user-profile vectors and use it to define *Similar-Interest based Activity-Partner Recommendation (SIAPR)*:

$$p([u_t, a_l] \rightarrow u_c) \propto SI(u_t, u_c) = Sim_{Cosine}(\overline{r_t}, \overline{r_c}) = \frac{\overline{r_t} \cdot \overline{r_c}}{||\overline{r_t}|| ||\overline{r_c}||}, \tag{2}$$

where the vectors r_t and r_c capture the interests (i.e., the sets of preferred items) of u_t and u_c, respectively.

Also-Like Hypothesis. Besides the above hypothesises, assuming that users prefer to attend an activity together with users who also prefer to attend the activity, we rank the activity-partner candidates by their attendance preference to the activity item:

$$pf([u_t, a_l] \rightarrow u_c) \propto AL(u_c, a_l) = pf(u_c \rightarrow a_l), \tag{3}$$

We call this method *Also-Like based Activity-Partner Recommendation (ALAPR)*. The attendance preference of the activity-partner candidates to the activity item can be estimated by any activity-item recommendation system. For example, we can use user-based collaborative filtering [5] (explained in detail in Section 3.2) to estimate the attendance preference.

3.2 Utilizing Training Together Preference

In this section, we propose an alternative method to the simple strategies introduced in Section 3.1. Our objective is to predict a user's together preference via his/her past together preference records. We first discuss about the possible sources of past together preference data for the target user. Then, we will show how known together preference data can be used to predict together preference for a new item.

Extracting Together Preference Data. Several methods can be used to retrieve together preference data. First, some domains own the together preference data already. For example, consider the case where the activity items are online games. The system that hosts the games can easily record whether two users have played some game together. Together preferences can also be derived from users' behavior at the activity-partner recommendation web service. For example, if we set up an activity-partner recommendation system with an interface similar to the one in the of Figure 1(c), users' clicking behavior on the invitation button is a indicator of activity-partner preference. One typical source of together-preference data are the check-in records of geo-social networks. Assume that we have access to the check-in data of users together with their social connections. If two users who are friends checked in at the same activity venue very close in time, we can infer that they attended the activity together. For example, two friends who checked in at the same Chinese restaurant at 8:00 pm and 8:15 pm on the same day, most probably had dinner together.

Using Together Preference Data. With the availability of past together-preference data, recommending activity partners seems to be a typical recommendation problem if we regard the combination of target user and activity (e.g. $[u_t, a_l]$) as a special "user". Up to now, two main classes of recommendation approaches exist: collaborative [3] or content-based filtering [4]. Content-based filtering methods extract features from the items and recommend to users items with similar features to past items chosen by them. In our problem, the "items" to be recommended are activity partners, which lack a generalized definition of content. Therefore, collaborative filtering (CF) appears to be a more suitable approach for activity-partner recommendation. Therefore, we propose a method, called *Collaborative Filtering based Activity-Partner Recommendation (CFAPR)*, which appropriately extends the idea of CF methods to solve our problem.

Before presenting *CFAPR*, we first explain how user-based CF [5] works. Since it can be used for predicting the attendance preferences in our APR problem, here we take the process of accessing $pf(u_t \rightarrow a_l)$ of single user u_t on item a_l as an example. The first step of the approach is to calculate for each other user u_i the vector similarity between rating profiles of u_t and u_i (denoted as $\overline{r_t}$ and $\overline{r_i}$). For example, we can use the similar interests Equation (2) to calculate the similarity $S_{t,i}^u$ between u_t and u_i. The second step is as follows: if $S_{t,i}^u$ satisfies some condition (e.g., larger than a threshold or in the set of top-k highest similarities), we regard u_i to be in the *neighborhood* of u_t. To predict the preference $p_{t,l}^A$ of user u_t to activity a_l, we aggregate (weighted sum) the (known) preferences $p_{i,l}^A$ to a_l of all users u_i in the neighborhood of u_t, as follows:

$$p_{t,l}^A \propto \frac{1}{\sum_{u_i \in N^t} S_{t,i}^u} \sum_{u_i \in N^t} S_{t,i}^u r_{i,l}, \tag{4}$$

where N^t denotes the set of u_t's neighbor users who have rated a_l.

Now, assume that we try to apply this conventional user-based CF approach to predict the together preference $pf([u_t, a_l] \rightarrow u_c)$. Similarly, we can regard each $[u_t, a_l]$ as a special user unit. We call such a "user" unit a *ua-pair*. First, we should try to find the neighborhood of ua-pair $[u_t, a_l]$. However, since a conventional activity-item recommender always recommends to users activity items they have not rated yet, there must not be any historical together preference of $[u_t, a_l]$. The above fact means that all the elements of the profile vector of $[u_t, a_l]$ are unknown, thus we are not able to find neighbor ua-pairs of $[u_t, a_l]$ by computing the vector similarity between the row of $[u_t, a_l]$ and those of other ua-pairs. This problem is not unique to user-based CF. It also occurs when we try to use item-based [6] or matrix-factorization-based CF [7] methods, since the profile row of $[u_t, a_l]$ does not contain any known values.

To solve the problem discussed above, we employ an alternative method to define the neighbors of $[u_t, a_l]$ and their similarity. We just consider all $[u_t, a_m]$ $(m \neq l)$ as candidate neighbor ua-pairs of $[u_t, a_l]$. In other words, we only take the ua-pairs for which the user element is same as the target user u_t as candidates of neighbor ua-pairs, since we found that the together-preference patterns of different users are very different (this will be demonstrated in the next section). Then, we regard the similarity between $[u_t, a_l]$ and $[u_t, a_m]$ as the similarity between a_l and a_m $(m \neq l)$. For example, we can use the similarity between the profile vectors of a_l and a_m (i.e., item similarity) to model the similarity between $[u_t, a_l]$ and $[u_t, a_m]$. Note that we can also use content similarity between the activity items if the activity item carry a rich description. After calculating the similarity between $[u_t, a_l]$ and all $[u_t, a_*]$, we select the most similar $[u_t, a_*]$ as the neighbors of $[u_t, a_l]$ (i.e., those with similarity larger than a threshold or those with the highest similarities). Finally, we can predict $p_{t,l,c}^T$ (i.e., $pf([u_t, a_l] \rightarrow u_c)$) by aggregating all together preferences $p_{t,m,c}^T$ (i.e. $pf([u_t, a_m] \rightarrow u_c)$) of $[u_t, a_m]$ $(m \neq l)$ on u_c as:

$$p_{t,l,c}^T \propto \frac{1}{\sum_{[u_t, a_m] \in \mathcal{N}^{t,l}} S_{l,m}^a} \sum_{[u_t, a_m] \in \mathcal{N}^{t,l}} S_{l,m}^a p_{t,m,c}^T, \tag{5}$$

where $\mathcal{N}^{t,l}$ denotes the neighbor ua-pairs of $[u_t, a_l]$. We denote the above extended CF method by *CFAPR*. From the above equation, we can see that *CFAPR* actually assumes that people have similar preferences for patterns on similar activities, which is a reasonable assumption. For example, John likes to watch football matches and play football with his sports buddies, but prefers to watch romantic movies and have dinner in a restaurant with his girlfriend.

Algorithm 1 summarizes the whole process of *CFAPR*. Note that the size of the candidate set $\mathcal{C}^{t,l}$ of activity partners is an important parameter, since the problem size is determined by it. For example, in our experiments, we restrict the candidate set of activity partners to include only the users who have a friendship connection with u_t to control the problem size and the cost of *CFAPR*. Studying the effect of alternative methods for restricting the candidate-set is an important direction of our future work.

Algorithm 1. *CFAPR*

Input: (i), $\mathcal{C}^{t,l}$: the candidate set of partners recommended to user u_t when recommended activity-related item is a_l; (ii), $S^a(a_l, a_m)$: similarity function between two activity-items (i.e., a_l and a_m); (iii), *neighbor_condition*: a threshold or a value k for defining the number of neighbor ua-pairs.
Output: K partners recommended for u_t to attend a_l together

---

Initial $\mathcal{N}^{t,l} = \emptyset$; $\mathcal{A}^t =$ the activity items previously preferred by u_t;
for all $a_m \in \mathcal{A}^t$ **do** $Sim([u_t, a_l], [u_t, a_m]) = S^a(a_l, a_m)$
for all $a_l \in \mathcal{A}^i$
  if $Sim([u_t, a_l], [u_i, a_m])$ satisfies *neighbor_condition*
  then Add $[u_t, a_m]$ into $\mathcal{N}^{t,l}$
for all $u_c \in \mathcal{C}^{t,l}$ **do** Compute $pf([u_t, a_l] \to u_c)$ using Eq.(5)
Return K users in $\mathcal{C}^{t,l}$ having the highest K $pf([u_t, a_l] \to u_*)$ values.

---

4 Experiments

Section 4.1 demonstrates the meaningfulness of activity-partner recommendation via feedback collected from real web users. Section 4.2 evaluates the activity partner recommendation strategies described in Section 3.

4.1 Users' Favor of Activity-Partner Recommendation

To confirm the practical value of our work, we conducted an electronic survey that involved real-world web-users. The objective is to find out whether users to whom activity items are recommended are also interested in activity-partner recommendation for these items. The designed questionnaire, which has the format shown in Figure 3 asks people whether they prefer to be also recommended by activity partners and was released to public Chinese web-users from November 21, 2014. Until the submission of this work, 57 web-users (from various provinces of China) returned their answers to us. Although we did not get a lot of feedback (there were very few web users willing to fill-in the on-line questionnaire without a reward), we believe that the sample is big enough to reflect the opinion of typical web-users. Finally, *about 93% of participating users expressed their preference to activity-partner recommendation, compared to recommending activity items alone.* This indicates that our study has good potential in improving the quality of current recommender systems.

4.2 Effectiveness of Activity-Partner Recommenders

Data. In our effectiveness evaluation, we used data from location-based social networks (LSBN) to simulate a real-world scenario for our problem. We regard

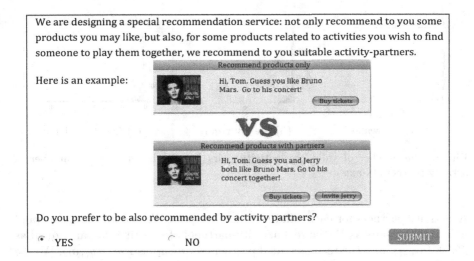

We are designing a special recommendation service: not only recommend to you some products you may like, but also, for some products related to activities you wish to find someone to play them together, we recommend to you suitable activity-partners.

Here is an example:

Fig. 3. The questionnaire used to assess the favor of web-users to activity-partner recommendation

locations in LSBN datasets as activity items. This is reasonable, since many activity items (e.g., tickets, dinner vouchers) refer to particular locations at particular time periods or moments. For example, location Han Dynasty (Chinese Restaurant, 4356 Main St, Philadelphia, PA 19127) can be regarded as activity-item "Coupon for eating Chinese food in Han Dynasty". Moreover, the activity partners with whom users attend (check-in) some activity items (location) can be inferred based on the check-in timestamps and the friendship relationships between users, as discussed in Section 3.2: if two users are friends and check-in at a same location at close timestamps (i.e., the time difference between their check-in timestamps is less than three hours), we regard that the two users attend the corresponding activity item (location) together and thus they are activity partners of each other with respect to the activity item. We used data[1] crawled from three popular LSBN websites: Gowalla (gewalla.com), Foursquare (foursquare.com) and Brightkite (brightkite.com). All these datasets have check-in timestamps and social links between users. Finally, we obtained 101400 (from Gowalla), 16220 (from Foursqure), and 1690 (from Brightkite) [user, activity] ua-pairs for testing (each such ua-pair is associated with at least one activity partners, e.g., [John, Han Dynasty]→ {Jerry, Bella, Nicole···})[2].

Evaluation Measures. After extracting the testing ua-pairs and their corresponding activity-partner knowledge, we use the tested methods to recommend

---

[1] Released on http://i.cs.hku.hk/~wttu/apr_project.html

[2] We discard the ua-pairs without any extracted activity partners. For example, if we find that none of John's friends checked in Han Dynasty at a close timestamp to John's, we infer that there are no activity partners of the ua-pair [John, Han Dynasty]; thus, we do not use [John, Han Dynasty] as a ua-pair in our experiments.

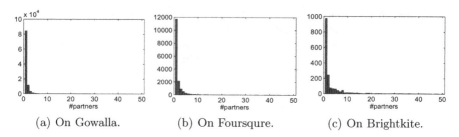

(a) On Gowalla.   (b) On Foursqure.   (c) On Brightkite.

Fig. 4. The number of valid user-activity pairs (y-axis) having a given number of activity partners (x-axis)

K activity partners for each valid user-activity pair. We denote the set of valid user-activity pairs as $\mathcal{V}$, the real activity-partners of a testing ua-pair $[u_t, a_l]$ as $Pa^{real}(u_t, a_l)$, and the recommended partners to $[u_t, a_l]$ as $Pa^{rec}(u_t, a_l)$. We use the classic precision and recall measures to evaluate the performance of recommending activity partners.

$$Precision = \frac{\sum_{(u_t, a_l) \in \mathcal{V}} |Pa^{rec}(u_t, a_l) \cap Pa^{real}(u_t, a_l)|}{\sum_{(u_t, a_l) \in \mathcal{V}} |Pa^{rec}(u_t, a_l)|}, \qquad (6)$$

$$Recall = \frac{\sum_{(u_t, a_l) \in \mathcal{V}} |Pa^{rec}(u_t, a_l) \cap Pa^{real}(u_t, a_l)|}{\sum_{(u_t, a_l) \in \mathcal{V}} |Pa^{real}(u_t, a_l)|}. \qquad (7)$$

Competitors. Besides methods *SCAPR, SIAPR, ALAPR* (introduced in Section 3.1), and *CFAPR* (introduced in Section 3.2), we include in the evaluation an additional strategy, which also employs together preference training. This method is called *Popular-Partner based APR (PPAPR)* and models the popularity of a activity partner candidate by the times s/he is preferred as an activity partner *by the target user only*. PPAPR is based on a partner consistency hypothesis, while *CFAPR* assumes that partners are sensitive to activities. In PPAPR, the popularity of a partner candidate u_c for a target user u_t is defined as follows:

$$pf([u_t, a_l] \to u_c) \propto Pop(u_t, u_c) = |\mathcal{V}_c^t|, \qquad (8)$$

where $\mathcal{V}_c^t$ is the set of valid user-activity pairs of user u_t whose activity partners include u_c.

While evaluating all methods, for each testing user-activity pair (e.g. $[u_t, a_l]$), we set the candidate set of activity partners as the friends set of u_t. Moreover, we use all neighbor candidates in *CFAPR* as the ua-pair neighbors (set *neighbor_condition* in Algorithm 1 initially as TRUE) and use the Cosine similarity between activities' rated vectors. Besides, for implementing *ALAPR*, we used user-based CF as a basis with the neighbors of a user being the 100 most similar users to the target user.

Results and Analysis. Before performing performance comparison, we analyze the number of activity partners users prefer when attending an activity. According to the check-in records of LBSN datasets, we found that most of (more than 95% in our experiments) the user-activity pairs have 1 to 5 activity partners. Therefore, we will test the performance of the methods introduced above on activity-partner recommendation when the size of recommendation list of activity partners is changing from 1 to 5. Figure 5 shows the results of all methods while K is varying from 1 to 5. Note that the precision of all methods falls and the recall increases as K increases, which indicates the predictions more close to the top are more accurate. When comparing performance of different methods, we can observe that:

- *CFAPR outperforms all other methods.*
  This indicates the suitability of *CFAPR* for activity-partner recommendation with training together-preference knowledge.
- *CFPAR outperforms PPAPR.*
  Both of *CFAPR* and *PPAPR* take use of past together preference. The difference between *CFAPR* and *PPAPR* is that *CFPAR* considers the influence of activity item together preferences. *CFAPR* assumes that the together preferences of a user on similar activity items are similar. The fact that *CFAPR* outperforms *PPAPR* has verified this assumption.
- *CFPAR and PPAPR outperform SIAPR, SCAPR, ALAPR.*
  In general, the methods which use past together preferences (i.e., *CFAPR* and *PPAPR*) of the target user perform better than methods which ignore this parameter (i.e., *SIAPR, SCAPR, ALAPR*). This fact shows that past together preferences play an important role in predicting activity partners.
- *SIAPR outperforms SCAPR, ALAPR.*
  SIAPR, ALAPR, SCAPR are three methods which uses information commonly seen in many current websites. Exploring their performance can pave the way toward constructing an initial activity-partner recommender for the case where there is no past partner knowledge about the target user. As the results show, *SIAPR* performs best among these three simple methods. Therefore, when there is no raining together-preference knowledge, *SIAPR* is a good choice to start up a activity-partner recommendation system.

Note that results in Figure 5 are on warm-start users; prior knowledge of together preferences is a requirement for methods such as *CFAPR* and *PPAPR*. The results show that *CFAPR* performs best for this set of users. In the case, where there are users with no past together preferences, we propose a hybrid strategy, where (i) *CFAPR* is used for recommending activity partners to users with past together preferences and (ii) *SIAPR* is used for the remaining users (recall that *SIAPR* performs best among the simple methods that do not rely on past together preference knowledge). We denote this hybrid method as *SICFAPR*. Figure 6 shows the result of *SICFAPR*, compared with using *SIAPR* to all users. Observe that *SICFAPR* exhibits constantly good performance on all tested cases.

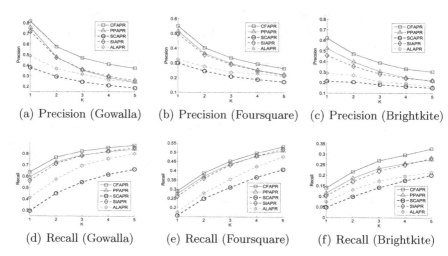

(a) Precision (Gowalla)   (b) Precision (Foursquare)   (c) Precision (Brightkite)

(d) Recall (Gowalla)   (e) Recall (Foursquare)   (f) Recall (Brightkite)

Fig. 5. Performance comparison of methods *CFAPR*, *PPFAPR*, *SCAPR*, *SIAPR* and *ALAPR*.

5   Related Work

The most related work to ours includes recommendation approaches that also utilize social or user-profile information (Section 3.1) and work on collaborative filtering (Section 3.2). We also discuss related work on problems that are similar to activity-partner recommendation.

Recommender Systems. Research on recommender systems in the previous years can be divided into two directions. The first is to improve existing models (e.g. collaborative filtering, content-based filtering, SVD based models) for recommendation. Another direction, which gains in popularity in the recent years, is to discover interesting applications of these models and extend base recommenders to domain-specific models and methods. Our work also falls in this direction. We study a new recommendation problem: recommend activity partners for the activity items suggested to a user.

Friend Recommendation. Recently, friend recommendation [11] became a popular research topic, assisting social networks to improve their service. Commonly to friend recommendation, the recommended object in our problem is also a user. However, the tasks of friend recommendation and activity-partner recommendation are very different. Friend recommendation systems predict user-user relationships (i.e., friendships) while our work explores (user, item)-user relationship (i.e., together preference from [user, activity item] to an activity partner). Friend recommendation estimates the likelihood that two non-friends will become friends in the future. Actually, the relatively bad performance of

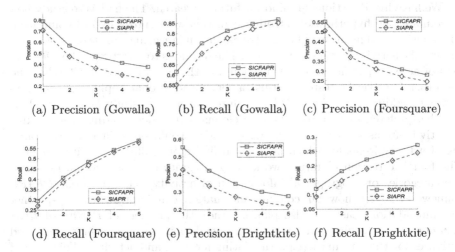

(a) Precision (Gowalla)   (b) Recall (Gowalla)   (c) Precision (Foursquare)

(d) Recall (Foursquare)   (e) Precision (Brightkite)   (f) Recall (Brightkite)

Fig. 6. Performance comparison of methods *SICFAPR* and *SIAPR* on all users (the ratio of #warm-start users to #cold-start users is about 2.0 (Gowalla), 1.5 (Foursqure) and 0.5 (Brightkite)).

the *SCAPR* method, which employs the social closeness between users to recommend activity partners, verifies the intrinsic difference between friendships and activity partners.

Group Recommendation. Group recommendation [12] is to explore the preference of a group of users to items. Currently, many services (e.g., Movielens, Tencent QQ) allow users to create groups that consist of several users. Then, a typical objective of group recommendation is to aggregate the preferences from group members to find relevant items for groups. The problem of activity-partner recommendation is different from the problem of group recommendation. Most works in group recommendation aim at selecting items for fixed groups, while activity-partner recommendation strives to find users as activity partners having as fixed variables a target user and an activity item (recommended by any activity-item recommendation system).

6   Conclusion and Future Work

In this paper, we have proposed and studied the problem of recommending activity partners to web-users for activity items suggested to them. Based on a questionnaire, we verify that real users have great interest in such a type of recommendation. We then show how to take advantage of different types of data and relationships, including attendance preference from users to activities, social context of users, and past together preference knowledge for activity-partner recommendation. Our experiments analyzed the strengths and weaknesses of the proposed activity partners recommendation models.

We have five directions in mind for future research. The first is to study how to combine the hypothesises introduced in this work into a hybrid component that considers all mentioned factors (social, attendance, and together preference) to rank the activity-partner candidates. The second is to investigate the effectiveness of activity-partner recommendation and the performance of *CFAPR* (and the other approaches tested in this paper) in additional application domains and with additional real-world data. Third, we plan to study how to combine together-preferences and attendance-preferences in order to adjust the ranking of activity items shown to the users. One idea would be to increase the ranking of activity items for which people can find more suitable activity partners. The fourth direction of future work is to integrate content information (e.g., the categories or geographical information of activity items) into our *CFAPR* framework, to see how far content information can improve activity-partner recommendation. Last, we also plan to study the problem of *Partner-Activity Recommendation*, where in friend recommendation we also include suggested activities for them to meet (e.g., the dating location and activity). This type of recommendation finds use in real-world applications, such as dating sites (e.g., www.jiayuan.com).

References

1. Zheng, V.W., Zheng, Y., Xie, X., Yang, Q.: Collaborative location and activity recommendations with gps history data. In: WWW, pp. 1029–1038 (2010)
2. Adamic, L.A., Adar, E.: Friends and neighbors on the web. Social Networks **25**(3), 211–230 (2003)
3. Su, X., Khoshgoftaar, T.M.: A survey of collaborative filtering techniques. Advances in Artificial Intelligence **2009**, 4 (2009)
4. Pazzani, M.J., Billsus, D.: Content-based recommendation systems. In: The Adaptive Web, pp. 325–341 (2007)
5. Schafer, J.B., Frankowski, D., Herlocker, J., Sen, S.: Collaborative filtering recommender systems. In: The Adaptive Web, pp. 291–324 (2007)
6. Sarwar, B., Karypis, G., Konstan, J., Riedl, J.: Item-based collaborative filtering recommendation algorithms. In: WWW, pp. 285–295 (2001)
7. Koren, Y., Bell, R., Volinsky, C.: Matrix factorization techniques for recommender systems. Computer **42**(8), 30–37 (2009)
8. He, Q., Pei, J., Kifer, D., Mitra, P., Giles, L.: Context-aware citation recommendation. In: WWW, pp. 421–430 (2010)
9. Linden, G., Smith, B., York, J.: Amazon. com recommendations: Item-to-item collaborative filtering. IEEE Internet Computing **7**(1), 76–80 (2003)
10. Wu, S., Sun, J., Tang, J.: Patent partner recommendation in enterprise social networks. In: WSDM, pp. 43–52 (2013)
11. Hannon, J., Bennett, M., Smyth, B.: Recommending twitter users to follow using content and collaborative filtering approaches. In: RecSys, pp. 199–206 (2010)
12. Gorla, J., Lathia, N., Robertson, S., Wang, J.: Probabilistic group recommendation via information matching. In: WWW, pp. 495–504 (2013)

Iterative Use of Weighted Voronoi Diagrams to Improve Scalability in Recommender Systems

Joydeep Das[1](✉), Subhashis Majumder[2], Debarshi Dutta[2],
and Prosenjit Gupta[2]

[1] The Heritage Academy, Kolkata, WB, India
joydeep.das@heritageit.edu
[2] Department of Computer Science and Engineering,
Heritage Institute of Technology, Kolkata, WB, India
subhashis.majumder@heritageit.edu,
debarshi301@gmail.com,
prosenjit_gupta@acm.org

Abstract. *Collaborative Filtering* (CF) technique is used by most of the *Recommender Systems* (RS) for formulating suggestions of item relevant to users' interest. It typically associates a user with a community of like minded users, and then recommend items to the user liked by others in the community. However, with the rapid growth of the Web in terms of users and items, majority of the RS using CF technique suffers from the *scalability* problem. In order to address this *scalability* issue, we propose a decomposition based Recommendation Algorithm using *Multiplicatively Weighted Voronoi Diagrams*. We divide the entire users' space into smaller regions based on the location, and then apply the Recommendation Algorithm separately to these regions. This helps us to avoid computations over the entire data. We measure Spatial Autocorrelation indices in the regions or cells formed by the Voronoi decomposition. One of the main objectives of our work is to reduce the running time without compromising the recommendation quality much. This ensures scalability, allowing us to tackle bigger datasets using the same resources. We have tested our algorithms on the MovieLens and Book-Crossing datasets. Our proposed decomposition scheme is oblivious of the underlying recommendation algorithm.

Keywords: Collaborative Filtering · Spatial autocorrelation · Weighted Voronoi Diagrams · Recommendation algorithms · Scalability

1 Introduction

Recommender Systems (RS) have been developed to address the problem of information overloading, where people face difficulties in finding their required information from an overwhelming set of choices. Collaborative Filtering (CF) [1, 13] is one of the most widely studied and widely adapted techniques behind recommendation algorithms. It tries to recommend items to users based on user-user or item-item similarities computed from existing data, often in the form of

© Springer International Publishing Switzerland 2015
T. Cao et al. (Eds.): PAKDD 2015, Part I, LNAI 9077, pp. 605–617, 2015.
DOI: 10.1007/978-3-319-18038-0_47

ratings given by users. Existing CF methods based on correlation criteria [9], non-negative matrix factorization (NNMF) [6] and singular value decomposition (SVD) [10] typically predict ratings accurately. However, these CF techniques suffer from high computational complexity. As a result, the methods do not scale well on very large datasets. To address this scalability problem, we propose a decomposition based CF algorithm. Our goal is to partition the entire users' space into smaller regions and apply the Recommendation Algorithm separately to the regions. However, without using any arbitrary partitioning method, we employ an intelligent partitioning technique using *Multiplicatively Weighted Voronoi Diagrams*. In this work, we partition the users' space according to the location of the user. The proposed work first find the weights (initial) associated with each of the voronoi cells formed by the ordinary (non-weighted) Voronoi Diagram, and then use these initial weights to construct the multiplicatively weighted Voronoi Diagram in an iterative manner. One of the objective of our work is to tessellate the users' space into clusters in such a way that the correlated users (users having similar preferences over items) end up in the same clusters. We find the Spatial Autocorrelation index (Geary's index) value in the regions (clusters) formed by the decomposition process to measure the correlation among the users in the regions.

In CF, finding similarity amongst N users is an $O(N^2)$ process. If N is large then similarity computation becomes quite expensive. Decomposition avoids this quadratic blowup and allows us to process bigger datasets even with limited computational resources. As for example, if we partition a region with n users into k partitions with nearly equal sizes, then the overall time required for performing collaborative filtering in all those k partitions will be proportional to $k.(n/k)^2 = (n^2/k^2).k = n^2/k$. So we can achieve a k order speed up by dividing the users' space into k partitions. The advantage of the proposed method is that less similarity computations are needed as we apply the CF algorithm to the partitions and not to the entire users' space. One disadvantage of this recommendation technique is that the recommendation quality may degrade, since we recommend only using the data of a particular region, and not the entire rating dataset. Our goal is to reduce the overall running time without sacrificing recommendation quality much. Experiments conducted indicate that our method is effective in reducing the running time, while maintaining an acceptable quality of recommendation. Moreover our proposed decomposition scheme is oblivious of the underlying recommendation algorithm.

The rest of the paper is organized as follows: In section 2, we provide background information about Weighted Voronoi Diagram, Spatial Autocorrelation and also review some of the past works related to Collaborative Filtering based RS. Section 3 outlines our contribution while sections 4 and 5 present our Decomposition and Recommendation Algorithms respectively. In section 6, we report and analyze the experimental results. Section 7 concludes discussing our future research directions.

2   Background and Related Work

2.1   Weighted Voronoi Diagram

Voronoi Diagrams[1] are used in computational geometry to decompose a metric space into regions based on distances from a specified finite set of points. These points are also called sites on the plane that provide certain services. Suppose that there is a person located at point x on the plane. Now, the question is which site he should access to get service from. Generally it is the site which is nearest from the point x, based on a suitable distance function. Voronoi Diagrams tessellate the plane into regions around each site, so that we know the service region for a particular site. Voronoi Diagrams can be generalized on the basis of the distance function that is used. One can define a distance function by assigning weights to the sites. The resultant diagrams are called Weighted Voronoi Diagrams. They are of two types - multiplicatively weighted and additively weighted Voronoi Diagrams. In a multiplicatively weighted Voronoi Diagram, we define the distance function as

$$d\left(p, s_i\right) = \| \, p - s_i \, \| \, / w_i \qquad (1)$$

where w_i is the weight of the site s_i. Here the effective distance is obtained by dividing the Euclidean distance by the weight or strength of the site, and therefore the region belonging to a site with higher weight will have a relatively larger area. Note that in the plane under the Euclidean metric, the edges of a multiplicatively weighted Voronoi Diagram can be circular arcs or straight line segments. Figure 1(a) shows an ordinary Voronoi Diagram consisting of 11 sites (from A to K), and the corresponding multiplicatively weighted Voronoi Diagram is shown in Figure 1(b). Applications of weighted Voronoi Diagrams can be found in various domains [2,3,8]. Dong developed a raster-based approach to generating and updating both ordinary and multiplicatively weighted Voronoi diagrams for point, line, and polygon features in a GIS environment [3]. Boots applied multiplicatively weighted Voronoi Diagrams to generate trade areas using store characteristics and assumptions about customer behaviour [2]. In this work, we use a multiplicatively weighted Voronoi Diagram to partition the users of our system with respect to location.

2.2   Spatial Autocorrelation

Spatial Autocorrelation [7] measures the co-variance of properties within a geographic space and it deals with both attributes and locations of spatial features. Spatial data tends to be highly self-correlated, i.e., people with similar characteristics, occupations and backgrounds tend to cluster in the same neighborhood. A commonly used measure of Spatial Autocorrelation is Geary's index (c) [4]. Geary's index measures the similarity of i's and j's attributes, c_{ij}, which can be calculated as follows:

---

[1] http://www.cs.utah.edu/suresh/compgeom/voronoi.pdf

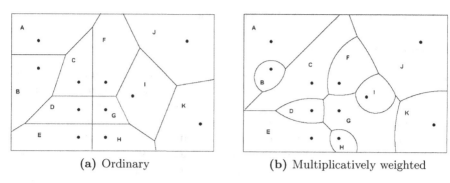

(a) Ordinary   (b) Multiplicatively weighted

Fig. 1. Voronoi diagram

$$c_{ij} = (z_i - z_j)^2$$

where z_i and z_j are the values of the attribute of interest for objects i and j.

A locational similarity w_{ij} is used in the calculation of Geary's index, where $w_{ij} = 1$ if i and j share a common boundary, and $w_{ij} = 0$ if not. Geary's index is expressed as follows:

$$c = \frac{\sum_i \sum_j w_{ij} c_{ij}}{2 \sum_i \sum_j w_{ij} \sigma^2} \quad where \quad \sigma^2 = \frac{\sum_i (z_i - \bar{z})^2}{(n-1)} \quad and \quad \bar{z} = \frac{\sum_i^n z_i}{n}$$

Here σ^2 is the variance of the attribute z values. If $c = 1$, the attributes are distributed independently of location. If $c < 1$, similar attributes coincide with similar locations, and if $c > 1$, attributes and locations are dissimilar.

2.3 Collaborative Filtering Based Recommender Systems

CF systems generate recommendations based on a subset of users that are most similar to the active user. However, each time a recommendation is requested, the algorithm needs to compute the similarity between the active user and all other users, based on their co-rated items. This similarity computation becomes very expensive with the growth of both the number of users and items in the database. We briefly discuss some of the past works that address this *scalability* problem in the remaining of this section.

Sarwar et al. [11] addressed the *scalability* issue associated with the CF task by clustering the complete user set on the basis of user-user similarity and used the cluster as the neighborhood. With the same perspective George and Merugu [5] use a collaborative filtering approach based on a weighted co-clustering algorithm that involves simultaneous clustering of users and items. Sarwat et al. [12] also proposed a scalable location-aware recommender system using location based rating. They employed user partitioning technique and produced recommendations twice as accurate compared to existing recommendation approaches. In this work, we address the scalability problem of the CF process by clustering the users' space on the basis of a Weighted Voronoi diagram.

3   Our Framework

The primary objective of our work is to deal with the computational complexity associated with the traditional CF algorithms. We propose a user partitioning technique using multiplicatively weighted Voronoi Diagrams to partition the entire users' space into smaller cells (regions) based on the location, and then apply the Recommendation Algorithm separately to these regions. In this work, we use two popular datasets MovieLens[2] and Book-Crossing[3] to test the effectiveness of our proposed algorithm. However, our core approach can be easily adapted to the other scenarios. One of our main goal is to partition the users' space in such a way that the correlated users are placed in the same regions. Our aim is to find the presence of spatial autocorrelation in the regions, and then recommend items with the idea that the suggested items will be liked by the user.

4   The Decomposition Algorithm

In this work, we use a multiplicatively weighted Voronoi Diagram based approach for space decomposition. Space partitioning is done on the basis of locations (zip-codes or cities) of the users. In order to construct the Voronoi Diagram, we represent each city or zip-code by 2D coordinates. Our work use the longitude-latitude of the centroids of the polygonal regions representing the zip-codes as the coordinates.

Generally, when weighted Voronoi Diagrams are used in a particular scenario, there are some weights associated with each voronoi site that we use in the distance function as described in Section 2.1. However, in our case, to start with we just have a flat set of user locations without any hierarchy. To identify some of them as voronoi sites, we introduce a threshold criterion described below. However, that is not enough to signify what weight should be associated with each of the sites. A natural expectation is that the weight associated with a particular site should be proportional to the number of users that will be present in the corresponding voronoi cell. With the above idea in mind, we proceed to construct the multiplicatively weighted Voronoi Diagram in two stages.

In stage I, the algorithm finds some cities having a minimum number of users (threshold), and use those cities as the voronoi sites (facilities). Each site S corresponds to a voronoi cell V(S) consisting of all the points (zip-codes or cities) closer to S than any other site. For the remaining points (cities) in the plane, we calculate the distance of a point (city) from each of the site points, and the point

---

[2] http://grouplens.org/datasets/movielens/

[3] http://www.informatik.uni-freiburg.de/cziegler/BX/

is allocated to the region represented by the site that has the minimum distance from that point. In this way, we map each point onto some voronoi cell. The Haversine distance formula[4] that computes great-circle distances between two points on a sphere from their longitudes and latitudes is being used to find the distances. Now we have a set of voronoi polygons consisting of user cities. These voronoi polygons correspond to an ordinary Voronoi Diagram, i.e., weights are all considered to be uniform and the boundaries between all cells are straight lines. Next we find how many users are located in each of the voronoi cells formed, and then use these numbers as the preliminary weights of those cells.

In stage II, we create the multiplicatively weighted Voronoi Diagram starting with the preliminary weights in an iterative manner. The preliminary weights of the cells are used to calculate the initial boundaries of the weighted Voronoi Diagram. As the boundaries change, due to the consideration of weights, the number of users belonging to each voronoi cell also change accordingly. We use these new numbers to modify the corresponding weights of each cell. Then we calculate the boundaries again on the basis of these new weights. We continue the above procedure for several iterations by which the boundaries get corrected again and again. Continuing this process for several iterations, we find that the weights of each voronoi cell reach a point of stability, when none of the weights change in two successive iterations. At this point, we accept the cell boundaries as the final boundaries of the weighted Voronoi Diagram. As we discuss later, the iterative process takes only a small part of the total time. We use these saturated weights as the final weights of the voronoi cells for further experimentations. The scheme is detailed in its algorithmic form as follows:

Algorithm Weighted Voronoi_Decomposition

Step 1: Select those zip-codes (or cities) as voronoi sites that satisfy the threshold criteria.

Step 2: Construct the initial Voronoi Diagram using these site points.

Step 3: Map the remaining cities to their destined voronoi cell (represented by the site closest to the city).

Step 4: Find the initial weights for the voronoi cells by calculating the total number of users in each cell.

Step 5: Create the weighted Voronoi Diagram in an iterative manner starting with the initial weights of the cells found in step 4.

   Step 5.1: The initial weights are used to define the initial boundaries of the weighted Voronoi Diagram.

   Step 5.2: Using the initial boundaries find the total number of users in each cell, and use these new numbers to modify the corresponding weights of each cell.

   Step 5.3: Calculate the boundaries again on the basis of these new weights.

Step 6: Continue steps 5.2 and 5.3 until the weights for the cells reach a stable point (saturated weights).

Note that for each user, we merely need to determine to which voronoi cell he or she belongs, which can be done in O(Number of cells) time. Therefore we

---

[4] http://www.movable-type.co.uk/scripts/latlong.htm

determine this information implicitly, avoiding the explicit construction of the cell boundaries of the weighted Voronoi Diagram, which is otherwise expensive. We next compute the value of Geary's index in the voronoi cells to verify whether users with similar tastes and preferences are grouped in the same neighborhood. In order to calculate the Geary's index, we use movie (or book) ratings as the parameter for computing attribute similarity (c_{ij}), while locational similarity (w_{ij}) is measured by calculating the distance between the pairs of user city and then use the inverse of that distance (as discussed in sub-section 2.2).

5 The Recommendation Approach

We investigate two popular CF methods - user-based and item-based and then combine these recommendation methods with our framework to verify whether their performance is improved. We use Pearson's correlation coefficient [13] as the similarity metric for finding user-user similarity while item-item similarity is captured using Cosine-Based similarity metric [13].

User-based *Top-N* recommendation algorithm first identify the K most similar users of the target user by computing the similarities between the target user and all other users in the region (cluster). Next, we form a set of top rated items (movies or books) by using the ratings of the K similar users. This set include only those items whose average rating from all the K similar users is more than a threshold value. Then the items in this set are again ranked in order of their rating frequency (no. of users rating the item). The system recommends to the target user the *top-N* items from the item set not rated by the user.

In item-based *Top-N* recommendation algorithm, we first compute the K most similar items for each item present in the cluster according to the similarity score. Then we form a set, *RC*, as recommendation candidates by taking the union of the K most similar items and then removing the items already rated by the target user. Let *RI* denote the set of items rated by the target user. Next, we calculate the similarities between each item of the set *RC* and the set *RI*. Then the items in the set *RC* are sorted in descending order of the similarity and the *Top-N* items from this set are recommended to the target user.

6 Experiments and Results

6.1 Data Description

Our experiments are performed on the MovieLens-1M and Book-Crossing datasets. MovieLens-1M data consists of 1,000,209 anonymous ratings (1-5) of approximately 3,900 movies made by 6,040 MovieLens users where each user has rated at least 20 movies. Book-Crossing dataset contains 278,858 users (anonymized but with demographic information) providing 1,149,780 ratings (explicit/implicit) on 271,379 books. Ratings are either explicit, expressed on an integral scale from 1-10 (higher values denoting higher appreciation), or implicit, expressed by 0.

6.2 Evaluation Metric Discussion

We use Mean Absolute Error (MAE) [13] and Root Mean Square Error (RMSE) [13] to evaluate the prediction accuracy while quality of the recommendation is measured using the *Precision, Recall* and *F1 score* metric. We have depicted the different combinations of recommendation that can be generated in a typical recommendation problem in Table 1. Note that a customer likes an item (movie or book) if he has given a rating of 4 or 5 to that item (in a scale of 1 to 5), otherwise dislikes it, i.e., his rating is 1, 2 or 3. A recommendation is positive if recommended rating coincides with the actual rating.

Table 1. Possible Recommendations

	Customer Likes (rating = 4 or 5)	Customer Dislikes (rating = 1, 2 or 3)
Recommend	True positives	False positives
Do not recommend	False negatives	True negatives

Precision: Precision measures the degree of accuracy of the recommendations produced by the algorithm. In our system, Precision measures what fraction of the recommended items are liked by the customers.

Recall: In our Recommender System, Recall measures what fraction of the items liked by the customers, has been recommended by the algorithm.

F1 score: F1 score is the harmonic mean of Precision and Recall.

$$Precision = \frac{True\ Positives}{True\ Positives + False\ Positives}$$

$$Recall = \frac{True\ Positives}{True\ Positives + False\ Negatives} \quad and \quad F1 = \frac{2 * Precision\ * Recall}{Precision + Recall}$$

6.3 Experimentation with Decomposition Algorithm

The Decomposition algorithm use threshold values (as discussed in section 4), which define the minimum number of users a city must have to be considered as a site in the Voronoi Diagram. For the MovieLens data, we use {5, 10, 15} as the threshold values, and for Book-Crossing we use {50, 100, 150} as threshold values. Note that, we used higher threshold values for the Book-Crossing data than the MovieLens data because the total number of users in the Book-Crossing dataset is significantly more than the MovieLens dataset. If lower threshold values (like 5, 10 and 15) are used then we will have a number of small cells with very few users (and ratings), which in turn may affect the recommendation quality. Table 2 shows the result of the weighted Voronoi decomposition using the thresholds. It also shows the total number of iterations required for the weights of the voronoi cells to reach a stable point. From Table 2, it can be noted that the time required to generate the weighted Voronoi Diagram using the iterative

Table 2. Results of Weighted Voronoi Decomposition

MovieLens-1M				Book-Crossing				
Threshold	No. of cells	No. of iterations	Time (sec)	% of total time	Threshold	No. of cells	No. of iterations	Time (sec)	% of total time
5	172	18	5.85	0.4	50	144	13	4.51	0.08
10	35	14	3.96	0.1	100	66	15	5.5	0.02
15	10	10	3.52	0.02	150	37	12	4.45	0.006

process is really negligible ($< 0.4\%$) compared to the total time required for recommendation as reported in Table 7.

We report the results of spatial autocorrelation performed on the MovieLens dataset in Table 3 and that of Book-Crossing dataset in Table 4. In the Tables, we compare the spatial autocorrelation values of the entire users' space with the corresponding values of the regions formed by the initial and weighted Voronoi Diagrams. We know that if the Geary's index value is less than 1, then spatial autocorrelation is present, otherwise absent. To measure correlation among the users in the regions we define two metrics - CI_1 and CI_2. CI_1 reports the percentage of items (movies or books) that falls below the correlation value 0.75, and CI_2 reports the percentage of items that falls below the correlation value 1.0. As for example, in Table 3, we notice that on an average 14.46% of the movies of the entire users' space fall below the correlation value 0.75 while 60.44% of the movies fall below 1.0. Similarly, for the Initial Voronoi Diagram, using Threshold = 5, 80.81% of the movies averaged across all the 172 regions fall below 0.75, and 94.53% of the movies fall below 1.0. We have the best results for spatial autocorrelation using the Weighted Voronoi approach ($CI_1 = 85.14$ and $CI_2 = 97.18$). Here, we achieve about 70.00% improvement in CI_1 value and about 37.00% in CI_2 over the entire space. In Table 3, we can also observe that the Weighted Voronoi approach also gives us better correlation values over the Initial (non-weighted) Voronoi for all the threshold values. Similar results can also observed in Table 4. In the tables, one can also observe that as the number of cells increases, or in other words sizes of the cells decrease, the percentage of spatial autocorrelation increases within each cell. However we cannot decrease the cell sizes beyond a certain point, since too few users in a cell will result in fewer number of ratings based on which the recommendations will be made and this in turn will affect the quality of recommendations.

6.4 Experimentation with Recommendation Algorithm

As we have already seen in the previous section that spatial autocorrelation exists in the regions, and therefore it seems very promising that if you recommended only using the users (or items) in the region of the target user, recommendation quality will improve. For both MovieLens and Book-Crossing datasets, we randomly split the user ratings into two sets - observed items (80%) for training and held-out items (20%) for testing. Ratings for the held-out items are to be predicted. We execute the algorithm with $K = 100$ and $N = 10$. That is we

Table 3. Correlation Comparisons on MovieLens Dataset

	No. of Cells	CI_1 (% < 0.75) (Average)	CI_2 (% < 1.0) (Average)
Entire Space	1	14.46	60.44
Threshold = 5			
Initial Voronoi	172	80.81	94.53
Weighted Voronoi	172	85.14	97.18
Threshold = 10			
Initial Voronoi	35	52.05	78.19
Weighted Voronoi	35	67.80	90.43
Threshold = 15			
Initial Voronoi	10	34.45	66.78
Weighted Voronoi	10	49.41	79.71

Table 4. Correlation Comparisons on Book-Crossing Dataset

	No. of Cells	CI_1 (% < 0.75) (Average)	CI_2 (% < 1.0) (Average)
Entire Space	1	72.97	88.10
Threshold = 50			
Initial Voronoi	144	94.43	97.51
Weighted Voronoi	144	96.05	99.38
Threshold = 100			
Initial Voronoi	66	91.21	95.31
Weighted Voronoi	66	93.20	98.58
Threshold = 150			
Initial Voronoi	37	88.52	94.14
Weighted Voronoi	37	89.41	97.34

Table 5. Performance on MovieLens Dataset

	No. of Cells	P@10 (Avg)	R@10 (Avg)	F1@10 (Avg)	MAE (Avg)	RMSE (Avg)
User-based						
Base	1	0.970	0.736	0.815	0.379	0.460
Th = 5	172	0.858	**0.830**	**0.828**	0.419	0.571
Th = 10	35	0.896	**0.829**	**0.842**	0.435	0.585
Th = 15	10	0.903	**0.766**	0.804	0.446	0.589
Item-based						
Base	1	0.882	0.735	0.797	0.412	0.501
Th = 5	172	0.812	**0.803**	**0.804**	0.421	0.532
Th = 10	35	0.824	**0.751**	0.783	**0.405**	0.511
Th = 15	10	**0.896**	**0.753**	**0.818**	0.442	0.562

Table 6. Performance on Book-Crossing Dataset

	No. of Cells	P@10 (Avg)	R@10 (Avg)	F1@10 (Avg)	MAE (Avg)	RMSE (Avg)
User-based						
Base	1	0.589	0.401	0.476	1.011	1.167
Th = 50	144	0.502	**0.501**	**0.495**	1.132	1.203
Th = 100	66	0.576	**0.570**	**0.565**	1.143	1.227
Th = 150	37	**0.630**	**0.616**	**0.613**	1.217	1.329
Item-based						
Base	1	0.576	0.417	0.484	1.121	1.23
Th = 50	144	0.491	**0.462**	0.475	1.17	1.272
Th = 100	66	0.536	**0.525**	**0.528**	1.142	**1.213**
Th = 150	37	0.563	**0.594**	**0.57**	1.18	1.331

consider a maximum of 100 similar users or items and recommend *top-10* items to the user.

We report the results of the Recommendation Algorithm performed on the MovieLens dataset in Table 5, and that of the Book-Crossing dataset in Table 6. Here term Th is abbreviation for threshold. In the Tables, we make a comparative analysis of the recommendation performance using different evaluation metrics. Here *base performance* indicates the performance of the algorithm using the entire users' space (without decomposition). We compare the overall performance in the regions formed by weighted Voronoi decomposition with the *base performance*. We use MAE and RMSE to evaluate the prediction accuracy and also use Precision@K, Recall@K and F1@K to evaluate the quality of the *top-K* recommended items. Note that, we present Precision (P@10), Recall (R@10) and F1 ($F1$@10) score on position 10. The bold numbers indicate that its value has an improvement over the base value.

In Tables 5 and 6, we have reported the performance of our recommendation algorithm averaged over all the regions. As for example, in Table 5, for threshold 5 (User-based case), we have an average Precision, Recall, F1, MAE and

Table 7. Running Time Comparisons on MovieLens and Book-Crossing Dataset

	MovieLens-1M				Book-Crossing		
	No. of Cells	Time(min) (WV)	Time(min) (MV)		No. of Cell	Time(min) (WV)	Time(min) (MV)
base	1	1233.3	1233.3	base	1	2755.56	2755.56
Th = 15	10	539.41	**257.36**	Th = 150	37	1355.50	**1106.22**
Th = 10	35	324.42	**45.18**	Th = 100	66	771.05	**373.42**
Th = 5	172	460.5	**20.45**	Th = 50	144	580.46	**93.5**

RMSE of 0.858, 0.830, 0.828, 0.419 and 0.571 respectively averaged across all the 172 regions. Here we can see that the algorithm performs better in terms of Recall and F1 score while in terms of Precision, MAE and RMSE, the base performance is slightly better. Similar results can also be observed in Table 6. Since we executed the algorithm only using the ratings of a particular region, it may sometimes compromise our recommendation quality as two users in two different regions may have similarity in the rating patterns. However, from the above tables, it is clear that our algorithm always performs better (in terms of Recall) than the base performance, while for the other evaluation metrics it has values which are nearly equal to the base.

6.5 Scalability

We report the running time of our algorithm for the MovieLens and Book-Crossing datasets in Table 7. In the 3rd and 7th column of Table 7, we record the overall time required for testing the algorithm (in minutes) in all the regions formed by Weighted Voronoi (WV) decomposition using the different thresholds. Note that, the running time comprises of both the spatial correlation calculation time and recommendation generation time for all the users of a region. Here base represents the entire dataset without decomposition. Our experiments are run on a computer with Core i3 - 2100 @ 3.10GHz x 4 CPU and 4 GB RAM.

From the 3rd and 7th column of Table 7, we can observe that the running time improves significantly when we divide the entire dataset into smaller cells and apply the algorithm independently to those cells. As for example, for Movie-Lens dataset, the overall time required for recommending all the users in the 172 regions is 460.5 minutes while that of the entire dataset is 1233.3 minutes. Thus the running time reduces by about 63% over the base performance. Similarly for the Threshold value of 10 and 15, the runtime reduces by about 73% and 57% respectively. However, to improve the runtime further, we analyzed our algorithm and found that the weights associated with some of the voronoi cells are significantly higher than the rest of the cells. For this reason, the recommendation algorithm spends considerable amount of time in recommending the users in those cells, which in turn affects the overall performance. In order to distribute the cell weights evenly, we then modified the distance function defined in equation 1 as follows.

616   J. Das et al.

$$d\left(p, s_i\right) = \| p - s_i \| / \sqrt{w_i} \qquad (2)$$

Then we again constructed the diagram using the distance function in equation 2. Note that using square root of weight is intuitively justified as weight of each cell is directly related to its area, whose dimension varies with the square of distance. We report the overall running time of our recommendation algorithm considering the cells as per this modified Voronoi diagram in Table 7 (4th and 8th column). Here the term MV is abbreviation for Modified Voronoi. Comparing the results of MV with WV approach, we can clearly observe that this small change produces significantly faster recommendations. The bold numbers indicate improvement over WV. As for example, for MovieLens dataset, using MV approach, the overall time required for recommending all the users in the 172 regions is 20.45 minutes while the recommendation time for the corresponding 172 regions using WV approach is 460.5 minutes. Thus the runtime reduces by an order of 2. Similar results can also be seen for the other threshold values. Thus we can conclude that our MV based technique is effective in reducing the running time further.

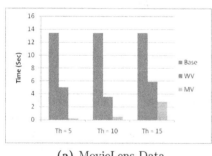

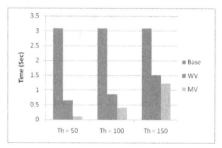

(a) MovieLens Data   (b) Book-Crossing Data

Fig. 2. Recommendation Time

In figure 2, we report the average recommendation time (in seconds) per user in the entire users' space (Base) and in the regions formed by both WV and MV based decomposition techniques. We can clearly observe that for both the MovieLens and Book-Crossing datasets, MV approach outperforms both the Base and WV based recommendation methods.

7 Conclusion and Future Work

In this paper, we have presented a scalable decomposition based Recommender System. We have implemented a decomposition technique that divides the users' space into some smaller regions with respect to location and then use spatial autocorrelation measures to capture the correlation among the users in a region. Experimental analysis using real datasets show that our model is efficient and scalable. Our proposed approach deals with the *Scalability* problem of the CF based recommendation methods by applying the Recommendation Algorithm separately to the regions. The focus of our future work is to use other metrics for finding the spatial correlation and similarities between users with the aim of optimizing the splitting technique and the Recommendation Algorithm. Finally as noted earlier, our proposed decomposition scheme is oblivious of the underlying recommendation algorithm and hence applicable with other recommendation algorithms as well. How this can be leveraged is a matter of future research.

References

1. Adomavicius, G., Tuzhilin, A.: Toward the next generation of recommender systems: A survey of the state-of-the-art and possible extensions. IEEE Transactions on Knowledge and Data Engineering **17**(6), 734–749 (2005)
2. Boots, B.: Modeling retail trade areas using higher-order, multiplicatively weighted voronoi diagrams. Journal of Retailing **20**, 519–536 (1997)
3. Dong, P.: Generating and updating multiplicatively weighted voronoi diagrams for point, line and polygon features in GIS. Comp. and Geosciences **34**, 411–421 (2008)
4. Geary, R.C.: The contiguity ratio and statistical mapping. The Incorporated Statistician **5**(3), 115–127 (1954)
5. George, T., Merugu, S.: A scalable collaborative filtering framework based on co-clustering. In: Proceedings of the Fifth IEEE International Conference on Data Mining. IEEE Computer Society (2005)
6. Hofmann, T.: Latent semantic models for collaborative filtering. ACM Transactions on Information Systems **22**(1), 89–115 (2004)
7. Lo, C.P., Yeung, A.K.W.: Concepts and Techniques of Geographic Information Systems. Prentice Hall (2007)
8. Majumder, S., Das, J.: Scalable recommendation technique using multiplicatively weighted voronoi diagrams. Indian patent pending, Ref. No. 966/KOL/2014, filed September 2014
9. Resnick, P., Iacovou, N., Suchak, M., Bergstorm, P., Riedl, J.: Grouplens: an open architecture for collaborative filtering of netnews. In: Proceedings of ACM Conference on Computer Supported Cooperative Work, pp. 175–186 (1994)
10. Sarwar, B., Karypis, G., Konstan, J., Riedl, J.: Application of dimensionality reduction in recommender systems - a case study. In: WebKDD Workshop (2000)
11. Sarwar, B., Karypis, G., Konstan, J., Riedl, J.: Recommender systems for large-scale e-commerce: Scalable neighborhood formation using clustering. In: Proceedings of the Fifth International Conference on Computer and Information Technology, pp. 158–167 (2002)
12. Sarwat, M., Levandoski, J.J., Eldawy, A., Mokbel, M.F.: LARS*: An efficient and scalable location-aware recommender system. IEEE Transactions on Knowledge and Data Engineering **26**(6), 1384–1399 (2014)
13. Su, X., Khoshgoftaar, T.: A survey of collaborative filtering techniques. Advances in Artificial Intelligence 2009 (2009)

Novel Methods and Algorithms

Principal Sensitivity Analysis

Sotetsu Koyamada[1,2]($\boxtimes$), Masanori Koyama[1], Ken Nakae[1], and Shin Ishii[1,2]

[1] Graduate School of Informatics, Kyoto University, Kyoto, Japan
[2] ATR Cognitive Mechanisms Laboratories, Kyoto, Japan
koyamada-s@sys.i.kyoto-u.ac.jp, ishii@i.kyoto-u.ac.jp

Abstract. We present a novel algorithm (Principal Sensitivity Analysis; PSA) to analyze the knowledge of the classifier obtained from supervised machine learning techniques. In particular, we define principal sensitivity map (PSM) as the direction on the input space to which the trained classifier is most sensitive, and use analogously defined k-th PSM to define a basis for the input space. We train neural networks with artificial data and real data, and apply the algorithm to the obtained supervised classifiers. We then visualize the PSMs to demonstrate the PSA's ability to decompose the knowledge acquired by the trained classifiers.

Keywords: Sensitivity analysis · Sensitivity map · PCA · Dark knowledge · Knowledge decomposition

1 Introduction

Machine learning is a powerful methodology to construct efficient and robust predictors and classifiers. Literature suggests its ability in the supervised context not only to reproduce "intuition and experience" based on human supervision [1], but also to successfully classify the objects that humans cannot sufficiently classify with inspection alone [2,3].

This work is motivated by the cases in which the machine classifier eclipses the human decisions. We may say that this is the case in which the classifier holds more knowledge about the classes than us, because our incompetence in the classification problems can be attributed solely to our lack of understanding about the class properties and/or the similarity metrics. The superiority of nonlinear machine learning techniques strongly suggests that the trained classifiers capture the "invisible" properties of the subject classes. Geoff Hinton solidified this into the philosophy of "dark knowledge" captured within the trained classifiers [4]. One might therefore be motivated to enhance understanding of subject classes by studying the way the trained machine acquires the information.

Unfortunately, trained classifiers are often so complex that they defy human interpretation. Although some efforts have been made to "visualize" the classifiers [5,6], there is still much room left for improvement. The machine learning techniques in neuroimaging, for example, prefer linear kernels to nonlinear kernels because of the lack of visualization techniques [7]. For the visualization

© Springer International Publishing Switzerland 2015
T. Cao et al. (Eds.): PAKDD 2015, Part I, LNAI 9077, pp. 621–632, 2015.
DOI: 10.1007/978-3-319-18038-0_48

of high-dimensional feature space of machine learners, Zurada et al. [8,9] and Kjems et al. [10] presented seminal works. Zurada et al. developed "sensitivity analysis" in order to "delete unimportant data components for feedforward neural networks." Kjems et al. visualized Zurada's idea as "sensitivity map" in the context of neuroimaging. In this study, we attempt to generalize the idea of sensitivity analysis, and develop a new framework that aids us in extracting the knowledge from classifiers that are trained in a supervised manner. Our framework is superior to the predecessors in that it can:

1. be used to identify a pair of discriminative input features that act oppositely in characterizing a class,
2. identify *combinations* of discriminative features that strongly characterize the subject classes,
3. provide platform for developing sparse, visually intuitive sensitivity maps.

The new framework gives rise to the algorithm that we refer to as "Principal Sensitivity Analysis (PSA)," which is analogous to the well-established Principal Component Analysis (PCA).

2   Methods

2.1   Conventional Sensitivity Analysis

Before introducing the PSA, we describe the original sensitivity map introduced in [10]. Let d be the dimension of the input space, and let $f : \mathbb{R}^d \to \mathbb{R}$ be the classifier function obtained from supervised training. In the case of SVM, f may be the discriminant function. In the case of nonlinear neural networks, f may represent the function (or log of the function) that maps the input to the output of a unit in the final layer. We are interested in the expected sensitivity of f with respect to the i-th input feature. This can be written as

$$s_i := \int \left(\frac{\partial f(\boldsymbol{x})}{\partial x_i} \right)^2 q(\boldsymbol{x}) d\boldsymbol{x}, \tag{1}$$

where q is the distribution over the input space. In actual implementation, the integral (1) is computed with the empirical distribution q of the test dataset. Now, the vector

$$\boldsymbol{s} := (s_1, \ldots, s_d) \tag{2}$$

of these values will give us an intuitive measure for the degree of importance that the classifier attaches to each input. Kjems et al. [10] defined $\boldsymbol{s}$ as **sensitivity map** over the set of input features.

2.2   Sensitivity in Arbitrary Direction

Here, we generalize the definition (1). We define $s(v)$ as the sensitivity of f in arbitrary direction $v := \sum_i^d v_i e_i$, where e_i denotes the i-th standard basis in $\mathbb{R}^d$:

$$s(v) := \int \left(\frac{\partial f(x)}{\partial v} \right)^2 q(x)\, dx. \tag{3}$$

Recall that the directional derivative is defined by

$$\frac{\partial f(x)}{\partial v} := \sum_{i=1}^d v_i \frac{\partial f(x)}{\partial x_i}.$$

Note that when we define the *sensitivity inner product*

$$\langle e_i, e_j \rangle_s := \int \left(\frac{\partial f(x)}{\partial x_i} \right) \left(\frac{\partial f(x)}{\partial x_j} \right) q(x)\, dx, \tag{4}$$

we can rewrite $s(v)$ with the corresponding *sensitivity norm*, as follows:

$$
\begin{aligned}
\|v\|_s^2 &:= \langle v, v \rangle_s \\
&= \left\langle \sum_i v_i e_i, \sum_j v_j e_j \right\rangle_s \\
&= \sum_{i,j} v_i v_j \langle e_i, e_j \rangle_s.
\end{aligned}
\tag{5}
$$

This inner product defines the kernel metric corresponding to the positive definite matrix K with ij-th entry given by $K_{ij} := \langle e_i, e_j \rangle_s$. This allows us to write

$$s(v) = v^{\mathrm{T}} K v. \tag{6}$$

2.3   Principal Sensitivity Map and PSA

The classical setting (2) was developed in order to quantify the sensitivity of f with respect to each individual input feature. We attempt to generalize this idea and seek the *combination of the input features* for which f is most sensitive, or the combination of the input features that is *"principal"* in the evaluation of the sensitivity of f. We can quantify such combination by the vector v, solving the following optimization problem about v:

$$
\begin{aligned}
\text{maximize} \quad & v^{\mathrm{T}} K v \\
\text{subject to} \quad & v^{\mathrm{T}} v = 1.
\end{aligned}
\tag{7}
$$

The solution to this problem is simply the maximal eigenvector $\pm v^*$ of K. Note that v_i represents the contribution of the i-th input feature to this principal combination, and this gives rise to the map over the set of all input features.

As such, we can say that v is the **principal sensitivity map (PSM)** over the set of input features. From now on, we call s in the classical definition (2) as the **standard sensitivity map** and make the distinction. The magnitude of v_i represents the extent to which f is sensitive to the i-th input feature, and the sign of v_i will tell us the relative direction to which the input feature influences f. The new map is thus richer in information than the standard sensitivity map. In Section 3.1 we will demonstrate the benefit of this extra information.

Principal Sensitivity Analysis (PSA). We can naturally extend our construction above and also consider other eigenvectors of K. We can find these vectors by solving the following optimization problem about V:

$$\begin{aligned} \text{maximize} \quad & \text{Tr}\left(V^{\mathrm{T}}KV\right) \\ \text{subject to} \quad & v_i^{\mathrm{T}}v_j = \delta_{ij}, \end{aligned} \tag{8}$$

where V is a $d \times d$ matrix. As is well known, such V is given by the invertible matrix with each column corresponding to K's eigenvector. We may define k-th dominant eigenvector v_k as the k-th **principal sensitivity map**. These sub-principal sensitivity maps grant us access to even richer information that underlies the dataset. We will show the benefits of these additional maps in Fig. 3. From now on, we will refer to the first PSM by just PSM, unless noted otherwise.

Recall that, in the ordinary PCA, K in (8) is given by the covariance $E\left[xx^{\mathrm{T}}\right]$, where x is the centered random variable. Note that in our particular case, if we put

$$r(x) := \left(\left(\frac{\partial f(x)}{\partial x_1}\right), \dots, \left(\frac{\partial f(x)}{\partial x_d}\right)\right)^{\mathrm{T}}, \tag{9}$$

then we may write $K = \int r(x)r(x)^{\mathrm{T}}q(x)\,dx = E\left[r(x)r(x)^{\mathrm{T}}\right]$. We see that our algorithm can thus be seen as the PCA applied to the covariance of $r(x)$ without centering.

Sparse PSA. One may use the new definition (8) as a starting point to develop sparse, visually intuitive sensitivity maps. For example, we may introduce the existing techniques in sparse PCA and sparse coding into our framework. We may do so [11] by replacing the covariance matrix in its derivation with our K. In particular, we can define an alternative optimization problem about V and α_i:

$$\begin{aligned} \text{minimize} \quad & \frac{1}{2}\sum_{i}^{N}\|r(x_i) - V\alpha_i\|_2^2 + \lambda \sum_{k}^{p}\|v_k\|_1 \\ \text{subject to} \quad & \|\alpha_i\|_2 = 1, \end{aligned} \tag{10}$$

where p is the number of sensitivity maps and N is the number of samples. For the implementation, we used scikit-learn [12].

2.4   Experiments

In order to demonstrate the effectiveness of the PSA, we applied the analysis to the classifiers that we trained with artificial data and MNIST data. Our artificial data is a simplified version of the MNIST data in which the object's *orientation* and *positioning* are registered from the beginning. All samples in the artificial data are constructed by adding noises to the common set of templates representing the numerics from 0 through 9 (Fig. 1). We then fabricated the artificial noise in three steps: we (1) flipped the bit of each pixel in the template picture with probability $p = 0.2$, (2) added Gaussian noise $\mathcal{N}(0, 0.1)$ to the intensity, and (3) truncated the negative intensities. The sample size was set to be equal to that of MNIST. Our training data, validation data, and test data consisted respectively of 50,000, 10,000, and 10,000 sample patterns. Using the

Fig. 1.   (a) Templates. (b) Noisy samples. Each figure is of 28×28 pixels.

artificial dataset above and the standard MNIST, we trained a feed forward neural network for the classification of ten numerics. In Table 1, we provide the structure of the neural network and its performance over each dataset. For either dataset, the training was conducted via stochastic gradient descent with constant learning rate. We also adopted a dropout method [13] only for the training on the MNIST dataset. The output from each unit in the final layer is given by the posterior probability of each class c. For computational purpose, we transform this output by log:

$$f_c(\boldsymbol{x}) := \log P(Y = c \mid \boldsymbol{x}), \tag{11}$$

where Y is, in the model governing the neural network, a random variable representing the class that the classifier assigns to the input $\boldsymbol{x}$. We then constructed the PSM and the standard sensitivity map for the f_c given above.

Table 1. Summary of training setups based on neural networks

Data set	Architecture	Unit type	Dropout	Learning rate	Error[%]
Digital data	784-500-10	Logistic	No	0.1	0.36
MNIST	784-500-500-10	ReLU	Yes	0.1	1.37

3   Results

3.1   PSA of Classifier Trained on Artificial Dataset

We will describe three ways in which the PSA can be superior to the analysis based on standard sensitivity map.

Fig. 2 compares the PSM and standard sensitivity map, which were both obtained from the common neural networks trained for the same 10-class classification problem. The color intensity of i-th pixel represents the magnitude of v_i. Both maps capture the characters that the "colorless" rims and likewise "colorless" regions enclosed by edges are insignificant in the classification. Note that the (1st) PSM distinguishes the types of sensitivities by their sign. For each numeral, the PSM assigns opposite signs to "the edges whose *presence* is crucial in the characterization of the very numeral" and "the edges whose *absence* is crucial in the numeral's characterization." This information is not featured in the standard sensitivity map. For instance, in the sensitivity map for the numeral 1, the two edges on the right and the rest of the edges have the opposite sensitivity. As a result, we can verify the red figure of 1 in its PSM. We are able to clearly identify the unbroken figures of 2, 4, 5 and 9 in their corresponding PSM as well. We see that, with the extra information regarding the sign of the sensitivity over each pixel, PSM can provide us with much richer information than the standard counterpart.

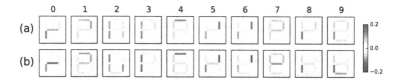

Fig. 2. (a) The standard sensitivity maps. (b) The PSMs.

Next, we will show the benefits of sub-principal sensitivity maps computed from PSA. Fig. 3(a) shows the 1st PSM through the 3rd PSM for the numerals 0 and 9.[1] In order to show how this extra information benefits us in visualization of the classification problem, we consider the following "local" sensitivity map integrated over the samples from a particular pair of classes:

$$s_{c,c'}(v) = \int \left(\frac{\partial f_c(x)}{\partial v} \right)^2 q_{c,c'}(x)dx, \tag{12}$$

where $q_{c,c'}$ is the empirical distribution over the set of samples generated from the classes c and c'. To get the intuition about this map, note that this value for $(c, c') = (9, 4)$ can also be pictorially written as

---

[1] We list the PSMs for all the numerals $(0, \ldots, 9)$ in the Appendix.

$$\lim_{\varepsilon \to 0} E_{\{9,4\}} \left[\left(\frac{\log P \left(Y = \boxed{\square} \mid \blacksquare + \varepsilon v \right) - \log P \left(Y = \boxed{\square} \mid \blacksquare \right)}{\varepsilon} \right)^2 \right], \quad (13)$$

where v can be the 3rd PSM of class 9, $\boxed{\square}$, for example. If v_k is the k-th PSM of the classifier, then $s_{c,c'}(v_k)$ quantifies *the sensitivity of the machine's answer to the binary classification problem of "c vs c'"* with respect to the perturbation of the input in the direction of v_k. By looking at this value for each k, we may visualize the ways that the classifier deals with the binary classification problem. Such visualization may aid us in learning from the classifiers the way to distinguish one class from another. Fig. 3(b) shows the values of $s_{c,c'}(v_k)$ for $c \in \{0,9\}$ and $k \in \{1,\dots,10\}$. We could see in the figure that, for the case of $(c,c') = (9,4)$, $s_{c,c'}(v_3)$ was larger than $s_{c,c'}(v_1)$. This suggests that the 3rd PSM is more helpful than the 1st PSM for distinguishing 4 from 9. We can actually verify this fact by observing that the 3rd PSM is especially intense at the top most edge, which can alone differentiate 4 from 9. We are able to confirm many other cases in which the sub-principal sensitivity maps were more helpful in capturing the characters in binary classification problems than the 1st PSM. Thus, PSA can provide us with the knowledge of the classifiers that was inaccessible with the previous method based on the standard sensitivity map.

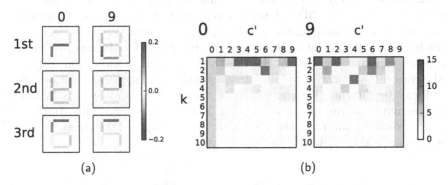

Fig. 3. (a) 1st $\sim$ 3rd PSMs of the classifier outputs f_c for the numerals 0 and 9. (b) $s_{c,c'}(v_k)$ for $c \in \{0,9\}$, $k \in \{1,\dots,10\}$, and $c' \in \{0,\dots,9\}\backslash\{c\}$.

Finally, we demonstrate the usefulness of formulation (8) in the construction of sparse and intuitive sensitivity map. Fig. 4 depicts the sensitivity maps obtained from the application of our sparse PSA in (10) to the data above. Note that the sparse PSA not only washes away rather irrelevant pixels from the canvas, but it also assigns very high intensity to essential pixels. With these "localized" maps, we can better understand the discriminative features utilized by the trained classifiers.

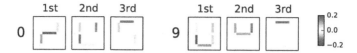

Fig. 4. Results of the sparse PSA on the classifiers f_c with $p = 3$ for the numerals 0 and 9. We ranked the 3 basis elements by the magnitude of $s(\boldsymbol{v})$. We selected the regularization term of $\lambda = 5$, and each PSM was normalized so that its L_2 norm was 1.

3.2   PSA of Classifier Trained on MNIST Dataset

We trained a nonlinear neural network-based classifier on the MNIST dataset, which consists of hand-written digits from 0 through 9. We then analyzed the trained classifier with our PSA. This dataset illuminates a particular challenge to be confronted in the application of the PSA. By default, hand-written objects do not share common displacement and orientation. Without an appropriate registration of input space, the meaning of each pixel can vary across the samples, making the visualization unintuitive. This is typical in some of the real-world classification problems. In the fields of applied science, standard registration procedure is often applied to the dataset before the construction of the classifiers. For example, in neuroimaing, one partitions the image data into anatomical regions after registration based on the standard brain, and represents each one of them by a group of voxels. In other areas of science, one does not necessarily have to face such problems. In genetics, data can be naturally partitioned into genes [14]. Likewise, in meteorology, 3D dataset is often translated into voxel structures, and a group of voxels may represent geographical region of specific terrain [15]. In this light, the digit recognition in unregistered MNIST data may not be an appropriate example for showing the effectiveness of our visualization method. For the reason that we will explain later, registration of multiclass dataset like MNIST can be difficult. We chose MNIST dataset here because it is familiar in the community of machine learning. Fig. 5 summarizes the results. Both the standard sensitivity map and the PSM were able to capture the character that outer rims are rather useless in the classification.

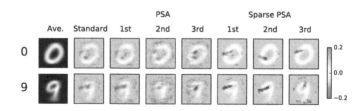

Fig. 5. Standard sensitivity map, PSA, and sparse PSA for $c \in \{0, 9\}$, $k \in \{1, 2, 3\}$, and $c' \in \{0, \ldots, 9\} \backslash \{c\}$. Ave. stands for the average of the testing dataset for the corresponding numerals.

Fig. 6 shows the values of $s_{c,c'}(\boldsymbol{v}_k)$. We can verify that small numbers of PSMs are complementing each other in their contributions to the binary classifications.

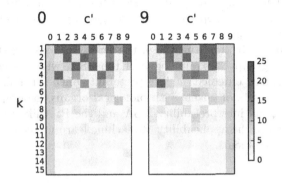

Fig. 6. $s_{c,c'}(\boldsymbol{v}_k)$ for $c \in \{0,9\}$, $k \in \{1,\ldots,15\}$, and $c' \in \{0,\ldots,9\}\backslash\{c\}$

We also applied sparse PSA to the classifier with $p = 3$ and $\lambda = 40$ (Fig. 5). We see that the sparse PSA highlights the essential pixels much more clearly than the normal PSA.

Since the orientation and position of each numeral pattern varies across the samples in this dataset, input dimensions hold different meanings in different samples. To perform more effective visualization, we would need registration to adjust each numeral pattern to a common standard template. This problem might not be straightforward, since one must prepare different templates for different numeral patterns. An elegant standardization suitable for our PSA-based visualization remains as a future study.

4   Discussion

We proposed a method to decompose the input space based on the sensitivity of classifiers. We assessed its performance on classifiers trained with artificial data and MNIST data. The visualization achieved with our PSA reveals at least two general aspects of the classifiers trained in this experiment. First, note in Fig. 3(b) and Fig. 6 that the first few (~ 10) PSMs of the trained classifier dominate the sensitivity for the binary classification problem. Second, we see that the classifier use these few PSMs out of 784 dimensions to solve different binary classification problems. We are thus able to see that the nonlinear classifiers of the neural network solve vast number of specific classification problems (such as binary classification problems) *simultaneously and efficiently* by tuning its sensitivity to the input in a data-driven manner. One cannot attain this information with the standard sensitivity map [8–10] alone. With PSA, one can visualize the decomposition of the knowledge about the input space learnt by the classifier. From the PSA of efficient classifier, one may obtain a meaningful decomposition of the input space that can possibly aid us in solving wide variety of problems. In medical science, for example, PSA might identify a combination of the biological regions that are helpful in diagnosis. PSA might also prove beneficial in sciences using voxel based approaches, such as geology, atmospheric science, and oceanography.

We may incorporate the principle of the PSA into existing standard statistical methods. A group Lasso analogue of the PSA, which is currently under our development, may enhance the interpretability of the visualization even further by identifying sets of voxels with biological organs, geographical location, etc. By improving its interpretability, PSA and the PSA-like techniques might significantly increase the applicability of machine learning techniques to various high-dimensional problems.

Appendix

In this section we list the figures that we omitted in the main text.

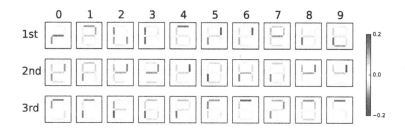

Fig. 7. 1st $\sim$ 3rd PSMs of the classifier trained on the artificial dataset

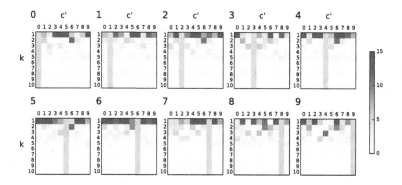

Fig. 8. $s_{c,c'}(\boldsymbol{v}_k)$ on the artificial dataset

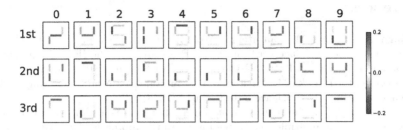

Fig. 9. Results of the sparse PSA on the classifiers trained on the artificial dataset

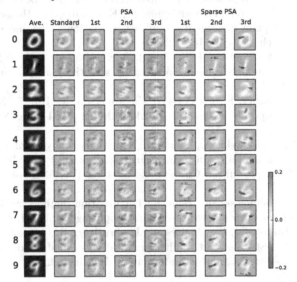

Fig. 10. Average, standard sensitivity map, PSA, and sparse PSA on MNIST data

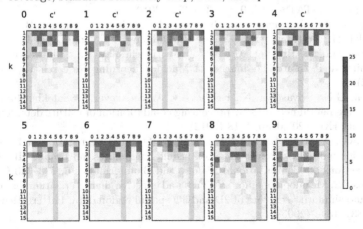

Fig. 11. $s_{c,c'}(\boldsymbol{v}_k)$ on MNSIT dataset

References

1. Taigman, Y., Yang, M., Ranzato, M., Wolf, L.: Deepface: closing the gap to human-level performance in face verification. In: Proceedings of the IEEE Conference on Computer Vision and Pattern Recognition (CVPR), pp. 1701–1708 (2014)
2. Horikawa, T., Tamaki, M., Miyawaki, Y., Kamitani, Y.: Neural decoding of visual imagery during sleep. Science **340**, 639–642 (2013)
3. Uberbacher, E.C., Mural, R.J.: Locating protein-coding regions in human DNA sequences by a multiple sensor-neural network approach. Proceedings of the National Academy of Sciences **88**, 11261–11265 (1991)
4. Hinton, G.E.: Dark knowledge. Presented as the keynote in BayLearn (2014)
5. Baehrens, D., Schroeter, T., Harmeling, S., Kawanabe, M., Hansen, K., Muller, K.R.: How to explain individual classification decisions. The Journal of Machine Learning Research **11**, 1803–1831 (2010)
6. Rasmussen, P.M., Madsen, K.H., Lund, T.E., Hansen, L.K.: Visualization of nonlinear kernel models in neuroimaging by sensitivity maps. NeuroImage **55**, 1120–1131 (2011)
7. LaConte, S., Strother, S., Cherkassky, V., Anderson, J., Hu, X.: Support vector machines for temporal classification of block design fMRI data. NeuroImage **26**, 317–329 (2005)
8. Zurada, J.M., Malinowski, A., Cloete, I.: Sensitivity analysis for minimization of input data dimension for feedforward neural network. In: Proceedings of the IEEE International Symposium on Circuits and Systems (ISCAS), vol. 6, pp. 447–450 (1994)
9. Zurada, J.M., Malinowski, A., Usui, S.: Perturbation method for deleting redundant inputs of perceptron networks. Neurocomputing **14**, 177–193 (1997)
10. Kjems, U., Hansen, L.K., Anderson, J., Frutiger, S., Muley, S., Sidtis, J., Rottenberg, D., Strother, S.C.: The quantitative evaluation of functional neuroimaging experiments: mutual information learning curves. NeuroImage **15**, 772–786 (2002)
11. Jenatton, R., Obozinski, G., Bach, F.: Structured sparse principal component analysis. In: Proceedings of the International Conference on Artificial Intelligence and Statistics (AISTATS), vol. 9, pp. 366–373 (2010)
12. Pedregosa, F., Varoquaux, G., Gramfort, A., Michel, V., Thirion, B., Grisel, O., Blondel, M., Prettenhofer, P., Weiss, R., Dubourg, V., Vanderplas, J., Passos, A., Cournapeau, D., Brucher, M., Perrot, M., Duchesnay, E.: Scikit-learn: Machine Learning in Python. The Journal of Machine Learning Research **12**, 2825–2830 (2012)
13. Hinton, G.E., Srivastava, N., Krizhevsky, A., Sutskever, I., Salakhutdinov, R.R.: Improving neural networks by preventing co-adaptation of feature detectors. arXiv preprint arXiv:1207.0580, 1–18 (2012)
14. Yukinawa, N., Oba, S., Kato, K., Ishii, S.: Optimal aggregation of binary classifiers for multiclass cancer diagnosis using gene expression profiles. IEEE/ACM Transactions on Computational Biology and Bioinformatics **6**, 333–343 (2009)
15. Kontos, D., Megalooikonomou, V.: Fast and effective characterization for classification and similarity searches of 2D and 3D spatial region data. Pattern Recognition **38**, 1831–1846 (2005)

SocNL: Bayesian Label Propagation
with Confidence

Yuto Yamaguchi[1]($\boxtimes$), Christos Faloutsos[2], and Hiroyuki Kitagawa[1]

[1] University of Tsukuba, Tsukuba, Japan
yuto_ymgc@kde.cs.tsukuba.ac.jp, kitagawa@cs.tsukuba.ac.jp
[2] Carnegie Mellon University, Pittsburgh, USA
christos@cs.cmu.edu

Abstract. How can we predict Smith's main hobby if we know the main hobby of Smith's friends? Can we measure the confidence in our prediction if we are given the main hobby of only a few of Smith's friends? In this paper, we focus on how to estimate the confidence on the node classification problem. Providing a confidence level for the classification problem is important because most nodes in real world networks tend to have few neighbors, and thus, a small amount of evidence. Our contributions are three-fold: (a) *novel algorithm*; we propose a semi-supervised learning algorithm that converges fast, and provides the confidence estimate (b) *theoretical analysis*; we show the solid theoretical foundation of our algorithm and the connections to label propagation and Bayesian inference (c) *empirical analysis*; we perform extensive experiments on three different real networks. Specifically, the experimental results demonstrate that our algorithm outperforms other algorithms on graphs with less smoothness and low label density.

1 Introduction

If we know that 5 out of 6 of Smith's friends love to play tennis, what would you say about Smith's main hobby? Same question, when we know that Johnson's 50 out of 60 friends love to play tennis - are we more confident about Smith, or about Johnson? Most people would be more confident in the latter case, despite the fact that the ratio of tennis-to-non-tennis friends, is the same in both cases. In this paper, we address the node classification problem on networks. Networks appear in numerous real-world applications, like social networks, citation networks, and biological networks. Often, the nodes of these networks have *labels*: E.g., users in social networks have demographic attributes (gender, age bracket, education level, e.t.c.) [10]. Although these labels are useful for a lot of practical applications, labels of a majority of nodes are often unavailable, which makes the node classification problem more important.

The node classification problem is informally as follows: **given** a partially labeled graph with labeled and unlabeled nodes, **find** correct labels of unlabeled nodes based on labeled nodes. Real-world applications of this setting are numerous, include research paper classification [2], personalized video suggestion [3], and anomaly detection [9].

© Springer International Publishing Switzerland 2015
T. Cao et al. (Eds.): PAKDD 2015, Part I, LNAI 9077, pp. 633–645, 2015.
DOI: 10.1007/978-3-319-18038-0_49

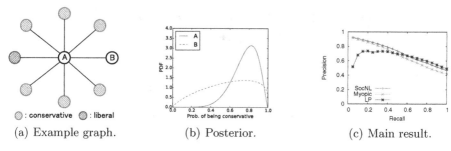

(a) Example graph.   (b) Posterior.   (c) Main result.

Fig. 1. (a)-(b) **Main idea**: An example where we want to classify nodes A and B. Our algorithm provides more confident posterior for node A because A has sufficient evidence (i.e., neighbors). (c): **Proposed SocNL wins or ties in the first place**: against other methods on POKEC network.

Here we propose SocNL, a *semi-supervised learning* (SSL) algorithm for the node classification problem. The SSL algorithm is one of the most promising algorithms for the node classification problem on sparsely labeled networks [4,14,15]. Similar to the other SSL algorithms, SocNL is also based on the *smoothness hypothesis* where a connected nodes tend to share a label. The main advantage of our algorithm is to provide the reliable confidence for each result[1]. From Bayesian inference perspective, the more evidence we observe, the more confident the estimate is, which is the principle we adopt in this paper.

The Main Idea and the Main Results. Figures 1(a) and 1(b) show our main idea. Suppose we have an example graph where nodes A and B are classified into conservative or liberal. As the intermediate result, SocNL can output the *posterior distribution* of the probability of being conservative for A and B shown in Figure 1(b), which have different shapes. The important point is that the posterior of A has more focused peak than B because A gets more evidence from its neighbors. By taking the posterior into account, the final result of the probability of being conservative is 0.76 for A while 0.59 for B, which agrees with our intuition that A is more confidently conservative. This result is quite differ from the result by label propagation [15] that does not consider the amount of evidence, where both nodes are assigned the identical probability 0.86. Figure 1(c) shows our main result illustrating SocNL wins or ties in the first place against other methods on POKEC network (see details in Section 6).

Contributions. Our contributions are summarized as follows:

1. **Novel Algorithm**: We propose a semi-supervised algorithm that is (a) *simple*; it only requires solving a linear system, (b) *fast*; each iteration of its recursive inference is linear on the input size and is proved to converge, and (c) *provides reliable confidence*; it takes into account the amount of the evidence to provide the reliable confidence.

---

[1] The name SocNL stands for *Socratic Node Labeling*, since it is self-aware, in the sense that it knows what it does not know, reminiscing of the Socratic principle "I know that I know nothing."

2. **Theoretical Analysis**: We show the solid theoretical foundation of our algorithm, indicating the convergence guarantee and the complexity. Also, we show that the special case of SocNL is equivalent to label propagation [15] and Bayesian inference over Dirichlet compound multinomial.

3. **Empirical Analysis**: We perform extensive experiments on three different real networks: a blog-citation network, a co-authorship network, and a social network with millions of nodes and edges. The experimental results demonstrate that our algorithm outperforms other algorithms on graphs with less smoothness and low label density.

Outline. The rest of the paper is organized as standard: problem definition, algorithm description, theoretical analysis, empirical analysis, and conclusion.

2   Related Work

In this section, we overview the semi-supervised learning, which makes use of unlabeled data in addition to labeled data to improve the performance. Most of SSL algorithms are classified into the generative models, the low-density separation, and the graph-based methods [5]. Here, we focus on the graph-based methods which our algorithm belongs to.

There have been proposed a lot of graph-based SSL algorithms, such as label propagation (LP) [15], label spreading [14], and manifold regularization [4]. Although these methods have achieved successful improvements, they do not consider the amount of evidence. This may cause some problems when we deal with the node classification. Because most of real-world networks have power law distributions [6] where a majority of nodes have a small number of neighbors, which means we cannot obtain sufficient evidence from neighbors.

Recently, a few algorithms aiming at providing the reliable confidence have been proposed [7,8,11]. Fang et al. [7] proposed DGR (Dirichlet-based Graph Regularization) that assumes every node has a Dirichlet prior and propagates it along edges. Although DGR provides the posterior like our algorithm, it has to solve a optimization problem numerically for each iteration, indicating it does not scale so much. Orbach et al. [11] devised an algorithm called TACO (Transduction Algorithm with Confidence). TACO infers the label probability and the uncertainty of it simultaneously. Chen et al. [8] also proposed an SSL algorithm called ReLISH (Reliable Label Inference via Smoothness Hypothesis). ReLISH is also formulated as a convex optimization problem and has a clear closed-form solution. However, it requires $O(n^3)$ complexity, meaning it does not fit large scale networks.

Table 1 shows the qualitative comparison between our algorithm and these algorithms. Note that LP can also incorporate the prior knowledge, but it does not consider the amount of evidence, meaning that LP does not provide the reliable confidence (evaluated in Section 6).

Table 1. Qualitative comparison

	LP [15]	DGR [7]	TACO [11]	ReLISH [8]	SocNL
Confidence		✓	✓	✓	✓
Large scale networks	✓		✓		✓
Closed-form solution	✓			✓	✓
Interpretable parameters	✓	✓			✓

3 Problem Definition

This section defines the terminologies and formulates the node classification problem. Table 2 gives the list of symbols. Let $G = (V, E)$ be a partially labeled graph where V is set of N nodes and E is set of M edges. The set of nodes is composed of two types of nodes. $V^L \subset V$ is a set of L *labeled nodes* whose labels are known, while $V^U = V \setminus V^L$ is a set of U *unlabeled nodes* whose labels are unavailable. Let $\mathcal{Y}$ be the set of K possible labels, and $Y_L = \{y_1, y_2, \cdots, y_L\}$ be the label assignments for the corresponding nodes in V^L. Using these terminologies, the node classification problem is formulated as follows:

Problem 1 (Node Classification)

- **Given**: a partially labeled graph $G = (V, E)$
- **Find**: label probability f_{ij} that node i has label j.

After obtaining the label probability for unlabeled nodes, we can develop a classification function $\mathcal{C}(v_i) = \arg\max_k f_{ik}$. Note that the maximum probability value $\max_k f_{ik}$ for each node indicates how confident the result is, which is used in the experiments.

Table 2. Symbols and Definitions

Symbols	Definitions
A	Adjacency matrix.
N, M, K	# of nodes, edges, and labels.
L, U	# of labeled and unlabeled nodes.
α_j	Prior belief that nodes have label j

4 Proposed Method

In this section, we propose SocNL, a novel semi-supervised node classification algorithm. Similar to the other SSL algorithms, SocNL is also based on the *smoothness hypothesis*: connected nodes are likely to share a label. Also, we adopt Bayesian principle that the estimate is inherently uncertain and becomes the more confident if we observe the mode evidence. This principle is suitable to our problem because if a node has many neighbors, we can obtain much evidence to infer the label probability of that node. On the other hand, if we cannot obtain sufficient evidence (i.e., a small number of neighbors), the inference result is unreliable. To formulate these ideas, SocNL adopts the followings:

- *Label propagation*: SocNL propagates labels from labeled nodes to unlabeled nodes based on the smoothness hypothesis.
- *Bayesian inference*: SocNL assigns each node with the prior label probability and then updates it by the evidence from its neighbors, which takes into account the amount of evidence and thus provides the reliable confidence.

4.1   The Model

In this section, we formulate the model. For now, let's ignore the unlabeled neighbors of target node i for simplicity. What we want to do here is to infer i's label probability $f_{ik} = P(\hat{y}_i = k|\tilde{N}_i)$ for all k given the set of i's labeled neighbors $\tilde{N}_i$, where $\hat{y}_i$ is the predicted label of i.

SocNL assumes that the label is a categorical random variable as $P(\hat{y}_i = k|\boldsymbol{\theta}) = \theta_k$ where $\boldsymbol{\theta}$ is the parameter of the categorical distribution. According to the smoothness hypothesis, we believe that a neighbor of node i shares the same parameter $\boldsymbol{\theta}$ as i. Then we get the multinomial likelihood function of labels of neighbors of i as follows:

$$P(\tilde{N}_i|\boldsymbol{\theta}) = Mul(\tilde{N}_i|\boldsymbol{\theta}) \propto \prod_{k=1}^{K} \prod_{j \in \tilde{N}_i} \theta_k^{\delta(y_j,k)} = \prod_{k=1}^{K} \theta_k^{n_{ik}}, \qquad (1)$$

where $\delta(y_j, k)$ takes 1 if $y_j = k$ otherwise 0, and $n_{ik} = \sum_{j \in \tilde{N}_i} \delta(y_j, k)$ is the number of i's neighbors whose label is k. Here we assume that labels of neighbors are i.i.d. As the conjugate prior of the multinomial distribution is Dirichlet distribution, let's think that the prior of parameter $\boldsymbol{\theta}$ is Dirichlet distribution:

$$P(\boldsymbol{\theta}) = Dir(\boldsymbol{\theta}|\boldsymbol{\alpha}) \propto \prod_{k=1}^{K} \theta_k^{\alpha_k-1}, \qquad (2)$$

where $\boldsymbol{\alpha} = (\alpha_1, \alpha_2, \cdots, \alpha_K)^T$ is the parameter of Dirichlet distribution. Putting these together, we get the posterior distribution as follows:

$$P(\boldsymbol{\theta}|\tilde{N}_i) \propto \prod_{k=1}^{K} \theta_k^{n_{ik}+\alpha_k-1}. \qquad (3)$$

After obtaining the posterior distribution, we can write the posterior predictive distribution for node i's label as follows:

$$P(\hat{y}_i = k|\tilde{N}_i, \boldsymbol{\alpha}) = \int_{\boldsymbol{\theta}} P(\hat{y}_i = k|\boldsymbol{\theta})P(\boldsymbol{\theta}|\tilde{N}_i, \boldsymbol{\alpha})d\boldsymbol{\theta} = \frac{n_{ik} + \alpha_k}{|\tilde{N}_i| + \alpha_0}, \qquad (4)$$

where $\alpha_0 = \sum_{k=1}^{K} \alpha_k$. By integrating out parameter $\boldsymbol{\theta}$ of the posterior distribution, SocNL takes into account all the possible value of $\boldsymbol{\theta}$ according to the posterior. We name this solution *Myopic baseline*, which only uses labeled neighbors $\tilde{N}_i$ of target node i.

Algorithm 1. Iterative Algorithm

Require: explicit labels Y_L, adjacency matrix $\boldsymbol{A}$, prior α
1: $\boldsymbol{F}^0 \leftarrow initialize F()$
2: $k \leftarrow 0$
3: **repeat**
4:    $\boldsymbol{F}_U^{k+1} \leftarrow (\boldsymbol{D}_U + \alpha_0 \boldsymbol{I})^{-1} \left(\boldsymbol{A}_U \boldsymbol{F}^k + \mathbf{1}\alpha^T \right)$
5:    $k \leftarrow k + 1$
6: **until** error between $\boldsymbol{F}_U^{k+1}$ and $\boldsymbol{F}_U^k$ becomes sufficiently small
7: **return** $\boldsymbol{F}_U^k$

4.2   Iterative Algorithm

In this section we develop our full algorithm, SocNL, which utilizes both labeled and unlabeled neighbors N_i. In this case, we do not know $\delta(y_j, k)$ for unlabeled nodes. Hence, instead of simply counting $\delta(y_j, k)$, we calculate n_{ik} as follows:

$$n_{ik} = \sum_{j=1}^{N} A_{ij} P(\hat{y}_j = k | N_j), \tag{5}$$

where we use the adjacency matrix $\boldsymbol{A}$. We can think that n_{ik} behaves as the expectation value of the number of i's neighbors with label k. For labeled nodes, we set $P(\hat{y}_j = k | N_j) = \delta(y_j, k)$. Plugging Eqn 5 into Eqn 4 and using $f_{ik} = P(\hat{y}_i = k | N_i)$, we get:

$$f_{ik} = \frac{\sum_{j=1}^{N} A_{ij} f_{jk} + \alpha_k}{\sum_{j=1}^{N} A_{ij} + \alpha_0}. \tag{6}$$

Since this equation is in the recursive fashion, we devise an iterative algorithm to solve it.

Hereafter, we formulate the matrix form of the iterative algorithm. Let $\boldsymbol{F}$ be row normalized $N \times K$ matrix. We write $\boldsymbol{F}_L$ and $\boldsymbol{F}_U$ as the upper $L \times K$ and lower $U \times K$ sub-matrices of $\boldsymbol{F}$, respectively. Also, we write $\boldsymbol{A}_L$ and $\boldsymbol{A}_U$ in the same way. The subscript L and U mean that the sub-matrices correspond to the labeled nodes in V^L and unlabeled nodes in V^U, respectively. Recall that each labeled node has $f_{ik} = \delta(y_i, k)$, which corresponds to the components of $\boldsymbol{F}_L$.

Using the matrices defined thus far, we write the following assignment formula:

$$\boldsymbol{F}_U \leftarrow (\boldsymbol{D}_U + \alpha_0 \boldsymbol{I})^{-1} \left(\boldsymbol{A}_U \boldsymbol{F} + \mathbf{1}\alpha^T \right), \tag{7}$$

where $\mathbf{1}$ is U dimensional column vector where each component is 1. $\boldsymbol{D}_U$ is $U \times U$ diagonal matrix with diagonal component $[D_U]_{ii} = \sum_j A_{ij}$. As we will prove in Section 5, the iterative algorithm repeating this assignment formula always converges to the solution if $\alpha_k > 0$ for all k, which corresponds to valid Dirichlet prior. The iterative algorithm of SocNL is shown in Algorithm 1. All f_{ik} of unlabeled nodes are initialized as arbitrary values. Note that SocNL is applicable to directed graphs without any modification of Algorithm 1, where the adjacency matrix A is asymmetric.

5   Theoretical Analysis

In this section, we show that SocNL has the solid theoretical foundation and connections to label propagation and Bayesian inference. All omitted proofs are shown in the appendix.

Convergence and Complexity. Here we show the convergence guarantee, the convergence speed, and the complexity of SocNL.

Theorem 1. *The iterative algorithm of SocNL always converges on arbitrary graphs if $\alpha_k > 0$ for all k.*

Corollary 1. *The fixed point solution of SocNL is written as:*

$$\boldsymbol{F}_U = (\boldsymbol{I} - \boldsymbol{P}_{UU})^{-1} \left(\boldsymbol{P}_{UL} \boldsymbol{F}_L + \boldsymbol{r} \boldsymbol{\alpha}^T \right), \tag{8}$$

where

$$\boldsymbol{P}_{UU} = (\boldsymbol{D}_U + \alpha_0 \boldsymbol{I})^{-1} \boldsymbol{A}_{UU}, \tag{9}$$
$$\boldsymbol{P}_{UL} = (\boldsymbol{D}_U + \alpha_0 \boldsymbol{I})^{-1} \boldsymbol{A}_{UL},$$
$$\boldsymbol{r} = (\boldsymbol{D}_U + \alpha_0 \boldsymbol{I})^{-1} \boldsymbol{1}.$$

Proof. *It follows directly from the proof of Theorem 1 (Appendix A).*   □

Theorem 2. *SocNL with prior strength α_0 converges faster than SocNL with another prior strength β_0 if $\alpha_0 > \beta_0$.*

Theorem 3. *The time complexity of SocNL is $O(hK(N + M))$ where h is the number of iterations.*

Connection to label propagation. Next, we show that LP is a special case of SocNL.

Theorem 4. *The special case of SocNL with parameter $\alpha_k = 0$ for all k is equivalent to LP.*

This means SocNL still works even if parameter $\alpha_k = 0$ for all k although it corresponds to invalid Dirichlet prior.

Corollary 2. *SocNL converges faster than LP.*

Proof. *It follows directly from Theorems 2 and 4.*   □

Connection to Bayesian inference. As mentioned in Section 4, SocNL is a natural extension of Bayesian inference.

Theorem 5. *SocNL is equivalent to Bayesian inference over Dirichlet compound multinomial if we ignore all the unlabeled neighbors of target node i.*

According to Theorems 4 and 5, SocNL behaves as the bridge between label propagation and Bayesian inference.

Table 3. Datasets

	N	M	K	Smoothness	Directed
POLBLOGS [1]	1,490	19,090	2	0.91 (0.49)	✓
COAUTHOR [12]	27,644	66,832	4	0.80 (0.23)	
POKEC [13]	1,632,803	30,622,564	187	0.45 (0.01)	✓

6 Empirical Analysis

In this section, we report the empirical analysis of our algorithm to answer the following questions:

- **Q1 - Prior**: How does the prior strength affect the performance of SocNL?
- **Q2 - Accuracy**: How accurate SocNL is compared to LP and Myopic?
- **Q3 - Convergence**: How fast does SocNL converge?

Datasets. Three network datasets described in Table 3 are used in our experiments. *Smoothness* is the probability that a connected pair has the same label. Values in parentheses are the smoothness after performing randomization of labels. POLBLOGS is a blog-citation network where the labels are political leanings of blogs. COAUTHOR is a co-authorship network where node i and j are connected if they co-write a paper. Labels on this network are the research field of authors (DB, DM, ML, and AI). POKEC is a social network in Slovakia where node i has an out-going edge to node j if i follows j. Labels are home locations of users.

Evaluation. We divide a set of labeled nodes into training nodes (30%), validation nodes (35%), and test nodes (35%), where labels of validation nodes and test nodes are hidden. Validation nodes are used in Section 6.1 to validate the prior strength, and test nodes are used in Section 6.2 to compare the performance. We perform node classification to infer hidden labels. Then we report the precision@p that is the precision of top p% of test (or validation) nodes ordered by the confidence value $\max_k f_{ik}$.

Reproducibility. The datasets we use in this paper are all available on the web as shown in Table 3. Also, our code is available on our website[2].

6.1 Q1 - Prior

In this section we study how the prior strength affects the performance of SocNL. Throughout the experiments in this paper, we use the class mass ratio as the prior $\alpha_k = \lambda L_k / L$ where L_k denotes the number of labeled nodes with label k and λ is the prior strength parameter which equals to α_0. Figure 2 shows the results where we vary the prior strength λ from 0.001 to 10.

We can see that larger λ results in well-calibrated confidence (i.e., higher precision at lower recall) but lower overall precision (right-most). For this reason

---

[2] https://github.com/yamaguchiyuto/socnl

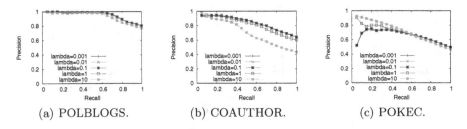

(a) POLBLOGS.   (b) COAUTHOR.   (c) POKEC.

Fig. 2. Larger prior strength is needed for less smoothness

we need to choose the correct value for λ to get a good trade-off between them. We can see that larger λ is needed to get well-calibrated confidence on POKEC network that has the relatively small smoothness. This is intuitive because if the smoothness is small, the data is not reliable and then we need more data to correctly update the prior, meaning that the large prior strength is needed.

Observation 1. *Strong prior is needed for graphs with weak smoothness.*

This agrees with intuition: the higher the value of α_0, the more emphasis it implies on the priors and the less emphasis on the evidence from the graph. Thus, if we know there is weak smoothness, then we should use the large prior. From the results, we choose 0.1 for POLBLOGS and COAUTHOR, and 10 for POKEC as the best prior strength for the experiments in the next section.

6.2 Q2 - Accuracy

Figure 3 shows the results, where we compare the accuracy of three algorithms. Myopic uses the same prior as SocNL. We can see that SocNL wins or ties in the first place on all networks. Specifically, LP shows low precision for "confident" results (left) because LP does not consider the amount of evidence. According to the results in this section and the last section, we can say that SocNL performs better than LP on graphs with less smoothness (e.g., POKEC) because the larger prior strength is needed.

Myopic shows low overall precision (right) because small-degree nodes do not have enough labeled neighbors. On the other hand, our algorithm achieves higher precision than Myopic because SocNL can propagate the evidence from more than 1-step away. This result means that similar to other SSL algorithms, SocNL tolerates low label density (i.e., small fraction of labeled nodes).

Observation 2. *SocNL outperforms LP and Myopic baseline on graphs with less smoothness and low label density.*

6.3 Q3 - Convergence

In this section, we compare the convergence speed of LP and SocNL with different λ. Figure 4 shows the results where x-axis indicates the number of iterations and the y-axis indicates the error between $\boldsymbol{F}_U^k$ and $\boldsymbol{F}_U^{k-1}$. The results confirm that SocNL converges faster if it uses the larger prior strength.

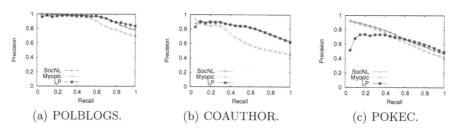

(a) POLBLOGS.   (b) COAUTHOR.   (c) POKEC.

Fig. 3. SocNL wins or ties in the first place on all networks

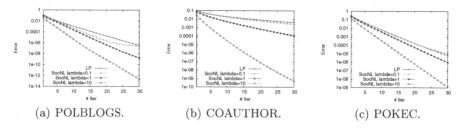

(a) POLBLOGS.   (b) COAUTHOR.   (c) POKEC.

Fig. 4. Faster convergence when using larger prior strength

Observation 3. *SocNL converges faster if it uses the larger prior strength.*

7   Conclusion

In this paper, we proposed SocNL, which addresses the node classification problem on networks. Specifically, we studied how to provide the reliable *confidence* of the classification result. Our contributions in this paper are:

- **Novel Algorithm**: we proposed a novel semi-supervised learning algorithm, called SocNL (Section 4).
- **Theoretical Analysis**: SocNL provably converges, and has connections to label propagation and Bayesian inference (Section 5).
- **Empirical Analysis**: experiments on three different real networks show that SocNL wins or ties in the first place (Section 6).

Our future work includes investigating how to address the ordinal or numerical labels such as movie ratings and user locations as coordinates. We plan to study the other distributions for the model.

Acknowledgments. This research was partly supported by the program *Research and Development on Real World Big Data Integration and Analysis* of the Ministry of Education, Culture, Sports, Science and Technology, Japan, and by JSPS KAKENHI, Grant-in-Aid for JSPS Fellows #242322. This material is also based upon work supported by the National Science Foundation (Grants No. IIS-1217559 and CNS-1314632), and by the Army Research Laboratory (under Cooperative Agreement Number W911NF-09-2-0053). Any opinions, findings, and conclusions or recommendations expressed in this material are those of the author(s) and do not necessarily reflect the views of the National

Science Foundation, or other funding parties. The U.S. Government is authorized to reproduce and distribute reprints for Government purposes notwithstanding any copyright notation here on.

Appendix

A. Proof of Theorem 1

Proof. *By rearranging Eqn 7, we get:*

$$F_U \leftarrow P_{UU} F_U + P_{UL} F_L + r\alpha^T.$$

It directly leads to

$$F_U = \lim_{n \to \infty} \left[(P_{UU})^n F_U^0 + \left\{ \sum_{i=0}^{n} (P_{UU})^i \right\} \left(P_{UL} F_L + r\alpha^T \right) \right].$$

Since $\alpha_0 > 0$, the ∞-norm $\|P_{UU}\|_\infty = \max_i \sum_j |[P_{UU}]_{ij}|$ is less than 1, which also means that the spectral radius $\rho(P_{UU})$ is less than 1. Therefore, the infinite series $\sum_{i=0}^{n}(P_{UU})^i$ converges to $(I - P_{UU})^{-1}$, and also $\lim_{n\to\infty}(P_{UU})^n = O$, which means SocNL converges. □

B. Proof of Theorem 2

Proof. *Here, let $\rho(\alpha_0)$ be the spectral radius of matrix P_{UU} when using prior strength α_0. According to Eqn 9, larger α_0 makes all the components of P_{UU} smaller, which means that all the eigenvalues of P_{UU} become smaller. Hence, the spectral radius satisfies $\rho(\alpha_0) < \rho(\beta_0)$. Also, the smaller spectral radius results in the faster convergence of the infinite series $\lim_{n\to\infty}(P_{UU})^n$. Consequently, SocNL with α_0 converges faster than SocNL with β_0.* □

C. Proof of Theorem 3

Proof. *In line 4 of Algorithm 1, the matrix multiplication needs $O(KM)$ and the matrix addition requires $O(KN)$ time. Putting together these operations, the time complexity of the iterative algorithm of SocNL is $O(hK(N + M))$.* □

D. Proof of Theorem 4

Proof. *By setting the prior $\alpha_k = 0$ for all k, we get:*

$$F_U = \left(I - D_U^{-1} A_{UU} \right)^{-1} D_U^{-1} A_{UL} F_L,$$

which is exactly the same fixed point solution of LP. □

E. Proof of Theorem 5

Proof. *Ignoring all the unlabeled neighbors means that we discard all edges among unlabeled nodes, leading to $A_{UU} = O$. Hence, the fixed point solution of SocNL becomes:*

$$F_U = P_{UL}F_L + r\alpha^T. \tag{10}$$

The element-wise form is as follows:

$$f_{ik} = \frac{\sum_{j=1}^N A_{ij}\delta(y_j, k) + \alpha_k}{\sum_{j=1}^N A_{ij} + \alpha_0} = \frac{n_{ik} + \alpha_k}{|\tilde{N}_i| + \alpha_0}, \tag{11}$$

where we use $f_{ik} = \delta(y_i, k)$ for labeled nodes. This equation is the same as Eqn 4, which is the solution for Bayesian inference. □

References

1. Adamic, L.A., Glance, N.: The political blogosphere and the 2004 us election: divided they blog. In: Proceedings of the 3rd International Workshop on Link Discovery, pp. 36–43. ACM (2005)
2. Aggarwal, C.C., Li, N.: On node classification in dynamic content-based networks. In: SDM, pp. 355–366. SIAM (2011)
3. Baluja, S., Seth, R., Sivakumar, D., Jing, Y., Yagnik, J., Kumar, S., Ravichandran, D., Aly, M.: Video suggestion and discovery for youtube: taking random walks through the view graph. In: WWW, pp. 895–904. ACM (2008)
4. Belkin, M., Niyogi, P., Sindhwani, V.: Manifold regularization: A geometric framework for learning from labeled and unlabeled examples. The Journal of Machine Learning Research 7, 2399–2434 (2006)
5. Chapelle, O., Schölkopf, B., Zien, A., et al.: Semi-supervised learning, vol. 2. MIT Press, Cambridge (2006)
6. Faloutsos, M., Faloutsos, P., Faloutsos, C.: On power-law relationships of the internet topology. In: SIGCOMM, pp. 251–262 (1999)
7. Fang, Y., Hsu, B.-J.P., Chang, K.C.-C.: Confidence-aware graph regularization with heterogeneous pairwise features. In: SIGIR, pp. 951–960. ACM (2012)
8. Gong, C., Tao, D., Fu, K., Yang, J.: Relish: Reliable label inference via smoothness hypothesis. In: AAAI (2014)
9. McGlohon, M., Bay, S., Anderle, M.G., Steier, D.M., Faloutsos, C.: Snare: a link analytic system for graph labeling and risk detection. In: KDD, pp. 1265–1274. ACM (2009)
10. Mislove, A., Viswanath, B., Gummadi, K.P., Druschel, P.: You are who you know: inferring user profiles in online social networks. In: WSDM, pp. 251–260. ACM (2010)
11. Orbach, M., Crammer, K.: Graph-based transduction with confidence. In: Flach, P.A., De Bie, T., Cristianini, N. (eds.) ECML PKDD 2012, Part II. LNCS, vol. 7524, pp. 323–338. Springer, Heidelberg (2012)

12. Sun, Y., Han, J., Gao, J., Yu, Y.: itopicmodel: Information network-integrated topic modeling. In: ICDM, pp. 493–502. IEEE (2009)
13. Takac, L., Zabovsky, M.: Data analysis in public social networks. In: International Scientific Conference and International Workshop Present Day Trends of Innovations (2012)
14. Zhou, D., Bousquet, O., Lal, T.N., Weston, J., Schölkopf, B.: Learning with local and global consistency. Advances in Neural Information Processing Systems **16**(16), 321–328 (2004)
15. Zhu, X., Ghahramani, Z., Lafferty, J., et al.: Semi-supervised learning using gaussian fields and harmonic functions. In: ICML, vol. 3, pp. 912–919 (2003)

An Incremental Local Distribution Network for Unsupervised Learning

Youlu Xing[1], Tongyi Cao[2], Ke Zhou[3], Furao Shen[1(✉)], and Jinxi Zhao[1]

[1] National Key Laboratory for Novel Software Technology, Department of Computer Science and Technology at Nanjing University, Nanjing, China
youluxing@sina.com, {frshen,jxzhao}@nju.edu.cn
[2] School of Physics at Nanjing University, Nanjing, China
caotongyi.is.tc@gmail.com
[3] School of Statistics at University of International Business and Economics, Bejing, China
02417@uibe.edu.cn

Abstract. We present an **I**ncremental **L**ocal **D**istribution **N**etwork (**ILDN**) for unsupervised learning, which combines the merits of matrix learning and incremental learning. It stores local distribution information in each node with covariant matrix and uses a vigilance parameter with statistical support to decide whether to extend the network. It has a statistics based merging mechanism and thus can obtain a precise and concise representation of the learning data called relaxation representation. Moreover, the denoising process based on data density makes ILDN robust to noise and practically useful. Experiments on artificial and real-world data in both "closed" and "open-ended" environment show the better accuracy, conciseness, and efficiency of ILDN over other methods.

Keywords: Incremental learning · Matrix learning · Relaxation representation

1 Introduction

In the field of unsupervised learning, many algorithms are designed to extract information from the distribution of data. Classic methods include k-means [1] and Neural Gas [2], which use fixed number of nodes to get different clusters. Self-Organizing Map [3] and Topology Representing Networks [4] represent the distribution and topological structure of the data with some given nodes.

Two drawbacks are obvious for these early methods. First, each node stores the mean feature vector of patterns belonged to the node and the metric is Euclidean. Correspondingly, each node is a simple unit with isotropic form and spherical class boundary, and thus has a poor description ability. Second, these methods need a predefined structure or number of nodes that requires additional knowledge of the data which is often hard to know. Also, the fixed structures render them unable to perform incremental learning or to handle the "*open-ended*" environment, i.e. data from new distributions may occur during learning.

© Springer International Publishing Switzerland 2015
T. Cao et al. (Eds.): PAKDD 2015, Part I, LNAI 9077, pp. 646–658, 2015.
DOI: 10.1007/978-3-319-18038-0_50

Two kinds of improvements have been made corresponding to these two problems, namely matrix learning and incremental networks. For the first problem, in order to obtain a precise yet concise representation of the data, a more expressive node is preferred, often based on mixture model, including PCASOM [5], Self-Organizing Mixture Network [6], localPCASOM [7], and MatrixNG and Matrix-SOM [8]. López-Rubio [9] gave a detailed review about these Mixture Model based Self-Organizing Maps which they called the Probabilistic Self-Organizing Map. For the second problem, many "growing networks" or "incremental networks" are proposed. Some grow after a fixed number of inputs learned such as GNG [11] and GSOM [12]; some use an adaptive threshold including GWR [13], Adjusted-SOINN [14], and TopoART [15]. Araujo and Rego [16] gave a detailed review about these incremental Self-Organizing Maps.

On one hand, though matrix learning methods record rich local distribution information and consider the anisotropy on different basis vectors, they have a common shortcoming - they cannot deal with the Stability-Plasticity Dilemma [10], i.e. many of these methods cannot learn data from new distributions after they are trained on the current data set; the other methods is able to learn data from new distributions but the previous learned knowledge will be forgotten, known as the "Catastrophic Forgetting". Thus, they can only work in a *"closed"* environment, with no new distributions occurring during learning. On the other hand, incremental networks process flexible structures that can adapt well for various data and environments, but they lose much useful information of the original learning data. Recently, an online Kernel Density Estimator (oKDE) [17] is proposed to introduce the Kernel Density Estimator to online learning. However, the learning environment (*"closed"* or *"open-ended"*) must be known in advance to set different parameters.

We propose an Incremental Local Distribution Network (**ILDN**), which, by combining the advantages of matrix learning and incremental learning, is able to obtain a precise and concise representation of the data as well as learning incrementally without forgetting previous knowledge. In summary, the characteristics of ILDN are:

(1) By storing in each node the covariant matrix to record rich local distribution information of the learning data, and adopting statistically supported node merging and denoising criterions, ILDN is able to obtain a precise and concise representation of the learning data, called a relaxation data representation.

(2) Through giving each node an adaptive vigilance parameter with statistics theoretical supporting, ILDN is able to learn new distributions data effectively without forgetting the previous learnt but still useful knowledge. That is, ILDN can handle the Stability-Plasticity Dilemma effectively.

2   Incremental Local Distribution Network

ILDN is an online incremental learning model which combines the advantages of matrix learning and incremental networks. The nodes in the network record not only the weight vector but also the data distribution information around its local region, i.e. the covariance matrix. Nodes which are close to each other in the feature space are connected. The connected nodes will be merged during learning if a concise data representation can be obtained, which is called a relaxation data representation.

In ILDN, each node i is associated with a 4-tuple $\langle c_i, M_i, n_i, H_i \rangle$: c_i, M_i and n_i are the mean vector, covariance matrix and number of input patterns belonged to node i. H_i is a vigilance parameter to decide whether an input pattern belongs to node i, it dynamically changes with the learning process. Assume that ILDN receives d-dimensional data $x \in \mathbb{R}^d$, the node can be described as a hyper-ellipsoid region using c_i, M_i and H_i:

$$i : \sqrt{(x - c_i)^T M_i^{-1}(x - c_i)} < H_i \qquad x \in \mathbb{R}^d \qquad (1)$$

The entire workflow of ILDN is as follows: when an input pattern comes, ILDN first conducts the Node Activation to find some activated nodes which are recorded in an activating node set S. Then Node Updating is conducted according to set S: If there is no activated node in S, a new node will be established for this input pattern; Else ILDN will find a winner among the activated nodes and update this winner node. Topology Maintaining module will create connections between the nodes in S and record these connections in the connection list set C. After that, ILDN will check the merging condition between the winner node and its neighbor nodes, if the merging condition is satisfied, Node Merging between the winner node and its neighbors will be executed to get a concise local representation. When all the steps above are done, ILDN will process the next input pattern. Denoising is implemented every λ patterns are learned. When the learning process is finished or users want the learning result, ILDN will Cluster the learned nodes and output the learning result.

2.1 Node Activation

When an input pattern x comes, ILDN first calculates the Mahalanobis distance between x and all node $i \in N$:

$$D_i(x) = \sqrt{(x - c_i)^T M_i^{-1}(x - c_i)}, \qquad i = 1, 2, ..., |N| \qquad (2)$$

where N is the set of nodes, $|N|$ represents the total number of the nodes. If $D_i(x) < H_i$, we say node i is activated. Then we put i in an activating set S and we get:

$$S = \{i | D_i(x) < H_i\} \qquad (3)$$

Then set S records all the activated nodes by the input pattern x.

2.2 Node Updating

If $S = \emptyset$, i.e. no node is activated by x, it means x is a new knowledge. A new node a is created for x as:

$$a : \langle c_a = x, M_a = \sigma I, n_a = 1, H_a = \varepsilon_{n_a} * \chi_{d,q}^2 \rangle \qquad (4)$$

To make M_a be nonsingular, we initialize it as σI, where I is the identity matrix and σ is a small positive parameter. This initialization ensures that the covariance matrix M_a is positive definite during learning. A small positive σ guarantees that the initial hyper-ellipsoid is compact convergence to the input pattern x, it decides the initial hyper-ellipsoid size of the new node. ε_{n_a} is a function of n_a to control the expansion trend of ellipsoid. $\chi^2_{d,q}$ is a value of χ^2 distribution with d degrees of freedom and q confidence, usually q is equal to 0.90 or 0.95. The details of such parameters will be discussed in Section 3.

If $S \neq \emptyset$, i.e. some nodes are activated by x, it means x is not new knowledge. ILDN will find a winner node i^* from set S:

$$i^* = \underset{i \in S}{\arg\min}\, D_i(x) \tag{5}$$

Then node i^*: $\langle c, M, n, H \rangle$ is updated in a recursive way as:

$$c_{new} = c + (x - c)/(n + 1); \quad n_{new} = n + 1; \quad H_{new} = \varepsilon_{n_{new}} \chi^2_{d,q}$$

$$M_{new} = M + [n(x - c)(x - c)^T - (n + 1)M]/(n + 1)^2 \tag{6}$$

2.3 Topology Maintaining

A topology preserving feature map is determined by a mapping Φ from a manifold $\mathcal{M}$ onto the vertices (or nodes) $i \in N$ of a graph (or network) $\mathcal{G}$. The mapping Φ is neighborhood preserving if similar feature vectors are mapped to vertices that are close within the graph. This requires that feature vectors v_i and v_j that are neighboring on the feature manifold $\mathcal{M}$ are assigned to vertices (or nodes) i and j that are adjacent (or connected) in the graph or network $\mathcal{G}$ [4].

Some methods achieve the topology preserving feature map through a predefined structure $\mathcal{G}$ such as SOM [3]. In this paper, the connections between nodes of $\mathcal{G}$ are built according to Hebbian learning rule: If two nodes are activated by one pattern, a connection between the two nodes is created. According to the definition of set S, we know that all nodes in S are activated by the current input pattern x. Thus, if no connection exists between node i and j in S, ILDN will add a new connection $\{\langle i, j \rangle | i \in S \land j \in S \land i \neq j\}$ into the connection list C. After a period of learning, these connections is able to organize the nodes into groups to represent different topology of the learning data.

2.4 Node Merging

As learning continues, there may be some nodes closing to each other and having similar principal components. Such nodes will be merged to obtain a concise local representation. At merging stage, two nodes will merge if the following two conditions are satisfied:

(i) two nodes i and j are connected by an edge; and

(ii) the volume of the combined node m is less than the sum volume of the two nodes i and j, i.e.

$$Volume(m) < Volume(i) + Volume(j) \tag{7}$$

If the above conditions are satisfied for node i and j, we merge i and j and let m represent the data in i and j collectively:

$$c_m = (n_i c_i + n_j c_j)/n_m; \quad n_m = n_i + n_j; \quad H_m = \varepsilon_{n_m} \chi^2_{d,q}$$
$$M_m = \frac{n_i}{n_m}(M_i + (c_m - c_i)(c_m - c_i)^T) + \frac{n_j}{n_m}(M_j + (c_m - c_j)(c_m - c_j)^T) \tag{8}$$

In practice, we only merge the winner node and its neighbors when a pattern is fed into ILDN. After merging, all the connections with original nodes in C are attached to the new node.

2.5 Denoising

The data from the learning environment may contain noise. Some nodes may be created by these noise data. Since ILDN records the distribution density of each node, we can use this information to judge whether a node is a noise node.

After every λ patterns learned, ILDN first calculates the mean value of the number of the input patterns belonged to each node as:

$$Mean = \sum_{i=1}^{|N|} n_i/|N| \tag{9}$$

where $|N|$ represents the total number of the nodes, n_i is the number of input patterns belonged to node i. We assume that the probability density of the noise is lower than the useful data. Based on this assumption, if n_i is smaller than a threshold $k * Mean$, we mark node i as a noise node and remove it. Where $0 \leq k \leq 1$ control the intensity of denoising. After denoising, all the connections with deleted nodes in C are also deleted.

2.6 Cluster

In [2], it is proved that the competitive Hebbian rule forms perfect topology preserving map of a manifold if the data is dense enough. Based on this opinion, we take each cluster as a manifold, therefore we can find different connected node domains as different clusters. Algorithm 1 shows the details of the clustering method.

2.7 Complete Algorithm of ILDN

As a summary for this section, we give the complete algorithm of ILDN (Algorithm 2).

Algorithm 1. Cluster Nodes

1: Initialize all nodes as unclassified.
2: Choose one unclassified node i from node set N. Mark i as classified and label it as class C_i.
3: Search node set N to find all unclassified nodes that are connected to node i with a "path". Mark these nodes as classified and label them as the same class as node i.
4: Go to Step 2 to continue the classification process until all nodes are classified.

Note: if two nodes can be linked with a series of connections, we say that a "path" exists between the two nodes.

Algorithm 2. Incremental Local Distribution Network

1: Initialize the network with $N = \emptyset$, $C = \emptyset$.
2: Input new sample $x \in \mathbb{R}^d$.
3: Determine set S using formula (3), where the elements of S is the nodes which activated by sample x.
4: If $S = \emptyset$, initialize a new node as formula (4) then goto Step 2.
5: If $S \neq \emptyset$, choose the winner node i^* by formula (5) and update i^* using formula (6).
6: Establish connections between the activated nodes in set S.
7: If the winner node i^* and its neighbors satisfy the merging conditions (i) and (ii), implement merging procedure as formula (8).
8: If the number of input patterns presented so far is an integer multiple of parameter λ. Denoising as in **Section 2.5**.
9: If the learning process is not finished, go to Step 2 to continue unsupervised online learning.

10: Cluster using Algorithm 1 and output the learning result.

---

3   Analysis

3.1   The Expansivity of the Nodes

As each node records the local regional distribution, we can assume that the patterns belonged to one node are generated by a Gaussian distribution $\mathcal{N}(c_X, \Sigma_X)$ where c_X is the center and Σ_X the covariance matrix. The hyper-ellipsoid boundary equation of each node is:

$$(x - c_X)^T \Sigma_X^{-1} (x - c_X) = H^2 \tag{10}$$

Let $K = H^2$, then K is a χ^2 distribution with d degrees of freedom. Giving a confidence q, the patterns from random variables X lie in the hyper-ellipsoid drawn by K can be described as:

$$P\{(x - c_X)^T \Sigma_X^{-1} (x - c_X) < K\} = q \tag{11}$$

Then K can be solved by $K = \chi^2_{d,q}$. For $K = H^2$, we can get the value of H. In practice, for node i we set $H_i = \varepsilon_{n_i} \sqrt{\chi^2_{n,q}}$, where $\varepsilon_{n_i} \geq 1$ and ε_{n_i} decreases when n_i increases. This strategy let the hyper-ellipsoid has a tendency of expansivity at the preliminary stage when $\varepsilon_{n_i} > 1$. With more patterns included in node i, H_i^2 approaches to $\chi^2_{d,q}$, then the hyper-ellipsoid arrives a stable state. We set $\varepsilon_{n_i} = (1 + 2 * 1.05^{1-n_i})$.

In ILDN, σ decides the initial hyper-ellipsoid size of new nodes. It can be understood as the initial size of the window we observe the learning data. Very big σ may lead to a new node cover several different clusters. Therefore, ILDN prefers small σ. Though small σ may make ILDN initially generate many nodes with small hyper-ellipsoid size, ILDN can merge these nodes following the learning process.

3.2   The Relaxation Data Representation

With the hyper-ellipsoid defined in formula (1), the volume of node n can be calculated:

$$Volume(n) = 2^{\lceil \frac{d+1}{2} \rceil} \pi^{\lceil \frac{d}{2} \rceil} \left(\prod_{i=0}^{\lceil d/2 \rceil - 1} \frac{1}{d - 2i} \right) \sqrt{|M_n|} H_n^d \tag{12}$$

where M_n and H_n are the covariance matrix and vigilance parameter of node n. $|M_n|$ represents the determinant of M_n, $[\cdot]$ represents the rounding operation. d is the dimension of the sample space. However, $|M|$ in formula (12) may be very close to 0 in some high dimensional task and thus not suitable to calculate the volume directly with it. To avoid directly calculate the volume with formula (12), for the covariance matrix M_i, M_j and M_m of node i, j and merging node m in node merging condition (ii), we do Singular Value Decomposition (SVD) as:

$$M = E^T diag(\lambda_1, \lambda_2, ..., \lambda_d) E \tag{13}$$

where $\lambda_1 \geq \lambda_2 \geq ... \geq \lambda_d$, and we get $\lambda_1^i, \lambda_2^i,..., \lambda_d^i$ for node i, $\lambda_1^j, \lambda_2^j,..., \lambda_d^j$ for node j and $\lambda_1^m, \lambda_2^m,..., \lambda_d^m$ for node m.

Then we find a truncated position t of all singular value with a predefined scaling factor ρ: p=$\underset{1 \leq p \leq d}{argmin} \sum_{i=1}^{t} \lambda_i \geq \rho \sum_{i=1}^{d} \lambda_i$. For node i, j and merging node m, we get t_i, t_j and t_m respectively. In this paper, we set ρ=0.95. Finally, we get a common truncated position t=max(p_i, p_j, p_{merge}) and calculate $|M_k|$ by using $\lambda_1^k \times \lambda_2^k \times ... \times \lambda_t^k$, $k = i, j, m$. Substituting $|M_k|$ (where $k = i, j, m$) into formula (12) and replace d with t, we get $Volume(i)$, $Volume(j)$ and $Volume(m)$. Substituting these three volumes and after a series of simplification, we get an equivalence merging condition:

$$\sqrt{\frac{\lambda_1^i H_i}{\lambda_1^m H_m} \cdot \frac{\lambda_2^i H_i}{\lambda_2^m H_m} \cdots \frac{\lambda_t^i H_i}{\lambda_t^m H_m}} + \sqrt{\frac{\lambda_1^j H_j}{\lambda_1^m H_m} \cdot \frac{\lambda_2^j H_j}{\lambda_2^m H_m} \cdots \frac{\lambda_t^j H_j}{\lambda_t^m H_m}} \geq 1 \tag{14}$$

In practice, we use formula (14) to judge the merging condition (ii). If formula (14) is not satisfied, node i and j will remain unchanged.

At the node merging step, we have two candidate data representation that are: (1) representation before merging and (2) representation after merging. Assume the domain of distribution $Q_i(x)$ of node i is $x \in R_i$, the domain of distribution $Q_j(x)$ of node j is $x \in R_j$. Then the data representation f_1 before node merging in domain $R_i \cup R_j$ is:

$$f_1 : \begin{cases} \sqrt{(x - c_i)^T M_i^{-1}(x - c_i)} < H_i & x \in R_i \\ \sqrt{(x - c_j)^T M_j^{-1}(x - c_j)} < H_j & x \in R_j \end{cases} \tag{15}$$

The data representation f_2 after node merging in domain $R_i \cup R_j$ is:

$$f_2 : \sqrt{(x - c_m)^T M_m^{-1}(x - c_m)} < H_m \qquad x \in R_i \cup R_j \tag{16}$$

When assume learning data in domain $R_i \cup R_j$ are generated by a Gaussian distribution, setting $H_m = \chi_{d,q}^2$ will guarantee the probability that the learning data in domain $R_i \cup R_j$ falls in f_2 equals to q (usually $q \geq 90\%$). Meanwhile, according to node merging condition (ii), the volume of node m is less than the total volume of node i and j. Thus, we get a much concise data representation on domain $R_i \cup R_j$.

Comparing of the two representations, f_1 uses more parameters than f_2, and the two regions in f_1 overlap each other. Thus, the representation f_1 is *tight* and we call f_1 as a *tight* data representation. On the other hand, the representation f_2 expresses the data distribution more concisely than f_1, it relaxes the requirement of the parameters, correspondingly, we call representation f_2 as a *relaxation* data representation.

4   Experiments

As ILDN aims to combine the advantages of incremental learning and matrix learning, we compare it with some classical and state-of-the-art methods of matrix learning and incremental learning. The matrix learning methods include localPCASOM [7], BatchMatrixNG [8] and oKDE (oKDE is also an incremental learning method) [17]. The incremental learning methods include TopoART [15] and Adjusted-SOINN (ASOINN) [14].

4.1   Artificial Data

Observe the Periodical Learning Results. We use the artificial data which is distributed in two belt areas (also used in [5]). Each belt area represents a

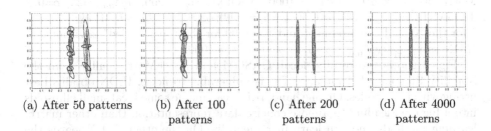

| (a) After 50 patterns | (b) After 100 patterns | (c) After 200 patterns | (d) After 4000 patterns |

Fig. 1. Periodical learning results of ILDN. The pink area represents the distribution of the learning data. Ellipses with black boundary-line are the learning result.

cluster and generates samples uniformly. Parameters of ILDN are $\chi^2_{d,q} = \chi^2_{(2,0.90)}$, $\sigma = 1e\text{-}5$, $k = 0.01$, $\lambda = 1000$. Fig. 1 illustrates the learning process of ILDN. At the early stage, many ellipsoids are generated to cover the learning data set (Fig. 1(a)). With the learning process continues, some ellipsoids are merged together, as at the 100 patterns stage (Fig. 1(b)). After 200 patterns, all ellipsoids are merged into 2 ellipsoids (Fig. 1(c)). Finally, ILDN gets 2 ellipsoids which fits the original data set very well. ILDN does not need to predetermine the number of the nodes, it automatically generates 2 nodes in this task.

Work in Complex Environment. In this section, we conduct our experiment on the data set shown in Fig. 2. The dataset is separated into five parts containing 20000 samples in total. Data sets A and B satisfy 2-D Gaussian distribution. C and D are concentric rings distribution. E is sinusoidal distribution. We also add 10% Gaussian noise and random noise to the dataset. Noise is distributed over the entire data set.

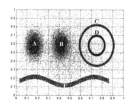

Fig. 2. Artificial data set used for the experiment. 10% noise is distributed over the entire data set.

The experiments are conducted in two environments. In the "closed" environment, patterns are randomly selected from the whole learning set. In the "open-ended" environments, five parts of the data are presented successively. In stage I, i.e. step 1 to 4000, patterns are chosen randomly from data set A. In stage II, i.e. step 4001 to 8000, the environment changes and patterns from B are chosen, etc.

We set the parameters of localPACSOM as $N = 20$, $\varepsilon=0.01$, $H=1.5$; Batch-Matrix as $N = 20$, $epoch = 1000$; ASOINN as $\lambda=200$, $age_{max}=25$, $c=0.5$; TopoART as $\beta_{sbm}=0.32$, $\rho=0.96$, $\varphi=5$, $\tau=100$; oKDE as $D_{th} = 0.01$, $N = 10$, $f = 1$ for the "closed" environment and $f = 0.99$ for the "open-ended" environment. ILDN as $\chi^2_{(d,q)}=\chi^2_{(2,0.90)}$, $\sigma=1e\text{-}5$, $k=0.5$, $\lambda = 1000$.

Fig. 3 shows the results. localPCASOM and BatchMatrix suffer from the Stability-Plasticity Dilemma. oKDE is vulnerable to noise. ILDN obtains best fitting with least nodes. On one hand, ILDN can merge some local small ellipsoids into a big one, leading to a more concise data representation than other matrix learning methods and can learn incrementally. On the other, ILDN learns the local distribution to describe original data while other incremental networks have to use a large number of nodes.

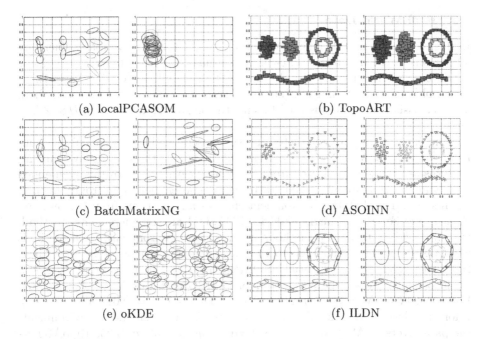

Fig. 3. Comparing results in the closed and open-ended environment. The left column of a method is the result in the closed environment, the right is the open-ended environment.

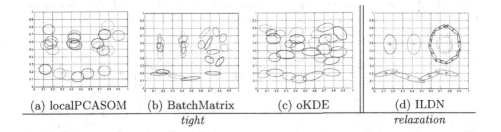

Fig. 4. The *relaxation* data representation vs. the *tight* data representation

In order to clearly observe the performance of the *relaxation* and the *tight* data representations, we use the dataset in Fig. 2 without noise to test localPCA-SOM, BatchMatrix, oKDE, and ILDN in the "closed" environment. The result is shown in Fig. 4. The *relaxation* representation can generate a more concise representation than the *tight* representation on the two Gaussian distributions. For the distribution which cannot be further simplified, the *relaxation* representation is able to get comparable result with the *tight* representation, such as the sinusoidal distribution. This experiment demonstrates that the *relaxation* data representation gets a much concise representation than the *tight* data representation while maintaining the same level of details and preciseness.

4.2   Real-World Data

In this section, we first do an experiment on the ATT_FACE database. There are 40 persons in the database and each person has 10 images differing in lighting conditions, facial expressions and details. The original image (size 92×112) is re-sampled to 23×28 image using the nearest neighbor interpolation method. Then Gaussian smoothing is used to smooth the 23×28 image with $HSIZE = [4; 4]$, $\sigma = 2$.

The experiments are conducted in the "closed" and "open-ended" environment for oKDE, ASOINN, TopoART and ILDN. In the closed environment, patterns are randomly selected from the whole learning set. In the open-ended environment, from step 1 to 200, patterns are chosen randomly from person 1. From step 201 to 400, the environment changes and patterns from person 2 are chosen, etc. For BatchMatrix and localPCASOM are not incremental method, we only conduct these two methods in the closed environment. Such methods need a predefined node number, to guarantee a good learning result, we set it as 200 which gives it a good initial condition: the initial nodes contain images of all 40 persons. We set the parameters of localPCASOM as $N = 200$, $\epsilon = 0.01$; BatchMatrix as $N = 200$, $epoch = 200$; oKDE as $D_{th} = 0.01$, $N = 10$, $f = 1$ for the closed environment and $f = 0.99$ for the open-ended environment; the parameters of ASOINN is set as $\lambda=100$, $age_{max}=50$, $c=0.25$; TopoART as $\beta_{sbm}=0.6$, $\rho=0.96$, $\varphi=3$, $\tau=100$; ILDN as $\chi^2_{(d,q)}=\chi^2_{(644,0.90)}$, $\sigma=1e\text{-}3$, $k=0.01$, $\lambda=1000$.

Table 1. Mean node number, missing person number and accuracy of 100 times learning results in closed and open-ended environment for ATT_FACE. The best performance is bolded.

	Environment	localPCASOM	BatchMatrix	oKDE	ASOINN	TopoART	ILDN
Node number	closed	200*	200*	3	276.24	16.65	247.29
	open-ended	—	—	3	317.33	260.82	247.31
Missing persons	closed	—	—	38	0	26.05	0
	open-ended	—	—	38	5.01	0	0
Accuracy	closed	23.4%	24.1%	—	96.7%	48.25%	**98.5%**
	open-ended	—	—	—	89.6%	96.3%	**98.5%**

We adopt 3 factors to evaluate the learning results: node number, missing person number and recognition ratio. Table 1 gives the learning results of 100 times learning results in both environments. oKDE only gets 3 nodes in the learning result in both environments, losing 38 persons. Though ILDN gets larger node number in the closed environment than TopoART, it does not lose any person. In the open-ended environment, ILDN gets the smallest node number (excluding oKDE) and does not lose any person. On the other hand, the ASOINN "forget" 5.01 persons. Moreover, the proposed ILDN has the least difference of number of node and do not "forget" any person in the two environments, which means ILDN is a more stable incremental method than others. At last, nearest neighbor is used to classify vectors in the original image vector. We do not

Table 2. Learning results on some UCI datasets. N/A represents that the methods do not give the learning result within 10 days. The best performance on each data set is bolded.

Dataset	localPCASOM	BatchMatrix	oKDE	ASOINN	TopoART	ILDN
Segment	465^*	465^*	24	376	110	465
	10.00%	15.16%	73.55%	85.01%	37.10%	**90.32%**
Shuttle	9^*	9^*	30	69	63	9
	79.16%	81.37%	90.66%	90.39%	86.48%	**92.96%**
Webspam	745^*	745^*	N/A	3531	4643	745
	65.78%	67.25%		85.45%	83.05%	**86.50%**
KDD99	N/A	N/A	N/A	127	149	32
				92.15%	92.10%	**92.81%**

test the oKDE because it is unsuitable for this task. Table 1 shows that the recognition accuracy of ILDN is higher than the other methods.

Next, we do the experiments on some UCI datasets including Segment, Shuttle, Webspam and KDD99 which differ in the length, dimensionality as well as the number of classes. The parameters of the comparison methods are set as they suggest. The parameters of ILDN are set as σ=1e-3, k=0.01, λ=1000. We define the node number of localPCASOM and BatchMatrix as same as ILDN, marked with * in the learning result.

The learning results are shown in Table 2. ILDN obtains highest accuracy in all datasets. Compared with three matrix learning methods localPCASOM, BatchMatrix and oKDE, ILDN uses far less nodes in three out of four datasets. Moreover, ILDN can handle very large scale dataset like KDD99, while localP-CASOM, BatchMatrix and oKDE cannot give a learning result within 10 days. We can also find that the incremental (or online) matrix learning method oKDE cannot give a learning result on the relatively small data set like Webspam within 10 days. Thus, ILDN is a more practical incremental (or online) matrix learning method than others.

5   Conclusion

This paper presents an incremental local distribution learning network (ILDN). It combines the advantages of matrix learning and incremental learning. The covariant matrix, statistical vigilance parameter and merging mechanism enable it to obtain precise and concise representation of the data and learn incrementally. Moreover, the denoising processing based on data density makes it robust to noise. The experiments on both artificial datasets and real-world datasets validate our claims and show that ILDN is more effective over other matrix learning methods and incremental networks, obtaining higher accuracy and being able to handle large-scale data.

658   Y. Xing et al.

Acknowledgments. We thank the anonymous reviewers for their time and effort. This work is supported in part by the National Science Foundation of China under Grant Nos. (61375064, 61373001) and Jiangsu NSF grant (BK2013-1279).

References

1. Lloyd, S.P.: Least squares quantization in PCM. IEEE Transactions on Information Theory **28**(2), 129–137 (1982)
2. Martinetz, T.M., Berkovich, S.G., Schulten, K.J.: Neural gas network for vector quantization and its application to timeseries prediction. IEEE Transactions on Neural Networks **4**(4), 556–558 (1993)
3. Kohonen, T.: Self-organized formation of topologically correct feature maps. Biological Cybernetics **43**(1), 59–69 (1982)
4. Martinetz, T., Schulten, K.: Topology representing networks. Neural Networks **7**(3), 507–552 (1994)
5. López-Rubio, E., Muñoz-Pérez, J., Gómez-Ruiz, J.A.: A pricipal components analysis self-organizing map. Neural Networks **17**, 261–270 (2004)
6. Yin, H., Allinson, N.M.: Self-organizing mixture networks for probability density estimation. IEEE Transactions on Neural Networks **12**(2), 405–411 (2001)
7. Huang, D., Yi, Z., Pu, X.: A new local PCA-SOM algorithm. Neurocomputing **71**, 3544–3552 (2008)
8. Arnonkijpanich, B., Hasenfuss, A., Hammer, B.: Local matrix adaptation in topographic neural maps. Neurocomputing **74**, 522–539 (2011)
9. López-Rubio, E.: Probabilistic self-organizing maps for continuous data. IEEE Transactions on Neural Networks **21**(10), 1543–1554 (2010)
10. Carpenter, G.A., Grossberg, S.: The ART of adaptive pattern recognition by self-organising neural network. IEEE Computer **21**(3), 77–88 (1988)
11. Fritzke, B.: A growing neural gas network learns topologies. In: Proceedings of the 1995 Advances in Neural Information Processing Systems, pp. 625–632 (1995)
12. Alahakoon, D., Halgamuge, S.K., Srinivasan, B.: Dynamic self-organizing maps with controlled growth for knowledge discovery. IEEE Transactions on Neural Networks **11**(3), 601–614 (2000)
13. Marsland, S., Shapiro, J., Nehmzow, U.: A self-organising network that grows when required. Neural Networks **15**(8–9), 1041–1058 (2002)
14. Shen, F., Hasegawa, O.: A fast nearest neighbor classifier based on self-organizing incremental neural network. Neural Networks **21**, 1537–1547 (2008)
15. Tscherepanow, M., Kortkamp, M., Kammer, M.: A hierarchical ART network for the stable incremental learning of topological structures and associations from noisy data. Neural Networks **24**(8), 906–916 (2011)
16. Araujo, A.F.R., Rego, R.L.M.E.: Self-organizing maps with a time-varying structure. ACM Computing Surveys **46**(1) Article No. 7 (2013)
17. Kristan, M., Leonardis, A., Skočaj, D.: Multivariate online kernel density estimation with Gaussian kernels. Pattern Recognition **44**(10–11), 2630–2642 (2011)

Trend-Based Citation Count Prediction
for Research Articles

Cheng-Te Li[1(✉)], Yu-Jen Lin[2], Rui Yan[3], Mi-Yen Yeh[2]

[1] Research Center for IT Innovation, Academia Sinica, Taipei, Taiwan
ctli@citi.sinica.edu.tw
[2] Institute of Information Science, Academia Sinica, Taipei, Taiwan
maxwellbiga@gmail.com, miyen@iis.sinica.edu.tw
[3] Natural Language Processing Department, Baidu Inc., Beijing, China
yanrui02@baidu.com

Abstract. This paper aims to predict the future impact, measured by the citation count, of any papers of interest. While existing studies utilized the features related to the paper content or publication information to do *Citation Count Prediction* (CCP), we propose to leverage the citation count trend of a paper and develop a *Trend-based Citation Count Prediction* (T-CCP) model. By observing the citation count fluctuation of a paper along with time, we identify five typical citation trends: early burst, middle burst, late burst, multi bursts, and no bursts. T-CCP first performs *Citation Trend Classification* (CTC) to detect the citation trend of a paper, and then learns the predictive function for each trend to predict the citation count. We investigate two categories of features for CCP, CTC, and T-CCP: the publication features, including author, venue, expertise, social, and reinforcement features, and the early citation behaviors, including citation statistical and structural features. Experiments conducted on the *Arnet-Miner* citation dataset exhibit promising results that T-CCP outperforms CCP and the proposed features are more effective than conventional ones.

Keywords: Citation count · Citation link · Citation category · Citation graph

1 Introduction

Nowadays, a large volume of research articles, in the order of hundreds per year if not in thousands, from diverse disciplines are continuously generated and published. When researchers kick off their research work, either during the seeking of emerging areas or after coming up with trending topics of interest, they would usually need to read a bundle of relevant and influential papers so that they can catch the trend and even be in the lead. However, due to the limitation of time, researchers might not be able to follow every paper and could be very difficult to determine which studies will possess high impact in the future. In addition, some relevant studies with different terms or keywords may be neglected due to limited human effort. Therefore, it would be useful if we can accurately measure the future impact or the influence of a paper at early stages after it gets published. As for funding agencies of government and industry, it is also essential to understand which projects are more potential based on either

© Springer International Publishing Switzerland 2015
T. Cao et al. (Eds.): PAKDD 2015, Part I, LNAI 9077, pp. 659–671, 2015.
DOI: 10.1007/978-3-319-18038-0_51

the paper publications of researchers' projects or the possible impact of a field. Hence being able to estimate the future impact of papers can support their decision to fairly and effectively distribute the resources.

To quantify the potential impact or influence of a particular paper, one of the most intuitive, objective, and commonly adopted measures is the number of citations, a.k.a. citation count, after the paper gets published [3]. The number of citations is the times a paper get cited by other articles. In fact, the citation count of a paper is also validated to be the major factor that affects the performance for the retrieval task of research articles [1], and also be the most influential factor in Google Scholar's ranking [6]. Nevertheless, regarding citation count to measure the impact suffers from a critical issue: it works only for papers that passed a long period of time after it gets published, say going beyond five years. Most scientific articles that have only few years after getting published tend to have a lower citation count, and thus the citation count would fail to estimate its impact. To deal with this problem, in this paper, we aim to predict the future number of citations for scientific papers. We believe that an accurate estimation of the long-term citation count will be beneficial for understanding the impact of a paper at early stages after it gets published (e.g., 1-3 years). Specifically, we propose to tackle three problems about citation count prediction. First, can we predict the number of citations of a paper given only its publication information, such as authors and venues? Second, can we accurately predict the citation count of a paper, given its citation information at early stages after it gets published? Third, what are the most important factors that determine whether a paper will get a large number of citation counts? To address these problems about citation behaviors, we develop a novel *Trend-based Citation Prediction* (T-CCP) framework, which can effectively model and predict the long-term citation count of a given paper.

Related Work. Castillo et al. [9] used the author reputation, measured by the number of papers, citations, and authority, with linear regression to estimate future citation counts. Yogatama et al. [10] mined textual features with graphical models to predict the download times of papers. Stern [13] identified the high-ranked papers using early citation information. Shi et al. [8] studied the structural properties of citation graphs in various disciplines. Pobiedina and Ichise [11] mined frequent graph patterns in a citation graph to estimate the future citation counts. Yan et al. [7][16], which is the most relevant work to this paper, combined content features, author features, and venues features, and used the regression models to learn the prediction function. In this work, we not only include the advantages of their proposed features, but also investigate more advanced features and propose the citation trends to boost the prediction accuracy. First, we investigate more content-independent features categorized into publication features and early citation behavioral features, in which the former not only contains the features used by Yan et al. [7][16] but also includes more new features (i.e., expertise, social, and reinforcement features). Second, analyzing and extracting features of paper content might be costly in terms of space and time as the number of research articles grows rapidly and drastically. We do not consider the paper contents for the citation count prediction. Third, instead of learning the prediction function directly from all the papers, we categorize the papers into several citation trends, train a predictive model for papers belonging to each trend, and find the most proper model for the test paper to do the task of citation count prediction.

We first analyze the behaviors of citation count over time and find that the evolution of citation count can be categorized into several patterns of trend. For example, some papers immediately burst to get lots of citation right after got published, but the citation count gradually decreased as time proceeds. On the contrary, some are inglorious at early stages but abruptly get a great number of citations. In addition, some articles never have bursts while others can obtain more than one bursts in their evolution of citation counts. Based on the stage that the bursts happen and the number of bursts, we divide the behaviors of citation count into five categories, termed *citation trends*. Given a scientific article, we first aim to classify which citation trend it should belong to, and then to predict its citation count after Δt years according to its classified citation trend. In other words, the proposed *Trend-based Citation Prediction* consists of two stages. The first is to classify which trend the target paper will behave. The second is to predict the exact citation count after n years. Moreover, to response the abovementioned questions, we divide the features into two sets: the publication features and the early citation features. The former is to capture the knowledge about authors, venues, and paper contents while the latter is the statistical and structural information of citation behaviors in the first k years of papers. We combine both feature sets and empirically derive accurate performance.

Contributions. We summarize the contributions of this paper as follows.

(a) We predict the future citation counts of papers based on the content-independent features of publication knowledge and/or the early citation information, to estimate the impact of papers in the long term. We also empirically investigate what factors are most influential on making a paper get a higher number of citations.

(b) We categorize the evolution of citation counts, termed *citation trends*, into five types, according to the stages of bursts of citation count as well as the number of bursts in papers' evolution of citation counts.

(c) We devise a *trend-based citation prediction* (T-CCP) model to estimate the future citation counts of papers. TCP is a two-stage prediction method, which first classifies the given paper into a potential citation trend, and predicts its exact citation count based the learned model of the corresponding trend.

(d) Experiments conducted on the well-known *ArnetMiner* citation data exhibit that the proposed T-CCP averagely outperforms CCP with 0.18 improvement in terms of R^2 measure, the early citation features can significantly boost the prediction accuracy with 0.34 improvement in average, and combining all of our features are averagely 0.43 better than conventional features in terms of R^2.

The structure of this paper is summarized as follows. We first introduce concrete definitions as well as the problem statements in Section 2. Then in Section 3, we describe the proposed method, followed by the elaboration of experimental results in Section 4. Finally we conclude this work in Section 5.

2   Problem Statements

Definition 1: Citation Count. For an archive of academic papers D, given a paper $d \in D$, its citation count $c(d)$ is the number of papers that cite d, denoted by $c(d) = |\{d' \in D : d' \, cites \, d\}|$, where $|S|$ is the number of elements in set S.

Definition 2: Citation Sequence. A citation sequence of a paper d, denoted by $s_{\Delta t}(d) = \langle c_1(d), c_2(d), ..., c_{\Delta t}(d) \rangle$, is a sequence of citation count $c_i(d)$ over a period of time $1, 2, ..., t$, where c_i is the citation count of the i-th year after d gets published.

Definition 3: Citation Trend. A citation trend p is a collection of citation sequences sharing a common pattern of evolution of citation count. Citation sequences of different citation trends demonstrate dissimilar evolutions of citation count.

Problem 1: Citation Count Prediction. Given a scientific article $d \in D$, the goal is to learn a predictive function f, and use the function $f(d, \Delta t)$ to predict the citation counts of d, i.e., $c_{\Delta t}(d)$, at a particular time period Δt after it gets published.

In order to effectively solve the citation count prediction problem, we propose a trend-based citation prediction model, which consists of two stages. The first is to classify the potential citation trend of the given paper while the second aims at predicting the citation count based on the model trained from papers belonging to the corresponding citation trend. We think for the model training, grouping papers into several citation trends is able to not only select the effective and representative data instances but also to reduce the noise. Hence the prediction performance is expected to be boosted. What follows gives the problem definitions for these two stages.

Problem 2: Citation Trend Classification. Given a scientific article $d \in D$, our goal is to accurately classify d into a certain citation trend p that it should belong to.

Problem 3: Trend-based Citation Count Prediction. Given a scientific article $d \in D$ and the citation trend p it belongs to, the goal is to learn a predictive function f_p, which is trained from papers whose citation sequences belong to trend p, and use the function $f_p(d, \Delta t)$ to predict the citation counts of d, i.e., $c_{\Delta t}(d)$, at a particular time period Δt after it gets published.

3   The Proposed Method

In this section, we first introduce five categories of citation trends, which are derived based on the time of getting bursts and the number of bursts. Then we describe a series of features used in this work, which can be divided into two main categories: publication features and early citation features. Finally, we present the predictive model that exploits the citation trends as well as the proposed features.

3.1   Categories of Citation Trends

To exploit the citation trends of paper for citation count prediction, we first investigate the evolution of citation count of papers over time. While existing study [15] has presented that the evolution of citation count is non-linear, we simplify the objective into identifying the patterns of citation count evolution, i.e., citation trends. The citation data collected by *Arnetminer* [5] is utilized for citation trend analysis. For each paper, we observe its citation counts within future Δt years after it gets published. Our goal is to exploit the idea of *burst* to categorize the citation count evolutions into several citation trends. A burst of citation count of a paper happens in a particular year t whose citation count is $H\%$ higher than the previous year (i.e., $t-1$). The evolution of citation count of a paper can have multiple bursts. We identify the citation trends according to two criteria. The first is **burst time**: when will the citation count of a paper gets burst, i.e., in which particular year the paper has the largest amount of citation count. We simply divide the time of getting burst into three stages: *early* stage, *middle* stage, and *late* stage within the first Δt years. The second is **burst number**: the number of bursts of citation count of a paper. We consider three kinds of burst number: *zero*, *single*, and *multiple*.

Based on burst time and burst number, we use *ArnetMiner* citation data to manually divide the citation trends into five categories, (a) *early burst*, (b) *middle burst*, (c) *late burst*, (d) *multi bursts*, and (e) *no bursts*, by using the *findpeaks* function in Matlab with the setting of $\Delta t = 8$ and $H\% = 75\%$. Four of the five identified citation trends are shown in Figure 1, except for *no bursts* which refers to the citation trends whose average citation counts over eight years is below 1 and represents papers having nearly no impact in the future. Note that the x-axis of Figure 1 is the year while the y-axis is the average *normalized citation count* (i.e., for a paper, the citation count of a particular year divided by the highest citation count within its eight years). Every paper with the early, middle, or late burst trend has a single citation burst and exhibits the highest citation counts at early, middle, or late stages respectively in their evolution of its citation count. The temporal positions of such three kinds of single burst are demonstrated in Figure 1. In addition, for the category of multi bursts, we do not consider burst time but burst number. Papers with more than one burst fall into this category. After averaging papers belonging to this category, in Figure 1, we can find the citation count goes up gradually and becomes stable in their evolution of citations.

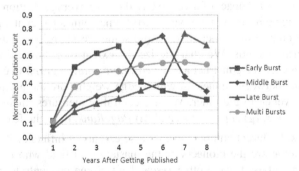

Fig. 1. Four of the five categories of citation trends

3.2   Publication Features

The publication features can be divided into five categories: (1) author, (2) venue, (3) social, (4) expertise, and (5) reinforcement. The extraction of publication features of a target paper is performed at the year that the paper gets published. Since there might be new authors or venues that have never appeared in the past with respect to the year of publication, we use zero values for such kind of features. In addition, since a paper may contain several authors, for each of author features, we use the *average* and the *maximum* scores among authors of a paper to be the feature values. We elaborate the detailed features of each category as follows. Note that some features in the first three categories have been explored in existing work [7][16].

Author Features. (**a**) An author with more papers published can be considered as having high productivity, and tends to get higher number of citations for his papers [7]. Therefore, we use the number of papers published to reflect the productivity of an author. (**b**) The *h-index* metric [3] has been widely used to evaluate the impact of an author. We regard the value of the *h-index* as one of the author features. (**c**) Famous authors usually ensure a certain amount of citations because their papers can permanently gain attention in the research communities. We measure an author's reputation as a feature, which is defined by his/her average number of citations (i.e., the citation count per paper). (**d**) The reputation of an author can be estimated in a relative manner as well. We calculate the *rank* of average citation count among those of all authors in a descending order. (**e**) If an author had ever co-worked with many other researchers, he/she has higher potential to be an influential scholar [1] because his/her expertise can attract the collaborations from others. We consider the number of coauthors to estimate the collaboration capability of an author.

Venue Features. (**a**) If a venue gets more citations, its prestige tends to be higher. We compute the average number of citations over the papers of a venue to capture its prestige. (**b**) The prestige can be also estimated relatively. Hereby we measure the relative venue prestige by calculating the *rank* of average number of citations over years. (**c**) The impact of a venue might change due to the growth of a research area. We measure the *recent* impact of a venue based on the average citation count over the papers of such venue in recent three years. (**d**) Some venues target at a particular field and obtain rare citations of other fields while some are interdisciplinary and thus attract citations across fields. We characterize the extent of cross-field or cross-venue citations for a venue. We construct a directed *venue citation graph* to capture the citation relationships between venues, in which edge weights are the normalized citation count from one venue to the other. We use centrality measures, including (**e**) *closeness*, (**f**) *betweenness*, (**g**) *eigenvector*, and (**h**) *PageRank*, to be the features.

Social Features. Different authors have diverse extents of influence on their research communities. Some are the pioneers or the masters of a field who had ever guided, collaborated, and advised many other researchers. Some concentrate on doing interdisciplinary studies, and play the role of mediators that connecting researchers from different fields. Some are junior researchers or passer-by individuals who might have less influence in their research communities. Therefore, we think it is critical to estimate the influence of an author considering the collaboration between authors.

We construct a *collaboration graph* to represent the co-authorships between authors. In addition, we further model the effectiveness of collaboration between authors into edge weights. Two authors with more co-authored papers are considered as having a good collaboration. We compute edge weights using *Jaccard coefficient*: $w(a_i, a_j) = |S_i \cap S_j|/|S_i \cup S_j|$, where S_i is the set of papers published by author a_i. To capture the influence of authors, based on the collaboration graph, we calculate several centrality measures, including (**a**) *betweenness*, (**b**) *closeness*, (**c**) *eigenvector*, and (**d**) *PageRank*, as the feature values. Based on the collaboration graph, we further investigate the membership of authors of a paper in terms of research communities. The collaboration between different research communities brings more audiences and leads to higher number of citations. We detect communities in the collaboration graph using *Louvain's algorithm* [14]. Then we measure the distribution of community membership of authors using the following scores: (**e**) the percentage of authors belong to the same community, (**f**) the percentage of new authors (i.e., never appear in the past), and (**g**) the number of communities for authors in the paper. (**h**) The number of authors is also an indicator that can reflect the visibility of a paper, and thereby affects the citations of a paper [17]. We regard the number of authors of a paper as a feature.

Expertise Features. Diverse expertise reflects different groups of audiences in either research topics or communities. An author who either has many distinct research skills or is involved in more research topics tends to have higher visibility and thus has higher potential to attract more audiences which lead to more citations. Likewise, a venue that includes more different topics of interest can gather more attention, which results in more readers and more potential citations. We propose to measure the *expertise diversity* for an author, a venue, and the reference of a paper. We model the expertise from two perspectives. The first is using the research *fields* provided by *ArnetMiner* [18] as the expertise. The second is exploiting the *communities* detected from the venue citation graph using *Louvain's algorithm* [14]. In other words, venues belonging to the same community are considered to be the same expertise. We take advantage of the entropy to measure (**a**) the *author expertise diversity* in terms of fields: $ExpDiv_{field}(a) = -\sum_{e \in E_{field}} P(e|a) \log (P(e|a))$, where $P(e|a)$ is the probability that the papers of author a belong to expertise e, and E_{field} is the set of fields. Similarly, we can derive, (**b**) the reference expertise diversity in terms of fields: $ExpDiv_{field}(r)$, (**c**) the author expertise diversity in terms of communities: $ExpDiv_{com}(a)$, and (**d**) the reference expertise diversity in terms of communities: $ExpDiv_{com}(r)$. Note that the author expertise diversity of fields had been proved to be useful [9][16] while the other five expertise diversity features are newly proposed in this paper. (**e**) The number of references of a paper is able to distinguish survey papers from regular, short, and poster papers. Since survey papers usually gain more citations than regular papers, we use the number of references as a feature. (**f**) If the research topic of a paper can either catch the *tendency* or be the lead of a potential area, the paper will have higher visibility and tend to have more citations. We measure the tendency of a paper by calculating the average year difference between the year of its publication and the years of its references.

Reinforcement Features. Important or influential authors tend to recognize representative and potential papers; hereby their papers usually cite the papers of important and renowned authors. On the other hand, influential authors also have higher potential to be cited by other essential papers due to their visibility and indicativity. In other words, the citation relationships between authors, i.e., citing and being cited, can be considered as a process of *mutual reinforcement*. We think the idea of reinforcement can benefit to the future citation count of a paper, and thus aim to characterize the reinforcement for both authors and papers. We construct an *author citation graph*, which is a directed and weighted graph, to represent the citation relationship between authors. Specifically, if the papers of author a_1 had ever cited the papers of author a_2, we construct an directed edge from a_1 to a_2, in which the edge weight is defined as $w(a_1, a_2) = |D(a_2|a_1)|/|D(a_1)|$, where $D(a_1)$ is the set of papers of author a_1 and $D(a_2|a_1)$ is the set of a_1's papers that cite a_2's papers. We exploit the graph-based ranking algorithms, *PageRank* and *HITS*, to measure the citation influence of authors. We denote the scores of *PageRank*, *hub*, and *authority* of author a as $inf_{pr}(a)$, $inf_{hub}(a)$, and $inf_{aut}(a)$. We compute the total/maximum/ average score for authors of paper d, and derive nine feature values: $\{sum, max, avg\}_{a \in A(d)} inf_x(a)$, where $x = \{pr, hub, aut\}$ and $A(d)$ is the set of authors of paper d.

3.3   Early Citation Features

Citation Statistical Features. (a) Existing study pointed out that the citation count at early stages as a paper gets published is highly correlated to its long-term citation count [2]. We directly utilize the citation sequence of the first Δt_e years of a paper d, termed $s_{\Delta t_e}(d) = \langle c_1(d), c_2(d), ..., c_{\Delta t_e}(d) \rangle$, to be the feature values, where $\Delta t_e < \Delta t$ and Δt is a long time (e.g., 10 years). (b) In addition to the citation counts at the first Δt_e years after getting published, we should also estimate the performance of authors of that paper. In other words, if an author is a rising star, his/her citation behavior of other papers within such t_e years might be also promising. Let $c_t(a)$ be the total number of citations, which are contributed from all the papers of author a before time t, where $t = t_s + \Delta t$ and t_s is the year that paper d gets published. Let $n_t(a)$ be the total number of papers published by author a before time t. Also let $m_t(a)$ be the total number of coauthors from all the papers of author a before time t. Then we can have the following nine features to characterize the early citation behaviors of paper d within the first t_e years: the total/maximum/average increased citation count of coauthors of d: $\{sum, max, avg\}_{a \in A(d)}(c_t(a) - c_{t_s}(a))$, the total, maximum, and average increased number of papers of coauthors of d, which is denoted by $\{sum, max, avg\}_{a \in A(d)}(n_t(a) - n_{t_s}(a))$, and the total, maximum, and average increased number of coauthors of d: $\{sum, max, avg\}_{a \in A(d)}(m_t(a) - m_{t_s}(a))$.

Citation Structural Features. The citation behavior of papers that are cited by paper d reflects the role about how d correlates with existing papers. The citation behavior of the set of papers that cite paper d within the first Δt_e years exhibits how d affects either the ecology of a certain field or the formation of multiple research fields. Papers that connect papers of different research fields or communities tend to have higher visibility and more citations. Papers that encourage the citation links within a field or a community might gain more attention and citations. Therefore, we think characterizing the citation behaviors of papers can benefit the prediction of citation count. We construct the *paper citation graph* to represent the citation behaviors of papers. Paper d_1 is connected to paper d_2 if d_1 has ever cited d_2. All the papers published before $t_s + \Delta t_e$ are used to construct the paper citation graph, where t_s is the year that paper d gets published. Based on the paper citation graph, we compute the scores of **(a)** *PageRank*, **(b)** *clustering coefficient*, **(c)** *hub*, and **(d)** *authority* of d to be feature values, denoted by $inf_{pr}(d)$, $inf_{cc}(d)$, $inf_{hub}(d)$, and $inf_{aut}(d)$. **(e)** To measure the citation behaviors of papers cites d, we further compute the average and maximum values of these scores over the set of papers that cite d within the first Δt_e years, denoted by $D_{\Delta t_e}(d)$: $\{avg, max\}_{d \in D_{\Delta t_e}(d)} inf_x(a)$, where $x = \{pr, cc, hub, aut\}$. **(f)** We consider the numbers of fields and communities over venues of papers in $D_{\Delta t_e}(d)$ as two features values. **(g)** The expertise diversity over venues of papers in $D_{\Delta t_e}(d)$: **ExpDiv**$_{field}(v)$ and **ExpDiv**$_{com}(v)$. **(h)** The average *venue rank* for papers in $D_{\Delta t_e}(d)$. **(i)** The average *author rank* for papers in $D_{\Delta t_e}(d)$.

3.4   The Prediction Models

Recall that we are tackling three problems based on a certain model trained by the proposed features. The first is to directly do the citation count prediction (CCP). The second is to the citation trend classification (CTC) of papers. Based on the trends derived by CTC, the third is the trend-based citation count prediction (T-CCP). For CCP and T-CCP, we leverage the technique of *Support Vector Regression* (SVR) [4] to learn the prediction function. Note that Yan et al. [16] have shown that SVR is one of the best methods to predict the citation count. For CTC, we use the technique of linear *Support Vector Machine* (LibLinear) [12] to classify the trend of a paper, in which the features used for CTC contain the publication features and early citations of the first three years after the paper gets published. It is worthwhile to note that for T-CCP, we train a separate predictive function $f_p()$ for each of the five trends p. As a paper d is classified to trend p, we utilize the corresponding function $f_p()$ to predict the future citation count.

4   Experiments

We conduct experiments to validate the effectiveness of our method. We aim to answer four questions: (1) can the proposed collection of features beat the state-of-the-art method by Yan et al. [9]? (2) To what extent can the citation information provided from the early stages after a paper gets published boost the performance? (3) Can the proposed model beat the direct prediction of citation counts? (4) Which category of features has the greatest impact on the accuracy of citation count prediction?

4.1   Evaluation Settings

We employ the *ArnetMiner* citation dataset [18], which contains major computer science publication data, for the experiments. The *ArnetMiner* citation data contains

1,383,158   papers, 855,629 authors, and 6,145 venues. We follow Yan et al. [7][16] to use the *coefficient of determination* (R^2), , to be the evaluation metric, $R^2 = \sum_{d \in D_{test}} (c_{pred}(d) - c_{mean}(D_{test}))^2 / \sum_{d \in D_{Test}} (c_{ground}(d) - c_{mean}(D_{test}))^2$, where $c_{pred}(d)$ is the predicted citation count for paper d in the test set D_{Test}, $c_{mean}(D_{test})$ is the mean of the ground-truth counts for papers in D_{Test}, and $c_{ground}(d)$ is the ground-truth citation count of paper d. $R^2 \in [0,1]$, and a larger R^2 value indicates better performance.

Table 1. Experimental results (R^2) by varying different future time period Δt, using different combinations of feature sets, and under CCP and T-CCP

		CCP			T-CCP		
Features		$\Delta t = 12$	$\Delta t = 11$	$\Delta t = 10$	$\Delta t = 12$	$\Delta t = 11$	$\Delta t = 10$
Yan et al. [7][16]		0.06	0.07	0.08	0.19	0.15	0.17
Publication (P)	Author	**0.18**	**0.27**	**0.20**	**0.33**	**0.27**	**0.30**
	Venue	0.07	0.09	0.11	0.17	0.13	0.13
	Expertise	0.07	0.07	0.08	0.10	0.09	0.10
	Social	0.06	0.07	0.08	0.10	0.09	0.14
	Reinforce	0.06	0.07	0.09	0.14	0.10	0.12
	Combined (P)	**0.19**	**0.28**	**0.23**	**0.38**	**0.46**	**0.34**
Early Citation (EC)	$\Delta t_e = 1$	0.32	0.18	0.24	0.43	0.33	0.32
	$\Delta t_e = 2$	0.44	0.29	0.29	0.41	0.44	0.43
	$\Delta t_e = 3$	0.41	0.43	0.40	0.48	0.57	0.49
	$\Delta t_e = 4$	**0.52**	**0.52**	**0.52**	**0.53**	**0.59**	**0.65**
Combined	P+EC($\Delta t_e = 1$)	0.39	0.33	0.30	0.63	0.53	0.46
	P+EC($\Delta t_e = 2$)	0.41	0.37	0.32	0.52	0.51	0.53
	P+EC($\Delta t_e = 3$)	0.45	0.47	0.41	0.69	0.63	0.63
	P+EC($\Delta t_e = 4$)	**0.53**	**0.53**	**0.52**	**0.67**	**0.68**	**0.67**

For papers in a particular year Y, we divide the papers into a training set and a testing set using five-fold cross validation. That says, the paper instances are divided into five parts, and each part is used for testing while the other parts are used for training. In addition, we consider the citation count accumulated up to 2013 as the ground-truth of a paper. The average R^2 over such five results is reported. We vary $Y = 2001, 2002, 2003$ so that we can see how the time period $\Delta t = 12, 11, 10$ after it gets published affects the performance. We compare the features we proposed with those utilized by Yan et al. [7][16], which has the most robust feature set to do the task of CCP. Those features used by Yan et al. is a small subset of ours, include author feature – {(a), (b), (c), (e)}, venue feature – {(c), (b), (h)}, social feature – {(d)}, and citation structural feature – {(a)}. Note that the settings of our experiments are different from those of Yan et al. [7][16]. Since their data contains the complete information of exact citation counts for every year, they are able to train the predictive model using papers before a particular year Y, and predict the citation counts of the future Δt years for papers published in Y. Because we do not know the citation counts for each year, we train and test on papers in the same year by regarding 80% as training and 20% as testing data. In addition, since we propose to use content-independent features, the content-related features of Yan et al. [7][16], including topic rank, topic diversity and versatility, will not be considered in our experiments. Due to different settings and using no content-related features, the prediction results of this paper are different from those of their work.

The evaluation consists of three parts. The first is the main experiment, which answers the four questions mentioned in the beginning of this section, by different combination of feature sets, various years, and those results under both CCP and T-CCP. The second is to report the performance of papers belonging to each citation trend over different future time periods. The third is to show the accuracy of CTC.

4.2   Experimental Results

The main results are shown in Table 1. First, T-CCP averagely outperforms CCP with 0.18 improvement among different feature combinations and future time periods. Such results reflect the usefulness of citation trends: predictive functions trained from separate trend can effectively capture the future citations, avoid learning the noise instances, and thus lead to the boost of accuracy. Second, the results of early citation features are better than those of publication features with 0.34 improvement in average, even using only the first or two years after getting published. This indicates that early citations tend to reveal the potential impact of a paper, and such clue is either more informative or at least equal to the knowledge provided by publication. In addition, it is natural to see that as more future time periods are used (e.g. up to $\Delta t_e = 4$), the performance significantly goes up, because more citation information is revealed. Third, as among the publication features, we can find the author feature has the most impact on the performance. We think it is due to the fact that researchers usually follow the research work of famous and outstanding authors, who might have high productivity or good reputation. Hence the author is the best indicator to distinguish the impact of a paper. Besides, though other publication features are not as effective as the author feature, they are competitive to those features used in Yan et al. [7][16], and combining all the publication features can improve the performance. Fourth, combining both publication and early citation features can further boost the performance with 0.43 improvement in average, compared to features used by Yan et al.

We report the performance of papers belonging to different citation trends using T-CCP with features P+EC ($\Delta t_e = 3$) under different future time periods. The results are shown in Table 2. The performance of papers possessing with at least one burst is at least 0.6 in general. Papers without bursts (i.e., "no bursts") are the most difficult to be accurately predicted. We think the reason is two-fold: (a) papers belonging to trends "no bursts", "middle burst", and "late burst" tend to have similar few citations at the early stage after getting published; (b) the publication features are able to distinguish whether or not a paper gets bursts.

Table 2. Experimental results (R^2) of different citation trends over different future time periods

T-CCP(P+EC3)	Early Burst	Middle Burst	Late Burst	Multi-Bursts	No Bursts
$\Delta t = 12$ (2001)	0.62	0.67	0.63	**0.68**	0.21
$\Delta t = 11$ (2002)	**0.84**	0.63	0.69	0.69	0.33
$\Delta t = 10$ (2003)	0.59	0.53	**0.84**	0.71	0.46

Finally, we report the accuracy of citation trend classification (CTC) using T-CCP with features P+EC($\Delta t_e = 3$). The confusion matrix of accuracy is shown in Table 3. We can find papers whose trends belonging to "no bursts" can be perfectly classified.

Sometimes papers with the remaining four trends have higher possibility to be classi-fied to "multi-bursts" than other three types of citation trends. We think it is because "multi-bursts" can be considered as a combination of "early burst", "middle burst", and "late burst", and thus papers belonging to trend "multi-bursts" might share similar features with those papers belonging to the other three trends.

Table 3. Confusion matrix of accuracy for CTC using Publication and 3-years Early Citations

CTC(P+EC3)	Early Burst	Middle Burst	Late Burst	Multi-Bursts	No Bursts
Early Burst	**0.479**	0.142	0.076	0.227	0.076
Middle Burst	0.080	**0.419**	0.145	0.295	0.062
Late Burst	0.080	0.123	0.353	**0.442**	0.001
Multi-Bursts	0.090	0.245	0.220	**0.446**	0.000
No Bursts	0.005	0.005	0.004	0.006	**0.980**

5   Conclusion

This paper considers citation trends to predict the future citation counts of papers using a robust set of features, consisting of publication and early citation features. Experimental results prove the effectiveness of T-CCP as well as the proposed fea-tures. In fact we can generalize the proposed model as a novel trend-based popularity predictor, and it can be exploited to predict the evolution of numbers in different types of time series data, such as the popularity of YouTube videos, the number of retweets of a post and the times of mentions of hashtag in Twitter. Nevertheless, the categories of trends might vary for different data. In the future, the mechanisms to automatically learn the trends such that the prediction accuracy is boosted need to be well tackled for not only citation count prediction, but also other types of time series data.

Acknowledge. This work is supported by the project of Ministry of Science and Technology in Taiwan, under the grant MOST 103-2221-E-001-006-MY2.

References

1. Bethard, S., Jurafsky, D.: Who should I cite: learning literature search models from citation behavior. In: Proc. of ACM International Conference on Information and Know-ledge Management (CIKM), pp. 609–618 (2010)
2. Adams, J.: Early citation counts correlate with accumulated impact. Scientometrics **63**(3), 567–581 (2005)
3. Hirsch, J.E.: An index to quantify an individual's scientific research output. Proc. of the National Academy of Sciences of the United States of America **102**(46), 16569 (2005)
4. Smola, A., Schölkopf, B.: A tutorial on support vector regression. Statistics and Compu-ting **14**(3), 199–222 (2004)
5. Tang, J., Zhang, J., Yao, L., Li, J., Zheng, L., Su, Z.: ArnetMiner: Extracting and mining of academic social networks. In: Proc. of ACM International Conference on Knowledge Discov-ery and Data Mining (KDD), pp. 990–998 (2008). (data: http://arnetminer.org/citation)

6. Beel, J., Gipp, B.: Google scholar's ranking algorithm: The impact of citation counts (an empirical study). In: Proc. of International Conference on Research Challenges in Information Science (RCIS), pp. 439–446 (2009)

7. Yan, R., Tang, J., Liu, X., Shan, D., Li, X.: Citation count prediction: Learning to estimate future citations for literature. In: Proc. of ACM International Conference on Information and Knowledge Management (CIKM), pp. 1247–1252 (2011)

8. Shi, X., Leskovec, J., McFarland, D.A.: Citing for high impact. In: Proc. of ACM/IEEE-CS Joint Conference on Digital Libraries (JCDL), pp. 49–58 (2010)

9. Castillo, C., Donato, D., Gionis, A.: Estimating Number of Citations Using Author Reputation. In: Ziviani, N., Baeza-Yates, R. (eds.) SPIRE 2007. LNCS, vol. 4726, pp. 107–117. Springer, Heidelberg (2007)

10. Yogatama, D., Heilman, M., O'Connor, B., Dyer, C, Routledge, B.R., Smith, N.A.: Predicting a scientific community's response to an article. In: Proc. of International Conference on Empirical Methods in Natural Language Processing (EMNLP), pp. 594–604, (2011)

11. Pobiedina, N., Ichise, R.: Predicting citation counts for academic literature using graph pattern mining. In: Ali, M., Pan, J.-S., Chen, S.-M., Horng, M.-F. (eds.) IEA/AIE 2014, Part II. LNCS, vol. 8482, pp. 109–119. Springer, Heidelberg (2014)

12. Fan, R.-E., Chang, K.-W., Hsieh, C.-J., Wang, X.-R., Lin, C.-J.: LIBLINEAR: A library for large linear classification. Journal of Machine Learning Research 9, 1871–1874 (2008)

13. Stern, D.I.: High-ranked social science journal articles can be identified from early citation information. PLoS ONE 9(11), e112520 (2014)

14. Blondel, V.D, Guillaume, J.-L., Lambiotte, R., Lefebvre, E.: Fast unfolding of communities in large networks. Journal of Statistical Mechanics: Theory and Experiment (10), P10008 (2008)

15. Chakraborty, T., Sikdar, S., Tammana, V., Ganguly, N., Mukherjee, A.: Computer science fields as ground-truth communities: Their impact, rise and fall. In: Proc. of IEEE/ACM International Conference on Advances in Social Networks Analysis and Mining (ASNOAM), pp. 426–433 (2013)

16. Yan, R., Huang, C., Tang, J., Zhang, Y., Li, X: To better stand on the shoulder of giants. In: Proc. of ACM/IEEE-CS Joint Conference on Digital Libraries (JCDL), pp. 51–60 (2012)

17. Katz, J.S., Hicks, D.: How much is a collaboration worth? A calibrated bibliometric model. Scientometrics 40(3), 541–554 (1997)

Mining Text Enriched Heterogeneous Citation Networks

Jan Kralj[1,2](✉), Anita Valmarska[1,2], Marko Robnik-Šikonja[3],
and Nada Lavrač[1,2,4]

[1] Jožef Stefan Institute, Jamova 39, 1000 Ljubljana, Slovenia
jan.kralj@ijs.si
[2] Jožef Stefan International Postgraduate School, Jamova 39,
1000 Ljubljana, Slovenia
[3] Faculty of Computer and Information Science, Večna pot 113,
1000 Ljubljana, Slovenia
[4] University of Nova Gorica, Vipavska 13, 5000 Nova Gorica, Slovenia

Abstract. The paper presents an approach to mining text enriched heterogeneous information networks, applied to a task of categorizing papers from a large citation network of scientific publications in the field of psychology. The methodology performs network propositionalization by calculating structural context vectors from homogeneous networks, extracted from the original network. The classifier is constructed from a table of structural context vectors, enriched with the bag-of-words vectors calculated from individual paper abstracts. A series of experiments was performed to examine the impact of increasing the number of publications in the network, and adding different types of structural context vectors. The results indicate that increasing the network size and combining both types of information is beneficial for improving the accuracy of paper categorization.

Keywords: Network analysis · Heterogeneous information networks · Text mining · Document categorization · Centroid classifier · PageRank

1 Introduction

The field of *network analysis* is a well established field which has existed as an independent research discipline since the late seventies [Zachary, 1977] and early eighties [Burt and Minor, 1983]. In recent years, analysis of *heterogeneous information networks* [Sun and Han, 2012] has gained popularity. In contrast to standard (homogeneous) information networks, heterogeneous networks describe heterogeneous types of entities and different types of relations. To encode even more information into the network, analysis of *enriched heterogeneous information networks*, where nodes of one type carry additional information in the form of experimental results or text documents, has arisen in recent years [Dutkowski and Ideker, 2011; Hofree et al., 2013].

This paper addresses the task of mining *text enriched heterogeneous information networks* [Grčar et al., 2013]. Compared to the original methodology, our

© Springer International Publishing Switzerland 2015
T. Cao et al. (Eds.): PAKDD 2015, Part I, LNAI 9077, pp. 672–683, 2015.
DOI: 10.1007/978-3-319-18038-0_52

implementation allows for the analysis of much larger heterogeneous networks given significantly decreased computation time. This was achieved by a modified PageRank computation, which takes into account only the parts of the network reachable from the given node, as explained in the methodology section. We showcase the utility of the improved approach on a large citation network in the field of psychology, where nodes—representing publications—are enriched with the publication abstracts. We analyze how the size of the network and the amount of structural information affect the accuracy of paper categorization.

The paper is structured as follows. Section 2 presents the related work. Section 3 presents the upgraded methodology used to analyze text enriched heterogeneous information networks. Section 4 presents the application of the methodology on a large data set of publications from the field of psychology. Section 5 presents the evaluation and analysis of information different components contribute to the quality of classifiers. Section 6 concludes the paper and presents the plans for further work.

2 Related Work

Network mining algorithms perform data analysis in a network setting, where each data instance is connected to other instances in a network of connections.

In ranking methods like Hubs and Authorities (HITS) [Kleinberg, 1999], PageRank [Page et al., 1999], SimRank [Jeh and Widom, 2002] and diffusion kernels [Kondor and Lafferty, 2002], authority is propagated via network edges to discover high ranking nodes in the network. Sun and Han [2012] introduced the concept of *authority* ranking for heterogeneous networks with two node types (bipartite networks) to simultaneously rank nodes of both types. Sun et al. [2009] address authority ranking of all nodes types in heterogeneous networks with a star network schema, while Grčar et al. [2013] apply the PageRank algorithm to only find PageRank values of nodes of a particular node type.

Classification is another popular network analysis task. Typically, the task is to find class labels for some of the nodes in the network using known class labels for a part of the network. A typical approach to solving this problem involves propagating the labels in the network, a concept used in [Zhou et al., 2004] and [Vanunu et al., 2010]. The concept of label propagation was expanded to heterogeneous networks by Hwang and Kuang [2010], performing label propagation to different node types with different diffusion parameters, similarly to the GNETMINE algorithm proposed by Ji et al. [2010]. Classification in heterogeneous networks can also be assisted by ranking, as shown by the ranking based classification approach described by Sun and Han [2012].

Another important concept, related to our work, is the concept of mining *enriched* information networks. While Dutkowski and Ideker [2011] and Hofree et al. [2013] explore biological experimental data using heterogeneous biological networks, Grčar et al. [2013] perform videolectures categorization in a heterogeneous information network of nodes enriched with text information.

Following the work of Grčar et al. [2013], our work is related also to text mining. The task addressed is text categorization in which one has to predict the category of a given document, based on a set of prelabeled documents. Most text mining approaches use the bag-of-words vector representation for each processed document. The resulting high dimensional vectors can be used by any machine learning algorithm capable of handling such vectors, such as a SVM classifier [D'Orazio et al., 2014; Kwok, 1998; Manevitz and Yousef, 2002], kNN classifier [Tan, 2006], Naive Bayes classifier [Wong, 2014], or a centroid classifier [Han and Karypis, 2000].

3   Methodology

This section presents the basics of the methodology of mining text enriched information networks, first introduced by Grčar et al. [2013]. The methodology combines text mining and network analysis on a text enriched heterogeneous information network (such as the citation network of scientific papers) to construct feature vectors which describe both the node content and its position in the network.

The information network is represented as a graph, a structure composed of a set of vertices V and a set of edges E. The edges may be either directed or undirected. Each edge may also have a weight assigned to it. The vertices (or nodes) of the graph in the information network are data instances. A *heterogeneous information network*, as introduced by Sun and Han [2012], is an information network with an additional structure which assigns a type to each node and edge of the network. The requirement is that all starting (or ending) points of edges of a certain type belong to the same type.

The data in a *text enriched heterogeneous information network* represents a fusion of two different data types: heterogeneous information networks and texts. Our data thus comprises of a heterogeneous information network with different node and edge types, where nodes of one designated type are text documents.

Network Decomposition. In the first step of the methodology, for the designated node type (i.e., text documents), the original heterogeneous information network is decomposed into a set of homogeneous networks. In each homogeneous network, two nodes are connected if they share a particular direct or indirect link in the original heterogeneous network. Take an example of a network containing two types of nodes, *P*apers and *A*uthors, and two edge types, *Cites* (linking papers to papers) and *Written_by* (linking papers to authors). From it, we can construct two homogeneous networks of papers: the first in which two papers are connected if one paper cites another, and the second in which they are connected if they share a common author[1]. The choice of links to be used in the network decomposition step is the only manual step of the methodology: taking into account the real-world meaning of links, the domain expert will select only the decompositions relevant for the given task.

---

[1] Depending on the application, any link between two papers, given by the heterogeneous network, may be used to construct either a directed or an undirected edge in the homogeneous network.

Feature Vector Construction. In the second step of the methodology, a set of feature vectors is calculated for each text in the original heterogeneous network: one bag-of-words vector constructed from the text document itself, and one feature vector constructed from every individual homogeneous network.

In bag-of-words (BOW) construction, each text is processed using traditional natural language processing techniques. Typically the following steps are performed: preprocessing using a tokenizer, stop-word removal, stemming, construction of N-grams of a certain length, and removal of infrequent words from the vocabulary.

For each homogeneous networks, obtained through network decomposition, the personalized PageRank (P-PR) algorithm [Page et al., 1999] is used to construct feature vectors for each text in the network.

The personalized PageRank of node v (P-PR$_v$) in a network is defined as the stationary distribution of the position of a random walker which starts its walk in node v and at either selects one of the outgoing connections or travels to his starting location. The probability (denoted p) of continuing the walk is a parameter of the personalized PageRank algorithm and is usually set to 0.85. The PageRank vector is calculated iteratively. In the first step, the rank of node v is set to 1 and the other ranks are set to 0. Then, at each step, the rank is spread along the connections of the network using the formula

$$r^{(k+1)} = p(A^T r^{(k)}) + (1 - p)r^{(0)} \tag{1}$$

where $r^{(k)}$ is the estimation of the PageRank vector after k iterations, and A is the coincidence matrix of the network, normalized so that the elements in each of its rows sum to 1.

Haveliwala and Kamvar [2003] have shown that the iteration, described by Equation 1, converges to the PageRank vector at a rate of p. In our experiments, the number of steps required ranged from 50 to 100, and since each step requires one matrix-vector multiplication, the calculation of a single P-PR vector may take several seconds for a large network, making the calculation of tens of thousands of P-PR vectors computationally very demanding.

Compared to Grčar et al. [2013], this work improves upon the original method by considerably decreasing the amount of computation for cases, where the size of the network taken into account during computation can be decreased. For each network node v, we can consider only the network G_v, composed of all the nodes and edges of the original homogeneous network that lie on paths leading from v. The P-PR$_v$ values, calculated on G_v, are equal to the P-PR values, calculated on the entire homogeneous network. If the network is strongly connected, G_v will be equal to the original network, yielding no change in the performance of the P-PR algorithm, However, if the network G_v is smaller, the calculation of the P-PR$_v$ algorithm will be faster as it is calculated on G_v instead of on the whole network. In our implementation we first estimate if the network G_v contains less than 50% of the original nodes. This is achieved by expanding all possible paths from node v and checking the number of visited nodes in each step. If the number of visited nodes stops increasing after a maximum of 15 steps, we know we have found the network G_v and can count its nodes. If the number of nodes is still increasing, we abort the calculation of G_v. We limit the maximum number of steps because each step of G_v is computationally comparable to one step in the PageRank iterative algorithm which converges in about 50 steps. Therefore we can considerably reduce the computational burden if we do not perform too many steps in the search for G_v.

Once calculated, the resulting PageRank and BOW vectors are normalized according to the Euclidean norm.

Data Fusion. The result of running both the text mining procedure and the personalized PageRank is a set of vectors $\{v_0, v_1, \ldots, v_n\}$ for each node v, where v_0 is the BOW vector, and where for each i ($1 \le i \le n$, where n is the number of network decompositions), v_i is the personalized PageRank vector of node v in the i-th homogeneous network. In the final step of the methodology, these vectors are combined to create one large feature vector. Using positive weights $\alpha_0, \alpha_1, \ldots, \alpha_n$ which sum to 1, a unified vector is constructed which fully describes the publication from which it was calculated. The vector is constructed as

$$v = \sqrt{\alpha_0}b \oplus \sqrt{\alpha_1}v_1 \oplus \cdots \oplus \sqrt{\alpha_n}v_n.$$

where the symbol $\oplus$ represents the concatenation of two vectors. The values of the weights α_i can either be set manually using a trial-and-error approach or can be determined automatically.

A simple way to automatically set weights is to use an optimization algorithm such as the multiple kernel learning (MKL), presented in [Rakotomamonjy et al., 2008] in which the feature vectors are viewed as linear kernels. For each i, the vector v_i corresponds to the linear mapping $\overline{v_i} : x \mapsto x \cdot v_i$. Another possibility is to determine the optimal weights using a general purpose optimization algorithm, e.g., differential evolution [Storn and Price, 1997].

4   Application and Experiment Description

In previous work, Grčar et al. [2013] used the described methodology to assist in the categorization of video lectures, hosted by the VideoLectures.net repository. The methodology turned out useful because of the rapid growth of the number of hosted lectures and the fact that there is a relatively large number of possible categories into which the lectures can be categorized. In this paper, the methodology is applied to a much larger network which allowed us to see 1) how the methodology scales up to big data and 2) if the information contained in the network structure is necessary at all when the textual data is abundant.

We collected data for almost one million scientific publications from the field of psychology. Like the video lectures, the publications belong to one or more categories from a large set of possible categories. The motivation is to construct a classifier which is capable to find appropriate categories for new publications with more probable categories listed first. Such a classifier can be used to assist in the classification of new psychology articles. The same methodology and data set could be exploited to form reading recommendations based on selected paper and to assist authors in submitting their papers to the most appropriate journal.

We first describe the structure and origin of the analyzed data set. Then, we describe creation of heterogeneous network of publications and authors and experiments performed on the data set.

Data Collection. The first step in the construction of a network is data collection. To the best of our knowledge, there is no freely available central database containing publications in the field of psychology. Because of this, we decided to crawl the pages connected with psychology on Wikipedia.

Wikipedia pages are grouped into categories which form a hierarchy. We visited the hierarchical tree of Wikipedia's subcategories of the category Psychology. We examined all categories up to level 5 in the hierarchy. The decision was based on the difference between the number of visited categories and the number of articles at depths 4, 5 and 6. We crawled through all Wikipedia pages, belonging to the visited categories, and extracted the DOIs (digital object identifiers) of all publications, referenced in the pages.

We queried Microsoft Academic Search (MAS) for each of the collected DOIs. If a publication was found on MAS, we collected the information about the title, authors, year of publication, the journal, ID of the publication, IDs of the authors, etc. Whenever possible, we also extracted the publication's abstract. Additionally, we collected the same information for all the publications that cite the queried publications.

Dataset. The result of our data collection process is a network consisting of 953,428 publications of which 63,862 "core publications" were obtained directly from Wikipedia pages. Other publications were citing the core publications. Each of the core publications was labelled with one or more Wikipedia categories from which it was collected. The categories at levels 3, 4 and 5 were transformed into higher level categories by climbing up the category hierarchy to level 2. This was done to decrease the total number of classes. We collected 93,977 abstracts of the publications, of which 4,551 belong to the core publications.

The heterogeneous network was decomposed into three homogeneous networks: the *paper-author-paper* (PAP) network, the *paper-cites-paper* (PP) network and a symmetric copy of the PP network in which directed edges are replaced by undirected edges (PPS).

Experiment Description. In all the experiments we used the same settings to obtain the feature vectors. As in [Grčar et al., 2013], n-grams of size up to 2 and a minimum term frequency of 0 was used to calculate the BOW vectors. For the calculation of personalized PageRank vectors the damping factor was set to 0.85 (the standard setting also used by Page et al. [1999]). In the experiments with more than one feature vector, the vectors were concatenated using weights determined by the differential evolution optimization [Storn and Price, 1997]. In all the experiments we used the centroid classifier using the cosine similarity distance. This classifier first calculates the centroid vector for each class (or category) by summing and normalizing all vectors belonging to instances of that class. For a new instance with feature vector w, it then calculates the cosine similarity distance

$$d(c_i, w) = 1 - c_i \cdot w$$

which represents the proximity of the instance to class i. The class (category) with the minimal distance is selected as the prediction outcome. We also use the "top n" classifier, where the classifier returns n classes with the minimal distances. As in [Grčar et al., 2013], we consider a classifier successful if it correctly predicts at least one label of an instance.

We use the centroid classifier for two reasons. First, Grčar et al. [2013] show that it performs just as well as the SVM and the k-nearest neighbor classifier. Second, for large networks calculating all the personalized PageRank vectors is computationally very expensive. As shown in [Grčar et al., 2013], the centroids of each class can be calculated in a single iteration of the PageRank algorithm.

We performed three sets of experiments using different number of papers and different homogenization of the heterogeneous network.

In the first set, we use the publications for which abstracts are available. Because most of the 93,977 qualifying papers are not core publications, we construct only two feature vectors for each publication: a bag-of-words (BOW) vector and a personalized PageRank vector obtained from the PAP network. We examine how the predictive power of the classifier increases as the number of publications used increases. We use 10,000, 20,000, 30,000, 40,000, 50,000, 70,000 and 93,977 publications.

In the second round of experiments all the collected papers are used (953,428 papers). Because the papers are labelled using citations the PP and PPS networks are not used because the links in this network were used to label the papers. Since the abstracts are not available for most of these papers, only the personalized PageRank vectors obtained from the PAP network are used in the classification.

In the third round of experiments, we use only the core publications for which an abstract is available (4,551 papers). While this is the smallest data set, it allows us to use all of the feature vectors the methodology provides: the BOW vectors and the personalized PageRank obtained from all three networks (PP, PPS and PAP).

5   Evaluation and Results

In each of the experiments, described in Sect. 4, we predicted the labels of the analyzed publications. The classification accuracy was measured for the top 1, 3, 5 and 10 labels, returned by the classifiers. For each experiment the data set was split into a training set, a validation set and a test set. In the first and third round of experiments, the sizes of the testing and validation set were fixed to 2500 instances, all the remaining instances were used for training. In the second round of experiments, the size of the validation and testing set was set to 1500 instances. The centroids of all classes were calculated using the training set and

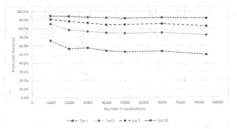

(a) The centroid classifier using BOW.

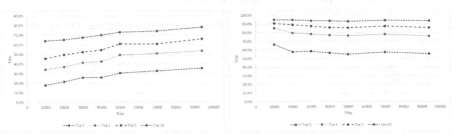

(b) The centroid classifier using PAP.   (c) The centroid classifier using both BOW and PAP.

Fig. 1. The classification accuracy of classifiers using different number of publications to predict labels

concatenated according to the weights optimized using the validation set. The performance of the algorithm (the percentage of papers for which the label is correctly predicted) was calculated using the test set.

The results of the first round of experiments are shown in Fig. 1. The performance of the classifier using BOW vectors does not increase with more instances, while the classifier using PAP vectors is steadily improving as we increase the number of publications. The classifier using both BOW and PAP vectors consistently outperforms the individual classifiers. This shows that combining the network structural information and the content of the publication is useful. As the performance of the PAP classifier increases, the gap between the BOW classifier and the classifier using both vectors also increases. The results obtained with all the 93,977 publications are shown also in Table 1.

Table 1. The classification accuracy of the centroid classifiers in the first and second round of experiments (the publications with abstracts)

top N	BOW+PAP	PAP	BOW	PAP
1	55.5	35.6	49.9	38.8
3	75.8	53.7	72.6	59.3
5	85.6	66.0	82.8	71.0
10	93.5	78.3	92.0	81.4

The classifier using the full PAP network (calculated in the experiment 2), also shown in Table 1, outperforms the classifiers using all other networks, showing that increasing the network size does help the classification. However, its performance is still lower than that of the BOW classifier for smaller networks. It appears that authors in the field of psychology are not strictly limited to one field of research, making prediction using co-authorship information difficult.

Table 2. The classification accuracy of the centroid classifiers in the second round of experiments, (the core publications with abstracts)

top N	All	noBOW	noPAP	noPP	noPPS	BOW+PAP	BOW+PP	BOW+PPS	PAP+PP	PAP+PPS	PP+PPS
1	64.9	49.5	64.7	61.3	65.9	57.7	59.1	62.8	50.3	49.0	44.3
3	84.3	64.6	82.5	74.3	83.5	80.0	78.4	82.0	65.6	63.7	56.7
5	90.2	72.5	90.0	88.6	90.6	88.1	86.4	89.6	72.7	72.0	64.0
10	95.4	81.7	95.4	94.7	95.9	94.9	94.4	95.1	81.5	81.4	73.2

top N	BOW	PP	PPS	PAP
1	55.4	43.5	42.9	30.6
3	78.8	55.8	54.2	47.5
5	87.4	62.4	61.5	58.9
10	93.8	72.1	72.8	72.7

Table 2 shows the accuracies, obtained in the third round of experiments. Because more information was extracted from the network, these results are the most comprehensive overview of the methodology. The results show that using a symmetric citation network (PPS), i.e. allowing the PageRank to use both directions of a citation yields better results than using the unidirectional citation network (PP). Combining both the PP and PPS vectors does not improve the performance of the classifier, which means that vectors, obtained from the PP network, carry no information that is not already contained in the PPS network. However, this is an exception and training classifiers with other vectors combinations increases the prediction accuracy over single vectors: using both BOW and PAP is better than using only BOW, and adding also PP increases the performance even further.

We also analyze the performance of classifiers for different class values. We analyze each class c in the following way. First, we obtain the ordered list of labels that the classifier returns for each test instance from class c. In this list, class values are ordered according to the distance between the instance and the (already computed) class centroids. The first element in the list is the class value whose centroid is closest to the given instance. We then find the rank of class c on this list. For each instance, we compute the minimum value of n for which the top-n classifier predicts class c for this instance. For each class value, we average the obtained ranks n over test instances with this class. This gives us an estimate of the ranking error.

We plot these average ranks versus number of instances with each class value. The results are shown in Fig. 2. The graphs are similar for classifiers using BOW, PPS and PAP vectors. We can see that classes containing a small number of instances have considerably higher average ranks than classes with many instances, meaning that prediction is much less successful for underrepresented class values. The classifier using PP vectors is the only classifier for which this

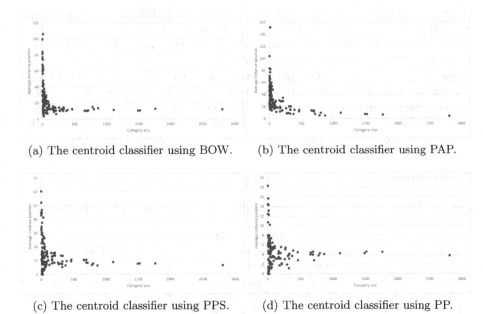

(a) The centroid classifier using BOW.   (b) The centroid classifier using PAP.

(c) The centroid classifier using PPS.   (d) The centroid classifier using PP.

Fig. 2. Graphs showing the average index of a class versus the class size

trend does not appear. For the PP classifier, the results for small classes show much more noise than for larger classifiers, but average ranks (i.e., error) does not decrease with increasing number of instances.

6   Conclusions and Further Work

While network analysis in general is an established field of research, analysis of heterogeneous networks is a much newer field. Methods taking the heterogeneous nature of the networks into account show an improved performance, as shown in Davis et al. [2011]. Some methods like RankClus and others presented in [Sun and Han, 2012] are capable of solving tasks that cannot even be defined on homogeneous information networks (like clustering two disjoint sets of entities). Another important novelty is joining network analysis with the analysis of data, either in the form of text documents or results obtained from various experiments [Dutkowski and Ideker, 2011], [Hofree et al., 2013] and [Grčar et al., 2013].

This paper presents a more efficient implementation of the methodology by [Grčar et al., 2013], which combines the information from heterogeneous networks with textual data. By improving the computational efficiency of the approach we were able to address a novel application, i.e. the analysis of a large citation network of psychology papers. Our contribution is also the analysis of performance with different number of instances and different types of network structures included. The results show that relational information hidden in the network structure is beneficial for classification, while the errors are shown to be mostly due to low number of instances for some categories.

In the work presented, we only use a part of the information we collected about the publications. In future, we will to examine how to incorporate the temporal information into our methodology; we have already collected the year of publication, which allows us to observe the dynamics of categories, aiming to improve the classification accuracy. In addition, we plan to use a combination of network analysis and data mining on PubMed and DBLP articles. We will also address biological networks enriched with experimental data and texts.

References

Burt, R., Minor, M.: Applied Network Analysis: A Methodological Introduction. Sage Publications (1983)

Davis, D., Lichtenwalter, R., Chawla, N.V.: Multi-relational link prediction in heterogeneous information networks. In: Proceedings of the 2011 International Conference on Advances in Social Networks Analysis and Mining, pp. 281–288 (2011)

D'Orazio, V., Landis, S.T., Palmer, G., Schrodt, P.: Separating the wheat from the chaff: Applications of automated document classification using support vector machines. Polytical Analysis 22(2), 224–242 (2014)

Dutkowski, J., Ideker, T.: Protein networks as logic functions in development and cancer. PLoS Computational Biology 7(9), (2011)

Grčar, M., Trdin, N., Lavrač, N.: A methodology for mining document-enriched heterogeneous information networks. The Computer Journal 56(3), 321–335 (2013)

Han, E.-H.S., Karypis, G.: Centroid-based document classification: analysis and experimental results. In: Zighed, D.A., Komorowski, J., Żytkow, J.M. (eds.) PKDD 2000. LNCS (LNAI), vol. 1910, pp. 424–431. Springer, Heidelberg (2000)

Haveliwala, T., Kamvar, S.: The second eigenvalue of the Google matrix. Technical report, Stanford InfoLab (2003)

Hofree, M., Shen, J.P., Carter, H., Gross, A., Ideker, T.: Network-based stratification of tumor mutations. Nature Methods 10(11), 1108–1115 (2013)

Hwang, T., Kuang, R.: A heterogeneous label propagation algorithm for disease gene discovery. In: Proceedings of SIAM International Conference on Data Mining, pp. 583–594 (2010)

Jeh, G., Widom, J.: SimRank: A measure of structural-context similarity. In: Proceedings of the 8th ACM SIGKDD International Conference on Knowledge Discovery and Data Mining, pp. 538–543. ACM (2002)

Ji, M., Sun, Y., Danilevsky, M., Han, J., Gao, J.: Graph regularized transductive classification on heterogeneous information networks. In: Balcázar, J.L., Bonchi, F., Gionis, A., Sebag, M. (eds.) ECML PKDD 2010, Part I. LNCS, vol. 6321, pp. 570–586. Springer, Heidelberg (2010)

Kleinberg, J.M.: Authoritative sources in a hyperlinked environment. Journal of the ACM 46(5), 604–632 (1999)

Kondor, R.I., Lafferty, J.D.: Diffusion kernels on graphs and other discrete input spaces. In: Proceedings of the 19th International Conference on Machine Learning, pp. 315–322 (2002)

Kwok, J.T.-Y.: Automated text categorization using support vector machine. In: Proceedings of the 5th International Conference on Neural Information Processing, pp. 347–351 (1998)

Manevitz, L.M., Yousef, M.: One-class SVMs for document classification. Journal of Machine Learning Research **2**, 139–154 (2002)

Page, L., Brin, S., Motwani, R., Winograd, T.: The PageRank citation ranking: Bringing order to the web. Technical report, Stanford InfoLab (1999)

Rakotomamonjy, A., Bach, F., Canu, S., Grandvalet, Y.: SimpleMKL. Journal of Machine Learning Research **9**, 2491–2521 (2008)

Storn, R., Price, K.: Differential evolution; a simple and efficient heuristic for global optimization over continuous spaces. Journal of Global Optimization **11**(4), 341–359 (1997)

Sun, Y., Han, J.: Mining Heterogeneous Information Networks: Principles and Methodologies. Morgan & Claypool Publishers (2012)

Sun, Y., Yu, Y., Han, J.: Ranking-based clustering of heterogeneous information networks with star network schema. In: Proceedings of the 15th ACM SIGKDD International Conference on Knowledge Discovery and Data Mining, pp. 797–806 (2009)

Tan, S.: An effective refinement strategy for KNN text classifier. Expert Syst. Appl. **30**(2), 290–298 (2006)

Vanunu, O., Magger, O., Ruppin, E., Shlomi, T., Sharan, R.: Associating genes and protein complexes with disease via network propagation. PLoS Computational Biology **6**(1) (2010)

Wong, T.-T.: Generalized Dirichlet priors for Naïve Bayesian classifiers with multinomial models in document classification. Data Mining and Knowledge Discovery **28**(1), 123–144 (2014)

Zachary, W.: An information flow model for conflict and fission in small groups. Journal of Anthropological Research **33**, 452–473 (1977)

Zhou, D., Bousquet, O., Lal, T.N., Weston, J., Schölkopf, B.: Learning with local and global consistency. Advances in Neural Information Processing Systems **16**(16), 321–328 (2004)

Boosting via Approaching Optimal Margin Distribution

Chuan Liu and Shizhong Liao[✉]

School of Computer Science and Technology,
Tianjin University, Tianjin 300072, China
szliao@tju.edu.cn

Abstract. Margin distribution is crucial to AdaBoost. In this paper, we propose a new boosting method by utilizing the Emargin bound to approach the optimal margin distribution. We first define the k^*-optimization margin distribution, which has a sharper Emargin bound than that of AdaBoost. Then we present two boosting algorithms, KM-Boosting and MD-Boosting, both of which approximately approach the k^*-optimization margin distribution using the relation between the kth margin bound and the Emargin bound. Finally, we show that boosting on the k^*-optimization margin distribution is sound and efficient. Especially, MD-Boosting almost surely has a sharper bound than that of AdaBoost, and just needs a little more computational cost than that of AdaBoost, which means that MD-Boosting is effective in redundancy reduction without losing much accuracy.

Keywords: Boosting · Emargin bound · k^*-optimization margin distribution

1 Introduction

Aiming to construct a "strong" classifier by combining a series of "weak" classifiers, boosting is currently one of the most successful techniques in classification. The "weak" classifier produced by weak learning algorithm only needs a classification ability better than a random guess. AdaBoost is the first practical boosting algorithm proposed by Freund and Schapire [5,6], and has exhibited excellent performance on benchmark datasets and real applications [4,19].

Behind the success, the working mechanism of AdaBoost has not been explained completely. Especially, the generalization error of AdaBoost keeps decreasing after a large number of base classifiers having been combined, which seems violating the Occams razor [1], intuitively.

In the statistics community, researchers have devoted much effort to study why and how boosting works, Breiman [2] and Friedman et al. [7] viewed boosting algorithm as gradient descent optimization in functional space. Mason et al. [13] developed AnyBoost for boosting arbitrary loss functions with a similar idea. In addition, some boosting algorithms have been proved to be consistent with

© Springer International Publishing Switzerland 2015
T. Cao et al. (Eds.): PAKDD 2015, Part I, LNAI 9077, pp. 684–695, 2015.
DOI: 10.1007/978-3-319-18038-0_53

Bayesian classifiers under some limitations [3,10]. However, these theories can't explain the resistance to overfitting of boosting. Margin theory is another popular explanation proposed by Schapire et al. [16], which argues that AdaBoost reduces the generalization error via improving the margin. Breiman [2] proved a minimum margin bound that is sharper than the bound given by Schapire et al. [16], thus they considered that minimum margin is more relevant to the boosting algorithms. Thus, much works [2,9] focused on maximizing the minimum margin. However, these algorithms do not always yield better performance. In fact, more often the opposite is true, and margin theory suffered serious doubt. Later, Koltchinskii et al. [11,12] showed the bound in Schapire et al. [16] can be improved based on Rademacher and Gaussian complexities, but these bounds can not be proved to be sharper than Breiman's minimum margin bound. Reyzin and Schapire [15] duplicated the experiments of Breiman's and they observed some flaws that lead to poor control in model complexity. They emphasized that margin distribution rather than minimum margin is crucial to boosting.

Recently, Wang et al. [21] proposed a new bound in terms of margin distribution called Emargin bound, which was proved to be sharper than previously well-known bounds [2,16]. In particular, they showed that if a boosting algorithm minimizes the Emargin bound, the learned classifier would converges to the optimal classifier in the hypothesis space. Gao et al. [8] presented the kth margin bound, from which they reformulated Emargin bound as the infimum of all the kth margin bound, that is, they proved that the Emargin bound is sharper than the minimum margin bound [2] from a new perspective. These results suggest a new boosting approach via optimizing Emargin bound. However, the Emargin bound can not be optimized easily. Although Wang et al. [20] designed the EEM algorithm to verify Emargin theory and obtained exciting results, the algorithm is not suitable in real applications due to the computational complexities.

Now the margin distribution of boosting algorithms is becoming more important [11,12,16]. Shen et al. [17] optimized the margin distribution through the expectation and variance, and they proved that AdaBoost approximately maximizes the unnormalized average margin and minimizes the margin variance. However, they provided no generalization error bound to support their method.

In this paper, we explicitly define an approximately optimal margin distribution called k^*-optimization margin distribution, and demonstrate that it would lead to a sharper generalization error bound than that of AdaBoost. From the definition we then develop two approximate algorithms, KM-Boosting and MD-Boosting. Both of the two algorithms present good results on benchmark datasets. In particular, the classifier generated by MD-Boosting empirically has a sharper generalization error bound than that of AdaBoost almost surely. Moreover, MD-Boosting can be viewed as an effective method to reduce redundancy because of its limited computational cost and less loss of accuracy.

2   Background and Related Work

Let the training set $S = \{(x_i, y_i)\}_1^n$ chosen independently from underlying distribution D over $X \times Y$. Here we focus on binary classification, that is, $Y = \{-1, 1\}$.

$\mathcal{H}$ is a given hypothesis space, and $\forall h \in \mathcal{H}$ is a mapping from X to Y. Let $\mathcal{C}(\mathcal{H})$ be the completion of the convex hull $\mathcal{H}$, i.e.,

$$\mathcal{C}(\mathcal{H}) = \overline{\{f | f = \sum \alpha_i h_i, \ \sum \alpha_i = 1 \text{ and } \alpha_i \geq 0\}}.$$

AdaBoost first initializes the weights of the training data uniformly. Then at each iteration, AdaBoost returns a base classifier $h \in \mathcal{H}$, and changes the distribution of the training data according to h. i.e., decrease the weights of the correctly classified examples and increase the weights of the misclassified examples simultaneously. The final classifier is a sign function of f, where $f = \sum \alpha_t h_t \in \mathcal{C}(\mathcal{H})$, and α_t is the corresponding weight of h_t.

For an example (x_i, y_i), the margin $y_i f(x_i)$ reflects the difference between the weights of correctly classified base classifiers and misclassified, that is

$$y_i f(x_i) = \sum_{t:y_i=h_t(x_i)} \alpha_t - \sum_{t:y_i \neq h_t(x_i)} \alpha_t. \tag{1}$$

$\Pr_S(yf(x) < \theta)$ can be regarded as a distribution over θ ($\theta \in [-1,1]$), called margin distribution, and denoted by $\mathrm{MD}(f)$. A "good" margin distribution means that most examples have large margins so that $\Pr_S(yf(x) < \theta)$ is small while the value of θ is not too small. The kth margin $\hat{y}_k f(\hat{x}_k)$ is defined as the kth smallest value in $\{y_i f(x_i), i = 1, 2, \ldots, n\}$. We define prediction matrix $H \in Q^{n \times T}$, where $H_{ij} = h_j(x_i)$ is the label of x_i given by $h_j(\cdot)$. We use $H_{i:}$ to denote the ith row of H, which describes all the outputs of the classifiers on x_i.

Denote Bernoulli relative entropy function by

$$\Delta(u; r) = u \log \frac{u}{r} + (1-u) \log \frac{1-u}{1-r}, \ 0 \leq u, r \leq 1.$$

It's easy to get that $\Delta(u; r)$ is a monotone increasing function of r for $u \leq r \leq 1$, thus the inverse of $\Delta(u; r)$ for a fixed u can be given

$$\Delta^{-1}(u; v) = \inf_w \{w : w \geq u \text{ and } \Delta(u; w) \geq v\}.$$

Theorem 1. (*The kth margin bound [8]*) *For any $\delta > 0$, if $\theta_k = \hat{y}_k f(\hat{x}_k) > \sqrt{\frac{8}{|\mathcal{H}|}}$, then with probability at least $1 - \delta$ over the random choice of sample with size n, every voting classifier $f \in \mathcal{C}(\mathcal{H})$ satisfies the following bound:*

$$\Pr_D[yf(x) < 0] \leq \frac{\ln |\mathcal{H}|}{n} + \Delta^{-1}\left(\frac{k-1}{n}; \frac{q}{n}\right), \tag{2}$$

where

$$q = \frac{8 \ln(2|\mathcal{H}|)}{\theta_k^2} \ln \frac{2n^2}{\ln |\mathcal{H}|} + \ln |\mathcal{H}| + \ln \frac{n}{\delta}. \tag{3}$$

The minimum margin bound is the trivial condition of kth margin bound with set $k = 1$, and the Emargin bound can be reformulated as:

$$\Pr_D[yf(x) < 0] \leq \frac{\ln |\mathcal{H}|}{n} + \inf_{k \in \{1,2,\ldots,n\}} \Delta^{-1}\left(\frac{k-1}{n}; \frac{q}{n}\right). \tag{4}$$

For convenience, we denote the Emargin bound and the kth bound of f by $\mathrm{EB}(f)$ and $\mathrm{KB}(f,k)$, respectively. Thus, we have

$$\mathrm{EB}(f) = \inf_{k \in \{1,2,\ldots,n\}} \mathrm{KB}(f,k).$$

3 k^*-optimization Margin Distribution

We first define a kind of margin distribution that has good property according the margin theory of boosting.

Definition 1. *For* $\forall f, g \in \mathcal{C}(\mathcal{H})$, *$f$ is produced by AdaBoost, $\mathrm{MD}(g)$ is the k^*-optimization margin distribution if the inequality:*

$$\hat{y}_{k^*} f(\hat{x}_{k^*}) \le \hat{y}_{k^*} g(\hat{x}_{k^*})$$

holds for $k^* = \arg\min_{k \in \{1,\ldots,n\}} \mathrm{KB}(f,k)$.

To describe the property of k^*-optimization margin distribution, we give the following lemma.

Lemma 1. $\Delta^{-1}(u;v)$ *ia a monotone increasing function of* v.

Proof. Since

$$\Delta^{-1}(u;v) = \inf_w \{w : w \ge u \text{ and } \Delta(u;w) \ge v\},$$

for $\forall\, u \le v_1 \le v_2 < 1$, we have

$$\begin{aligned}
\Delta^{-1}(u;v_1) &= \inf_w \{w : w \ge u \text{ and } \Delta(u;w) \ge v_1\} \\
&\le \inf_w \{w : w \ge u \text{ and } \Delta(u;w) \ge v_2\} \\
&= \Delta^{-1}(u;v_2),
\end{aligned}$$

where the inequality holds from the relation

$$\{w : w \ge u \text{ and } \Delta(u;w) \ge v_2\} \subset \{w : w \ge u \text{ and } \Delta(u;w) \ge v_1\}.$$

This completes the proof of the lemma.

Theorem 2. *If* $\mathrm{MD}(g)$ *is k^*-optimization margin distribution, g has sharper Emargin bound than f, i.e., $\mathrm{EB}(g) \le \mathrm{EB}(f)$.*

Proof. Since $\hat{y}_{k^*} f(\hat{x}_{k^*}) \le \hat{y}_{k^*} g(\hat{x}_{k^*})$ holds for $k^* = \arg\min_{k \in \{1,\ldots,n\}} \mathrm{KB}(f,k)$, and from Lemma 1 and formulae (2) and (3) we can easily get

$$\mathrm{KB}(g,k^*) \le \mathrm{KB}(f,k^*),$$

then

$$\mathrm{EB}(g) = \inf_{k \in \{1,\ldots,n\}} \mathrm{KB}(g,k) \le \mathrm{KB}(g,k^*) \le \mathrm{KB}(f,k^*) = \mathrm{EB}(f). \tag{5}$$

This completes the proof.

Theorem 2 guarantees the superiority of the k^*-optimization margin distribution, and the purpose of this paper is to develop a new boosting algorithm with the k^*-optimization margin distribution.

4   Two Optimization Strategies

This section introduces two approximate strategies to obtain k^*-optimization margin distribution. For simplicity, we assume $g(x) = \sum \beta_t h_t(x)$, where $h_t(x)$ is the hypothesis returned by the tth step of AdaBoost.

4.1   KM-Boosting

Definition 1 directly suggests a way to generate a voting classifier that has k^*-optimization margin distribution, which consists of two key steps, the first is to find the k^* corresponding to the Emargin bound, and the second is to improve the kth margin for a fixed $k = k^*$.

We first discuss the second step. First run the standard AdaBoost T steps, then a hypothesis set $\{h_1(x), h_2(x), \ldots, h_T(x)\}$ and their corresponding coefficients $\{\alpha_1, \alpha_2, \ldots, \alpha_T\}$ can be obtained. Compute the margin of each sample and sort them in ascending order. Then drop out the points with margin ordered in the first $k - 1$ and maximize the minimum margin for the rest points, where the rest points are marked $\{(x_{s_1}, y_{s_1}), (x_{s_2}, y_{s_2}), \ldots, (x_{s_{n-k+1}}, y_{s_{n-k+1}})\}$, i.e., improving kth margin can be reduced to the following linear program.

$$\max_{\beta, m} m$$

$$\text{s.t. } y_i \sum_{t=1}^{T} \beta_t h_t(x_i) \geq m \ (i = s_1, s_2, \ldots, s_{n-k+1}), \tag{6}$$

$$\beta_t \geq 0, \quad \sum_{t=1}^{T} \beta_t = 1.$$

Theorem 3. *The classifier f' with coefficients solved from formula (6) has a larger kth margin than f.*

Proof. Obviously $(\alpha, \hat{y}_k f(\hat{x}_k))$ is a feasible solution of formula (6), then

$$y_i f'(x_i) \geq \hat{y}_k f(\hat{x}_k)$$

holds for $\forall i \in \{s_1, s_2, \ldots, s_{n-k+1}\}$. That is, there are at most $k - 1$ samples that have margin less than $\hat{y}_k f(\hat{x}_k)$. Therefore, we have

$$\hat{y}_k f'(\hat{x}_k) \geq \hat{y}_k f(\hat{x}_k).$$

This completes the proof of the theorem.

Actually, formula (6) can be understood to maximize the minimum margin but ignores the outliers, that is, the dropped $k - 1$ samples are interpreted as outliers. In particular, the LP-AdaBoost is the trivial condition with $k = 1$. This is why LP-AdaBoost performs rather poor when much noise exists.

Algorithm 1. KM-Boosting

Input :
Sample set $S = \{(x_1, y_1), (x_1, y_2), \ldots, (x_n, y_n)\}$,
the number of iterations T.

Procedure:
Run AdaBoost T steps to obtain $f(x) = \sum_{t=1}^{T} \alpha_t h_t(x)$.
Compute the margin of each sample as (1).
Sort the margins in ascending order.
Confirm k^* through validation error of AdaBoost as (7).
According to k^* drop outliers and get new training set.
Solve new coefficients $\{\beta_1, \beta_2, \ldots, \beta_T\}$ from the linear program (6).
Output:
The final classifier $G(x) = \text{sgn}(g(x))$, where $g(x) = \sum_{t=1}^{T} \beta_t h_t(x)$.

Now we talk about how to choose k^*. Since we interpret the first $k^* - 1$ samples as outliers, choosing k^* means confirming the percentage of outliers. Here we consider the percentage of outliers in training set as follows:

$$e' = \Pr_S[\Pr(\hat{y} = y|x) < 0.5] \approx \Pr_D[\Pr(\hat{y} = y|x) < 0.5],$$

i.e., the Bayes error, where $\hat{y}$ is the prediction of y given by Bayes classifier, and the approximation comes from the fact that S drawn independently from D. Further more, we approximate it by the validation error ve' of AdaBoost. That is, we take

$$k^* = \lfloor ve' * n \rfloor + 1. \tag{7}$$

Algorithm 1 is the formal description of KM-Boosting. In fact, KM-Boosting is highly consistent with margin theory. Remind that "good" margin distribution can be simply described as the distribution in which most examples has large margins. This idea is completely reflected in KM-Boosting, namely, the $n - k^* + 1$ examples are viewed as the "most" examples. We maximize the minimum margin of them and don't care the margins of the other $k^* - 1$ points.

4.2 MD-Boosting

KM-Boosting directly optimize the k^*th margin, however, it must save some points as validation sets. Here we present another method that could obtain k^*-optimization margin distribution simply.

First look at the fact in Figure 1, it shows the kth margin increase along with the growth of iteration when k is small. However, when k reaches a certain value, the kth margin always decrease. We have mentioned that larger kth margin would produce smaller kth margin bound for a fixed k. This means, it's the increased kth margin that really help to improve the Emargin bound. Based on

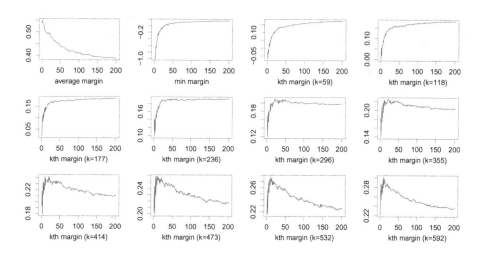

Fig. 1. The change of minimum margin, average margin and the kth margin as the iterations increase in "twonorm". In each subgraph, the horizontal axis is the iterations and the vertical axis is the corresponding kth margin.

these results, we define set K^* as follows

$$K^* = \{k \mid \hat{y}_k f(\hat{x}_k) > \sqrt{\frac{8}{|\mathcal{H}|}} \wedge$$

the kth margin increase with the iteration$\}$,

where $\hat{y}_k f(\hat{x}_k) > \sqrt{\frac{8}{|\mathcal{H}|}}$ comes from Theorem 1, and then approximate the original Emargin bound by

$$\Pr_D \left(yf(x) < 0 \right) \leq EB(f)$$
$$= \min_{k \in \{1,2,\ldots,n\}} KB(f,k) \approx \min_{k \in K^*} KB(f,k) \tag{8}$$

Proposition 1. *For $\forall g \in \mathcal{C}(\mathcal{H})$, MD($g$) is k^*-optimization margin distribution if the inequality*

$$\hat{y}_k g(\hat{x}_k) \geq \hat{y}_k f(\hat{x}_k)$$

holds for all $k \in K^*$.

This proposition is easy to get from formula (8). Therefore, the object turns to finding a learner satisfying Proposition 1:

$$\max_{g \in \mathcal{C}(\mathcal{H}), \boldsymbol{\xi}} \sum \xi_k$$
$$\text{s.t. } \hat{y}_k g(\hat{x}_k) \geq \hat{y}_k f(\hat{x}_k) + \xi_k \quad (k \in K^*), \tag{9}$$
$$\xi_k \geq 0.$$

Solving equation (9) is intractable, and we approximately formulate it as

$$\max_{\boldsymbol{\beta},\boldsymbol{\xi}} \quad \sum \xi_i$$

$$\text{s.t.} \quad y_k H_{i:}\boldsymbol{\beta} \geq y_k H_{i:}\boldsymbol{\alpha} + \xi_i \quad ((x_i, y_i) \in S^*),$$

$$\beta_t \geq 0, \quad \sum \beta_t = 1, \tag{10}$$

$$\xi_i \geq 0,$$

where

$$S^* = \{ (x_i, y_i) \mid \text{od}(y_i f(x_i)) \in K^* \},$$

and $\text{od}(y_i f(x_i))$ is the order of $y_i f(x_i)$ in all the margins.

Many researchers have discussed that it's the hard points that really influence the result of AdaBoost, while the weights of easy points quickly vanish and give no (asymptotic) contribution. Actually, formula (10) can be viewed as a method to improve the margins of the hard points.

Then we turn to the question how to build the set K^*. Notice that Figure 1 shows that the average margin decreases while the iteration increases. In fact, this phenomenon appears in all datasets we have tested. So empirically we have for $\forall k \in K^*$,

$$\hat{y}_k f(\hat{x}_k) < \mu_S(Yf(X))$$

where $\mu_S(Yf(X))$ is the average margin of f in S. Shen et al. [18] proved the margin of AdaBoost follows the Gaussian distribution, thus $|K^*| \leq \frac{n}{2}$, approximately.

Tracking the change of the kth margin to build K^* would lead to additional computation and storage costs. Actually, it's unnecessary since we find K^* is not very sensitive to the result, that is, as long as $|K^*|$ reaches some certain value, the result will be good. So we replace the original K^* with

$$K^* = \{k \mid n * q_0 \leq k \leq \max\{500, 0.1 * n\} + nq_0 \land$$
$$\hat{y}_k f(\hat{x}_k) < \mu_S(Yf(X))\}, \tag{11}$$

where $q_0 = \text{Pr}_S[yf(x) \leq 0]$. Therefore, the problem size of formula (10) is at most $0.1 * n + T$ when n is very large while other modified boosting algorithms usually require to solve a linear program with size $n + T$.

5 Experimental Results and Analysis

The goal of this paper is not to find a boosting algorithm that outperforms all the other variants of boosting, but to show a way to approach the optimal margin distribution. So we only compare the proposed algorithms with AdaBoost. We verify the proposed algorithms on 13 benchmark datasets with two types of base classifiers, decision stumps and three-layer decision trees. We train on a randomly drawn subset of 40% of the examples in a data sets and validate on a subset of 20% of the examples, which is disjointed with the above training set,

---

Algorithm 2. MD-Boosting

Input :
Sample set $S = \{(x_1, y_1), (x_1, y_2), \ldots, (x_n, y_n)\}$,
the number of iterations T.

Procedure:
Run AdaBoost T steps
to obtain base classifiers $h_t(x)$ and corresponding coefficients α_t.
Compute the margin of each sample as (1).
Sort the margin in ascending order and confirm K^* as (11).
Solve the new coefficients $\{\beta_1, \beta_2, \ldots, \beta_T\}$ by (10).
Output:
The final classifier $G(x) = \text{sgn}(g(x))$, where $g(x) = \sum_{t=1}^{T} \beta_t h_t(x)$.

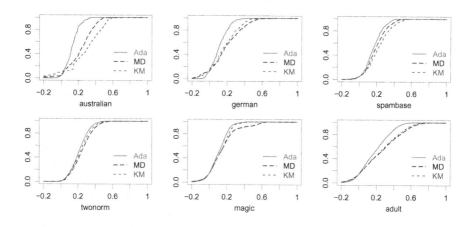

Fig. 2. Comparison of the margin distributions on different datasets with decision stumps, where the horizontal axis is θ, and the vertical axis is the percentage of the samples with margin less than θ.

the other 40% is used as the test set. For comparison, the iteration steps in each experiment is fixed at 200, and we repeat 10 times for each settings.

Figure 2 and 3 present the comparison of the margin distributions on different datasets with decision stumps and three-layer decision trees, respectively. From the figures we can see that the black line representing the margin distribution of MD-Boosting and the blue line representing the margin distribution of KM-Boosting usually appear bellow the red line representing the margin distribution of AdaBoost. We assert that both MD-Boosting and KM-Boosting can improve the margin distribution, and that for the classifier g generated by MD-Boosting,

$$\forall \theta > 0 \left(\Pr_S[y_i g(x_i) < \theta] \leq \Pr_S[y_i f(x_i) < \theta] \right)$$

holds on all the datasets. This means that the classifier yielded by MD-Boosting has a sharper generalization bound than that of AdaBoost according the mar-

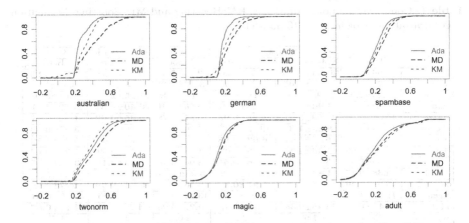

Fig. 3. Comparison of the margin distributions on different datasets with decision trees, where the horizontal axis is θ, and the vertical axis is the percentage of the samples with margin less than θ.

gin theory of boosting. As far as we know, MD-Boosting is the first boosting algorithm that has this property. For all the other modified boosting algorithms, such as LP$_{Reg}$-AdaBoost [14], MDBoost [17], and the KM-Boosting, the margin distribution curve usually appear above that of AdaBoost for some $\theta > 0$.

Table 1 shows the comparison among AdaBoost, KM-Boosting and MD-Boosting, where the best performances are marked in bold face. The last row is the average ranking of the three methods on all 13 datasets. We can see both MD-Boosting and KM-Boosting have a better performance than that of AdaBoost on decision stumps, while KM-Boosting performs a little worse than AdaBoost does on decision trees. This may be due to the fact that some leaf nodes have no sample, causing uncontrolled complexities. In consideration of both of accuracy and performance, it seems that decision stumps are more suitable for KM-Boosting.

Figure 4 shows the size of the base classifier in the final classifier generated by MD-Boosting, which is a decision stump and a decision tree respectively. The sizes of both base classifiers could be reduced, especially the decision stump, the size does not exceed 50 on ten datasets. That is, MD-Boosting with decision stump is more effective for feature selection where small ensembles are needed.

6   Conclusion

It is widely accepted that margin distribution plays an important role in the success of AdaBoost. Following this line, we first define the k^*-optimization margin distribution, which is closer to the optimal margin distribution than that of AdaBoost. Then we propose two boosting algorithms: KM-Boosting and MD-Boosting, both of which can approximate the k^*-optimization margin distribution and improve the accuracy of AdaBoost. Especially, MD-Boosting almost surely has a sharper generalization error bound than that of AdaBoost, and can

Table 1. Comparison among AdaBoost, KM-Boosting and MD-Boosting: Estimation of generalization error in % on 13 datasets

Alg\Data	decision stumps			decision trees (deep = 3)		
	AdaBoost	KM-Boosting	MD-Boosting	AdaBoost	KM-Boosting	MD-Boosting
heart	23.06±5.65	**20.09±7.13**	22.50±8.06	22.13±7.50	22.87±5.93	**21.94±5.83**
wisconsin	4.74±2.19	**4.65±2.65**	4.93±1.64	**3.47±1.28**	3.68±1.50	3.87±1.31
german	27.25±2.95	26.95±6.00	**26.18±6.08**	28.08±4.18	**27.15±2.38**	27.63±5.20
diabetes	**24.90±4.45**	**24.90±3.99**	25.18±4.29	26.56±5.91	**26.40±4.32**	27.79±4.74
australian	15.82±2.64	15.69±3.50	**15.25±3.95**	15.87±4.78	**15.25±4.02**	16.52±3.48
phoneme	18.88±1.07	**18.83±1.34**	19.08±0.96	14.20±1.16	14.19±0.94	**13.97±0.93**
spambase	6.11±1.11	6.44±1.05	**6.03±0.98**	5.50±0.77	5.66±1.09	**5.48±1.19**
chess	4.42±0.90	**3.96±1.15**	4.29±1.03	1.05±0.54	1.13±0.69	**1.02±0.42**
twonorm	3.30±0.49	3.49±0.48	**3.28±0.51**	3.04±0.46	3.23±0.41	**3.01±0.36**
breast	4.38±0.17	**3.89±2.32**	4.27±1.82	**3.21±2.26**	3.32±1.79	3.39±1.72
coil2000	5.98±0.40	5.98±0.38	**5.97±0.39**	6.56±0.88	**6.24±0.69**	6.37±0.78
magic	15.92±1.02	**15.91±0.97**	15.92±1.09	13.44±0.53	13.48±0.56	**13.42±0.61**
adult	**14.41±0.38**	14.45±0.38	14.42±0.37	14.21±4.96	14.25±0.53	**14.19±0.57**
avg-rank	2.46	1.69	1.92	2.08	2.15	1.77

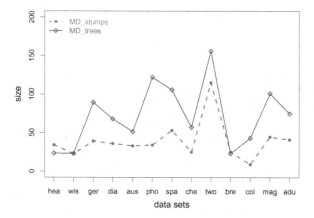

Fig. 4. The size of base classifiers in the final classifier generated by MD-Boosting. MD-stumps denotes MD-Boosting + decision stumps, and MD-trees denotes MD-Boosting + decision trees.

reduce redundancy without accuracy loss. As far as we know, MD-Boosting is the first boosting algorithm that has these properties.

Acknowledgments. The work is supported in part by the National Natural Science Foundation of China under grant No. 61170019.

References

1. Blumer, A., Ehrenfeucht, A., Haussler, D., Warmuth, M.K.: Occam's razor. Information Processing Letters **24**(6), 377–380 (1987)
2. Breiman, L.: Prediction games and arcing algorithms. Neural Computation **11**(7), 1493–1517 (1999)

3. Bühlmann, P.L.: Consistency for L2Boosting and matching pursuit with trees and tree-type basis functions. Tech. Rep. 109, ETH Zürich (2002)
4. Dietterich, T.G.: An experimental comparison of three methods for constructing ensembles of decision trees: Bagging, boosting, and randomization. Machine Learning **40**(2), 139–157 (2000)
5. Freund, Y., Schapire, R.E.: Experiments with a new boosting algorithm. In: Proceedings of the 13th International Conference on Machine Learning, vol. 96, pp. 148–156 (1996)
6. Freund, Y., Schapire, R.E.: A decision-theoretic generalization of on-line learning and an application to boosting. Journal of Computer and System Sciences **55**(1), 119–139 (1997)
7. Friedman, J., Hastie, T., Tibshirani, R.: Additive logistic regression: A statistical view of boosting. Annals of Statistics **28**(2), 337–407 (2000)
8. Gao, W., Zhou, Z.H.: On the doubt about margin explanation of boosting. Artificial Intelligence **203**, 1–18 (2013)
9. Grove, A.J., Schuurmans, D.: Boosting in the limit: maximizing the margin of learned ensembles. In: Proceedings of the 15th National Conference on Artificial Intelligence, pp. 692–699 (1998)
10. Jiang, W.: Process consistency for Adaboost. Annals of Statistics **32**(1), 13–29 (2004)
11. Koltchinskii, V., Panchenko, D.: Empirical margin distributions and bounding the generalization error of combined classifiers. Annals of Statistics **30**(1), 1–50 (2002)
12. Koltchinskii, V., Panchenko, D.: Complexities of convex combinations and bounding the generalization error in classification. Annals of Statistics **33**(4), 1455–1496 (2005)
13. Mason, L., Baxter, J., Bartlett, P., Frean, M.: Boosting algorithms as gradient descent. In: Advances in Neural Information Processing Systems, pp. 512–518 (1999)
14. Rätsch, G., Onoda, T., Müller, K.R.: Soft margins for AdaBoost. Machine Learning **42**(3), 287–320 (2001)
15. Reyzin, L., Schapire, R.E.: How boosting the margin can also boost classifier complexity. In: Proceedings of the 23rd International Conference on Machine learning, pp. 753–760 (2006)
16. Schapire, R.E., Freund, Y., Bartlett, P., Lee, W.S.: Boosting the margin: A new explanation for the effectiveness of voting methods. Annals of Statistics **26**(5), 1651–1686 (1998)
17. Shen, C., Li, H.: Boosting through optimization of margin distributions. IEEE Transactions on Neural Networks **21**(4), 659–666 (2010)
18. Shen, C., Li, H.: On the dual formulation of boosting algorithms. IEEE Transactions on Pattern Analysis and Machine Intelligence **32**(12), 2216–2231 (2010)
19. Viola, P.A., Jones, M.J.: Rapid object detection using a boosted cascade of simple features. In: IEEE Computer Society Conference on Computer Vision and Pattern Recognition, pp. 511–518 (2001)
20. Wang, L., Deng, X., Jing, Z., Feng, J.: Further results on the margin explanation of boosting: New algorithm and experiments. Science China Information Sciences **55**(7), 1551–1562 (2012)
21. Wang, L., Sugiyama, M., Jing, Z., Yang, C., Zhou, Z.H., Feng, J.: A refined margin analysis for boosting algorithms via equilibrium margin. Journal of Machine Learning Research **12**, 1835–1863 (2011)

o-HETM: An Online Hierarchical Entity Topic Model for News Streams

Linmei Hu[(✉)], Juanzi Li, Jing Zhang, and Chao Shao

Tsinghua National Laboratory for Information Science and Technology,
Department of Computer Science and Technology, Tsinghua University,
Beijing 100084, China
{hulinmei1991,lijuanzi2008,zhangjinglavender,birdlinux}@gmail.com

Abstract. Nowadays, with the development of the Internet, large amount of continuous streaming news has become overwhelming to the public. Constructing a dynamic topic hierarchy which organizes the news articles according to multi-grain topics can enable the users to catch whatever they are interested in as soon as possible. However, it is nontrivial due to the streaming and time-sensitive characteristics of news data. In this paper, to address the challenges, we propose a Hierarchical Entity Topic Model (HETM) which considers the timeliness of news data and the importance of named entities in conveying information of who/when/where in news articles. In addition, we propose online HETM (o-HETM) by presenting a fast online inference algorithm for HETM to adapt it to streaming news. For better understanding of topics, we extract key sentences for each topic to form a summary. Extensive experimental results demonstrate that our model HETM significantly improves the topic quality and time efficiency, compared to state-of-the-art method HLDA (Hierarchical Latent Dirichlet Allocation). In addition, our proposed o-HETM with an online inference algorithm further greatly improves the time efficiency and thus can be applicable to the streaming news.

Keywords: News streams · Topic hierarchy · Hierarchical entity topic model · Online inference

1 Introduction

Recently, once a thing of great concern such as "The Missing Flight MH370" happens, massive news articles [1] covering a wide spectrum of aspects ranging from the missing situation to possible causes, to investigation and searching will be continually published. Though many web sites have created a "Special Topic" web page for organizing all the related news article URLs together, it is still impossible for users to go through all the contents and manually identify the newly emerging and important sub-topics. This necessitates a dynamic structured summarization for

---

[1] e.g., http://news.sohu.com/s2014/jilongpofeiji_gd/

© Springer International Publishing Switzerland 2015
T. Cao et al. (Eds.): PAKDD 2015, Part I, LNAI 9077, pp. 696–707, 2015.
DOI: 10.1007/978-3-319-18038-0_54

the streaming news. Using a hierarchical structure which presents topics at different levels of granularity can enable the users to feel free to find whatever they are interested as soon as possible.

Considerable studies have been done on streaming text. One of the most popular researches is topic detection which aims to discover the topics reported in news articles and group them in terms of their topics [1–3]. Most of the state-of-art topic detection systems are based on probabilistic generative models, clustering techniques and Vector Space Model (VSM) model. Under the frameworks, researchers have proposed several practical methods such as LDA (Latent Dirichlet Allocation), single-pass clustering and incremental K-means clustering etc. However, most of them focused on flat topical structures. Some researches [4] constructed hierarchical document-level clusters instead of hierarchical theme-level topics. Hierarchical topic models have been successfully applied for topic hierarchy construction in text mining [5,6]. Considering that the news is time sensitive and named entities are critical in conveying when, where, who in news articles, we present a new hierarchical entity topic model as well as an online inference method for streaming news.

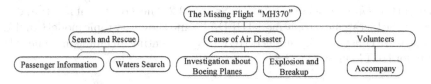

Fig. 1. The topic hierarchy about The Missing Flight "MH370"

In this paper, we propose to dynamically construct a hierarchy of topics and subtopics from the streaming news articles about a special topic. For example, Fig. 1 shows part of the topic hierarchy about "*The Missing Flight MH370*". Each topic is represented as ranked lists of words and entities. The task is not trivial due to various challenges in analyzing streaming news data. First, the whole data cannot be fit into memory at once and has to be processed incrementally. Second, the algorithm need to be efficient due to real-time response rate requirements. In addition, news data is time-sensitive and puts emphasis on named entities (persons, locations, organizations and time) which convey the information of when, where and who [7]. These challenges necessitate algorithms that should meet the following demands: 1) be incremental; 2) run in real time; 3) take the arriving time of news documents into consideration; 4) model the relationship between topics and entities. To this end, we develop an online Hierarchical Entity Topic Model (o-HETM) to automatically construct topic hierarchies from fast-coming news streams. Specifically, we incorporate the time factor into the well-known nCRP (nested Chinese Restaurant Process) and construct a non-parametric hierarchical topic model based on it. The topic model distinguishes entities from topics and models the relationship between topics and entities. To adapt the model to online news streams, we present an online Gibbs sampler for fast inference. The main contributions of this work can be summarized as follows:

(1) We propose a nonparametric hierarchical entity topic model HETM for topic hierarchy construction from news data. The model considers the timeliness of news data and the relationship between topics and entities.

(2) We propose online HETM (o-HETM) by presenting a fast online inference algorithm for HETM to adapt it to streaming news data.

(3) Experimental results on real datasets demonstrate that our model HETM significantly improves the topic quality and time efficiency, compared to HLDA (Hierarchical Latent Dirichlet Allocation). With the online inference algorithm based on Gibbs sampling, o-HETM further improves the time efficiency dramatically and can be used for streaming news.

2 Our Model

In this section, we detail our proposed online nonparametric hierarchical entity topic model o-HETM for dynamic topic hierarchy construction from news streams. We first introduce the *Time-Dependent nCRP* which considers the timeliness of news data for constructing nonparametric hierarchical topic models in Section 2.1. Then we present the graph representation and document generative process of HETM based on the *Time-Dependent nCRP* in Section 2.2. The model combines the advantages of the hierarchical topic model HLDA [8] and the entity topic model CorrLDA [9]. Last, we construct the model o-HETM by presenting an online inference algorithm for HETM in Section 2.3.

2.1 Time-Dependent nCRP

As described in HLDA [8], the *nCRP* places priors over trees and does not limit the branching factors or the depths of the trees. It supposes that there are an infinite number of infinite-table Chinese restaurants. A customer enters the root restaurant with infinite tables where each refers to another restaurant and each restaurant is referred to exactly once. He chooses a table according to *CRP* (Chinese Restaurant Process) which is a "preferential attachment" way. Specifically, the $(n+1)^{th}$ customer chooses a new table with probability $\frac{\gamma}{n+\gamma}$ and chooses an already existing i^{th} table with probability proportional to the number of people sitting there, namely $\frac{n_i}{n+\gamma}$. Then he reaches another restaurant which is referred to by this table. And so forth, the structure repeated infinite times will reach an infinitely branched and deep tree.

In this paper, considering that news data is time-sensitive, we present a variant of *nCRP*, *Time-Dependent nCRP*. It supposes that the much former customers will have less influence to the current customer [10]. Therefore, in our *Time-Dependent nCRP*, we use a common time discount function in drawing a table for a customer. We use t to denote time (period). The $(n+1)^{th}$ customer at time t chooses a new table with probability $\frac{\gamma}{n'+\gamma}$ where n' denotes the discounted number of customers before time t. He chooses an already occupied table with probability $\frac{n_i'}{n'+\gamma}$ where $n_i' = \sum_{\delta=0}^{\Delta} e^{-\frac{\delta}{\lambda}} n_{i,\delta}$ is the discounted number of customers sitting at table i at time t and n' is the total discounted number of customers at time t. This expression defines a time-decaying function parameterized by Δ (width) and λ (decay factor). The *Time-Dependent nCRP* supposes that only the previous customers in the time period from $t - \Delta$ to t will influence the current customer at t. The parameter λ controls the decay rate of the influence of previous customers with time. When $\Delta = 0$, the *Time-Dependent nCRP* only considers the previous customers at current time t. When $\Delta = t$ and $\lambda = \infty$, it degenerates to a common nCRP without considering the factor of time.

2.2 Online Hierarchical Entity Topic Model

In this section, based on the *Time-Dependent nCRP*, we can construct our hierarchical entity topic model HETM. Considering a document as a customer, the model first chooses a table for the customer in the root restaurant. Then the customer enters another restaurant at deeper level referred to by the table. And so forth, we can finally get a path for the document and each node on the path represents a topic. The depth of the infinite tree is controlled by *Stick-Breaking Process* parameterized by (m, π) in which $m \in (0, 1)$ controls the mean of the stick lengths and $\pi > 0$ determines the variance of the stick lengths [8]. The process supposes that there is a stick whose length equals to 1. We sample stick lengths ranging in $(0, 1)$ according to $V_i Beta(m\pi, (1 - m)\pi)$. In most applications, we fix the depth of a tree as L. Therefore, for the last level L, the stick length is $1 - \prod_{i=1}^{L-1}(1 - V_i)$. These lengths correspond to the probabilities of topics on the path and form the prior distribution of the document over the topics. Then words are sampled from the L topics along that path. After sampling all the words of the document, entities are drawn according to the topics of words.

Table 1. Notations

Sym	Definition	Sym	Definition
S	news stream	M	the number of documents in S
L	the number of levels	K	the number of topics
θ	topic distribution	β	word distribution
$\tilde{\beta}$	entity distribution	γ, δ	parameters of *Time-Dependent nCRP*
m, π	level distribution hyperparameter	η	word distribution hyperparameter
$\tilde{\eta}$	entity distribution hyperparameter	N	the number of words in a document
$\tilde{N}$	the number of entities in a document	$\mathbf{c}$	a document path containing $c_1, ..., c_L$

Table 1 summarizes the notations that will be used throughout this paper. The graph representation of our hierarchical entity topic model HETM is illustrated in Figure 2. The model assumes a tree with infinite branches but fixed depth of L levels. It combines the advantages of the hierarchical topic model HLDA [8] and the entity topic model CorrLDA [9][7], namely, it not only generates a structured topic hierarchy for the news data but also models the relationship between topics and entities. Thus, it can better fit the news data.

The document generative process of the model is as follows:

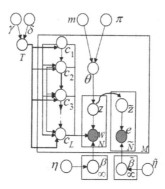

Fig. 2. Graph Representation of HETM

1. For each table $k \in T$ in the infinite tree, draw a word distribution $\beta_k \sim Dir(\eta)$ and entity distribution $\tilde{\beta}_k \sim Dir(\tilde{\eta})$
2. For each document $d \in S$
   - draw $\mathbf{c}_d \sim Time - Dependent\ nCRP(\gamma, \delta)$
   - draw a distribution over levels in the tree, $\theta_d | m, \pi \sim stick - breaking(m, \pi)$
3. For each word $w \in d$
   - choose a level $z_{d,n} | \theta_d \sim Mult(\theta_d)$
   - choose a word $w_{d,n} | z, \mathbf{c}_d, \beta \sim Discrete(\beta_{\mathbf{c}_d}[z_{d,n}])$
4. For each entity $e \in d$
   - choose a level $\tilde{z}_{d,n} \sim Unif(z_{w_1}, z_{w_2}, ..., z_{w_{N_d}})$
   - choose an entity $e_{d,n} \sim Discrete(\tilde{\beta}_{\tilde{z}_{d,n}})$

As shown in the generative process, first, we associate each table (topic) in the tree with a prior word distribution and entity distribution. When generating a document d, we first determine its path according to the *Time-Dependent nCRP* process parameterized by γ and δ. Then the topic distribution of the document is drawn from a stick-breaking process parameterized by (m, π). The following process of generating words and entities is similar to the entity topic model CorrLDA [9]. First, we sample topics for words **w** and then sample topics for entities **e** based on the sampling results of word-topic assignments. Therefore, we can learn the relationship between topics and entities.

2.3 Online Inference Algorithm

In this section, we introduce the online inference algorithm for our model to fit the streaming news. We apply Gibbs sampling which uses $p(z_i | \mathbf{z}_{\neg i}, \mathbf{w})$ to simulate the intractable posterior distribution $p(\mathbf{z} | \mathbf{w})$. The desired Gibbs sampler runs a Markov chain for enough iterations and then it converges to the desired posterior distribution. Inspired by the online LDA [11], we have the following extension for our o-HETM:

In Algorithm 1, we first apply batch Gibbs Sampler [8] on the first 10% data because its content can cover most content of the later coming data [11] (Line 1).

Algorithm 1. Online Gibbs Sampler for o-HETM

---

1 Use batch Gibbs Sampler on the first 10% data;
2 **foreach** *document $d \in S$* **do**
3   sample a path c_d according to $p(c_d|\mathbf{w}, \mathbf{c}_{-d}, \mathbf{z})$;
4   **foreach** *word $w_i \in \mathbf{w}$* **do**
5   sample a topic according to $p(z_i|\mathbf{z}_{\neg i}, \mathbf{w}, \mathbf{c})$;
6   **foreach** *entity $e_i \in \mathbf{e}$* **do**
7   sample a topic according to $p(\tilde{z}_i|\tilde{\mathbf{z}}_{\neg i}, \mathbf{e}, \mathbf{c})$;
8   **if** *the index of d can be divided by Count (we set Count = 100 in this work)* **then**
9   **foreach** *document in the previous Count documents before d* **do**
10   repeat 3-7;

Afterwards, we sample topics of each new word by conditioning words of previous documents observed so far (Lines 2-7). To improve the accuracy of topics, we resample the topics of some previous words (Lines 8-10). The probability distribution $p(c_d|\mathbf{w}, \mathbf{c}_{-d}, \mathbf{z})$ is used for path sampling, where $\mathbf{c}_{-d}$ denotes paths of documents before time of document d , $p(z_i|\mathbf{z}_{\neg i}, \mathbf{w})$, $p(\tilde{z}_i|\tilde{\mathbf{z}}_{\neg i}, \mathbf{e})$ is used for level (topic) sampling where $\mathbf{z}_{\neg i}$ denotes all the topics of all words except the ith word, so dose $\tilde{\mathbf{z}}_{\neg i}$. They can be derived as:

$$p(c_d|\mathbf{w}, \mathbf{c}_{-d}, \mathbf{z}, \eta, \gamma, \delta) \propto p(c_d|\mathbf{c}_{-d}, \gamma, \delta)p(\mathbf{w}_d|\mathbf{c}, \mathbf{w}_{-d}, \mathbf{z}, \eta) , \tag{1}$$

where the first term is the prior on paths implied by the *Time-Dependent nCRP* and the second term denotes the probability of the data given a particular choice of path. We can refer to [8] for details. For sampling topics for words:

$$p(z_{d,i}|\mathbf{z}_{-(d,i)}, \mathbf{c}, \mathbf{w}, m, \pi, \eta) \propto p(z_{d,i}|\mathbf{z}_{d,-i}, m, \pi)p(w_{d,i}|\mathbf{z}, \mathbf{c}, \mathbf{w}_{-(d,i)}, \eta) , \tag{2}$$

where the first term is the conditional topic distribution given all other words' topics, $p(z_{d,i} = k|\mathbf{z}_{d,-i}, m, \pi) = \frac{m\pi + \#[\mathbf{z}_{d,-i}=k]}{\pi + \#[\mathbf{z}_{d,-i}\geq k]} \prod_{j=1}^{k-1} \frac{(1-m)\pi + \#[\mathbf{z}_{d,-i}>j]}{\pi + \#[\mathbf{z}_{d,-i}\geq j]}$. It is determined by the prior distribution and the word-topic assignments. The prior distribution is the truncated stick breaking process parameterized by m, π where larger m indicates a larger probability of higher levels. The second term is the word distribution given all the other variables.

$$p(w_{d,i}|\mathbf{z}, \mathbf{c}, \mathbf{w}_{-(d,i)}, \eta) \propto \#[\mathbf{z}_{-(d,i)} = z_{d,i}, \mathbf{c}_{z_{d,i}} = c_{d,z_{d,i}}, \mathbf{w}_{-(d,i)} = w_{d,i}] + \eta .$$

After sampling all the words, we sample levels for entities according to the already known topic distribution among words and the distributions of the topics over entities. Formally,

$$p(\tilde{z}_{d,i}|\mathbf{z}, \mathbf{c}, \mathbf{e}, m, \pi, \tilde{\eta}) \propto p(\tilde{z}_{d,i}|\mathbf{z}_d)p(e_{d,i}|\mathbf{z}, \mathbf{c}, \mathbf{e}_{-(d,i)}, \tilde{\eta}) \tag{3}$$

Due to space limitation, we don't present how to commutate the distributions in detail. For more details, we can refer to HLDA [8].

3   Topic Summary

Key sentences are selected to form a summary for each topic. We note that the title of a news article briefly summarizes the content of the article. Therefore, we utilize the titles of the news articles belonging to the same topic to construct the candidate sentences for the topic's summary. The representative sentences are selected as follows. First, a topic signature word set TW_z and entity set TE_z are generated by extracting top 10 words and 10 entities with highest $p(w|z)$ and $p(e|z)$. Second, for each sentence s, a word set TW_s and an entity set TE_s are formed by extracting informative words (i.e., noun, verb, adj, and adv.) and entities from the the sentence [12]. Third, each sentence is ranked by measuring the weighted average score of the similarity between its word set and the topic signature word set and the similarity between its entity set and the topic signature entity set. Jaccard Similarity[2] is used as the similarity metric. Formally, $Sim(s,z) = \lambda_1 \cdot Sim_{Jac}(TW_s, TW_z) + \lambda_2 \cdot Sim_{Jac}(TE_s, TE_z)$. The higher the similarity, the higher the rank. The weight parameters λ_1 and λ_2 allow the users to freely control the importance of entities compared to words. In this work, they are empirically set as 0.4 and 0.6 respectively to emphasize entities more.

4   Experiments

4.1   Datasets

Due to restrictions of data crawling on many websites, it is difficult to collect data for our experiments. We collect three news datasets about different topics. The first dataset (in Chinese) and the third dataset (in English) were respectively crawled from the news agency **Sohu** and **Sina**, while the second dataset (in English) was collected from the well-known news agency **The Guardian**. For each news document, we keep the publication time, title and body content. For all the datasets, we sort the documents by their publication time and perform preprocessing as follows: 1) word segmentation (for only Chinese dataset) and entity recognition with ICT-CLAS or StandfordNER [3]; 2) removal of stop words (e.g.,"a", "the", "of", etc.). Statistics of the three datasets including the number of documents, vocabulary size and entity vocabulary size after preprocessing are summarized in Table 2.

4.2   Experimental Setup

Our evaluation of the efficacy of our proposed online hierarchical entity topic model is threefold: 1) comparison with state-of-the-art method HLDA implemented by Chua et al. [13]; 2) comparison with gold standards constructed according to the manually created tables of contents in related Wikipedia articles; 3) a case study.

Comparison with State-of-the-art Methods. For fair comparison, on one hand, the common hyper-parameters shared by these methods are set as the same,

---

[2] Defined as $Sim_{Jac} = \frac{|A \cap B|}{|A \cup B|}$, A and B are two sets

[3] http://nlp.stanford.edu/software/CRF-NER.shtml

Table 2. Statistics of the collected datasets

Datasets	♯ Docs	Vocabulary	Entity Vocabulary
The Missing Flight MH370	798	8,125	2,344
2012 US Election	940	31,240	7,184
2010 Chile Earthquake	170	5,918	1,277

i.e., $m = 0.25, \pi = 500, \eta = \langle 1.0, 0.5, 0.25 \rangle, \gamma = 1$ according to the previous studies [8]. For additional hyper-parameters $\tilde{\eta}$ of HETM and o-HETM, we set them as same as η. We set the maximum level of all the models to 3. On the other hand, as HLDA didn't consider the factor of time, we set the time width as $\Delta = t$ and the decay factor as $\lambda = \infty$ which makes Time-dependent nCRP degenerate to nCRP. We compare our models HETM and o-HETM with HLDA in terms of time efficiency and topic coherence.

Comparison with Gold Standards. We construct a gold standard for each topic by leveraging the Contents table in Wikipedia articles (e.g., "*2010 Chile earthquake*" [4]). The Contents tables summarize the topics about the thing that the title stands for. We ask five college students to work together to pick up the effective topics which forms the gold standard hierarchy. Only when all of them reach a consensus, we will consider a topic in the tables of contents as effective. Then we in our results with the the gold standards manually.

Case Study. For qualitative analysis of topic hierarchies generated by our method, we present the result of the largest dataset "*2012 US Election*" generated by our method as a case study.

4.3 Evaluation Metrics

Topic Coherence. We use topic coherence to evaluate the topic quality [14]. Given a list of words, the more often the words co-occur, the larger the topic coherence is and the list is more likely to represent a topic. Formally, it is computed as $C(t, \mathbf{W}^{(t)}) = \sum_{n=2}^{N} \sum_{l=2}^{n-1} log((D(w_n^{(t)}, w_l^{(t)}) + 1)/D(w_l^{(t)}))$ where a topic is described using top N words with the largest probabilities, $\mathbf{W}^{(t)} = \{w_1^{(t)}, ..., w_N^{(t)}\}$, $D(w_i)$ is the document frequency of word w_i, $D(w_i, w_j)$ is the co-document frequency of word w_i and w_j. suppose .

Recall. The recall of our topic hierarchy is defined as $R(H) = (|T_H \cap T_{H_S}|)/|T_{H_S}|$ where T_H and T_{H_S} are respectively the topic sets of our topic hierarchy and the gold standard.

4.4 Results and Analysis

Comparison with State-of-the-art Methods. Table 3 shows the average topic coherence scores and running time of different methods. As we can see, our model HETM significantly outperforms HLDA in terms of both topic quality and time efficiency. It shows that distinguishing entities from words can not only discover the

---

[4] http://en.wikipedia.org/wiki/2010_Chile_earthquake

Table 3. Topic Coherence and Time Efficiency

	MH370		Election		Earthquake	
	TC	Time	TC	Time	TC	Time
HLDA	-68.17	9h	-88.09	121h	-44.2	2.13h
HETM	**-66.15**	7h	**-48.46**	65h	**-36.89**	1.53h
o-HETM	-75.36	**0.17h**	-92.84	**4.26h**	-52.5	**0.09h**

relations between topics and entities but also improve the time efficiency. With the online inference algorithm, o-HETM further improves the time efficiency by more than 20-50 times. The time of dealing with a document reaches 2-50 milliseconds, which meets the demand of real-time news processing. However, without surprise, the topic quality of o-HETM is of inferior quality, compared to HLDA in terms of topic coherence. The significant improvement of the time efficiency is made at the expense of the topic quality.

Comparison with the Gold Standards. The recall values of our topic hierarchies about the three events "The Missing Flight MH370", "2012 US Election" and "2010 Chile Earthquake" are respectively 71.4%, 62.5% and 90.9%. Considering that our topic hierarchies are generated from the real news data while the gold standard are constructed manually without referring to the news, the recall of 62.5%-90.9% has demonstrated the effectiveness of our method. For example, the gold standard about the *"2012 US presidential election"* [5] include topics of *"primaries"*, *"campaigns"* and *"races"*, most of which can be found in our topic hierarchy as shown in Fig. 3. In addition to that, our method discovers many popular topics such as *"tax"* and *"scandals"* and provides a more complete view. The gold standard of *"The Missing MH370"* contains some specific topics that don't occur a lot in news articles such as *"electrical fire speculation"* and thus cannot be discovered by our method. In future work, we can leverage the *"table of contents"* for semi-supervised topic modeling of news to improve our results. The recall vale on the dataset of "2010 Chile earthquake" is the highest. There are 10 aligned topics (e.g., tsunami, damage, government response and so on). Only one topic, i.e., *"prison escape"* is not in our topic hierarchy. However, our model discovers hot topics such as *"copper"*.

Case Study. Due to space limitation, we present only the main part of the topic hierarchy about the *"2012 US election"* in Fig. 3. As we can see, the topic of *"economy"* is the most hot topic, which includes more than 2/3 of the documents and has subtopics of *"job"* and *"primaries"*. Another small but hot topic is "Affair about sex". It has subtopics of *"sexual harassment"* and *"Corruption"*. Other topics which are not shown in the figure are in smaller size. In terms of topic summary, we can see an example that the most related news title to the topic of "job" is *"US politics live blog: Rick Perry's jobs policy, New Hampshire v Nevada, Herman Cain's 9-9-9 tax plan"*. Overall, our results accord well with our common sense. However,

---

[5] http://en.wikipedia.org/wiki/United_States_presidential_election,_2012

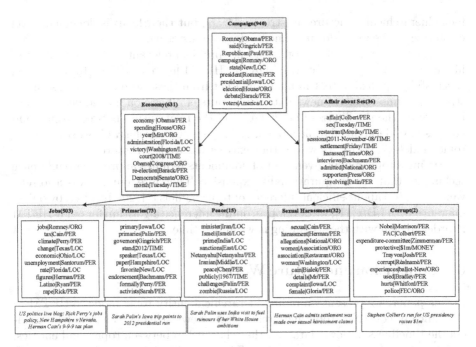

Fig. 3. The main part of the topic hierarchy about the *"2012 US election"*. We show top 10 words and entities separated by "|" in each topic and manually label the topic for better understanding. The number in the bracket is the number of documents belonging to the topic. The bottom shows the topics' most relevant news titles.

some expected topics such as *"debate"* and "voting" are mixed with the root topic of "Campaign" as we can see the top 10 words contain *"debate"* and *"voters"*. All the models have the problems. The reason may be that these topics are too related to each other and thus cannot be separated.

Parameter Analysis. We also test the model with different values of the parameters Δ and λ, and find that the quality of the topic hierarchy is highly insensitive to the parameters. However, smaller Δ and λ are likely to result in more specific topics. A user can choose different parameters according to the datasets and practical demands. In terms of the parameter of online-algorithm, if *Count* is larger, the model will be more approximate to the algorithm of batch sampling, and will get better results at the expense of time.

5 Related Work

Topic discovery from streaming data has been studied a lot. For example, [2] presented general probabilistic methods for discovering and summarizing the evolutionary patterns of themes in a text stream. Topic detection and tracking aims to discover topics and group the streaming documents in terms of their topics [1–3]. However, most of the studies focused on flat structures of the topics. Some studies focus hierarchical structures in streams [4,15], but their focus is document level clustering, while we perform theme level word clustering.

Topic hierarchy construction is another problem relevant to our work. A lot of hierarchical topic models, e.g., HLDA [5], HPAM [6], and hHDP [16] have been successfully applied in text mining. Different from these studies, we focus on the streaming time-sensitive news data which puts emphasis on named entities, and incorporate the timeliness of news data and the relationship between topics and entities into hierarchical topic modeling. There is also considerable work on entity topic models [7,17–19]. However, they extract flat topic structures.

Fitting a topic model given a set of documents requires approximate inference techniques that are computationally expensive. Therefore, our work is also relevant to studies about efficient inference of topic models [11,20]. Inspired by online inference with LDA [11], we develop an online inference algorithm for our o-HETM in order to deal with the streaming news in real time.

6   Conclusion and Future Work

In this paper, we present an online hierarchical entity topic model o-HETM to dynamically construct topic hierarchies from news streams. The model considers the timeliness of news data and the relationship between topics and entities, which are very important for news data. The fast online inference algorithm significantly improves the time efficiency of the model and thus adapt it to the streaming news. Extensive experiments have verified the effectiveness and efficiency of the proposed model, compared to the baseline model HLDA. In future work, we can investigate and visualize the hierarchical topic evolutionary patterns based on the current work.

Acknowledgments. The work is supported by 973 Program (No. 2014CB340504), NSFC-ANR (No. 61261130588), Tsinghua University Initiative Scientific Research Program (No. 20131089256), Science and Technology Support Program (No. 2014BAK04B00) and THU-NUS NExT Co-Lab.

References

1. Allan, J., Papka, R., Lavrenko, V.: On-line new event detection and tracking. In: Proceedings of the 21st Annual International ACM SIGIR, pp. 37–45. ACM (1998)
2. Mei, Q., Zhai, C.: Discovering evolutionary theme patterns from text: an exploration of temporal text mining. In: KDD, pp. 198–207. ACM (2005)
3. Banerjee, A., Basu, S.: Topic models over text streams: a study of batch and online unsupervised learning. In: SDM, vol. 7, pp. 437–442. SIAM (2007)
4. Trieschnigg, D., Kraaij, W.: Hierarchical topic detection in large digital news archives. In: Proceedings of the 5th Dutch Belgian Information Retrieval Workshop, pp. 55–62 (2005)
5. Griffiths, D., Tenenbaum, M.: Hierarchical topic models and the nested chinese restaurant process. Advances in Neural Information Processing Systems **16**, 17 (2004)

6. Mimno, D., Li, W., McCallum, A.: Mixtures of hierarchical topics with pachinko allocation. In: ICML, pp. 633–640. ACM (2007)
7. Newman, D., Chemudugunta, C., Smyth, P.: Statistical entity-topic models. In: KDD, pp. 680–686 (2006)
8. Blei, D.M., Griffiths, T.L., Jordan, M.I.: The nested chinese restaurant process and bayesian nonparametric inference of topic hierarchies. Journal of the ACM (JACM) **57**(2), 7 (2010)
9. Blei, D.M., Jordan, M.I.: Modeling annotated data. In: Proceedings of the 26th Annual International ACM SIGIR, pp. 127–134. ACM (2003)
10. Ahmed, A., Xing, E.P.: Dynamic non-parametric mixture models and the recurrent chinese restaurant process: with applications to evolutionary clustering. In: SDM, pp. 219–230. SIAM (2008)
11. Canini, K.R., Shi, L., Griffiths, T.L.: Online inference of topics with latent dirichlet allocation. Journal of Machine Learning Research - Proceedings Track, 65–72 (2009)
12. Hu, P., Huang, M., Xu, P., Li, W., Usadi, A.K., Zhu, X.: Generating breakpoint-based timeline overview for news topic retrospection. In: ICDM, pp. 260–269. IEEE (2011)
13. Chua, F.C.T.: Summarizing amazon reviews using hierarchical clustering. Technical report, Technical report (2009)
14. Mimno, D., Wallach, H.M., Talley, E., Leenders, M., McCallum, A.: Optimizing semantic coherence in topic models. In: Proceedings of EMNLP, pp. 262–272. Association for Computational Linguistics (2011)
15. Kleinberg, J.: Bursty and hierarchical structure in streams. Data Mining and Knowledge Discovery **7**(4), 373–397 (2003)
16. Zavitsanos, E., Paliouras, G., Vouros, G.A.: Non-parametric estimation of topic hierarchies from texts with hierarchical dirichlet processes. The Journal of Machine Learning Research **12**, 2749–2775 (2011)
17. Agrawal, P., Tekumalla, L.S., Bhattacharya, I.: Nested hierarchical dirichlet process for nonparametric entity-topic analysis. In: Blockeel, H., Kersting, K., Nijssen, S., Železný, F. (eds.) ECML PKDD 2013, Part II. LNCS, vol. 8189, pp. 564–579. Springer, Heidelberg (2013)
18. Hu, L., Li, J., Li, Z., Shao, C., Li, Z.: Incorporating entities in news topic modeling. In: Zhou, G., Li, J., Zhao, D., Feng, Y. (eds.) NLPCC 2013. CCIS, vol. 400, pp. 139–150. Springer, Heidelberg (2013)
19. Kim, H., Sun, Y., Hockenmaier, J., Han, J.: Etm: entity topic models for mining documents associated with entities. In: ICDM, pp. 349–358 (2012)
20. Yao, L., Mimno, D., McCallum, A.: Efficient methods for topic model inference on streaming document collections. In: Proceedings of the 15th ACM SIGKDD, pp. 937–946. ACM (2009)

Modeling User Interest and Community Interest in Microbloggings: An Integrated Approach

Tuan-Anh Hoang[(✉)]

Living Analytics Research Centre,
Singapore Management University, Singapore, Singapore
tahoang.2011@smu.edu.sg

Abstract. To explain why a user generates some observed content and behaviors, one has to determine the user's topical interests as well as that of her community. Most existing works on modeling microblogging users and their communities however are based on either user generated content or user behaviors, but not both. In this paper, we propose the *Community and Personal Interest* (**CPI**) model, for modeling interest of microblogging users jointly with that of their communities using both the content and behaviors. The **CPI** model also provides a common framework to accommodate multiple types of user behaviors. Unlike the other models, **CPI** does not assume a hierarchical relationship between personal interest and community interest, i.e., one is determined purely based on the other. We build the **CPI** model based on the principle that a user's personal interest is different from that of her community. We further develop a regularization technique to bias the model to learn more socially meaningful topics for each community. Our experiments on a Twitter dataset show that the **CPI** model outperforms other state-of-the-art models in topic learning and user classification tasks. We also demonstrate that the **CPI** model can effectively mine community interest through some representative case examples.

Keywords: Microbloggings · Topic modeling · Behavior mining

1 Introduction

In microblogging sites, users can publish short messages (called *tweets*), as well as adopt a wide range of behaviors spontaneously. The behaviors include following other users, mentioning hashtags or other users in tweets, and forwarding (or *retweeting*) messages received from other users, etc.. Empirical and user studies have shown that both the tweets and behaviors of a microblogging user are determined by her personal interest or that of her community [14,29,34,38]. However, most existing works on modeling user personal interests and community interest in microbloggings consider only either user generated content or user behaviors but not both (e.g.,[16,22,28,44]). This existing approach neglects the relationship between user generated content and behavior, and thus learning

© Springer International Publishing Switzerland 2015
T. Cao et al. (Eds.): PAKDD 2015, Part I, LNAI 9077, pp. 708–721, 2015.
DOI: 10.1007/978-3-319-18038-0_55

the two type of interests can be inaccurate. It also cannot leverage user generated content to provide a semantic interpretation of the behaviors.

Moreover, most works on modeling microblogging content and behavior assume that there is a dependent relationship between user and community interests. These works either determine a user's interests purely based on her communities (e.g., [32,33,43]), or determine a community's interests purely by aggregating it's members' interests ([21,22]). This approach suffers from two drawbacks. First, it ignores the fact that a user's personal interests may be different from that of the communities she belongs to. For example, a user can belong to a political community at the same time expressing interest in entertainment topics. Second, it suffers from *trivial topics*. These are popular but not socially meaningful topics. One such topic may be about food and drinks, and another about daily activities. These trivial topics are shared by many microblogging users ([17]). These topics are also likely be modeled by existing models as community interest leading to multiple communities sharing these common topics. While sharing trivial topics is reasonable for overlapping communities ([15,40]), it is not practical for mutually exclusive communities (e.g., political communities, and professional communities, etc.). These communities should be characterized by clear topics.

In this work, we therefore aim to model topical interest of microblogging users and that of their *mutually exclusive communities* considering both user generated content and behavior. Moreover, we want to differentiate between topical interests of each user and that of each community. For each user, we also want to learn the bias of the user towards her community in generating both content and behavior. Lastly, for each community, we want the community to be clearly distinguished by socially meaningful topics.

Our main contributions in this work consist of the following.

- We propose a generative model, called *Community and Personal Interest* model (abbreviated as **CPI**), for modeling topics and user topical interest as well as modeling user community in microbloggings. Our model is designed to work with data consists of user communities that are mutually exclusive. The **CPI** model encapsulates different types of user behaviors in a common framework, and associates the behaviors with the user generated content through a set of latent topics.
- We develop a sampling method to infer the model's parameters. We also develop a technique to regularize the sampling process so that: (1) trivial topics are less likely to be assigned as community interest topics; while (2) non-trivial topics shared mostly by users within a community are more likely be assigned to be interest of the community.
- We apply **CPI** model on a Twitter dataset and show that it outperforms other state-of-the-art models in modeling content topics and user classification tasks. We also conduct an empirical analysis of personal interest and community topics found in the dataset to demonstrate the efficacy of the **CPI** model.

The rest of the paper is organized as follows. We first discuss the related works in Section 2. We then present the **CPI** model in detail in Section 3. The algorithm for learning parameters of the **CPI** model and the regularization technique are presented in Section 4. Next, we describe the experimental dataset and report the results of experiments of applying the proposed model on the dataset in Section 5. Finally, we give our conclusions and discuss future work in Section 6.

2   Related Work

2.1   Topic and Community Analyis

Michelson *et. al.* first examined topical interests of Twitter users by analyzing the named entities mentioned in their tweets [25]. Hong *et. al.* [16], Mehrotra *et. al.* [24], and Ramage *et. al* [29] conducted empirical studies on different ways of performing topic modeling on tweets using the original LDA model [16], Author-topic model [31], and Supervised LDA model [30]. Later, Zhao *et. al.* [44] proposed TwitterLDA model in which: (a) each user has a topic distribution and they share a common background topic; and (b) a topic is assigned to each tweet (instead of to each word). Recently, Qiu *et. al.* proposed to use TwitterLDA for jointly modeling topics of tweets and their associated posting behaviors (i.e., tweet, retweet, or reply) [28]. These works however only consider user generated content in modeling user interest. Our work, in the other hand, considers both user generated content and user behaviors.

Early works on community mining in social networks are purely based on either social links among the users (e.g., [1, 27]) or user generated content (e.g., [39, 45]). Ding *et. al.* conducted an empirical studies showing that both social links and user generated content should be considered in community mining in order to find coherent user communities [10]. Our work considers the social links among users as part of user behaviors, e.g., users follow and retweet other users, and users mention other users in tweets. Moreover, most of existing works that consider both social link and user content are based on the assumption that users/documents within a community have similar interest and are densely connected (e.g., [2, 4, 5, 26, 43]). This assumption is not practical in microblogging context where users express interest in a vast variety of topics, and their interest is therefore not always determined by their communities only. Our model, on the other hand, seeks to differentiate between a user's personal interest from that of her community. It is also important to note that, unlike works on mining overlapping communities (e.g., [2, 15, 43]), our work aims to mine mutually exclusive communities where each user belongs to only one of the communities.

Lastly, there are also existing works on finding community interest. However these works either (a) determine a community's interest by aggregating interest of users within the communitiy (e.g., [21, 22]), or (b) determine a user's interests purely based on interest of communities the user belongs to (e.g., [13, 33, 43]). The first approach suffers from trivial topics which are shared by many users, and hence more likely be assigned to community interests. The second approach is not able to differentiate between user personal interest and community interest. In contrast, our model differentiates between users' personal interests and communities' interests. We learn the two interests simultaneously with a regularization so that socially meaningful topics are more likely be assigned to community interests.

2.2   User Behavior Analyis

There has been a number of works analyzing user behavior in microblogging data. For example, Kwak *et. al.* [19,20], Wu *et. al.* [38], and Feller *et. al.* [11] studied the patterns of following behavior; Conover *et. al.* [7], Wu *et. al.* [38], and Suh *et. al.* [35] examined retweet behaviors; Hannon *et. al.* [12] and Yin *et. al.* [42] proposed models for recommending following behavior; and Yang *et. al.* [41], Dabeer *et. al.* [9], and Cui *et. al.* [8] proposed models for modeling retweet behavior. However, most of these works (i) only consider a single type of behaviors, or (ii) do not consider the user generated content when studying user behaviors. Our model, on the other hand, allows different types of user behaviors to be modeled simultaneously when modeling user generated content.

3   Community and Personal Interest (CPI) Model

In the **CPI** model, each tweet is a bags-of-words chosen from a vocabulary denoted by $\mathcal{V}_t$, and each behavior belongs to one of L behavior types. Each type-l behavior is drawn from a set of all possible values denoted by $\mathcal{V}_{bl}$. **CPI** model has K latent topics, where each topic k has a multinomial distribution ϕ_k over the vocabulary $\mathcal{V}_t$ and a multinomial distribution λ_{lk} over the vocabulary $\mathcal{V}_{bl}$ for each behavior type-l.

The **CPI** model assumes that there are C mutually exclusive communities and U users. Each community c has a multinomial distribution σ_c over the K topics that represents interest of the community. The personal interest of each user u is also represented by a topic distribution θ_u over the K topics. Each user belongs to one of the C communities following a multinomial distribution π. We denote the community of user u by c_u. Moreover, each user u has a dependence distribution μ_u which is a Bernoulli distribution indicating how likely the user behaves according to her own personal interest (μ_u^0) or according to interest of her community ($\mu_u^1 = 1 - \mu_u^0$). Lastly, we assume that θ_u, π_u, σ, λ_l, and ϕ have Dirichlet priors α, τ, η, γ_l, and β respectively, while μ_u has Beta prior ρ.

The generative process of **CPI** model is shown in Figure 1. To generate a tweet t for user u, we first flip a biased coin y_t (whose bias is μ_u) to decide if the tweet will be based on u's personal interest or that of her community. If the coin is head up, (i.e., $y_t = 0$), we then choose the topic z_t for the tweet according to u's topic distribution θ_u. Otherwise, (i.e., $y_t = 1$), we choose z_t according to her community's topic distribution σ_{c_u}. As tweets are short with a limited number of charac-

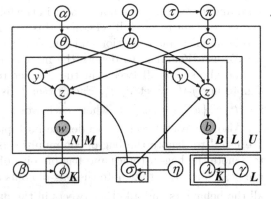

Fig. 1. Plate notation for the **CPI** model

ters, we assume that each tweet has only one topic. Once the topic z_t is chosen, words in t are then chosen according to the topic's word distribution ϕ_{z_t}. Similarly, we assume the same process for all adopted behaviors, except that, for a behavior b of type l, once the topic z_b is chosen, the behavior is then sampled according to the topic's behavior distribution λ_{lz_b}. In summary, the generative process is as follows.

– Sample the community distribution $\pi \sim Dirichlet(\tau)$
– For each $k = 1, \cdots, K$, sample the k-th topic $\phi_k \sim Dirichlet(\beta)$
– For each $c = 1, \cdots, C$, sample c-th community's topic distribution $\sigma_c \sim Dirichlet(\eta_c)$
– For each type of behavior l ($l = 1, \cdots, L$), and each topic k, sample type-l's k-th behavior distribution $\lambda_{lk} \sim Dirichlet(\gamma_l)$
– For each user u: sample the user's community $c_u \sim Dirichlet(\pi)$, topic distribution $\theta_u \sim Dirichlet(\alpha)$, and dependence distribution $\mu_u \sim Beta(\rho)$
– Generate tweets for the user u: for each tweet t that u posts:
  1. Sample coin $y_t \sim Bernoulli(\mu_u)$
  2. Sample topic for the tweet: if $y_t = 0$, sample $z_t \sim Multinomial(\theta_u)$; otherwise ($y_t = 1$), sample $z_t \sim Multinomial(\sigma_{c_u})$
  3. Sample the tweet's words: for each word slot n, sample the word $w_{t,n} \sim Multinomial(\phi_{z_t})$
– Generate behaviors for the user u: for each behavior b of type-l that u adopts:
  1. Sample $y_b \sim Bernoulli(\mu_u)$
  2. Sample topic for the behavior: if $y_b = 0$, sample $z_b \sim Multinomial(\theta_u)$; otherwise ($y_u = 1$), sample $z_b \sim Multinomial(\sigma_{c_u})$
  3. Sample behavior instance $b \sim Multinomial(\lambda_{lz_b})$

4   Model Learning

4.1   Gibbs Sampling

In learning the parameters of the **CPI** model, we use collapsed Gibbs sampler ([23]) to iteratively sample the latent variables (i.e., coins, topics, and communities) for tweets, behaviors, and users. Due to the space limitation and its similarity to the sampling for tweet, we do not present in the following the sampling for behaviors and leave it out to the full version of this paper [37]. We also describe in [37] the details about implementation and complexity of the learning procedure.

We use W and $\mathcal{T}$ to denote the number of words in the tweet vocabulary V_t and the set of all tweets in the dataset respectively. For each user u_i, we denote her j-th tweet by t_j^i. Each tweet t_j^i is a bag-of-words with length N_{ij}, i.e., $t_j^i = \{w_1^{ij}, \cdots, w_{N_{ij}}^{ij}\}$, where each word w_n^{ij} is drawn from the vocabulary $\mathcal{V}_t$. Also, for each tweet t_j^i, we denote its topic and coin by z_j^i, y_j^i respectively. We use $\mathcal{C}$ to denote the bag-of-communities of all the users; and use $\mathcal{Z}$ and $\mathcal{Y}$ to denote the bag-of-topics and bag-of-coins of all the tweets and behaviors. We use $\mathcal{Y}_{-t_j^i}$ and $\mathcal{Z}_{-t_j^i}$ to denote the bag-of-coins and bag-of-topics, respectively, of all the behaviors and all other tweets in the dataset except the tweet t_j^i. Lastly, we use $\mathcal{C}_{-c_{u_i}}$ to denote the bag-of-communities of all the users except u_i, and use $\mathcal{Z}_{-u_i}$ to denote the bag-of-topics of the tweets behaviors posted/adopted by all other users except u_i

Sampling for Tweet t_j^i. The coin y_j^i is sampled according to Equations 1 and 2, while the topic z_j^i is sampled according to Equations 3 and 4. In these equations, $\mathbf{n_y}(y, u, \mathcal{Y})$ records the number of times the coin y is observed in the set of tweets and behaviors of user u. Similarly, $\mathbf{n_{zu}}(z, u, \mathcal{Z})$ records the number of times the topic z is observed in the set of tweets and behaviors of user u (i.e., those tweets and behaviors currently have coins 0); $\mathbf{n_{zc}}(z, c, \mathcal{Z}, \mathcal{C})$ records the number of times the topic z is observed in the set of tweets and behaviors that are tweeted/adopted based on interest of community c and by any user; and $\mathbf{n_w}(w, z, \mathcal{T}, \mathcal{Z})$ records the number of times the word w is observed in the topic z for the set of tweets $\mathcal{T}$ and the bag-of-topics $\mathcal{Z}$.

$$p(y_j^i = 0 | \text{rest}) \propto \frac{\mathbf{n_y}(0, u_i, \mathcal{Y}_{-t_j^i}) + \rho_0}{\sum_{y=0}^{1} \left(\mathbf{n_y}(y, u_i, \mathcal{Y}_{-t_j^i}) + \rho_y \right)} \frac{\mathbf{n_{zu}}(z_j^i, u_i, \mathcal{Z}_{-t_j^i}) + \alpha_{z_j^i}}{\sum_{k=1}^{K} \left(\mathbf{n_{zu}}(k, u_i, \mathcal{Z}_{-t_j^i}) + \alpha_k \right)} \quad (1)$$

$$p(y_j^i = 1 | \text{rest}) \propto \frac{\mathbf{n_y}(1, u_i, \mathcal{Y}_{-t_j^i}) + \rho_1}{\sum_{y=0}^{1} \left(\mathbf{n_y}(y, u_i, \mathcal{Y}_{-t_j^i}) + \rho_y \right)} \frac{\mathbf{n_{zc}}(z_j^i, c_{u_i}, \mathcal{Z}_{-t_j^i}, \mathcal{C}) + \eta_{c_{u_i} z_j^i}}{\sum_{k=1}^{K} \left(\mathbf{n_{zc}}(k, c_{u_i}, \mathcal{Z}_{-t_j^i}, \mathcal{C}) + \eta_{c_{u_i} k} \right)} \quad (2)$$

$$p(z_j^i = z | y_j^i = 0, \text{rest}) \propto \frac{\mathbf{n_{zu}}(z, u_i, \mathcal{Z}_{-t_j^i}) + \alpha_z}{\sum_{k=1}^{K} \left(\mathbf{n_{zu}}(k, u_i, \mathcal{Z}_{-t_j^i}) + \alpha_k \right)} \prod_{n=1}^{N_{ij}} \frac{\mathbf{n_w}(w_n^{ij}, z, \mathcal{Z}_{-t_j^i}) + \beta_{z w_n^{ij}}}{\sum_{v=1}^{W} \left(\mathbf{n_w}(v, z, \mathcal{Z}_{-t_j^i}) + \beta_{zv} \right)} \quad (3)$$

$$p(z_j^i = z | y_j^i = 1, \text{rest}) \propto \frac{\mathbf{n_{zc}}(z, c_{u_i}, \mathcal{Z}_{-t_j^i}, \mathcal{C}) + \eta_{c_{u_i} z}}{\sum_{k=1}^{K} \left(\mathbf{n_{zc}}(k, c_{u_i}, \mathcal{Z}_{-t_j^i}, \mathcal{C}) + \eta_{c_{u_i} k} \right)} \prod_{n=1}^{N_{ij}} \frac{\mathbf{n_w}(w_n^{ij}, z, \mathcal{T}_{-t_j^i}, \mathcal{Z}_{-t_j^i}) + \beta_{z w_n^{ij}}}{\sum_{v=1}^{W} \left(\mathbf{n_w}(v, z, \mathcal{T}_{-t_j^i}, \mathcal{Z}_{-t_j^i}) + \beta_{zv} \right)}$$

$$(4)$$

Sampling for User u_i. The community c_{u_i} is sampled according to Equation 5. In the equation, $\mathbf{n_c}(c, \mathcal{C})$ records the number of times the community c is observed in the bag-of-communities $\mathcal{C}$, and $\mathbf{n_z}(z, u)$ records the number of tweets/ behaviors of u are observed in the topic z and has coin 1.

$$p(c_{u_i} = c | \text{rest}) \propto \frac{\mathbf{n_c}(c, \mathcal{C}_{-c_{u_i}}) + \tau_{c_{u_i}}}{\sum_{g=1}^{C} \left(\mathbf{n_c}(g, \mathcal{C}_{-c_{u_i}}) + \tau_g \right)} \prod_{z=1}^{K} \left[\frac{\mathbf{n_{zc}}(z, c, \mathcal{Z}_{-u_i}, \mathcal{C}_{-c_{u_i}}) + \eta_{cz}}{\sum_{k=1}^{K} \left(\mathbf{n_{zc}}(k, c, \mathcal{Z}_{-u_i}, \mathcal{C}_{-c_{u_i}}) + \eta_{ck} \right)} \right]^{\mathbf{n_z}(z, u_i, \mathcal{Y}, \mathcal{Z}, \mathcal{B})}$$

$$(5)$$

4.2 Semi-supervised Learning

The **CPI** model presented as above is totally unsupervised with two parameters, i.e., number of topics K and number of communities C. In some settings, however, we may have known the community labels for some users but not the others. For example, a subset of users may explicitly share their political and professional labels. By assigning users within the same known community labels with the same community label (i.e., a value of c), and by fixing their community label assignments during the sampling process (i.e., do not sample community for those users), we can use **CPI** model as a semi-supervised model. On one hand, this helps to bias the **CPI** model to more socially meaningful communities. On the other hand, this also helps to overcome the weakness of supervised methods that require large number of labeled users in user classification task [6].

4.3 Sparsity Regularization

Community Topic Regularization. To avoid learning trivial community top-ics, community topic regularization aims to make every topic covered by mostly one community. Trivial topics (see Section 1) are usually shared by almost all users and hence are likely covered by multiple communities. Such topics are less likely be clear community topics. In contrast, a community topic is pre-ferred to be more unique among users within the community. We thus apply the *entropy based regularization* technique [3] to obtain the sparsity in the distribu-tion $p(c|z)$. We implement this regularization in each coin and topic sampling steps for tweets and behaviors since they are main steps to determine whether a topic is community topic or personal interest topic. Again, due to the space limitation, we do not present in the following the regularization in sampling for behaviors and leave it out to [37].

When sampling coin for the tweet t_j^i, we multiply the right hand side of Equa-tions 1 and 2 with a corresponding regularization term $\mathcal{R}_{coin}(y|c_{u_i}, z_j^i)$ which is defined by Equation 6. Similarly, when sampling topic for the tweet t_j^i, we multi-ply the right hand side Equation 4 with regularization term $\mathcal{R}_{topicComm}(z|c_{u_i}, t_j^i)$ which is defined by Equation 7. Lastly, when sampling community for user u_i, we multiply the right hand side of Equation 5 with a corresponding regularization term $\mathcal{R}(c)$ which is defined by Equation 8.

$$\mathcal{R}_{coin}(y|c_{u_i}, z_j^i) = exp\left(-\frac{\left(H_{y_j^i=y}(p(c_{u_i}|z_j^i)) - \mathcal{E}_{topicComm}\right)^2}{2\sigma_{topicComm}^2}\right) \tag{6}$$

$$\mathcal{R}_{topicComm}(z|c_{u_i}, t_j^i) = \prod_{z'=1}^{K} exp\left(-\frac{\left(H_{z_j^i=z}(p(c_{u_i}|z')) - \mathcal{E}_{topicComm}\right)^2}{2\sigma_{topicComm}^2}\right) \tag{7}$$

$$\mathcal{R}_{topicComm}(c|u_i) = \prod_{z=1}^{K} exp\left(-\frac{\left(H_{c_{u_i}=c}(p(c|z)) - \mathcal{E}_{topiComm}\right)^2}{2\sigma_{topicComm}^2}\right) \tag{8}$$

In Equations 6, 7, and 8, $H_{y_j^i=y}(p(c_{u_i}|z_j^i))$ is the empirical entropy of $p(c_{u_i}|z_j^i)$ when $y_j^i = y$; and $H_{z_j^i=z}(p(c_{u_i}|z')$ and $H_{c_{u_i}=c}(p(c|z))$ has similar meaning with respectively regards to $p(c_{u_i}|z')$ and $p(c|z)$. The parameters $\mathcal{E}_{topicComm}$ and $\sigma_{topicComm}$ are the expected mean and variance of the entropy of $p(c|z)$ respec-tively. These are pre-defined parameters. Obviously, with a small expected mean $\mathcal{E}_{topComm}$ (which is corresponding to a skewed distribution), these regulariza-tion terms (1) increase weight for values of y and z that give lower empirical entropy of $p(c_{u_i}|z_j^i)$ (or $p(c_{u_i}|z_j^{i,l})$), hence increasing the sparsity of these distri-butions; and (2) decrease weight for values of y and z that give higher empirical entropy of $p(c_{u_i}|z_j^i)$ (or $p(c_{u_i}|z_j^{i,l})$), hence decreasing the sparsity of these distri-butions. The expected variance $\sigma_{topicComm}$ can be used to adjust the strictness of the regularization: smaller $\sigma_{topicComm}$ imposes stricter regularization. When $\sigma_{topicComm} = \infty$, the model has no regularization on $p(c|z)$.

Community Distribution Regularization. Even with the above community topic regularization, we may still have an extreme case where there is a com-munity that (1) includes all if not most of the users, and (2) covers largely trivial topics. To avoid this extreme case, we need to achieve a balance of user populations among the communities, i.e., we need to regularize the community distribution so that it is not too skewed to a certain community. To achieve this, we again use *entropy based regularization* technique [3] to facilitate a balanced community distribution $p(c)$. We implement this regularization in each community sampling step for users since it is the main step to determine the community distribution. That is, when sampling community for user u_i, we also multiply the right hand side of Equation 5 with the regularization term defined by the Equation 9.

$$\mathcal{R}_{comm}(c|u_i) = exp\left(-\frac{\left(H_{c_{u_i}=c}(p(c)) - \mathcal{E}_{comm}\right)^2}{2\sigma_{comm}^2}\right) \tag{9}$$

In Equation 9, $H_{c_{u_i}=c}(p(c))$ is the empirical entropy of $p(c)$ when $c_{u_i} = c$. Similar to above, the pre-defined parameters $\mathcal{E}_{comm}$ and σ_{comm} are the expected mean and variance of the entropy of $p(c)$ respectively. With a high enough expected mean value of $\mathcal{E}_{comm}$ (which corresponds to a balanced distribution), this regularization term (1) decreases the weight for values of c that give lower empirical entropies of $p(c)$ (and hence increases the balance of the distribution); while (2) increases weight for values of c, that give higher empirical entropies of $p(c)$ (and hence decreases the balance of these distributions). Similarly, the expected variance σ_{comm} can be used to adjust the strictness of the regularization: smaller $\sigma_{topicComm}$ imposes stricter regularization. When $\sigma_{comm} = \infty$, the model has no regularization on $p(c)$.

In our experiments, we set $\mathcal{E}_{topicComm} = 0$ (this is corresponding to the case where each topic is assigned to at most one community) and $\sigma_{topicComm} = 0.2$; and set $\mathcal{E}_{comm} = \ln(C)$ where C is the number of the communities (this is corresponding to the case where the communities are perfectly balanced), and $\sigma_{comm} = 0.3$. We also used symmetric Dirichlet hyperparameters with $\alpha = 50/K$, $\beta = 0.01$, $\rho = 2$, $\tau = 1/C$, $\eta = 50/K$, and $\gamma_l = 0.01$ for all $l = 1, \cdots, L$. Given the input dataset, we train the model with 600 iterations of Gibbs sampling. We took 25 samples with a gap of 20 iterations in the last 500 iterations to estimate all the hidden variables.

5 Experimental Evaluation

5.1 Dataset

We collected tweets from a set of Twitter users who are interested in software engineering for evaluating the **CPI** model. To construct this dataset, we first utilized the list of 100 most influential software developers in Twitter provided in [18] as seed users. These are highly-followed users who actively tweet about software engineering topics, e.g., *Jeff Atwood*[1], *Jason Fried*[2], and *John Resig*[3]. We further expanded the user set by adding all users following at least five seed users so as to get more technology savvy users. Lastly, we took all tweets posted

---

[1] http://en.wikipedia.org/wiki/Jeff_Atwood
[2] http://www.hanselman.com/blog/AboutMe.aspx
[3] http://en.wikipedia.org/wiki/John_Resig

by these users in August to October 2011 to form the experimental dataset. In this work, we consider the following behavior types: (1) *mention*, and (2) *hashtag*, and (3) *retweet*. These are messaging behaviors beyond content generation that users may adopt multiple times.

We employed the following preprocessing steps to clean the dataset. We first removed stopwords from the tweets. Then, we filtered out tweets with less than 3 non-stopwords. Next, we excluded users with less than 50 (remaining) tweets. Lastly, for each behavior, we filtered away the behaviors with less than 10 adopting users; and for each user and each type of behaviors, we filtered out all the user's behaviors if the user adopted less than 50 behaviors of the type. These minimum thesholds are necessary so that, for each behavior and each user, we have enough number of adoption observations for learning both influence of the user's personal interest and that of her community on behavior adoption.

Based on the biographies of the users, we were able to manually label 3,023 users, including 2,503 **Developers** and 520 **Marketers**. The labeling work is mostly unambiguous as the biographies are quite short and clear, and only users with explicit declaration of their professionals were labeled. We therefore used these labels as ground truth community labels in our experiments.

Table 1. Statistics of the experimental dataset

#user	14,595
#labeled users	3,023
#tweets	3,030,734
#mention adoptions	354,463
#hashtag adoptions	894,619
#retweet adoptions	909,272

Table 1 shows the statistics of the experimental dataset after the preprocessing steps. The statistics show that the dataset after the filtering is still large. This allows us to learn the parameters accurately.

5.2 Experimental Tasks

Content Modeling. In this task, we compare **CPI** against **TwitterLDA** model [44] in modeling topics in the content. **TwitterLDA** is among state-of-the-art modeling methods for microblogging content. To evaluate the performance, we run both models with the number of topics varied from 10 to 100.

User Classification. In this task, we evaluate the performance of the **CPI** model as a semi-supervised learner (see Section 4.2). The task is chosen since: (1) we have ground truth community labels (**Developer** and **Marketer**) for only a small fraction of users the dataset (20.7%); and (2) the supervised learning approach for user classification in microbloggings may not practical as shown in [6]. We compare **CPI** model against the state-of-the-art semi-supervised learning (**SSL**) methods provided in [36]. Those are *label propagation* based methods which iteratively update label for each (unknown label) user u based on labels of the other users who are most similar to u. Here, we use cosine similarity between pairs of users. We represent each user as a vector of features, which include: (a) tweet-based features, and (b) bags-of-behaviors of the users. The tweet-based features for each user are the components in topic distribution of the user's tweets discovered by **TwitterLDA** model. For the **CPI** model, we set the communities to 3 since: (a) it is reasonable to have one more community in each of the two datasets since there are users who do not belong to any of the two manually identified communities; and (b) this is to ensure that the **CPI** model run with the same settings as the **SSL** baseline methods.

5.3 Evaluation Metrics

We adopt *likelihood* and *perplexity* for evaluating the content modeling task. To do this, for each user, we randomly selected 90% of tweets of the user to form a training set, and use the remaining 10% of the tweets as the test set. Then for each method, we compute the likelihood of the training set and perplexity of the test set. The method with a higher likelihood, or lower perplexity is considered better for the task.

For user classification task, we adopt *average $F1$ score* as the performance metric. We first evenly distributed the set of labeled users in each dataset into 10 folds such that, for each user label, every fold has the same proportion of users having the label. Then, for each method, we run 10-fold cross validation. More precisely, for each method and each time, we chose 1 fold of labeled users as test set. We hide label of user in this fold and consider them as unlabeled users. Then, we use 9 remaining folds of labeled users and all unlabeled users as the (semi-) training set. We then compute the average $F1$ score obtained by each method in both label classes (i.e., **Developer** and **Marketer**). The method with a higher score is the winner in the task.

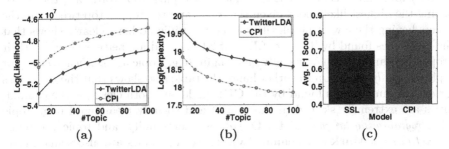

Fig. 2. (a) Likelihood and (b) Perplexity of **TwitterLDA** and **CPI** models in the content modeling task; and (c) Average $F1$ scores of **SSL** and **CPI** models in the user classification task

5.4 Results

Content Modeling. Figures 2 (a) and (b) show the performance of **TwitterLDA** model and **CPI** model in content modeling task when varying the number of topics K. As expected, larger number of topics K gives larger likelihood and smaller perplexity, and the amount of improvement diminishes as K increases. The figures show that **CPI** model significantly outperforms **TwitterLDA** model in the task. Considering both time and space complexities, we set the number of topics to 80 for the remaining experiments.

User Classification. Figure 2 (c) shows the performance of **SSL** methods and the **CPI** model in the user classification task. In the figure, the **SSL** bar shows

Table 2. Top topics of each community found by different models

Community	TwitterLDA+SSL			CPI		
	Topic	Topic Label	Prob	Topic	Topic Label	Prob
Developer	32	Daily activities	0.072	46	Programming languages	0.57
	77	Programming languages	0.052	36	Project hosting services	0.34
	64	Daily life	0.036	71	Operating systems	0.03
Marketer	57	Online marketing	0.142	7	Online marketing	0.987
	72	Business	0.098	78	Mobile business	0.009
	4	Social networks	0.056	59	Technology business	0.003

the best performance obtained by methods provided in [36]. The figure clearly shows that the **CPI** model significantly outperforms the **SSL** baseline methods in the task.

5.5 Topic Analysis

Community Topics. We now examine the representative topics for each community as found by the **CPI** model and **TwitterLDA** in both the two datasets. As the **TwitterLDA** model does not identify community for each user, we first use the best user classifier among the learnt **SSL** classifiers to determine community for all the users. We then compute topic distribution of each community by aggregating topic distributions of all users within the community.

Table 2 shows the top topics for each ground truth community in the experimental dataset found by **TwitterLDA+SSL** method and **CPI** model. Note that the topic labels are manually assigned after examining the topics' top words[4]) and top tweets. For each topic, the topic's top words are the words having the highest likelihoods given the topic, and the topic's top tweets are the tweets having the lowest perplexities given the topic. Table 2 clearly shows that the top topics found by **TwitterLDA+SSL** method are neither clear (as their proportions are small) nor socially meaningful (e.g., topic 32 (*Daily activities*) or topic 64 (*Daily life*)). On the other hand, the table also shows that the top topics for each community as found by the **CPI** model are both clear (as the communities are extremely skewed to the topics) and socially meaningful (e.g., topic 46 (*Programming languages*) for **Developer** community; and topic 7 (*Online marketing*) for **Marketer** community). These top topics are also semantically reasonable. It is expected that the **Developer** community are mainly interested in programming related topics, and the **Marketer** community are mainly interested in marketing related topics.

Personal Interest Topics. Next, we examine the representative personal interest topics found by **CPI** model. Table 3 shows the top topics in aggregated personal topic distributions of all users in the dataset. The table clearly shows that these representative topics are reasonable. It is expected that the top personal interest topics include *Entertainment* (topic 34) and a trivial topic (*Daily*

Table 3. Top personal interest topics found by **CPI**

Topic	Topic Label	Probability
34	Entertainment	0.054
33	Daily life	0.041
39	Smartphone	0.031

---

[4] The top words of topics found by the models are not shown here due to the space limitation.

Table 4. Top behaviors of representative topics found by **CPI** model

Topic	Top hashtags	Top mentions	Top retweeted
7	#seo,#socialmedia,#marketing #sm,#marketin,#facebook	@jeffbullas,@leaderswest @markwschaefer,@smexamine	mashable,sengineland marketingland,jeffbullas
34	#debat,#debate,#debate201 #vpdebat,#breakingbad	@twitter,@mike,@nytimes @mat,@medium,@branch	robdelaney,pourmecoffee anildash,theonion
36	#fail,#ruby,#nodejs #github,#mongodb,#android	@twitter,@github,@dropbox @kickstarter,@newsycombinator	codinghorror,oatmeal rickygervais,github
46	#javascript,#programming #java,#ruby,#python,#php	@github,@skillsmatter,@twitter @rubyrogues,@steveklabnik	codinghorror,garybernhardt steveklabnik,dhh,mfeathers
78	#mobile,#mobil,#facebook #app,#retail,#advertising	@techcrunc,@sa,@mashabl @fastcompan,@mediapos	techmeme,gigaom,mashable allthingsd,sai,techcrunch

life - topic 33). It is also expected that a technology related topic (*Smartphone* - topic 39) is among the top personal interest topics of users in the experimental dataset as most of its users are working in IT industry. This also shows the effectiveness of our regularization technique in differentiating between trivially popular topics and socially meaningful ones so that to assign the formers to user personal interest, and assign the latter to community interest.

5.6 User Behaviors Analysis

Lastly, we examine the user behaviors associated with the result topics. Table 4 show some of representative topics (shown in Tables 2 and 3) together with the topics' top behaviors. For each topic, the topic's top behaviors are the behaviors having the highest likelihoods given the topic. The table show that the extreme behaviors for each of the topics are reasonable. For example, it is expected that people use marketing and social media related hashtags (*#seo, #socialmedia, #marketing*, etc.), mention online marketers and bloggers (*@jeffbullas, @leaderswest, @markwschaefer*, etc.), and retweet from marketing magazines (*mashable, sengineland, marketingland*) for topic *Online marketing* (topic 7); people also use programming related hashtags (*#javascript, #programming, #java, ruby*, etc.), mention big IT companies and hosting services (*@twitter, @github*, etc.), and retweet from influential developers (*codinghorror, garybernhardt, steveklabnik*, etc.) for topic *Programming languages* (topic 46). A qualitatively similar result holds for the remaining topics as well as topics that are not shown in the two tables. We leave out these analysis due to the space limitation.

6   Conclusion

In this paper, we propose a novel topic model for simultaneously modeling mutually exclusive community and user topical interest in microblogging data. Our model is able to integrate both user generated content and multiple types of behaviors to determine user and community interests, as well as to derive the influence of each user's community on her generated content and behaviors. We also report experiments on a Twitter dataset showing the improvement of the proposed model over other state-of-the-art models in content modeling and user classification tasks.

In the future, we would like to extend the proposed model to incorporate social factors in studying user generate content and behavior. These factors include the users' interaction, their social communities, and the temporal and spatial dynamics of the users and the communities.

Acknowledgments. This research is supported by the Singapore National Research Foundation under its International Research Centre @ Singapore Funding Initiative and administered by the IDM Programme Office, Media Development Authority (MDA).

References

1. Airoldi, E.M., Blei, D.M., Fienberg, S.E., Xing, E.P.: Mixed membership stochastic blockmodels. J. Mach. Learn. Res. **9** (2008)
2. Balasubramanyan, R., Cohen, W.W.: Block-LDA: jointly modeling entity-annotated text and entity-entity links. In: SDM (2011)
3. Balasubramanyan, R., Cohen, W.W.: Regularization of latent variable models to obtain sparsity. In: SDM13 (2013)
4. Chang, J., Blei, D.M.: Relational topic models for document networks. In: AIS-TATS (2009)
5. Chang, J., Boyd-Graber, J., Blei, D.M.: Connections between the lines: augmenting social networks with text. In: KDD (2009)
6. Cohen, R., Ruths, D.: Classifying political orientation on twitter: it's not easy! In: ICWSM (2013)
7. Conover, M., Ratkiewicz, J., Francisco, M., Gonçalves, B., Flammini, A., Menczer, F.: Political polarization on twitter. In: 5th ICWSM (2011)
8. Cui, P., Wang, F., Liu, S., Ou, M., Yang, S., Sun, L.: Who should share what?: item-level social influence prediction for users and posts ranking. In: SIGIR (2011)
9. Dabeer, O., Mehendale, P., Karnik, A., Saroop, A.: Timing tweets to increase effectiveness of information campaigns. In: 5th ICWSM (2011)
10. Ding, Y.: Community detection: Topological vs. topical. J. Informetrics (2011)
11. Feller, A., Kuhnert, M., Sprenger, T., Welpe, I.: Divided they tweet: the network structure of political microbloggers and discussion topics. In: ICWSM (2011)
12. Hannon, J., Bennett, M., Smyth, B.: Recommending twitter users to follow using content and collaborative filtering approaches. In: RecSys 2010 (2010)
13. Hoang, T.A., Cohen, W.W., Lim, E.P.: On modeling community behaviors and sentiments in microblogging. In: SDM14 (2014)
14. Hoang, T.A., Cohen, W.W., Lim, E.P., Pierce, D., Redlawsk, D.P.: Politics, sharing and emotion in microblogs. In: ASONAM (2013)
15. Hoang, T.-A., Lim, E.-P.: On joint modeling of topical communities and personal interest in microblogs. In: Aiello, L.M., McFarland, D. (eds.) SocInfo 2014. LNCS, vol. 8851, pp. 1–16. Springer, Heidelberg (2014)
16. Hong, L., Davison, B.: Empirical study of topic modeling in twitter. In: SOMA (2010)
17. Java, A., Song, X., Finin, T., Tseng, B.: Why we twitter: understanding microblogging usage and communities. In: WebKDD/SNA-KDD 2007 (2007)
18. Jurgen, A.: Twitter top 100 for software developers. http://www.noop.nl/2009/02/twitter-top-100-for-software-developers.html
19. Kwak, H., Chun, H., Moon, S.: Fragile online relationship: a first look at unfollow dynamics in twitter. In: CHI (2011)
20. Kwak, H., Lee, C., Park, H., Moon, S.: What is twitter, a social network or a news media? In: WWW (2010)
21. Li, D., He, B., Ding, Y., Tang, J., Sugimoto, C., Qin, Z., Yan, E., Li, J., Dong, T.: Community-based topic modeling for social tagging. In: CIKM 2010 (2010)
22. Lim, K.H., Datta, A.: Following the follower: detecting communities with common interests on twitter. In: HT (2012)

23. Liu, J.S.: The collapsed gibbs sampler in bayesian computations with applications to a gene regulation problem. J. Amer. Stat. Assoc (1994)
24. Mehrotra, R., Sanner, S., Buntine, W., Xie, L.: Improving LDA topic models for microblogs via tweet pooling and automatic labeling. In: SIGIR (2013)
25. Michelson, M., Macskassy, S.A.: Discovering users' topics of interest on twitter: a first look. In: AND 2010 (2010)
26. Nallapati, R.M., Ahmed, A., Xing, E.P., Cohen, W.W.: Joint latent topic models for text and citations. In: KDD (2008)
27. Newman, M.E.J.: Modularity and community structure in networks. PNAS (2006)
28. Qiu, M., Jiang, J., Zhu, F.: It is not just what we say, but how we say them: LDA-based behavior-topic model. In: SDM (2013)
29. Ramage, D., Dumais, S.T., Liebling, D.J.: Characterizing microblogs with topic models. In: ICWSM (2010)
30. Ramage, D., Hall, D., Nallapati, R., Manning, C.D.: Labeled LDA: a supervised topic model for credit attribution in multi-labeled corpora. In: EMNLP (2009)
31. Rosen-Zvi, M., Griffiths, T., Steyvers, M., Smyth, P.: The author-topic model for authors and documents. In: UAI (2004)
32. Sachan, M., Contractor, D., Faruquie, T.A., Subramaniam, L.V.: Using content and interactions for discovering communities in social networks. In: WWW (2012)
33. Sachan, M., Xing, E., et. al.: Spatial compactness meets topical consistency: jointly modeling links and content for community detection. In: WSDM (2014)
34. Schantl, J., Kaiser, R., Wagner, C., Strohmaier, M.: The utility of social and topical factors in anticipating repliers in twitter conversations. In: WebSci (2013)
35. Suh, B., Hong, L., Pirolli, P., Chi, E.H.: Want to be retweeted? large scale analytics on factors impacting retweet in twitter network. In: SocialCom (2010)
36. Talukdar, P.P., Pereira, F.: Experiments in graph-based semi-supervised learning methods for class-instance acquisition. ACL (2010)
37. Tuan-Anh, H.: Modeling user interest and community interest in microbloggings: an integrated approach. https://www.dropbox.com/s/h0o7dca1i83qkck/CPI.pdf
38. Wu, S., Hofman, J.M., Mason, W.A., Watts, D.J.: Who says what to whom on twitter. In: WWW (2011)
39. Xu, W., Liu, X., Gong, Y.: Document clustering based on non-negative matrix factorization. In: SIGIR (2003)
40. Yang, J., McAuley, J., Leskovec, J.: Community detection in networks with node attributes. In: ICDM (2013)
41. Yang, J., Counts, S.: Predicting the speed, scale, and range of information diffusion in twitter. In: ICWSM (2010)
42. Yin, D., Hong, L., Davison, B.D.: Structural link analysis and prediction in microblogs. In: CIKM (2011)
43. Yin, Z., Cao, L., Gu, Q., Han, J.: Latent community topic analysis: integration of community discovery with topic modeling. ACM TIST (2012)
44. Zhao, W.X., Jiang, J., Weng, J., He, J., Lim, E.-P., Yan, H., Li, X.: Comparing twitter and traditional media using topic models. In: Clough, P., Foley, C., Gurrin, C., Jones, G.J.F., Kraaij, W., Lee, H., Mudoch, V. (eds.) ECIR 2011. LNCS, vol. 6611, pp. 338–349. Springer, Heidelberg (2011)
45. Zhou, D., Manavoglu, E., Li, J., Giles, C.L., Zha, H.: Probabilistic models for discovering e-communities. In: WWW 2006 (2006)

Minimal Jumping Emerging Patterns: Computation and Practical Assessment

Bamba Kane, Bertrand Cuissart[(✉)], and Bruno Crémilleux

GREYC - CNRS UMR 6072, University of Caen Basse-Normandie,
14032 Caen Cedex 5, France
{bamba.kane,bertrand.cuissart,bruno.cremilleux}@unicaen.fr

Abstract. Jumping Emerging Patterns (JEP) are patterns that only occur in objects of a single class, a *minimal JEP* is a JEP where none of its proper subsets is a JEP. In this paper, an efficient method to mine the *whole set* of the minimal JEPs is detailed and fully proven. Moreover, our method has a larger scope since it is able to compute the essential JEPs and the top-k minimal JEPs. We also extract minimal JEPs where the absence of attributes is stated, and we show that this leads to the discovery of new valuable pieces of information. A performance study is reported to evaluate our approach and the practical efficiency of minimal JEPs in the design of rules to express correlations is shown.

Keywords: Pattern mining · Emerging patterns · Minimal jumping emerging patterns · Ruled-based classification

1 Introduction

Contrast set mining is a well established data mining area [14] which aims at discovering conjunctions of attributes and values that differ meaningfully in their distributions across groups. This area gathers many techniques such as subgroup discovery [17] and emerging patterns [2]. Because of their discriminative power, contrast sets are highly useful in supervised tasks to solve real world problems in many domains [1,7,12].

Let us consider a dataset of objects partitioned into several classes, each object being described by binary attributes. Initially introduced in [2], emerging patterns (EPs) are patterns whose frequency strongly varies between two datasets. A *Jumping Emerging Pattern* (JEP) is an EP which has the notable property to occur only in a single class. JEPs are greatly valuable to obtain highly accurate rule-based classifiers [8,9]. They are used in many domains like chemistry [12], knowledge discovery from a database of images [7], predicting or understanding diseases [3], or DNA sequences [1]. A *minimal JEP* designates a JEP where none of its proper subsets is a JEP. Minimal JEPs are of great interest because they capture the vital information that cannot be skipped to characterize a class. Using more attributes may not help and even add noise in a classification purpose. Mining minimal JEPs is a challenging task because it is

© Springer International Publishing Switzerland 2015
T. Cao et al. (Eds.): PAKDD 2015, Part I, LNAI 9077, pp. 722–733, 2015.
DOI: 10.1007/978-3-319-18038-0_56

a time consuming process. Current methods require either a frequency threshold [4] or a given number of expected patterns [16]. On the contrary, one of the results of this paper is to be able to compute the *whole set* of minimal JEPs.

The contribution of this paper can be summarized as follows. First, we introduce an efficient method to obtain *all minimal JEPs*. A key idea of our method is to introduce an alternative definition of a minimal JEP which stems from the differences between pairs of objects, each of a different class. A backtrack algorithm for computing all minimal JEPs is detailed and the related proofs are provided. Our method does not require either a frequency threshold or a number of patterns to extract. It provides a general approach and its scope encompasses the essential JEPs [4] (i.e., JEPs satisfying a given minimal frequency threshold) and the k most supported minimal JEPs [16] which constitute the state of the art in this field. Second, taking into account the absence of attributes may provide interesting pieces of knowledge to build more accurate classifiers as experimentally shown by Terlecki and Walczak [15]. We address this issue. Our method integrates the absence of attributes in the process by adding their negation. It produces the whole set of minimal JEPs both with the present and absent attributes. Practical results advocate in favor of this addition of negated attributes in the description of the objects. Third, the results of an experimental study are given. We analyze the computation of the minimal JEPs, including the absence of attributes and comparisons with essential JEPs and top-k minimal JEPs. Finally, we experimentally assess the quality of minimal JEPs, essential JEPs and top-k minimal JEPs as correlations between a pattern and a class.

Section 2 gives the preliminaries. The description of our method is provided in Section 3. Section 4 presents the experiments. We review related work in Section 5 and we round up with conclusions and perspectives in Section 6.

2   Preliminaries

Let $\mathcal{G}$ be a *dataset*, a multiset consisting of n elements, an element of $\mathcal{G}$ is named an *object*. The *description* of an object is given by a set of attributes, an *attribute* being an atomic proposition which may hold or not for an object. The finite *set of all the attributes occurring in $\mathcal{G}$* is denoted by $\mathcal{M}$. In the remainder of this text, for the sake of simplicity, the word "object" is also used to designate the description of an object.

A *pattern* denominates a set of attributes, an element of the power set $\mathcal{M}$, denoted $\mathcal{P}(\mathcal{M})$. A pattern is *included in* the object g if p is a subset of the description of g: $p \subseteq g$. The *extent of a pattern* p in $\mathcal{G}$, denoted $p'_{\mathcal{G}}$, corresponds to the set of the objects that include p: $p'_{\mathcal{G}} = \{g \in \mathcal{G} : p \subseteq g\}$. A pattern is *supported* if it is included in at least one object of the dataset. Moreover, we define a relation, I, on $\mathcal{G} \times \mathcal{P}(\mathcal{M})$ as follows: for any object g and any pattern p, $gIp \iff p \subseteq g$.

Usual data mining methods only consider the presence of attributes. With binary descriptions, the absence of an attribute can be explicitly denoted by adding the negation of this attribute in order to build patterns conveying this

Table 1. A dataset of 6 objects

Objects \ Attributes		1	¬1	2	¬2	3	¬3	4	¬4
$\mathcal{G}_+$	g_1	X			X	X		X	
	g_2		X	X		X			X
$\mathcal{G}_-$	g_3		X	X			X	X	
	g_4		X	X		X		X	
	g_5		X		X	X		X	
	g_6	X		X		X			X

Table 2. Differences from the dataset in Table 1

	g_3	g_4	g_5	g_6
g_1	1,3,¬2	1,¬2	1	¬2,4
g_2	3,¬4	¬4	2,¬4	¬1
$\mathcal{D}_{\bullet j}$	1,3,¬2,¬4	1,¬2,¬4	1,2,¬4	¬1,¬2,4

information. We integrate this idea in this paper by adding the negation of absent attributes and thus the description of an object always mentions every attribute either positively or negatively. In other words, $\mathcal{M}$ explicitly contains the negation of any of its attributes, the symbol ¬ is used to denote the negation of an attribute (cf. Table 1 as an example).

Minimal Jumping Emerging Pattern. We now suppose that the dataset $\mathcal{G}$ is partitioned into two subsets $\mathcal{G}_+$ and $\mathcal{G}_-$, every subset of such a partition is usually named a *class* of the dataset. We call an object of $\mathcal{G}_+$ a *positive object* and an object of $\mathcal{G}_-$ a *negative object*. We say that a supported pattern p is a *JEP* if it is never included in any negative object: $p'_{\mathcal{G}} \neq \emptyset$ and $p'_{\mathcal{G}} \subseteq \mathcal{G}_+$.

A JEP is *minimal* if it does not contain another JEP as a proper subset. The set of the minimal JEPs is a subset of the set of the JEPs which groups all the most general JEPs. As a JEP contains at least one minimal JEP, when an object includes a JEP then it includes a minimal JEP.

Table 1 displays a dataset of 6 objects partitioned in two datasets: $\mathcal{G}_+ = \{g_1, g_2\}$ and $\mathcal{G}_- = \{g_3, g_4, g_5, g_6\}$. The pattern $p = \{1, ¬2\}$ is a JEP as $p'_{\mathcal{G}_+} = \{g_1\}$ and $p'_{\mathcal{G}_-} = \emptyset$ and $\{1\}$ and $\{¬2\}$ are not JEPs, p is thus a minimal JEP.

3 Contribution

Section 3.1 introduces the key notion of a *difference* between two objects, it provides a new definition of a minimal JEP. The latter is the support of our algorithm for extracting minimal JEPs which is detailed and proven in Section 3.2.

3.1 A Relation Between the Minimal JEPs and the Differences Between Objects

Let $\mathcal{G}$ be a dataset partitioned into two subsets $\mathcal{G}_+$ and $\mathcal{G}_-$. The *difference between an object i and an object j* groups the attributes of i that are not satisfied by j: $\mathcal{D}_{i,j} = i \setminus j = \{m \in \mathcal{M} : i \ I \ m \text{ and } ¬j \ I \ m\}$. When one focuses on a negative object j, the *gathering of the differences for a negative object j* corresponds to the union of the differences between i and j, for any *positive object i*: $\mathcal{D}_{\bullet j} = \cup_{i \in \mathcal{G}_+} \mathcal{D}_{i,j}$. In Table 2, the *gathering of the differences* for the negative object 4 is $\mathcal{D}_{\bullet 4} = \mathcal{D}_{1,4} \cup \mathcal{D}_{2,4} = \{1, ¬2\} \cup \{¬4\} = \{1, ¬2, ¬4\}$.

The following lemma is a direct consequence of the definition of the gathering of the differences for a negative object.

Lemma 1. *Let j be a negative object and p be a pattern. If $\mathcal{D}_{\bullet j} \cap p \neq \emptyset$ then p is not included in j : $\neg(j\ I\ p)$.*

It follows that, if a supported pattern p intersects with every gathering of the differences for a negative object and, thanks to Lemma 1, p cannot be included in any negative object, thus p is a JEP. We now reason by contraposition and we suppose that a supported pattern p does not intersect with the gathering of the differences for one negative object j_0: $\mathcal{D}_{\bullet j_0} \cap p = \emptyset$. If p is supported by a positive object i_0, as $\mathcal{D}_{\bullet j_0} \cap p = \emptyset$ implies $\mathcal{D}_{i_0, j_0} \cap p = \emptyset$, then p is supported by j_0. Thus p cannot be a JEP.

A JEP corresponds to a supported pattern which has at least one attribute in every $\mathcal{D}_{\bullet j}$, for j a negative object. Proposition 1 follows:

Proposition 1. *A supported pattern p is a JEP if $\mathcal{D}_{\bullet j} \cap p \neq \emptyset$, $\forall j \in \mathcal{G}_-$*

On the example, the JEP $p = \{1, \neg 2\}$ intersects with every $\mathcal{D}_{\bullet j}$ (see Table 2): $\mathcal{D}_{\bullet g_3} \cap p = \{1, \neg 2\}, \mathcal{D}_{\bullet g_4} \cap p = \{1, \neg 2\}$, $\mathcal{D}_{\bullet g_5} \cap p = \{1\}$ and $\mathcal{D}_{\bullet g_6} \cap p = \{\neg 2\}$.

We now establish a relation between the gathering of the differences and the minimal JEPs.

Proposition 2. *A JEP p is a minimal JEP if, for every attribute a of p, $\exists j \in \mathcal{G}_-$ such that $p \cap \mathcal{D}_{\bullet j} = \{a\}$.*

On the example, the JEP $p = \{3, 1, \neg 2\}$ is not a minimal JEP since it contains the JEP $\{1, \neg 2\}$. Proposition 2 gives another point of view: since no intersection between p and a $\mathcal{D}_{\bullet j}$ (for j a negative object) corresponds to $\{3\}$, the attribute $\{3\}$ does not play a necessary part in the discriminative power of p, thus p is not a minimal JEP.

Proof (of Proposition 2). Let p be a JEP.

Suppose p is not minimal: there exists a JEP q, different from p, such that $q \subsetneq p$. Consider an attribute a such that $a \in p \backslash q$. As q is a JEP, Prop. 1 imposes that $\forall j \in \mathcal{G}_-$, $q \cap \mathcal{D}_{\bullet j} \neq \emptyset$, it ensues that $\forall j \in \mathcal{G}_-$, $p \cap \mathcal{D}_{\bullet j} \neq \{a\}$. One now can state that, if p is not minimal, then p contains one attribute a such that $\forall j \in \mathcal{G}_-$, $p \cap \mathcal{D}_{\bullet j} \neq \{a\}$.

Conversely, suppose there exists an attribute a in p such that $\forall j \in \mathcal{G}_-$, $p \cap \mathcal{D}_{\bullet j} \neq \{a\}$. As p is a JEP, Prop. 1 ensures that $\mathcal{D}_{\bullet j} \cap p \neq \emptyset$, $\forall j \in \mathcal{G}_-$. It follows that, $\forall j \in \mathcal{G}_-, \mathcal{D}_{\bullet j} \cap p \backslash \{a\} \neq \emptyset$. By applying Prop. 1, $p \backslash \{a\}$ is a JEP and p cannot be minimal. □

Prop. 2 states that a minimal JEP is a supported pattern that excludes all the negative objects and where every attribute is necessary to exclude (at least one) object. It follows:

Consequence of Prop. 2. Let p be a minimal JEP for the dataset $\mathcal{G}_+ \cup \mathcal{G}_-$ and $g_- \in \mathcal{G}_-$. If p is not a minimal JEP for the dataset $\mathcal{G}_+ \cup \mathcal{G}_- \setminus \{g_-\}$ then there exists a unique attribute a, $a \in p$, such that $p \backslash \{a\}$ is a minimal JEP for the dataset $\mathcal{G}_+ \cup \mathcal{G}_- \setminus \{g_-\}$.

3.2   Calculation of the Minimal JEPs

We now introduce a structure designed to generate all the minimal JEPs for a dataset: a rooted tree whose "valid" leaves are in a one-to-one correspondence with the minimal JEPs. We suppose here that for $\forall j \in \mathcal{G}_-$, $\mathcal{D}_{\bullet j} \neq \emptyset$, as it follows from Prop. 1 that this condition is a necessity for the existence of at least one minimal JEP. We also assume that an arbitrary order is given on the negative objects: for two negative objects j and j', $j \prec j'$ if j is accounted before j'.

Rooted Tree. A *rooted tree* (T, r) is a tree in which one node, the root r, is distinguished. In a rooted tree, any node of degree one, unless it is the root, is called a *leaf*. If $\{u, v\}$ is an edge of a rooted tree such that u lies on the path from the root to v, then v is a *child* of u. An *ancestor* of u is any node of the path from the root to u. If u is an ancestor of v, then v is a *descendant* of u, and we write $u \leqslant v$; if $u \neq v$, we write $u < v$.

A Tree of the Minimal JEPs. We create the tree (T, r) as a rooted tree in which each node x, except the root r, holds two labels: an attribute, $l_{attr}(x) \in \mathcal{M}$, and a negative object $l_{obj}(x) \in \mathcal{G}_-$. For a node x of (T, r), $Br(x)$ gathers the attributes that occur along the path from the root to x: $Br(x) = \{l_{attr}(y), y \leqslant x\}$; $Br(x)$ indicates the pattern considered at x. For any node x of T and any attribute a, $a \in Br(x)$, $crit(a, x)$ gathers the negative objects already considered at the level of x and whose exclusion is due to the sole presence of a in $Br(x)$: $crit(a, x) = \{j \preceq l_{obj}(x) : \mathcal{D}_{\bullet j} \cap Br(x) = \{a\}\}$.

Definition 1 (A tree of the minimal JEPs (ToMJEPs)). *A rooted tree (T, r) is a tree of the minimal JEPs for $\mathcal{G}$ if:*

  i) *any node x, except the root r, holds two labels: an attribute label, $l_{attr}(x) \in \mathcal{M}$, and a negative object label, $l_{obj}(x) \in \mathcal{G}_-$.*
  ii) *if x is an internal node then:*
    a) *the children of x hold the same negative object label: $l_{obj}(y) = min\{j \in \mathcal{G}_- : \mathcal{D}_{\bullet j} \cap Br(x) = \emptyset\}$, $\forall y$ a child of x,*
    b) *every child of x holds a different attribute label,*
    c) *the union of the attribute labels of the children y of x corresponds to $\mathcal{D}_{\bullet l_{obj}(y)}$.*
  iii) *x is a leaf if it satisfies one of the following conditions:*
    a) *$\exists z \preceq x$ such that $crit(l_{attr}(z), x) = \emptyset$,*
    b) *$\forall j \in \mathcal{G}_-$, $\mathcal{D}_{\bullet j} \cap Br(x) \neq \emptyset$.*

A leaf which satisfies the criteria iii)a) is named *dead-end leaf*, otherwise it is named a *candidate leaf*.

Figure 1 depicts a ToMJEPs for the dataset of Tables 1 and 2. The nodes with a dashed line are the dead-end leaves, the nodes surrounded by a solid line the candidate leaves. A candidate leaf surrounded by a bold plain line is associated to a *supported* pattern: it represents a minimal JEP. For example, the node x such that $Br(x) = \{1, \neg 2\}$ is associated to a minimal JEP while the node

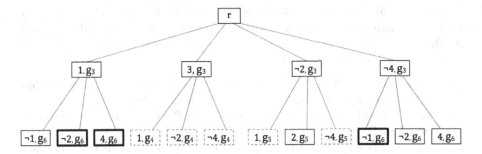

Fig. 1. Example of a tree for minimal JEPs

y such that $Br(y) = \{\neg 4, \neg 2\}$ is associated to a pattern which is not supported by the dataset. The node z such that $Br(z) = \{3, \neg 2\}$ is a dead-end leaf: since $\forall j \in \{g_3, g_4\}, \{3, \neg 2\} \cap \mathcal{D}_{\bullet j} \neq \{3\}$, the attribute 3 does not fulfill the constraint raised by Prop. 2, thus $crit(3, z) = \emptyset$.

We will now demonstrate that there is a one-to-one mapping between the "supported" candidate leaves of a ToMJEPs and the minimal JEPs. The following lemma is an immediate consequence of the definition of a ToMJEPs, together with the application of Prop. 1 and 2.

Lemma 2. *Let (T, r) be a ToMJEPs and x be a node of T, different from a dead-end leaf. If there exists $i \in \mathcal{G}_+$ such that $i \mathcal{I} Br(x)$ then $Br(x)$ is a minimal JEP for the dataset $\mathcal{G}' = \mathcal{G}_+ \cup \{j \leq l_{obj}(x)\}$.*

Proof. By definition of a ToMJEPs, for a node x, we have $Br(x) \cap \mathcal{D}_{\bullet j} \neq \emptyset, \forall j \leq l \leq l_{obj}(x)$. Thanks to Prop. 1, it follows that $Br(x)$ is a JEP for $\mathcal{G}_+ \cup \{j \leq l_{obj}(x)\}$.

If x is not a dead-end leaf, by definition of a ToMJEPs, we have $\forall z \leq x$, $crit(l_{attr}(z), x) \neq \emptyset$, thus $\forall a \in Br(x)$, $\exists j \in \cup \{j \leq l_{obj}(x)\}$ such that $Br(x) \cap \mathcal{D}_{\bullet j} = \{a\}$. Prop. 2 ensures that $Br(x)$ is a minimal JEP for the dataset $\mathcal{G}_+ \cup \{j \leq l_{obj}(x)\}$. $\qquad\square$

Lemma 3. *Let (T, r) be a ToMJEPs. Let p be pattern. If p is a minimal JEP for the dataset $\mathcal{G}_+ \cup \mathcal{G}_-$ then there exists a unique candidate leaf x such that $Br(x) = p$.*

Proof. The proof reasons inductively on $\mathcal{G}_-$. For a sake of simplicity, we denote here the set of the negative objects as $\{1, \ldots, k\}$ with $k = |\mathcal{G}_-|$ and $\forall 1 \leq j \leq k - 1$, $j \prec j + 1$.

Definition 1 implies that the children of the root r deal with 1 (the first negative object), we have $\mathcal{D}_{\bullet 1} = \{l_{attr}(x) : x \text{ is a child of } r\}$. Moreover, as by definition of a ToMJEPs, $crit(l_{attr}(x), x) \neq \emptyset$, no child of r is a dead-end leaf. Thus, associated to any pattern p which is a minimal JEP for the dataset $\mathcal{G}_+ \cup \{1\}$, there is a unique node x, different from a dead-end leaf such that $Br(x) = p$.

Let us now suppose that, considering any minimal JEP p for $\mathcal{G}_+ \cup \{1,\ldots,l\}$ with $l < k$, there exists a unique node x, different from a dead-end leaf, such that $Br(x) = p$. When we consider a pattern q, minimal JEP for the dataset $\mathcal{G}_+ \cup \{1,\ldots,l,l+1\}$, two cases arise:

- If q is a minimal JEP for $\mathcal{G}_+ \cup \{1,\ldots,l\}$, then, thanks to the induction hypothesis, there exists a unique node x_q such that $Br(x_q) = q$.
- Otherwise, thanks to the consequence of Prop. 2, there exists one attribute a such that $\mathcal{D}_{\bullet l+1} \cap q = \{a\}$ and $\mathcal{D}_{\bullet j} \cap a \neq \{a\}, \forall j \preceq l$. Prop 2 ensures that $q \setminus \{a\}$ is minimal JEP for $\mathcal{G}_+ \cup \{1,\ldots,l\}$. Thanks to the induction hypothesis, there exists a unique node x, different from a dead-end leaf, such that $Br(x) = q\setminus\{a\}$. By definition of a ToMJEPs, there exists a unique child of x, such that $Br(q) = x$. As q is a *minimal* JEP, x is not a dead-end leaf. □

Prop. 3 is a consequence of Lemmas 2 and 3:

Proposition 3 (One-To-One correspondence). *Let (T,r) be a ToMJEPs. There is a one-to-one correspondence between the set of the candidate leaves x such that $Br(x)$ is a supported pattern and the set of the minimal JEPs.*

Prop. 3 ensures that we can generate the minimal JEPs by simply performing a depth first traversal of a ToMJEPs and output the candidate leaves such that $Br(x)$ is a supported pattern. Note that it is not necessary to compute and store the entire ToMJEPs. A depth first traversal only requires to store the path from the root to the node currently visited.

The sketch of implementation provided in Section 4.1 gives information about the calculation of the extent, the calculation of the essential JEPs and the top-k minimal JEPs that are inferred from a ToMJEPs.

4   Experimental Evaluation

This section provides and comments results from a study conducted on 13 benchmark datasets. We investigate the computation of the JEPs according to running time, setting a minimum frequency threshold. It also indicates the reliability of correlation between a JEP and a class. In the following, a *JEP* denominates a supported pattern with respect to any class.

4.1   Material and Methods

The datasets. The study is conducted on 13 usual datasets described in Table 3. All the datasets are available from the UCI Machine Learning repository [10]. We selected these datasets because they have been used, at least once, in an experimental assessment of JEPs [3,4,16]. Non binary attributes were converted into a binary valued format by applying a sanctioned method [6,11] which is available at Frans Coenen's website[1].

---

[1] http://cgi.csc.liv.ac.uk/~frans/KDD/Software/LUCS-KDD-DN/exmpleDNnotes.html

Table 3. The datasets and their characteristics

Datasets	Objects	Attributes	Classes	Datasets	Objects	Attributes	Classes
breast	699	20	2	mushroom	8124	90	2
congres	435	34	2	pima	768	38	2
ecoli	336	34	8	tic-tac-toe	958	29	2
glass	214	48	7	waveform	5000	101	3
heart	303	52	5	wine	178	68	3
hepatitis	155	56	2	zoo	101	42	7
iris	150	19	3				

Implementation. Our algorithm partially explores a ToMJEPs in a depth first manner, it outputs every candidate leaf whose associated pattern is a supported one. We implemented two solutions to ensure to only output supported patterns. The first one, called *post-filtering* solution, generates all the candidate leaves and then checks whether their extent is empty or not. The second one, named *maintaining_extent* solution, integrates the computation of the extents with the calculation of the child of an internal node of a ToMPJEPs. It enables to backtrack as soon as the extent is empty.

Moreover, when a minimum frequency threshold is provided, the *maintaining_extent* solution is straightforwardly adapted to improve the computing of the essential JEPs. Indeed, the frequency of candidate essential JEPs [4] is directly derived from the cardinality of the extent. For the same reason, this solution also enables to compute the top-k minimal JEPs [16] when a value for k is provided. Moreover, the pruning strategy becomes more and more efficient during the mining step because the minimal frequency threshold to belong to the top-k minimal JEP only increases during the mining.

Protocol. In order to compute *all* the minimal JEPs whatever the positive class is, we successively consider each class (of the dataset) as the positive class while the union of the others classes constitutes the negative class. Computations were performed on a server using Ubuntu 12.04 with 2 processors Intel Xeon 2.80 GHz and 512 gigabytes of RAM.

4.2   Results and Discussions

Computation of the Minimal JEPs. We computed all the minimal JEPs on the 13 selected datasets, by using the *post-filtering* and *maintaining_extent* solutions. Moreover, essential JEPs are computed with two minimum frequency thresholds (1% and 5%), and the top-k JEPs with $k = 10$ and $k = 20$. Table 4 gives the cardinalities of the sets of the minimal JEPs and the running times. For computing all the minimal JEPs, the *maintaining_extent* solution always operates faster than the *post-filtering* solution, by a factor varying from 1.6 to 3. By observing the results for the essential JEPs and top-k minimal JEPs, one notes that the running time decreases significantly when a minimal threshold is set for the cardinality of the extent. The use of a frequency constraint related to the cardinality of the extent is efficient, obviously there is the risk to miss interesting patterns.

Minimal JEPs as Rules to Express Correlations. A JEP expresses a correlation between the occurrence of a pattern and one class of objects. This part provides

Table 4. Computation of minimal JEP including negation of attributes

		All minimal JEPs		Essential JEPs		Top-K minimal JEPs	
		post-filtering	*maintaining_extent*	1%	5%	10	20
Datasets	Min.JEPs	Time	Time	Time	Time	Time	Time
iris	40	70.564 ms	24.348 ms	14.316 ms	9.783 ms	13.043 ms	17.303 ms
breast	38	924.998 ms	347.572 ms	190.432 ms	79.198 ms	95.212 ms	119.213 ms
ecoli	200	842.345 ms	353.734 ms	173.658 ms	98.982 ms	134.314 ms	136.712 ms
zoo	3323	1339.008 ms	579.208 ms	232.023 ms	101.032 ms	67.178 ms	79.032 ms
pima	1443	7.323 s	3.093 s	895.053 ms	532.123 ms	1.009 s	1.694 s
glass	59747	27.172 s	12.418 s	6.927 s	3.241 s	1.439 s	2.081 s
congres	55449	89.396 s	38.077 s	19.145 s	8.380 s	3.107 s	4.929 s
hepatitis	410404	123.520 s	53.706 s	25.576 s	14.419 s	2.978 s	3.097 s
heart	122865	3.351 mn	1.194 mn	29.560 s	15.201 s	9.432 s	8.921 s
tic-tac-toe	109949	5.664 mn	2.797 mn	55.860 s	13.182 s	4.541 s	6.325 s
wine	1353996	200.321 mn	99.366 mn	58.053 mn	36.324 mn	8.342 mn	11.821 mn
mushroom	17345228	673.563 mn	423.116 mn	192.743 mn	101.765 mn	27.545 mn	50.325 mn
waveform	23895434	1845.431 mn	954.190 mn	421.813 mn	238.425 mn	47.342 mn	59.175 mn

experimental results to assess the interest of such rules: do these rules cover a large part of the objects? Are they confident enough? We have also performed experiments to evaluate the usefulness of the explicit description of the absent attributes by adding their negations.

The study has been conducted by using a leave-one-out framework: every object has been successively discarded from the dataset. For every object g, the minimal JEPs have been extracted by considering $\mathcal{G} \setminus \{g\}$ as the dataset and the resulting rules have been applied on g.

Table 5 provides results obtained by applying minimal JEPs, essential JEPs, or top-k minimal JEPs as association rules. *No Negated attributes* designates the descriptions which do not explicitly take into account the absence of attributes whereas *With Negated attributes* points the descriptions that explicitly consider the absence of attributes. The column *Cov* denotes the *coverage* of the set of association rules (the part of the objects for which at least one association rule has applied). The column *Con* refers to the *average confidence* (i.e., the ratio between the number of correct applications of the rules over the whole number of applications of the rules). For example, if we consider the dataset named *breast*, whith the *No Negated attributes* description, 47.78% of the objects contain at least one minimal JEP, this coverage raises to 49.33% of the objects when the descriptions *With Negated attributes* are accounted. With the same dataset, by using the *No Negated attributes* description, 98.19% of the rules resulting from a minimal JEP apply on an object of the proper class ; this average confidence slightly decreases to 96.13% when the *No Negated attributes* description is used.

First of all, the JEPs often apply on a large portion of the objects: for 7 datasets among the 13 datasets, more than 80% of the objects contain at least one JEP. Note that this coverage increases when the description turns from *No Negated attributes* to *With Negated attributes*, up to 8% for the *hepatitis* dataset.

The average confidences indicate that minimal JEPs often point a reliable association between a pattern and a class, even when no frequency constraint is set.

Table 5. Evaluation of minimal JEPs as rules to express correlations

Datasets	No negated attributes										With Negated attributes									
	All Min. JEPs		eJEP 1%		eJEP 5%		top-k 10		top-k 20		All Min. JEPs		eJEP 1%		eJEP 5%		top-k 10		top-k 20	
	Cov	Con	Cov	Con	Cov	Con	Cov	Con	Cov	Con	Cov	Con	Cov	Con	Cov	Con	Cov	Con	Cov	Con
breast	47.78	98.19	45.2	98.32	44.81	99.10	39.16	98.5	43.29	97.01	49.33	96.13	47.42	97.49	45.25	98.51	38.21	97.92	40.21	95.85
congres	99.08	89.7	91.98	80.37	91.98	92.31	93.72	78.53	91.67	78.32	99.00	88.10	92.28	90.31	85.42	92.13	60.43	95.23	70.14	94.23
ecoli	34.52	63.79	30.71	66.38	71.63	20.32	75.73	25.92	68.97	37.32	37.32	71.23	32.12	73.98	26.43	74.21	25.92	75.43	27.13	74.01
glass	90.18	62.17	70.38	77.90	78.12	70.45	75.27	70.29	68.43	94.13	64.15	87.12	57.82	80.23	68.45	75.43	70.29	80.32	70.47	74.78
heart	97.02	58.5	61.90	80.32	64.87	70.32	71.43	74.23	69.25	98.02	56.42	86.32	57.82	82.55	60.43	61.21	75.43	80.32	67.01	78.43
hepatitis	82.42	78.39	81.30	69.32	84.01	69.91	85.90	72.72	83.94	91.03	75.38	73.23	79.17	70.31	81.83	62.62	80.72	70.39	77.38	
iris	88.67	90.98	92.23	75.43	94.61	71.60	96.10	74.38	72.72	83.94	89.40	85.32	95.39	83.21	96.37	78.62	97.23	82.76	96.37	
mushroom	70.15	77.12	79.44	47.53	88.43	58.18	81.38	60.06	79.92	94.67	78.43	78.62	85.35	83.21	82.54	63.74	80.32	66.34	77.98	
pima	17.05	54.2	14.47	13.54	60.01	13.01	62.43	14.43	58.91	18.54	61.20	82.34	16.03	63.01	57.92	80.01	62.91	80.32	66.34	59.54
tic-tac-toe	81.62	86.57	58.12	25.74	89.01	60.43	80.21	62.32	62.32	78.32	82.12	82.34	70.64	83.21	65.53	84.24	72.31	69.43	75.32	67.30
waveform	45.28	72.19	32.93	74.91	75.10	33.19	73.10	35.92	35.92	53.42	74.76	78.43	40.67	78.01	39.52	82.94	37.91	79.63	40.47	74.78
wine	72.91	67.43	56.72	41.73	83.01	51.84	70.84	53.42	62.81	62.81	73.54	65.89	61.92	77.81	59.02	69.67	55.81	69.51	61.90	67.39
zoo	89.11	92.22	72.19	94.44	95.10	68.92	73.78	73.24	72.01	87.32	91.43	91.43	79.32	92.47	77.19	93.93	60.33	92.48	67.39	90.03

By paying the price of a lower coverage, setting a minimum frequency threshold – as it is done for the essential JEPs or, indirectly, for the top-k minimal JEPs – causes an increase of the average confidence, depending on the dataset. The average confidence levels reached by the two descriptions, *No Negated attributes* and *With Negated attributes*, are very comparable.

As a conclusion, both description families, *With Negated attributes* and *No Negated attributes*, lead to minimal JEPs reaching a similar level of confidence. However, the minimal JEPs extracted with the *With Negated attributes* descriptions cover a wider range of objects than the minimal JEPs extracted with the *No Negated attributes* descriptions, but with a longer running time.

5 Related Work

Since the key paper of Dong and Li [2], subsequent research has focused on mining emerging patterns and contrast sets. However, there are very few attempts to tackle the discovery of minimal JEPs. Fan and Ramamohanarao have proposed an algorithm extracting the minimal JEPs whose frequency of occurrence is greater than a given threshold, such JEPs are called *essential JEPs* [4]. Terlecki and Walczak have designed a computational method based on a CP-Tree to get the k most supported minimal JEPs, named *top-k minimal JEPs* [16]. These methods require either a frequency threshold or a given number of expected patterns. On the contrary, our method is free from these parameters and computes the whole set of minimal JEPs. Terlecki and Walczak [15] have experimentally shown that taking into account the absence of attributes may provide interesting pieces of knowledge to build more accurate classifiers. We have dealt with this issue since our method extracts minimal JEPs including the negation of the attributes which are absent.

In addition, JEPs can be associated to *version space* [13]. A *version space* gathers the descriptions that match all objects of one class and no object of the other class. Therefore a version space corresponds to the JEPs that match all objects of one class. JEPs are also related to the concept of *disjunctive version space* since a JEP corresponds to all descriptions of objects that match at least one object of one class and no object for the other classes. In Formal Concept Analysis, a JEP is also named "hypothesis" [5] (a *hypothesis* brings together the descriptions of objects that match at least one object in one class and no object in others).

6 Conclusion

We have introduced an efficient method to extract the whole set of minimal JEPs. To the best of our knowledge, it is the first method which does not require either a frequency threshold or a given number of expected patterns. Our method is also able to straightforwardly extract the essential JEPs and the k most supported minimal JEPs. Moreover it enables the integration of negated attributes that can be precious for a classification purpose. We have experimentally analyzed the computation of these JEPs, together with the reliability of the correlations between a JEP and a class.

The structure of tree of the minimal JEPs constitutes a framework for designing and expressing algorithms to compute the minimal JEPs from a dataset. In order to speed up the calculation, this framework will be used to seek for efficient orderings on the attributes or on the objects. Another direction is to produce patterns correlated to one class to a lesser extent and mine emerging patterns with high growth-rate values. Beyond this work, we plan to use minimal JEPs in the design of an advanced rule-based classifier.

References

1. Chen, X., Chen, J.: Emerging patterns and classification algorithms for dna sequence. JSW **6**(6), 985–992 (2011)
2. Dong, G., Li, J.: Efficient mining of emerging patterns: discovering trends and differences. In: KDD, pp. 43–52 (1999)
3. Dong, G., Li, J.: Mining border descriptions of emerging patterns from dataset pairs. Knowl. Inf. Syst. **8**(2), 178–202 (2005)
4. Fan, H., Ramamohanarao, K.: An efficient single-scan algorithm for mining essential jumping emerging patterns for classification. In: Chen, M.-S., Liu, B., Yu, P.S. (eds.) PAKDD 2002. LNCS (LNAI), vol. 2336, pp. 456–462. Springer, Heidelberg (2002)
5. Ganter, B., Kuznetsov, S.O.: Hypotheses and version spaces. In: Ganter, B., de Moor, A., Lex, W. (eds.) ICCS 2003. LNCS (LNAI), vol. 2746, pp. 83–95. Springer, Heidelberg (2003)
6. Kerber, R.: Chimerge: discretization of numeric attributes. In: Swartout, W.R. (ed.) AAAI, pp. 123–128. AAAI Press / The MIT Press (1992)
7. Kobyliński, Ł., Walczak, K.: Spatial emerging patterns for scene classification. In: Rutkowski, L., Scherer, R., Tadeusiewicz, R., Zadeh, L.A., Zurada, J.M. (eds.) ICAISC 2010, Part I. LNCS (LNAI), vol. 6113, pp. 515–522. Springer, Heidelberg (2010)
8. Kobylinski, L., Walczak, K.: Efficient mining of jumping emerging patterns with occurrence counts for classification. T. Rough Sets **13**, 73–88 (2011)
9. Li, J., Dong, G., Ramamohanarao, K.: Making use of the most expressive jumping emerging patterns for classification. In: Terano, T., Liu, H., Chen, A.L.P. (eds.) PAKDD 2000. LNCS (LNAI), vol. 1805, pp. 220–232. Springer, Heidelberg (2000)
10. Lichman, M.: UCI machine learning repository (2013). http://archive.ics.uci.edu/ml
11. Liu, H., Setiono, R.: Feature selection via discretization. IEEE Trans. Knowl. Data Eng. **9**(4), 642–645 (1997)
12. Lozano, S., Poezevara, G., Halm-Lemeille, M.P., Lescot-Fontaine, E., Lepailleur, A., Bissell-Siders, R., Cremilleux, B., Rault, S., Cuissart, B., Bureau, R.: Introduction of Jumping Fragments in Combination with QSARs for the Assessment of Classification in Ecotoxicology. J. Chem. Inf. Model. **50**(8), 1330–1339 (2010)
13. Mitchell, T.M.: Generalization as search. Artif. Intell. **18**(2), 203–226 (1982)
14. Novak, P.K., Lavrac, N., Webb, G.I.: Supervised descriptive rule discovery: A unifying survey of contrast set, emerging pattern and subgroup mining. Journal of Machine Learning Research **10**, 377–403 (2009)
15. Terlecki, P., Walczak, K.: Jumping emerging patterns with negation in transaction databases classification and discovery. Inf. Sci. **177**(24), 5675–5690 (2007)
16. Terlecki, P., Walczak, K.: Efficient discovery of top-K minimal jumping emerging patterns. In: Chan, C.-C., Grzymala-Busse, J.W., Ziarko, W.P. (eds.) RSCTC 2008. LNCS (LNAI), vol. 5306, pp. 438–447. Springer, Heidelberg (2008)
17. Wrobel, S.: An algorithm for multi-relational discovery of subgroups. In: Komorowski, J., Żytkow, J.M. (eds.) PKDD 1997. LNCS, vol. 1263, pp. 78–87. Springer, Heidelberg (1997)

Rank Matrix Factorisation

Thanh Le Van[1]([✉]), Matthijs van Leeuwen[1],
Siegfried Nijssen[1,2], and Luc De Raedt[1]

[1] Department of Computer Science, KU Leuven, Leuven, Belgium
{thanh.levan,matthijs.vanleeuwen,siegfried.nijssen,luc.deRaedt}@cs.kuleuven.be
[2] Leiden Institute for Advanced Computer Science,
Universiteit Leiden, Leiden, The Netherlands

Abstract. We introduce the problem of *rank matrix factorisation* (RMF). That is, we consider the decomposition of a rank matrix, in which each row is a (partial or complete) ranking of all columns. Rank matrices naturally appear in many applications of interest, such as sports competitions. Summarising such a rank matrix by two smaller matrices, in which one contains partial rankings that can be interpreted as local patterns, is therefore an important problem.

After introducing the general problem, we consider a specific instance called Sparse RMF, in which we enforce the rank profiles to be sparse, i.e., to contain many zeroes. We propose a greedy algorithm for this problem based on integer linear programming. Experiments on both synthetic and real data demonstrate the potential of rank matrix factorisation.

Keywords: Matrix factorisation · Rank data · Integer linear programming

1 Introduction

In this paper, we study a specific type of matrix called *rank matrices*, in which each row is a (partial or complete) ranking of all columns. This type of data naturally occurs in many situations of interest. Consider, for instance, sailing competitions where the columns could be sailors and each row would correspond to a race, or consider a business context, where the columns could be companies and the rows specify the rank of their quotation for a particular service. Rankings are also a natural abstraction of numeric data, which often arises in practice and may be noisy or imprecise. Especially when the rows are incomparable, e.g., when they contain measurements on different scales, transforming the data to rankings may result in a more informative representation.

Given a rank matrix, we are interested in discovering a set of rankings that repeatedly occur in the data. Such sets of rankings can be used to succinctly summarise the given rank matrix. With this aim, we introduce the problem of *rank matrix factorisation* (RMF). That is, we consider the decomposition of a rank matrix into two smaller matrices.

T. Cao et al. (Eds.): PAKDD 2015, Part I, LNAI 9077, pp. 734–746, 2015.
DOI: 10.1007/978-3-319-18038-0_57

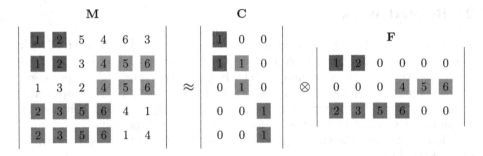

Fig. 1. Rank matrix factorisation toy example. Rank matrix **M** is approximated by the product of indicator matrix **C** and sparse profile matrix **F** ($k = 3$).

To illustrate the problem of rank matrix factorisation, let us consider the toy example in Figure 1. It depicts a rank matrix that is approximated by the product of two smaller matrices. Rank matrix **M** consists of five rows and six columns. Assuming no ties and complete rankings, each row contains each of the numbers one to six exactly once. Now, the task is to decompose a $n \times m$ rank matrix **M** into a $n \times k$ matrix **C** and a $k \times m$ matrix **F**, where **C** is a binary indicator matrix, **F** consists of *rank profiles*, and k is a user-specified parameter. Intuitively the rank profiles in **F** are (partial) rankings and can be interpreted as *local patterns*. For example, together **C** and **F** show that the first two columns are ranked first and second in the first row.

In this paper we focus on a specific rank matrix factorisation problem: the problem of finding sparse rank profiles where rows of **F** contain zeroes. This allows us to discover recurrent structure that occurs in the rankings of **M**, and not to focus on any noise that may be present. Within this setting we do not necessarily aim at finding a factorisation that approximates the original matrix as closely as possible; the *reconstructed rank matrix* $\mathbf{C} \otimes \mathbf{F}$ may deviate from **M**, as long as its overall structure is captured. Hence, here we focus on one specific of choices within the RMF framework; we would like to stress that within the generic framework many other choices are possible. The same can be said with regard to the choices made for, e.g., rank profile aggregation and quantification of the reconstruction error. RMF is a general framework with numerous possibilities, and we propose and solve a first instance to demonstrate its potential.

The key contributions of our paper are 1) the introduction of the problem of rank matrix factorisation (RMF), 2) the introduction of a scoring function and an algorithm, based on integer linear programming, for Sparse RMF, an instance of rank matrix factorisation, and 3) an empirical evaluation on synthetic and real-life datasets that demonstrates the potential of RMF. It is shown that rank matrix factorisations can provide useful insights by revealing the rankings that underlie the data.

2 Related Work

To the best of our knowledge, we are the first to investigate the problem of rank matrix factorisation. Mining rank data, although a very new topic, has attracted some attention by the community lately. In our earlier work [1] we proposed to mine *ranked tiles*, e.g., rectangles with high ranks, and we will empirically compare to them in the experiments. Furthermore, Henzgen and Hüllermeier [2] proposed to mine frequent subrankings. The latter approach aims to mine individual patterns, whereas we aim to find a *set of patterns* that together covers most of the data.

RMF is clearly related to matrix factorisation approaches such as NMF [3,4], BMF [5,6], and positive integer matrix factorisation (PIMF) [7]. NMF, PIMF, and RMF have in common that the values in the factorisation are constrained to be positive, but are quite different otherwise. RMF specifically targets rank data, which requires integer values, making the results easier to interpret, a different scoring function, and a different algebra. RMF considers rank matrices instead of Boolean matrices and is therefore clearly different from BMF.

3 Rank Matrix Factorisation

In this section we formally define rank matrices and introduce the rank matrix factorisation problem that we consider.

Definition 1 (Rank matrix). *Let M be a matrix consisting of m rows and n columns. Let $\mathcal{R} = \{1, ..., m\}$, $\mathcal{C} = \{1, ..., n\}$ be index sets for rows and for columns respectively. The matrix M is a rank matrix iff:*

$$\forall r \in \mathcal{R} : \cup_{c \in \mathcal{C}} M_{r,c} \subseteq \sigma, \tag{1}$$

where $\sigma = \{1, 2, ..., n\} \cup \{0\}$.

In our setting, columns are items or products that need to be ranked; rows are rankings of items. Here, the rank value 0 has a special meaning. It denotes *unknown* rankings. For example, in rating datasets, it might happen that there are items that are not rated. Such items will have rank value 0.

Given a rank matrix, we would like to find a short description of the rank matrix in terms of a fixed number of rank profiles, or patterns, consisting of partial rankings. We formalise this problem as a matrix factorisation problem.

Problem 1 (Rank matrix factorisation). *Given a rank matrix $M \in \sigma^{m \times n}$ and an integer k, find a matrix $C^* \in \{0,1\}^{m \times k}$ and a matrix $F^* \in \sigma^{k \times n}$ such that:*

$$(C^*, F^*) \equiv \underset{C,F}{\text{argmax}} \ d(M, C \otimes F). \tag{2}$$

where $d(,)$ is a scoring function that measures how similar the rankings in the two matrices are, and $\otimes$ is an operator that creates a data matrix based on two factor matrices. Rows $F_{i,:}$ of matrix F indicate partial rankings, columns $C_{:,i}$ of matrix C indicate in which rows a partial ranking appears.

Within our generic problem statement we first need to specify the operator $\otimes$. If multiple patterns are present in one row, this operator essentially needs to combine the different partial rankings into a single ranking. This problem is well-known in the literature as the problem of *rank aggregation*. In this first study, we use a very simple aggregation operator, namely, we use normal matrix multiplication to combine the matrices. More complex types of aggregation are left for future work.

An important drawback of normal matrix multiplication is that the product $\mathbf{CF}$ is not necessarily a rank matrix even if $\mathbf{C}$ is binary and $\mathbf{F}$ contains partial rankings. We address this here by restricting the set of acceptable matrices to those for which $(\mathbf{CF})_{ij} \leq n$ for all $i \in \mathcal{R}$ and $j \in \mathcal{C}$.

Next, we need to define the scoring function d. In the definition of this function we first need the concept of a *cover* for a rank matrix factorisation. The cover of a factorisation is the set of cells in the reconstructed matrix where at least one pattern occurs, i.e., where the reconstructed matrix is non-zero.

Definition 2 (Ranked factorisation cover)

$$cover(\mathbf{C}, \mathbf{F}) \equiv \{(i,j) | i \in \mathcal{R}, j \in \mathcal{C}, (\mathbf{CF})_{i,j} \neq 0\}. \tag{3}$$

Coverage is the size of the cover, i.e., $coverage(\mathbf{C}, \mathbf{F}) = |cover(\mathbf{C}, \mathbf{F})|$.

To support the aim of mining patterns in rank matrices, the scoring function $d(,)$ in Equation 2 needs to be designed in such a way that it: 1) rewards patterns that have a high coverage, 2) penalises patterns that make a large error within the cover of the factorisation.

To penalise patterns that make a large error, we define an error term that quantifies the disagreements between the reconstructed and the original rank matrix. We first define notation for the data matrix identified by the cover of a factorization.

Definition 3 (Ranked data cover). *The ranked data cover matrix* $\mathbf{U}(\mathbf{M}, \mathbf{C}, \mathbf{F})$ *is a matrix with cells* u_{ij}, *where:*

$$u_{ij} = \begin{cases} \mathbf{M}_{i,j} & if\ (i,j) \in coverage(\mathbf{C}, \mathbf{F}) \\ 0 & otherwise. \end{cases} \tag{4}$$

Now the ranked factorisation error is defined as follows.

Definition 4 (Ranked factorisation error)

$$error(\mathbf{M}, \mathbf{C}, \mathbf{F}) = \sum_{i=1}^{m} d(\mathbf{U}(\mathbf{M}, \mathbf{C}, \mathbf{F})_{i,:}, \mathbf{C}_{i,:}\mathbf{F}) \tag{5}$$

Here, $d(\cdot, \cdot)$ *is a function that measures the disagreement between two rankings over the same items.*

Hence, the ranked factorisation error is the total of rank disagreements between the reconstructed rank matrix and the true ranks in the original rank matrix. The score is calculated row by row.

Many scoring functions can be used to measure the disagreement between rows, for instance, Kendall's tau or Spearman's Footrule (see [8] for a survey). For an efficient computation, we choose the Footrule scoring function.

Definition 5 (Footrule scoring function). *Given two rank vectors, $u = (u_1,\ldots,u_n)$ and $v = (v_1,\ldots,v_n)$, the Footrule scoring function is defined as $d_F(u,v) = \sum_{i=1}^n |u_i - v_i|$.*

Having defined the ranked factorisation coverage and ranked factorisation error, we now can completely define the Sparse Rank Matrix Factorisation (Sparse RMF) problem as solving the following maximisation problem:

$$(\mathbf{C}^*, \mathbf{F}^*) \equiv \underset{\mathbf{C},\mathbf{F}}{\mathrm{argmax}}\ d(\mathbf{M}, \mathbf{CF}) \tag{6}$$

$$\equiv \underset{\mathbf{C},\mathbf{F}}{\mathrm{argmax}}\ \alpha * coverage(\mathbf{C}, \mathbf{F}) - error(\mathbf{M}, \mathbf{C}, \mathbf{F}) \tag{7}$$

$$= \underset{\mathbf{C},\mathbf{F}}{\mathrm{argmax}}\ \sum_{i=1}^m \sum_{j=1}^n (\alpha[(i,j) \in coverage(\mathbf{C}, \mathbf{F})] -$$

$$|\mathbf{U}(\mathbf{M}, \mathbf{C}, \mathbf{F})_{ij} - \sum_{t=1}^k \mathbf{C}_{i,t}\mathbf{F}_{t,j}|) \tag{8}$$

where α is a threshold and [.] are the Iverson brackets.

Note that in this scoring function, for each cell we have a positive term if the error is smaller than α; we have a negative term if the error is larger than α. In practice, we often use a relative instead of an absolute threshold. We denote such a threshold as a percentage, i.e., $\alpha = a\%$ implies $\alpha = a\% \times n$.

4   Sparse RMF Using Integer Linear Programming

We propose a greedy algorithm that uses integer linear programming (ILP). First, we present two theorems that can be used to calculate the ranked factorisation coverage and ranked factorisation error. Then, we present the algorithm.

Theorem 1. *Let CF be a decomposition of a rank matrix M. Let $A \in \{0,1\}^{m \times n}$ satisfy the following two properties:*

$$A_{i,j} \leq \sum_{t=1}^k C_{i,t}F_{t,j} \tag{9}$$

$$nA_{i,j} \geq \sum_{t=1}^k C_{i,t}F_{t,i} \tag{10}$$

then

$$\mathbf{A}_{ij} = 1 \leftrightarrow (i,j) \in cover(C,F) \tag{11}$$

Theorem 2. *Let A be a binary matrix that satisfies Theorem 1, then*

$$error(M, C, F) = \sum_{i=1}^{m} \sum_{j=1}^{n} |M_{i,j} A_{i,j} - \sum_{t=1}^{k} C_{i,t} F_{t,j}| \tag{12}$$

Given a binary matrix A satisfying Theorem 1, the ranked factorisation in Equation 8 can be formulated as:

$$\arg\max_{C,F,Y} \sum_{i=1}^{m} \sum_{j=1}^{n} \alpha A_{i,j} - Y_{i,j} \tag{13}$$

subject to

$$M_{i,j} - \sum_{t=1}^{k} C_{i,t} F_{t,j} \leq Y_{i,j} \qquad \text{for } i = 1, \ldots, m, \ j = 1, \ldots, n \tag{14}$$

$$-M_{i,j} + \sum_{t=1}^{k} C_{i,t} F_{t,j} \leq Y_{i,j} \qquad \text{for } i = 1, \ldots, m, \ j = 1, \ldots, n \tag{15}$$

$$A_{i,j} \leq \sum_{t=1}^{k} C_{i,t} F_{t,j} \qquad \text{for } i = 1, \ldots, m, \ j = 1, \ldots, n \tag{16}$$

$$n A_{i,j} \geq \sum_{t=1}^{k} C_{i,t} F_{t,i} \qquad \text{for } i = 1, \ldots, m, \ j = 1, \ldots, n \tag{17}$$

$$\sum_{t=1}^{k} C_{i,t} F_{t,j} \leq n \qquad \text{for } i = 1, \ldots, m, \ j = 1, \ldots, n \tag{18}$$

$$C_{i,t} \in \{0, 1\} \qquad \text{for } i = 1, \ldots, m, \ t = 1, \ldots, k \tag{19}$$

$$F_{t,j} \in \sigma \qquad \text{for } j = 1, \ldots, n, \ t = 1, \ldots, k \tag{20}$$

where $Y_{i,j}$ is the upper bound of $|M_{i,j} - \sum_{t=1}^{k} C_{i,t} F_{t,j}|, i = 1, \ldots, m, \ j = 1, \ldots, n$.

Inequalities (14) and (15) are introduced to remove the absolute operator of the summations in Equation (12). Inequalities (16) and (17) are due to Theorem 1. Inequality (18) ensures that the reconstructed matrix is a rank matrix.

Note that the newly introduced optimisation problem in (13) - (20) is an ILP problem if either C or F is known. This makes it possible to apply an EM-style algorithm as shown in Algorithm 1, in which the matrix F is optimised given matrix C, and matrix C is optimised given matrix F, and we repeat the iterative optimisation till the optimal score cannot be improved any more.

To avoid local maxima, we need to initialise the iterative process in a reasonable way, i.e., smarter than random. The solution we choose is to initialise the matrix C using the well-known K-means algorithm. To compute the similarities of rank vectors in K-means, we use the Footrule scoring function. The K-means algorithm clusters the rows in k groups, which can be used to initialise the k

columns of **C**. Note that this results in initially disjoint patterns, in terms of their covers, but the iterative optimisation approach may introduce overlap.

We implemented the algorithm in OscaR[1], which is an open source Scala toolkit for solving Operations Research problems. OscaR supports a modelling language for ILP. We configured OscaR to use Gurobi[2] as the back-end solver. Source code can be downloaded from our website, http://dtai.cs.kuleuven.be/CP4IM/RMF.

Algorithm 1. *Sparse RMF algorithm*

Require: Rank matrix **M**, integer k, threshold α
Ensure: Factorisation **C**, **F**
 1: Initialise **C** using K-means algorithm
 2: **while** not converged **do**
 3:     **F** ← Optimise (13) - (20) given **C**
 4:     **C** ← Optimise (13) - (20) given **F**
 5: **end while**

---

5   Experiments on Synthetic Datasets

The goal of Sparse RMF is to find a set of rank profiles (local patterns), which can be used to summarise a given rank matrix. Alternative methods for summarising matrices are bi-clustering [9] and ranked tiling [1]. While ranked tiling and Sparse RMF work on ranked data, bi-clustering algorithms are mostly applied to numeric data. Hence, to compare to all of these algorithms, we first generate continuous data and then convert them to ranked data as in [1]. The main idea is to benchmark the performance of the considered algorithms in terms of recovering implanted tiles in the synthetic data. Different from ranked tiling [1], where implanted tiles only have high average values, we now implant tiles that have both low and high average values.

Data Generation. We use the generative model that we introduced in [1] to generate continuous data. First, we generate background data whose values are sampled from normal distributions having mid-ranged mean values. Second, we implant a number of constant-row tiles whose values are sampled from normal distributions having low/high mean values. Finally, we perform a complete ranking of columns in every row to obtain a rank matrix.

Formally, background data is generated by this generative model:

$$\forall r \in \mathcal{R}, \forall c \in \mathcal{C}, \ \mathbf{M}_{r,c} \sim \begin{cases} N(\mu_r^1, 1) & \text{if } x_1 = 1 \\ N(\mu_r^2, 1) & \text{if } x_2 = 1 \\ N(\mu_r^3, 1) & \text{if } x_3 = 1 \end{cases} \tag{21}$$

---

[1] https://bitbucket.org/oscarlib/oscar/wiki/Home
[2] http://www.gurobi.com/

where $\mu_r^1 \sim U(3,5)$, $\mu_r^2 \sim U(-5,-3), \mu_r^3 \sim U(-3,3)$; $x = (x_1, x_2, x_3), x_i \in \{0,1\}, \sum_i x_i = 1, x$ has mass probability function $\mu = (p, p, 1 - 2p), 0 \le p \le 0.5$.

A constant-row tile $\mathbf{M}_{R,C}$ having high average values is generated as:

$$\forall r \in R, \quad \mu_r \sim U(3,5) \tag{22}$$

$$\forall r \in R, \forall c \in C, \quad \mathbf{M}_{r,c} \sim N(\mu_r, 1) \tag{23}$$

Tiles having low average values are generated in a similar way. However their mean values are sampled from a different uniform distribution: $U(-5, -3)$.

Setup. We generate four 500 rows × 100 columns datasets for different p, i.e., $p \in \{0.05, 0.10, 0.15, 0.20\}$. In each dataset, we implant seven constant-row tiles. Three tiles have low average values, the other four have high average values.

We evaluate the ability of the algorithms to recover the implanted set of tiles. We do this by measuring *recall* and *precision*, using the implanted tiles as ground truth. Overall performance is quantified by the *F1* measure, which is the average of the two scores.

Varying the Parameters. We varied the parameters k and α in Equation 8 and then applied the Sparse RMF algorithm on the four synthetic datasets. For each parameter combination, the algorithm was executed ten times and the result maximising the score was used. This is to get rid of effects that are due to differences in the initialization based on K-means.

Precision and recall are calculated based on the union of the coverage area, which has non-zero values, by the k components $\mathbf{C}(:, i)$ $\mathbf{F}(i, :), i = 1 \ldots k$. The average performance of the algorithm on the four datasets is summarised in Figure 2. The figure shows that the Sparse RMF can recover implanted tiles and

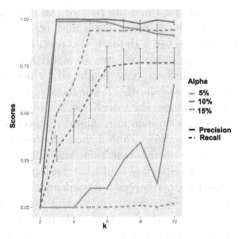

Fig. 2. Average precision and recall on the four synthetic datasets with varying parameters α and k

remove noise outside when $k = 5$ and $\alpha \sim 15\%$. When α is too small, i.e., 5%, the algorithm cannot recover the tiles. In general, the algorithm has high performances when k is large. This matches our expectation though, since a higher α results in higher tolerance to noise and a larger k results in more patterns and hence a more detailed description of the data.

Comparison to Other Algorithms. In this experiment, we compare our approach to ranked tiling [1] and several bi-clustering algorithms. These include CC

Table 1. Comparison to ranked tiling and bi-clustering. Precision, recall and F1 quantify how accurately the methods recover the seven implanted tiles

Algorithm	Data type	Pattern type	Prec.	Recall	F1
Sparse RMF	Ranks	Rank profile	99%	94%	96%
Ranked tiling [1]	Ranks	Ranked tile	37%	41%	39%
CoreNode [16]	Numerical	Coherent values bicluster	30%	8%	19%
FABIA [13]	Numerical	Coherent values bicluster	99%	51%	75%
Plaid [12]	Numerical	Coherent values bicluster	91%	46%	67%
SAMBA [14]	Numerical	Coherent evolution bicluster	52%	9%	31%
ISA [15]	Numerical	Coherent values bicluster	43%	17%	30%
CC [10]	Numerical	Coherent values bicluster	7%	5%	6%
Spectral [11]	Numerical	Coherent values bicluster	-	-	-

[10], Spectral [11], Plaid [12], FABIA[3] [13], SAMBA[4] [14]and ISA[5] [15]. CC, Spectral and Plaid are part of the R biclust[6] package.

Since large noise levels may conversely affect the performance of the algorithms, we use a dataset also used for the previous experiments, with $p = 0.05$ (low noise level). We ran all algorithms on this dataset and took the first seven tiles/bi-clusters they produced, which have the highest scores (SAMBA) or largest sizes (all other). For most of the benchmarked algorithms, we used their default values. For CoreNode, we use $msr = 1.0$ and $overlap = 0.5$. For ISA, we applied its built-in normalised method before running the algorithm itself.

The results in Table 1 show that our algorithm achieves much higher precision and recall on this task than any of the ranked tiling and bi-clustering methods. Note that Spectral could not find any patterns. Ranked tiling only finds highly ranked tiles, whereas our rank matrix factorisation is more general and allows to capture any recurrent partial rankings in the rank matrix. Some of the bi-clustering methods attain quite high precision, e.g., FABIA and Plaid, but their recall is much lower than for Sparse RMF. The reason is that the synthetic data contains incomparable rows, with values on different scales. These results confirm that converting such data to rank matrices is likely to lead to better results.

6   Real World Experiments

This section presents results on three real world datasets: 1) Eurovision Song Contest voting data, 2) Sushi preferences, and 3) NBA basketball team rankings.

We previously collected the European Song Contest (ESC) dataset [1]. This dataset contains aggregated voting scores that participating countries gave to

---

[3] http://www.bioinf.jku.at/software/fabia/fabia.html

[4] http://acgt.cs.tau.ac.il/expander/

[5] http://cran.r-project.org/web/packages/isa2/

[6] http://cran.r-project.org/web/packages/biclust/

Table 2. Dataset properties, parameter settings, and performance statistics of Sparse RMF on three datasets

Dataset	EU Song Contests [1]	NBA team rankings [17]	Sushi dataset [18]
Size	44x37	34x30	5000x10
0s in data	40%	60.4%	0%
#runs	200	200	10
k	10	9	8
α	10%	5%	20%
Coverage	30%	86%	78.2%
Average error	1.59	1.0	1.31
0s/pattern	59.7%	12.6%	13.8%
Overlapping	2%	0%	0%
Convergence	6.1±1.7	3.2±1.6	6.2±1.1
Time/run	3s	1.2s	53min

competing countries during the period from 2010 to 2013. We aggregated the data by calculating average scores that voting countries award to competing countries and transformed it to ranked data. The NBA basketball team ranking dataset was collected by the authors of [17]. It consists of rankings of 30 NBA basketball teams by a group of professional agencies and groups of students. The Sushi dataset was collected by the authors of [18]. It contains preferences of five thousand people over ten different sushi types.

We initially applied sparse rank matrix factorisation on these datasets with varying α and k. Based on these preliminary experiments, we used the following heuristics to choose reasonable parameter values to report on. We choose α such that it results in high coverage and low error. Given the chosen α, for k we choose the largest value such that each resulting pattern is used in at least two rows. When k is further increased, patterns are introduced that are used in only one row of the rank matrix, or even in none. This would clearly result in redundancy, which we would like to avoid.

Table 2 presents a summary of the results obtained by the Sparse RMF algorithm on all datasets. The upper five rows describe dataset properties and the used parameter values. For each dataset, the algorithm is executed a number of times (#runs) and the highest-scoring result is used for the remaining statistics (except for convergence and time/run, for which all runs are used).

The coverages and average errors (per covered cell) show that the algorithm can achieve high coverage with low error. With the Sushi dataset, for example, 78% of the matrix can be covered by just 8 rank profiles, and on average the ranks in the reconstructed rank matrix differs just 1.3 from those in the covered part of the matrix. The numbers of zeroes per pattern demonstrate that the algorithm successfully finds local patterns in the three studied datasets: partial rankings are used to cover the matrix. The overlapping statistic indicates that only the ESC dataset needs multiple patterns per row to cover a large part of the matrix: 2% of the rows are covered by more than one pattern.

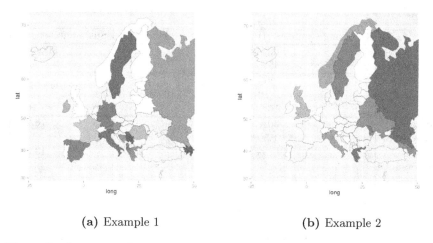

(a) Example 1   (b) Example 2

Fig. 3. Rank patterns discovered on the ESC dataset. Voters are painted green. Competitors are painted by their ranks: the darker the red, the higher the score.

Convergence indicates the average number of iterations in which the run converges, with standard deviation. This shows that the algorithm needs only few iterations to converge, typically 3 to 6. Finally, the average time per run shows that our algorithm runs very efficiently on modestly sized rank matrices, but making it more efficient for larger datasets is left for future work.

To show how rank patterns can provide insight into the data, we visualise two typical rank profiles obtained on the European Song Contest data in Figure 3. Both depict a set of voting countries (in green) and their typical voting behaviour (in red). For example, countries in Eastern Europe tend to give higher scores to Russia and nordic countries than to other countries.

7   Conclusions

We introduced the novel problem of rank matrix factorisation (RMF), which concerns the decomposition of rank data. RMF is a generic problem and we therefore introduced Sparse RMF, a concrete instance with the goal to discover a set of local patterns that capture the recurrent rankings in the data.

We formalised Sparse RMF as an optimisation problem and proposed a greedy, alternate optimisation algorithm to solve it using integer linear programming. Experiments on both synthetic and real datasets demonstrate that our proposed approach can successfully summarise rank matrices by a small number of rank profiles with high coverage and low error.

Acknowledgments. This research was supported by the DBOF 10/044 Project, the Natural and Artifical Genetic Variation in Microbes project, Post-doctoral Fellowships of the Research Foundation Flanders (FWO) for Siegfried Nijssen and Matthijs van Leeuwen, and the EU FET Open project Inductive Constraint Programming.

References

1. Le Van, T., van Leeuwen, M., Nijssen, S., Fierro, A.C., Marchal, K., De Raedt, L.: Ranked tiling. In: Calders, T., Esposito, F., Hüllermeier, E., Meo, R. (eds.) ECML PKDD 2014, Part II. LNCS, vol. 8725, pp. 98–113. Springer, Heidelberg (2014)
2. Henzgen, S., Hüllermeier, E.: Mining rank data. In: Džeroski, S., Panov, P., Kocev, D., Todorovski, L. (eds.) DS 2014. LNCS (LNAI), vol. 8777, pp. 123–134. Springer, Heidelberg (2014)
3. Paatero, P., Tapper, U.: Positive matrix factorization: A non-negative factor model with optimal utilization of error estimates of data values. Environmetrics **5**(2), 111–126 (1994)
4. Lee, D.D., Seung, H.S.: Learning the parts of objects by non-negative matrix factorization. Nature **401**(6755), 788–791 (1999)
5. Monson, S., Pullman, N., Rees, R.: A survey of clique and biclique coverings and factorizations of (0, 1)-matrices. Bull. Inst. Combinatorics and Its Applications **14**, 17–86 (1995)
6. Miettinen, P., Mielikainen, T., Gionis, A., Das, G., Mannila, H.: The discrete basis problem. IEEE Transactions on Knowledge and Data Engineering **20**(10), 1348–1362 (2008)
7. Lust, T., Teghem, J.: Multiobjective decomposition of positive integer matrix: application to radiotherapy. In: Ehrgott, M., Fonseca, C.M., Gandibleux, X., Hao, J.-K., Sevaux, M. (eds.) EMO 2009. LNCS, vol. 5467, pp. 335–349. Springer, Heidelberg (2009)
8. Marden, J.I.: Analyzing and Modeling Rank Data. Chapman & Hall (1995)
9. Madeira, S.C., Oliveira, A.L.: Biclustering algorithms for biological data analysis: a survey. IEEE/ACM Transactions on Computational Biology and Bioinformatics **1**(1), 24–45 (2004)
10. Cheng, Y., Church, G.M.: Biclustering of expression data. In: Proc. of the 8th International Conference on Intelligent Systems for Molecular Biology, vol. 8, pp. 93–103 (2000)
11. Kluger, Y., Basri, R., Chang, J.T., Gerstein, M.: Spectral Biclustering of Microarray Data : Coclustering Genes and Conditions. Genome Research **13**, 703–716 (2003)
12. Turner, H., Bailey, T., Krzanowski, W.: Improved biclustering of microarray data demonstrated through systematic performance tests. Computational Statistics & Data Analysis **48**(2), 235–254 (2005)
13. Hochreiter, S., Bodenhofer, U., Heusel, M., Mayr, A., Mitterecker, A., Kasim, A., Khamiakova, T., Van Sanden, S., Lin, D., Talloen, W., Bijnens, L., Göhlmann, H.W.H., Shkedy, Z., Clevert, D.A.: FABIA: factor analysis for bicluster acquisition. Bioinformatics **26**(12), 1520–1527 (2010)
14. Tanay, A., Sharan, R., Shamir, R.: Discovering statistically significant biclusters in gene expression data. Bioinformatics **18**(Suppl. 1), S136–S144 (2002)

15. Ihmels, J., Friedlander, G., Bergmann, S., Sarig, O., Ziv, Y., Barkai, N.: Revealing modular organization in the yeast transcriptional network. Nature Genetics **31**(4), 370–377 (2002)
16. Truong, D.T., Battiti, R., Brunato, M.: Discovering non-redundant overlapping biclusters on gene expression data. In: ICDM 2013, pp. 747–756. IEEE (2013)
17. Deng, K., Han, S., Li, K.J., Liu, J.S.: Bayesian Aggregation of Order-Based Rank Data. Journal of the American Statistical Association **109**(507), 1023–1039 (2014)
18. Kamishima, T.: Nantonac collaborative filtering: recommendation based on order responses. In: Proceedings of the Ninth ACM SIGKDD International Conference on Knowledge Discovery and Data Mining, KDD 2003, New York, NY, USA, pp. 583–588. ACM (2003)

An Empirical Study of Personal Factors and Social Effects on Rating Prediction

Zhijin Wang[1,2], Yan Yang[1,2](✉), Qinmin Hu[1,2], and Liang He[1,2]

[1] Shanghai Key Laboratory of Multidimensional Information Processing,
East China Normal University, Shanghai 200241, China
[2] Department of Computer Science and Technology,
East China Normal University, Shanghai 200241, China
zhijin@ecnu.cn, {yangyan,qmhu,lhe}@cs.ecnu.edu.cn

Abstract. In social networks, the link between a pair of friends has been reported effective in improving recommendation accuracy. Previous studies mainly based on the assumption that any pair of friends shall have similar interests, via minimizing the gap between user's taste and the average (or similar) taste of this user's friends to reduce the error of rating prediction. However, these methods ignore the diversity of user's taste. In this paper, we focus on learning the diversity of user's taste and effects from this user's friends in terms of rating behavior. We propose a novel recommendation approach, namely <u>P</u>ersonal factors with <u>W</u>eighted <u>S</u>ocial effects Matrix Factorization (PWS), which utilities both user's taste and social effects to provide recommendations. Experimental results carried out on 3 datasets, show the effectiveness of the proposed approach.

Keywords: Personal factors · Social effects · Rating prediction

1 Introduction

In recent years, due to the rapid growth and increasing popularity of social networks, social recommendation [20] receives much attention. The task of social recommendation is to provide recommendations by systematically leveraging the social links [21] between users as well as their past behavior.

The link between a pair of friends has been reported effective in improving recommendation accuracy (e.g., [1,5,9,22]). However, the diversity of user's taste is one of the most challengeable problems which decreases the improvement of accuracy. Informally, users at both ends of a link may rate common items, but may rate different items at most occasions. As Figure 1(a) shows, given user u_1 and u_3 is a pair of friends, they both rated item i_2, but u_1 preferred item i_1 while u_3 do not. Hence, a prediction of preference from u_3 to i_1 directly via the preference of u_1 to i_1 will be a slip.

We consider that user's ratings are affected by (1) *personal factors*: user's personal taste and interests, as well as (2) *social effects*: the effects from the interests

© Springer International Publishing Switzerland 2015
T. Cao et al. (Eds.): PAKDD 2015, Part I, LNAI 9077, pp. 747–758, 2015.
DOI: 10.1007/978-3-319-18038-0_58

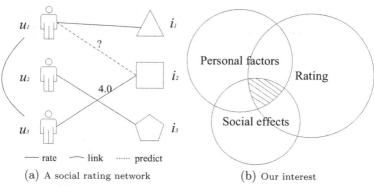

(a) A social rating network   (b) Our interest

Fig. 1. A sample social network and our interest

of this user's friends, as shown in Figure 1(b). The later case is illustrated with the following example. Given user u_1 and user u_3 are friends, and the attitude of user u_1 on item i_2 is between neutral and good (Let's say 3.5 stars), but she has to choose either the neutral (3 stars) or the good (4 stars). Probably, she has chosen the rating 4 because of her friend u_3 rated 4 on item i_2.

We aim to improve the recommender performance by capturing both personal factors and social effects. There are two challenges: (1) How to build up the relation between personal factors and social effects? (2) How to observe personal factors and social effects solely on the basis of rating logs?

To address these challenges, we propose Personal factors with Weighted Social effects Model (PWS), which is based on matrix factorization technique [8] and incorporates personal factors with weighted social effects.

To summarize, our main contributions are as follows:

- We consider that a rating is affected by personal interests and interests of friends. In particular, we formulate the relation between personal factors and social effects in terms of rating prediction.
- Based on the consideration, we develop PWS to learn personal factors and social effects and provide recommendations.
- Finally, We demonstrate how the model can be applied to improve recommendation. We systematically compared our approaches with other algorithms on 3 public real datasets.

The rest of the paper is organized as follows. Section 2 provides a brief review of related work on social recommendation methods. Section 3 presents our proposed recommendation approach. Section 4 presents the experimental setup. Section 5 shows the experimental results and analyses. Finally, we conclude in Section 6.

2   Related Work

In this section, we review several popular approaches for social recommendation.

According to the way the trust links are used in each method, social recommendation techniques can be categorized into two types: memory-based (e.g., [1,5,14,15]) and model-based (e.g., [4,6,10,11,13,24]).

Typically, memory-based methods generate predictions via similar users or items which are usually calculated by predefined similarity functions (e.g., Cosine or Pearson similarity measurement). In trust networks, methods in this category use social trust metric to represent the similarity between two users, and the degrees of trust are estimated by means of propagation over the trust network. In [16], Massa et al. propose a trust-aware method for recommender systems. In this work, the weight for prediction function is measured by the combination of estimated degree of trust and user similarity. The experiments on Epinions dataset show that the enhancement of precision while preserving the coverage (number of ratings that are predictable). In [1,5], trust-aware methods are proposed to improve standard collaborative filtering methods. The experimental results reveal that the social trust information can help enhance recommendation performance.

In contrast, the model-based approaches use the observed ratings to train a designed learning model. User social information is fused into traditional matrix factorization framework to improve recommendation accuracy, since the effectiveness and efficiency of this framework. In [12], social relations are integrated into probabilistic matrix factorization [18] to reduce prediction error. In [11], social trust ensemble (STE) is proposed to linearly combine user rating and friends' rating on an item in matrix factorization framework. In [13], Ma et al. propose social regularization terms to constraint matrix factorization objective functions. In cases of explicit social relations are not available, [10] proposes implicit social recommendation methods, and the implicit social relation is built on the top-N similar users which are calculated by Pearson Correlation Coefficient.

The aforementioned matrix factorization based social recommendation methods move a nice step forward in the research of recommender systems by fusing (explicit or implicit) social relations into a regularizer term to constraint learning process or embedding trust links into.

However, regularizer methods often ignore the diversity of user's taste, and embedding methods can not be directly used for learning the personal factors and social effects, In this paper, we present an empirical study on PWS which provides insights for learning personal factors and social effects on rating prediction.

3   Personal Factors and Social Effects Modeling

In this section, we present our approaches PWS, to incorporate personal factors and social effects into a matrix factorization model for rating prediction.

3.1   Problem Definition

In social rating networks (See Figure 1(a)), we can get two major data sources: 1) the user-item rating matrix R which records users' past behavior, and 2) the binary social trust matrix T which denotes trusts among users.

Let $\mathcal{U}, \mathcal{I}$ be the set of users and items, respectively, and $\mathcal{V}$ be the set of values users can assign to items. Let r_{ui} present the entry of matrix R, which indicates the rating of user u on item i. The ratings are explicitly defined as:

$$R = \{(u, i, r_{ui}) | u \in \mathcal{U}, i \in \mathcal{I}, r_{ui} \in \mathcal{V}\}$$

where $\mathcal{V}$ is set of integers usually in the range $[1, 5]$. Let t_{uv} denote the value of social trust u has on v as a real number in $[0, 1]$, 0 means no trust and 1 means full trust. Therefore, the trust relations among users are formulated as:

$$\mathcal{T} = \{(u, v, t_{uv}) | u \in \mathcal{U}, v \in \mathcal{U}, u \neq v, t_{ui} \in [0, 1]\}$$

note that $\mathcal{T}$ is asymmetric in general.

The task of rating prediction is as follows: Given a user $u \in \mathcal{U}$ and an item $i \in \mathcal{I}$ for which $\hat{r}_{ui}$ is unknown, predict the rating for u on i using R and $\mathcal{T}$.

3.2   Matrix Factorization (MF)

In this subsecition, we review the matrix factorization method that is widely studied in the literature.

Given a $m \times n$ rating matrix R describing m users' ratings on n items. the low-rank approach builds a rank-d representation of R, decomposing it a user-factor matrix $P \in \mathbb{R}^{d \times m}$ and an item-factor matrix $Q \in \mathbb{R}^{d \times n}$ with $d \ll \min(m, n)$, such that $R \approx P^T Q$. The predicted rating for a user u and an item i is calculated as follows:

$$\hat{r}_{ui} = q_i^T \cdot p_u \tag{1}$$

where p_u denotes the u-th column of P, and q_i denotes the i-th column of Q.

The basic MF model can be enhanced to include user and item biases [7], e.g., the tendency of users and items to deviate from the global rating mean. when biases are included, Equation 1 becomes:

$$\hat{r}_{ui} = \mu + b_u' + b_i'' + q_i^T \cdot p_u \tag{2}$$

where μ is the global mean rating, b_u' and b_i'' indicate the observed deviations of user u and item i respectively, p_u denotes the u-th column of P, and q_i denotes the i-th column of Q.

Typically, R is very sparse. This poses a challenge for training the model, which is addressed by learning P, Q, b_u', b_i'' from observed ratings by minimising the following objective function [8]:

$$L = \min_{P,Q,b_u',b_i''} \frac{1}{2} \sum_{r_{ui} \in R} (r_{ui} - (\mu + b_u' + b_i'' + q_i^T \cdot p_u))^2$$
$$+ \frac{\lambda_1}{2} \|P\|^2 + \frac{\lambda_2}{2} \|Q\|^2 + \frac{\lambda_3}{2} \sum_{u \in \mathcal{U}} {b_u'}^2 + \frac{\lambda_4}{2} \sum_{i \in \mathcal{I}} {b_i''}^2 \tag{3}$$

where $\lambda_1, ..., \lambda_4$ are parameters to the regularisation part of the objective function, which avoid overfitting.

The optimization problem in Equation 3 is minimized by implementing a stochastic gradient descent method[1].

3.3 PWS

In this subsection, we present our approach PWS to fuse social links into the matrix factorization.

Social Interests. We use social interests to present the taste of user's friends. As the example shown in Figure 2(a), user's taste are positively (or negatively) affected by this user's friends. The pattern of these interests is formulated as:

$$\bar{\mathcal{X}}_u = \frac{\sum_{v \in \mathcal{T}(u)} t_{uv} x_v}{\sum_{v \in \mathcal{T}(u)} t_{uv}} = \frac{\sum_{v \in \mathcal{T}(u)} x_v}{|\mathcal{T}(u)|} \tag{4}$$

where $\bar{\mathcal{X}}_u$ is the estimated latent social interests vector of user u, $x_v \in \mathbb{R}^{d \times 1}$ is latent factors of user v, t_{uv} is a binary variable which indicates a trust link between user u and v, and $|\mathcal{T}(u)|$ is the number of neighbors of user u.

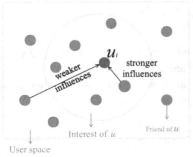

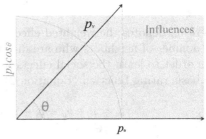

(a) The social effects in user space. The assumption is that the more similar (or opposite) interests between a pair of friends, the more stronger influences they affect each other.

(b) The degree of influences from user $v \rightarrow u$ is presented in a inner product manner. The similar interests will have a positive effect, and the dissimilar taste leads to a weak effect.

Fig. 2. Presenting social effects in user space

In order to reduce the complexity of parameter learning, we use the user latent factor vector p_v to represent the pattern of social interests $x_v, v \in \mathcal{T}(u)$. By this means, the interacted rating opinions between user and her friends will be directly connected. Hence, equation 4 is transformed to:

---

[1] The stochastic gradient descent is chosen due to its speed and ease of implementation. An alternative strategy is alternating least square (ALS) [2,3]. While ALS can be parallelised (Parallel threads or processes) [17,19], these advantages are irrelevant in our case.

$$\bar{\mathcal{X}}_u = \frac{\sum_{v \in \mathcal{T}(u)} p_v}{|\mathcal{T}(u)|} \qquad (5)$$

where $\bar{\mathcal{X}}_u$ means the total interests of u's friends.

Weighted Social Effects. Due to the diversity of user's taste, as Figure 2(a) shows, the social effects to a target user contain both relative stronger (or weaker) influences. More realistically, for a pair of friends, the influences tend to be strengthened if their interests are more similar. By benefiting from inner product (as shown in Figure 2(b)), we formulate the social effects as:

$$\bar{S}_u = \bar{\mathcal{X}}_u^T \cdot p_u = \frac{\sum_{v \in \mathcal{T}(u)} p_v^T}{|\mathcal{T}(u)|} \cdot p_u \qquad (6)$$

where $\bar{S}_u$ denotes the effects received from friends of user u, $|\mathcal{T}(u)|$ is the number of neighbors who are directly linked to user u, and p_u, p_v denote the latent factor of user u, v respectively.

We use the constant parameter w to control the degree of social influences from neighbors to a user. Therefore, we have:

$$\bar{S}_u = \left(\frac{w}{|\mathcal{T}(u)|} \sum_{v \in \mathcal{T}(u)} p_v^T\right) \cdot p_u \qquad (7)$$

where $\bar{S}_u$ denotes the weighted effects received from friends of user u, $|\mathcal{T}(u)|$ is the number of neighbors who are directly linked to user u.

In order to learn the social effects from ratings, we fuse the social effects into the basic rating function (Equation 2) and get:

$$\begin{aligned}
\hat{r}_{ui} &= \mu + b_u + b_i + q_i^T p_u + \left(\frac{w}{|\mathcal{T}(u)|} \sum_{j \in \mathcal{T}(u)} p_j^T\right) p_u \\
&= \mu + b_u + b_i + \left(\frac{w}{|\mathcal{T}(u)|} \sum_{j \in \mathcal{T}(u)} p_j^T + q_i^T\right) p_u
\end{aligned} \qquad (8)$$

where $\hat{r}_{ui}$ is the predicted rating of user u to item i, μ is the global mean rating, b_u and b_i indicate the observed deviations of user u and item i respectively, p_u denotes the u-th column of P, and q_i denotes the i-th column of Q, $\mathcal{T}(u)$ is the friends of user u.

Therefore, we substitute Equation 8 into Equation 3, and get the objective function:

$$\begin{aligned}
L = \min_{P,Q,b_u,b_i} \sum_{r_{ui} \in R} \left(r_{ui} - \mu - b_u - b_i - \left(\frac{w \sum_{j \in \mathcal{T}(u)} p_j^T}{|\mathcal{T}(u)|} + q_i^T\right) p_u\right)^2 \\
+ \frac{\lambda_1}{2} \|P\|^2 + \frac{\lambda_2}{2} \|Q\|^2 + \frac{\lambda_3}{2} \sum_{u \in \mathcal{U}} b_u'^2 + \frac{\lambda_4}{2} \sum_{i \in \mathcal{I}} b_i''^2
\end{aligned} \qquad (9)$$

where $\lambda_1, ..., \lambda_4$ are parameters to the regularisation part of the objective function, which avoid overfitting.

As shown in Algorithm 1, we exploit stochastic gradient descent to learn the proposed model. In order to update the parameters more easily in training process, we scan the datasets according to each user (Line 4-13). The updating rules are as follows:

$$p_u \leftarrow p_u + \gamma(e_{ui} \cdot q_i - \lambda_1(p_u + \frac{w}{|\mathcal{T}(u)|} \sum_{j \in \mathcal{T}(u)} p_j^T))$$

$$q_i \leftarrow q_i + \gamma(e_{ui} \cdot p_u - \lambda_2 q_i)$$

$$b_u' \leftarrow b_u' + \gamma(e_{ui} - \lambda_3 b_u')$$

$$b_i'' \leftarrow b_i'' + \gamma(e_{ui} - \lambda_4 b_i'')$$

where

$$e_{ij} = r_{ij} - (\mu + b_u' + b_i'' + q_i^T \cdot p_u + (\frac{w \sum_{j \in \mathcal{T}(u)} p_j^T}{|\mathcal{T}(u)|} + q_i^T) p_u)$$

and γ is the learning rate.

Algorithm 1. Pseudo code for minimising Equation 9 by stochastic gradient descent

---

Input:
    R, user-item rating matrix; $\mathcal{T}$, social trust links;
    D, dimensionality of latent vectors
    $\lambda_1, \lambda_2, \lambda_3, \lambda_4, \gamma, MaxEpoch, w$
Output:
    P, user-factor matrix; Q, item-factor matrix
    b_u, biases of users; b_i, biases of items
1: Initialise P, Q, b_u, b_i with random values in [0,1] and D;
2: $\mu \leftarrow \sum_{r_{ui} \in R} r_{ui}/|R|$;
3: **for** $epoch = 1$ to $MaxEpoch$ **do**
4:     **for** all user $u \in \mathcal{U}$ **do**
5:       **for** all rating $r_{ui} \in I(u)$ **do**
6:         $\bar{\mathcal{X}}_u \leftarrow \frac{w}{|\mathcal{T}(u)|} \sum_{j \in \mathcal{T}(u)} p_j^T$ // Social interests
7:         $e_{ui} \leftarrow r_{ui} - (\mu + b_u + b_i + q_i^T \cdot p_u + \bar{\mathcal{X}}_u \cdot p_u)$
8:         $b_u \leftarrow b_u + \gamma(e_{ui} - \lambda_3 b_u)$
9:         $b_i \leftarrow b_i + \gamma(e_{ui} - \lambda_4 b_i)$
10:        $p_u \leftarrow p_u + \gamma(e_{ui} q_i - \lambda_1(p_u + \bar{\mathcal{X}}_u))$
11:        $q_i \leftarrow q_i + \gamma(e_{ui} p_u - \lambda_2 q_i)$
12:       **end for**
13:     **end for**
14: **end for**

Algorithm 1 exhibits the steps for learning the proposed PWS.

4 Experimental Setup

In this section, we illustrate dataset collections, evaluation metrics, algorithm configurations and comparable methods.

4.1   Datasets

Epinions Dataset. Epinions[2] is a consumer review site which allows visitors read reviews about a variety of items to help them decide on a purchase. The social relationships in epinions are directed. The Epinions dataset[3] we use was published by authors of [15]. Each user has on average 13.5 expressed ratings and 9.9 neighbors.

Flixster Dataset. Flixster[4] is a social networking service in which user can rate movies[5]. Users can also add other users to their friend list and create a social network. Unlike epinions, the social relations in Flixster are undirected [6]. Possible rating values in Flixster are 10 discrete numbers in the range [0.5, 5] with step size 0.5. On average each user has 8.9 friends and each users has rated 10.4 movies. However, if we ignore the many users who have not rated any movies and only consider users with at least one rating, each user has rated 55.5 movies on average.

DouBan Dataset. DouBan[6] is a Chinese social website providing user rating, review and recommendation services for movie, book and music. Users can make friends with each other through the emails. DouBan dataset[7] is crawled and shared by Ma [13]. In this dataset, users can rate movies, books and songs in a 5-start numerical rating scale.

Table 1. General Statistics of Epinions, Flixster and DouBan Datasets

Statistics	Epinions	Flixster	DouBan
Users	49,290	787,213	129,490
Items	139,738	48,794	58,541
Ratings	664,824	8,196,077	16,830,839
Social Relations	487,183	7,058,819	1,692,952
Users with Rating	40,163	147,612	129,490
Users with Friend	49,288	786,936	111,210

The general statistics of the Epinions, Flixster and DouBan dataset are shown in Table 1.

4.2   Evaluation Metrics

We adopt two metrics, the Mean Absolute Error (MAE) and the Root Mean Square Error (RMSE), to evaluate the performance of our proposed in comparison with traditional methods.

---

[2] www.epinions.com
[3] http://www.trustlet.org/wiki/Epinions_dataset/
[4] www.flixster.com
[5] http://www.sfu.ca/~sja25/datasets/
[6] www.douban.com
[7] http://dl.dropbox.com/u/17517913/Douban.zip

The metric MAE is defined as:

$$MAE = \frac{1}{|R'|} \sum_{r_{ij} \in R'} |r_{ij} - \hat{r}_{ij}| \tag{10}$$

where r_{ij} denotes the rating user i gave to item j in the test dataset R', $\hat{r}_{ij}$ denotes the rating user i gave to item j as predicted by a method, and $|R'|$ denotes the number of tested ratings. The metric RMSE is defined as:

$$RMSE = \sqrt{\frac{1}{|R'|} \sum_{r_{ij} \in R'} (r_{ij} - \hat{r}_{ij})^2} \tag{11}$$

We can see that a smaller MAE or RMSE value means a better performance.

4.3 Comparable Methods

SR-MF. Social regularization based matrix factorization (SR-MF)[13] explicitly utilizes the social relationships to regulate the latent user factors.

Note that to make a better reading, Equations in comparable methods are represented by the symbols used in this paper.

ASS-MF. Adaptive Social Similarity based matrix factorization [23] is proposed to alleviate the zero similarity problem among friends who are without common items.

SWS. To study how the social influences $\bar{S}_u$ affects user rating behavior and performance without the interference of biases (Equation 2), we introduce Simple Weighted Social Effect Model (SWS) by fusing the social effects (Equation 6) into the basic rating function (Equation 1). Hence, we have:

$$L = \min_{P,Q} \frac{1}{2} \sum_{r_{ui} \in R} (r_{ui} - (\frac{w \sum_{j \in T(u)} p_j^T}{|T(u)|} + q_i^T)p_u)^2 + \frac{\lambda_1}{2}\|P\|^2 + \frac{\lambda_2}{2}\|Q\|^2 \tag{12}$$

where λ_1, λ_2 are parameters which avoid overfitting, $T(u)$ is the friends of user u, p_u denotes the u-th column of P, and q_i denotes the i-th column of Q.

In all experiments, the learning rate is set to 0.01 and the dimensionality D is set to 10. In the experiments conducted on Epinions dataset, we set $\lambda_1 = \lambda_2 = 0.01$. In the experiments conducted on Flixster and DouBan datasets, we set $\lambda_1 = 0.005, \lambda_2 = 0.02$. And λ_3 is configured as 0.01 with respect to SR-MF or ASS-MF. For the proposed PWS, we configure $\lambda_3 = \lambda_4 = 0.01$.

5   Experimental Results

In this section, we use the real world user rating data and their corresponding social networks to empirically validate the proposed Personal factors with Weighted Social effects Matrix Factorization (PWS). Our experiments are intended to address the following questions:

- How the weight of social influence w affects the learning of social effects $\bar{S}_u$ and recommendation performance?
- Can personal factors and social effects be learned by the designed rating pattern?
- Can the recommendation performance benefit from the captured personal factors and social effects?

5.1   Effects on Parameter w

To study how the social effects $\bar{S}_u$ affects the learning process of proposed model and recommendation, we measure the performance in terms of MAE and RMSE as w changes.

In the experiments based on SWS, we configure the step size of w as 0.2 to observed the recommendation performance, and w is still increased until the model became hard to be learned. As shown in Figure 3, (1) the performance is greatly improved when fusing the social effect $\bar{S}_u$ into. (2) the optimal recommendation performance can be obtained when w are set to 2.6, 5.0 and 0.4 respectively in these datasets. According to the obtained optimal w values, we can find that users tend to communicate 2.6 times (in average) on an item in Epinions.com. But users tend to deliver 0.4 messages in DouBan.com, a possible reason is that users in DouBan.com are inconvenient to communicate with each other. For users in Flixster.com, each user communicates with 5.0 friend averagely, the possible reason is that item categories in Flixster.com are mainly about movies, while items are multiple in both DouBan.com and Epinions.com. (3) as w increases to a given value, the error of output of this model become larger. This phenomenon reveals that a user is affected by some of this user's

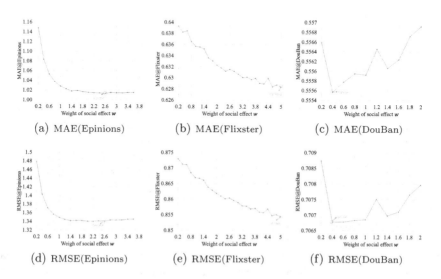

(a) MAE(Epinions)   (b) MAE(Flixster)   (c) MAE(DouBan)

(d) RMSE(Epinions)   (e) RMSE(Flixster)   (f) RMSE(DouBan)

Fig. 3. The effects of parameter w on results

friend. And if w is so large, the rating noisy from this user's friends (e.g., some friends tend to rate higher) would be intolerable according to this model.

Therefore, the more realistic model PWS is proposed by leveraging personal biases (See Equation 8). Experimental results carried on PWS show that the optimal values can be found when w is set to 0.01. We illustrate the experimental results of PWS in the following subsection.

5.2 Comparing with Regularization

We perform 5-fold cross validation in our experiments. For each dataset used in this paper, we randomly select 80% ratings as training set and other 20% as test set. For comparison purpose, the dimensionality D is set to 10 in all the experiments conducted in this paper.

Table 2. Performance comparisons with MAE and RMSE on the provide datasets

Models	Epinions		Flixster		DouBan	
	MAE	RMSE	MAE	RMSE	MAE	RMSE
SR-MF	1.1204 ± .0005	1.4775 ± .0007	0.6410 ± .0005	0.8770 ± .0008	0.5576 ± .0008	0.7125 ± .0010
ASS-MF	1.1105 ± .0004	1.4570 ± .0006	0.6404 ± .0005	0.8772 ± .0009	0.5566 ± .0007	0.7112 ± .0010
SWS	1.0166 ± .0007	1.3422 ± .0010	0.6330 ± .0007	0.8620 ± .0010	0.5560 ± .0006	0.7075 ± .0012
PWS	0.8016±.0005	1.0489 ± .0007	0.6207 ± .0003	0.8365 ± .0006	0.5519 ± .0005	0.7009 ± .0010

Table 2 reports the MAE and RMSE values of all comparison partners on the three datasets. The parameter w is set to 2.0 for experiments on SWS, and 0.01 for PWS. As Table 2 reveals, (1) SWS slightly improves the state-of-the-art social regularizer methods SR-MF and ASS-MF, (2) PWS outperforms the SWS on the provided three datasets, and (3) the improvement of PWS on Epinions dataset is more significant (by 28%) than other 2 datasets (by 1%-4%), which shows the effectiveness of PWS in dealing the with diversity of item, since the item classes are multiple in Epinions [6].

It should be noted that the datasets we use in this paper are the raw data. Hence, there is a slip between the duplicated experimental results (See Table 2) and the result in literatures (SR-MF [13] and ASS-MF [23]), since some items are filtered out in data preprocess stage.

6 Conclusions and Future Work

In this paper, we focus on learning the behavior of personal factors and social effects when a user rates an item. We consider the social effects as a variation of ratings. Based on this consideration, we propose PWS to model the connection between ratings and social links, which extends matrix factorization framework. Experimental results conducted on 3 public available datasets show the effectiveness of rating prediction.

The study of common but differentiated weight of social effects for each user is a potential future direction of this work.

Acknowledgments. We would like to thank the anonymous reviewers for their valuable comments and suggestions to improve the quality of this paper. This work is supported by the National Key Technology R&D Program (No.2012BAH93F03) and the Shanghai Science and Technology Commission Foundation (No.13511506201).

References

1. Bedi, P., Kaur, H., Marwaha, S.: Trust based recommender system for semantic web. In: IJCAI 2007, pp. 2677–2682 (2007)
2. Bell, R.M., Koren, Y.: Scalable collaborative filtering with jointly derived neighborhood interpolation weights. In: ICDM 2007, pp. 43–52 (2007)
3. Hu, Y., Koren, Y., Volinsky, C.: Collaborative filtering for implicit feedback datasets. In: ICDM 2008, pp. 263–272 (2008)
4. Huang, J., Cheng, X., Guo, J., Shen, H., Yang, K.: Social recommendation with interpersonal influence. In: ECAI 2010, pp. 601–606 (2010)
5. Jamali, M., Ester, M.: TrustWalker: a random walk model for combining trust-based and item-based recommendation. In: KDD 2009, pp. 397–406 (2009)
6. Jamali, M., Ester, M.: A matrix factorization technique with trust propagation for recommendation in social networks. In: Recsys 2010, pp. 135–142 (2010)
7. Koren, Y.: Collaborative filtering with temporal dynamics. In: KDD 2009, pp. 447–456 (2009)
8. Koren, Y., Bell, R.M., Volinsky, C.: Matrix factorization techniques for recommender systems. IEEE Computer 42(8), 30–37 (2009)
9. Lu, W., Ioannidis, S., Bhagat, S., Lakshmanan, L.V.S.: Optimal recommendations under attraction, aversion, and social influence. In: KDD 2014 (2004)
10. Ma, H.: An experimental study on implicit social recommendation. In: SIGIR 2013, pp. 73–82 (2013)
11. Ma, H., King, I., Lyu, M.R.: Learning to recommend with social trust ensemble. In: SIGIR 2009, pp. 203–210 (2009)
12. Ma, H., Yang, H., Lyu, M.R., King, I.: Sorec: social recommendation using probabilistic matrix factorization. In: CIKM 2008, pp. 931–940 (2008)
13. Ma, H., Zhou, D., Liu, C., Lyu, M.R., King, I.: Recommender systems with social regularization. In: WSDM 2011, pp. 287–296 (2011)
14. Massa, P., Avesani, P.: Trust-aware collaborative filtering for recommender systems. In: Meersman, R. (ed.) OTM 2004. LNCS, vol. 3290, pp. 492–508. Springer, Heidelberg (2004)
15. Massa, P., Avesani, P.: Trust-aware bootstrapping of recommender systems. In: ECAI 2006 Workshop on Recommender Systems, pp. 29–33 (2006)
16. Massa, P., Avesani, P.: Trust-aware recommender systems. In: Recsys 2007, pp. 17–24 (2007)
17. Pilászy, I., Zibriczky, D., Tikk, D.: Fast als-based matrix factorization for explicit and implicit feedback datasets. In: Recsys 2010, pp. 71–78 (2010)
18. Salakhutdinov, R., Mnih, A.: Probabilistic matrix factorization. In: NIPS 2007 (2007)
19. Schelter, S., Boden, C., Schenck, M., Alexandrov, A., Markl, V.: Distributed matrix factorization with mapreduce using a series of broadcast-joins. In: Recsys 2013, pp. 281–284 (2013)
20. Shen, Y., Jin, R.: Learning personal + social latent factor model for social recommendation. In: KDD 2012, pp. 1303–1311 (2012)
21. Victor, P., Cock, M.D., Cornelis, C.: Trust and recommendations. In: Recommender Systems Handbook, pp. 645–675 (2011)
22. Yao, Y., Tong, H., Yan, G., Xu, F., Zhang, X., Szymanski, B.K., Lu, J.: Dual-regularized one-class collaborative filtering. In: CIKM 2014, pp. 759–768 (2014)
23. Yu, L., Pan, R., Li, Z.: Adaptive social similarities for recommender systems. In: Recsys 2011, pp. 257–260 (2011)
24. Yuan, Q., Chen, L., Zhao, S.: Factorization vs. regularization: fusing heterogeneous social relationships in top-n recommendation. In: Recsys 2011, pp. 245–252 (2011)

Author Index

Ai, Xusheng I-251
Al-Hamoudi, Asmaa I-525
Al-Hassani, Shamma I-525
Al-Shamsi, Eiman I-525
Al-Shehhi, Ameera I-525
An, Yuan II-598
Arandjelović, Ognjen I-550
Azzag, Hanene II-134

Baba, Yukino II-255
Bailey, James II-422
Baron, Michael II-383
Bar-Zev, Yedidya II-27
Bayer, Immanuel II-447
Beutel, Alex II-201
Beykikhoshk, Adham I-550
Bui, Hung H. I-330, I-343

Cai, Wandong I-96
Cao, Jinli II-280
Cao, Longbing I-176, II-707, II-732
Cao, Tongyi I-646
Cardell-Oliver, Rachel II-522
Chakraborty, Abhijnan I-108
Chan, Jeffrey II-422
Chan, Keith C.C. II-409
Chatzakou, Despoina I-122, II-201
Chawla, Nitesh V. I-264
Chawla, Sanjay II-319
Chen, Enhong I-472, I-576, II-534
Chen, Fang I-459
Chen, Hsuan-Hsu I-199
Chen, Meng II-344, II-357
Chen, Ming-Syan I-3, I-45, I-70
Chen, Ting I-96
Cheung, David W. I-591
Chin, Wei-Sheng I-442, II-690
Chou, Chung-Kuang I-70
Christen, Peter I-380, II-549, II-562, II-574
Crémilleux, Bruno I-722
Cui, Zhiming I-251
Cuissart, Bertrand I-722

Cule, Boris II-637
Cuzzocrea, Alfredo II-146

Das, Joydeep I-605
Das, Shreyasi I-108
De Raedt, Luc I-734
Desai, Kalpit I-151
Desrosiers, Christian II-165
Dutta, Debarshi I-605

Elovici, Yuval II-119
Estephan, Joël II-319

Faloutsos, Christos I-122, I-633, II-201
Fang, Binxing II-79
Fang, Chen I-317
Feng, Weiwei II-79
Fournier-Viger, Philippe II-625
Fu, Bin II-732
Fukui, Ken-ichi II-293

Ganguly, Niloy I-108
Gao, Kai II-3
Gao, Yang II-675
Ge, Jiaqi II-243, II-268
Ge, Yong I-563
Ghesmoune, Mohammed II-134
Ghosh, Saptarshi I-108
Giatsoglou, Maria I-122, II-201
Goethals, Bart II-637
Gopakumar, Shivapratap II-331
Gudmundsson, Joachim II-319
Gueniche, Ted II-625
Guo, Danhuai I-563
Guo, Li II-79
Guo, Yunchang I-563
Gupta, Prosenjit I-605
Gupta, Saurabh II-720
Gupta, Sunil Kumar I-226, I-303

Han, Shuguo II-305
Hao, Hongwei II-473

Haque, Ahsanul II-383
Harris, Greg I-538
He, Liang I-188, I-747, II-534
He, Lifang II-485
He, Saike II-15
He, Yulin I-405
He, Zengyou II-177
Heiser, John I-511
Hendrickx, Tayena II-637
Hoang, Tuan-Anh I-708
Horton, Michael II-319
Hu, Bin I-239
Hu, Jun I-16
Hu, Linmei I-696
Hu, Qingbo I-82
Hu, Qinmin I-188, I-747
Hu, Xiaohua II-598
Huang, Dong I-511, II-305
Huang, Hao I-511
Huang, Jiaji I-429
Huang, Joshua Zhexue I-405, II-459
Huynh, Viet I-343

Ifada, Noor II-510
Ishii, Shin I-621

Jabbour, Said II-662
Jia, Xiuyi II-52
Jin, Bo I-164
Johnson, Reid A. I-264
Juan, Yu-Chin I-442, II-690

Kajimura, Shunsuke II-255
Kajino, Hiroshi II-255
Kalb, Paul I-511
Kane, Bamba I-722
Kang, Ying II-106
Kashima, Hisashi II-255
Katz, Gilad II-27
Khan, Latifur I-525, II-383
Khoa, Nguyen Lu Dang I-459
Kimura, Masahiro I-135
Kitagawa, Hiroyuki I-633
Kitamoto, Asanobu II-64
Koyama, Masanori I-621
Koyamada, Sotetsu I-621
Kralj, Jan I-672

Lau, Francis C.M. I-164
Laukens, Kris II-637

Lavrač, Nada I-672
Le Van, Thanh I-734
Le, Trung II-189
Lebbah, Mustapha II-134
Leckie, Christopher I-486, II-215, II-422
Lee, Wang-Chien I-3
Li, Changliang II-15
Li, Cheng II-92
Li, Cheng-Te I-659
Li, Fangfang II-707
Li, Jianhui I-563
Li, Juanzi I-696
Li, Mark Junjie II-459
Li, Rumeng I-212
Li, Xiaoli II-305
Li, Zhaoxing II-177
Liang, Wenxin II-177
Liao, Shizhong I-684
Lin, Chih-Jen I-442, II-690
Lin, Shuyang I-82
Lin, Ying Chun II-649
Lin, Yu-Feng I-199
Lin, Yu-Jen I-659
Liu, Chengchen II-228
Liu, Chuan I-684
Liu, Chunming I-176
Liu, Fangbing I-33
Liu, Qi I-576, II-534
Liu, Rujie I-369
Liu, Wei I-459, II-522
Liu, Xiaosheng II-675
Liu, Yang II-344, II-357
Lu, Jian II-756
Lu, Ziyu I-591
Luo, Jiawei II-228
Luo, Jun II-370
Luo, Zhilin I-96
Lv, Guangyi II-534
Lv, Lei II-357

Ma, Guowei I-576
Majumder, Subhashis I-605
Mamoulis, Nikos I-591
Manwani, Naresh I-151
Mao, Chengsheng I-239
Masud, Mohammad M. I-525
Meng, Dan II-106
Meysman, Pieter II-637
Miao, Jiansong II-744
Mirsky, Yisroel II-119

Mitra, Bivas   II-720
Miwa, Makoto   I-289
Moore, Philip   I-239
Moriyama, Koichi   II-293
Motoda, Hiroshi   I-135
Mustapha, Samir   I-459

Naaman, Mor   I-16
Nagata, Masaaki   II-39
Nagel, Uwe   II-447
Nakae, Ken   I-621
Nakagawa, Hiroshi   II-498
Naulaerts, Stefan   II-637
Nayak, Richi   II-510
Ng, Wee Siong   I-498
Nguyen, Ha Thanh Huynh   II-280
Nguyen, Hoang Tu   II-228
Nguyen, Khanh   II-189
Nguyen, Long   I-343
Nguyen, Minh-Tien   II-64
Nguyen, Phuoc   I-277
Nguyen, Thanh-Tung   II-459
Nguyen, Thuy Thi   II-459
Nguyen, Tri-Thanh   II-64
Nguyen, Tu Dinh   II-331
Nguyen, Vu   I-330
Nijssen, Siegfried   I-734
Ning, Xia   I-429
Numao, Masayuki   II-293

Ofek, Nir   II-27
Ohara, Kouzou   I-135
Okada, Yoshiyuki   II-293
Orimaye, Sylvester Olubolu   II-610

Panangadan, Anand   I-538
Pathak, Sayan   II-720
Pei, Jian   I-199, II-422
Phan, Duy Nhat   II-435
Phung, Dinh   I-303, I-330, I-343, I-550,
    II-92, II-189, II-331
Prasanna, Viktor K.   I-538

Qian, Hongze   II-586
Qian, Wenbin   II-397
Qiu, Huimin   II-744
Qiu, Tianyu   I-418

Raeder, Troy   I-264
Rahimi, Seyyed Mohammadreza   II-370

Rajasegarar, Sutharshan   II-215
Ramadan, Banda   II-574
Ramamohanarao, Kotagiri   II-422
Raman, Rajeev   II-625
Rana, Santu   I-303, II-92
Ranbaduge, Thilina   II-549
Rashidi, Lida   II-215
Raza, Rana Aamir   I-405
Rendle, Steffen   II-447
Robnik-Šikonja, Marko   I-672
Rockmore, Daniel N.   I-317
Rokach, Lior   II-119
Rudra, Koustav   I-108
Rui, Yong   I-472
Runcie, Peter   I-459

Sabourin, Robert   II-165
Saha, Budhaditya   I-226
Sais, Lakhdar   II-662
Saito, Kazumi   I-135
Salhi, Yakoub   II-662
Sanner, Scott   I-380
Sasaki, Yutaka   I-289
Sasidharan, Sanand   I-151
Sato, Issei   II-498
Sebastian, Yakub   II-610
Sethi, Manav   I-108
Shah, Alpa Jayesh   II-165
Shah, Neil   I-122, II-201
Shang, Lin   II-52
Shao, Chao   I-696
Shao, Dongxu   I-498
Shao, Weixiang   II-485
Shapira, Bracha   II-27, II-119
Shen, Chih-Ya   I-3
Shen, Furao   I-418, I-646
Shen, Hua   II-177
Shen, Yi   I-563
Sheng, Victor S.   I-251
Shi, Ziqiang   I-369
Shindo, Hiroyuki   I-212
Shu, Wenhao   II-397
Shuai, Hong-Han   I-45
Siew, Eu-Gene   II-610
Slutsky, Anton   II-598
Sohrab, Mohammad Golam   I-289
Su, Yun   I-239
Sudoh, Katsuhito   II-39
Sun, Chonglin   I-164
Sun, Zhengya   II-473

Sun, Zhenlong   II-675
Sundararajan, Ramasubramanian   I-151

Thi, Hoai An Le   II-435
Tian, Guanhua   II-15
Tian, Tian   I-392
Tokui, Seiya   II-498
Tran, Dat   I-277
Tran, Khoi-Nguyen   I-380
Tran, Truyen   II-331
Tseng, Vincent S.   I-199, II-625, II-649
Tu, Wenting   I-591

Vaithianathan, Tharshan   I-486
Vakali, Athena   I-122, II-201
Valmarska, Anita   I-672
van Leeuwen, Matthijs   I-734
Vatsalan, Dinusha   II-549, II-562
Venkatesh, Svetha   I-226, I-303, I-330,
    I-343, I-550, II-92, II-189, II-331
Vinh, Nguyen Xuan   II-422

Wang, Guan   I-82
Wang, Houfeng   I-355
Wang, Jian   II-243, II-268
Wang, Jin   II-522
Wang, Jiushuo   II-3
Wang, Manman   I-239
Wang, Peng   II-79
Wang, Qing   II-562
Wang, Senzhang   I-58
Wang, Weiping   II-106
Wang, Xiaoting   I-486
Wang, Xin   II-370
Wang, Xizhao   I-405
Wang, Xun   II-39
Wang, Yang   I-459
Wang, Yi   II-675
Wang, Yibing   II-586
Wang, Yinglong   II-397
Wang, Yue   I-96
Wang, Zhihai   II-732
Wang, Zhijin   I-747
Wei, Baogang   II-586
Wu, Cheng-Wei   II-649
Wu, Gaowei   II-15
Wu, Huayu   I-498
Wu, Jian   I-251
Wu, Le   I-472, I-576

Wu, Xintao   I-96
Wu, Zhiang   II-732

Xia, Chaolun   I-16
Xia, Shu-Tao   I-33
Xia, Yuni   II-243, II-268
Xiao, Xiao   I-563
Xie, Hairuo   I-486
Xie, Junyuan   I-58
Xie, Lexing   I-380
Xie, Xing   I-472
Xing, Youlu   I-646
Xu, Bo   II-15
Xu, Changsheng   II-473
Xu, Feng   II-756
Xu, Guandong   II-707, II-732
Xu, Hua   II-3
Xu, Jin   I-511

Yamaguchi, Yuto   I-633
Yan, Guo   II-756
Yan, Jianfeng   II-675
Yan, Rui   I-659
Yang, De-Nian   I-3, I-45
Yang, Hebin   I-188
Yang, Jun   II-397
Yang, Min   I-591
Yang, Wenxin   II-370
Yang, Yan   I-747
Yao, Liang   II-586
Yao, Yuan   II-756
Yao, Yufeng   I-251
Yeh, Mi-Yen   I-659
Yoo, Shinjae   I-511
Yu, Bo   II-106
Yu, Dantong   I-511
Yu, Philip S.   I-45, I-58, I-82, II-485
Yu, Xiaohui   II-344, II-357
Yuan, Nicholas Jing   I-472

Zeng, Jia   II-675
Zhan, Qianyi   I-58
Zhang, Bang   I-459
Zhang, Chunhong   II-744
Zhang, Jiawei   I-58
Zhang, Jing   I-696
Zhang, Peng   II-79
Zhang, Qing   I-355
Zhang, Xianchao   II-177

Zhang, Yaowen   II-52
Zhang, Yin   II-586
Zhao, He   II-459
Zhao, J. Leon   II-534
Zhao, Jinxi   I-418, I-646
Zhao, Pengpeng   I-251
Zhao, Yangyang   II-473
Zheng, Yi   II-534
Zhou, Chuan   II-79

Zhou, Chunting   I-164
Zhou, Ke   I-646
Zhou, Pei-Yuan   II-409
Zhou, Yuanchun   I-563
Zhou, Yujun   II-15
Zhu, Jun   I-392
Zhu, Yan   I-16
Zhu, Yin   I-472
Zhuang, Yong   I-442, II-690

Printed in the United States
By Bookmasters